Birkhäuser

Modern Birkhäuser Classics

Many of the original research and survey monographs in pure and applied mathematics published by Birkhäuser in recent decades have been groundbreaking and have come to be regarded as foundational to the subject. Through the MBC Series, a select number of these modern classics, entirely uncorrected, are being re-released in paperback (and as eBooks) to ensure that these treasures remain accessible to new generations of students, scholars, and researchers.

Klaus Hentschel
Editor

Physics and National Socialism

An Anthology of Primary Sources
Translated by Ann M. Hentschel

Reprint of the 1996 Edition

Birkhäuser

Editor
Klaus Hentschel
Head of Section for History of Science
and Technology
University of Stuttgart
Keplerstrasse 17
70174 Stuttgart
Germany
klaus.hentschel@po.hi.uni-stuttgart.de

2010 Mathematics Subject Classification: 01A75, 01A60, 01A80, 00A79

ISBN 978-3-0348-0202-4 e-ISBN 978-3-0348-0203-1
DOI 10.1007/978-3-0348-0203-1

Library of Congress Control Number: 2011937343

Originally published under the same title as volume 18 in the Science Networks. Historical Studies series by Birkhäuser Verlag, Switzerland, ISBN 978-3-7643-5312-4
Reprint 2011 by Springer Basel AG

Printed on acid-free paper

Springer Basel AG is part of Springer Science+Business Media

www.birkhauser-science.com

Contents

Introduction

I Controversies Prior to 1933

II After the Nazi Seizure of Power

Appendix

List of Figures

List of Tables

Introduction

1 Aim and General Description of the Anthology

The purpose of this anthology is to introduce the English speaking public to the wide spectrum of texts authored predominently by physicists portraying the actual and perceived role of physics in the Nazi state. Up to now no broad and well-balanced documentation of German physics during this time has been available in English, despite the significant role physics has played both politically (e.g., in weaponry planning) and ideologically (e.g., in the controversy over the value of theoretical ('Jewish') vs. experimental ('Aryan') physics), and even though prominent figures like the scientist-philosopher and emigré Albert Einstein and the controversial nuclear physicist Werner Heisenberg have become household names. This anthology will attempt to bridge this gap by presenting contemporary documents and eye-witness accounts by the physicists themselves. Authors were chosen to represent the various political opinions and specialties within the physics community, omitting some of the more readily accessible texts by leading physicists (e.g., Einstein, Heisenberg, Lenard) in favor of those by less well-known but nonetheless important figures (e.g., Finkelnburg, Max Wien, Ramsauer). In this way we hope not only to circumvent the constricted 'Great Men' approach to history but also to offer a broader picture of the activities and conflicts within the field and the effects of the political forces exerted upon them.

Recent publications indicate that the theme of physics under National Socialism is still very much in the public eye, attracting attention well beyond the academic community.[1] In the secondary literature primary sources are often referred to but only sometimes summarized or quoted and then only cursorily. Thus there is a deficiency of complete English translations of such texts, whether they appeared in the notorious Nazi journal *Zeitschrift für die gesamte Naturwissenschaft* or the non-conformist *Die Naturwissenschaften*, whether private correspondence or public lectures. Such an anthology enables a wider audience to read these primary documents without the filter of the historian. Thus the envisaged audience includes not only scholars and students of science, history, politics, and sociology, but also the general reader. For this reason semipopular accounts, such as Heisenberg's article on the *Uranverein*'s activities referencing the original literature, were preferred over highly technical texts. A small number of texts originally written and published in English are included to represent the viewpoint of contemporary British and American physicists as well as their German

[1] Cf., e.g., the recent controversy between Jeremy Bernstein and Martin Heisenberg (a son of Werner Heisenberg) in *The New York Review of Books*, **39** (issues no. 1 and 2, Jan. 16, 1992), pp. 56–57, or between Jensen, Rechenberg and von Weizsäcker in *Die Zeit*, April 24, 1992, p. 41, and June 8, 1992, p. 84; Goldberg & Powers [1992], Simonsohn [1992], and the commentary in subsequent issues. There was another wave of interest when the Farm Hall transcripts were published in 1993; cf. here footnote 28.

colleagues in exile. The documents presented range from pamphlets by Nazi fanatics to very critical rebuttals by non-conformists like the Nobelist Max von Laue and such eminent exiles as Albert Einstein and James Franck. In addition, letters, extracts from legal documents, and newspaper articles are included as well as some book reviews, which are otherwise generally neglected in the scholarly literature. The selection had to be substantial enough to achieve sufficient breadth and to document the wide spectrum of positions at a specific moment along with the changes of these positions over time. Extensive cross-referencing and the thematic introduction are also designed to preserve the readability of each document for readers interested only in specific aspects. The anthology can thus also serve as a course reader; and since a considerable number of the selected texts were previously unpublished, though some are frequently referred to in the standard literature, it is equally well suited as a source companion to books on Nazi science, in particular those by Beyerchen, Cassidy, Heilbron, Macrakis, and Walker.

1.1 Purpose of the Introduction

The introduction covers the general context of the period and is only intended as a summary of the current historiography of science in the Third Reich, concentrating on physics. It addresses Nazi science policy and its impact on physical instruction and research, the social history and statistics of academic appointments, student enrollment, the emigration of physicists and scientists from related fields after 1933 and, of course, the physical research conducted between 1933 and 1945 in Germany, especially in areas related to weapons technology. The changing attitude of scientists in Germany toward Nazi science policy, is also discussed as well as the policies of specific German scientific institutions in the Nazi state, such as the *Kaiser-Wilhelm-Gesellschaft*, the Kaiser Wilhelm Institute of Physics, and the *Physikalisch-Technische Reichsanstalt*. We also touch upon the history of important scientific societies in Germany such as the *Deutsche Physikalische Gesellschaft* and the *Deutsche Gesellschaft für technische Physik* and trace the origins of the 'Aryan Physics' movement to conflicts in Weimar physics, such as the controversy between theorists and experimentalists, and the rivalry between Berlin (Prussian) and provincial (Southern German) physicists, as well as the sudden resurfacing of vitriolic anti-Semitism in 1920 from its more implicit form prevalent in academic circles during the Weimar Republic and earlier.[2] After discussing briefly the legacy of National Socialism to postwar Germany, the introduction ends with an analysis of style and rhetoric in the selected texts.

By reading these primary materials a fairer evaluation can be made of the complex position of individuals like Heisenberg, Ramsauer, or Prandtl who un-

[2]Cf. doc. 1 and the annotation there; on the cultural and intellectual roots of anti-Semitism see also, e.g., Poliakov [1974-85], and Klamperer (Ed.) [1995]; on academic anti-Semitism see footnote 244 below.

like many of their colleagues in exile, chose to manoeuver in the dangerous waters between the scylla of opportunism and the charybdis of dismissal or deportation. These texts reveal the exceptionality of clear-cut cases of either the 'Nazi' or the 'resistance fighter' among German physicists—most hovered in the gray area in-between. It is hoped that this collection will inspire much-needed indepth studies on this aspect.

1.2 Organization of the Anthology

The texts presented in this anthology are ordered strictly chronologically, using the date of publication for published documents, the date of composition in case of letters, and the delivery date for public talks. The few texts quoted extensively in other selected documents are cross-referenced to avoid duplication. The detailed annotation provides the specific context and information on the actors and institutions involved, referencing similar or related texts not included in the anthology.

The selection is based on the view that historical trends are not contained within convenient time units but rather grow out of the immediately preceding, and continue into the following stages of historical development. For this reason the volume is not restricted to the years 1933-45 when National Socialists monopolized German parliament, but is divided into the following five parts: Controversies Prior to 1933; After the Nazi Seizure of Power; The Ideological Schism 1936–1939; Physics at War 1939–1945; The Legacy of National Socialism.

Part 1 retraces the development of arguments raised against major areas of modern physics: Relativity and quantum theory. Many conflicts that became volatile in the Nazi era such as, for instance, the confrontation between experimental and theoretical physicists, have their origins well within the Weimar Republic. Max von Laue's book review of Johannes Stark's 1923 pamphlet on what the latter perceived to be a 'crisis' in German physics (doc. 2) reveals these two Nobelists and leading figures from the two rival groups, as personal enemies long before the National Socialists' rise to power. The succeeding documents show that these conflicts only became more entrenched in the final years of the Weimar Republic, especially following the onset of the global economic crisis in 1929.

Part 2 covering the years 1933 to 1936 documents the rapid encroachment of National Socialism into science through jurisdiction. Excerpts are taken from the Law for the Restoration of the Professional Civil Service, for example (see doc. 7), which was the legal basis for the persecution of Jewish university employees as well as of the few liberal and left-wing professors, forcing them into exile, though not without some public scandal, as is demonstrated by James Franck's and Wolfgang Köhler's articles from 1933 (docs. 9 and 13). We also document Nazi attempts at achieving political streamlining, the opportunistic applause by some, and the ongoing opposition among the majority of physicists against the loss of academic freedom which often went hand in hand with toleration and accommodation to the changed working conditions. But we also encounter flagrant cases of self-mobilization for the new regime among engineers and scientists.

Part 3 covers the heated ideological disputes in the period 1937–1939 over the 'Aryan Physics' issue as conceived by the 'old guard' Nazis Lenard and Stark. This controversy conceals much more mundane matters of contention, however, such as decision-making influence in science policy and appointments to new professorships and key positions. We find premeditated attacks in Nazi organs on Heisenberg and other proponents of the style of physics that had been so successful in the Weimar years, but we also encounter some carefully worded answers as well as protests and petitions to the authorities. The documents in this part, including the *REM*'s decree concerning the taking of a doctorate by Jews (doc. 53), must be seen against the increasingly constrained living conditions under which the Jewish citizens were subjected. As their rights were plucked away one by one, violence against them increased, culminating in the pogrom of November 1938 which triggered a more systematic politically condoned exploitation and exclusion of Jews from the economy. With the annexation of Austria in March 1938, Austrian Jews like Lise Meitner also fell under German jurisdiction. The country's great-power policy and aspirations toward becoming a world might temporarily concealed the growing domestic and social problems.[3]

Part 4 deals with the war years of the Nazi dictatorship when ideological controversies fell into the background as the political focus moved toward a well-functioning weapons industry that ultimately depended on a sufficient number of physicists and technicians in industrial and academic research laboratories (see, for instance, the articles by Feitl and Karlson from 1943 on 'war physics', docs. 100 and 101). Realizing the necessity for the unimpeded operation of science in the interest of the war economy, the powerful Nazi officials Himmler as head of the *SS* and Göring as head of the *Luftwaffe*, reined in their activists who had been agitating against relativity and quantum theory. The pragmatists Prandtl, Ramsauer, Finkelnburg and others gained influence, many working in private industry as opposed to the public sector as university staff. Repeated efforts on the side of the Party to centralize scientific research effectively, such as the establishment and reorganization of the Reich Research Council (*Reichsforschungsrat*), failed (see, e.g., docs. 52 and 98 regarding the Reich Research Council, and Carl Ramsauer's essay of 1943 (doc. 102) on the crucial role of physics). In this part we can also trace the radical policy changes for research and development in a war economy as the initial successes of the 'lightning war' (*Blitzkrieg*) of the first two years turned into defeat in the winter of 1941. The last documents in this section illustrate the ferocious competition over scarce supplies (docs. 103ff.), and the desperate efforts in the final years of the war by the head of the Reich Research Council's Planning Office Werner Osenberg, and the Plenipotentiary of Nuclear Physics Walther Gerlach, to prevent their scientists and engineers from being

[3]Cf., e.g., Bracher [1969], chaps. 6–7; N. Frei and M. Funke in Broszat & Frei (Eds.) [1989].

enlisted to die futilely in defense of the collapsing homefront (doc. 107).

Part 5 documents the legacy of National Socialism to German physics, throwing light on ideological changes and at the same time on the personal continuity of the postwar era. Lise Meitner's two letters to Otto Hahn from 1945 and 1948 (docs. 108 and 120) illustrate how strained and complicated their formerly extremely close scientific relationship had become as a result of their geographic separation since her flight into exile in the summer of 1938. Planck's report on his audience with Adolf Hitler and Finkelnburg's boasted victory over 'Party physics' provide two retrospective accounts by leading actors (cf. doc. 110 and doc. 114 as well as the annotation on the problematics of source evaluation), and the final document exemplifies that physicists were no exception in signing the necessary statements on oath to exonerate their colleagues during the denazification process (informally referred to as 'white-wash certificates' (*Persilscheine*), see doc. 121). Finally, we also sample the postwar controversy over the wartime activities of German scientists and their moral responsibility fought out primarily between von Laue and Heisenberg against Goudsmit and Morrison in the American *Bulletin of the Atomic Scientists* (see docs. 115–118).[4]

The **appendix** does not aspire to encyclopedic completeness but is designed to help the reader through the maze of often unfamiliar names and German establishments mentioned in the texts. It is divided into the following categories: professional institutions and associations (A), state and military institutions (B), National Socialist organizations (C), major industrial firms (D), journals and periodicals (E), and biographical profiles of authors of the selected texts, prominent German scientists and some important officials referred to frequently (F). In each section the entries are listed in alphabetical order.

The **bibliography** provides full references to frequently cited or particularly relevant literature. Short titles are used in the annotation consisting of the author's name, the year of publication within brackets, and where necessary, a lowercase letter. Infrequently cited specialized publications appear in full in the footnotes. A **name index** at the end of the volume lists individuals referred to in the introduction and the main annotated text.

2 Notes on the Historiography of National Socialism

The National Socialist regime in Germany is the subject of many[5] historical, political, and sociological studies,[6] including a plethora of biographical or psy-

[4]Cf., e.g., Walker [1989]a, chap. 6, or [1990]b. For Heisenberg's later role as science adviser in the Adenauer era, see Eckert [1990].

[5]According to Michaelis & Schmid [1983], the British Library in London and the Library of Congress in Washington registered more than 55,000 publications on Hitler, Nazi Germany and World War II after 1945.

[6]For useful general overviews of the historiography of the Nazi period, see Kershaw [1985] and Kühnl [1990]; on a recent controversy in Germany on the historization of this period

chological studies on some of its prominent political figures, from the pinnacle of government, the *Führer* Adolf Hitler, to the lowest echelons, the lackey *SS-Obersturmbannführer* Eichmann.[7] Some thorough studies on institutions also exist, though mostly in German, such as on the Reich Education Ministry,[8] the National Socialist Party (*NSDAP*), Storm Detachment (*Sturmabteilung*: *SA*), the Defense Squadron (*Schutzstaffel*: *SS*) and the Secret Police (*Gestapo*).[9]

The *NSDAP*'s efforts to create a National Socialist professoriate through their affiliated organizations such as the University Lecturers League (*NS-Dozentenbund*) founded in 1935 and through the reorganization of universities have also been studied in depth.[10] Furthermore, social historians have studied many aspects of National Socialist Germany including mass organizations and membership, the function of certain social groups such as workers, professors or students;[11] and more recently, the historiographic trends of writing a 'history from below', *histoire des mentalités*, and a history of everyday life has led to the move away from the traditional focus on political leaders and selected individuals toward popular opinion and the behavior of social groups.[12] Two useful anthologies in this respect are, on the one hand, the reports written by workers throughout Germany that were smuggled abroad, where they were assembled and edited by the Social Democratic Party in exile, and on the other hand, the formerly secret reports by the Security Service of the *SS* on public opinion in Germany.[13]

---

known, see, e.g., the anthology *Historikerstreit. Die Dokumentation der Kontroverse um die Einzigartigkeit der nationalsozialistischen Judenvernichtung*, Munich: Beck, 1987. Noteworthy early attempts at a broad structural description include Marx [1936], Loewenstein [1939], and Neumann [1942]; from the many of later date see, e.g., Bracher [1969], and Broszat [1969]; cf. also Bracher, Funke & Jacobsen (Eds.) [1986], as well as footnote 18. On science see the surveys mentioned in footnote 19 and further references in secs. 3–6 of this introduction.

7 On Hitler see, for instance, Fox [1979] and Wolf-Rüdinger Hartmann: 'Adolf Hitler: Möglichkeiten seiner Deutung', *Archiv für Sozialgeschichte* **15** [1979], pp. 521–535; cf. als Loewenstein [1939], pp. 28–32, and the overview in Kershaw [1985]c, chaps. 4–5. On Eichmann, see Arendt [1977], which together with Waite [1977] exemplifies the psycho-historical approach; cf. Wehler [1980].

8 See, in particular, Seier [1964], [1988], pp. 273ff.; cf. Graf zu Rantzau [1939].

9 See, for instance, Orlow [1969], Bollmus [1970], and Kater [1983] on the *NSDAP*, Longerich [1989] on the *SA*, and Gellately [1990] or Delarue [1964/79] on the *Gestapo*. See also, e.g., Hüttenberger [1976] on their interrelations and, ibid., pp. 427ff. for the complex substructure of the *NSDAP* itself.

10 See, in particular, the Ph.D. thesis by Kelly [1973], as well as Heinemann (Ed.) [1980], Heiber [1991/92].

11 For a prominent example, see Schoenbaum [1967]; cf. Kershaw [1985]c, chap. 7, the following footnote, as well as secs. 3.3–3.4, and 4 of this introduction for further references on lecturers, students, and emigrés.

12 See for instance, Kershaw's study on the popular reception of Hitler in *The Hitler Myth. Image and Reality in the Third Reich*, Oxford, 1987; or the studies by Martin Broszat and others on the Bavaria project, published in 6 Volumes as *Bayern in der NS-Zeit*, Munich, 1977–83, cf. also Benz [1990] for a recent attempt to combine structural analysis with history of mentality focussing on the mechanisms of National Socialist power.

13 See *Deutschland-Berichte der Sopade 1934–1940*, Frankfurt am Main: Zweitausendeins,

Among the social groups of relevance to this anthology, German engineers have been studied quite thoroughly in this regard,[14] certainly better than physicists and physical chemists, where a detailed exploration of mentality and social history still needs to be done.[15] Many engineers were affected seriously by the economic crisis of the late 1920's and by the severe restraints imposed by the Versailles Treaty on research and development in sectors like airplane production. When the Nazis came to power, these restrictions were largely ignored in areas of military relevance (see, e.g., doc. 75); naturally engineers welcomed the resulting boost to these fields and were thus easily caught up in the Nazi rhetoric on 'Aryan technology' with its anti-Semitic overtones, and were lured by the promise of increased social prestige and the prospect of new jobs. It has been argued that the new appreciation for *Deutsche Technik*, the analog to 'Aryan Physics' (cf. sec. 5.3), also had a distinctly anti-theoretical flavor, so that engineers saw an opportunity to gain prestige at the expense of the previously elevated researchers in fundamental science.[16]

Some anthologies of primary texts on the general history of the period are available both in English and German.[17] Through these historical and sociological studies, the earlier image drawn by the left-wing analyzers of fascism Borkenau, Franz Neumann or Hannah Arendt of the Nazi regime as a totalitarian system has given way to a much more complex, polycentric portrayal of the government as a "polycracy".[18] In this interpretation the individual subsystems fought amongst themselves at times at least as vigorously as against the Allied forces (for instance: *SS* versus *SA* in the 1934 *Röhmputsch* affair, Air Force (*Luftwaffe*) against the Army Ordnance Office (*Heereswaffenamt*), the Party main office (*Braunes Haus*) against the Reich Education Ministry (*REM*) in the daily bureacratic bickering over the implementation of Nazi legislation, the *Reichsforschungsrat* against the

1980, and *Meldungen aus dem Reich. Die geheimen Lageberichte des Sicherheitsdienstes der SS 1938–1945*, edited by Heinz Boberach, Herrsching: Pawlak, 1984.

[14]See the references in footnote 273 below and Gerd Hortleder: *Das Gesellschaftsbild des Ingenieurs. Zum politischen Verhalten der technischen Intelligenz in Deutschland*, Frankfurt: Suhrkamp, 1970.

[15]See, however, Eckert [1993] for a social history of the Sommerfeld School, as well as Paul Forman's and Steffen Richter's contributions on physicists and their funding in the Weimar era and beyond.

[16]See, in particular, A. Heinemann-Grüder in: Renneberg & Walker (Eds.) [1994]; it would be worthwhile to investigate whether the electrical engineers employed at the big German electrical trusts *Siemens*, *Telefunken* and *AEG* were an exception (as suggested to me by J. Lemmerich).

[17]See Noakes & Pridham (Eds.) [1990]a, Hofer (Ed.) [1957] or Poliakov & Wulf (Eds.) [1959/83], the latter two to be taken *cum grano salis*, however. Cf. also footnote 38 for German anthologies on scientific subjects.

[18]See, e.g., Broszat [1969], pp. 363ff., 423ff., Bollmus [1970], Klaus Hildebrand in: Bracher et al. (Eds.) [1986], pp. 73–96. Hüttenberger [1976], pp. 420ff., explains the emergence of this myriad of conflicting accountabilities as the result of simultaneous efforts to expand power, penetrate into foreign spheres, differentiation within these spheres of influence and resulting compromises.

Air Force's *Forschungsführung*, the chemical and electrical trusts versus the heavy metals trusts, and eventually even some of the military leaders against Hitler in several failed assassination attempts). Political streamlining or coordination (*Gleichschaltung*) was thus clearly much less efficient than the Party (*NSDAP*) intended and the *Führer's* directives were often supplemented by other, sometimes contradictory decisions by other centers of the complicated hierarchy of power. Since the sources assembled here also suggest this multidimensional power structure, with party institutions as *one* but by no means the *only* decision-making element in the network of rapidly changing alliances, we adopt this 'structuralist' or 'functionalist' view in our interpretation of the documents to describe physics within Nazi Germany.

History of science is still far behind history proper in this regard,[19] partly because the community of historians of science is much smaller, but also because of a certain regrettable tendency in the older literature to dismiss this period as an 'unfruitful interlude' in an otherwise unbroken line of 'continuous progress'. We have personal accounts by scientists from these tumultuous times, of course, such as Samuel Goudsmit's *Alsos*,[20] or the reminiscences of Born, Frisch, Szilard, and Weisskopf, to name but a few physicist exiles on the one hand,[21] and by von Ardenne, Hahn, Heckmann, and Stark, who stayed in Germany at that time, on the other.[22] Most of these were written retrospectively, however, or are subjectively blurred, usually either out of defensive motives and as postwar apologia in the case of suspected collaborators of the system, or in outrage in the case of its victims.[23]

The collected works of some of the most important physicists of the period have either already been published (Bohr, Debye, Heisenberg, von Laue, Pauli, Sommerfeld) or are beginning to appear (Einstein);[24] but so far only in the case of Pauli is the scientific correspondence of the crucial years after 1933 now available in print (up to 1949);[25] in most other cases, however, only extracts have been published.[26] Some contemporary physicists have been interviewed, and in

[19]See the overviews provided in Mehrtens [1979], in Mehrtens & Richter (Eds.) [1980], pp. 15–87, as well as Beyerchen [1992] for a more recent account of "What we now know about Nazism and science".

[20]See Goudsmit [1947].

[21]Born [1968], Frisch [1979], Leo Szilard, 'Reminiscences' in: Fleming & Bailyn (Eds.) [1969], and Weisskopf [1991].

[22]Characteristically, the latter biographies have not yet been translated into English, as is true of von Ardenne [1972], Heckmann [1976], and Stark [1987].

[23]Cf., e.g., Walker [1989]a, chap. 6 on the Heisenberg-Goudsmit controversy in which both sides charged the other of onesidedness.

[24]The Heisenberg edition in particular has made available some new material of relevance to the theme under discussion here. It includes some of Heisenberg's speeches for the Army Ordnance Office and some of his classified reports for the *Uranverein* in Vols. A II and C IV; see also Kleint & Wiemers (Eds.) [1993], part I.

[25]See Meyenn et al. (Eds.) [1978–93].

[26]See, for instance, the selections in Nathan & Norden (Eds.) [1960], the appendices to Bohr's

some cases the transcripts have appeared;[27] and the notorious Farm Hall tape recordings by the British secret service during the internment of 10 top German physicists in Godmanchester (near Cambridge, England) between July and December 1945, have also recently been published.[28] As the latter are readily available now in English, we have decided not to incorporate any passages from these conversations, but rather to concentrate on less easily accessible materials, a considerable portion of which was hitherto unpublished.

Furthermore, a few carefully researched biographies of important physicists and physical chemists exist, most notably on Werner Heisenberg, Max Planck, Arnold Sommerfeld, and Fritz Haber,[29] while others (such as on Lise Meitner or James Franck) are still in preparation.[30] There are also autobiographies, obituaries or biographical sketches by close colleagues on several other scientists and science policy-makers.[31] Autobiographical accounts and obituaries are often undocumented, frequently apologetic in character and usually written in retrospect and thus unfortunately cannot always be accepted at face value. Other studies have focussed on specific episodes in the life and work of individuals such as, for example, the theoretical physicist Pascual Jordan,[32] the natural philosophers Hugo Dingler and Bernhard Bavink,[33] and the astronomer Otto Heckmann.[34]

Helpful texts about the fate of several institutions of major importance to physical research have also been published, such as on the *Kaiser-Wilhelm-Gesell-*

Collected Works, the selections from the Arnold Sommerfeld Papers presented in: Eckert et al. (Eds.) [1984] and from the Walther Gerlach Papers in: Heinrich & Bachmann (Eds.) [1989] respectively. Selected Heisenberg and Sommerfeld correspondence editions are underway.

[27]See, e.g., Ermenc (Ed.) [1989] as well as the interviews recorded for the Archive for the History of Quantum Physics by T. S. Kuhn, John Heilbron and Paul Forman in the 1960's, transcripts on microfilm in several research libraries and available on loan at the American Institute of Physics, College Park, Maryland.

[28]See Frank (Ed.) [1993]; cf. also Hoffmann (Ed.) [1993]b for the German retranslation of the preserved transcript excerpts, which with one exception have only survived in the official English translation. See also Goldberg & Powers [1992], Walker [1993]b.

[29]See Cassidy [1992]a, Heilbron [1986]b, Benz [1975], and Stoltzenberg [1994]; cf. also Mason [1992] on the physical geochemist Victor Moritz Goldschmidt who taught at Göttingen up to his emigration in 1935.

[30]On Meitner, see Sime [in press], cf. also Krafft [1978]a, [1988] and Kerner [1986]; cf. also Kamp et al. [1982] and Lemmerich (Ed.) [1982] on Born and Franck.

[31]See, e.g., Born [1968], von Ardenne [1972], Steenbeck [1977], and Stark [1987] as four very different examples of autobiographies by German physicists; Schmidt-Ott [1952], and Glum [1964] for autobiographies by science administrators. For other obituaries and biographical sketches see the bibliographical references in Appendix F. See also Nachmansohn [1979] and Volkov [1987] for prosopographic studies on Jewish scientists in Weimar Germany, and Eckert [1993] on the members of the Sommerfeld School.

[32]See the contribution by Norton Wise in Renneberg & Walker (Eds.) [1994], pp. 224–254; cf. also Beyler [1994] on Jordan.

[33]See Wolters [1992], Hentschel [1993].

[34]As, e.g., Hentschel & Renneberg in: Meinel & Voswinckel (Eds.) [1994], pp. 201–207, Renneberg & Hentschel [1995]. On other astronomers cf., e.g., Kuiper [1946], Litten [1992], and G. Wolfschmidt in: Meinel & Voswinckel (Eds.) [1994], pp. 152–159.

schaft, the principle scientific academies, the *Kaiser-Wilhelm-Institut für Physik* in Berlin, or the Einstein Tower at the Astrophysical Observatory in Potsdam,[35] but again the majority only in German. David Cahan's important study on the *Physikalisch-Technische Reichsanstalt* unfortunately only covers the years 1871–1918, and the sequel for the years 1918–1948 has only appeared in German.[36] While earlier official histories of universities and research institutions tended only to touch upon the period 1933–45 (if at all), in the last 10 to 15 years, considerable efforts have been made to write the history of German universities during the Nazi period—those on the Technical University of Berlin, and on the universities in Giessen, Heidelberg, Tübingen, Göttingen, Cologne, Kiel, and recently, Hamburg and Munich are especially noteworthy;[37] but none of these studies is likely to be translated into English. The same is true of the anthologies of primary texts compiled by German historians of science and education on German aerodynamics and aviation research, and on physics instruction during the Third Reich.[38] The latter underscores how the ideological and militaristic tendencies krept into the schoolteacher's pedagogical realm—a sphere which is only touched upon briefly in this anthology (cf. esp. doc. 59).

University professors as a group or the student body have been studied in some detail both concerning the typical 'apolitical' self-image of the former before the rise of National Socialism and their performance in the Third Reich.[39] On the

[35]See, in particular, Burchardt [1975], Macrakis [1993], the chapter written by H. Albrecht and A. Hermann in: Vierhaus & vom Brocke (Eds.) [1990], pp. 356–406, and Oexle [1994] on the *KWG* in general; Scriba (Ed.) [1995] for a comparison of the academies in Berlin, Munich, Heidelberg, Vienna and the Leopoldina, Wennemuth [1994] on the Heidelberg Academy; Grau, Schlicker & Zeil [1979] and Walker [1995], chaps. 4–5, on the Prussian Academy in Berlin; Hentschel [1996] on the Einstein Tower; Heisenberg [1971], Macrakis [1986], [1993], chap. 8, and Kant [1987], [1992] on the *KWI* of Physics (cf. also sec. 5.1).

[36]See Cahan [1989] and Kern [1994] respectively; compare the latter with a one-sided history of the *PTR* up to 1937 by its president Stark [1937], or with Bortfeldt et al. (Eds.) [1987], where the Nazi period is only covered briefly on pp. 106–112, symptomatic of the general tendency to avoid discussion of the events during the Third Reich in jubilee volumes and in-house histories (see also sec. 5.1 here on the *PTR*).

[37]See Böhles et al. [1982] on Giessen (cf. Moraw [1982], pp. 210–225); Adam [1977] on Tübingen, Rürup (Ed.) [1979] on the Berlin Polytechnic (especially the contributions by Cassidy and Mehrtens); Becker, Dahms & Wegeler (Eds.) [1987]; Buselmeier et al. (Eds.) [1985], Wolgast [1986], pp. 142–166, Mußgnug [1988] and Jansen [1992] on Heidelberg; Golczewski [1988] on Cologne; Uhlig (Ed.) [1991] on Kiel; Krause et al. (Eds.) [1991] on Hamburg, and Wengenroth [1993] on the Technical University at Munich in the Weimar and Nazi periods. See also the relevant sections in the broader histories of the universities of Stuttgart by Voigt [1981], chap. 4, and Steinmetz et al. [1958], chap. 9, on Jena (with a heavy Marxist flavor).

[38]See Trischler (Ed.) [1992], and Brämer & Kremer (Eds.) [1980]; cf. also Bäumer(-Schleinkofer) [1990] which contains rich material on Nazi biology, and Weinreich [1946] or Eilers [1963], pp. 13ff., on various other disciplines such as history or geography.

[39]See, for instance, Nolte [1965], Stern [1972], Döring [1974], Kater [1981], Gerda Freise in: Brämer (Ed.) [1983], pp. 31–58, Funke [1986], Olszewski [1989], Mehrtens in: Siegele-Wenschkewitz & Stuchlik (Eds.) [1990], Vogel [1991], Heiber [1991/92], and Jansen [1992],

ideological backdrop to the rise of Nazism, see also Fritz Ringer's famous study on the *Decline of the German Mandarins*, and Kurt Sontheimer's work which also focusses on the time preceding 1933.[40] Frederic Lilge made an early attempt to explain the failure of the Germany university system in terms of the corrosive influences of Nietzsche's satire of academic culture, and the growing irrationalism fostered in the writings of Oswald Spengler, Stefan George, Martin Heidegger and other thinkers of the time.[41] There has been some interest early on in the role played by German scholarship from fields ranging from geography to history, from anthropology to political science, in the crimes against the Jewish people.[42] Jeffrey Herf's study analyzes the ambiguous position of National Socialist ideology toward technology, which has its roots in the 'reactionary modernism' of the neo-romantic, irrationalist, and radically subjectivist ideas in German nationalism.[43] His prime witnesses include the intellectuals Ernst Jünger, Carl Schmitt, and Werner Sombart who prepared the way for an irrational embrace of technology which Joseph Goebbels later termed the 'century of steely romanticism' (*stählerne Romantik*). Indeed, this strange coexistence between political irrationalism and a rearmament-based industrial rationalization is precisely what was so typical of the National Socialist state. The open conflict between these two diverging ideologies provides another important reason for the ultimate failure of the Nazi state.

Apart from these studies by general historians, some topics of special interest to the general public have also been dealt with quite extensively, notably the German atomic bomb project,[44] the massive emigration wave of Jews and other persecuted persons, which later proved so important to American science,[45] or the surreptitious importation of Nazi scientists by Allied governments after World War II.[46] However, only very few texts of broader perspective dealing with physicists during the Third Reich are available in English: the classic study *Scientists under Hitler* by Alan D. Beyerchen is one;[47] the other is a very recent book by

chap. VI on the change in attitude by professors after 1933. On students see sec. 3.4 below.

[40]See Ringer [1969], Sontheimer [1969], and the contributions by the social historians Ringer, Burchardt and Sontheimer in: Schwabe (Ed.) [1988] on the German professoriate prior to 1933.

[41]See Lilge [1948], p. 170: "The German universities failed, while there was still time, to oppose publicly with all their power the destruction of *Wissenschaft* and of the democratic state. They failed to keep the beacon of freedom and right burning during the night of tyranny so that it could be seen by the entire world." For a different, more political perception see Sontheimer [1962]. On Heidegger's role see, e.g., Jürg Altwegg (Ed.) *Die Heidegger Kontroverse*, Frankfurt: Athenäum, 1988.

[42]One of the earliest studies is Weinreich [1946], research director of the Yiddish Scientific Institute, who backed his severe attacks on 'Hitler's professors' with documentary evidence.

[43]See Herf [1984].

[44]Cf. the literature mentioned in sec. 5.4 below.

[45]Cf. the literature mentioned in sec. 4 below.

[46]Cf. the end of sec. 5.5 below.

[47]Published by Yale University Press in 1977, reprinted several times and also translated into German.

Mark Walker entitled *Nazi Science*, on Stark, Heisenberg, the Prussian Academy of Science, German nuclear research and the myths formed about the latter in particular.[48] Moving outside of physics, the role of mathematicians and in particular the efforts of Ludwig Bieberbach and others to legitimate their concept of 'Aryan mathematics'[49] based on pseudo-anthropological typology developed by the psychologist Erich Rudolf Jaensch[50] has also been studied intensively. Contributions on the history of other sciences during the Third Reich soon followed.[51]

In the related field of technology, particularly engineering research for the military, Leslie Simon's pioneering study is complemented by Karl-Heinz Ludwig's survey of technology and engineers in the Third Reich[52] as well as by studies on the interaction between industry and technology, science and the state.[53]

3 The Impact of Nazi Science Policy on Physics Instruction and Research

The political realignment (*Gleichschaltung*) of all state-controlled institutions came on the heels of Hitler's rise to power on January 31, 1933. University employees were thus not immune to this policy which involved the expulsion of all political and ideological opponents to the ruling party, most importantly activists of left-wing political organizations and Jews, the latter defined on a purely 'racial' basis regardless of their actual convictions. The infamous 'Law for the Restora-

[48]See Walker [1995] and [1989]a, chap. 2; cf. also the papers by Richter [1978/79], Walker [1989]b and Wolff [1993] in German, or Guérout [1992] in French.

[49]On mathematicians see, e.g., Pinl [1969-76], Pinl & Furtmüller [1973], Schappacher & Kneser [1990]; cf. also Lindner [1980],Mehrtens [1987]a, [1989], as well as his contribution in Renneberg & Walker (Eds.) [1994]; Siegmund-Schultze [1986], [1989], [1993], and in Olff-Nathan (Eds.) [1993] on mathematics. On individual mathematicians' careers see, e.g., Reid [1976] on Courant, Siegmund-Schultze [1984] on Vahlen, Mehrtens [1987]b, [1990], pp. 308–315 on Bieberbach, Schappacher & Scholz (Eds.) [1992] on Teichmüller, Litten [1994]b on Perron.

[50]See Jaensch [1931]. For a broader analysis of the intellectual origins of the 'Aryan science' concept, see Guérout [1992], pp. 79–87, and further references there.

[51]For instance, the collections of historical analyses in Mehrtens & Richter (Eds.) [1980] (aside from physics, also dealing with mathematics, chemistry, pharmacology, ethology, and architecture), Lundgren (Ed.) [1985] (focussing on the humanities and social sciences), Erickson [1985] on theologians, Muller [1986] on social scientists, Leske [1990] on philosophers, Deichmann [1991] on biologists, Rössler [1990] on geography, Olff-Nathan (Ed.) [1993] with contributions in French mainly on mathematics, physics, the bio-sciences and geography, Renneberg & Walker (Eds.) [1994] (mainly on physics, mathematics and technology), and Meinel & Voswinckel (Eds.) [1994], for a broad spectrum of subjects and approaches. For references to studies in other fields such as history, linguistics, etc., see, e.g., Seier [1988], p. 249.

[52]See Simon [1948] and Ludwig [1974]; cf. also Guérout [1992], pp. 36–41 on the German Engineers Association (*Verein Deutscher Ingenieure*).

[53]See, e.g., Borkin [1978] and Hayes [1987] on the *IG Farben* trust, Braun [1992] on the aviation industry, Neufeld [1995] on Peenemünde rocket research, and Hans Mommsen on *Volkswagen* (*VW*), in preparation. See also, e.g., Thomas Trumpp as well as Hans-Erich Volkmann, in Bracher et al. (Eds.) [1986] on the *NSDAP*'s dependency on big industrial companies.

tion of the Professional Civil Service', for example, gave this purge legal backing and had a direct impact on teaching and research. The title of this law is itself a prime example of the euphemistic language used by the Nazis in an attempt to attach positive connotations, such as here the civil service, commonly associated with the Prussian values of efficiency, correctness and parsimony, to their drastic measures whose purpose was to destroy the former democratic system. The word 'restoration' implies that during the Weimar period the civil service had lost these virtues through party patronage. The reception of this law in the daily press was mixed. In several newspapers the new law was welcomed as an end to 'Communist' party favoritism and politically motivated hiring of officials (*Parteibuchbeamten*); others, however, feared the 'ideological realignment'.[54] As a result of this law, at some universities up to a quarter of the physics faculty was forced into early retirement or emigration (cf. here sec. 4 and references there). Because this and other related laws issued after April 7, 1933 are referred to frequently, they have been included in this anthology (see docs. 7, 8, 12, 17, 36 and 52f. and the annotation with further references to the secondary literature). That these laws had to be bolstered by a whole series of implementation regulations and legislative measures is symbolic of the improvisational character of Nazi science policy.[55] We also include a rare immediate public protest to the issuance of this law of April 1933 (doc. 13), another commentary in an anonymous article (doc. 14), and a British article to represent its reception abroad (doc. 21).[56] Finally, the consequences of these laws on the lives of physicists are demonstrated with James Franck's voluntary resignation (doc. 9), for example, and related correspondence (docs. 10 and 16), and material related to Haber's resignation (docs. 23f.) and Jaffé's dismissal (docs. 25f.). The annotation refers readers to the existing secondary literature on National Socialist legislation and related issues and indicates some of the legal options provided by the new laws.[57]

[54]Cf. *Göttinger Tageblatt*, Vol. 45, No. 895, April 10, 1933, p. 1, 'Das Ende des roten Bonzentums'. See also the rather critical commentaries in the non-realigned *Berliner Tageblatt* **62**, issues no. 165, Saturday, April 8, 1933, p. 1, No. 168 A, Tuesday, April 11, 1933, p. 4: 'Das neue Beamtenrecht', and No. 172 A, Thursday, April 13, 1933, p. 1–2: 'Geistige Gleichschaltung'. On M. Planck's, M. von Laue's and W. Heisenberg's reactions to this law see Wolff [1993], pp. 268f. The fullest commentary on the background and consequences of this law for the civil service is provided in Mommsen [1966], esp. pp. 39ff.

[55]See, e.g., Walk (Ed.) [1981] on the labyrinthine sequence of anti-Semitic legislation passed after April 1933 which also includes the revocation of the famous exemption clause protecting Jewish veterans who had served "at the front for the German Reich or whose fathers or sons had been casualties in the World War" (doc. 7 §3 and footnote 8 there). Cf. also Seier [1984], p. 145, where the National Socialist policy is portrayed as a "chain of improvisations", and pp. 145–149, for an interesting periodization.

[56]Cf. also N.N. [1933]e-f, [1936]b, Gumbel [1938], Notgemeinschaft (Ed.) [1936] and N.N. [1938]c-d.

[57]See in particular the classic study by Hans Mommsen [1966], the commented anthology by Walk (Ed.) [1981], as well as the contributions in Heinemann (Ed.) [1980].

3.1 The Changing Attitude of Scientists within Germany

1933 was a pivotal year also for German science. Contemporary evaluations of the political situation show the extent to which intellectuals living in Germany misjudged Hitler and the *NSDAP*. Lise Meitner's letter describing Hitler's Potsdam speech in March 1933 (doc. 5) is an example. The intense propaganda in the two new and effective means of mass communication, radio and film, under the control of Goebbel's Ministry of Public Information and Propaganda, probably played a significant role.[58] In March 1933, 301 university teachers signed a declaration in support of Hitler that appeared in the *Völkischer Beobachter*, the backbone of the National Socialist press. In the following autumn the National Socialist Teachers League (*NSLB*) circulated another declaration of confidence in Hitler and the National Socialist government among the German professoriate on the occasion of Germany's withdrawal from the League of Nations, collecting a list of 675 names.[59] However, despite this initial wave of enthusiasm, the private correspondence of physicists only one year later reveals the extent to which the situation had changed (cf. docs. 33f.).

Since our anthology is ordered strictly chronologically, the reader will be able to trace the changes in attitude, the hopes and fears of those who chose not to emigrate, and the options that remained open to them at various stages. There were many different motives behind the decision to stay in Germany. Max Planck, for instance, wanted to form an 'island of stability' to prepare for the rebuilding of his field of physics after the hopefully brief interlude of Nazi domination in parliament. According to Werner Heisenberg, early in the summer of 1933 Planck was of the opinion "that at that time large portions of the German population were in a state of complete hysteria and that the emigration of many German professors would only turn things for the worse. He had the definite hope that this hysteria would not last longer than one or two years and that after this time it would be important to have as many reasonable people as possible inside Germany and at German universities".[60] Heisenberg was of the same opinion; moreover, he was far too attached to the Alps and his native country to leave. Carl Friedrich von Weizsäcker banked on changes in government policy after the initial wave of radicalism had subsided.[61] On the other hand, advanced students

[58]See in particular Welch [1983].

[59]See *Völkischer Beobachter*, Berlin Ed., March 4, 1933; and *Bekenntnis der Professoren an den deutschen Universitäten und Hochschulen zu Adolf Hitler und dem nationalsozialistischen Staat*, Dresden, 1934, pp. 129–135; statements of loyalty by world-famous scholars, which were also translated into French and English, preceded this list. These authors included the philosopher Martin Heidegger and the physician Ferdinand Sauerbruch. See also Kater [1981], pp. 64ff., for a statistical analysis of the academic groups who signed this document (472 were professors). It must be noted, however, that the circumstances of the collection of these signatures are not clear; cf., e.g., Zneimer [1978].

[60]Heisenberg in a letter to James O'Flaherty, Nov. 28, 1956, in O'Flaherty [1992], p. 490.

[61]Both Heisenberg and von Weizsäcker later admitted that these hopes had been naive: Cf.,

and scientists, particularly in lower or middle ranking academic positions, took advantage of the changed political circumstances to improve their careers under much less competitive conditions.[62]

The new political setting thus caused serious strain in the education and training of the next generation of scientists. Petitions issued by alarmed academic physicists and prominent industrial researchers indicate the growing frustration with the mismanagement and faulty allocation of ever-shrinking resources. The petitions by the most senior professor of physics at that time, Max Wien, of 1934 (doc. 35), and the 1936 one co-signed with Heisenberg and Geiger (doc. 49), are examples along with Prandtl's letter to Göring in 1941 (cf. docs. 62ff.), and Ramsauer's submissions to Education Minister Rust in 1942 (cf. docs. 84ff.). Wien's petition and Ramsauer's attachment (see in particular docs. 35 and 91) provide informative statistics on unoccupied chairs and on research output. Section 3.4 below provides additional figures, especially on declining student enrollment after 1933 and on the physics curriculum in particular. Carl Krauch's article 'Youth to the front line' (cf. doc. 58) is particularly revealing in this context. Employing an unmistakably martialistic tone, this influential *IG Farben* lobbyist felt the need to emphasize the value of science in the increasingly militarized state which tended to underestimate the importance of research and development, and neglected university education, which ultimately hampered the upcoming generation of scientists.

3.2 Science Policy Measures after 1934

It is arguable that the Nazis did not have a carefully planned and comprehensive longterm science policy.[63] On the other hand, disregarding the relentless power struggle among the various agencies, the in-fighting and the disfunctionality of newly created centralized institutions like the Reich Research Council (*Reichsforschungsrat*) (see below), the following features were quite widespread:[64]

e.g., Heisenberg [1967], von Weizsäcker [1991], p. 234, and Hoffmann (Ed.) [1993], pp. 338f.

[62]See the next subsection as well as, e.g., Glum [1962], p. 294, on the 'struggle of the left-over lecturers' (*sitzengebliebene Privatdozenten*), "that is what unpaid university lecturers still seeking chairs were called who had not succeeded in becoming full professors for lack of exceptional achievement and who were frequently willing to use any means."

[63]Cf. Seier [1964], [1984], pp. 149ff., Maier (Ed.) [1966], pp. 71–102, Ludwig [1974], chap. 6, Müller [1978], Mehrtens and Schottlaender in: Rürup (Ed.) [1979], pp. 427ff., 444ff., Kelly and Kleinberger in: Heinemann (Ed.) [1980], Vol. II, pp. 9–30, 61–76, Lundgren (Ed.) [1985], pp. 10ff., Titze [1989], Guérout [1992], p. 9: "non-politique de la science", Macrakis [1993], chap. 4, Siegmund-Schultze [1993], Lundgren and Mehrtens in: Fischer et al. (Eds.) [1994], Rudolf Vierhaus in: Scriba (Ed.) [1995], p. 252: "there was no uniform and consistent science policy in the Third Reich", as well as Achim Thom's critique, ibid., pp. 252f.

[64]See, e.g., Rosenberg [1934], Goebbels [1934], Hitler [1934], Krieck [1936], Goering [1937], and Mentzel [1940] which enunciate the many partly conflicting aims of the following list.

1. The adaptation of the university system to follow the authoritarian *Führerprinzip*;
2. The transformation of academic faculties through 'purges' and politically-based recruitment policies;
3. The politicalization of scientific disciplines through placing a new focus on racially based nationalistic (*völkische*) aspects;
4. The professionalization and differentiation of new research areas, with a strong emphasis on applied research;
5. The militarization and deployment of research and development toward the 'ultimate victory' (*Endsieg*);
6. The transferral of areas of research from universities to industry and to state-owned or semiprivately-run research establishments.

As a result although clearly not intended, in some cases industry profited from the increased availability of highly qualified scientists (e.g., the case of Gustav Hertz, mentioned here on p. lxxxviii). However, only large companies could venture such hirings on the strength of their own economic standing, and thus industry did not become a major haven for dismissed academic scientists. The almost haphazard efforts to cull all unwanted individuals from the faculties and to centralize German research had the effect of seriously reducing existing research and development capacities at universities and polytechnics. Theoretical physics and mathematics suffered particularly high emigration losses (see the next section). On the other hand, Nazi science policy did succeed in establishing some new specialized branches. The military science department (*Wehrwissenschaft*) at the Polytechnic in Charlottenburg (Berlin), for example, was organized under the directorship of General Karl Becker.[65] Other areas such as, e.g., metallurgy, geopolitics and anthropology or 'racial hygiene', were strengthened because of their obvious links to National Socialist expansionist and racist policies. The new police state also fostered the field of psychology, for example, because of its relevance to the assessment of an individual's fitness for military service.[66]

Despite these isolated successes in the assignment of research priorities, altogether university research departments proved to have much more inertia in protecting the *status quo* than the new rulers had anticipated. The government made successive attempts to gain control over the selection and prioritization of research topics. One of these was the foundation of the Reich Research Council (*Reichsforschungsrat*) in March 1937. General Karl Becker was again chosen as its

[65]See, in particular, K. Becker [1939] and here doc. 75; cf. also Ebert and Rupieper in: Rürup (Ed.) [1979], Vol. 1, pp. 469–491. On the institutionalization of *Wehrwissenschaft* in general, see Guérout [1992], pp. 96–99 and further references there.

[66]Cf. in particular Geuter [1992] on the professionalization of psychology at that time.

chairman. This research planning agency was originally placed under the supervision of the Reich Education Ministry (*REM*) and its purpose was to coordinate centrally research conducted at various locations.[67] It consisted essentially of 13 separate specialty divisions mostly directed by politically 'reliable' scientists who were in fact often longtime *NSDAP* members and not infrequently even *SS* or *SA* officials. Examples include the high-frequency specialist Abraham Esau for the physics and engineering division, Peter Adolf Thiessen for chemistry and organic materials, and Erwin Marx for electrical engineering.[68] However, many fields of research remained beyond the reach of this council from its very inception. Aeronautics was one of these, along with industrial research conducted in the private laboratories of big companies like the *IG Farben AG* chemical trust, which received major subsidies within the context of the Four-Year Plan rearmament program of 1937.[69] The research and development department (*Abteilung Forschung und Entwicklung*) of Göring's German Resources and Materials Office (*Amt für deutsche Roh- und Werkstoffe*, founded in the spring of 1936), was under the autocratic control of Karl Krauch, a chemist and *IG Farben* board member. The close cooperation between the state and this concern was so close that the agency established to promote trade and industry, *Reichsamt für Wirtschaftsausbau*, acquired the unofficial nickname: 'Office for the Expansion of *IG Farben*' (*Amt für IG-Farben Ausbau*).

From the institutional point of view the Reich Research Council was ill-conceived from the start. The German national science foundation (*Deutsche Forschungsgemeinschaft: DFG*), which had been founded in 1920 under the name *Notgemeinschaft der Deutschen Wissenschaft* on the initiative of Fritz Haber and Friedrich Schmidt-Ott, continued to co-exist with the Reich Research Council and practically dominated research planning and funding. It was presided over by Johannes Stark between 1934 and 1936, later by the senior battalion commander (*Obersturmbannführer*) of the *SS* Rudolf Mentzel. Total funding for the Council and the science foundation for 1937 and 1938 rose to about twice the 1934 and 1935 levels; no major budget increases occurred during the war years, however.[70] At the outbreak of World War II, the two rival institutions were formally inte-

---

67 See here doc. 52, dated Mar. 16, 1937, as well as Ludwig [1974], pp. 217ff., Laitko et al. (Eds.) [1987], pp. 565f.

68 On the long-distance power transmission expert E. Marx, party member since 1937 and a convinced National Socialist, see Maier [1993]. The appointment of General Karl Becker as president and Peter Thiessen, who had entered the party in 1925, as head of organic chemistry, even though his specialty was rather in physical chemistry, illustrates that political reliability clearly outweighed professional competency.

69 Cf. in particular Petzina [1968], Borkin [1978], Zilbert [1981], chap. 3, and Hayes [1987], as well as Ludwig [1974], pp. 216–233, and Dannenfeldt in Brämer (Ed.) [1983]; see also Janssen [1968], chap. 1, and Guérout [1992], pp. 102–107.

70 Cf. Zierold [1968], Nipperdey & Schmugge [1970], and Ludwig [1974], p. 218.

grated into Rust's *REM* and came under the control of Rudolf Mentzel.[71]

The powerful head of the German *Luftwaffe*, Hermann Göring, managed to maintain complete control over aeronautical research. He built up his own extended network of research organizations, including the *Lilienthal-Gesellschaft* and the German Aeronautical Research Academy (*Deutsche Akademie der Luftfahrtforschung*), both founded in 1936, the latter later renamed Air Warfare Academy (*Luftkriegsakademie*); the German Glider Research Institution (*Deutsche Forschungsanstalt für Segelflug*) in Darmstadt (founded 1934); the Air-to-Ground Communications Research Institution (*Flugfunkforschungsanstalt*) in Oberpfaffenhofen close to Munich (1937); the Munich Aeronautical Research Institution (*Luftfahrtforschungsanstalt München*) (1939); the Research Institution Graf Zeppelin in Stuttgart (1941); and other aeronautical and aviation research institutions in Adlershof (Berlin), Göttingen, Braunschweig, etc.[72] Even the Reich Postal Ministry (*RPM*) under Wilhelm Ohnesorge had its own research center in Miersdorf near Berlin where television and high-frequency technologies were developed along with research in metrology and acoustics. A cyclotron was also built there between 1940 and 1945. The *RPM* also employed independent researchers like Manfred von Ardenne to work on isotope separation, for example. These institutions were all separate from university research departments or privately funded research centers.[73] Some of the most infamous research conducted during the Third Reich, such as low-temperature, low-pressure and typhus experiments on human beings was performed as so-called "applied military medicine research" (*Wehrmedizinische Zweckforschung*) by the *SS Ahnenerbe* organization.[74]

On Dec. 3, 1941 Hitler decreed that all production under military contracts be rationalized, and in the following January he also issued the 'Armament 1942' order. In February Albert Speer was named Minister of Ammunition; and in May of the same year, central planning (*Zentrale Planung*) for raw materials allocation was placed in Speer's hands.[75] Hitler's decree of June 9, 1942 (doc. 98) attempted to rectify the Reich Research Council's failure to centralize and coordi-

[71]On the role played by the German élite in the preparation for World War II see, e.g., Eichholtz & Schumann (Eds.) [1969], or Broszat & Schwabe (Eds.) [1989] and references there.

[72]On aviation research in Germany 1933–45, see particularly Trischler [1992], his paper in Renneberg & Walker (Eds.) [1994], and Tollmien in: Becker, Dahms & Wegeler (Eds.) [1987], Tollmien and Groehler in: Tschirner & Göbel (Eds.) [1990]; cf. also Simon [1948].

[73]According to Laitko et al. (Eds.) [1987], p. 566, airplane research and development expenses and production costs rose from 400 million reichsmarks in 1935 to 12 billion in 1944. On the expansion of aeronautics under the *Luftwaffe*'s protective wing see, e.g., Simon [1948], Ludwig [1974], and Trischler [1992]a, part II; on the *RPM*'s research institution, see, e.g., Lundgren et al. [1986], pp. 122ff., Leclerc [1988], and von Ardenne [1972].

[74]See in particular Kater [1974]; cf. also Proctor [1988]; Deichmann [1992], chap. 8, and Lifton [1986].

[75]Cf. Braun [1992], pp. 5ff.: Between January and October 1942 productivity in some key fields such as aero-engine production did in fact rise after the introduction of some centralization measures combined with the rationalization of production with the introduction of assembly lines.

nate research planning. The Council was transferred from the *REM*'s control and placed under Göring's command. To ensure the independence of his own aerodynamics research network from this Council, Göring created another agency, the *Reichsstelle Forschungsführung des Reichsministers der Luftfahrt und Oberbefehlshabers der Luftwaffe* (*'FoFü'*), and assigned the Göttingen aerodynamics specialist Ludwig Prandtl and three other experts to direct it.[76] The German Navy also established its own research planning organization. The *Vermittlungsstelle der Kriegsmarine für Entwicklungs und Forschungsaufgaben* was founded in 1942 with the head of the machine tools department at the Hanover Polytechnic Werner Osenberg as its acting director.[77] Thus the Nazi regime's basic polycratic structure was reproduced on a smaller scale: The three physical research policy organizations competed directly for funds, personnel, and increasingly limited supplies and raw materials.

Although Nazi science administrators did bring about changes in the local research profile at universities and polytechnics, such as, e.g., the installation of defense research (*Wehrforschung*) at the Polytechnic in Charlottenburg (Berlin), they did not succeed in creating the envisioned 'Aryan science' on a large scale. Between 1936 and 1940 some of the most outspoken ideologues among the physicists such as Ludwig Glaser and Wilhelm Müller complained bitterly about the continued influence of representatives of the old scientific élite, including the theoretical physicists Arnold Sommerfeld, Max Planck, and Max von Laue.[78] The anti-Semitic and antitheoretical 'Aryan Physics' movement,[79] led by the experimentalists Philipp Lenard and Johannes Stark, may have won some battles over appointments of professors in theoretical physics,[80] but it certainly failed to win over the majority in the physics community. Furthermore, especially after 1940 when war demands reoriented the entire field of physics research away from ideological disputes about the political correctness of relativity theory and quantum mechanics toward military applications (cf. doc. 82), this reactionary movement became politically obsolete. How far the pragmatists had come in their fight against the fanatic ideologues in 1942 is perhaps best illustrated by the protocol of the secret meeting between Göring, Speer and a number of other Nazi officials and military leaders. On this occasion Göring did not mince his words in expressing his annoyance about the inflated influence of the Lecturers League leader Walther Schultze in university affairs (cf., e.g., doc. 99).

At various academic institutions there were indications of obstinacy, dissent, or at least of non-conformism. Despite the political pressures, there was a remark-

[76]Cf., e.g., Ludwig [1974], p. 327, and Trischler (Ed.) [1992], 173ff., 185ff.

[77]Cf. Ludwig [1974], pp. 237f.

[78]Cf. docs. 77 and 83, as well as Thüring [1937/38], Müller [1936], [1939], (Ed.) [1941].

[79]Cf., e.g., docs. 3, 18, 40 and sec. 5.3 below.

[80]On the Sommerfeld chair at the University of Munich which ended up in the hands of the scientifically insignificant Nazi ideologue Wilhelm Müller instead of going to Werner Heisenberg, the faculty's unanimously favored candidate, see footnote 267 below and references there.

ably persistent attempt to uphold university autonomy in appointment decisions and other matters traditionally linked to the university, even the autonomy of the individual faculties. Furthermore, university staff members also expressed their solidarity with colleagues subjected to Nazi intimidation and propaganda measures. Friedrich Hund's immediate courageous response to a nasty politically motivated smear campaign against Heisenberg in the the *SS* journal *Das Schwarze Korps* (cf. docs. 55–57), and the Munich science faculty's insistence on their candidate to the Sommerfeld chair are two such examples.[81] Many cases are also recorded in which administrative delay tactics were employed successfully, and projected measures diluted or sometimes even practically eliminated.[82] This reticence was because traditionally German universities were virtually a state within a state: They had their own laws, their own juries, their own decision procedures and their own presidents. Any attempt to encroach upon the autonomy of the academic world was thus automatically met with 'resistance'—though hardly comparable with the resolve of the French *Résistance*, for instance, or the plotters of the assassination attempt on July 20, 1944. Recent historiography, in particular Martin Broszat's analyses of the various forms of resistance found among the population in Bavaria, has attempted to differentiate between various degrees of opposition, or 'levels of non-conformity' ranging from the most potent but extremely rare form of fundamental opposition which truly deserves the label 'resistance', to the weaker variants of 'recalcitrance', 'renitence' and 'civil disobedience' down to the weakest forms of 'partial resistance' and 'ideological dissidence'.[83] At universities there certainly never was any coordinated opposition to Nazi attempts to gain control of their spheres of influence, nor was there any mentionable public protest. The rare exceptions of the Munich and Hamburg student 'White Rose' groups, whose members were rounded up for distributing hectograph handouts critical of Hitler's policy, rather confirm the general rule.[84] 'Resistance' at universities and state institutions remained at the local level and was restricted to specific points of debate. These single-issue opponents always

[81]For other examples see, e.g., Litten [1994]b on Perron, Swinne [1992] on a joint effort in support of Richard Gans, etc.

[82]See, for instance, Wolgast [1986], pp. 147f., 156f., or Golczewski [1988], pp. 128ff., 248ff. and 317ff., for examples of such evasive and stalling tactics behind the scenes at the Heidelberg and Cologne Universities. Cf. also Hirsch [1975] for a sociological analysis of this phenomenon.

[83]See in particular Martin Broszat: 'Resistenz und Widerstand', in: Broszat, Fröhlich & Großmann (Eds.), *Bayern in der NS-Zeit*, Munich, 1981, Vol. IV, pp. 691–709; cf. also sec. II,3 in Bracher et al. (Eds.) [1986], esp. pp. 621ff., Seier [1988], pp. 250f., and Trischler in Renneberg & Walker (Eds.) [1994], pp. 72f. and compare this with the older literature such as Perron [1946], Wallach [1946], Haberditzl [1963] or Scherzer [1965] based on a misleading monolithic concept of 'resistance'.

[84]On this organization see, e.g., Inge Scholl, *White Rose. Students against Tyranny: The Resistance of the White Rose, Munich 1942–1943*, translated by A. R. Schultz, Middleton, Conn.: Wesleyan Univ. Press, 1978 (reprinted 1983); Annette E. Dumbach & Jud Newborn, *Shattering the German Night. The Story of the White Rose*, Boston: Little, Brown & Co., 1986.

made it very clear, however, that they were otherwise basically in agreement with the overall policy so as not to lose their voice in the decision-making process. The declaration by the Association of German Universities (*Verband Deutscher Hochschulen*) of April 22, 1933 is a case in point: Only after stating their basic sympathy with the National Socialist revolution as the "rebirth of the German people and the rise of the new German Reich" and "in fulfillment and confirmation of their most ardent wishes and hopes", did the deans of altogether ten universites politely but firmly plea for a continuation of the same basic power structures at universities under the new regime.[85]

Nonetheless, the '*Führer* principle' was quickly installed by decree.[86] The Reich Education Ministry (*REM*) was established in May 1934 with the aim of centralizing control of the education sector, which had previously been under the authority of the separate states (*Länder*).[87] The rector (or president) of each university was now nominated by the *REM* under Minister Rust and no longer by the university academic council (*Senat*). The minister assumed the powers previously assigned to advisory faculty committees or the academic council. Nevertheless, many of the structures and procedures within universities remained basically intact. Thus, for instance, the faculty continued to compile the candidate lists for vacant chairs. However, these names were subjected to the additional scrutiny of among others the head of the local University Lecturers League and the party spokesman (*Vertrauensmann*), nominated by the Party's University Commission. Only then could they be submitted to the Ministry.[88] Not infrequently the faculty's evaluations of professorship candidates based on scholarly merit were at odds with the political evaluations issued by the Nazi officials. Additional delays were quite often caused by conflicting political evaluations among the party authorities involved (the Nazi Lecturers and Student Leagues, the *NSDAP* the *REM* and occasionally even the *Gestapo*). Years often elapsed before a single candidate could pass through all the political filters and receive a call from the Reich Chancellery. Only then could negotiations start with the local state ministry, later

[85]See the *Erklärung des Vorstandes des Verbandes Deutscher Hochschulen*, April 22, 1933: "We defend our ancient, venerable forms of organization: The autonomous administration by rector, academic council and the faculties, [...] independent replacement of the teaching staff"; copy among the von Mises Papers, Harvard University.

[86]For a nationwide law see the guidelines issued by the *REM*, 'Richtlinien zur Vereinheitlichung der Hochschulverwaltung', on the 1st of April 1935. Several states had already passed similar laws. Prussia's law, for instance, was issued on Oct. 28, 1933; cf., e.g., Huber et al. (Eds.) [1942], Vol. 1, pp. 33ff., Lundgren (Ed.) [1985], pp. 10ff., Wolgast [1986], p. 150, and Wendehorst [1993], p. 189.

[87]On the *REM*'s policy with which it became possible for the first time in Germany's history to direct a uniform education policy throughout the Reich, see in particular Seier [1964].

[88]Cf., e.g., Kelly [1973], p. 331, for a listing of the many steps necessary in carrying out an appointment, and van den Bussche et al. in: Krause et al. (Eds.) [1991], Vol. III, pp. 1274–1275, for a diagram of the quite complex appointment process both before and during World War II when the Party's University Commission under Rudolf Heß was finally eliminated.

directly with the *REM*.[89] In the interim many chairs remained vacant, only aggravating the problems arising from the brain drain triggered by the ruling party's intolerant and racist policy. A lack of courage by individuals and the isolation of the few dissident thinkers prevented the formation of any coordinated opposition at universities, which were governed instead by nervous 'self-realignment' (*Selbstgleichschaltung*) policies, if not by outright voluntary mobilization to advance the aims of the new regime.[90]

Another government interference in higher education involved a nationwide regulation on the qualification of university lecturers (*Reichshabilitationsordnung*) decreed by Minister Rust on Dec. 13, 1934.[91] Henceforth the 'habilitation' procedure involved not only a formal evaluation of the scholarly qualifications and teaching skills of candidates for professorships, but also explicitly incorporated political considerations: "Professors in the National Socialist State must satisfy particularly strict requirements of professional suitability, personality and character to qualify as educators, lecturers and researchers. Utmost care must be taken in the selection and shaping of their successors in the academic teaching profession."[92] Under the new regulation, acceptance of a habilitation thesis by the university commission was only the first step in the evaluation procedure. A separate "thorough and rigorous assessment of didactic abilities and especially the suitability of character and person for teaching at a university in the National Socialist State" was also called for.[93] In order to obtain a teaching certificate, the candidate also had to participate in a political 'community camp' or retreat (*Gemeinschaftslager*) and in a 'University Lecturers Academy' over a period of four to six weeks, which were usually located in rural areas.[94] If a candidate refused to take part in one of these *Dozentenlager*, his chances for a university career were nil. Sommerfeld's assistant Heinrich Welker is an example. Because

[89]For detailed descriptions of two such appointment incidents see, e.g., Litten [1994]a, and Renneberg & Hentschel [1995].

[90]For a provocative portrayal of this timerous mentality see in particular Mehrtens's papers, for instance, in Renneberg & Walker (Eds.) [1994] (in English translation), in Füllgraf & Falter (Eds.) [1990], and in Meinel & Voswinckel (Eds.) [1994] where he describes the "basic compromising" and ensuing "collaborative conditions". Rainer Brämer even speaks of "secret complicity" with regard to the attempts to retain academic autonomy. All too often it had to be recompensed with *de facto* collaboration with the system: Brämer (Ed.) [1983], pp. 7, 28. According to Mitchell Ash, this period is characterized by "increasing cooperation of scientists with the state", the "technologization of basic research", and the regrouping of resource constellations: See Ash [1995], pp. 6ff.

[91]See *Deutsche Wissenschaft* **1** [1935], Amtlicher Teil, pp. 12–14; cf. ibid. **5** [1939], pp. 126–135 (Feb. 17, 1939) for a modified version of this decree.

[92]Ibid., p. 12.

[93]Ibid., p. 13. In the later version the submission of proof of 'Aryan' origin of both the candidate and his spouse was also expressly required (*Deutsche Wissenschaft* [1939], p. 126).

[94]Cf., e.g., 'Nichtwissenschaftliche Leistungen zur Erlangung der Dozentur', *Deutsche Wissenschaft* **3** [1937], pp. 266–267, as well as Juilfs [1939] for a contemporaneous report on the atmosphere and the indoctrination exercised at these camps.

the habilitation degree was separate from the permission to teach (*venia legendi*), Welker was not prevented from submitting his habilitation thesis for acceptance by the faculty, however, so that he was able to continue his research and even to prolong his position as teaching assistant.[95] Another example is Fritz Straßmann, who despite his outstanding work in radiochemistry together with Otto Hahn at the *KWI* of Chemistry could not habilitate because of his refusal to join the Nazi Lecturers League.[96] However, statistically speaking, the new rulers succeeded neither in changing the predominance of newly habilitated assistants from a particular social background (75% were from academic, artistic or middleclass families of property), nor in lowering the average age of habilitation candidates—it even rose slightly.[97] They did not even manage to establish the desired new image of "the leader-educator, the youthful yet hardened soldierlike ideological lectern hero to replace the bookish armchair scholar".[98]

The University Lecturers League (*NSDDB*), the self-assigned "shock troops of the movement" at universities founded by hardcore National Socialist professors in 1934, had a crucial voice in the political evaluation of the next generation of university teachers. This was in keeping with Nazi university policy. The Party realized that it was not realistic to decapitate universities completely by replacing entire faculties, which were generally conservative but not overly sympathetic to the National Socialist cause. Instead, it opted for a gradual rebuilding from the bottom upward. The next generation of professors had to pass strict regulations, and any lecturers who appeared to be too openly opposed to the regime or who were considered 'racially inferior' were dispensed with.[99]

The effectiveness of this strategy is disputable, however. In her thesis on 'National Socialism and German University Teachers', Reece Conn Kelly came to the conclusion that "the party's *Dozentenbund* failed to gain full control over the selection of the future generation of university teachers, [...] failed to attain an effective control over 'personnel politics' and over the government of the universities", and that "the *Dozentenbund* was frustrated by intra-party conflicts in its efforts to overcome its ineffectiveness in controlling and transforming German

[95]See Eckert [1993], p. 216. Max Delbrück's (1906–1981) application for habilitation was also rejected twice because of his unwillingness to join the Nazi Lecturers League, and when he finally was habilitated in August 1936, he was refused permission to teach: See Fischer [1985], p. 67. According to Seier [1984], p. 156, twice as many habilitations were granted in Thuringia between 1934 and 1938 than *venia legendi*.

[96]Cf., e.g., Krafft [1981], secs. 1.2.1–4, especially 1.2.2, and Hahn [1962]b, p. 127.

[97]See the statistical average of the habilitations in the period 1920–33 against 1933–44 in Ferber [1956], pp. 149f., and p. 177; cf. also Seier [1988], pp. 255f.

[98]On these profiles which underscore the naivité and primitivity of the Nazi view of science see, e.g., Seier [1988], p. 261, and references there.

[99]There was a general mistrust of university professors among Nazi officials. Minister Rust hinted at it in his addresses to academic audiences, suggesting that the German professoriate had betrayed the Nazi movement before 1933. On the "never quite quenchable suspicion of the collective unreliability of professors", cf., e.g., Seier [1964] and Wendehorst [1993], pp. 180f.

learning and the universities".[100] The task of the local leader of the University Lecturers League was a difficult one. A fragile balance had to be reached between loyalty to the internal standards of the discipline on the one hand, and to the political standards of the new rulers on the other. Consequently the newly founded *NSDDB* often had problems in finding willing qualified persons for that position.

The corresponding organization for schoolteachers, the Nazi Teachers League (*NSLB*) initially had similar problems in spreading its ideology among teachers, especially at high schools, because since its foundation in 1927, it had attracted ultraright-wing elementary schoolteachers,[101] and this 'lower-class' reputation, together with the well-known anti-intellectualism of many Nazi leaders, led to a fairly strong reluctance to join, if not to outright opposition in more conservative high-school teachers unions such as, e.g., the German Philologists Union (*Deutscher Philologenverband*). Nevertheless, a sharp rise in membership after March 1933 as well as strong support from leading party officials, such as Education Minister Bernhard Rust (himself a former elementary schoolteacher) led the way toward the coordination of the teachers associations within the German Education Association (*Deutsche Erziehungsgemeinschaft*) in mid-1933. With the issuance of the law on civil servant associations of May 27, 1937, the last independent teachers associations were dissolved, so that the *NSLB* was able to claim a 97% membership rate of all teachers.[102] Conceptually National Socialist pedagogy placed primacy on exercising the body as opposed to the mind. It laid great emphasis on character development, encouraging conscienciousness, thoroughness, and a sense of order and cleanliness, etc.,[103] and awakened "pride in the results of German science", with frequent references to mythology, attempting to embed the teaching goals in the pupils' daily routine. Other important trends include the fostering of biology, especially genetics, 'racial hygiene' and the racist pseudo-anthropology favored by the National Socialists,[104] as well as aviation,[105] with the obvious intention of generating increased numbers of aspiring pilots. In the sciences physics was also stressed, as is shown, for instance, in Brämer's and Kremer's detailed investigation where the total number of hours per week devoted to science instruction increased steadily between 1890 and 1939 at

[100]See Kelly [1973], pp. ii–iii, as well as her contribution in Heinemann (Ed.) [1980], Vol. II.

[101]According to Brämer & Kremer (Eds.) [1980], the *NSLB* listed less than 6,000 members around 1932.

[102]See particularly Eilers [1963] and Brämer & Kremer (Eds.) [1980], pp. 18–29.

[103]Cf., e.g., H.J. Fischer [1937/38], Requard [1937/38], [1938/39], [1940], and Evola [1940] on the link between practicing physics and the so-called 'Erbcharakter'.

[104]A decree issued by the *REM* in early 1935 required the introduction of genetics and race studies in the class curriculum and its inclusion as an obligatory examination subject for school leavers; see particularly Bäumer(-Schleinkofer) [1990] for primary material, as well as Eilers [1963], or Weingart et al. [1988] for analysis.

[105]On Nov. 17, 1934, the *REM* issued a edict incorporating a new course on aeronautics and aviation into the curriculum to "foster the aviation spirit at schools"; cf. also Brämer & Kremer (Eds.) [1980], pp. 55, 62f., 169, 184.

most types of schools in Germany.[106] One effective strategy to convince National Socialist leaders of the usefulness of physics instruction was to emphasize its role in improving the nation's military preparedness. This new emphasis on 'defense physics' (*Wehrphysik*), featuring such themes as motor-vehicle engineering, firearms and ballistics, airplanes and air-raid protection, led to a heavy militarization of physics education and ultimately conditioned the younger generations to the coming war.[107] In 1944 and early 1945 the extreme was reached when mobile training units in trucks or even horse-drawn carts moved between frontlines and anti-aircraft stations.to teach physics to the 16 yearold Peoples Army conscripts.[108] Thus by the end of World War II the 'total unity' of didactic subject matter and daily routine clamored after since 1933 had perversely been achieved.

3.3 Faculty Membership in Nazi Organizations

One indicator of the degree to which National Socialist science policy succeeded in changing the intellectual climate at universities are statistics on membership in Nazi organizations. These membership rates must be evaluated carefully though, particularly because only a few investigations by social historians exist so far on membership rates of specific groups such as biologists and psychologists; physicists and chemists are still unexamined. Also, the documentation on which these investigations rely, namely the membership card catalogs of many Nazi organizations preserved at the Berlin Document Center (BDC), which was under Allied control until the German reunification in 1989, are not complete.[109] Similar statistics on the student body are discussed in sec. 3.4 below.

We start with the National Socialist German Workers Party (*NSDAP*), which had existed since 1919 as an ultraright-wing splinter group under the name German Workers Party (*Deutsche Arbeiterpartei*). Following the failed Hitler *putsch* in Munich it was outlawed from 1923 to 1925. It only gained real political significance in the last years of the Weimar Republic when the global economic crisis in 1929 led to growing dissatisfaction by many social groups with the other Weimar parties that were incapable of forming stable coalitions. Though the *NSDAP* advanced to becoming the dominant party in the two elections of 1932, it did not obtain an absolute majority in the last fair elections of November 1932; there was even a slight decline in the number of supporting votes. Nevertheless, on Jan. 30, 1933 Hindenburg nominated Hitler as chancellor in a well-staged impressive

[106]See Brämer & Kremer (Eds.) [1980], p. 52. The apparent decrease sometimes claimed is due to not taking into consideration a reduction in the total amount of hours per week after 1938, caused by the elimination of the last year at high schools (*Oberprima*) ibid., p. 53.

[107]See esp. doc. 59 for a more detailed listing of the contents of one school textbook; cf. also Banse [1933], Günther [1933], [1936], and Brämer & Kremer (Eds.) [1980], pp. 164ff.

[108]See, e.g., E. Dehn and Karl Hahn (1944), both reprinted in Brämer & Kremer (Eds.) [1980], pp. 240–244 on the physical education of Air Force and Navy assistants.

[109]The *NSDAP* membership catalog, for instance, only accounts for about 80% of party membership, according to Ash & Geuter [1985], p. 265.

state ceremony (cf. Lise Meitner's account, doc. 5). After the parliament building (*Reichstag*) fire and Hitler's 'seizure of power' in March 1933, the other right-wing parties and the few remaining left opposition parties (including, in particular, the Communist and Socialist Parties) were driven underground. In June 1933 the Socialist Party (*SPD*) was declared illegal and the dissolution of other parties soon followed. By July 14, 1933, the *NSDAP* had established a monopoly.[110]

In considering membership in the *NSDAP*, various statistical factors must be taken into account that are not always immediately obvious. For one, admittance to the party was suddenly barred in May 1933 in response to a surge in membership applications in the wake of Hitler's 'seizure of power' (*Machtergreifung*) in March 1933. The 'old guard' who held low membership card numbers suspected these so-called 'March casualties' (*Märzgefallene*) of opportunistic motives, in many cases justifiably.[111] It was only in the middle of 1937 that new admissions became possible again, and then only with the personal support of the local party leader. These restrictions were finally lifted in 1939. Furthermore, it should also be noted that prior to 1933, a considerable number of Nazi sympathizers decided *not* to join the party for tactical reasons. This was not just based on the disadvantages attached to being associated with the *NSDAP* during the Weimar Republic; some individuals knew that they were thus able to work more effectively within the Weimar system for the Nazi cause.[112] However, we also find cases of scientists who openly embraced and supported the National Socialism long before 1933;[113] however, statistically speaking, this was a small minority of 1.2% of university teachers.[114] Things are complicated further by the fact that it is also not justified simply to assume that every party member was a hardcore Nazi. The motives of those who joined are as varied as the degrees of participation in the regime. The decision to become a member of the *NSDAP* cannot always be seen as an effort to ingratiate oneself with the new powers-that-be but must also be seen in light of the social pressure to conform, reinforced by intense propaganda campaigns; and this pressure was typically stronger in smaller towns

[110] For a description of the political transition between 1932 and 1934 see, e.g., Bracher [1969], chaps. 4–5, Broszat [1969], chaps. 3–6, or Günther Plum in: Broszat & Frei (Eds.) [1989].

[111] This term originally referred to the fallen of the March Revolution of 1848 and was later applied sarcasticly to such instant Nazi converts.

[112] See Kater [1981], p. 62; cf. Hentschel [1993], p. 5, on Bavink for an example. According to Zneimer [1978], p. 148, the Nazi philosopher Ernst Krieck was penalized with a four-month suspension from his teaching duties by the Prussian Education Ministry for his outspoken support of the *NSDAP*.

[113] See, e.g., Lenard's and Stark's public declaration of 1924, doc. 3; Stark in particular, an academic outsider after leaving Würzburg University in 1922, was no longer impeded by professional considerations from joining the *NSDAP* in 1930; and he felt free to act henceforth as a Nazi activist in his southern Bavarian home territory at *NSDAP* election raids: cf., e.g., Walker [1995], pp. 15f.

[114] This figure from Seier [1984], p. 145, is even more telling when contrasted against the considerable number of Marburg professors (60.8%) who had openly supported Hindenburg (then Hitler's opponent) in the parliamentary elections of 1932 (ibid.).

than in the big cities, and varied dramatically among different social groups even within academia.[115] Total membership in the *NSDAP* rose from 850,000 in January 1933 to 2.5 million in May. By 1939 the number had reached 5.3 million and by 1945 it came to 8.5 million.[116] Moreover, by no means all applications for party membership were accepted for reasons of race (Jewish ancestry reaching as far back as four generations) or physical disability, as well as due to a lack of contacts (an applicant needed character witnesses and the intercession of two party members); and such applicants are not reflected in the statistics. Finally it must also be mentioned that sometimes persons were automatically filed as party members as a result of membership in another organization that was engulfed by the *NSDAP* or by one of its many suborganizations, such as the Nazi Motorist Corps (*NS Kraftfahrerkorps*) or Pilot Corps (*NS Fliegerkorps*), although all too often this was merely a convenient but inapplicable excuse used during postwar interrogations about party involvement (cf. sec. 5.5 on denazification).

There are only a few studies on party membership among the German professoriate by disciplines as a whole during the regime, such as in psychology, medicine and biology. Some interdisciplinary overviews of membership rates also exist for individual universities, such as Berlin, Hamburg and Frankfurt; and statistics are also available on German nuclear physicists and physical chemists involved in the German uranium project.[117] At the University of Hamburg, for instance, membership rates in the four major faculties were as follows: in mathematics and science 66.3%, that is, 63 out of a total of 95 non-Jewish faculty members were members of the *NSDAP*, compared to 72.7% in medicine, 54.7% in philosophy, and 50% in law, with an overall average of 64% of all 280 non-Jewish Hamburg faculty members in all subjects. The figures for other universities are of the same order of magnitude (Frankfurt 38%, Berlin 44.5%, Giessen 56%) and also confirm the basic pattern of higher membership rates in medicine, and lower ones in law (as compared to the sciences).[118] Deichmann's perusal of the BDC files on 440 biologists yielded a total of 234 *NSDAP* memberships, amounting to 57.6% of the total of 404 biologists remaining in Germany. The comparative figure for 117 psychologists by Ash and Geuter is 54.7%, and for medical doctors by Kater, 44.8%.[119] From a total of 71 members of the German *Uranverein*, which was composed of a number of branches (cf. sec. 5.4) and is thus a large subset of

[115]On the much stronger social pressures in small university towns see in particular Kater [1981], pp. 66, 68, 74.

[116]Figures according to Broszat & Frei (Eds.) [1989], p. 195; cf. also the Nazi Party publication: Reichsorganisationsleiter der NSDAP (Ed.): *Partei-Statistik*, Munich, 1935, 3 Vols., and Kater [1983] as well as Broszat [1969], p. 51, for the social structure of *NSDAP* membership.

[117]See Ash & Geuter [1985], Kater [1989], and Deichmann [1992], chap. 10, as well as the analysis of a sample group of engineers in Ludwig [1974]; and Giles in: Krause et al. (Eds.) [1991], Walker [1989]a, pp. 193f.

[118]Giles in: Krause et al. (Eds.) [1991], p. 115, on Hamburg; Zneimer [1978] on Frankfurt; Seier [1984], p. 157, on the University of Berlin; Böhles et al. [1982] on Giessen.

[119]See Deichmann [1992], p. 225; Ash & Geuter [1985], p. 265; Kater [1989], p. 56.

German nuclear physicists, Mark Walker has found that 56% were *NSDAP* members. 8% of these 40 physicists and physical chemists were from the 'old guard' who had joined the ranks of the *NSDAP* prior to 1933, 68% joined the party between 1933 and 1938, and 13% during the war years.[120] 18 (thus nearly one half) of the Party members joined upon the lifting of the membership restrictions in 1937. According to Walker, "this flood of scientists into the National Socialist German Workers Party in the years before the war cannot be explained merely by opportunism or by fear of political retribution alone. Many of these physicists and chemists sincerely wanted to take part in what contemporaries called the 'National Socialist Revolution', a return to traditional values after the brief aberrant Weimar Republic."[121]

The figures on technicians and engineers should be compared against Ludwig's results on Nazi party membership among a sample group of members in the National Socialist Technology League (*Bund Deutscher Technik: NSBDT*). This organization had the imposing total membership of 81,000 engineers, which is partly due to the circumstance that an apolitical trade organization was merged with the Nazi organization after 1933 and its members were admitted automatically. Prior to 1933 party membership had totalled approximately 7,000 and another 14,000 engineers joined the *NSDAP* between 1933 and 1937. Thus membership rates increased among engineers by 200% within 4 years, compared to the population average of 230% within 2 years. Engineers who had not already done so before 1933 were less willing to become Party members than the national average. The economic crisis of 1929 and the subsequent years of rising unemployment provided the main motive for those who did so.[122]

These relatively high membership rates for scientists in the *NSDAP* after 1933 contrasts sharply with their fairly minimal political involvement during the Weimar Republic. Only 26.6% of altogether 271 Hamburg faculty members and only 16.5% of the faculties in mathematics and science there had had any party affiliation whatsoever prior to 1933, and if so mainly in the monarchist *Deutsch-Nationale Volkspartei* (*DNVP*), the conservative *Deutsche Volkspartei* (*DVP*), or the liberal *Deutsche Demokratische Partei* (*DDP*), later renamed *Deutsche Staatspartei.*[123] In predominantly Catholic regions like Bavaria, the Center Party also had a small share of academic members.[124] However, prior to 1933 the great majority of university teachers abstained from party membership, considering it inappropriate. They shied away from making open political statements, speeches

[120]See Walker [1989]a, pp. 193 and 262, for a full list; when the remaining 9% joined the *NSDAP* is not clear.

[121]Walker [1989]a, p. 194.

[122]Figures according to Ludwig [1974], pp. 107ff.

[123]Cf., e.g., Hering in: Krause et al. (Eds.) [1991], Vol. 1, pp. 89ff., 100–104, for statistics.

[124]Cf., e.g., Wengenroth [1993], p. 223, for pre-1933 membership from the Munich Polytechnic faculty, of which 61 professors were then politically active: 22 in the local *Bayerische Volkspartei*, 11 in the *DNVP*, 9 in the *DVP*, 4 in the *DDP* and 6 in the *NSDAP*.

or public declarations of sympathy to any of the many competing parties in the Weimar Republic, quite in accord with the scholar's apolitical self-image. How widespread this attitude was even among those who later acted in the most courageous and impeccable manner is revealed, for example, in the exchange of letters between von Laue and Einstein in May 1933, after the latter had withdrawn his membership in the Prussian Academy to forestall being expelled for purportedly making anti-German pronouncements in the United States (cf. doc. 6). Max von Laue warned: "The political struggle requires different methods and different characters to scientific research. As a rule the scholar comes to grief." Einstein's response was, however: "I do not share your view that men in the sciences should keep silent about political matters, i.e., about matters of humanity in the widest sense. From the situation in Germany you can see just where such self-restraint leads. It means surrendering leadership to the blind and irresponsible. Does not a lack of a sense of responsibility lie behind this? Where would we be now if people like Giordano Bruno, Spinoza, Voltaire, and Humboldt had thought and acted in this way?"[125]

However right Einstein may have been, the majority of German scientists and scholars did not agree with him. By remaining politically noncommittal, university teachers left the way open to obtaining influential government advisory positions. Nevertheless, the numerous 'March casualties' had to overcome this traditional abhorrence of political involvement.[126] It is instructive to look at the age aspect of membership in the *NSDAP*: In Hamburg the vast majority of the youngest lecturers were party members, 84.4% of those between ages 20 and 29 and nearly 80% of those between ages 30 and 39. 62.4% of all university teachers between ages 40 and 49 were members, 54% between the ages of 50 and 59, and only 18.2% between ages 60 and 69. These figures show that those who wished to advance their careers within the university system, up the academic ladder of 'habilitation', lectureship and finally professorial appointment, were more inclined to join the *NSDAP* as an outward sign of political conformity.[127] The corresponding rates for *SA* and *SS* memberships were considerably lower,

[125]Max von Laue to Einstein, May 14, 1933, and Einstein to von Laue, May 26, 1933 (DMM), quoted, e.g., in Seier [1988], p. 289; see also Stern [1972]b.

[126]Cf., e.g., Wendehorst [1993], pp. 199f., on one such case, and further literature mentioned in footnote 39 above on the attitudes and self-image of the German professoriate. See in particular Jansen [1992] for an analysis of the political activities of the Heidelberg professoriate between 1914 and 1935: More than the 10–15% estimated in other studies were involved in some kind of political activity, though often not linked to political parties.

[127]See, e.g., Kater [1981], p. 55, and the references in footnotes 95f. above. This contemporaneous evaluation of the potential professional advantage from such membership should not be confused with historical retrospective analyses of the gains actually achieved from this step and the upward mobility of university teachers, which was not as great as could be assumed (see below). Cf. Zneimer [1978] on the correlation between social origin, age and party activity, or Deichmann [1992], chap. 1, on the weak correlation between party membership and funding by the *DFG*.

starting at 33.8% *SA* members among those between 20 and 29, dropping down to 5% for those between 50 and 59 years old in 1933, with the *SS* figures as low as 9% and 4% for these two age groups.[128]

Another relevant Nazi organization was the *Reichsfachschaft Hochschullehrer*, the national professional organization of university teachers founded by Erich Seidl as a branch organization within the National Socialist Teachers League (*Nationalsozialistischer Lehrerbund: NSLB*), which organized all teachers at schools and institutions in higher education. Membership was obligatory for all lower-level untenured professors (*Nichtordinarien*) at German universities and polytechnics until the close of 1935. According to Giles's study on the Hamburg faculty in particular, membership in the Hamburg *NSLB* in 1933 was about 45% of all university teachers between the ages of 20 and 29, 58% between ages 30 and 39, 63% between ages 40 and 49, 40% between ages 50 and 59, and only 30% between ages 60 and 69.[129] Walker has determined that 72% of all physicists and physical chemists involved in the extended German uranium project were members of either the *NSDAP* or the *NSLB*.[130] Compared with 56% membership in the *NSDAP*, this shows that about one third of this research collective had decided against joining the *NSDAP* and had opted for the professional association as the weaker form of Nazi affiliation. This changed, however, when a decree by Minister Rust in mid-1935 made it no longer obligatory. According to Kelly, most university teachers stopped paying their dues to the League, and it soon ran into financial and organizational trouble, leading finally to its dissolution as one of the economizing measures taken in 1943 "as a simplification measure determined by the war".[131]

While the *Reichsfachschaft Hochschullehrer* of the *NSLB* sank into political oblivion, the University Lecturers League (*Dozentenbund, NSDDB*) gained correspondingly in importance. In 1935 the *NSDDB* was reorganized as a full-fledged party agency on lines similar to the Student League (*NSDStB*) (see the next section), the Hitler Youth, or the *SS*. Its central headquarters was in Munich,

[128]Figures taken from Giles in: Krause et al. (Eds.) [1991], pp. 116f. The comparative figures for German non-emigré biologists are: A total of 22.5% *SA* members and 5.4% *SS* members (according to Deichmann [1992], p. 225); and 17.6% *SA* members among Ash & Geuter's [1985] reference group of psychologists. Kater [1989], p. 70, found circa 7% *SS* members among male physicians as defined by a list of the Reich General Medical Council (*Reichsärztekammer*) of 1935, and Brämer & Kremer (Eds.) [1980], p. 29, where an 11% membership in the *SA* among the member teachers is mentioned.

[129]Unfortunately, Giles does not give the average membership figure for full professors in the *NSLB*, but it must have been significantly lower than the above numbers because of the obligatory nature of membership for lower-ranking professors. Cf., Giles in: Krause et al. (Eds.) [1991], pp. 117f.

[130]Cf. Walker [1989]a, p. 193.

[131]See Kelly [1973], pp. 221ff., esp. p. 239, for the very low membership rates in 1936 (for instance, two members out of 557 teachers in the Baden district), and Brämer & Kremer (Eds.) [1980], pp. 25–29.

and it had a hierarchical organization with subordinate district and local groups, and a fairly influential local representative, the *Dozentenbundsführer*. Membership numbers again reveal that only about one fifth of all university teachers were official members of this Nazi organization. Giles gives a total of 21 members in the Hamburg *NSDDB* within the mathematics and physics sector, thus totaling up to 22% of all faculty members in these disciplines (membership from the field of medicine was slightly lower at 21.9%, philosophy at 18.9%, and law at 16.6%).[132]

How much did the decision to join one or several of these Nazi organizations actually further the professional careers of these individuals? Several incompatible answers have been proposed. Richard Zneimer's study on the Frankfurt faculty came to the conclusion that although a couple of accelerated appointments were made on the basis of a candidate's political correctness and at what point he had entered the party, after 1936 it "ceased to be a vehicle of upward professional mobility" for academics.[133] This conclusion is also basically confirmed in later studies linking university careers to *NSDAP* membership.[134] The implication that Germany's war preparations from 1936 based on the Four-Year Plan increased the demand for highly qualified experts and thus strengthened the hand of pragmatists and reduced the importance of lip-service to party ideology is flatly contradicted in Ulrich Adam's study on the University of Tübingen, for example. According to this study from 1938 on, academic calls were decided primarily on the basis of membership in Nazi organizations and other open demonstrations of political conformity.[135] Though we will encounter some examples of these politically motivated appointments within the field of physics—most notably Wilhelm Müller's call to Sommerfeld's chair in theoretical physics (cf. doc. 85, p. 265, as well as footnote 267 below)—I do not think that they are representative of appointment policy during the second half of the Third Reich.[136]

---

[132]Giles in: Krause et al. (Eds.) [1991], Vol. 1, p. 119. For revealing quotes on the *NSDDB*'s bad reputation among university officials see, e.g., Seier [1988], pp. 260f.

[133]After 1936 "an inverse relationship between party activism and mobility emerges": See Zneimer [1978], p. 155; and, for instance, Steinmetz [1958], pp. 634–641, for other examples of politically 'reliable' candidates installed in the faculty in the early years of the regime; cf. Mehrtens in: Mehrtens & Richter (Eds.) [1980], p. 48, and Ash & Geuter [1985], pp. 272f.

[134]For instance, Ebert in Rürup (Ed.) [1979], Vol. 1, pp. 462ff., Kater [1981], pp. 55ff. According to Ash & Geuter [1985], p. 271, 80% of all persons who received calls in psychology between 1933 and 1935 were party members, while afterwards, this rate drops to 56%. From Seier [1988], p. 157, we learn that from among the 25 associate professors at Berlin University who had supported Hitler and the *NSDAP* prior to 1932, only eight acquired a full professorship three years later. According to Kater's study on the *SS-Ahnenerbe*, its members encountered problems in submitting their habilitations, and only very few of these actually succeeded in obtaining professorships afterwards: See Kater [1974], pp. 287–290; cf. also Seier [1988], p. 285.

[135]See Adam [1977], p. 207.

[136]For case studies see, e.g., Kater [1981], pp. 55ff., Ash & Geuter [1985], p. 274; Becker, Dahms & Wegeler (Eds.) [1987], pp. 38, 66, 81, 107ff., 187, 234f., 315f., 422; Golczewski [1988], pp. 317ff., 350ff., Walker [1989]a, pp. 195ff.; Deichmann [1992], pp. 228f. 235ff.; Litten [1994], Renneberg & Hentschel [1995]. Cf. also Seier [1964], [1988], pp. 263ff.

Recent studies have also shown that the percentage of *NSDAP* membership was significantly higher among associate (*Extraordinariate*) and untenured professors than among full professors, and that the party membership percentage for German and Austrian biologists who habilitated after 1933 is also much higher (80% in the case of biologists) when compared against those who acquired their professorship qualifications before that date.[137] The pressure to conform within the new Nazi system was thus clearly strongest among young assistants and unsalaried university lecturers (*Privatdozenten*) whose futures were uncertain. However, it is not correct to say categorically that without membership in a Nazi organization an academic career would have been impossible. Examples to the contrary exist, but the above numbers suggest that it was not attempted very often (in less than 20% of all cases), since other forms of political indoctrination such as participation in 'lecturers camps' (*Dozentenlager*) were obligatory in any case (see sec. 3.2 above). Social pressure at this level was fairly high, and it was occasionally exerted not only by Nazi institutions but also by well-meaning colleagues and seniors who often downplayed this step as a formality.[138]

3.4 Student Enrollment in the Sciences

The political revolution that had led to the Weimar Republic was known as the 'revolution without students'. In contrast, in the early stages at least, students were among the most outspoken supporters of the National Socialist movement.[139] Many of the youngest World War I veterans who had returned with broken self-esteem and were generally disorientated, entered university, and throughout the entire Weimar Republic right-wing student fraternities and antidemocratic paramilitary *Freikorps* units flourished.[140] The other major force in student government was the German Student League (*Deutscher Studentenbund*, called *Akademischer Bund* until 1925). Originally it was an alliance of liberal and leftist student organizations, which split into socialist and democratic wings in 1929. In 1926 the National Socialist Student League (*NSDStB*) was founded by Wilhelm Tempel. Under Baldur von Schirach's direction from 1928 on it took up the *NSDAP*'s party line and was affiliated officially. The Student League (under the

[137]See Ash & Geuter [1985], pp. 267f., and Deichmann [1992], pp. 231f.

[138]See, for instance, the astronomer Otto Heckmann's remark on Werner Heisenberg's advice in 1937 to join the party, to the effect that it would open the way for him to get a chair that would otherwise be occupied by some Nazi crank (quoted in Renneberg & Hentschel [1995]). Cf. also Deichmann [1992], p. 232, and footnote 95 above.

[139]For general overviews see Bleuel & Klinnert [1967], Faust [1971], Steinberg [1973], Kater [1975] and Giles [1985], as well as Böhles et al. [1982], chaps. 2–5, Jarausch [1984], secs. III–IV, Norbert Giovanni in: Buselmeier et al. (Eds.) [1985], pp. 293–306, Golczewski [1988], chaps. 2–3, Wengenroth [1993], pp. 215–223, and in particular Grüttner [1995].

[140]See, for instance, Fieberg in Böhles et al. [1982], pp. 38ff., and Bleuel & Klinnert [1967] on the nationalistic conservative mentality of the vast majority of students after 1918, and ibid., pp. 261f., for membership statistics in student associations in the summer term of 1930.

name *Völkischer Bund* in some regions until 1929) won around 10% of student votes in the 1928 and 1929 elections but then steadily gained strength, so that by 1931 it had become the majority voice at almost all student assemblies.[141]

It is also noteworthy that the available comparative statistics on male and female voters reveal that the Student League was able to attract an equal percentage of female voters, while the chauvinistic student associations were less successful.[142] An important impetus to joining or at least supporting the League was, of course, the bad employment situation for academics since the great depression of the late 1920's. In 1931 the number of enrolled students was twice as high as the estimated demand for graduated scientists. Thus the 'Law Against the Overcrowding of Universities' which came into effect on April 25, 1933 (doc. 12) was more than just yet another anti-Semitically motivated piece of legislation (limiting the number of newly matriculating Jewish students to 1.5% and the total number of Jewish students to 5%).[143] It also reflected a fairly widespread concern about a growing 'academic proletariate'.[144]

Although the Student League already dominated in student parliaments and in the German Student Association (*Deutsche Studentenschaft*)[145] when the Nazis took over government in 1933, it still competed against the latter to represent the overall student body. A new law that came into force in April 1933 prescribed that members of the Student Association, i.e., those students eligible to vote in the student parliament, be of 'German origin' and of 'German tongue', but German nationality was not required, thus fostering the expansionist ambitions of the Weimar right toward a Greater Germany.[146] In fact, it has been claimed

[141]The best overview is now available in Grüttner [1995], pp. 496ff. For statistics on student government elections in Hamburg, see Giles [1985], p. 54, as well as their contributions in: Krause et al. (Eds.) [1991], Vol. 1, pp. 204f. On the Berlin Polytechnic, where the *NSDStB* gained more than 60% of the votes in 1931/32, see Laitko et al. (Eds.) [1987], pp. 514f.; and Rürup (Ed.) [1979]. On Giessen (with 55% in 1931) see Böhles et al. [1982], p. 67. On Göttingen (where the League already had an absolute majority by 1931), see Becker, Dahms & Wegeler (Eds.) [1987], pp. 22–24, and Popplow [1977], pp. 159, 170–172. On the Munich Polytechnic (with 15 of 30 seats for the *NSDStB* in November 1932), see Wengenroth [1993], pp. 215–223. On Jena (with 50% in February 1933) see Steinmetz et al. [1958], p. 649. On Stuttgart (with 88% for *völkische* and Nazi student groups in November 1932), see Voigt [1981], p. 32.

[142]See the Hamburg statistics cited above, and Grüttner [1995], p. 498.

[143]Cf., e.g., Wengenroth [1993], pp. 234ff., where it is asserted that the Munich Polytechnic was not very affected by the law because they only had few Jewish students in 1929 (35), owing to particularly strong existing anti-Semitism which had caused such students to prefer more liberal universities such as Berlin or Göttingen. In Cologne, for instance, the number of Jewish students was slashed by 60% or more within half a year, while the total number of students only decreased by 7.8% within this short period: See Golczewski [1988], pp. 95f.

[144]On the growing pauperization of students after 1929 and the problematic job market situation in the late Weimar Republic see, e.g., Jarausch [1984], pp. 141f., 183ff.

[145]Founded in July 1919 as the first national student parliament; cf., e.g., Jarausch [1984], pp. 120ff., 160f., 167f.

[146]See, e.g., 'Neues Studentenrecht' in *Berliner Tageblatt* **62**, No. 172, ed. A, Thursday, April 13, 1933, p. 1 , cols. 1–2, as well as the critical comment in the *Vossische Zeitung*, evening edition

that the activities that led to the infamous book-burning in May 1933 had their origin in an attempt by the representatives of the Student Association to prove their willingness and ability to mobilize students for the National Socialist cause independently of the Student League[147]—in itself a prime example of the polycratic power struggles inherent in the Nazi system. While Communist, Socialist, Jewish and pacifist student organizations were dissolved between April and July 1933, the Student League co-existed with the nationalistic and conservative, often anti-Semitic but not necessarily National Socialist student fraternities and *Freikorps* units until May 1935, at which time they were also either dissolved or forcibly incorporated into the Student League.[148] In 1937 even the few remaining Jewish students of German nationality still tolerated at German universities were prohibited from obtaining a doctorate, in 1938 Jewish students of German nationality could no longer matriculate at German universities, and soon they were not even allowed to enter public libraries.[149]

Government grant applicants or those interested in a career in the civil service now felt increasing pressure to join the Student League. The League's membership rates in Hamburg, for example, were reportedly about 33% in 1938, 50% in 1939, and 35% in 1942/43. By comparison, membership in the National Socialist Party of the same group was 40%, 33% and 11% in the same years.[150] As membership in the Student League became increasingly a matter of course, motivation among its members fell. Their enthusiasm was taxed by the introduction of obligatory labor service which often cost many hours in the term breaks during the harvest season,[151] in addition to numerous political training evenings (*Schulungsabende*), intense indoctrination at political 'camps',[152] and the imposition of a rigorous

of the same day, p. 1, col. 3. The corresponding laws for non-Prussian states were passed by the Reich Cabinet on April 22, and the Bavarian laws were released one week later.

[147]See Krackow [1983], Schöne [1983], Strätz [1986], Jan Hans in: Krause et al. (Eds.) [1991], Vol. 1, p. 239; Golczewski [1988], pp. 76ff. On the various student organizations and the relevant legislation, see also Steinmetz et al. [1958], pp. 650f., and Grüttner [1995], sect. II.2.

[148]Cf., e.g., Steinmetz et al. [1958], pp. 652–659, Jarausch [1984], pp. 123f., 157f., 171f.

[149]See 'Erwerb der Doktorwürde durch Juden deutscher Staatsangehörigkeit', *Deutsche Wissenschaft* **3** [1937], pp. 224–225; cf. also Götz von Olenhusen [1966] for other Draconian measures against 'non-Aryan' students.

[150]For these average figures see again Giles [1985], pp. 104, 226, 245–247; Jarausch [1984], pp. 156f., 166, 173; Michael Grüttner in: Krause et al. (Eds.) [1991], Vol. 1, p. 220. and Grüttner [1995], pp. 500f. There is a striking difference in membership between the sexes: While male students were more often members of the Nazi Party, female students tended to prefer membership in the Student League. The ratios were 60.6% male to 32.8% female Party members as opposed to 23.2% male to 77.7% female Student League members.

[151]For depictions of the daily life of students during this time see in particular Böhles et al. [1982], pp. 77–113, and Giles [1985], chaps. 3–5; see also the *Deutsche Allgemeine Zeitung*, Nov. 12, 1934, p. 1 : "Die Pflichten der Studierenden im neuen Semester". On the 'student deployment' (from 1937 on) in the form of agricultural and industrial labor see also Jarausch [1984], pp. 194f., and Grüttner [1995], pp. 227ff.

[152]For detailed accounts of the program in two such *Reichslager*, see Willi Menzel's internal report of the Darmstadt meeting in February 1936, in carbon copy at the Bundesarchiv Koblenz,

daily routine in fellowship club houses (*Kameradschaftshäuser*).[153]

This decline in appeal went hand in hand with increased marginalization of the Student League politically. Its influence was reduced more or less to denunciation of politically and racially undesirable lecturers and classmates to Party and secret police agencies. But the Student League, along with its counterpart for academic teachers, the German University Lecturers League (*NSDDB*), maintained considerable influence in faculty appointments. It is in this context that the Student League activists Bruno Thüring and Fritz Kubach appear in this anthology: Thüring, who had already brought about the dismissal of the astronomer Alexander Wilkens at the Munich Observatory, viciously attacked Heisenberg and the 'formalistic' trend in modern physics. Kubach on the other hand supplied the readers of the *Zeitschrift für die gesamte Naturwissenschaft* (coedited by him and Thüring) with articles in which the hagiographic approach was fused with an imperialistic effort to 'Aryanize' all major figures in science: Copernicus, Kepler and Lenard are portrayed as 'Great German scientists'.[154] Interesting contradictions arose whenever it was attempted to combine this heroic solitary genius image with the *Volksgemeinschaft* message that stressed the necessity for efficient teamwork.

Nearly all fields suffered from an abrupt decline in enrollments by new students and from the consequences that became apparent a couple of years later. According to the official 10-year statistics compiled by the Reich Education Ministry in 1943, the total number of enrolled students, which had peaked at around 150,000 in 1931, declined drastically to under 100,000 in 1933, below 80,000 in 1934, and less than 60,000 in 1936. Up to 1937 the trend was the same both at universities and polytechnics.[155] In 1937 matriculations at polytechnics, which had sunk below the 10,000 mark, was again on the rise, peaking around 12,300 in 1939. With the outbreak of World War II when trimesters were introduced

---

Rep. 15.19, Mappe 68, pp. 131–142, and Juilfs's account of the first German Mathematicians Camp in July 1938, published in Juilfs [1939]. See also: *Die Reichsstudentenführung: Arbeit und Organisation des deutschen Studententums*, ed. by Gustav Adolf Scheel, Berlin: Junker & Dünnhaupt, 1938, as well as the issue on the *Reichsberufswettkampf: Die berufliche Aufrüstung der deutschen Jugend* which appeared by the same publisher in 1935, edited by Günter Kaufmann. Cf. also Steinmetz [1958], pp. 659ff.

[153]See, e.g, the 1943 *NSDStB* brochure edited by the Political Education Office of the Reich Student Leadership (*Amt für politische Erziehung der Reichsstudentenführung*): *Dienstanweisung für die Kameradschaft vom 20. 4. 1943*, which provided full instructions on how to establish 'discipline and order'. See also Giles [1985], p. 213, for statistics revealing the underutilization of these fellowship club houses in 1937, and Grüttner [1995], p. 502 for membership rates.

[154]See, e.g., docs. 40, 42, and 83; Kubach [1942], Thüring [1936]a–b, [1941/43]; cf. also Lenard [1929], Stark [1936]a, Stark & Müller [1942], Weigel (Ed.) [1937], May [1940], Toepler [1941], Teichmann [1942], and Wesch [1942]. On Thüring see, e.g., Litten [1992]. Nothing has been published yet on Kubach or May; see, however, e.g., Grüttner [1995], p. 509, and Meinel & Voswinckel (Eds.) [1994], pp. 265–294, on the contemporary 'Aryan' history of science.

[155]See here fig. 1, p. li (the logarithmic scale of student numbers disguises the abrupt decline somewhat); cf. also Lorenz [1943], Vol. I, N.N. (Ed.) [1949], pp. 622f., and Quetsch [1960], Jarausch [1984], pp. 129f., 178f., 208f., for overall statistics.

to shorten the necessary period of study, the number of students plummeted again (many male students were drafted); after the first military successes in the Benelux countries and in France, student numbers rose slightly to 5,600 in the summer term of 1941, and then dropped again.[156] A certain amount of this drop is also due to very low birth rates in the German Reich after 1900, especially between circa 1911 and 1933, giving rise to a lot of speculation about the reasons for this decline during the Weimar Republic itself.[157] The enormous drop in student numbers referred to above was significantly greater than this, however, especially if the overall tendency to open universities to new social groups and the global increase in student numbers in other countries is taken into account. In this respect it is instructive to compare the student numbers against the total population. While in the USA a rate of 104 students per 10,000 inhabitants had been reached by 1938 (including junior colleges), the comparable numbers in Germany were 20 (in the 1929/30 academic year), 15 (in 1933/34), and 8 per 10,000 inhabitants (in 1937/38).[158]

In physics the number of new students fell quite dramatically to around 15% of the 1933 level in 1936/37 (in the same year as the Heisenberg-Geiger-Wien petition, doc. 49), and it only recuperated slowly to around 50% of the 1933 level in 1940/41. In contrast, chemistry received favored treatment in the Four-Year Plan as a result of the dominant influence of the *IG Farben* chemical concern. New enrollments already matched the 1932/33 level in 1937/38 and continued to make gains, peaking around 330% of the 1933 reference level in 1939/40.[159] The percentage of matriculated students in the sciences fell from 11% in the academic year 1929/30 to 8% in 1937/38, while medicine rose from 17% to 37% in the same period.[160] During the war, the number of physics students at universities declined further from about 500 in 1939 to 250 in 1942, then rose slightly to about 300 in 1943. This means that compared with 1932, the total number of students in physics sank from 2,230 to about 550 in 1939 (including polytechnics), that is, by around 75%, while mathematics sank by 95% (from 4,160 to 186) and even chemistry lost 35% (from 4,000 to roughly 2,500).[161]

This dramatic decline in student attendance was, of course, correlated with

[156]These figures from Ludwig [1974], p. 276; cf. also Trischler [1992]a, pp. 229, on the acute lack of airplane engineers, and Aly [1936] for an early warning by the head of the engineering group of the Nazi Student League about the imminent shortage in young engineers.

[157]This factor is too often ignored in evaluations of student statistics. According to the figures given in John E. Knodel: *The Decline of Fertility in Germany, 1871–1939*, Princeton Univ. Press, 1974, pp. 5ff., during 1916–20 the gross birth rate was less than half of the previous values before the turn of the century. See also the *Beiträge zum deutschen Bevölkerungsproblem*, Berlin, Statistisches Reichsamt, 1929, 1933 and 1935 as one example taken from the copious contemporary literature on the *Geburtenrückgang*.

[158]Numbers from Quetsch [1960], pp. 4–7, 13, and 42; cf. also Eckert [1993], p. 198.

[159]Figures according to Lorenz [1943], Vol. I, diagram 21b.

[160]Figures from Quetsch [1960], p. 43.

[161]Figures according to Lorenz [1943], Vol. I, App., pp. 12–15, and Fischer [1988], pp. 95f.

a diminished, or at least stagnating, output of graduate students: 99 physics diplomas were issued at polytechnics both in 1936 and 1937, while in 1938 the number fell to 75; and after an artificial peak in 1939 of 113 (due to special arrangements for students who then went to the front), it fell to 61 in 1940, with the same tendency in other fields as well.[162] Prior to 1939 there had been some discussion about the need for a fundamental reform of education in physics; with the outbreak of World War II it became even more critical.[163] It was resolved to shrink the physics curriculum to only 7 terms; independently of this, trimesters were introduced between 1939 and 1941 to shorten the length of each term, which was certainly not conducive to the quality of the training.

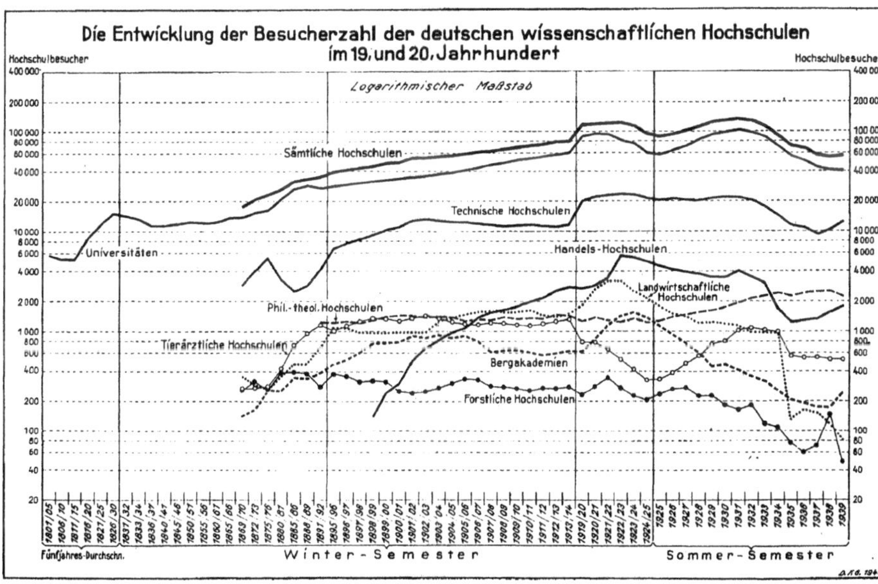

Fig. 1: The development of attendance at German universities in the 19th and 20th Centuries. Captions from the top: All, universities, polytechnics, business colleges, agricultural colleges, liberal arts colleges, veterinarian colleges, mining academies, forestry colleges, five-year average; from Lorenz [1943], Vol. I, diagram 2 (note that the logarithmic scale downplays the actual decline after 1933).

[162]According to Ludwig [1974], p. 281. Cf. also Seier [1988], p. 259, for quotes from Heydrich's Security Headquarters secret reports, based extensively on information collected by Secret Police informants placed at all levels of society. In 1938 the Security Service (*SD*) warned of a virtually catastrophic lack of young replacements in all fields of the arts and sciences.

[163]See, e.g., Feigl & Hamel [1939], Orthmann [1939], Stuart [1939]; cf. also Ramsauer [1938].

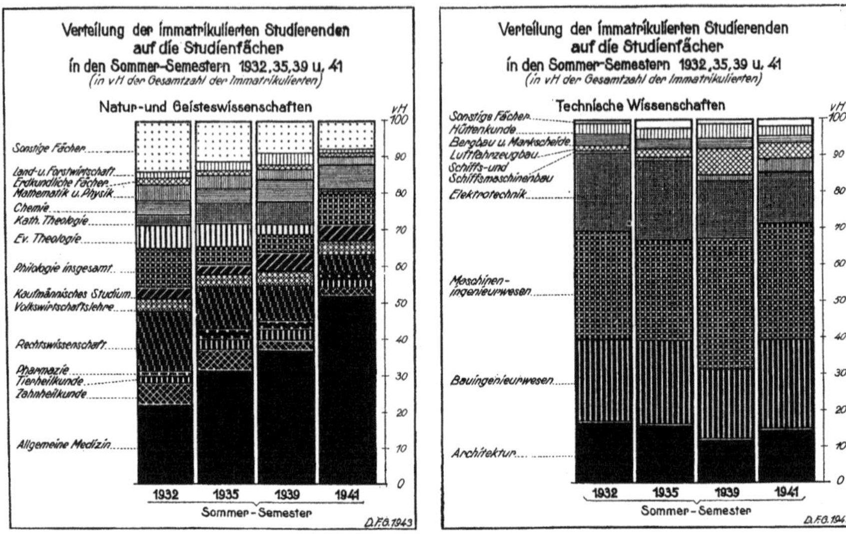

Fig. 2: Distribution of enrolled students by subject in the 1932, '35, '39 and '41 summer terms (percentages of total number of enrolled students). **Left:** Sciences and Humanities: Other subjects, agriculture and forestry, geosciences, mathematics and physics, chemistry, Catholic theology, Protestant theology, philology, business studies, economics, law, pharmacology, veterinary science, dentistry, general medicine. **Right:** Technical Sciences: Other subjects, metallurgy, mining engineering, aviation engineering, ship building and engine construction, electrical engineering, mechanical engineering, architecture; from Lorenz [1943], Vol. I, diagram 11b–c.

Ironically, although National Socialist ideology strictly favored the traditional role of women as homemakers, after 1939 female students assumed an increasing share in the shrunken student body: Starting at about 200 female students at polytechnics around 1938/39, their numbers rose to more than 700 in 1941, over 1,000 in 1942, and over 1,500 in 1943.[164] World War II thus led to an astonishing reversal in female education: The total number of female students, which had sunk to 6,342 in 1939, rose to 42,210 in 1944, so that in the summer of 1944, they constituted nearly half (49.3%) of all students. 65% of newly matriculated students were female (against 9.4% before the war); they dominated particularly in the sciences (63.5%) and humanities (83.4%).[165]

164See again Ludwig [1974], table 7, p. 276; cf. also Lorenz [1943], Vol. I, diagrams 3ff., which demonstrate that female students came to nearly 30% of all students in the winter term of 1941 while they had never been more than 15% prior to 1939.

165Figures from Jarausch [1984], pp. 177, 202ff.; cf. also Quetsch [1960], pp. 16f., 50f. These numbers are confirmed in local studies: Giles [1985], chap. 7, and Michael Grüttner in: Krause et al. (Eds.) [1991], Vol. 1, p. 213, report that the 1,009 women studying at Hamburg University in 1944 constituted 46.8% of all students in Hamburg, and according to Wendehorst [1993], p. 211, the percentage of female students rose from 11% to a spectacular 61% at the University of Erlangen between 1939 and 1944. For a detailed study on female students during the Nazi regime, see Pauwels [1984].

4 Emigration Research

The huge emigration wave of professionals from Germany during the Third Reich has only come to the attention of the general public relatively recently. Systematic studies focussing on the subject have been appearing sporadically, however, and interest has been on the increase since the 1980's.[166] In the last five years, the German national research foundation (*DFG*) has been funding a major program on the history of emigration in the sciences, and several publications have already appeared in print.[167]

4.1 General Emigration Statistics

Before going into the specifics of the exodus of scientists from Germany, let us first have a brief look at the overall emigration statistics to get an idea of scale. On acceding to power, the Nazi Party immediately started implementing its discriminatory and anti-Semitic agenda, forcing politically or 'racially undesirable' persons into exile.[168] By 1935, for instance, around 65,000 persons had left Germany as a result of racial discrimination, 5–6,000 were Social Democrats, 6–9,000 Communists, and 5,000 political dissidents from other parties or social groups; around 1,600 were scientists.[169]

Within the German Reich's 1937 borders there were approximately 562,000 people of Jewish extraction,[170] which amounted to a mere 0.7% of the German population. Altogether, in the 12 years of growing terrorization in the Nazi regime, around 330,000 persons of Jewish origin emigrated, 220,000 of these between 1933

[166]On emigration in general see, for instance, the special issue of *Berichte zur Wissenschaftsgeschichte* **7** [1984], as well as Fermi [1968], Fleming & Bailyn (Eds.) [1969], Jackman & Borden (Eds.) [1983], Fischer [1987], [1991]b, Sven Papcke in: Koebner et al. (Eds.) [1988], pp. 13–27, Roeder in: Stadler (Ed.) [1989], pp. 102–114, Herbert A. Strauss: in Strauss et al. (Eds.) [1991], pp. 9–23, and the invaluable reference book by Roeder & Strauss (Eds.) [1980/83], Vol. 2. On physics and mathematics in particular, see Weiner [1969], Holton [1983], Reingold [1983], Hoch [1983], Rider [1984], Stuewer [1984], Eckert [1993], chap. 7, as well as Klaus Fischer's and Stefan Wolff's contributions mentioned in the following section.

[167]See, for instance, Deichmann [1991] on biology (she is currently working on a similar prosopographic work covering the field of chemistry in the same period); Wolff [1991], [1993] and Fischer [1988], [1991]a on physics.

[168]See, e.g., the *NSDAP*'s program of 1930, points 18 and 24 translated in Noakes & Pridham (Eds.) [1990], Vol. 1, pp. 15f. For a detailed chronology and analysis of the anti-Semitic measures that ultimately led to a policy of systematic extermination of the European Jewry, see in particular, Hilberg [1961]; for a collection of anti-Semitic legislation, see Walk (Ed.) [1981].

[169]All preceding estimates according to Broszat & Frei (Eds.) [1989], pp. 117f., and from the *List of Displaced Scholars*, issued by the Notgemeinschaft deutscher Wissenschaftler im Ausland (Ed.) [1936], which lists 1,628 persons.

[170]As defined by the Nuremberg laws of 1935 based on ancestry; only about 503,000 of these were practising Jews. All figures in this section are taken from Broszat & Frei (Eds.) [1989], p. 220. Hilberg cites slightly different figures, however: 520,000 Jews in 1933 reduced to 350,000 in 1938. See Hilberg [1961], Vol. 2, p. 412, and the introduction to Roeder & Strauss (Eds.) [1980/83], Vol. 2, pp. xv ff.

and 1938 and another 100,000 in the period between 1939 and 1941.[171] As the German Reich expanded, successive emigration waves were set off, starting with the *Anschluß* of Austria in 1938, when 150,000 Austrian Jews (out of the total Jewish population of about 190,000) left their native country. The occupation of the ethnic German Sudeten region of former Czechoslovakia caused another 25,000 people to flee; and by 1941 6,000 more refugees had left Fascist Italy (which was however only about 27% of the Italian Jewish population, due to a much more relaxed enforcement of the discriminatory laws there).[172] Another 10,000 people managed to escape during the holocaust years of 1942–45. Later historiography has shown that in an attempt to 'solve' the 'Jew problem' increasingly extreme measures were adopted as the burden of detaining such huge numbers of people began to weigh heavily on the already strained economy. From 1942 on, these concentration camps developed into sites of holocaust, of systematic mass-murder at the hands of *SS* cadres implementing the notorious 'final solution' that had been resolved in secret at the Wannsee Conference on January 20, 1942.[173]

By the time the Nazi regime finally came to an end, only 25,000 Jews were still alive in Germany. Although the German population had been well aware of the systematic mass deportation of Jews to 'the East' and of the existence of concentration camps, the full scale of the holocaust in all its heinous brutality had been concealed as best as possible from the general public. Many Germans only discovered the truth of the rumors after 1945 when photographs and eye-witness accounts by the few survivors were published and when the war tribunal proceedings against some of the major offenders were underway (see sec. 5.5 below on denazification). Despite a great deal of effort especially since the 1970's, we cannot say that Germans have completely succeeded in coming to terms with their recent history.

4.2 The Emigration of Scholars, Scientists and Engineers

Scientific emigration must be seen as a subset of overall emigration from central Europe. Science was particularly vulnerable to political intervention, since a lot of research is conducted at universities. Those holding upper-level positions there, including university teachers, were mostly civil servants and thus directly affected

[171]These Jewish refugees thus constituted the vast majority of the approximately 500,000 German-speaking migrants (estimate by Strauss in: Strauss et al. (Eds.) [1991], p. 10). According to Werner Roeder (ibid.), only 6% of these refugees, that is 30,000, were political refugees not of Jewish origin, with the percentage much higher for specific groups such as intellectuals, scientists and artists.

[172]See Hilberg [1961], pp. 421ff.

[173]On the incremental development of the Holocaust see, e.g., Hilberg [1961], Kogon [1973], Broszat [1977] and in: Bracher et al.(Eds.) [1986], pp. 85f., sec. II,2, as well as Mommsen [1991], pp. 184–232, and further literature cited in Kershaw [1985]b, chaps. 5 and 8.

by the Nazi government's attempts to align the official apparatus politically.[174] The 'Law for the Restoration of the Professional Civil Service' of April 7, 1933, along with its many implementation regulations (see here docs. 7–8, 12, 17 and 36) was one purging measure aimed at removing all political opponents, such as Communists, Socialists and liberal activists, as well as specifically persons of Jewish descent, from all publicly funded permanent positions. The first wave of dismissals in 1933/34 affected 1,145 university teachers (including 313 full professors, 109 assistant professors, and 184 untenured professors).[175] By comparison, the staffing numbers for the 1930/31 winter term were: 2,741 full professors, 1,741 associate professors, 1,779 non-tenured university lecturers; thus altogether 6,744 active university teachers, not counting another 10,584 persons employed in academic research. As a result something like 15% of all university teachers (of whom 2% were women) were suspended, with the precise percentages depending on the specific reference group.[176] The sinister dynamics of this process soon became apparent as the dismissal lists published in British newspapers by newly established emergency aid organizations (see below sec. 4.6, p. lxi) lengthened. The dismissals were also usually announced in local German newspapers, though rarely with any commentary.[177] It is difficult to assess the quantitative estimates of these dismissals, because the figures given are not uniform and not even always consistent with each other. According to one source, by 1939 the dismissals had increased to a total of about 1,684 university employees, but according to another, in 1936 (thus prior to the emigration wave from Austria and Czechoslovakia) 1,617 university teachers, of whom 124 were physicists, had already been dismissed from their posts.[178] In 1956 von Ferber initially estimated out of the total university personnel an average of 39% emigrés, with 30% of these coming from the experimental sciences. His figures came from a comparison against university staffing in 1931 and 1938, but he did not take into account the heterogenous age distribution, natural departures, and that on average only 60–65% of those dismissed also chose to emigrate.[179] Surveys have shown that

[174]For a study on the civil service in the Third Reich, see esp. Mommsen [1966]; for an anthology of the legislation affecting universities, see Huber et al. (Ed.) [1942].

[175]Figures from Hartshorne [1937], p. 112; cf. Hartshorne [1938], pp. 13–14, Beyerchen [1977], pp. 44f., 221, and Seier [1988], pp. 252f., all based on a comparison of course catalogs and files of the *Academic Assistance Council* in London (see below). The official figures collected by the *REM* in 1934 were published recently in Gerstengarbe [1994].

[176]Strauss et al. (Eds.) [1991], p. 10, also estimates an emigration quota of 15% (thus a total of 1,100–1,500 persons) among university teachers of all ranks (unpaid lecturers to full professors) by 1940; Seier [1988] gives an estimate of 11% of all full professors.

[177]See, for instance, the list of 200 dismissals in the *Manchester Guardian Weekly* of May 19, 1933: N.N. [1933]c. For a rare public protestation in Germany against the new discriminatory science policy, see Wolfgang Köhler's statement, doc. 13.

[178]See Notgemeinschaft deutscher Wissenschaftler im Ausland (Ed.) [1936], Pross [1955], Kröner [1983], p. 13, Eckert [1993], p. 147; cf. also Fischer [1988].

[179]For a thorough critique of von Ferber's [1956] data, see esp. Beyerchen [1977], p. 221, Seier [1988], p. 253, and Fischer [1991]b, pp. 536f.

the average age of emigrés from several fields was distinctly lower than that of the average university teacher: For instance, two thirds of the emigrés to Great Britain about whom age information is available, were younger than 40 at the time of their dismissal and the median age was just above 30.[180] More recent estimates based on the listing in the *International Biographical Dictionary of Central European Emigrés 1933–1945*[181] are that 14–17% of all German university teachers emigrated, 950 of whom were either physicists, mathematicians, chemists or biochemists, 200–300 being physicists.

4.3 Emigration Statistics by Region and Discipline

The above quantitative estimates must be differentiated both according to region and to discipline. In the following we will confine ourselves to the figures related to physics and its subdisciplines.[182] In the first place, only 40% of all 36 German universities were affected at all by dismissals in the fields of physics and mathematics. 209 out of a total of 322 physicists, that is, two thirds of all physicists in the German civil service with permission to teach, were employed at these 15 universities. Over 40% of all emigrés in these fields left either Göttingen or Berlin.[183] As pointed out by Fischer [1988], the bigger departments usually faced the biggest relative losses, with interdisciplinary averages of nearly a third for Frankfurt and Berlin, roughly 25% in Heidelberg, 20% in Breslau, Göttingen, Freiburg in Breisgau and Hamburg, only 7% in Erlangen, and even less for Tübingen and

---

[180] According to Rider [1984], pp. 157f.; compare with the average age compiled by Roeder & Strauss (Eds.) [1980/83], Vol. 2, pp. lxxviii ff. On the medium-term consequences of the dismissals and the resulting generation gap in Germany see, e.g., Richard Hinton Thomas & R. H. Samuel: *Education and Society in Modern Germany*, Westport, Conn.: Greenwood Press, 1971 (reprint of 1949 ed.). They also note that by 1939 about 45% of all academic positions in Germany had either been eliminated, the occupant replaced or were otherwise altered.

[181] Co-edited by the Institut für Zeitgeschichte at Munich University and by the Research Foundation for Jewish Emigration in New York—see Roeder & Strauss (Eds.) [1980/83], Vol. II. See also the listing of 1,522 scientists, social scientists and scholars from other fields in the handbook compiled by Kröner [1983].

[182] For comparative numbers referring to all dismissals effectuated shortly after April 1933, see Gerstengarbe [1994], pp. 33f. She concludes that only 0.5% of all dismissals were based on § 2a (directed against Communists) and 11.5% on the basis of § 4 (directed against other 'politically unreliable' persons) of the civil service law, while nearly 70% were directed against persons of Jewish descent (§ 3), and another 17.5% were based on § 6 for reasons of administrative efficiency.

[183] The previous figures according to Fischer [1988], p. 87. On the University of Göttingen see in particular Beyerchen [1977], chap. 2, esp. pp. 34f., and Rosenow in: Becker, Dahms & Wegeler (Eds.) [1987], of which the appendix includes a comprehensive list of dismissed faculty, pp. 490–499. Beyerchen estimated a loss of 67% in mathematics and physics at Göttingen, excluding Prandtl's aerodynamics research institute. On the exodus of scientists from Berlin see in particular W. Fischer et al. (Eds.) [1994] and Schottlaender [1988]; on the emigration of theoretical physicists from Munich, see Eckert [1993], pp. 149ff.

Rostock.[184] All the German polytechnics, at Charlottenburg (Berlin), Munich, Stuttgart and Dresden, were much less affected because they had a lower initial percentage of Jewish employees, stemming from anti-Semitic appointment policies, while the newly founded city universities at Hamburg, Cologne and Frankfort were known for their liberalism (comparatively speaking).[185] These regional discrepancies are confirmed in an analysis of membership in the German Physical Society (*Deutsche Physikalische Gesellschaft*). In January 1933 there were about 1,227 members, 960 of whom were living in Germany at the time. The highest numbers of members living abroad were registered in the districts of Berlin (29 = 11.5% of the local membership), Lower Saxony (18 = 18%), Austria (10 = 14.7%), Thuringia, Saxony and Silesia (10 = 8.5%), and Prague (6 = 18%).

The emigration statistics are equally unbalanced among the various subdisciplines within physics. Stefan Wolff's recent study on the theoretical physics community composed of 60 university teachers finds that 26 chose exile.[186] This high proportion (43% which is much higher than the overall average of 15% cited above) is, of course, correlated with the drastic deterioration in the political climate toward theoretical physics, and in particular toward quantum mechanics and relativity theory, from 1933 onwards.[187] Klaus Fischer's thorough scientiometric study recorded the number of publications written in German in the period 1925–33, distinguishing between authors who remained in Germany and those who later emigrated. His results were: 25% of all such publications concerning quantum mechanics in this period were authored by future emigrés. The corresponding numbers for nuclear physics were about 18.8%, spectrum analysis 12.7%, electricity and magnetism 8.2% and acoustics only 3.8%.[188] Future expatriates had a 14.5% share of articles published between 1926–33 in the scientifically more adventurous journal *Zeitschrift für Physik*, while they contributed only 5.9% of the papers published between 1930–33 in the more conservative *Annalen der Physik*. Fischer thus concludes that there is a positive correlation between contributions by future emigrants to a journal specifically orientated toward basic, fundamental research, and a negative correlation with regard to the paper's degree of practical application. The number of papers by future emigrés

[184]These figures (not limited to the sciences) from Wendehorst [1993], p. 188; cf. also Fischer [1988], p. 88; for dismissals at Heidelberg Univ. in particular, see Mußgnug [1988], for Kiel see Uhlig (Ed.) [1991], for the Berlin Polytechnic: Ebert [1979], for other Berlin institutions see W. Fischer et al. (Eds.) [1994], for Göttingen: Dahms [1986], and in: Becker et al. (Eds.) [1987].

[185]On the correlation of emigré percentage with physics department employee totals, see Fischer [1988], p. 88; on anti-Semitism at German universities prior to 1933 see in particular Hammerstein [1995], pp. 76ff. on the historical roots, as well as the references in footnote 244 below.

[186]See Wolff [1993]; cf. also Eckert [1993], chap. 7.

[187]Cf. here sec. 5.3. This high emigration rate is even higher than the figures for some other particularly heavily affected fields, such as economics and the social sciences (41% of full professors), and law (36%)—figures according to Seier [1988], p. 253.

[188]See Fischer [1988], p. 97.

is positively correlated with the novelty of the topics and negatively with the conservativeness of a journal. Thus physicists destined to seek asylum abroad typically came from the youngest, most up-to-date disciplines, such as quantum mechanics and nuclear physics, and much fewer came from well-established traditional disciplines like acoustics. This is corroborated by the number of times individual papers are cited elsewhere, which can be looked upon as an indicator of their scientific value. Among the most frequently cited German-speaking nuclear physicists, emigrants are represented more than three times over what the statistics would lead us to expect: Although German-speaking authors (excluding non-exiles) came to 118 individuals (4.1%) within the total nuclear physics community of roughly 2,861 members (defined as having written a paper on this subject between 1921 and 1947), emigrants took 14.3% of the citations within this period. "Out of the 26 German nuclear physicists most often cited throughout the world before 1933, not one or two—as would be expected statistically—but 13 emigrated or moved abroad for job market reasons, that is, no less than 50 percent. Thus it was primarily the intellectual leaders, sociologically speaking, the chieftains, who left."[189]

4.4 Reaction to the Dismissals and to the First Wave of Emigrations

Few prominent scientists not specifically affected by the new discriminatory legislation left their positions in protest against the new Nazi government. The decision by the theoretical physicist and Nobel laureate Erwin Schrödinger to move to Oxford in 1933 (where he only stayed until 1936, remigrating to Graz and then to Dublin in 1939) left a gaping hole in the department for theoretical physics at the University of Berlin. Some scientists of Jewish origin, initially immune to dismissal because of their military service during World War I, did the same.[190] Thus, for instance, the Göttingen atomic physicist James Franck and the world-famous physical chemist and discoverer of ammonia synthesis Fritz Haber both resigned in protest, the latter in explicit response to the dismissal of some of his most qualified colleagues at the Kaiser Wilhelm Institute of Physical Chemistry in Dahlem (Berlin).[191] But open protest against the new legislation or other forms of intervention remained the exception and was only aimed at

[189]Fischer [1991]b, p. 543; cf. also Fischer [1993], table 37, p. 63, for a ranking of top nuclear physicists according to the number of citations between 1920 and 1947; compare with Stuewer's analysis using a more qualitative method of historical analysis: Stuewer [1984]. Cf., e.g., Beyerchen [1977], table 2, p. 48, or Guérout [1992], p. 31, for a list of 20 Nobel laureates who emigrated from Germany after 1933 or from Austria after the *Anschluß* in 1938.

[190]For this provision and its background see doc. 7, § 3, No. 2, and the annotation there.

[191]On Franck's resignation see, e.g., here docs. 9f., and *Göttinger Zeitung*, April 18, 1933; on Haber see docs. 15f., as well as Nipperdey & Schmugge [1970], pp. 50ff., and Stoltzenberg [1994].

preventing the application of Nazi measures to specific leading scientists.[192] Few dared to express public sympathy with those who had been dismissed (see doc. 13 for an exception). Such comments can usually only be found in private correspondence between trusted friends (see, e.g., doc. 10). As a result this first wave of dismissals encountered little resistance. Many hoped that the Nazi regime had reached its peak and that it was only a matter of time before it would go the way of the many previous weak and short-lived Weimar coalitions.[193] Those who had relied on the war veterans exemption clause, which had been inserted on Hindenburg's insistence and exempted all Jewish citizens who had fought on the front in World War I as well as their children,[194] were soon disappointed. In 1935 the Nuremburg Laws removed their exemption status and categorically excluded all Jews from employment in the civil service, also cutting deeply into the personal freedom of the remaining Jews in Germany and the subsequently annexed lands. Thus, while the total loss of scientists to emigration is around 15%, the actual losses sustained within certain specialties, such as quantum mechanics or nuclear physics, were significantly larger and could have amounted to as much as 50% when one considers actual scientific output.[195]

4.5 Host and Transit Countries

On the whole, emigrants first moved to neighboring areas such as the Saar region, which was under French occupation since the Treaty of Versailles.[196] 37,000 people chose this region as their first refuge until a plebiscite there determined its reannexation to Germany in 1935. In 1933 alone French territory was the first choice for approximately 65,000 emigrants, the Soviet Union for ca. 6,000, Sweden for 5,000 and Switzerland for 10,000. The extent of continued migration after departure from the country of origin is revealed when we examine Great Britain's emigration totals, for example. Around 1940 there were about 63,000 individuals in exile there, while in 1943 this number had reduced to ca. 25,000. The majority migrated on to the United States, some for fear of a German invasion of the British Isles, but the majority sought the much wider variety of opportunities

[192] See, e.g., Schottlaender [1988], pp. 85–98, as well as Planck's intercession for Haber in early 1933 in an audience with Hitler (doc. 114), at which time he distinguished between 'good' and 'bad' Jews—quite in the tone of the times. Cf. also Albrecht [1993] and Mehrtens in: Meinel & Voswinckel (Eds.) [1994], pp. 20f.

[193] See, e.g., Kopfermann's report to Bohr, doc. 20, for a perceptive contemporary portrayal of the feelings and attitudes commonly held by scientists toward the political events.

[194] See, e.g., Walle [1982] on German Jewish soldiers, and footnote 8 in doc. 7 for literature on the motives behind this legal provision.

[195] See Fischer [1991], [1987] for a discussion of the problems attached to the use of quantitative measures to describe such qualitative aspects as emigration loss to a mother country against the corresponding gains to the host countries.

[196] On the distribution of Jewish refugees from Germany and Austria in selected host countries see, e.g., Roeder & Strauss (Eds.) [1980/83], Vol. 2, pp. xx ff.

in America, particularly during the war-economy boom years. Large-scale military research programs such as those at Los Alamos, Hanford or Oak Ridge, where $400 million were spent on uranium separation facilities and a plutonium-producing pile alone, provided jobs for thousands of qualified persons in science and engineering. The annual federal budget investment in R & D rose sharply from $48 million to $500 million, increasing from 18% before the war to 83% of the total national expenditures for this purpose.[197]

The emigration of scholars, scientists and engineers must be set against the backdrop of streams of uprooted people seeking a fresh start in completely foreign surroundings. Most economies were still struggling to overcome the global crisis of 1929, and the host countries were staggering under high unemployment rates even before the huge influx of Europeans. They were thus not always welcomed. According to recent statistical studies, the United States was by far the most important destination country.[198] Approximately 48% of all scientists and engineers ended up there, followed by England (10%), Palestine (8%) and Switzerland as well as Latin America (ca. 4% each). The corresponding percentages of those who chose these countries as their first refuge are quite different: USA only 35%, England 37%, Palestine and Switzerland 8%. This is because it was very difficult to find long-term positions in countries like England where there was a very limited number of permanent jobs in science.[199] Only about one half of the 139 foreign physicists and mathematicians seeking refuge in Britain stayed permanently. 45 or so moved on to the USA, which a total of 192 physicists and mathematicians chose as their permanent home.[200] Strained job market conditions after 1933 even for an American-trained post-doctoral graduate, as well as a considerable anti-Semitic tendency among higher administrative officials at some American universities dating back to the twenties,[201] were not conducive to the placement

[197]See, e.g., Kevles [1987], chaps. 21–22, esp. pp. 341f., on the impact of the USA's entry in World War II on its economy, and for the foregoing figures. In the spring of 1945 there were 4,000 staff employees at Los Alamos alone.

[198]See Möller [1984], based on a collection of 950 entries on emigré scientists as listed in Roeder & Strauss (Eds.) [1980/83].

[199]According to Moseley [1977] and Rider [1984], p. 155, there were between 800 and 1,000 physicists in Britain around 1920 (based on J.J. Thomson's estimate); whereas in 1940 the British Ministry of Labor listed nearly 2,800 physicists.

[200]For the preceding figures and for more details on the absorption of these professionals in the American system of higher education, as well as their incorporation into the physics community in general, see Rider [1984], pp. 155f. According to Kevles [1987], pp. 202, 275, the American Physical Society had about 3,600 members in 1939.

[201]For instance, in 1922 the Harvard president A. Lawrence Lowell insisted on establishing a quota system for Jewish undergraduates. *The Nation* commented: "Harvard is not the first university to attempt to limit the proportion of Jews in its midst. It is merely the frankest" (see N.N. [1922]). Arthur Holly Compton's suggestion that it was unethical to give positions in the U.S. to Europeans while many American students were unemployed only lost general acceptance after 1939 when the first temporary visiting fellow named Wilhelmy was sent back to Germany and died soon afterwards. Cf., e.g., Hoch [1983], pp. 239–243, Jones [1984], and

of so many predominantly 'Jewish' scientists at these institutions.[202] This discriminatory trend weakened gradually as economic conditions improved and the dramatic events in Europe from 1939 onward became generally known.

Approximately 100 scholars and scientists found refuge in Turkey, where the head of the Turkish Republic Kemal Atatürk closed the existing 'house of science' (*Dar-ül-fünun*) and established the reformed Universities of Istanbul and Ankara. The availability of specialists expulsed from Germany was thus providential; but not many of these new professors stayed there for more than a couple of years. Adaptation to the cultural setting was not easy: For instance, it was requested officially, that the Turkish language be used for delivering lectures.[203] It must be emphasized that it was quite typical for emigrés to find only temporary positions initially, and it was only after a series of moves that they were finally able to settle down. This was the case particularly for professionals who had not yet made a name for themselves in their fields, and their futures were often very uncertain.[204] It should also be noted that these less successful emigré scientists left the fewest traces behind for today's historians—thus we have no reliable estimates on the number of those who chose to emigrate and were cut off from their original professional communities as a result or dropped out of their fields altogether, some even committing suicide in despair. For lack of documentation, their history will never be written.

4.6 Emigré Relief Organizations

Several relief organizations were established in and after 1933 to help alleviate the situation for refugee scholars and scientists by providing institutional support and

Reingold [1983], pp. 209ff., for further examples of anti-Semitism at American universities, as well as Saul S. Friedman: *No Haven for the Oppressed. United States Policy toward Jewish Refugees 1938-1945*, Detroit: Wayne State Univ. Press, 1973. See also Rider [1984], tables 2 and 3, for a comparison of the varying degrees of regional absorption of emigré scientists in the U.S., and according to the ranking of schools in mathematics and physics; cf. also footnote 244 below for literature on the historical roots of anti-Semitism in Europe and Germany in particular.

[202]For instance, Eugene Feenberg applied in vain for a position in North Carolina despite excellent recommendations, including E.C. Kemble's assurance that "he is a tall rangy Texan and neither looks nor acts like a New York Hebrew". He received the reply: "it is practically impossible for us to appoint a man of Hebrew birth [...] in a southern institution." (quoted from Hoch [1983], p. 241).

[203]See in particular the studies by Widmann [1973] and Neumark [1980] on emigration to Turkey. These included the experimentalist Arthur von Hippel, the astronomer Erwin Freundlich, and the philosopher Hans Reichenbach. Cf. also the contemporary internal *REM* report in Grothusen (Ed.) [1989].

[204]See, for instance, docs. 25–26 as well as Rider [1984], pp. 109ff. on the fate of Georg Jaffé and Alfred Tarski (both already very well-known within their fields). See also Hentschel [1996], chap. 6 on E. F. Freundlich's migration from Istanbul to Prague, and then onwards to Scotland to flee the German occupation forces. At each of these destinations he established new observatories without being able to reap the fruits of his labors.

at least temporary employment. The earliest and most important of these was the Academic Assistance Council (AAC), founded in April 1933 in London by the director of the London School of Economics and Political Sciences, Sir William Beveridge. The physicist Rutherford was chairman until 1937, the astrophysicist Walter Adams general secretary, and Esther Simpson its dedicated assistant secretary.[205] This Council, renamed Society for the Protection of Science and Learning (SPSL) in 1936, soon had more than 2,000 paying members who made regular voluntary contributions to it. Within the first two years of its existence the AAC succeeded in placing 62 professors in tenure positions and helped another 148 emigrés to find temporary positions, 54 of these in the many branches of London University, 31 at Cambridge, 13 at Oxford, 7 at Scottish universities and the rest at other universities and colleges throughout Great Britain.[206] The fellowships offered included a grant of around £250 for married and £182 for unmarried scholars and scientists, while the regular lecturer's salary at the same universities was typically between £400–500 sterling. The Imperial Chemical Industries also offered 18 fellowships; however, until 1935 the trust refused support to scientists who had any connections to the *IG Farben* German chemical concern. At the end of April 1935, the German exiles Fritz Demuth, Philipp Schwartz and Moritz Bonn founded the emergency organization for German scientists abroad, the *Notgemeinschaft Deutscher Wissenschaftler im Ausland* in Zurich (its headquarters were relocated subsequently to London in 1936). They succeeded in placing about 2,000 emigrés from Germany and Austria mainly in the USA and England, frequently applying the organization's leading figures' (such as the physicist Max Born) good connections with local colleagues at the host locations. In May 1933 the *Comité des Savants*, later renamed *Comité pour l'Accueil et l'Organisation du Travail des Savants Étrangers*, was founded in France. Later that same year Philipp Hartog formed and presided over the *Comité International pour le Placement des Intellectuels Réfugiés* in Geneva, and the League of Nations installed Norman Bentwich as first high commissioner of refugees from Germany.[207] In New York City, the Emergency Committee in Aid of Displaced Scholars was formed in 1933 by American academics, with Stephen Duggan as chairman and Betty Drury as executive secretary. They compiled extensive dismissal lists and were quite successful in finding positions for refugee researchers at many American

[205]See, e.g., Rutherford [1934] for a fund-raising article in *Science* which portrays the AAC's activities up to that point; cf. also Hirschfeld [1988], and Cooper (Ed.) [1992] chap. 3 for the history of the AAC, and the Council's files in Bodleian Library at Oxford.

[206]These figures are quoted from Hirschfeld [1992], p. 78. Somewhat different figures are given in Bentwich [1953], p. 13: London 56, Cambridge 30, Oxford 15, Manchester 16; and Rutherford [1934], pp. 533f.: London 67, Cambridge 31, Oxford 17. According to Eric Ashby in: Cooper (Ed.) [1992], p. 7, by 1939, the SPSL had aided about 1,000 scholars and scientists.

[207]Cf. Bentwich [1953] as well as K. Düwell in: Strauss et al. (Eds.) [1987], pp. VIII–X for surveys of most of these and other aid organizations.

colleges and universities based on the local need.[208] Finally, several prominent individuals took great pains especially to help personal acquaintances or people referred by them. Rudolf Ladenburg, who emigrated to the USA in 1932, Eugene Wigner, Albert Einstein and Hermann Weyl all went out of their way to help such persecuted individuals find appropriate positions in their new surroundings.[209]

While earlier emigration studies have focussed mostly either on providing statistical estimates for various scientific fields or on documenting individual stories, more recent studies have also begun to study the qualitative changes in science in the host countries.[210] It would be erroneous to assert that modern physics only took off in the USA as a result of the huge influx of refugees. Americans had already made much progress in new research fields like astrophysics and nuclear physics beforehand. Large research facilities like the Mt. Wilson Observatory near Pasadena were built, which became a model in its time; and the first cyclotrons (spiral-pathed particle accelerators) transformed Berkeley into the world center for nuclear physics in the 1930's. Emigrés like Hans Bethe joined an already viable program.[211] It is certainly justified to state, however, that American research benefited considerably from the many bright young minds both qualitatively and quantitatively. Even before the Japanese attack on Pearl Harbor, circa 1,700 physicists were working on American defense research, and by the fall of 1942 the demand for physicists had risen to three to four times the annual output from universities.[212] Thus at least a part of this need could be filled by emigrés. In general, physicists able to adapt to the very different style of conducting physical research in the USA were most successful. But many were unable to adjust to the closer cooperation between theorists and experimenters or else they clung to the traditionally distant and authoritarian role of the German professor.[213]

It is well-known that especially after the USA entered the war against Japan and Germany in December 1941, many top emigré scientists who had passed the rather strict security screenings[214] joined weapons development programs. Aside

[208]See in particular Duggan & Drury [1948]; cf. also the finding aid of the Emergency Committee in the relevant collection stored at the Research Libraries of the New York Public Library, Accession no. 46 M 43, and the 48 boxes of files containing this organization's files.

[209]Cf., for instance, the list of refugees attached to the Wigner and Ladenburg letters.

[210]Cf., e.g., Fischer [1987], Hoch [1987].

[211]See in particular Heilbron & Seidel [1989] for Lawrence's program in experimental nuclear physics in the USA; cf. also Osietzki [1993].

[212]Figures according to Kevles [1987], p. 320.

[213]The files of the above relief organizations provide full reports on the successes and failures, confirming these two patterns in the careers of emigrants: See, e.g., the sample reports by emigrés as well as by host colleges and universities in Duggan & Drury [1948], chap. ix: 'Scholars speak for themselves', and chap. x: 'Opinions of administrators'; cf. also Thomas B. Appleget: 'The foundation's experience with refugee scholars', dated March 5, 1946, in the files of the Rockefeller Foundation's Archive, call no. 1.1.200, Box 47, Folder 545a, for a fairly positive report on the long-term placement of their fellows. For a detailed study on Schrödinger's failure to adjust to the Oxford environment, see Hoch & Yoxen [1987].

[214]Rudolf Peierls, for instance, was not permitted to participate in British radar research, even

from the Radiation Laboratory at the Massachusetts Institute of Technology (MIT), the many nuclear research sites included:

- the Chicago metallurgical laboratory (administered by Arthur Holly Compton, but led by the Italian emigré Enrico Fermi) where the first stabile nuclear chain reaction was observed in a nuclear reactor on Dec. 2, 1942,
- research labs at Columbia University in New York where gaseous diffusion techniques were studied,
- electromagnetic separation and plutonium research labs at Berkeley, where Lawrence worked on cyclotron development,
- the huge gaseous diffusion and electromagnetic facilities at Oak Ridge, Tennessee, with its $400 million uranium separation plant,
- the plutonium works near Columbia River in Hanford, Washington, designed by the Hungarian-German emigré Eugene Paul Wigner and realized by the major industrial contractor DuPont Chemical Co.,
- and finally, from March 1943 on, the Los Alamos complex in New Mexico, where some of the best physicists and engineers from other war research laboratories and universities as well as several thousand other staff members collaborated under the tightest secrecy to produce the uranium and plutonium bombs.

These large-scale research projects in radar and nuclear physics ultimately determined the outcome of the war against Japan—Germany had already surrendered before the first atomic bomb was tested in the New Mexican desert and before other prototypes were dropped on Hiroshima and Nagasaki in August 1945.[215] Molecular biology was another research area to profit from the influx of emigré physicists interested in biological applications.[216]

---

after acquiring British citizenship in the spring of 1940; however, he later participated in British nuclear research. Hans Bethe was only permitted to contribute to American war research after his naturalization in December 1941. On the other hand, the screening process was far from perfect. Klaus Fuchs who later spied for the Soviet Union for political reasons encountered no problems in getting employed at Los Alamos; cf. Norman Moss, *Klaus Fuchs, the Man Who Stole the Atom Bomb*, New York: St. Martins Press, 1987; Robert Chadwell Williams, *Klaus Fuchs. Atom Spy*, Cambridge, Mass.: Harvard Univ. Press, 1987.

[215]See, e.g., Guerlac [1987], or Kevles [1987], chap. 20, on the MIT radiation lab; Johnson [1978], chap. 1–2, and Fisher [1988] on radar development in general; Heilbron & Seidel [1989] on Berkeley; Gowing [1964] on the British nuclear research program; Smyth [1945], Groves [1962], Hawkins [1986], Kevles [1987], chap. 21, Rhodes [1986], and Hoddeson et al. [1993] for a technical history of the Los Alamos Project; cf. also Gowing & Arnold [1979] for more references to the older literature, and Goldberg [1995] for a recent reassessment of the political background to the decision to drop two atomic bombs.

[216]See, e.g., Fleming [1969], and Evelyn Fox Keller: 'Physics and the emergence of molecular biology: A history of cognitive and political synergy', *Journal of the History of Biology* **23** [1990], pp. 389-409. On chemistry see, e.g., Carroll [1983].

5 Physics in Nazi Germany

Before providing a brief overview of some of the more important institutions in physical research (sec. 5.1) as well as of major research programs (sec. 5.4), let us start by looking at the overall statistics of physicists in academia. Between 1931 and 1938, the total number of physicists at universities fell from 175 to 157, while there was a small gain at polytechnics from 139 to 151 physicists in the two reference years respectively. Thus, even with the great qualitative losses incurred by the emigration of leading scientists after 1933 (see part 4 of this introduction), in purely quantitative terms physics in Germany essentially stagnated during this period, while in the USA the boom of preceding decades continued on despite the Great Depression.[217] Another general feature in Germany was the strict institutional barrier between theoretical and experimental physics. At universities there was typically one institute for experimental physics or more, and a separate institute for theoretical physics. This was quite unlike the American physics department which brought experimentalists and theoreticians into a much closer collaborative setting, and it was drastically different to the novel teamwork-based form of big science (such as the Los Alamos complex) to emerge in the USA at that time.[218]

5.1 Institutions Connected to Physical Research

The Nazi government was quick to ensure that state-run institutions were directed by politically sympathetic individuals. For instance, the president of the *Physikalisch-Technische Reichsanstalt* (*PTR*) Friedrich Paschen was forced to retire; and one of the very few early vociferous members of the *NSDAP* in the sciences, Johannes Stark,[219] succeeded him in May 1933.[220] One of Stark's first actions was to cancel all contracts with "Jews and leading figures of the previous regime" with the specific aim of removing Max von Laue from the *PTR*'s advisory committee. The few Jews among the research staff had already been fired in April 1933 on the basis of the Law for the Restoration of the Professional Civil Service (docs. 7f. and 17). Soon afterwards Stark began to indulge in a plan to expand the *PTR* on a grand scale. It was to become Germany's central physical research institution with several hundred laboratories and thousands of scientific employees. This conception fit right in with Hitler's gigantomanic visions of redesigning the German capital; it never materialized, however, owing to bureaucratic bickering, a shortage of funds and resistance by physicists who were

[217]See Ferber [1956], p. 197; cf. Eckert [1993], p. 199, Kevles [1987], chaps. 13–21, on the USA.

[218]This difference is emphasized in Eckert [1993], p. 199, where some of the later failures of German science against Allied research are attributed to this disciplinary isolation (p. 211).

[219]See, e.g, his declaration in support of Hitler of 1924, doc. 3, as well as Stark [1930]a-c, [1934]c and sec. 5.3 below.

[220]See Lenard's jubilant newspaper article on Stark's nomination, doc. 18, as well as von Laue's retrospective account of the background to the change in the *PTR*'s presidency, doc. 19.

loath to being subjected to Stark's dictatorial discipline.[221] Nevertheless, Stark did succeed in reorganizing the Institute's research branches and expanding the staff from 292 employees in 1932 to a total of 444 in 1938, 138 of whom were scientists. He reoriented research to an extent toward the physics of the atomic shell (his own subject of interest) and turned the focus to matters of national and military importance. For instance, research was conducted on low-temperature carbonization in a rather unsuccessful attempt to obtain hydrocarbons from the low-temperature distillation of coal and peat; and he established a laboratory for the analysis of the hardness and strength of materials.[222] However, other areas of research conducted at the *PTR* continued as before. One particularly successful branch was the development of quartz clocks under the high-frequency physicist Adolf Scheibe.[223] In 1934 Stark became furthermore president of the *Notgemeinschaft der Deutschen Wissenschaft* (founded in 1920 and one of the two major German funding organizations for scientific research, similar to the American postwar National Science Foundation). It is curious that Max von Laue, one of Stark's most resolute opponents, regarded his nomination to these positions as the lesser evil (cf. Gustav Mie's letter, doc. 33), noting that the nominee was at least a qualified physicist as opposed to some Nazi bureaucrat. However, Stark soon had to relinquish his presidency of the *Notgemeinschaft* in 1936, had been officially renamed *Deutsche Forschungsgemeinschaft* (*DFG*) under his ægis. He got embroiled in administrative quarrels with the *REM* as it was taking over from the *RIM* supervision of scientific research in August 1934. Stark's incompetence in setting research priorities and his emphasis on questionable research projects like the fiasco of attempting to regain gold from German swampland were the principal reasons for his replacement, however.[224] Stark was succeeded by his arch-rival the senior *SS* battalion commander and *REM* official Rudolf Mentzel, but he retained his seat as president of the *PTR* until his regular retirement in 1939. His successor was the politically influential high-frequency specialist Councillor of State Abraham Esau,[225] who was later also appointed Göring's Plenipotentiary of Nuclear Physics (*Bevollmächtigter des Reichsfeldmarschalls für Kernphysik*), thus becoming the scientific head of the German *Uranverein* for the brief interlude

[221]On the foregoing see in particular Hoffmann [1993]a.

[222]See, e.g., Stark [1937]; cf. Hoffmann in: Albrecht (Ed.) [1993], pp. 122–124, Kern [1994], pp. 217–255, and Walker [1995], pp. 20f.

[223]A. Scheibe (1895–1958) was a former student of Max Wien. By the early 1930's quartz clocks developed at the *PTR* had reached the precision of 0.00025 sec/month; cf., e.g., A. Scheibe & U. Adelsberger, 'Die Gangleistung und technischen Einrichtungen der Quarz-Uhren der PTR in den Jahren 1932–1944' (distributed manuscript) 1950, and Lemmerich [1987], pp. 90–93.

[224]Cf., e.g., Zierold [1968], pp. 188–190, 208–212; Beyerchen [1977], pp. 121f., Walker [1995], pp. 37f. On the Reich Interior Ministry (*RIM*), where Johannes Stark enjoyed good relations with the minister, Wilhelm Frick, for instance, but which lost its authority over the *PTR* and similar institutions in August 1934 and consequently its importance, see Pfundtner [1937] as well as the references in the appendix.

[225]On the *PTR* under his leadership, see Kern [1994], chap. 7.

of a little more than one year (see below).

Thus staff changes reveal the shifting weight between the various competing factions within the Nazi polycracy. Stark had aligned himself with the Rosenberg faction in the *NSDAP* and with the Interior Ministry under Wilhelm Frick, but both were quite weak players in the political game. Mentzel, on the other hand, was supported by the powerful *SS* controlled by Heinrich Himmler, who also continually expanded his own research organization, the *Ahnenerbe.* Historical research has not yet clarified sufficiently whether the remarkable shifts in power that occurred after 1937 away from ideologues like Stark and toward hard-core pragmatists like Mentzel and Thiessen can be correlated with the increased dominance of the Nazi right wing (*SS*) over the leftist and social revolutionary branch of the National Socialist State since the Röhm *putsch* affair in 1934. Stark's uncareful choice of allies could thus explain this sudden loss of power and political marginalization after 1937. At any rate, emphasis on concepts like 'community' (*Volksgemeinschaft*) is ubiquitous in Stark's political writings.

Another interesting case is the *Kaiser-Wilhelm-Gesellschaft*, founded in 1911. When the the Nazis came to power it was presided over by the theoretical physicist Max Planck, who had succeeded Adolf von Harnack at his death in 1930. Although the society itself was a private registered association and thus legally independent of the new rulers, many of its numerous major research institutions were supported by public funds, such as the *KWI* of Physical Chemistry, and were thus affected by the new Nazi legislation (see here secs. 3.1–3.2). In order to appease the new government the society's governing board, including in particular the science administrators General Secretary Friedrich Glum and the president Max Planck, decided to steer a course of self-imposed political realignment. When the head of the *KWI* of Physical Chemistry Fritz Haber was forced to dismiss his renowned department heads Herbert Freundlich and Michael Polányi at the end of April 1933 for reasons of race, he submitted his resignation in protest (cf. doc. 15), much to the consternation of Planck who had hoped to prevent such an escalation with his low-key policy of compliance.[226]

In accordance with the so-called Harnack principle, the society's many research institutes were formed around prominent personalities, the one of most interest here being the Kaiser Wilhelm Institute of Physics, created for Einstein in 1917, which functioned initially as a funding organization since it lacked a building of its own and employed very few researchers. Its premises were finally completed in 1937, well after those for chemistry and physical chemistry, for instance, on funds provided by the Rockefeller Foundation, which had decided to support this project in April 1930 and was then faced with the difficult decision

[226]On Planck's role as 'spokesman for German science' and as one of the main advocates of this 'self-realignment' politics by the *KWG*, see in particular Heilbron [1986]b; on the history of the society during the Third Reich, see also Glum [1964], part 3, Albrecht & Hermann in: Vierhaus & vom Brocke (Eds.) [1990], and Macrakis [1993]; cf. also the research reports by its many diverse institutes in Hartmann (Ed.) [1936].

of whether to keep its committment despite the subsequent political changes in Germany.[227] The minutes of a meeting during the planning stages of the *KWI* of Physics in June 1934 (doc. 30) reflect how the society's president and influential scientist tried to use the pending status of the funding for this costly enterprise as a means to moderate Nazi science policy. At this meeting Planck subtly warned the German government against such radical measures as repelling valuable scientists through anti-Semitic legislation, pointing to the risk of losing the Rockefeller grant of over 2 million reichsmarks. The success of this strategy was limited at best. In the long run it did not prevent the situation from worsening for Jewish citizens as a result of the infamous Nuremberg Laws of 1935 and later; the academic degree restrictions (doc. 53) are just one example. The Rockefeller money arrived nonetheless, and the newly erected *KWI* of Physics was inaugurated in 1937 by its director, the Dutch theoretician Peter Debye (cf. doc. 54 for his description of the building and its research divisions). When World War II broke out and the military took an interest in the research being conducted at his institute, Debye was given the alternative of either assuming German nationality or taking a leave of absence. He chose the latter course and went to the USA. The institute was reorganized under the control of the Army Ordnance Office (*Heereswaffenamt, HWA*) in 1940 (cf. the contract between the *KWG* and the *HWA*, doc. 79, and Rust's approval, doc. 80) and became one of the research centers exploring nuclear fission applications (see sec. 5.4 below on the *Uranverein*). At the beginning of 1942, however, when the Army Ordnance Office reached the conclusion that this research could not yield results in the immediate future for military applications, such as developing an atomic bomb or a compact energy source for submarines and aircraft and returned its authority over the Institute to the Reich Research Council. In July 1942 the Research Council, and thus the *KWI* of Physics, was subordinated to the head of the German Air Force, *Reichsfeldmarschall* Hermann Göring. He in turn commissioned first Abraham Esau (in November 1942) and later Walther Gerlach (officially on Jan. 1, 1944) to guide nuclear research at the numerous research sites subsumed under the *Uranverein* (see also pp. lxxxii ff., as well as doc. 103 on the many institutions enlisted for this nuclear research project).

Compared with the British and American war research efforts united in the Manhattan Project, to this day the prime example of 'big science', the *Uranverein* was only a loosely knit, decentralized network of researchers with quite different research agendas. Rather than teamwork as on the American end, on the German side we find cut-throat competition, personal rivalries, and fighting over the limited resources.[228] Overall, the *Uranverein*'s total expenses amounted to no more

[227] See, e.g., Macrakis [1986] on the *KWI* of Physics, and [1993] on other *KWG* institutes.

[228] It is interesting that Bagge and Diebner both praised the American institutionalization of teamwork highly in their postwar booklet on the peaceful uses of nuclear energy: Bagge, Diebner & Jay [1957], pp. 13–15. Their categorization of the German *Uranverein* as such an "up-to-date research organization" (ibid., p. 15) seems to be based more on national pride than

than about 8 million reichsmarks, equivalent to about $2 million at the time.[229] The only German research and development project in any way comparable in scale to the Manhattan Project is the Peenemünde rocket research facility, which employed 1,960 scientists and engineers and some 3,852 workers by 1942, and at its peak had a total of nearly 6,000 operational staff,[230] set against a total of nearly 5,000 employees at Los Alamos.[231] While the German *Vergeltungswaffen* (reprisal weapons) effort, which produced a pilotless jet-propelled plane (V-1) and a supersonic ballistic missile (V-2), is estimated to have cost between a half and three billion wartime dollars,[232] the U.S. atomic bomb project cost approximately two billion dollars (in the 1940 valuation). However, merely comparing costs does not give the full picture. As the curator of World War II history at the National Air and Space Museum, Michael Neufeld put it: "Since the German war economy was significantly smaller than the American one at its peak, the Army rocket program imposed a burden on the Third Reich roughly equivalent to that of [the] Manhattan [Project] on the United States."[233] Despite the comparable economic burden of these two big research and development projects, they were vastly different in terms of their short-term effectiveness: Whereas ca. 15,500 deaths in England and Belgium are directly related to successful hits of rockets and unmanned flight-bombs built at Peenemünde and in the Nordhausen underground *Mittelwerke* mine shafts,[234] there was a much heavier death toll of about 20,000 from the hazardous production process itself, involving approximately 60,000 slave laborers. The missiles were assembled in camps under appalling living conditions—a well-known fact among the German scientists who were employed at Peenemünde.[235] By contrast, the bombing of Hiroshima and Nagasaki alone caused a death toll of 340,000. The fusion of these two technologies is still with us in the form of the intercontinental nuclear missile.

---

on actual fact, however.

[229]See, e.g., von Weizsäcker [1991] who emphasizes that this is a mere one thousandth of the cost of the Manhattan Project.

[230]Prior to the Allied bombing of the site on Aug. 17, 1943 resulting in 735 deaths. The missile production facilities were rebuilt in an old gypsum mine in the Harz mountains—figures from Tarter [1992], pp. 155, 160; cf. also Neufeld [1995], p. 206, who arrives at a total staff of 12,000 including the foreign workers and prison laborers.

[231]Ibid.; cf. Gowing [1964], Hawkins et al. [1983], Rhodes [1986], Hoddeson et al. [1993].

[232]Tarter [1992], p. 161, estimates three billion wartime dollars, while Neufeld [1995], p. 273, estimates only 2 billion marks, that is half a billion US dollars of World War II vintage (about 5 billion dollars in early 1990).

[233]Neufeld [1995], p. 273; on the German war economy cf. Milward [1965], Janssen [1968] pp. 325ff., 391ff., and Zilbert [1981].

[234]The V-1 killed about 5,500 London inhabitants and injured 16,000. The V-2 launched against London caused up to 2,700 deaths and 6,000 injuries; and casualties in Antwerp numbered 2,200 with 7,000 seriously injured. See Bode & Kaiser [1995], pp. 112–118, and Neufeld [1995], p. 264, for the figures as well as Johnson [1978], chap. 3 for the perspective of the British intelligence on these weapons.

[235]On the slave labor camps and eye-witness accounts, see Neufeld [1995], pp. 208ff., 264f.

5.2 German Physical Associations

The history of the two main professional organizations of the period between 1933 and 1945, the German Physical Society (*Deutsche Physikalische Gesellschaft: DPG*) on the one hand,[236] and the German Society of Technical Physics (*Deutsche Gesellschaft für technische Physik: DGtP*) on the other,[237] captures another important dimension of physics and National Socialism. Unlike other major organizations, such as the chemical trusts and non-government research institutions like the *KWG* (see the previous section), these two societies did not opt for rapid voluntary political realignment after 1933, choosing rather to take a relatively independent course. This worked surprisingly well until the Nazi-activist *DPG* members Herbert Stuart and Wilhelm Orthmann filed a petition demanding the expulsion of all remaining Jewish members. Since this petition could not be rejected without causing a major scandal, the president of the society Peter Debye requested the 'voluntary resignation' of all Jewish members in late 1938 (cf. docs. 67f.). I have compared the *DPG*'s membership lists for 1938 and 1939 and find that 121 names do not reappear in the second year (it can be assumed that only a small number of these resignations were for normal reasons); on the other hand, 84 new memberships are listed. The increased political pressure on the society (of roughly 1,200 members) as well as on the *DGtP* (with a membership of 3,000 shortly after its foundation in 1920 and ca. 2,500 in 1940, after which membership plummeted) is also reflected in the internal memorandum to the Science Ministry (doc. 66). There the technical physicist Karl Mey, who was president of the *DGtP* throughout the entire Nazi period up to 1945, is denounced as unreliable on both political and personal grounds, backed primarily by the argument that at a meeting of the society he failed to toast the *Führer*.

5.3 The Protracted Conflict between Experimental and Theoretical Physics and the Emergence of the 'Aryan Physics' Movement

Some of the conflicts between different factions during the Third Reich have their origins well within the Weimar period. An example in the physics community is the ferocious fight between theorists and experimentalists, i.e., von Laue and Sommerfeld versus Lenard and Stark.[238] It is important to see this historical continuity behind the apparently drastic changes that took place in 1933.[239] For this reason, some pre-1933 documents have been included in this anthology dating as

[236]On the early history of the *DPG*, dating back to the famous *Berliner Physikalische Gesellschaft* founded in 1845 see, e.g., Mayer-Kuckuck (Ed.) [1995] and Heinicke [1985].

[237]See in particular Hoffmann & Swinne [1994].

[238]For a general description of Weimar Germany see, e.g., Gay [1968]; on the rise of antidemocratic thinking at that time see, e.g., Sontheimer [1962], Stern [1972]a, and Ringer [1983]; on science in particular, see the classic studies by Forman [1971], [1973], and [1974] as well as Nachmansohn [1979] or Volkov [1987] on the many outstanding contributions by Jewish scientists.

[239]See, e.g., Kleinert [1978], Stern [1982] and [1986] for the prehistory of 'Aryan Physics'.

early as the 1920's. In 1923 the theoretician and close friend of Albert Einstein, Max von Laue, reviewed Stark's recent book on the supposed 'crisis' in German physics (doc. 2). Although Stark had supported Einstein's revolutionary concept of light quanta of 1905, he turned later into a sharp opponent of modern physics in general,[240] decrying relativity and quantum theory as 'typically Jewish' fabrications without any empirical justification. Max von Laue's review of Stark's book illustrates the superior level of von Laue's rebuttal to arguments that had degenerated from naive common-sense objections to unfounded and polemical attacks by experimentalists unable to keep up with the developments in modern physics, which required considerable proficiency in fairly complex mathematics. The first document in this anthology shows how heated the discussion had become in 1920 with Einstein's angry response to two lectures delivered under the ægis of a fictitious 'antirelativist' association organized by one of the obscure protagonists of this propaganda campaign, who soon admitted openly the anti-Semitic motives behind his actions.[241]

The experimentalists Stark and Lenard, both Nobel Prize recipients as Einstein was, felt underrated. Public interest in confirmations of Einstein's theories went beyond anything they themselves had experienced.[242] The publicity surrounding the famous light-deflection measurements of November 1919 completely overshadowed Stark's discovery in late 1913 of the influence of electric fields on spectral line patterns, for example. Many experimentalists felt they could no longer understand recent developments in theoretical physics and as a result suspected theoreticians of maliciously working on a purely formalistic level. Aside from relativity theory this also included Planck's and Sommerfeld's contributions to quantum theory, as well as Heisenberg's and Born's contributions to quantum mechanics since 1925, doing away with the last remnants of intuitiveness (*Anschaulichkeit*) in favor of an algebraic manipulation of terms and a probabilistic interpretation that consciously repudiated all former attempts at creating a causal world view.[243] The physics, and to a greater extent the mathematics on which these new theories were based went beyond what the classically prepared old school led by Stark and Lenard could follow. But their 'culture shock' was more than a simple lack of comprehension of 'modern' physics. It also incorporated old fears of a sinister Jewish conspiracy to create mass-hysteria in the population, reinforcing their general distrust of theory, and reawakening

---

[240]Cf. Hermann [1966] and [1967] on Stark's relations with Einstein and Arnold Sommerfeld; cf. also Kleinert & Schönbeck [1978] on Lenard's relations with Einstein prior to 1920.

[241]See Weyland (Ed.) [1921] and the annotation to doc. 1. On the development of arguments against the theory of relativity and the emergence of flagrant anti-Semitism after 1920 see, e.g., Grundmann [1967], Hermann [1977], Kleinert [1979]a, and Hentschel [1990]a, sec. 3.2.

[242]On Einstein's reception in the press see, e.g., Elton [1986], Hentschel [1990]a, chap. 2, and further references given there.

[243]See in particular Forman [1971] for a description of the intellectual climate under which these massive changes in the conceptual foundations took place.

repressed inferiority complexes.[244] These elements had been seething under the surface throughout the Weimar Republic and had only occasionally spilled over into scientific publications, because the protagonists of modern theories holding some of the most influential posts in scientific societies, journal publishing houses, and other institutions and were able to stem the tide. The theoreticians Albert Einstein, Max von Laue, Arnold Sommerfeld, Peter Debye, and Max Planck, held key positions in central research institutions such as the *KWI* of Physics, the committees of the *Notgemeinschaft*, and the Prussian Academy, etc., and thus controlled the largest share of financial support for research projects in Weimar Germany.[245] Furthermore, it was they who were asked for expert opinions when decisions were made by university committees on physics appointments within the German-speaking countries. The Munich theoretician Arnold Sommerfeld, for instance, was extremely successful in placing his many pupils and became so influential that he had the reputation of a gray eminence.[246] Lenard and Stark, on the other hand, were much less successful in terms of science policy and career management: In 1920 Stark had begun to solicit members for a newly founded organization called *Fachgemeinschaft Deutscher Physiker* (German Professional Society of Physicists) that somehow did not manage to compete seriously with the traditional German Physical Society. In 1922 he left Würzburg University to concentrate on porcelain research and development.[247] This move shut him out of academia during the Weimar period, however; his applications for vacant chairs all fell through, partly because of his choleric disposition, but partly also as a result of his conflicts with Sommerfeld, who barred his receiving the call to fill Willy Wien's chair for experimental physics at the University of Munich in 1929.[248] In 1925 Lenard decided to resign his membership in the *DPG* after a quarrel over citation policy in German physics periodicals, and he allegedly even went so far as to post outside of his Heidelberg lab the following notice: "No Entry to Members of the So-Called German Physical Society!"[249] Other antirelativists,

[244]On the history of anti-Semitism in Europe and the global conspiracy theory see, e.g., Poliakov [1955/68], [1968], Kampmann [1979], and Klamper (Ed.) [1995], esp. pp. 264ff.; on academic anti-Semitism in Germany see, e.g., Schottländer [1979], Ringer [1983], Kampe [1985], Grab (Ed.) [1986], Volkov [1987], Hammerstein [1995], and other references there. For a pictorial synopsis of anti-Semitic clichés in their most venomous form see Hans Diebow: *Der ewige Jude*, Munich and Berlin: Zentralverlag der NSDAP, 1937.

[245]See, in particular, Forman [1974], and Richter [1971], [1972], [1973].

[246]See, e.g., Benz [1975], pp. 126f.; Eckert et al. (Eds.) [1984], pp. 115ff.; Eckert [1993], pp. 97ff.; cf. also Stark [1930]a-b whose complaints about Sommerfeld's importance are indirect evidence of this point. The selected edition of Sommerfeld's correspondence being prepared by Michael Eckert at the DMM will also provide ample examples of Sommerfeld's key role in professorship appointments in physics.

[247]See the Stark file in: Akten des Rektorats und Senats der Universität Würzburg, No. 837.

[248]Cf. Hermann [1967], Eckert [1985], pp. 88f., [1993], p. 200; Benz [1975], pp. 171f. Walther Gerlach was chosen instead; cf. also Heinrich & Bachmann (Eds.) [1989], pp. 65ff.

[249]See, e.g., Steenbeck [1977], p. 71, Beyerchen [1977], pp. 111ff., 240, Stark file in the Würzburg Univ. archive (see footnote 247 above); Walker [1995], chaps. 2–3. On the increasing

such as the *PTR* physicist Ernst Gehrcke and Stark's pupil Ludwig Glaser were also marginalized and driven to ever closer alliances with radical elements and suspect outsiders, such as the right-wing activist and trickster Paul Weyland or the editors of the obscure pamphlet entitled *100 Authors against Einstein*, which appeared in 1931.[250] These badly chosen alliances in turn destroyed their reputations altogether so that they were simply ignored by the powerful scientific élite.[251] It did not help that the scientific ideals propagated by Lenard and Stark were hopelessly outmoded. Mechanistic models for the atom and the prerequisite of intuitiveness in theories were part of the mental set of classical physics, which had been superceded by relativity theory and quantum mechanics after several arduous efforts to uphold their premises had ultimately failed.[252] Their distrust in mathematical formulations in theories went even beyond the degree reached in say, Maxwell's theory of electromagnetism; and their naive inductivism and animosity against the modern form of scientific activities actually only impeded their efforts to produce any new results that may have convinced others of the fruitfulness of their approach.[253] Lenard's speculations in 1910 on the ether and its interaction with matter were altogether obscure, because he was unable to address the many problems posed by the conception of a stationary medium.[254] Lenard's postulation of an *Uräther* that was supposedly modified in the presence of matter may have been consistent with his unshakable belief in layers of reality forever shielded from human discovery, but they did not correlate with any empirical findings such as, for instance, his collaborator's repetition of the Michelson-Morley interferometer experiment using the light of fixed stars. Disappointingly, this experiment by Rudolf Tomaschek yielded a negative result, turning this effort to detect an ether wind into failure once again.[255]

---

importance of English-speaking scientific journals see, e.g., the statistics in Hooker [1935], and Siegmund-Schultze [1993].

[250]Cf. the very critical review of this book by the astronomer Albert von Brunn (doc. 4); see also Goenner [1994]. On Weyland see Kleinert [1993], Goenner [1993], and Hentschel [1990]b. On Gehrcke see Goenner [1993], and on Glaser see Walker [1995], pp. 11, 55–57, and doc. 77 where Glaser's anti-Semitic resentment is given free rein.

[251]Cf. von Laue's sarcastic tone in his review of Stark [1922] (doc. 2) or Weizel's review of Müller's pamphlet in 1942 (doc. 89). Stark's frustation is evident in his polemical article against the 'Jewification' (*Verjudung*) of German universities, Stark [1930]a.

[252]On the development of quantum theory and quantum mechanics see, e.g., Jammer [1966], and the numerous other secondary sources mentioned in part there. See also doc. 31. On Stark's vision of the alternatives see, e.g., Stark [1927], [1928], [1931]. See also the controversies between Sommerfeld and Stark: Sommerfeld [1930] and Stark [1930]b.

[253]For a broader analysis of this antagonism between modernists and antimodernists on the basis of the debate over the foundations of mathematics (formalists versus intuitionists), see Mehrtens [1990].

[254]See Lenard [1910]; cf., for instance, Whittaker [1951/53], Vol. 1, chaps. 4–5, 9–10.

[255]See Lenard [1920/21], and Tomaschek [1924], as well as [1942], Teichmann [1942] and the correspondence between Lenard and Tomaschek, copies of which are in the possession of Andreas Kleinert, Univ. of Halle. For a history of experiments of the Michelson-Morley type, see Lloyd

With the Nazi *Machtergreifung* in 1933 the anti-modernists now sought their revenge and lunged for the key positions formerly held by their adversaries. As members of the 'old guard' in the National Socialist movement, Lenard and Stark were free from the suspicion of being political opportunists;[256] they held good cards indeed in the early days of the National Socialist 'renewal'. Both became top-level consultants in the Nazi bureaucracy and even had access to Hitler himself. They also meddled in professorship appointments.[257] Max von Laue points out this change in the élite in a letter to Walther Gerlach in July 1933: "Besides, the way to obtain a position now is not the one via Planck and yours truly, but via Lenard and Stark. [...] Planck and me are so ill-famed—you know as what—that I would be more of a hindrance than a help if I were to recommend anyone in Germany."[258] Not surprisingly, in May 1933 Stark was nominated as president of the *PTR* despite the unanimous advice to the contrary of several scientists who had been consulted.[259] Although he did not succeed in his bid for the presidency of the German Physical Society (*DPG*) as well, Stark was appointed president of the *Notgemeinschaft* (subsequently renamed *Deutsche Forschungsgemeinschaft*) in June 1934.[260]

Once these prestigious positions had been taken over, moves were made toward dominating science policy or at least determining who was to refill the many vacant chairs and academic positions since the recent dismissals. These efforts were less successful, however. Only about a dozen scientists, many of these being pupils of Stark or Lenard, decided to join the bandwagon of 'Aryan Physics' at that time. The overwhelming majority of the ca. 1,000 members of the *DPG* held privately to their conviction that science is a supranational enterprise and that it made no sense to distinguish between 'Jewish' and 'Aryan' contributions; yet

Swenson: *The Etherial Æther*, Austin, Univ. of Texas Press, 1972.

[256]As early as 1924 Lenard and Stark had published in a newspaper a declaration of loyalty to Hitler—see doc. 3; cf. also Stark [1930]c for his peculiar account of Hitler's aims and personality, and Stark [1987] for his retrospective remarks about his relations to Hitler. Stark had even been a political activist in the Traunstein region for the 1932 elections—cf. Walker [1995], p. 16, and the sources from Stark's party file from the BDC cited there.

[257]See the correspondence between Stark and the Reich Minister of the Interior Wilhelm Frick, November 1933, cited in Walker [1995], p. 18 and footnote 45, as well as Kleinert [1980], p. 43. Cf. also Voigt [1981], pp. 42f., and Deichmann [1992], pp. 234f.

[258]July 25, 1933, DMM, Gerlach Papers, in Heinrich & Bachmann (Eds.) [1984], p. 73, fig. 44.

[259]See Lenard's emphatic praise of this decision in doc. 18 as well as von Laue's later account, doc. 19. Cf. also Beyerchen [1977], pp. 100, 114f., 241, Kleinert [1980], pp. 35f., 43f., Walker [1995], pp. 16ff.

[260]See the talks presented at the Würzburg conference of the German Physical Society in September 1933 (docs. 27 and 28), the correspondence between von Laue and Mie of 1934 (docs. 33f.), and Stark's contemporary declarations on German science policy in Stark [1934]a–d. See also Kleinert [1980] on Stark's and Lenard's strained relations with the *KWG*, Eckert [1985], pp. 89f. on his and Glaser's criticism of the physics collection, esp. the Sommerfeld atom models in the *Deutsches Museum*, Munich, as well as Walker [1995], pp. 18ff., for Stark's imposing but strategically inept presentation as the dictator (*Führer*) of physics at the Würzburg meeting. Cf. Stark's postwar account in Stark [1947] and von Laue's [1947]b rebuttal.

only a few dared to put it in writing.[261] Lenard's textbook on 'Aryan Physics', completed in August 1935 with the aim of providing German universities with a politically correct textbook incorporating only proper (that is classical) physics was welcomed enthusiastically only by the few known supporters of the movement. The vast majority of physicists continued to shun the latter, however. One future member of the Peenemünde research team, at that time still a student of physics, reports:

> When Lenard's book *Deutsche Physik* was published, it met with head-shaking and amazement among colleagues. We young physicists read a few pages out of curiosity, and then put it aside. I remember that Hans Geiger once said to a group of students, 'This is all very strange. One cannot do away with the facts of physics just like that. I'm so surprised that Lenard should have digressed so far; he used to be a very good experimenter.' Under the circumstances, it was very courageous of Geiger to say that much. We students got the message. I remember that I was glad to have this assurance and confirmation of my own thoughts.[262]

A similar impression is gained from a survey of physics textbooks and problem collections published for use in schools. Teachers were reluctant to indulge in Lenard's obscure ether models, nor did they show much inclination to adopt the recommendations for 'Germanized' physics terminology published in a science pedagogy journal in 1935.[263] Walther Gerlach's and Max von Laue's articles from the mid-1930's on the relation between theoretical and experimental physics should also be interpreted as an indirect refutation of the one-sided opinions raised by representatives of the 'Aryan Physics' movement on this issue.[264] Thus the 'Aryan Physics' movement did not really leave much trace in textbooks, quite in contrast to the new vogue for topics like 'military' physics, which transformed the character of physics education after 1933.[265] Thus around 1936 the Lenard-Stark circle heightened the ideological pressure with a new wave of publications severely attacking 'undesirable' tendencies in physics. In their earlier publications Stark and his cohorts concentrated on 'racially Jewish' targets, with Einstein as the prime example of a "ruthless formalist". Now they broadened their line of attack and openly criticized 'racially Aryan' scientists such as Heisenberg, von Laue and

[261] See, e.g., Bavink [1934]a, [1938], von Laue [1936], Weizel [1942], Heisenberg [1943]; cf. Hentschel [1993] on the unsuccessful efforts to censor Bavink's later textbooks.

[262] Ernst Stuhlinger in a letter to Donald E. Tarter, February 1989, quoted in Tarter [1992], p. 159; cf., ibid., p. 160, for a quote from another physicist, Georg von Tisenhausen, who reportedly never heard of 'Aryan Physics' throughout his entire physics training at that time. Cf. doc. 39 for the very polemical foreword to this textbook, and von Laue's review, doc. 45.

[263] See Hillers [1935]; cf. also Haberditzl [1963]. Even Lenard had his own policy of only refraining from using words that required a 'foreign' pronunciation: See doc. 39, footnote ††.

[264] See Gerlach [1936] and von Laue [1937].

[265] On the emergence of *Wehrphysik* see sec. 3.2 and references there; on the failure of 'Aryan Physics' to penetrate school textbooks, see also the complaints to this effect by Stark [1939] and Wesch [1939], as well as Brämer & Kremer (Eds.) [1980], pp. 68–75.

Sommerfeld, for whom they coined the term 'White Jews', for continuing to promote this supposedly formalistic, counter-intuitive and unrealistic style of physics (see in particular doc. 55). This move only exposed the complete uselessness of their pseudo-racial classification of physics which, to quote Prandtl, basically boiled down to: "Whatever's beyond my comprehension, I regard as Jewish pretension" (see doc. 85). The theoreticians decided nevertheless to counter these attacks, which they perceived as character assassination campaigns. A vicious battle over the proper ideological setting of physics ensued: See, for instance, the attack on Heisenberg and other theoreticians by Lenard's student Willi Menzel in the Party paper *Völkischer Beobachter* in January 1936, Heisenberg's reply in the same newspaper in February 1936 (docs. 42 and 43), as well as the even more biting attacks on Heisenberg in the *SS* journal *Das Schwarze Korps* in July 1937, the subsequent petitions by his defenders (see docs. 55ff.), and Heisenberg's own moves, which eventually led to Himmler's personal intervention in the matter (see doc. 64). A further setback came with the so-called 'Munich synod' (*Münchener Religionsgespräche*) arranged by Wolfgang Finkelnburg, the new vice president of the *DPG* since 1941, at which an equal number of proponents and opponents of the modern theories of relativity laid out their cases (cf. doc. 93 for the full text of the conciliatory compromise agreed to and doc. 110 for Finkelnburg's version of the event and its background).

Although this debate by no means stopped the flood of antirelativistic pamphlets,[266] its outcome indicated that the heyday of the openly ideological wing of the Nazi movement was over. With the onset of World War II, and at latest with the American involvement after the Japanese attack on Pearl Harbor in 1941, hotheaded demagogues were supplanted by cold pragmatists like Mentzel, Thiessen, Harteck and Vögler, who were much more interested in an efficiently running research and development sector in science and industry than in any politically correct programmatics. The success of the aerodynamics specialist Wilhelm Müller in succeeding Sommerfeld in the prestigeous chair for theoretical physics at Munich University in 1939, even though he had never even published in a physics journal, was a last Pyrrhic victory for the 'Aryan Physics' movement,[267] since his incompetence in theoretical physics was so obvious that it gave influential physicists and engineers at the pragmatic end of the political spectrum, such as Ludwig Prandtl and Carl Ramsauer, ample cause to complain officially about the worrisome state of physics in general (see, e.g., docs. 62, 90ff.) and of theoretical physics in particular (see, e.g., doc. 85).

All in all not more than perhaps two dozen physicists adopted a flagrantly pro-

[266]For later treatises see, for instance, May [1941]a, Müller [1941]a, [1944], as well as Müller's article, doc. 83.

[267]On this appointment which was made against the unanimous wishes of the faculty see, e.g., Beyerchen [1977], pp. 164–167, Cassidy [1992]a, chaps. 18ff., Eckert [1993], pp. 200f., Walker [1995], pp. 51–55. Compare this case also with the astronomer Otto Heckmann's drawn out appointment proceedings at Hamburg, described in Renneberg & Hentschel [1995].

Nazi stance within their field. These included as noted above Wilhelm Müller, August Becker who became Lenard's successor to the chair for experimental physics at Heidelberg, Ludwig Wesch, also from Heidelberg University where he held the chair in theoretical physics, Alfons Bühl who was professor at the Karlsruhe Polytechnic, and finally Ferdinand Schmidt, another pupil of Lenard's and professor of experimental physics at Stuttgart Polytechnic. According to Beyerchen, by the end of 1939, outspoken representatives of the 'Aryan Physics' movement occupied six of the 81 professorships available in Germany and Austria, that is less than 10% of them.[268] The notorious *NSDStB* activists Fritz Kubach and Bruno Thüring, Stark's doctoral student Ludwig Glaser, and Willi Menzel, are added to this select list, along with several other fringe figures such as the opportunist philosophers Hugo Dingler, Eduard May and Friedrich Requard.[269]

It is nevertheless an interesting facet of the social dynamics within the National Socialist system that *de facto* collaboration with the Nazi regime was not only tolerated but in fact widely practiced among German physicists,[270] while explicit *ideological* commitment to the Nazi phraseology was repudiated and invariably led to isolation within the physics community.[271] Scientists as a whole were not prepared to accept that the value of scientific results depended on the origin of the scientist, a position which also only seemed to gain any logical foothold when seen within the framework of the interrelation between science and nationality, and society and race as suggested in the Nazi ideology.[272] However, this same concern for the well-being of the discipline, which thus impeded a stronger outward ideologization within the National Socialist period, also made

[268]See Beyerchen [1977], pp. 171f., and Walker [1995], p. 51; on Heidelberg, where the physics institute was renamed 'Philipp-Lenard Institut' in 1936, see N.N.[1936], P.F.F. [1936], A. Becker [1942], Wolf [1962], and Neumann [1986]; for contemporary statistics of occupied and vacant positions in 1934 cf. here doc. 35.

[269]On Dingler, see Wolters [1992]; on E. May see Fischer et al. (Eds.) [1994], p. 94. For a listing of most of the other anti-Einstein activists see Hentschel [1989] and [1990], sec. 3.2.

[270]On the wider problematics of *de facto* collaboration through "loyalty to the state and carefully conformed agitation" by scientists who nevertheless saw themselves as non-conformists, see in particular Mehrtens [1989], and Mehrtens in: Renneberg & Walker (Eds.) [1994], and in: Meinel & Voswinckel (Eds.) [1994], pp. 19ff. On the commonplace self-image of "resistance through collaboration", see also Walker [1995], p. 63; on Heisenberg's interpretation of 'active resistance' within a totalitarian state as opposed to 'passive resistance' from outside, formulated in November 1947 in a memorandum in defense of Carl Friedrich von Weizsäcker's father, who was secretary of state in the German Foreign Ministry, see Walker [1989]a, pp. 201–203, for excerpts and a critique.

[271]Cf., e.g., Tarter [1992], p. 159, for a quote from an interview with the engineer Ernst Stuhlinger: "Lenard, Stark and Tomaschek were really ostracized. Physics was taught as usual, with Einstein's relativity, Bohr's atom model, Heisenberg's and Schrödinger's quantum mechanics, Pauli's principle, etc."

[272]The scientific community's resistance to precisely this tenet of the ideology is comparable to today's repudiation of advocates of social constructivism, who also try to link scientific results to the specific social groups that produce (rather than discover) the results and thereby threaten the Mertonian ideal of scientific universalism.

it vulnerable to *de facto* conformity of research activities to the needs of the Behemoth, quite often even anticipating it by ceding self-imposed realignment or even self-mobilization.[273] The good intentions of attempting to prevent worse from happening, of avoiding having less-qualified persons assume important positions in science and administration, were a guiding motive behind the decision by leading scientists such as Heisenberg, and others to stay in Germany rather than choose exile.[274] It was too late when much later some realized that this initial decision, followed by the many small concessions along the way so as not to lose ground, had transformed the vast majority of scientists and engineers into easy targets and convenient servants of a criminal state.[275]

5.4 Physical Research in Germany between 1933 and 1945

This selection obviously cannot encompass the entire breadth of physics pursued in Nazi Germany.[276] Early quantum field theory and quantum electrodynamics have to be omitted, for example, because of a lack of sufficiently popular texts.[277] Solid state physics had already become much too specialized, and a complete explanation would require a considerable amount of space.[278] Topics that can be

[273]See Ludwig [1974], p. 241, Trischler [1988], pp. 14ff., as well as C. Tollmien and A. Heinemann-Grüder in: Tschirner & Göbel (Eds.) [1990], Maier [1993] pp. 28ff., the 7 biographical studies in Böndel et al. [1995], as well as Heinemann-Grüder and Trischler in: Renneberg & Walker (Eds.) [1994] on the mentality of German engineers; self-mobilization is defined by Helmuth Trischler as "voluntary involvement, the free devotion of an individual's ability, above and beyond the professional call of duty, to advance the objectives of the regime" (p. 73). On the motives behind 'self-mobilization', see Mehrtens in: Meinel & Voswinckel (Eds.) [1994], pp. 27–29.

[274]Cf., e.g., Cassidy [1992]a, p. 307, on Planck's advice to Heisenberg to stay "in order to form islands of stability", and Heisenberg's letter to Born, dated June 2, 1933, in which he tried to convince Born to return to Göttingen for the sake of theoretical physics (doc. 22; cf. also Cassidy [1992], pp. 307f.).

[275]In his history of National Socialism, Friedrich Glum, general director and manager of the administrative section of the *KWG* throughout the 1930's, used the metaphor of an escalator on which persons were swept away, once they had taken the decision not to emigrate, never having a moment to take their bearings: See Glum [1962], p. VII.

[276]Cf., for instance, Bothe & Flügge (Eds.) [1948] on nuclear physics and cosmic rays, as well as other issues in this Fiat Review series such as Betz (Ed.) [1948] on hydrodynamics and aerodynamics, Goubau & Zenneck (Eds.) [1948] on electronics, Kappler (Ed.) [1948] on the physics of liquids and gases, Kopfermann (Ed.) [1948] on the physics of electron shells, Joos (Ed.) [1948] on the physics of solids, and ten Bruggencate (Ed.) [1948] on astronomy. See also technical progress reports for individual fields, e.g., Euler & Heisenberg [1938] on cosmic rays, or the *Uranverein*'s secret progress reports listed in David & Warheit (Eds.) [1952].

[277]Cf., however, the anthologies edited by Julian Schwinger, *Quantum Electrodynamics*, New York: Dover, 1958; and Arthur I. Miller (Ed.), *Early Quantum Electrodynamics: A Source Book*, Cambridge: Cambridge Univ. Press, 1994.

[278]See, however, the remarks in Eckert [1993], pp. 212–216, Schubert [1987] and the following pages for the links to the development of semiconducting radar detectors at *Siemens* and elsewhere. See also the contributions by P.K. Hoch, J. Teichmann, K. Szymborski and S. Weart in:

addressed in relatively short texts have been preferred. However, other ideologically relevant topics such as relativity theory and quantum mechanics, which bore the brunt of the attacks by the *Deutsche Physik* movement, are represented, as well as technological applications of physics in the weapons industry. Nuclear fission and radar technology have been of some interest to the public in both the past and the present (see below). The newspaper report on the 1940 physics war conference (doc. 82) is an example of the extent to which research activities in physics were molded by the needs of the military after the outbreak of World War II. This is underscored by two more specific texts on the physics of submarine warfare and the aerodynamics of bomb dropping (docs. 100 and 101).

It should be pointed out, though, that despite occasionally successful research and development work especially in the laboratories of the German Air Force, all in all the National Socialist polycracy and the disfunctionality of organizations established specifically to centralize and coordinate research such as the *RFR* (cf. sec. 3.2 above) prevented these technological innovations from coming to full fruition—fortunately we might add, since the war would otherwise have been extended. The 'reactionary modernism' of top officials, such as the powerful but incompetent president of the *RFR* and High Commander of the Air Force Göring, did imply a hightech orientation, fostering great hopes in the saving power of new 'wonder weapons' in the final years of the war. However, it was not combined with a farsighted allocation of the scarce resources. They were squandered on a few hypertrophic projects that had caught the fancy of the vengeful *Führer*. The high-speed jet fighter so crucial to aerial combat above German territory came too late to the serial production stage, for example, even though the basic technology of jet engines had already been developed prior to the war,[279] and delta-wing jet fighter designs never left the drawing board. During the last apocalyptic years of World War II, increasingly hectic efforts were made to compensate for earlier mistakes in research and development planning.[280] For instance, in May 1944 the aircraft producer *Heinkel A.G.* confiscated an Austrian gypsum mine in Hinterbrühl (Vienna), and turned it into an underground production plant employing 2,000 forced laborers to construct the jet fighter He

Lillian Hoddeson et al. (Eds.) *Out of the Crystal Maze. Chapters from the History of Solid-State Physics*, Oxford: Oxford Univ. Press, 1992.

[279]On the jet engine, which was developed both in Britain at the Thomson Houston Company from 1936 on, and in Germany where *Heinkel* built the first flying prototype in August 1939, nearly two years before the first British jet flight see, e.g., John Ziman: *The Force of Knowledge*, Cambridge: Cambridge Univ. Press, 1976, pp. 182–188, esp. p. 186: "The failure of the *Luftwaffe* to win air superiority with a fast jet fighter during the War was mainly due to incompetent organization at the top. The German Air Ministry believed in piston engines and failed to give adequate support to the brilliant engineering teams of their aircraft companies."

[280]See, e.g., Trischler [1992]a, pp. 245ff., and pp. 225f., on the delta-wing design at the *Deutsches Forschungsinstitut für Segelflug* branch headed by the flight pioneer Alexander Lippisch; cf. also the latter's autobiographical account in: *Ein Dreieck fliegt. Die Entwicklung der Delta-Flugzeuge bis 1945*, Stuttgart, 1976.

162. When a shortage of metallic materials developed, it was resolved to build suicide aircraft and unpowered combat gliders (in some cases economizing by using wooden parts and other materials to replace metal) as part of the last great effort of the 'People's Army'[281] against Allied bomber formations.[282]

Another key area of research was the detection and ranging of radio waves (radar) which was carried out both by American and British researchers in specialized research centers such as the Royal Navy's research labs at the experimental department of HM Signal School in Portsmouth, and from the summer of 1940 on, in the newly created 'radiation laboratory' at MIT in Cambridge, Mass., which already had a staff of 90 physicists and engineers, 45 technicians, and 6 Canadian guest researchers by November 1940.[283] Radar direction finding experiments with long waves in the meter and decimeter range were also conducted in Germany,[284] but with the development of the top-secret air-to-surface vessel radar (ASV), the Allies gained a huge advantage. These devices could be installed in airplanes, thus making it possible to trace surfacing German submarines (submarines could not avoid being in this vulnerable position at regular intervals to recharge their batteries and replenish their air supply). This advance in turn only became possible with the development of shortwave radar in the microwave (cm) range.[285] German submarines were not equipped with sensors to detect Allied shortwave radar, which proved decisive in Germany's defeat in the U-boat war; by the summer of 1942 there was a sharp increase in Allied successful hits.[286]

On the whole, there was a considerable reduction in new inventions in Germany as indicated, e.g., by the patent application statistics between 1930 and 1940, down from 78,748 to 43,479.[287] Furthermore, as Ludwig, Herf and others

---

[281]The *Volkssturm* was the German home guard to serve as a desperate last defense. Hitler's decree of Sep. 25, 1944 mobilized all males between 16 and 60 not already enlisted in the Armed Forces in this militia for the protection of local districts, but many units were sent to the rapidly receding front.

[282]On the increasingly absurd features in hightech airplane development in the Third Reich's final stages which had distinct similarities to Japanese *kamikaze* tactics, see in particular Ulrich Albrecht in: Renneberg & Walker (Eds.) [1994] and in: Tschirner & Göbel (Eds.) [1990].

[283]According to Eckert [1993], p. 227, its staff increased to a total of 3,897 in 1945, divided into over one hundred highly specialized teams.

[284]See in particular Brandt [1962], pp. 9–112, Reuter [1971], and Trenkle [1986]; the first patent application in this direction was issued to the German engineer Hülsmeyer in 1904, but was not followed up for nearly three decades.

[285]See, e.g., Guerlac [1987], Fisher [1988], Howse [1993], and Eckert [1993], pp. 224ff. For a popular exposition of the contemporary technology see, e.g., Hallows [1946].

[286]See, e.g., Guerlac [1987], p. xxii, Kevles [1987], p. 308, and Eckert [1993], p. 232: "The atom bomb only ended the war. Radar won it." For a detailed military history, see Fisher [1988] and Howse [1993], on Allied anti-submarine technology in particular see Johnson [1978], chap. 4.

[287]See Ludwig [1974], p. 227; the granted patents numbered 26,737 compared to 14,647 in these two years; for later regulations restricting the development of 'useless' inventions to prevent the wasteful dissipation of national resources and funds see, e.g., 'Prüfung von Erfindungsvorschlägen', published, e.g., in *Elektrotechnische Zeitschrift* **63** [1942], Aug. 27 issue, p. 402. On Hitler's inventors edict of April 17, 1942, which allowed each soldier to address the Army Major General

document, Germany's technology was inferior to that of its enemies in many crucial areas including shortwave radar development, operations research, communications technology, air defense, sea mine and torpedo technology, as well as in its sheer production capacity.[288] German aero-engine production at 15,000 in 1940, for instance, had a 1:1 ratio to American production, and a 3:4 ratio to Great Britain. In 1942 the respective numbers were 37,000 engines in Germany, 54,000 in Great Britain, and 138,000 in the USA.[289] German steel production in 1939 totalled 27 million tons per year, compared with over 100 million tons produced by Great Britain, the Soviet Union and the USA. The disadvantage in tank production rose from a 1:1.5 ratio to 1:4 between 1941 and 1944, disregarding the drop in quality. In 1943 German airplane production amounted to only about 20% of that of the Allies, and 10,757 new tanks against about 30,000 on the side of the Allies. On the basis of these figures Herf comes to the following conclusion: "In short, the German dictatorship simply lacked the technical means to win the war, except, of course, its war against the European Jews."[290] From 1933 on armament expenses rose steadily against the national income: In 1933 they came to 3 billion RM against 46.5 billion receipts. In 1935 the proportion was 10:58.5 billion, in 1938 37:79.5 billion and in 1942 91:127 billion RM. Thus the German war economy was essentially being driven by military expenditures.[291]

The German project on the use of nuclear fission for military and civilian purposes (in short: the *Uranverein*) is a classic example of fundamental research with military implications. It originated in probably the most important scientific discovery made in Nazi Germany between 1933 and 1945: the discovery of nuclear fission by Otto Hahn and Fritz Straßmann in late Dec. 1939 at the Kaiser Wilhelm Institute of Chemistry in Berlin Dahlem. In late autumn of 1939 they found barium instead of the expected radium in their radiochemical analysis of the decay product of a uranium sample that had been bombarded with slow neutrons.[292] Hahn informed Lise Meitner in Sweden, where she had been

---

directly about proposed improvements in weapons design, see Boelcke [1969], p. 94, and Guérout [1992], p. 17.

[288]On the scientific aspects of mine and sea-mine sweepers see, e.g., Steenbeck [1977], pp. 94ff., and Johnson [1978], chap. 5, on torpedos, see Rössler [1984]; on the submarine industry, see Zilbert [1981], chap. 5.

[289]Numbers according to Braun [1992], p. 5; cf. also Zilbert [1981], chap. 6. Leading industrialists at a *Daimler-Benz* board meeting on Dec. 11, 1941 are quoted there as attributing this inability to keep up with the competition to unsuccessful coordination of production.

[290]For the preceding figures and the quote, see Herf [1984], p. 20; cf., e.g., Ludwig [1974]. However, the numbers given by Trischler [1992]a, p. 176 on airplane production between 1931 and 1944 show that their absolute numbers had increased from a mere 13 to nearly 40,000.

[291]Figures from Bode & Kaiser [1995], p. 123. See also the figures in the official statistics N.N. (Ed.) [1949], as well as Milward [1965], Janssen [1968], pp. 325ff., 391ff., Speer [1969], pp. 549–580, and Zilbert [1981], pp. 260–268.

[292]See Hahn & Straßmann [1939], Meitner & Frisch [1939], Frisch [1939], and Wohlfarth (Ed.) [1979]; cf. also, e.g., Hahn [1962]b, chap. VI B, [1968/86], the chapter on Berlin 1919–44, Herrmann [1964], [1990], Frisch [1979], Krafft [1978]b, [1981], chap. 3, Rhodes [1986], chap. 9, and

living in exile since July 1938, about these findings,[293] and she interpreted this phenomenon together with her nephew Otto Robert Frisch who was visiting her from Cambridge, England, as the splitting of a nucleus into two halves, releasing energy and neutrons in the process. Physicists around the world reasoned (several of them in published articles),[294] that if these neutrons could be used to induce further fissions of uranium atoms, a chain reaction might result releasing a large amount of energy. In Germany, a concerted research program was initiated after two groups of physicists pointed independently to the potential military application of nuclear fission and neutron multiplication in letters addressed to the *REM* and the *HWA* in April 1939.[295] While the *HWA* showed little interest initially, the *REM* passed the information on to the Reich Research Council (*RFR*) on whose behalf a first conference was organized on April 29, 1939. Erich Bagge, Walther Bothe, Peter Debye, Robert Döpel, Hans Geiger, Wolfgang Gentner, Wilhelm Hanle and Gerhard Hoffmann participated in this first *Uranverein* meeting. Soon afterwards several of these members were drafted at the beginning of World War II. A second meeting was arranged on Sep. 16, 1939, this time under the ægis of the *HWA*. It was decided then to defer several nuclear physicists from military service (these included Bothe, Siegfried Flügge, Geiger, Paul Harteck, Hoffmann, Josef Mattauch and Georg Stetter) to research the viability of building an atomic bomb based on nuclear fission, or at least to find a new compact source of energy to drive submarines or airplanes and thereby significantly extend their radius of action.[296] After extending this research group, referred to as the second *Uranverein*, to include Klaus Clusius, Döpel, Otto Hahn, Werner Heisenberg, Georg Joos and Carl Friedrich von Weizsäcker, it was resolved in October 1939 at another conference at the *HWA* to proceed with research on nuclear fission and chain reactions at several research sites throughout Germany.

Lemmerich [1988].

[293]Cf., e.g. Sime [1990], Kerner [1986] pp. 83ff., Lemmerich [1988], pp. 155ff., or Hahn [1962]b, pp. 126f.

[294]See, e.g., Flügge [1939] as well as here doc. 74, where the enormous increase in explosive potential of such nuclear product is pointed out. Cf. also P.K. [1942], Flechtner [1942], Krbek [1942]b for other popular texts along this vein, and Pollard [1942] or Heisenberg [1943]b for some of the last published textbooks on nuclear physics before it fell under the secrecy restrictions.

[295]The Göttingen experimental physicist Georg Joos wrote to the *REM* on April 22, 1939, and Paul Harteck and Wilhelm Groth wrote to the *HWA* on April 24, 1939; cf., e.g., Irving [1967], pp. 32ff., Walker [1989]a, p. 17. Scientists also took the initiative on the Allied side with Albert Einstein, prompted by the three Hungarian emigrés Leo Szilard, Edward Teller and Eugene Wigner, informing the U.S. President of the existence of a German nuclear program in a letter dated Aug. 2, 1939.

[296]Bagge quoted Diebner's superior H. Basche's opening statement at the meeting as follows: "It would naturally be very nice if it were possible to tap a new energy source; it would very probably also have military significance. However, a negative determination would be just as important, because one could be sure that the enemy would also not be able to make use of it." See Bagge in: Bagge, Diebner & Jay [1957], p. 23; cf. also Rechenberg [1988]b, p. 457, and Walker [1989]a.

A research team at the University of Leipzig (under Werner Heisenberg until 1942, including Robert and Klara Döpel and others), worked mainly on various possible arrangements of the uranium material in the reactor.[297] It was in Leipzig that the first experimental verification of neutron multiplication in a uranium machine was made in February 1942, four months earlier than the corresponding experiments by Fermi in Chicago, in June. Another of the German uranium project's many research branches was directed by Kurt Diebner, and from 1942 on by Werner Heisenberg at the *KWI* of Physics in Dahlem (Berlin) (see p. lxvii above). This group also included Siegfried Flügge, Fritz Bopp, Erich Fischer, von Weizsäcker and Karl Wirtz. In 1939 they started to work primarily on the theory behind nuclear fission and chain reactions.[298] From late 1940 on they also experimented with a horizontal arrangement of uranium oxide and paraffin layers, using deuterium as the moderator in a specialized research lab built into a small shed that bore the code name 'Virus House'. In addition, the *KWI* of Physics had a high-tension facility that could be used to induce nuclear fission artificially, and a bomb-proof underground laboratory was built in 1943.[299] Aside from the *KWI* of Physics in Dahlem (Berlin) and the research team at the University of Leipzig, the following institutions were also involved in nuclear research under the ægis of the *Uranverein*:

- the physics section of the *KWI* of Medical Research in Heidelberg (under Walther Bothe) with a half dozen physicists working on determining nuclear constants and using one of the very few European cyclotrons that was operational from the autumn of 1943;[300]

- a similarly sized research group of nuclear chemists and physicists at the *KWI* of Chemistry (under Otto Hahn) studying the physical and chemical characteristics of fission products,

- the research installations of the *HWA* in Gottow close to Berlin (under Kurt Diebner) where at least five physicists measured properties of fission products and conducted nuclear chain reaction experiments,

- the department of physical chemistry at the University of Hamburg (Paul Harteck and five other physical chemists, physicists and chemists) specializing in isotope

[297]See, e.g., Kleint & Wiemers (Ed.) [1993] part I, and Kleint in: Geyer et al. (Eds.) [1993].

[298]See, e.g., Flügge's survey article, Flügge [1939] and the more popular version (here doc. 74) and compare with Bothe & Flügge (Eds.) [1948], esp. part I, pp. 107f., and part II, pp. 143–147, 174ff. Cf., e.g., Irving [1967], [1968], Rechenberg [1988]b, and Walker [1989]a for historical descriptions of this research program.

[299]Cf., e.g., Heisenberg & Wirtz in: Bothe & Flügge (Eds.) [1948], part. II, pp. 142ff., 152ff., and doc. 116; Irving [1967], pp. 132ff.; Walker [1989]a, pp. 150ff., and Macrakis [1993], pp. 177ff.

[300]See Bothe [1938], W. Gentner in: Bothe & Flügge (Eds.) [1948], part. II, pp. 28–31; see also H. Salow (ibid.), pp. 32–49 on the cyclotron at the research lab of the *Reichspost* Ministry in Miersdorf. Cf. also Heilbron [1986]a, Walker [1989]a, p. 173, and Weiss in: Meinel & Voswinckel (Eds.) [1994], on the importance of this key technology.

separation, and production procedures for heavy water, the moderator substance used in the uranium machine,

- as well as several other teams at Göttingen (2 physicists), Vienna (6 physicists under Georg Stetter), Munich (4 physical chemists and physicists under Klaus Clusius), etc.[301]

The progress reports[302] on this project from 1942, 1943, and 1947 corroborated by Samuel Goudsmit's accounts of 1946 and 1947 (cf. docs. 95, 111, 115, and 116), imply that especially after 1942 and once the *HWA* lost immediate interest in this project due to its slow progress, the German project changed its focus toward developing a uranium pile, a mini reactor that the participants called a 'uranium machine'. However, the internal progress reports drawn up by these researchers and other sources such as the Farm Hall transcripts indicate clearly that at least some were aware of the dual nature of this technology for both civilian and military applications. Moreover, during the Third Reich Heisenberg, Harteck and von Weizsäcker repeatedly emphasized the potential usefulness of this research in weapons development.[303] Although it was assumed that this was part of a lobbying effort to convince Nazi officials to prolong the *uk* priority status of their research and thus ensure its continuation in the midst of World War II, we have evidence that its potential military applications played a considerable role in the *Uranverein*'s research. Though undoubtedly not of central concern, the duality of nuclear research prevented a clean separation of the two aspects. For instance, a study by Carl Friedrich von Weizsäcker of July 1940 pointed to the possibility of breeding a new type of radioactive material, element 93 (or *Eka-Rhenium*, as he called it), inside a uranium machine that, unlike the uranium isotope U 235, could also serve both as a nuclear fuel and as an explosive.[304] Thus the uranium machine

[301]See, e.g., doc. 103 for a more complete listing; cf. also Walker [1989]a, pp. 52f., for estimates on the number of team members, and Irving [1967], p. 214, for a map showing the locations of all of these research groups before and after their relocation to bomb-proof evacuation sites in 1944–45.

[302]These reports called *Deutsche Kernphysikalische Forschungsberichte* were top-secret at the time and only distributed to a handful of the parties involved. In 1945, they were confiscated by the Allies, and listed in David & Warheit [1952]; see also Diebner (alias Werner Tautorus) [1956] for another listing.

[303]See, e.g., W. Heisenberg's report dated Dec. 6, 1939 in *Kernphysikalische Forschungsberichte* **G-39**: "The greater the enrichment, the smaller the machine can be designed. Enrichment [...] is furthermore the only method by which an explosive can be produced that supercedes the explosiveness of the most powerful conventional explosive by several powers of ten." Von Weizsäcker (1940) cited in the following footnote, p. 5; cf. also Walker [1989], pp. 21ff. and the document of summer 1942 quoted in footnote 13 of doc. 95.

[304]This new element, now called plutonium, is formed when the isotope U 238 captures thermal neutrons. It transforms into U 239 and decays into plutonium: See von Weizsäcker, 'Eine Möglichkeit der Energiegewinnung aus U 238' (dated July 17, 1940), *Kernphysikalische Forschungsberichte* **G-59** and **DJ 29** (see also Cassidy [1992]a, p. 424; Kevles [1987], p. 328, and references to American research on the same topic there). This mechanism was used by

was not simply a 'peaceful' new source of energy, but also a potential deliverer of nuclear explosives. One report on experiments conducted by a *HWA* research group in Gottow in 1944 to induce chain reactions using cylindrical explosives is a striking example of evidently bomb-related research.[305] Furthermore, when after 1942 the central concern was to produce a working uranium machine in which a controlled chain reaction can take place, by no means were military applications excluded completely. The Navy hoped to acquire a concentrated energy source, for instance, that would extend the periods on duty of submarines away from harbor where they were most vulnerable.[306]

The research on nuclear chain reactions and on the physical properties of nuclear fission products conducted by the *Uranverein* between 1939 and 1945 is described extensively (though by no means exhaustively) in Werner Heisenberg's postwar overview article, which has been annotated here in detail (see doc. 115). However, Heisenberg's report is typical of all retrospective accounts by German scientists involved in this nuclear research program insofar as they all portray their roles in the development of the 'uranium combuster' (*Uranbrenner*), as it was also called (nowadays known as a reactor) in an innocent light. Heisenberg and von Weizsäcker in particular implied repeatedly, if not stating so explicitly, that they had decided consciously against working on the military aspects with the specific aim of preventing Hitler from acquiring the bomb, by arguing consistently against its feasibility in the strained German wartime economy.[307] However, in later interviews and published statements, both Heisenberg and in particular von Weizsäcker retracted this at least in part and conceded that the decision on

the Allies to produce the plutonium for the second atomic bomb dropped over Nagasaki on Aug. 9, 1945. See also doc. 95: "once such an engine is operational, the matter of producing an explosive also takes on new significance", cf. Cassidy [1992], p. 445.

[305]See W. Herrmann, G. Hartwig, H. Rackwitz, W. Trinks & H. Schaub, 'Versuche über die Einleitung von Kernreaktionen durch die Wirkung explodierender Stoffe', *Kernphysikalische Forschungsberichte* **G-303**; see also Irving [1967], p. 194, and compare with the Allied research in the 'Explosives Division' of the Manhattan Project as described by David Hawkins in: Brown & MacDonald [1977], pp. 490ff.

[306]This type of military application is emphasized in publicity articles appearing as late as 1940 (see, e.g., P.K. [1940]) and also in Heisenberg's talk in 1942 in which he mentioned that a working uranium machine would provide enough power to drive battleships and submarines (cf. here doc. 95, footnote 13).

[307]See, e.g., the joint declaration drafted by the Farm Hall internees on Aug. 8, 1945, in Frank (Ed.) [1993], pp. 102ff., or von Weizsäcker's remark (ibid., pp. 76f.) "I believe the reason we didn't do it was because all the physicists didn't want to do it, on principle. If we had all wanted Germany to win the war we would have succeeded", whereupon Hahn replied: "I don't believe that, but I am thankful we didn't succeed". See also Heisenberg [1948]b, [1967]a; or Jungk [1958] who based his apologetic account on interviews and materials provided by von Weizsäcker, Heisenberg and others. (Cf. Jungk's foreword to the German translation of Walker [1989]a, p. 7 where he retracts his earlier account and complains about his source as having "not only spoken with two separate tongues but also having thought and felt with two minds": See also Robert Jungk, *Trotzdem. Mein Leben für die Zukunft*, Munich: Hanser, 1993, pp. 298–300; however, cf. also Powers [1993], chaps. 12, 14, 36f. for a recent reappearance of this version.

whether or not to build the bomb had actually been out of their hands.

> Obviously we were not *fully* aware of the extent of the danger, but then in the first two years the following became evident: It was possible to build nuclear reactors relatively easily with moderate means, i.e., certainly in the period of a few years available to us, that is, *that* particular reactor we knew would work, namely a reactor using natural uranium and heavy water. It was *also* clear that an explosive is produced in such a reactor from which atom bombs could be made; but luckily it was *also* clear then that this would involve an enormous technological investment lasting many years. Therefore we could report these results with complete honesty and a good conscience to the government agencies, and the consequence was—just as we had hoped—that the government decided to make *no* effort to construct atom bombs, but that we received certain—albeit modest funds—to continue work on the design of a reactor, precisely on a heavy-water reactor."[308]

At Farm Hall, Heisenberg even admitted: "We wouldn't have had the moral courage to recommend to the Government in the spring of 1942 that they should employ 120,000 men just for building the thing up" whereby he implied that the fear of what would have happened if it had not worked in the end had been too great. Indeed, Heisenberg only requested a budget of 200,000 reichsmarks for the uranium research at his Kaiser Wilhelm Institute (an increase of 75,000 reichsmarks with a total budget of 350,000 reichsmarks), and the flow of money was in fact so sluggish that in the autumn of 1942 a reminder notice had to be sent to the *HWA* for the committed amount of 50,000 reichsmarks. It also took until the end of 1943 for Walther Bothe in Heidelberg finally to receive the cyclotron that had been ordered in 1937.[309] Thus, to a large extent, the decision fell on the *HWA* in early 1942 when it turned the focus away from this research. However, the myths formed in the early postwar setting on both sides of the former frontline are still very much alive today.[310]

However careful a reading we subject these retrospective accounts of the German nuclear research effort to (such as Heisenberg's quote above or in doc. 115), it is clear that the decentralized organization of the research in so many widely dispersed laboratories had serious drawbacks. The limited supply of uranium, which was provided mostly by the *Auer* Company under the direction of Nikolaus Riehl, and by *Degussa* in Frankfurt, was a particular problem.[311] The raw

[308]Quoted from Heisenberg's speech in honor of Wolfgang Finkelnburg, Feb. 24, 1968 (publ. in Rechenberg (Ed.) [1992]a, pp. 205f.). Cf. also Heisenberg [1967]b: "Thank God we could not build it".

[309]On the contemporary budget see, e.g., Rechenberg [1992]b, on the complications in the delivery of the cyclotron see Osietzki [1988], pp. 35ff., and Eckert & Osietzki [1989], pp. 45ff.

[310]On these myths see Walker [1989]a, [1990], [1993]a or [1995], chaps. 8–10.

[311]On the stages of chemical purification of the raw material pitchblende using fractional crystallization, and where necessary the reduction of uranium oxides to their metallic forms

material for these chemical processes came from foreign sources: the Joachimsthal mines in Sudetenland in Czechoslovakia annexed in 1938, and from Belgium; and in 1940 the deuterium (D_2O, also called heavy water) that was needed as a moderator substance could only be produced at a single electrolysis factory in southern Norway. This indicates how reliant this type of research was on the German expansionist foreign policy. Especially toward the end of the war, material shortages became a primary concern. The Plenipotentiary of Nuclear Physics since the close of 1943, Walther Gerlach, was assigned the task of distributing the limited number of uranium blocks fairly among the rival research groups.[312] According to Heisenberg, the supply of uranium blocks and heavy water was critical for the last experiment performed jointly by the Berlin and Heidelberg research teams at the bomb-proof Haigerloch redoubt in March and April 1945. Only the lack of sufficient amounts of these materials—they had been allotted 664 blocks of uranium and approximately 1.5 tons of heavy water—prevented them from reproducing the results Fermi and his collaborators had obtained in Chicago in December 1942. In conclusion, until mid-1942 the German *Uranverein* produced results comparable with and in some crucial cases even ahead of American efforts. But thereafter they fell behind so considerably that by the end of the war they had barely reached the level of sophistication necessary for the construction of a nuclear reactor.[313] Thus when the select group of 10 German atomic scientists interned at Farm Hall in southern England heard about the dropping of an atomic bomb over Hiroshima on Aug. 6, 1945, they were incredulous (Heisenberg in particular refused to believe the news at first).[314] On the other hand, it did not take them very long to figure out fairly accurately how much enriched uranium was needed to build such a bomb (a calculation that Heisenberg had apparently never attempted before) and how it worked.[315] Thus if the *HWA* or other powerful Nazi state agencies had realized the importance of this type of weapons development and had pressed hard enough, the German scientists would certainly have been able to provide the Nazi rulers with the weapon; and very likely they were also willing to do so. It is another merely practical matter whether it would have been possible to actually produce the bomb before the Americans and to obtain the considerable quantities of enriched uranium 235 or plutonium 239 needed under the strained wartime conditions and continuous air raids along with the sporadic sabotage of key sites like the heavy water production plant *Norsk-Hydro*

---

with calcium and the subsequent precipitation of the resulting calcium oxides using acid see, e.g., Riehl [1988], pp. 17–20, 24–25, 27f., and Irving [1967], pp. 75f.

[312]See in this respect, in particular, Gerlach's correspondence in Irving (Ed.) [undated], reel no. 29 (DMM), and here docs. 103f.

[313]See also the 1945 report to Compton by A. W. Weinberg & L. W. Nordheim, here doc. 109.

[314]See Frank (Ed.) [1993], pp. 70f.

[315]See again Frank (Ed.) [1993], pp. 125–164, for the full text of Heisenberg's lecture of Aug. 14, 1945; on the issue of the critical mass see ibid., pp. 4–7, 73–77, 83, 86, 126; Walker [1989]a, pp. 48, 155, 171f., Eckert [1993], p. 234.

at Vemork near Rjukan.[316]

It has been claimed that the National Socialist regime inadvertently fostered a closer interaction between science and industry.[317] The Nazis did not dare to interfere with and reorganize the industrial sector of society. On the other hand, they meddled extensively in the state-run universities (see sec. 3.2 above). Thus industry appeared to be a good alternative for many physicists. Georg Joos, for example, became fed up with the political interference and resigned his professorship at the Göttingen department of experimental physics in 1941 to become chief physicist at the *Carl Zeiss* Optical Company in Jena, which developed into the main optical supplier for the military.[318] Gustav Hertz, who was dismissed from his professorship at the Polytechnic in Charlottenburg (Berlin) on racial grounds, was employed as director of a second newly founded *Siemens* research lab in 1935. This electrotechnical concern was powerful enough through its sheer size to be able to employ such politically 'undesirable' scientists. Fundamental scientific research, which had hitherto mostly been conducted at universities and other state institutions, shifted to industrial laboratories. Industrialists in turn became leading figures on science policy boards and science organizations. For example, the *IG Farben* chemist Carl Bosch was the president of the Kaiser Wilhelm Society (cf. here p. lxvii) between 1937 and 1940. After his death in 1940 he was succeeded by Albert Vögler, president of a leading German steel concern. The director of the *AEG* research labs Carl Ramsauer was president of the German Physical Society (see sec. 5.2) from 1940 to 1945. There does not seem to be any broad overview of industrial research in physics and related fields during this period, but at least some sectors such as the development of betatrons and cyclotrons at the *AEG* and *Siemens* companies, for instance, have been studied in closer detail.[319] Although one of the fundamental ideas behind multiple acceleration was published by the German engineer Rolf Wideröe at the early date of 1928, it was an American, Ernest O. Lawrence, and his doctoral student Stanley M. Livingston, who first built a working 80,000 eV cyclotron in 1931 in which charged particles are accelerated on a spiral path by powerful homogeneous magnetic fields. In 1939 Lawrence received the Nobel Prize in physics for this achievement that provided nuclear physicists with a viable source of particles

[316]This Norwegian plant produced deuterium using the method developed by Bonhoeffer, and optimized by Harteck, Groth and Clusius. It was partially destroyed in an act of sabotage in February 1943 and bombed in November 1943. The remaining heavy water supply was sunk on board the ferry 'Hydro' in February 1944; see, e.g., Irving [1967], pp. 126ff., 152ff.; Walker [1989]a, pp. 27f., 118ff., 137ff.

[317]See, e.g., Ash [1995], Eckert [1993], pp. 203ff., and Eckert & Osietzki [1989], pp. 43ff.

[318]According to Hermann [1989], p. 151, in 1943 sales reached a record 236 million RM, of which almost 194 million (or 84%) were to the military; cf. also Ley [1942] and Münster [1946].

[319]See, e.g., Osietzki [1988], pp. 32ff., [1993], pp. 207ff., and her paper in: Eckert & Osietzki [1989], pp. 43ff., and references there. On this and other research at *Siemens* see also Trendelenburg [1975], esp. pp. 68ff., and Schubert [1987].

at predetermined charges and ushered in a new phase of research practice.[320] Another important area of research was electron microscopy conducted at *Siemens* by Ernst Ruska.[321] One problem with historical research in the industrial sector is that in the past some companies have been reluctant to open their archives to scholars interested in studying their activities during the Nazi regime. There are scattered remarks by physicists who had worked in industry, published in autobiographical accounts.[322] By contrast, the field of chemistry has been studied much more intensively in this respect.[323] On the research conducted at the *PTR* with a heavy emphasis on meteorology see here sec. 5.1.

5.5 The Legacy of National Socialism

Although the 'thousand-year Reich' had lasted but a dozen years, it cost up to 26 million lives. When the war came to an end, there was a sweeping social reorganization: Some of those who had held key positions now faced war tribunals; the majority, however, had to go through the process of 'denazification'. According to Allied Control Council Directive No. 24 of January 1946 and No. 38 of September 1946, and the U.S. Directive of April 26, 1945, the population had to undergo political screening in an effort to eradicate Nazism. The intent was to disqualify a reasonably large segment of the German upper class and parts of the middle class from leading positions in public life and thus simultaneously force them to accept responsibility for what had happened.[324]

---

[320]On the research by Ernest Orlando Lawrence and his many colleagues, see Heilbron & Seidel [1989] and as well as footnote 8 of doc. 82. For a concise contemporary description of the physical principle behind cyclotrons, see Watzlawek [1942]a, Osietzki in: Renneberg & Walker (Eds.) [1994], p. 260, highlights the extremely competitive atmosphere that pervaded the Radiation Laboratory at Berkeley which led to the attitude: 'bigger is better'. She continues, "Lawrence thereby presented a fundamental challenge to the theoretically accented European leading role in physics, which had been founded on the quantum mechanics that had been developed there."

[321]On the early history of electron microscopy, see Ernst Ruska, *Die Frühentwicklung der Elektronenlinsen und der Elektronenmikroskopie*, Halle: Deutsche Akademie der Naturforscher Leopoldina, 1979, and Trendelenburg [1975], pp. 76f.

[322]See, e.g., Brandt [1962], pp. 9–112 on ultrashortwave (radar) research at *Telefunken*, Riehl [1988] for the *Auer* Company, Ramsauer [1949] on the *AEG*, von Ardenne [1972] on his own institute at Lichterfelde (Berlin) (supported financially by the Reich Postal Ministry), Casimir [1983] on the *Philips* Company in Eindhoven, or Steenbeck [1977] on the *Siemens-Schuckert* Research Lab. There are also interviews with solid-state physicists and quantum theorists available at the American Institute of Physics.

[323]On the *IG-Farben* trust, see the literature mentioned in footnote 69 above as well as *OMGUS (Office of Military Government for Germany, United States, Finance Division – Financial Investigation Section). Ermittlungen gegen die I.G. Farben*, edited by Hans-Magnus Enzensberger, Nördlingen: Greno, 1986, Petzina [1968], and Teltschik [1992] for a general overview on the chemical industry.

[324]From Wolfgang Krüger: *Entnazifiziert! Zur Praxis der politischen Säuberung in Nordrhein-Westfalen*, Wuppertal, 1982, p. 22; cf. also Golczewski [1988], p. 386 on Cologne, Niethammer

According to the Law for the Liberation from National Socialism and Militarism of March 5, 1946, which placed denazification in the hands of German officials in the American zone (from 1947 also in the French and British zones), every German citizen had to fill out a questionnaire concerning his former membership in Nazi organizations, which was first checked by Allied military personnel against the confiscated membership lists of the *NSDAP*, *SA*, *SS* and other Nazi organizations (until recently filed at the Berlin Document Center). It was then passed on to a public prosecutor who initiated a hearing at which the defendent could present his exonerating evidence or witnesses. The proceedings ended with a formal notice that classified each person under one of the following 5 categories:

1. Major offender (1st degree)
2. Incriminated activist, militant or profiteer
3. Lesser offender
4. Fellow traveller (i.e., nominal supporter), and
5. Exonerated individual.

Membership in Nazi organizations and especially in the *NSDAP*, the *SA* and the *SS* was certainly an important indicator in this mandatory evaluation, but an effort was made also to take other evidence into consideration, because it was clear that various motives existed for joining these organizations.[325] Major offenders convicted at the Nuremberg trials faced a maximum of life imprisonment or the death sentence. Others assigned by German tribunals within the first two categories had to face severe consequences, such as being forbidden to exercise their profession within the public sector, and also the loss of their voting rights; the maximum sentence by German tribunals for major offenders was 10 years of forced labor and the confiscation of their entire personal wealth, and for incriminated persons 5 years labor camp and partial confiscation. By Sept. 30, 1946, however, the German tribunals in Hessen had decided only 9,628 cases, convicting only 20 major offenders (2%), 215 incriminated, and 686 lesser offenders. Owing to several amnesties, by 1949 a total of 2.8 million cases were simply closed.[326]

[1982] on Bavaria, and in particular Vollnhals [1991] for a general overview of denazification in all four German occupation zones.

[325] See, in particular, Vollnhals [1991], pp. 11, 18, Golczewski [1988], pp. 380f. and Frank R. Pfetsch in: Buselmeier et al. (Eds.) [1985], pp. 367f., on this issue.

[326] For these statistics see, e.g., Vollnhals [1991], pp. 21ff., 332ff.: By 1949/50 out of 3.66 million denazification cases processed in the Western zones, 1.1 million (approximately one third) were closed, 1.2 million defendents were exonerated, 1 million classified as fellow travellers, 150,000 as minor offenders, 23,000 as incriminated and less than 1,600 (0.05%) as major offenders. Cf. also Walker [1989]a, p. 192, where slightly different figures are provided based on a total of 6 million denazification cases.

Regarding the denazification of scientists in particular, a couple of chemists and engineers who had exploited slave laborers loaned from concentration camps were involved in the *IG Farben* trial.[327] However, very few physicists were similarly tried. Philipp Lenard, for instance, died before his denazification hearings could be held.[328] Apparently, Johannes Stark was one of the very few physicists to be brought before a tribunal after 1945. The tribunal of the local Traunstein county court classified him as a major offender for his active role as a protagonist of the 'Aryan Physics' movement and his leadership at the *PTR* and the *DFG* after 1933. He was sentenced to 6 years of forced labor, but a Munich appeals court preferred to categorize him as a minor offender and reduced his sentence accordingly to a one-thousand mark fine.[329]

The urgent need for 'cleared' staff for the many vacant positions and tasks in postwar Germany accelerated the denazification process and this despite the problematic latitude allowed in the classification scheme. Further complications arose because well-meaning but often insufficiently informed persons with otherwise good records issued 'whitewash certificates' (*Persilscheine*, derived from the name of a well-known washing powder).[330] These complicating factors resulted in a considerable number of persons being falsely 'cleared', some of whom even acquired important positions later on. Thus the tribunals earned the reputation of being a fellow traveller production line ("*Mitläuferfabrik*").[331] Occasionally big scandals broke when the truth about past Nazi involvement resurfaced.[332]

At universities, essentially only the most outspoken representatives of the Nazi system were dismissed without compensation, while those of the lesser categories 2 or 3 were often just forced into early retirement. Loyalty toward fellow colleagues governed most internal university decisions on ambiguous cases,

---

[327]See, e.g., Borkin [1978] and Hayes [1987]; cf. Gerhardt [1948].

[328]On June 19, 1947 the Heidelberg tribunal requested Lenard's personnel file as a civil servant from the Baden State to prepare for his hearing, only to learn at the beginning of 1948 after finishing their review that the individual in question had died on May 20th of the preceding year: See Lenard's personnel file, sheets 653–656, Badisches Generallandesarchiv Karlsruhe. (I am indebted to Prof. Andreas Kleinert for this information).

[329]See Kleinert [1983] and Walker [1989]a, pp. 199f., as well as ibid., pp. 197f., for a few other examples of the whitewashing of fairly tarnished figures after 1945. A systematic study on the hitherto neglected tribunal proceedings and denazification trials of scientists in the Munich area is currently being prepared by Freddy Litten at the Research Institute at the Deutsches Museum, Munich; cf. also Litten [1994]b, p. 12, where we are informed that Bruno Thüring and Wilhelm Müller were also classified as 'fellow travellers' in their appeals despite their involvement in all kinds of suspect activities and denunciations.

[330]For examples of such letters see, e.g, the special folder concerning denazification affairs in the Walther Gerlach Papers, DMM, as well as here doc. 121. Cf., e.g., Walker [1989]a, pp. 198f., for Heisenberg's exoneration certificates.

[331]See Niethammer [1982] for an exemplary study of denazification in Bavaria.

[332]One such case is Hans Globke (1898–1973), who worked at the *RIM* from 1932–45 and co-authored an infamous commentary to the Nuremberg Laws in 1936 (see Stuckart & Globke [1936]). He nevertheless later succeeded in obtaining a position in the public service as state secretary and head of the German Federal Chancellery from 1953 to 1963.

and many preferred to look to the future. In May 1946, out of the 102 former Göttingen university teachers in the civil service, 14 were still undergoing denazification, 16 had been dismissed upon being assigned to categories 1–3, and 13 had already succeeded in returning to the positions they had held prior to 1945 as a result of a successful appeal. By the time the denazification procedure had ended, only three had been assigned to category 3 (which entailed an automatic vocational ban) and five to category 4. However, according to the Law on the Termination of Denazification in Lower Saxony of Dec. 18, 1951, all category 3 and 4 convictions were transformed into category 5 without further trial—thus all previously convicted and disqualified university teachers could hope to regain their former positions and along with it their right to a state pension.[333] Students matriculating at a German university or polytechnic (which were reopened soon after the war) also had to submit a clearance certificate issued by the Allied Military Government of Germany, and from the beginning of 1946 on this was extended to all persons who registered for examinations, and to all submissions of masters, doctoral, and habilitation theses, even if these individuals had never enrolled as students previously. However, the principle concern then was not so much to confront the past as to re-establish functioning research and teaching institutions—ideal conditions for repressing the many problematic issues of Germany's past.[334]

But this tendency to cover up the past was by no means confined to the Germans. Soon after the war the Western Allies initiated a top secret military 'Project Overcast' (also known by its later name 'Operation Paperclip') which amounted to hiring about 700 former enemy experts in aerodynamics, aircraft construction, jet propulsion, rocket technology, metallurgy, chemistry, and medicine, irrespective of their former involvement in Nazi affairs, thus explicitly violating U.S. immigration and naturalization restrictions, which prohibit the entry of former *SS* and *NSDAP* members into the country, for instance.[335] A corresponding transfer to the USSR under the acronym *OSOAWIACHIM* of at least 2,370 (but probably more like 2,500) 'specialists' in similar fields, together with approximately

[333]All figures from Becker, Dahms & Wegeler (Eds.) [1987], pp. 46–48; cf. also Voigt [1981], pp. 53f., on Stuttgart; Golczewski [1988], pp. 376ff., on the universities of Cologne; Pfetsch in: Buselmeier et al. (Eds.) [1985], pp. 366ff., on Heidelberg University (according to which every fourth scientist from a total of 359 university members had originally been classified as category 1 or 2); and Sywottek in: Krause et al. (Eds.) [1991], pp. 1391ff., on Hamburg University. These figures contrast with a rate of 57% of civil servant dismissals and ca. 26% dismissals in the private sector in the state of Hessen (Vollnhals [1991], pp. 14f.).

[334]On the rebuilding of science institutions and its problems see, e.g., N.N.[1946], Hahn [1947], Heisenberg [1948]a, Laue [1949]. Cf. also with the many references in: Michaelis & Schmid [1983], Osietzki [1984], Lundgren et al.[1986], M. Heinemann and T. Stamm-Kuhlmann in: Vierhaus & vom Brocke (Eds.) [1990], pp. 407ff. and 886ff., Trischler [1992]a, chap. 3, Hermann [1993], and Oexle [1994].

[335]See, e.g., Lasby [1971], Gimbel [1986], Bower [1990], Hunt [1991], and Bode & Kaiser [1995], pp. 141ff., as well as docs. 112f.

4,600 family members, took place. They worked for the Soviet Union between 1945 and 1959;[336] 84% of these were 'persuaded' or forced to make this move in October 1946.[337] One example of postwar plundering involved the *Carl Zeiss* optical factory and the Schott glass works in Jena. When the city of Jena was occupied by U.S. Forces on April 13, 1945, American officers immediately started to assess the largely intact factory complexes and prepared plans to remove a significant portion of it to the West, since Jena was designated to the Eastern Soviet zone of Germany. Because of the time constraints determined by the hand-over of the territory to the Soviet Army, the American Forces eventually only took a small portion of the material originally estimated at about 600 freight car loads. Around 80,000 technical drawings and laboratory equipment parts as well as approximately 2,000 employees, 400 of whom were scientists and engineers (together with their relatives coming to about 1,300 people) were transferred to Heidenheim near Stuttgart, 14 of these went on to the USA under 'Operation Paperclip'. Then the Soviets arrived and surveyed the whole factory, initially giving the impression of supporting its being rebuilt, but then suddenly, in the night of Oct. 21, 1946, they shipped out train loads of dismantled machinery to the Soviet Union and transferred 274 scientists, engineers and other specialists, some against their will.[338] Later, rival branches of the former *Zeiss* works were rebuilt in Oberkochen and Jena on both sides of the German iron curtain, and there were numerous disputes in court over patents and even over who was entitled to the well-known Zeiss trademark.[339]

Several of the postwar documents selected for this anthology illustrate the problem of moral evaluation within the Cold War environment. The bitter debate between Goudsmit and Heisenberg after the war about what the goals and achievements of the German *Uranverein* had been, can only be fully understood in the light of this postwar background (see the annotation to docs. 111f. and 116ff.). Goudsmit found it politically important to 'prove' the failure of science in Nazi Germany to convince American authorities of the necessity of freedom of research, uninhibited by secrecy regulations and nationalistic and ideological impediments in the increasingly tense Cold War atmosphere.[340] Another controversy arose between Goudsmit vs. Bethe and Sack about importing German scientists and engineers into the U.S. to work on military research programs under 'Operation Paperclip' (cf. docs. 112f.).

[336]See Albrecht et al. [1992], Ciesla [1993], and Bode & Kaiser [1995], pp. 146–154.

[337]For autobiographical accounts of scientists transferred to the Soviet Union after World War II see, e.g., von Ardenne [1972], Steenbeck [1977], Riehl [1988].

[338]See, e.g., Brüche [1946], pp. 198ff., 223f., Hermann [1989], chap. 13, and [1993], pp. 77ff.

[339]See again Hermann [1989], chaps. 14ff. The Western branch *Zeiss Oberkochen* later received a compensation payment of 23.1 million DM from the US Office of Occupation Damages (ibid., p. 166). For other aspects of "exploitation and plunder in Post-World War II Germany" see Gimbel [1990].

[340]His personal loss at the hands of the Nazis is another explanation for his tendency to belittle German scientific achievements (see footnote 3 of doc. 117).

That so few scientists were actively involved in the 'Aryan Physics' movement made it a simple matter to use them as scapegoats after the fall of the Nazi regime. Until recently it was a quite common strategy to limit the influence of Nazi ideology to this small group of outspoken ideologues (see here p. lxxvii), and thus absolve the vast majority of scientists and engineers who had supposedly just continued their everyday business and otherwise did what they could to combat these few evil exceptions. One of the aims of this anthology is to take up the new historiographic trend led by Herbert Mehrtens, Mark Walker and others in combating this simplistic woodcut view and to demonstrate through a selection from a variety of texts how many shades of gray there are between the many journalistic and tendentious black-and-white accounts.

6 Style and Rhetoric

The style and rhetoric in some of the texts selected for this anthology require special comment. All texts published in Germany, particularly after 1934 were written under the conditions of a totalitarian state, when the danger of being denounced by *Gestapo* agents or informants had become obvious to everyone with the aim of controlling the mentality of the populace using all possible means: positively through propaganda in the censured media, including public radio broadcasts (*Volksempfänger*), and filtered weekly news reports at the cinema; and negatively through intimidation.[341] Only a few publishing houses, such as the *Lehmanns Verlag*, actively supported the Nazi movement prior to 1933; and because 'Aryanization' of the private sector of society took time, other publishers like Julius Springer could proceed as usual for some time.[342] Nevertheless, as most emigré scientists published in Allied journals, German scientific journals soon lost their leading position to their English-language counterparts.[343] However, newspapers were brought more quickly into line. In the aftermath of the *Reichstag* fire, the new government quickly disposed of the entire Marxist and leftist opposition press (about 50 Communist and 130 Social Democratic Party-owned papers, compared with about 120 National Socialist newspapers existing then), and in October 1933, the 'Editor Act' elevated the press to a 'public institution', placing more control over the "execution of the editorial function" into the hands of the Nazi Propaganda Ministry.[344]

[341] On Nazi propaganda in the German cinema, see in particular Welch [1983].

[342] On the importance of the publishing house owned by Julius F. Lehmann (1864–1935) as a "patron of the racial movement" see G. Stark [1976/77]; cf. with the inhouse publications in 1940 and 1965 to celebrate the 50th and 75th anniversaries. On Springer, the publisher of *Die Naturwissenschaften*, *Zeitschrift für Physik* and other journals that regularly carried contributions on quantum theory and other modern sciences, see Frank Holl in *Archiv für die Geschichte der Buchwesens* (forthcoming).

[343] See Michaelis & Schmid [1983], pp. 69f., where postwar journals are ranked according to the number of citations, and the number of publications registered in the *Chemical Abstracts*.

[344] On the 'coordination' of the press and its consequences see, e.g., Marx [1936], pp. 99ff.

As often happens under such oppression, a new language develops: Not everything is said directly; many things (especially criticism of official doctrine and Nazi issues) are camouflaged, or are left unsaid, appearing between the lines in an apparently innocuous text. Some contemporary writers used the form of parables, legends or historic subjects. For instance, Max von Laue officially spoke about Galileo in his Würzburg talk of 1933, praising his resistance to the Catholic church's stance on science, but everyone present immediately understood the implied parallel with the National Socialist crusade against Einstein (see doc. 27). In his obituary on Fritz Haber, von Laue again employed the form of comparison with a historical figure (in this case Themistocles) to bring his message across in a thinly veiled critique of the policy of dismissing excellent scientists on the basis of racial origin (cf. doc. 31). Others chose lyrical expression to conceal their opposition to National Socialist slogans, but in other cases it was also an indication of their apolitical escapism; many were caught up in the wave of sentimental subjectivity that flourished throughout the Nazi period.[345] Analyzing such texts often requires recognizing and deciphering these hidden messages, which were so obvious to an initiated contemporary audience.

Thus, the translated documents require a careful reading without *a priori* assumptions. It is by no means granted, for instance, that everyone who signed a letter with the officially prescribed closing 'Heil Hitler' was a hardened Nazi. Though this might well be the case, it is advisable to note who the addressee is and the date it was written. (In official letters, this closing was less circumventable than in private correspondence. In the years 1933 or 1934, and then again from 1942–45, its use signals a higher degree of ideological commitment.) These rather standardized features were often adopted by persons criticial of the Nazi state so as not to become all too easy targets for their enemies. Subtleties in the wording took on a greater importance. The discussions about the precise phrasing of the letter reluctantly sent out by the president of the *DPG* Peter Debye in late 1938 requesting the withdrawal of its few remaining Jewish members are a case in point. In his draft, Debye had indicated that "under the prevailing circumstances" he was "forced" to this take this measure, but the two pro-Nazi instigators Herbert Stuart and Wilhelm Orthmann were quite unhappy with this rather apologetic tone. The final version dated December 9, 1938, managed to retain its objectionable ambiguity: "Under the prevailing compelling circumstances the membership of German Reich Jews in the sense of the Nuremberg Laws in the German Physical Society can no longer be upheld." (See doc. 67 and the subsequent discussion during the meeting of the society five days later, doc. 68.)

The many petitions and memoranda reproduced in this anthology addressed to ministries, military leaders and top Nazi officials also have to be seen from this double perspective, since they were clearly molded according to the expectations

[345]See, in particular, Ritchie [1983], pp. 124f., 126ff., on poetry, and pp. 133ff., on literary forms of resistance; cf. also Welch [1983] on the Nazi propaganda techniques.

and the language of the intended recipient. To achieve anything in this time as members of the intellectual élite, leading physicists had to assume the politically correct tone that was not necessarily their own. Thus in a sense, the language of these documents already reflects one of the central dilemmas faced by scientists at this time: To stay within the system meant making compromises, not only in their choice of words but also in their actions. Some form of linguistic adaptation was unavoidable, though the degree depended very much on the audience and on the type of publication. Hellmut Seier has spoken of 'semantic mimicry' in this context to indicate the common use of certain "tributes to the spirit of the day that exhibited an ideological, militaristic, racially hygienic or otherwise tailored matter of consequence in order to court students, officials, censors, editors, publishers", etc.[346] A prime example is Wolfgang Köhler's 'Conversations in Germany' (here doc. 13), which appeals to the conscience of his Christian readers by invoking terms like 'moral duty', quoting from the Psalms, and reminding his readers of Lessing's *Nathan* as the symbol of the great tradition of tolerance in the German enlightenment.

The Göttingen university lecturers' public declaration that Prof. Franck's resignation "is tantamount to an act of sabotage" (see doc. 11) was specifically aimed at exciting the interest of those in power. This same aim of soliciting the support of military agencies also lay behind Ludwig Prandtl's warnings about the "serious threat" to physical instruction and the "sabotage to technological development" in the context of his counterattacks against the 'Aryan Physics' movement (cf. doc. 84 and Attachment V). Göring and Himmler were not interested in the least in some internal academic dispute between rival groups of physicists, but accusing their opponents of "terroristic behavior" was a strategy that had at least some chance of attracting the attention of the targeted audience. Although Ramsauer and Prandtl at first had difficulty reaching the decision-makers in the complex Nazi polycracy, their manoeuvers finally met with success (cf. the increased frankness of Prandtl's petition in 1938 against the petition to Himmler in 1941 (docs. 62 and 84). Another interesting example is Pascual Jordan's martialistic tone (see, e.g., doc. 87). The abundance of belligerent metaphors, allusions to death and struggle, brute force and victory, culminate in Jordan's illustration of the pragmatic meaning of objectivity in terms of the superiority of one country's machine guns over another's.[347] The same can be traced in the less graphic texts by Schumann (cf. doc. 75), Stark (cf. doc. 56), or Heisenberg (cf. doc. 95), for instance.

A rather amusing chapter in the history of National Socialist efforts to change their *Lebenswelt* are their endeavors to reform terminology in physics and engineering. In 1935 the provisional results of a commission's work on proposals to

[346]See Seier [1988], p. 290.

[347]See Jordan [1935], pp. 56–57. For an analysis of Jordan's style and traces of Nazi ideology in his writings, cf. in particular N. Wise in: Renneberg & Walker [1994], p. 226.

Germanize physical terms were published in a science teacher's journal. Though they had to admit that on first reading, changes such as *veil* for *violet*, *kress* for *orange*, *Abirrung* (digression) for *Aberration*, *Schauteil* ('viewing piece') and *Dingteil* ('object piece') for *Okular* and *Objektiv* would sound quite 'presumtuous', in fact in the first instance perhaps even 'funny', they hoped nonetheless that these new terms would become very attractive "once they are really used seriously."[348] Not surprisingly, these proposals found very few supporters even among the orthodox.

Extreme cases of Nazi rhetoric can be found in Ludwig Glaser's or Philipp Lenard's anti-Semitic tirades, which seem to have completely lost touch with their audience. It was not an easy task to translate the heavy, pseudo-poetic style of Lenard's and Stark's eulogy of the Nazi cause, which is rather atypical for a newspaper article. The language is deliberately intended to convey the spirit of 'Aryan-Germanic blood' in all its cumbersome stiffness (see footnote 7 of doc. 3). Empty phrases, reminiscences of the great days of yore, and superlatives (such as 'highest and holiest') abound. The choice of vocabulary (e.g., 'spirit', 'holy', 'Christ') and phrases like "God's gifts from times of old when races were purer, people were greater, and minds were less deluded", gives it the feel of a sermon or a religious treatise. This document condenses the new dogma as:[349]

- A profession of faith of a numerically small sect within the physics community;
- a celebration of Hitler as the redeemer to lead Germany out of the dark Weimar period into a new era; and
- the installation of Hitler as the new '*Führer*' to be (blindly) followed as the new god, with National Socialism as the new religion.

Lenard's text also reminds us that the use of rhetoric is by no means confined to outspoken or hidden critiques of the Third Reich. It was used on an even grander scale by Nazi propagandists.[350] Some of the more interesting examples occur in passages where leading Nazi functionaries have to conceal certain weaknesses or defects in the new system, or where they downplay the break with deeply entrenched ideals to make the transition to the new norms appear smoother and more consistent. One such case occurs in Stark's Würzburg speech in which he tries—unsuccessfully, thanks to von Laue's intervention—to sell the new research policy (which aimed at limiting the traditional conception of freedom of research that allowed academic researchers to choose their own topic of inquiry), describing his Stark's own vision of the centralized organization of scientific research

[348]See Hillers [1935], p. 304, or Brämer & Kremer (Eds.) [1980], pp. 209ff.

[349]The pseudo-religious character of Nazi pamphlets has also been pointed out in Müller [1961].

[350]For an analysis of Hitler's rhetorics in *Mein Kampf* see Kenneth Burke: 'The rhetorics of Hitler's 'Battle", *Southern Review* **5** [1939], pp. 1–21, and Schnauber [1972].

under his authoritarian control. An even more blatent example is Göring's remark at a meeting in 1942 (cf. doc. 99) in which he ridicules the ideal of the internationality of science with the scatological simile of the scientist who cannot contain himself but must recklessly publish his results. The new regime wanted, instead of free accessibility to research results, a nationalized research with strict secrecy constraints on scientists. In areas of military relevancy such conditions were imposed to a degree unknown previously: The researchers of the *Uranverein* were officially not even allowed to retain a personal copy of their own research reports issued in the *Kernphysikalische Forschungsberichte.* Only a handful of copies were distributed to a very limited number of officials.

How idiosyncratic the semantics of contemporary documents can be is illustrated by an anecdote Carl Friedrich von Weizsäcker recently related in an interview: In his correspondence with his father Ernst von Weizsäcker, secretary of state at the Foreign Ministry, a dash in the text indicated that the contrary was actually meant.[351] This verbal cat-and-mouse game did not end with the demise of the Third Reich. It continued well into the postwar years. Heisenberg's report of 1947 on the aims and results of the German *Uranverein*, for instance, is as noteworthy for what it does not mention as for its content. He devotes not a single word to the military applications of this research, in striking contrast to the frequent references to this aspect of nuclear research in previous internal reports (see footnote 306 above). Does this mean that German nuclear physicists had managed throughout the years to swindle the Nazi regime into assigning the highest priority rating to a research project that they knew had no realistic military applications? Though this was implied in some of these texts, and understood often enough in precisely this misleading way,[352] it is counterfactual nonetheless. In a way it is tragic that the newly regained freedom of speech after 1945 was not used to conduct a free and open discussion of what had happened, but the conditions of the postwar days were different ones. Some of these scientists were facing denazification and did not want to incriminate themselves, others who had nothing to hide themselves still felt compelled to conceal the mistakes of their colleagues for the sake of the discipline. Most people simply did not want to think back in time, but rather look ahead to the future and to the future of their discipline.

No reading of any historical account, fair and multifacetted though it might be, can replace a close study of original documents with all their emotional engagement, one-sidedness and occasional fanaticism. Nothing can replace the actual comparison of styles as different as *Geheimrat* Planck's dry bureaucratic

[351]See C.F. von Weizsäcker in: Hoffmann (Ed.) [1993], p. 331.

[352]For instance, by Robert Jungk, who had been led to assume that this had indeed been the clear intention of Heisenberg and others at that time—a myth that he later unwittingly spread in his book Jungk [1958], though later research has clearly proven this to be wrong: cf. Walker [1989]a and Jungk's foreword to the German translation of Walker's book (p. 7) and Jungk [1993] (ft. 307), pp. 298–300.

prose with its carefully worded hints at disagreement with the political authorities (cf. doc. 38) and Professor Stark's agitative verbosity against the proponents of what he considered 'Jewish Physics' (cf., e.g., doc. 56).

Signs of disrespect for the Nazi ideology and phraseology are even more fascinating, be they as frank as in von Laue's writings, or as carefully concealed as in Ludwig Prandtl's petitions. The analysis of a coded vocabulary used by scientists to communicate with each other when voicing their discontents, to avoid the attention of the perceived omnipresent *Gestapo* with its feared devastating consequences, still remains to be done. Many of the selected letters from this period will hopefully awaken new interest in the many available and unexplored resources—we are fortunately now entering a time when estates are coming within the public domain.[353]

7 The Text

The sources of the texts selected for this anthology was either physicists' estates, institutional archives, or German journals and newspapers published during the period in question (see the Table of Contents). Short texts (of around three pages length) were preferred and translated in full. In some cases (i.e. longer or repetitive texts) the most relevant passages were excerpted, but we have tried to minimize this wherever possible. The sources used appear under the title of each document.

7.1 Tips for the User

The documents are self-contained with cross-references to other related texts within the anthology to facilitate use of this collection by readers interested in specific topics. Appendix entries are marked in the annotation by an arrow (→). The annotation is based on the latest secondary literature to place the documents in context. Typographical errors and emendations are indicated with square brackets, and in the case of institutional documents, bureaucratic notes are included in the annotation where they are significant to the content. Omissions are kept to a minimum, and in all cases are indicated by ellipses enclosed in square brackets to indicate that the omission is ours. Emphasis in the original texts is indicated in boldface, omitting purely aesthetic emphasis following the conventions of certain publications (such as surnames in spaced type), however. Original footnotes are clearly distinguished from the editorial annotation through

[353]In 1991, for instance, the von Laue papers were opened at the *Archiv der Kaiser-Wilhelm/Max-Planck-Gesellschaft* in Berlin. Restrictive terms can range from 30 to 60 years, depending on the country and the individual archives. Thanks to Dipl.phys. Jost Lemmerich, the Meitner papers have finally been organized and inventorized at Churchill College, Cambridge University (cf. Lemmerich [1993]).

the use of non-numerical footnote markers (∗, †, etc.). In the original texts italics is reserved for publication titles and for German terms.

7.2 A Note on the Translation

As already indicated in the rhetoric section, the selected texts include a challenging variety of styles. Words that pose particular translation problems include the specialized vocabulary adopted by the National Socialists,[354] e.g., '*Ahnenerbe*' (ancestral heritage) or '*Welteislehre*' (cosmic-ice theory), to package their peculiar ideologies.[355] These terms, institution names and titles or occupations that cannot be translated unambiguously, appear in the original language in the annotation along with a brief explanation where necessary (see also the note preceding the biographical appendix). German terms that appear frequently in English texts, such as 'Reich' and 'heil', have been retained and are not emphasized. The appendix also provides translations of institution names and journal titles for the convenience of the reader (though the latter are referred to consistently in the original language). English spelling of the translations is based upon the 1989 edition of *Webster's Encyclopedic Unabridged Dictionary of the English Language*, Portland House, New York. British spelling is retained in the few documents originally published in British publications. Geographic names appear in their common English form.

7.3 Acknowledgments

Our special thanks go to Prof. Mark Walker for his inspiration to compile this volume and for his continued support. We are also indebted to Prof. Erwin Hiebert, Prof. Dr. Andreas Kleinert, Dipl. phys. Jost Lemmerich, Prof. Roger Stuewer and Prof. Mark Walker for their constructive comments on draft versions of this book. The editor assumes sole responsibility for any remaining errors. During the archival research we have received generous assistance from: Prof. Dr. Eckart Henning and Christel Wegeleben, Archive for the History of the Max Planck Society, Berlin; Dr. Füßl, Deutsches Museum Special Collections; Dr. Ulrich Hunger, Göttingen University Archive; Prof. Dr. E.-A. Müller, Max-Planck-Institut für Strömungsforschung, Göttingen; Dr. Wolfgang Heinicke, Deutsche Physikalische Gesellschaft, Bad Honnef; Dr. Helmut Rechenberg, Max-Planck-Institut für Physik, Heisenberg Archive, Munich; Prof. Dr. Roswitha Schmid,

[354]For an indepth study on the Nazi phraseology see: Victor Klemperer, *LTI. Lingua Tertii Imperii. Die Sprache des Dritten Reiches* [1946], reprinted by Leipzig: Reclam, 1975, or: Kornelia Berning, *Vom 'Abstammungsnachweis' zum 'Zuchtwart'. Das Vokabular des Nationalsozialismus*, Berlin: De Gruyter, 1964 or S. Bork: *Mißbrauch der Sprache. Tendenzen nationalsozialistischer Sprachregelung*, Bern, 1970; see also the references in footnote 350 as well as Hiller's [1935] strange recommendations for 'Germanized' physics terminology.

[355]On the *Welteislehre* as a prime example of pseudoscientific discourse, see in particular Nagel [1991] and Schröder [1991].

Naturwissenschaftliche Rundschau, Munich; Prof. Dr. F. Baumgart and Peter A. Süß M.A. of the Kommission für die Geschichte der Julius-Maximilians-Universität Würzburg; Dr. A. Miller, Kernforschungszentrum Karlsruhe; and Michele Zeccardi at the Center for History of Physics, American Institute of Physics, College Park, MD. We also thank the librarians from the following institutions: Staats- und Universitätsbibliothek Göttingen; Staatsbibliothek Preußischer Kulturbesitz, Berlin (Haus 1 & 2); Hahn-Meitner-Institut, Berlin; Library of the Deutsches Museum, Munich; Boston Public Library; and the Library of Congress in Washington, D.C.

We are particularly grateful for the copyright permissions provided by Prof. Dr. Friedrich Hund, Göttingen, Dr. med. Helgard Krietsch, Munich, Prof. Dr. Theodore von Laue, Worcester, Mass., PD Dr. med. J. Peter Nordmeyer (Hagen), and Prof. Alvin M. Weinberg, Oak Ridge, Tennessee. Other copyright permissions have been obtained from the following archives and journals:

Archiv zur Geschichte der Max-Planck-Gesellschaft, Berlin
Bundesministerium der Justiz, Bonn (legal texts)
Deutsches Museum, Munich (DMM)
Educational Foundation for Nuclear Science (*Bulletin of the Atomic Scientists*), Chicago
Frankfurter Allgemeine Zeitung, Frankfurt
Geheimes Staatsarchiv Preussischer Kulturbesitz, Berlin
Kernforschungszentrum Karlsruhe (*Kernphysikalische Forschungsberichte*)
Max Planck Institut für Strömungsforschung, Göttingen (Prandtl)
Nature, London
Die Naturwissenschaften (Springer), Berlin
Review of Scientific Instruments (AIP), New York
Science, Washington, DC
Ullstein Bilderdienst, Berlin

In some cases the journals were discontinued in 1945, or the articles and authors are no longer covered by copyright restrictions.

Finally, we would also like to thank the publisher, in particular Dr. Annette A'Campo and Ms. Doris Wörner at the Birkhäuser mathematics readers' department for their kind assistance in the production of this book as well as the editors for the decision to include it in this series.

Part I
Controversies Prior to 1933

1 Albert Einstein: My Reply. On the Anti-Relativity Theoretical Co., Ltd. [August 27, 1920]

Source: Albert Einstein, 'Meine Antwort. Ueber die anti-relativitätstheoretische G. m. b. H.' in *Berliner Tageblatt*, Ser. 49, No. 402, Morning Edition A, August 27, 1920, p. 1.

A motley group has joined together to form a company under the pretentious name "Syndicate of German Scientists" currently with the single purpose of denigrating the theory of relativity and me as its author in the eyes of non-physicists.[1] Recently Messrs. Weyland and Gehrke held a lecture with this intent at Philharmonic Hall which I attended personally.[2] I am fully aware of the fact that both speakers are unworthy of a reply from my pen; for, I have good reason to believe that there are other motives behind this undertaking than the search for truth. (Were I a German national, whether bearing a swastika or not, rather than a Jew of liberal international bent ...)[3] I only respond because I have received repeated requests from well-meaning quarters to have my view made known.[4]

First of all I observe that to my knowledge hardly any scientists today who have made any substantial contribution to theoretical physics do not concede that the entire theory of relativity is logically and consistently structured and that it

[1] For analyses as well as historical background on the political climate of interwar Germany and the informally established 'Syndicate of German Scientists for the Preservation of Pure Science, reg. assoc.' (*Arbeitsgemeinschaft deutscher Naturforscher zur Erhaltung reiner Wissenschaft, e. V.*), see in particular Grundmann [1967], Hermann [1977], Kleinert [1978], [1979]a, Goenner [1993], and Hentschel [1990]a, sec. 3.2, as well as Kleinert [1993], who notes on pp. 203f. that the syndicate was never officially registered.

[2] On the anti-Semitic radical activist Paul → Weyland, who publicly denounced relativity theory as an "enormous bluff" and A. → Einstein as a plagiarizer, see in particular Kleinert [1993] and Goenner [1993], pp. 120–123, where Weyland is reported to have accumulated a criminal record as an impostor; cf. also Hentschel [1990]b. For his lecture see Weyland [1920]a. A synopsis of Ernst → Gehrcke's arguments is published in Gehrcke [1920]b.

[3] By articulating the clearly anti-Semitic motivations behind the attacks, Einstein inadvertently introduced this subject into the scientific debate. Gehrcke, for instance, who had avoided resorting publicly to such arguments, was now prompted to respond on this level. On the resulting escalation of the debate, see Hentschel [1990]a, sec. 3.2.

[4] However, Einstein later regretted publishing this retort in the daily press, since it only contributed to the public polemics. On Sept. 6, 1920 he wrote to Arnold Sommerfeld: "Perhaps I ought not to have written the article. But I wanted to prevent that my unbroken silence to the systematically reiterated objections and accusations be taken as an endorsement"; and on a similar vein to Max Born on Sept. 9, 1920: "Everyone has to offer his sacrifice at the altar of stupidity from time to time, to humor God and man. And I did a thorough job of it with my article." See Hermann (Ed.) [1968], p. 69; Born [1968]b, p. 59.

agrees with the verified experimental data now available. Prominent theoretical physicists—I name H. A. Lorentz, M. Planck, Sommerfeld, Laue, Born, Larmor, Eddington, Debye, Langevin, Levi-Civita—use the theory as a basis, and most have made valuable contributions to it. From among physicists of international renown I can only name Lenard as an outspoken critic of relativity theory. I admire Lenard as a master of experimental physics;[5] however, he has yet to accomplish something in theoretical physics, and his objections to the general theory of relativity are so superficial that I had not deemed it necessary until now to reply to them in detail. I intend to return to this.[6]

I am accused of conducting a tasteless campaign for the theory of relativity.[7] I can certainly say that all my life I have been a friend of well-considered, tempered words and concise description. High-flown phrases and catchwords give me goose flesh, be they about relativity theory or anything else. I have often been amused by effusions for which I am now ultimately being called to account. By the way, I gladly cede this pleasure to the gentlemen of the Co., Ltd.

Now to the talks. Mr. Weyland, who does not seem to be a specialist at all (is he a doctor? engineer? politician? I could not find this out), presented nothing of pertinence.[8] He broke out in coarse abuse and base accusations. The second speaker, Mr. Gehrke,[9] partly made directly inaccurate statements and partly attempted to create a false impression for uninformed laymen through a biassed data selection and through distortion of the facts. The following examples ought to illustrate this:

Mr. Gehrke alleges that the theory of relativity would lead to – solipsism, an allegation that any specialist would take for a joke. He supports this with the familiar example of the two clocks (or twins), of whom **one** takes a round trip in relation to the inertial system, the other does not. He asserts—despite having been refuted frequently both orally and in writing by the most qualified specialists in the theory—that in this case the theory leads to the really nonsensical result that two clocks at rest next to each other imitate one another in relation to each

[5]Philipp → Lenard had been awarded the Nobel Prize in 1905 for his experimental work on cathode rays and the photoelectric effect, which Einstein provided with a theoretical explanation in the same year.

[6]Einstein [1918]b does address Lenard's common sense arguments, to which Lenard [1918/21] was the reply. Einstein [1918]a also discusses Gehrcke's conception of the physics of the ether. Around this time, however, Einstein gradually withdrew from the public debate.

[7]The unprecedented wave of publicity on the theory particularly after 1919 was the result of numerous popular science publications and coverage in the media: For statistics on books written on the subject at that time, see Hentschel [1990]a, sec. 2.2 and references there.

[8]Weyland had indeed never studied physics. In 1953 he claimed to have studied chemistry, but this also is uncertain. He appeared in the public eye as a political activist, e.g., as editor of a blatantly anti-Semitic periodical *Deutsch-Völkische Monatshefte*, which anticipated later Nazi publications like the infamous *Stürmer*; see in particular Kleinert [1993].

[9]On the experimental physicist Ernst Gehrcke, an optics specialist at the *Physikalisch-Technische Reichsanstalt* (→ *PTR*), see in particular Goenner [1993], pp. 114–116.

other. I can only interpret this as a deliberate attempt at misleading the lay public.[10]

Furthermore, Mr. Gehrke alludes to Mr. Lenard's objections, many of which are linked to examples in mechanics from everyday life.[11] These are already invalidated on the basis of my general proof that on first approximation the statements of the general theory of relativity conform with those of classical mechanics.[12]

What Mr. Gehrke said about the experimental confirmation of the theory is to me the most striking proof that he was not concerned with revealing the true facts.

Mr. Gehrke wants to give credence to the idea that the motion of Mercury's perihelion is explanable without relativity theory.[13] There are two options. On the one hand, suitable interplanetary masses can be postulated that are large enough and distributed in such a way as to cause the amount of motion observed at Mercury's perihelion; this is naturally a highly unsatisfactory alternative to the one offered by the theory of relativity, which yields Mercury's perihelion motion without making any special assumptions. Or else you base yourself on a paper by Gerber in which he had anticipated my correct formula for Mercury's perhelion motion.[14] But specialists in the field agree not only that Gerber's derivation is thoroughly incorrect, but that the formula cannot even be obtained as a consequence of Gerber's leading assumptions.[15] Mr. Gerber's paper is therefore utterly worthless, a failed and irreparable attempt at a theory. I establish

[10]On the history and explanation of this so-called clock or twin paradox see, e.g., Leslie Marder: *Time and the Space-Traveller*, London: Allen & Unwin, 1971.

[11]Lenard questioned, for example, Einstein's principle of relativity, because accelerations as those in a train gathering speed could certainly be used to distinguish particular reference systems.

[12]Einstein's opponents were incapable of accepting even these very relevant arguments, because the latter were already on a too abstract level.

[13]The shifting of the point in Mercury's orbit nearest to the sun by an amount of 43″ per century, first observed and calculated by LeVerrier and others in the mid-19th Century, could not be explained by classical celestial mechanics; cf. N. T. Roseveare, *Mercury's Perihelion: From Le Verrier to Einstein*, Oxford: Oxford Univ. Press, 1983.

[14]In 1902 a paper appeared by the schoolteacher Paul Gerber (1854–?) on the propagation velocity of gravitation entitled 'Über die Fortpflanzungsgeschwindigkeit der Gravitation' in the obscure *Programmabhandlungen des städtischen Realgymnasiums Stargard in Pommern.* There the motion of Mercury's perihelion was derived from the assumption of a velocity-dependent gravitation potential, thus using an entirely different method to Einstein's in his paper 'Erklärung der Perihelbewegung des Merkur aus der allgemeinen Relativitätstheorie' published at the end of 1915 in *Sitzungsberichte der Preußischen Akademie der Wissenschaften*, pp. 831–839. Gehrcke rediscovered Gerber's paper in 1916 and published excerpts from it in the *Annalen der Physik* (4) **52** [1916], pp. 415–441. In his own commentary he challenged Einstein's priority; see also Gehrcke [1916].

[15]See in this regard Laue [1917], S. Oppenheim: 'Zur Frage nach der Fortpflanzungsgeschwindigkeit der Gravitation', *Annalen der Physik* (4) **53** [1917], pp. 163–168 or Hugo von Seeliger's contribution, ibid., pp. 31–32.

that the general theory of relativity has provided the first realistic explanation for the motion of Mercury's perihelion. I did not mention Gerber's article originally, because I did not know about it when I wrote my paper on the motion of Mercury's perihelion; but even if I had been acquainted with it, I would have had no cause to mention it. The personal attack Messrs. Gehrke and Lenard have launched against me based on these circumstances has been generally regarded as unfair by real specialists in the field; I had considered it beneath my dignity to waste a word on it.

In his lecture Mr. Gehrke has placed in a distorted light the reliability of the masterfully conducted English measurements on the sun's deflection of passing lightrays, by only mentioning **one of the three** separate groups of exposures which as a result of warping had to produce faulty results. He concealed the fact that in their official reports the English astronomers themselves called their results a brilliant confirmation of the general theory of relativity.[16]

With regard to the problem of the redshift of spectral lines, Mr. Gehrke did not mention that current determinations contradict one another, and that a final decision on this matter has not yet been reached.[17] He only produced evidence speaking **against** the existence of the line shifts predicted by relativity theory but omitted mentioning that the conclusiveness of those earlier results has been shaken by more recent investigations by Grebe and Bachem and by Perot.[18]

Finally I would like to note that on my initiative arrangements are being made for a discussion to be held on relativity theory at the scientific conference in Nauheim.[19] Anyone willing to confront a professional forum can present his objections there.

---

[16]See Frank W. Dyson, A. S. Eddington & C. Davidson, 'A determination of the deflection of light by the sun's gravitational field from observations made at the total eclipse of May 29, 1919', *Philosophical Transactions of the Royal Society of London* **A 220**, pp. 291–334.

[17]Compare, e.g., Gehrcke's note: 'Was beweisen die Beobachtungen über die Richtigkeit der Relativitätstheorie?', in *Zeitschrift für technische Physik* **1** [1920], p. 123, and the contemporary overview article by F. Croze: 'Les raies du spectre solaire et la théorie d'Einstein', *Annales de Chimie et de Physique* (9) **19** [1923], pp. 93–229, and further references there to studies on redshift in the gravitational field, which Einstein had postulated in 1907.

[18]See the paper by the Bonn experimental physicists Leonhard Grebe (1883–1967) and Albert Bachem (1888–1957): 'Über die Einsteinverschiebung im Gravitationsfeld der Sonne', *Zeitschrift für Physik* **1** [1920], pp. 51–54 and **2** [1920], pp. 415–422, as well as the contributions by the French astrophysicist Alfred Pérot (1863–1925) in the *Comptes Rendus des Sciences Hebdomadaires* of the Paris Academy of Sciences, **171** [1920], pp. 223–232, **172** [1921], pp. 578–581; however, cf. also the critical papers by Ludwig → Glaser, a student of J. Stark's, in *Physikalische Zeitschrift* **23** [1922], pp. 100–102 and in *Jahrbuch der Radioaktivität und Elektronik* **20** [1923], pp. 277–352.

[19]On the session on relativity theory at the 86th convention of the scientific society *Gesellschaft Deutscher Naturforscher und Ärzte* at the spa town of Bad Nauheim on Sep. 20, 1920 see, e.g., Hermann Weyl's eyewitness account published in *Umschau* **24** [1920], pp. 609–611 as well as Gehrcke and Weyl, ibid. **25** [1921], pp. 99, 123, 227, Pincussen [1920], and Weyland [1920]c on the supposed 'stifling' of Einstein's opponents; see also Goenner [1993], pp. 123–127 and references there.

A peculiar impression will be left abroad, particularly to my Dutch and English colleagues H. A. Lorentz and Eddington,[20] who have both worked closely on relativity theory and have lectured on the subject repeatedly, when they see the theory and its author being disparaged in such a way in Germany itself.[21]

Fig. 3: Group photograph of Berlin physicists and chemists (1920). From left to right standing: Walther Grotrian, Wilhelm Westphal, Otto von Baeyer, Peter Pringsheim, Gustav Hertz. Sitting (from left to right): Hertha Sponer, Albert Einstein, Ingrid Franck, James Franck, Lise Meitner and Otto Hahn. (Sponer and Franck worked at the Second Institute of Physics at Göttingen University from 1921–1933). From Lemmerich [1982] p. 49 with kind permission.

[20]On Hendrik Antoon Lorentz's (1853–1928) research on relativity theory see, e.g., A.J. Kox in: Don Howard & John Stachel (Eds.) *Einstein and the History of General Relativity*, Boston: Birkhäuser, 1989, pp. 201–212. For a unique interpretation of the theory by the astronomer Sir Arthur Stanley Eddington (1882–1944) see, e.g., Hentschel [1990]a, sec. 4.12.1.

[21]A short notice in the same issue on p. 3 carried the headline 'Albert Einstein wants to leave Berlin!' and objected to the "Pan-German provocations and pseudo-scientific methods" used by the assailants. In anticipation of this, an article appeared in four separate newspapers on August 26 and 27 signed by Max → von Laue, Walther → Nernst and Heinrich Rubens rejecting the accusations and defending Einstein's reputation: See Laue, Nernst & Rubens [1920]. Other expressions of solidarity by Walther Nernst, Max Reinhardt, Stefan Zweig and many others soon followed: See Bie et al. [1920], and N.N. [1920]a-b; cf., e.g., Goenner [1993], pp. 116ff.

2 Max von Laue: Review of Johannes Stark's 'The Current Crisis in German Physics' [January 12, 1923]

Source: Max von Laue, review of Johannes Stark: *Die gegenwärtige Krise in der deutschen Physik*, in *Die Naturwissenschaften*, Vol. 11, Issue No. 2, January 12, 1923, pp. 29–30.

Stark, Johannes, *The Current Crisis in German Physics*. Leipzig, J. A. Barth, 1922. VI, 32 pp., 15 × 23 cm. Price 40.– M[arks].

This work has 4 chapters: The role of theory in physics; the position of Einstein's general theory of relativity in physics and its propaganda; the dogmatism of quantum theory; and the one-sidedness of modern physics and the move away from application.

Not much can be said against the first chapter; its content is not exactly new. In the other three the author pours his heart out and pontificates on everything he currently does not like in German physics. Unfortunately, this is quite a lot. He would like to banish the general theory of relativity completely from physics to mathematics and philosophy. In his opinion quantum theory damages experimental research, in that some scientists point in its favor instead of verifying whether certain singularities in their series of measurements are not simply the result of observational error. Finally, he believes that theoretical physics circles do not sufficiently acknowledge technical achievements out of haughtiness.[1]

We will not go into the attacks against individuals which the book allows itself.[2] Tact, above all, forbids us from expressing an opinion about claims that some researchers today are too old.[3] But to the point, we would like to say: Never before have we heard of such haughtiness by physicists towards engineers. On the contrary, discerning people from all walks of life, including engineering, know very well the essential difference between an intellectual activity directed purely towards knowledge and one directed towards the application of this knowledge; and they occasionally also draw practical consequences from it, e.g., in habilitations.[4]

1 See also, e.g., doc. 56 for J. → Stark's invectives against modern physics as well as the Introduction, sec. 5.3, on the 'Aryan Physics' movement.

2 Specifically, Stark accused Albert → Einstein of being a shallow propagandist, plagiarist, and a traitor to his country, and Arnold → Sommerfeld of being a dogmatist, see Stark [1922], pp. 6ff., 12–13, 15, 20.

3 Stark criticized Walther → Nernst as vain and power-hungry, ibid., pp. 31f., because of the latter's appointment as president of the *Physikalisch-Technische Reichsanstalt* (→ *PTR*) in 1922 at the age of 58. Stark had protested openly against this appointment and was probably envious. He finally obtained this title through his Nazi Party connections in 1933, replacing Nernst's aged successor, F. → Paschen: Cf. footnote 6 and Max → von Laue's version of the events, doc. 19, as well as Beyerchen [1977], pp. 114–115, Schröder [1993], p. 331.

4 An academic requirement similar to a dissertation or second doctoral thesis qualifying the author for professorship at German universities. Laue is alluding to an incident in 1929 when one of Stark's pupils, the antirelativist Ludwig → Glaser, had difficulty getting his habilitation thesis accepted at the University of Würzburg, where Stark was professor of physics from 1920; see also the following footnote. Cf Stark [1987], pp. 60f.

But this **essential** difference between the two approaches is certainly not a **value** difference as Mr. Stark seems to conclude—though in the contrary sense—when he writes: "If physics had been conducted exclusively as a 'pure' science, ignoring any application whatsoever ...., then it would have shriveled up as a ridiculous intellectual game."

Mr. Stark left a teaching and research position in order to devote himself to technology. That is his due right; since an individual, notably an important one, is free to seek his occupation at any time and in any field he pleases. This severance surely did not take place without some conflict.[5] But Mr. Stark should really have preserved enough respect for his own former activity to not debase it publicly.

Otherwise, even the title of the work is an unfortunate choice. The crisis, which undoubtedly exists in physics and which is equally undoubtedly attributable above all to quantum theory, is not at all limited to **German** science. It manifests itself in the same way in all countries where physical research is being conducted; and it can only be overcome when science has found the answer to the quantum puzzle. There is no other remedy. We would like to agree with the author, however, that Germany should try to increase the amount of experimental research at the expense of theoretical work. But surely we all know that the current disproportion is mainly due to the predicament which our entire nation is in and that physicists are unfortunately not in the position to change it much.

All in all, we would have wished that this book had remained unwritten, that is, in the interest of science in general, of German science in particular, and not least of all in the interest of the author himself.[6]

3 Philipp Lenard & Johannes Stark: The Hitler Spirit and Science [May 8, 1924]

Source: Philipp Lenard and Johannes Stark,[1] 'Hitlergeist und Wissenschaft', *Großdeutsche Zeitung. Tageszeitung für nationale und soziale Politik und Wirtschaft*, Vol. 1, Issue No. 81, Thurs-

[5] Annoyed by the faculty's objections to the form of his expert opinion on Glaser's habilitation and incensed by the faculty's lack of trust, Stark resigned his professorship, left the university, and took up research in a private laboratory in 1922; cf. Stark [1987], pp. 64ff. He had other motives for this career change as well, however. See footnotes 8ff. of doc. 18.

[6] The reviewer proved to be right. After his porcelain factory failed, Stark was unable to obtain another university position during the Weimar period. It was only after A. → Hitler's seizure of power that Stark was again able to acquire important positions by virtue of his early loyalty to the → *NSDAP*: Cf. doc. 3 for P. → Lenard's and Stark's declaration of solidarity with Hitler in 1924. He obtained the presidency of the *PTR* in 1933 (cf. doc. 18) and also of the *Deutsche Forschungs-Gemeinschaft* (→ *DFG*) in 1934 (cf. docs. 33f.).

[1] In Stark [1934]d and in the second to last paragraph of doc. 40, J. → Stark claims that P. → Lenard (of Heidelberg) had written this proclamation on his own and that Stark himself (then in Ullersricht) had only provided his signature. It was in any case unusual for German physicists to publish declarations of this type in newspapers; it was not without effect: Cf., e.g., Beyerchen [1977], sec. 5, pp. 95ff.

day, May 8, 1924, pp. 1–2;[2] also reprinted in full with minor changes in *Nationalsozialistische Monatshefte*, Vol. 7, Issue No. 71, February 1936, pp. 110–111.

We have had the pleasure of receiving general recognition as scientists and would like to speak here in this capacity. Following our innermost feelings we would like to make known herewith our recognition as scientists of Hitler and his comrades. We recognize—in the way they expressed themselves before and during the recently concluded legal proceedings[[3]] and in their whole thinking and sentiment—the **very same** spirit [*Geist*] that we ourselves have constantly looked for, striven toward, and expressed in our own work to make it profound and successful: That spirit of total clarity, of honesty towards the outer world, and at the same time of inner uniformity; that spirit which hates any compromising activity because it is insincere. But we have already recognized early on and revered this—to us exemplary—spirit in the great scientists of the past as well: in Galileo, Kepler, Newton, and Faraday.[[4]] We admire and revere it in the same way also in Hitler, Ludendorff, Pöhner, and their comrades.[[5]] We recognize in them the closest affinity of ideas with our own.

Consider what it means to be privileged to have this kind of genius living among us in the flesh. They were rare at all times, these culture-bearing intellects. But all the inner aspirations of humanity and all the resulting achievements making our life on earth worthwhile and beautiful are based on their efforts. Experience reveals that the incarnations of this spirit are only of Aryan-Germanic blood, just as the great men of science named above were also of this blood. But blood also can die out; Romans and Greeks died out along with their culture. Do not be deceived, the Aryan-Germanic blood, the carrier of its unique spirit,

[2]The newspaper editors inserted the following introductory comment under the authors' names: 'We are exceptionally pleased to be able to offer to two noteworthy representatives of German science the chance to present their views in the following article. Both are recipients of the Nobel Prize who, as they have informed us, wish to strengthen the national movement with their joint declaration and to support the efforts to restore the lives and occupations of our freedom fighters.'

[3]After the failure of Adolf → Hitler's attempted *putsch* in Munich on Nov. 8–9, 1923, Hitler, the ex-general Erich von Ludendorff, and the leaders of the *Kampfbund*, who had coordinated the activities of the National Socialist Party and other paramilitary groups in Bavaria were put on trial for high treason in Munich. The trial opened on Feb. 26, 1924, received enormous publicity, and ended on April 1 with the minimum sentence of five years imprisonment for Hitler and three other defendants; von Ludendorff was acquitted. For contemporary documents about the *putsch* attempt and the succeeding trial see, e.g., Noakes & Pridham (Eds.) [1990]b, Vol. 1, pp. 26–35.

[4]Cf. Lenard [1929] for his hagiographical description of the life and work of 'great men' of science throughout history, especially pp. 31–48 for Galileo and Kepler, pp. 76–99 for Newton, and pp. 211–222 for Faraday.

[5]General Erich Friedrich Wilhelm von Ludendorff (1865–1937) had served in World War I and supported Hitler's plans for a 'national revolution'. He was to become head of the national army. Ernst Pöhner (1870–1925), Munich's police chief, sympathized with the Nazis and was appointed by Hitler as the new 'Prime Minister' with dictatorial powers during the *putsch*.

is already in the process of disappearing rapidly; a racially alien spirit has been busily at work for over 2,000 years. The exact same force is at work, always with the same Asian people behind it that had brought Christ to the cross, Jordanus Brunus[6] to the stake, and that shoots at Hitler and Ludendorff with machine guns and confines them within fortress walls: It is the fight of the dark spirits against the torchbearers in an endeavor to eliminate their influence on earth. Each time this was done through 'legal' means. But it is not the means but the end result that matters. This awareness and understanding of the threat to the ever rarer incarnations of the spirit of light among us is, of course, beyond the ability of many to see, even among the leadership. But we nationalists [*Völkische*] have begun to pour over the pages of history, guided by an inner feeling that had instructed us: The Germanic spirit of honesty, which we have always held since our youth to be the highest and holiest in man and which had guided us in our life's work as well, is increasingly at stake here and has been so already for years. With its extermination, we are also sawing off the branch on which science sits. And just because we felt this and felt it increasingly keenly, we became 'nationally minded' [*völkisch*], which means: We now place the highest emphasis on protecting that which is inherited in our blood, because we have learned to recognize it as the blessing of all of mankind. But we will not simply protect our own intellectual identity, we will first start digging it out again from under the alien-spirited rubble within ourselves—primarily in order to find ourselves again.[7] We need lucid minds not only as scientists—no, people should not be segregated according to their occupation (a frequently used Semitic deception!). We want lucid, well-rounded characters, just as Hitler is one. He and his comrades in the struggle appear to us as God's gifts from times of old when races were purer, people were greater, and minds were less deluded. This we feel; and these divine gifts should not be taken from us. This thought alone should already be a solid enough basis to hold the nationally-minded together toward their great goal: Founding a new Germany, with Hitler 'beating the drum', in which the German spirit is not just tolerated again to a certain extent and released from imprisonment, no, but in which the German spirit is protected, nursed, and assisted so that it can then finally thrive again and develop itself further for the vindication of the honor of life on our planet which is now dominated by an inferior spirit.[8] Universities and

[6]Giordano Bruno (1548–1600), Italian philosopher and poet, believed in an infinitude of worlds with living beings. Since this was in open conflict with church doctrine, Bruno was arrested in Venice in 1591 and was handed over to the Inquisition in Rome in 1593. After 7 years of imprisonment he was burned at stake at Campo de'Fiori.

[7]This sentence is an example of the ambiguous language and pseudo-religious tone often found in Nazi texts: 'Nicht nur verteidigen, sondern sogar erst wieder ausgraben aus fremdgeistiger Verschüttung in uns müssen wir unseres Geistes Eigen – um vor allem uns selbst wieder gefunden zu haben.'

[8]On stylistic similarities to Hitler's *Mein Kampf* (of which Vol. 1 appeared one year later), and to Alfred → Rosenberg's *Mythus des 20. Jahrhunderts* (written at about the same time, but published only in 1930), see Beyerchen [1977], sec. 5.5. These texts share the contemporary

their students have failed, most of all precisely in those subjects that should have set the pace long ago. But it is also much better that 'the man of the people' is doing it. He is here. He has revealed himself as the '*Führer*' of the sincere. We shall follow him.[9]

4 Albert von Brunn: Review of '100 Authors against Einstein' [March 13, 1931]

Source: Albert von Brunn,[1] review of the collection of essays: H. Israel et. al. (Eds), *100 Autoren gegen Einstein* in *Die Naturwissenschaften*, Vol. 19, Issue No. 11, March 13, 1931, pp. 254–256.

100 Authors against Einstein. Edited by K. Israel, Erich Ruckhaber, and R. Weinmann. Leipzig: R. Voigtländer 1931. 104 pp. 15 × 23 cm. Price 2.40 R[eich] M[arks].[2]

---

style of *völkisch* popular literature, which conveys a romantic mystic notion of blood and soil and the recurrent image of a death struggle between the 'Jewish forces of darkness' and the 'Aryan-Germanic forces of enlightenment'. See also the Introduction, sec. 6.

[9]This text demonstrates how Hitler turned the ignominious failure of the *putsch* into a propaganda victory. The outcome of his 'patriotric act' was blamed on the pusallinamity of the national leaders and on others (such as here universities and students). Hitler emerged as a hero, as someone with the courage and energy to act on his convictions, not only for the approximately 55,000 members of the subsequently banned Nazi Party but also for a much larger segment of the nationalistically and anti-Semitically inclined population: The *Führer* image had been created.

[1]The astrophysicist [Friedrich Julius Witholm Lothar] Albert von Brunn (1880–1940) studied in Leipzig, Würzburg and Göttingen 1899–1904 and took his doctorate under K. Schwarzschild in 1905. 1906–08 von Brunn conducted planetary observations. In 1922 he was named honorary professor at the Polytechnic in Danzig and 1923–26 was director of the Danzig Observatory. 1926–28 he was a volunteer staff member first at the Breslau Observatory, then 1929–40 at the Einstein Institute in Potsdam. 1929–31 von Brunn worked on light deflection in the sun's gravitational field, 1930 also on the solar center-limb effect, and in 1935 on the distribution of extragalactic nebulae.

[2]Hans Israel (1881–?), erroneously identified here as K. Israel, was an engineer who had earned his degree in chemistry in 1905 and was a self-proclaimed philosopher. His contribution to this collection of essays is entitled: 'Mathematische Widerlegung der Relativitätstheorie', pp. 14–15. Erich Ruckhaber (1876–1956) was a professional translator and interpreter in Berlin with various disparate interests including philosophy (he earned his degree in philosophy in 1927) and relativity theory. Ruckhaber's essay is entitled: 'Die völlige Unlogik der Relativitätstheorie', pp. 47–49. The senior editor Rudolf Weinmann (1870–?) also held a degree in philosophy from 1895 in Munich, started a career in drama in 1899 and subsequently also published a number of articles and pamphlets on relativity theory. His contribution is entitled: 'Die spezielle Relativitätstheorie. Der unzureichende Anlaß—Die falsche Voraussetzung—Die absurden Konsequenzen—Die anderen Möglichkeiten', pp. 60–64. For more information on the editors and other antirelativist contributors, see a recent paper examining the background to this book and its authors by Hubert Goenner, who writes: "Obviously, these three men were united not only by their common interest in philosophy and opposition to relativity theory but also by their incompetence in the fields of mathematics and physics." Goenner [1994], p. 254. Albert → Einstein's revolutionary theory of relativity also encountered a lot of resistance within

In principle it is not astonishing that many have formed an unfavorable prejudice against the theory of relativity after having witnessed the development of only its most outward manifestations. This is the work of over-zealous but less well-informed enthusiasts of this theory, who have made serious tactical errors and gross blunders without its author himself being in any way to blame.[3] Thank goodness knowledgeable supporters of the new concept have stifled in time attempts to let the *vox populi* decide on the theories. Even individual fanatic scientific advocates of the Einsteinian theory seem to have finally abandoned their tactic of cutting off any discussion about it with the threat that every criticism, even the most moderate and scrupulous ones, must be discredited as an obvious effluence of stupidity and malice. But even if these monstrous products of the 'Einstein frenzy' [*Einstein-Taumel*] now belong to history and are thus eliminated from consideration, thoroughly respectable reasons for a certain discomfort with relativity theory still do remain: The special theory of relativity already demands certain *sacrificia intellectus*: Primarily, ††† renunciating the strict **determinability** of simultaneity—note, by no means the **concept** of simultaneity itself![4] Naturally, for some philosophers, this is equivalent to an unexpiable crime against Kant's eternal infallibility, because they just do not understand the inevitability of Einstein's objectives.[5] Scientists will not find this matter so tragic, simply because they know that both relativity theories obviously leave completely untouched '1st order effects', like general, planetary and secular aberration, Doppler effect, and the Olaf Römer effect.[6] Compared to the view held up to now, uncertainty only appears in order of magnitude $\frac{v^2}{c^2}$ (a clear conventional symbol to all interested persons today).[7] This means that within the entire solar system,

the physics community itself. On the Voigtländer publishing house, see footnote 20 below.

[3]See Hentschel [1990]a, chap. 2, for an account of some misinterpretations of the theory of relativity by laymen.

[4]The three daggers stand for trice making the sign of the cross with the connotation: "God forbid!" upon the mere thought of relinquishing the uniqueness of simultaneity. In paragraph 1 of Albert Einstein's 1905 paper 'On the electrodynamics of moving bodies' ('Zur Elektrodynamik bewegter Körper'), *Annalen der Physik* (4) **17** [1905], pp. 891–921, reprinted in *The Collected Papers of Albert Einstein*, Vol. 2 *The Swiss Years: Writings, 1900–1909*, Princton: Princeton Univ. Press, 1989, pp. 275–310, and translated by Anna Beck, ibid., pp. 140–170, it is pointed out that simultaneity is not given *per se*, but has to be defined for distant events by means of light signal exchange.

[5]On the interpretation of relativity theory by neo-Kantian adherents to Immanuel Kant's (1724–1804) philosophy see Hentschel [1990]a, sec. 4.1.

[6]Aberration effects refer to the fact that the apparent angle of observation θ of a star image changes with the velocity of the observer v according to the relation $\sin(\theta) = v/c$, with c as the velocity of light in a vacuum. In secular aberration, v is the earth's velocity on its orbit around the sun. The Doppler effect describes a change in the frequency ν of an emitted wave with the speed of the observer and emitter in the medium. The 'Römer effect' refers to the Danish astronomer Ole Rømer's (1644–1710) determination of the velocity of light in 1676 by measuring the duration of eclipses of Jupiter's moons, which varied with varying distances between the earth and Jupiter.

[7]Einstein's theory of relativity predicts changes in measurements of physical magnitudes

Pluto included, the 'margin of error' for establishing simultaneity objectively does not exceed the order of magnitude of 2^s.[8] The 'general' theory of relativity is in a worse position though. Here it is necessary to assume that the mysterious arena of nature's workings is a four-dimensional manifold, in which space and time appear fused together in a mathematically clearly formulated, thus by no means obscure, manner.[9] But here also there is no reason whatsoever for the guardians of the Kantian sanctuary to be upset, since naturally, relativists also can only observe the effect of the mysterious four-dimensional workings of nature within the empirical world, that is, within the empirical space and time perceptions (of Euclid, Galileo/Newton, and Kant) which are imposed upon us once and for all by nature.[10]

Thus the present book under review could conceivably be tolerated, if need be, provided it owed its existence merely to these in principle justified, though also very exaggerated, fears of an "evaporation of the concept of reality" in modern science. Yet, when an author who lacks sufficiently objective judgment collects outside criticism with a certain bias against a scientific branch of study, he must put up with the fact that his approach will be dismissed as pamphleteering, irrespective of the moral evaluation of his aim.

However, even the most tolerant critic will not be able to ferret out any mitigating circumstances whatsoever for this '100 authors book'. For, of what elements is this Areopagus composed?! 90% of the 'authors' are raving Kantians,[11]

---

proportionate to v^2/c^2 (that is, in the second order of v/c). With the speed of light at $c \simeq 3 \cdot 10^8$m/s and most speeds of macroscopic bodies like the motion of planets around the sun at $v \leq 3 \cdot 10^5$m/s, the relativistic effect of $v^2/c^2 \leq 10^{-6}$ is well beyond detection for most phenomena examined in classical physics, with the motion of mercury's perihelion as the msot notable exception.

[8]This estimate of 2 seconds as the maximum relativistic time correction comes from multiplying the sidereal periods of the planets with the relativistic correction factor v^2/c^2, with v of order of magnitude 10–50 km/sec as the average speed of the planets along their orbits.

[9]This 'union of space and time' was first described in a talk on 'space and time' by the Göttingen mathematician Hermann Minkowski (1864–1909). Minkowski took his doctorate in Königsberg in 1885, became university lecturer in 1887, assistant professor in 1893, associate professor in 1884, and full professor in 1885 in Königsberg. In 1896 he became professor of higher mathematics at the Zurich Polytechnical Institute and from 1902 was full professor of mathematics at Göttingen. Minkowski specialized in number theory, diophantine approximations, hydrodynamics and electrodynamics. See H. Minkowski: 'Raum und Zeit', *Verhandlungen der deutschen Naturforscher und Ärzte* **80** [1908], pp. 4–9; 'Das Relativitätsprinzip', *Annalen der Physik* (4) **47** [1915] (= posthumous publication of a talk on Nov. 5, 1907). Cf. also, Peter Galison: 'Minkowski's space-time: From visual thinking to the absolute world', *Historical Studies in the Physical Sciences* **10** [1979], pp. 85–131, and Lewis Pyenson: 'Hermann Minkowski and Einstein's special theory of relativity', *Archive for History of Exact Science* **A** [1977], pp. 71–95.

[10]The Greek mathematician Euclid (c. 280 BC), the Italian physicist and astronomer Galileo Galilei (1564–1642), and the English natural philosopher, astronomer and mathematician Sir Isaac Newton (1642–1727) together form the basis of what is now referred to as 'classical' physics.

[11]See footnote 5 above.

who have not the faintest notion of the epistemological predicament in which modern physicists had found themselves as a result of the failure of all attempts to prove absolute motion by optical means and as a result of the absolute proportionality between 'inertial' and 'gravitational' mass. Therefore, their racket and railing has weight zero. What, for example, are we to do with such sententious statements—to quote only a few at least reasonably sensible comments—as:

> Einsteinianism asserts the equivalence of acceleration and gravitation. In other words: It teaches that an effect (acceleration) is equivalent to its cause (gravitation). This thesis is sheer nonsense (Dr. A. Reuterdahl).[12]

Or:

> To me Einstein's theory is a functional recasting of reality. His reference system: variable space and time scales, invariable light velocity (despite a variable refractive index) is not to my taste (Prof. Dr. Strehl).[13]

And this is by no means the worst of this genre. It is out of the question in a review to go into more detail on these 'arguments', which are repeated to the point of satiety. Therefore, in a word: Since zero always yields zero when multiplied by any finite number, the compilers might just as well have presented 1,000 rather than 100 of such authors without even the quintessence of their remarks being able to yield any weight other than zero. They should have realized that just as it is true that a majority of votes at a ladies' teaparty can scarcely confirm Einstein's theories, in the same way, the accumulation of 'verdicts' by authors who command a little of the phraseology of Kant's critical philosophy but who have not felt even a whisper of his genius can hardly present a case against relativity theory.

[12]On Arvid Reuterdahl (1876–?), head of the engineering department at St. Thomas College, St. Paul, Minnesota and an active opponent of Einstein see, e.g., Hentschel [1990]a, p. 158; cf. ibid., footnote 34 for references to some of Reuterdahl's articles in American local newspapers. His contribution is entitled: 'Der Einsteinismus/Seine Trugschlüsse und Täuschungen', pp. 40–45.

[13]Karl Wilhelm Andreas Strehl (1864–1940) studied in Erlangen and Munich 1882–87, taught at a commerce school in Augsburg in 1889 and at high schools (*Gymnasien*) in Augsburg from 1890, in Weissenburg/Saale from 1894, and in Erlangen from 1897 until 1905, where he submitted his Ph.D. thesis in 1900. 1905–23 he taught at a *Gymnasium* in Hof, Bavaria. Despite his marginal role in astronomy and optics, one of the editions of *Poggendorff's Handwörterbuch* contains his pretentious self-description as a 'contemporary of Gauß'. For further information on Strehl's work on the theory of refraction in optical instruments see: *Zeitschrift für ophthalmologische Optik* **29** [1941], pp. 170–172, and the *Zentralzeitung für Optik und Mechanik* **48** [1927], pp. 75–77, and **61** [1940], p. 118. Strehl's comment above is his entire contribution, entitled 'Relativität der Relativitätstheorie', p. 49. Strehl refers to the equality between the velocity of light in a medium c_m with refracting index n and $c_m = c_0/n$, which thus ceases to be a constant; but he neglects the fact that the theory of relativity only postulates the constancy of the velocity of light c_0 in a vacuum.

Some reasonable critics in philosophy and physics have allowed themselves to be called in among these 'authors', with whom relativist scientists need not, and actually also do not consider it beneath their dignity to cross swords. (Although Einstein himself, by nature a pure scientist, is uninterested in such academic disputes!)[14] But even here a prominent name unfortunately does not yet ensure an equal match of weapons. If it is known, for example, that one of these 'prominent persons' even quite recently on another occasion called 'empirical astrology' (sic!) a science *κατ εξοχην*,[15] then one could hardly expect an exact scientist to engage himself in a conversation on the *raison d'être* of the deductive methods of relativity theory with someone who does not even know the limits between science and idle pastimes.[16] And indeed, there are also sentences by him like:

["]The idea that motion, which is supposed to be only relative, of course, has an absolute real effect (gauge contraction, etc.), is discussed in a completely inacceptable manner.["] These are Pythian oracles, but no argument against Einstein's crystal-clear logic. But even the serious critics that finally remain (philosophers and a few physicists—the reviewer doubts whether all of the latter are edified by their names being misused in such a way!) cannot understand the great theoretical predicament from which Einstein attempted to free the world. It is clearly impossible to detect absolute motion through optical means (optical, understood in the widest sense!) and to comprehend the proportionality of inertial and gravitational mass, which has been proven to the utmost limit of the most sophisticated art of observation. The first is resolved by special relativity theory, though not without demanding sensitive sacrifices; the second riddle is solved by the 'general' theory of relativity (more appropriately called gravitation theory), in which it is shown that gravitation in the 'four-dimensional space' of the space-time manifold is merely an 'pseudo-force' [*Scheinkraft*] approximately analogous to centrifugal and Coriolis force in classical mechanics.[17] Whoever does not know about, or is not willing to know about the predicament in which physical epistemology was when it turned to Einstein, should not at all presume to be

[14]Einstein withdrew from such disputes in the twenties, having previously shown a keen interest in the theory's reception. Cf. Hentschel [1990]a, sec. 3.4 for a description of the protective flank of relativity theory supporters, who took up the fight on behalf of and sometimes even without the consent of Einstein.

[15]*Kat exochen*, Greek for 'in the truest sense'.

[16]This alludes to Hans Driesch (1867–1941), professor of philosophy at the University of Leipzig, who wrote a book discussing occult phenomena: *Parapsychologie. Die Wissenschaft von den okkulten Erscheinungen*, Munich: Bruckmann, 1932; see Goenner [1994], p. 255. His contribution to the book under review is entitled: 'Meine Haupteinwände gegen die Relativitätstheoretiker', pp. 7–8.

[17]The Coriolis effect is the apparent deflection (Coriolis acceleration) of a body in motion with respect to Earth, as seen by an observer on Earth, attributed to a hypothetical force (Coriolis force) but actually caused by the rotation of the Earth and appearing as a deflection to the right in the Northern Hemisphere and a deflection to the left in the Southern Hemisphere. Also called compound centrifugal force or deflecting force. Named after the French engineer and mathematician Gaspard Gustave de Coriolis (1792–1843).

competent to sit in judgment upon the General Theory of Relativity; and only this ignorance protects philosophers from the accusation of deliberate deception, when they falsely insinuate again and again that the motive in Einstein's analyses is to fool the world through eccentricities [*Bizarrerien*] that are enigmatic to the mathematical layman. And—shall I repeat myself once again?—since the relativistic 'margin of uncertainty of simultaneity' is practically infinite, why in the world should this theory therefore be 'absurd', 'unintuitive', or even 'stupid', when the incorrigible philosopher still has the option, as a last resort, to read the statements of relativity theory in the sense of Vaihinger's 'as if' philosophy![18] This interpretation might not fully agree with Einstein's—I do not know for sure—but under no circumstance does it contradict with the fact that Einstein has always only denied that simultaneity can be determined precisely but has never claimed that the 'concept' of simultaneity itself was nonsensical.

Thus of the '100 authors' criticism nothing remains but a handful of in themselves certainly quite considerable objections raised by acute philosophers and physicists. Relativity theory ought obviously to address these—had it not already done so successfully long ago. When, for instance, Prof. Hartog of Amsterdam gives carefully phrased warnings against also extending 'relativization' to natural occurrences as an **inward** experience, or even into the area of moral values, the author of relativity theory would certainly be the first to sharply reject such an improper interpretation of his ideas.[19] Thus even this most praiseworthy contribution actually only batters down open doors. On the whole, this book is in any case a product of such pitiful impotence (only in politics has such a low point been reached; could so-called 'ideological' [*weltanschaulich*] antipathies perhaps be the sole motive of this pamphlet anyway?) that this regression into the sixteenth and seventeenth century can only be coldly regretted. Whether it was necessary that R. Voigtländer's publishing house provide proof that he could compromise himself **even** more, in defiance of all human odds, outside of his most peculiar domain of cosmic-ice theory, is for him to judge.[20] I have never felt, for either aesthetical or

[18]See Hentschel [1990]a, chap. 4.4, for the interpretation of relativity theory by adherents to the 'philosophy of as if' (also called fictionalism), founded by Hans Vaihinger (1852–1933) in his book *Philosophie des Als Ob*, which first appeared in 1911. According to the philosophy of as if, all postulates of relativity theory, such as, for instance, the constancy of the velocity of light in a vacuum, were introduced into physical theory merely as convenient fictions.

[19]The essay by the Dutch philosopher and theologian Hendrick de Hartog (1869–1938), who had published on Hegel, Fichte, Schelling, Schoppenhauer as well as on church doctrine, Calvin, Wilhelm Wundt and Luther, is entitled 'Philosophische Grundgesichtspunkte', pp. 13–14.

[20]See Nagel [1991] on the cosmic-ice theory, proposed by Hanns Hörbiger (1860–1931) around 1894, and its fate during the 'Third Reich'. The Voigtländer publishing house, founded by [Fürchtegott Leberecht] Robert Voigtländer (1821–1887) in 1847 in Bad Kreuznach, then relocated by his son Robert Voigtländer (1849–1931) to Leipzig in 1888, had published most of the books on cosmic-ice theory by 1931. In 1931 the company came under the direction of the founder's great-grandson [Robert Paul] Otto Voigtländer (born in 1911), who was also one of the founding members of the *Verein für kosmotechnische Forschung* established in 1923 to promote Hörbiger's theory: See Nagel [1991], pp. 51ff., 105, 108–112.

epistemological reasons, that the reasoning of the relativity theories immediately agrees with my own thinking in general. I have rather come to accept it step by step through experience. I also demand, of course, along with all knowledgeable supporters, that relativity theory settle all objections genuinely aimed at investigating into the truth. But anyone in the same position as I with respect to relativity theory will therefore be all the more against such mindless pamphleteering, which if taken seriously, would gravely upset the platform of discussion between antagonists of equally high standing on the **real** difficulties left open by relativity theory. Finally, objection must be raised against presenting authors in a bibliography of opponents to relativity theory, who possibly once voiced reservations in one sense or another, but who are or were all in all clearly supporters of relativity theory (Bottlinger, Poincaré, Prey!).[[21]] It can only be hoped that German science will not be compromised again by such pathetic contrivances.

[21]Kurt [Felix Ernst] Bottlinger (1888–1934) studied astronomy under von Seeliger in Munich and took his Ph.D. in 1912. 1913–14 he was an engineer at the Swiss Geodesical Commission in Basel; 1919 he became teaching assistant, 1921 an astronomical observer, from 1927–34 professor at the university observatory in Babelsberg (Berlin). Bottlinger published supporting surveys of experimental tests of the general theory of relativity from 1920. The *100 Authors* cites the following papers by Bottlinger: 'Die astronomischen Prüfungsmöglichkeiten der Relativitätstheorie', *Jahrbuch der Radioaktivität und Elektronik* **17** [1920], pp. 146–161; 'Über eine astronomische Prüfungsmöglichkeit des Relativitätsprinzips', *Astronomische Nachrichten* **211** [1920], Issue No. 5051, cols. 239–240. [Jules] Henri Poincaré (1854–1912) acquired his title *Dr. ès-sc.* in Paris in 1879, then his doctorate at Caen University. From 1881 he was professor of mathematics at the Sorbonne. His specialty was in the theory of autemorphous functions, algebraic geometry and topology. Philosophically, Poincaré advocated a varient of conventionalism. His considerations on space and time conventions had brought him close to Einstein's special theory of relativity, which he nevertheless never supported publically. Poincaré's book *Dernières Pensées*, Paris: Flammarion, 1913, is listed as against relativity. Adalbert Prey (1873–1949) took his doctorate in philosophy in Vienna in 1896. Until 1909 he was employed as an assistant (*Adjunkt*) in the Austrian bureau of standards (*Gradmessungsbüro*), after which he became associate professor of astronomy at the University of Innsbruck. In 1917 he was professor of astronomy and director of the observatory at the German University in Prague and 1930–46 professor of theoretical astronomy at Vienna University. Prey published nothing on relativity theory: Only his name appears in the listing of antirelativists.

Part II
After the Nazi Seizure of Power

5 Lise Meitner: Letter to Otto Hahn [March 21, 1933]

Source: Archive of the Max Planck Society, Berlin: III, Rep. 14B, No. 21. Handwritten signed letter on letterhead: 'Kaiser Wilhelm-Institut für Chemie, Professor Dr. Lise Meitner. Berlin-Dahlem, Thiel-Allee 63'. Salutation: 'Lieber O. H.!' See also the excerpted transcription in: Lemmerich [1988], pp. 112–113, doc. no. 118.

Dear Otto Hahn,[[1]]

Yesterday your 1st letter of the 10th of this month arrived and a couple of days ago the one from New York. Cordial thanks for both. I am glad that you feel comfortable on the whole in America, and I have read the beginning of your first course of lectures with pleasure, which is adapted so splendidly to American tastes.[[2]] All due respect to you!

Here naturally everything and everyone is affected by the radical political changes. Today is the ceremonious *Reichstag* inauguration in Potsdam.[[3]] Last week already we had received instructions from the K[aiser]. W[ilhelm]. S[ociety]. to hoist the swastika flag beside the Black-White-Red [imperial flag]; the K[aiser]. W[ilhelm]. S[ociety]. is covering the cost of the flags.[[4]]

The wife of Privy C[ouncillor] Schiemann and Edith [Hahn[5]] were just here to listen to the radio broadcast of the Potsdam ceremony. It was thoroughly amicable and dignified. Hindenburg said a few short sentences and then allowed Hitler to speak, who spoke very moderately, tactfully, and personally.[[6]] Hopefully

[1] For earlier correspondence between Lise → Meitner and Otto → Hahn, and on their relationship in the first stage of their scientific collaboration at the Kaiser Wilhelm Institute of Chemistry in Berlin (→ *KWIPC*), which lasted until Meitner's exile from Germany in 1938, see Ernst (Ed.) [1992]; for their correspondence in 1938–39 see D. Hahn (Ed.) [1975], pp. 75–129.

[2] Otto Hahn was guest professor at the George-Fisher Baker Institute for Chemistry at Cornell University in Ithaca, New York, from March 7, 1933. The events in Germany eventually forced him to return earlier than scheduled in June 1933. See Hahn [1962]b, pp. 90f., Hahn [1968/86], pp. 142–145, and D. Hahn (Ed.) [1988], p. 146, for Hahn's description of his stay in the USA.

[3] After the burning of the old *Reichstag* building on Feb. 27, 1933, a provisional parliamentary building was opened officially in Potsdam's *Garnisonskirche* with A. → Hitler and Hindenburg as main speakers during the inauguration ceremony.

[4] Cf., for example, the brief note on the obligatory raising of both flags in the *Göttinger Zeitung* of March 21, 1933, **71**, no. 23 350, p. 2.

[5] Elisabeth Schiemann (1881–1972) was a botanist who researched plant breeding and was a lifelong friend of Lise Meitner. Her extensive correspondence with Meitner, spanning 1911–68, is preserved at the Meitner papers, Churchill College, Cambridge, England; some excerpts are published in Kerner [1986]. Edith Junghans (1887–1968) was married to Otto Hahn in March 1913. She did not accompany her husband on his trip to America but stayed in Berlin.

[6] See, for instance, the *Göttinger Zeitung*, **71**, Tuesday, March 21, 1933, no. 23 350, pp. 1–2

things will continue in this vein. If the level-headed leaders can prevail, among whom Papen can now be counted primarily, then there is hope for developments turning out well in the end.[7] Periods of transition almost inevitably produce all kinds of blunders, of course. Everything now depends on rational moderation.

It must surely have been difficult for Haber to have to raise the swastika flag. I was glad to hear by chance that he personally gave Kühn the directions in hoisting the brand-new flag; that is so much more dignified for him than if this requirement had been forced upon him.[...[8]]

Yours, Lise

6 Albert Einstein: Letters to the Prussian Academy of Sciences and the Academy's Response [March 28–April 5, 1933]

Source: Archives of the Berlin Academy of Sciences: AAW Berlin, II-IIIa, Vol. 28b, Inventory A, Sheets 6, 16, 44, Nos. 783, 799, 815. Published in: Frühauf et al. [1959], p. 9, and Kirsten & Treder (Eds.) [1979], pp. 246, 248, 254–255: Document No. 169: Handwritten signed letter dated March 28, 1933 (received March 30, 1933), written on letterhead of the vessel: 'S. S. Belgenland, Red Star Line'. Document No. 173: Typed draft of the Academy's press release of April 1, 1933. Document No. 181: Handwritten letter to the Academy dated April 5, 1933 (received April 10, 1933) from Le Coq sur Mer on the Belgian coast.

---

or the *Berliner Tageblatt* **62**, no. 132 A, March 21, 1933, p. 1, for extensive coverage of 'The day of Potsdam', the inauguration ceremony of the new *Reichstag*, which was also transmitted by all German radio stations from 9 a.m. until after 10 p.m.; and *Göttinger Zeitung*, Wednesday, March 22, 1933, no. 23 352, pp. 1–2, or *Berliner Tageblatt* **62**, no. 133, pp. 1ff., for the full text of the welcoming speech by General Paul von Hindenburg (1847–1934), the second German president between 1925–34, and Hitler's statement of policy. See also Noakes & Pridham (Eds.) [1990]b, Vol. 1, Doc. 102, pp. 153–154, for another eyewitness account: "In the morning a broadcast of the ceremonies in Potsdam. All cleverly done, impressive, spell-binding even, at any rate for the masses. But we too cannot and must not shut our eyes in the face of what is going on. Today and here, the marriage took place [...] between the masses led by Hitler and the 'Spirit of Potsdam', Prussian values, represented by Hindenburg".

[7]Franz von Papen (1879–1969), a former member of the Center Party, represented national conservative groups. Between June 1 and Nov. 17, 1932, he was Reich Chancellor, and from July 20, 1932 also Prussian *Reichskommissar*. Between Jan. 30–April 7, 1933, von Papen was Vice Chancellor and Prussian *Reichskommissar* in a coalition government of the National Socialists (→ *NSDAP*), the German National People's Party (*DNVP*) and the *Stahlhelm*, with Adolf Hitler as Reich Chancellor. Backed by Reich President Hindenburg, von Papen and others hoped to be able to use Hitler and the *NSDAP* as a weapon against leftist parties, labor unions and the Weimar system, to install a conservative and authoritarian government with a program supporting the interests of big industry.

[8]Kühn was the building superintendent. On Fritz → Haber, who soon resigned from his position as director of the *KWIPC* in Dahlem (Berlin) for political reasons, see Hahn [1962]b, p. 91 and Hahn [1968/86], pp. 144–146, as well as docs. 15ff., 24. The subsequent text and postscript contain news about colleagues and a report on the progress in her research.

Letter of Resignation, Antwerp, March 28, 1933

To the Prussian Academy of Sciences, Berlin

The current state of affairs in Germany compel me to resign herewith from my position at the Prussian Academy of Sciences. For 19 years the Academy has given me the opportunity to devote my time to scientific research, free from all professional obligations.[1] I know how very much I am obliged to her. I withdraw reluctantly from this circle, also because of the intellectual stimulation and the fine human relationships which I have enjoyed throughout this long period as a member and have always valued highly.

But under the present circumstances I consider my position's inherent dependence upon the Prussian government intolerable.[2]

Respectfully yours Albert Einstein

Prussian Academy of Sciences Press Release. Berlin, April 1, 1933

The Prussian Academy of Sciences was shocked to learn from newspaper reports[3] about Albert Einstein's participation in the loathesome [anti-German] campaign in America and France. It has demanded an immediate explanation from him. Einstein has since given notice of his withdrawal from the Prussian Academy of Sciences on the grounds that he could no longer serve the Prussian State under the current government. Because he is a Swiss citizen, he apparently also intends to give up his Prussian citizenship, which he had acquired in 1913 by virtue of his acceptance in the Academy as a regular full-time member.

[1] A. → Einstein had been holding this position since July 1913 as regular member of the Prussian Academy of Sciences with a yearly salary of 12,000 marks, partly financed by a donation from the financier Leopold Koppel (ca. 1854–1933). Einstein had no teaching duties to allow him to concentrate on research with the option of giving talks and seminars at the University of Berlin, which he only chose to do occasionally.

[2] Einstein had already left Germany in December 1932, that is before → Hitler's seizure of power in January 1933: In July 1932 the → *NSDAP* already controlled 230 seats in the *Reichstag*. The Einsteins left Germany on a steamer from Bremerhaven to the USA, Einstein supposedly telling his wife that they would never again see their summer house in Caputh. He was formally on leave of absence to give lectures in California and at Princeton. For his opinion on Nazi Germany see the subsequent letter to the Prussian Academy of April 5, 1933; cf. also, Einstein [1933]a,b; Clark [1971], chap. 16, Kirsten & Treder [1979].

[3] According to Frühauf et al. [1959], pp. 8f., on Mar. 29, 1933 the Prussian Minister for Sciences, Arts and Education requested verification of relevant newspaper reports and if appropriate the initiation of disciplinary proceedings against Einstein. The presiding secretary von Ficker sent a letter to M. → Planck on the same day asking him to write to Einstein and to suggest he withdraw membership in the academy. Planck complied, but in his reply to von Ficker dated Mar. 31, 1933 he also emphasized that he was perfectly sure that "in the history of the coming century the name Einstein will be celebrated as one of the most brilliant stars ever to have shone at our Academy." However, Einstein's letter of resignation was already on its way.

Einstein's agitatorial behavior abroad is particularly offensive to the Prussian Academy of Sciences, because it and its members have felt intimately attached to the Prussian State since times past; and for all its strict restraint in political matters, it has always emphasized and preserved the national idea.[4] For this reason it has no cause to regret Einstein's resignation.

For the Prussian Academy of Sciences Heymann[5]

Einstein's Response, Le Coq sur Mer, April 5, 1933

To the Prussian Academy of Sciences,

I have been informed by a thoroughly reliable source that in an official statement the Academy of Sciences had alluded to "Albert Einstein's participation in the loathesome [anti-German] campaign [*Greuelhetze*] in America and France".[6]

I declare herewith that I have never participated in a loathesome campaign; and I must add that I have never seen any sign whatsoever of such incitement anywhere. On the whole, reports were confined to describing and commenting on the official announcements and orders by the responsible members of the German government as well as on the plans regarding the destruction of German Jews through economical means.[7]

---

[4]On the early history of the Prussian Academy founded in 1700 see, e.g., Adolf von Harnack: *Geschichte des königlichen preußischen Akademie der Wissenschaften zu Berlin*, Berlin, 1900 (reprinted Hildesheim: Olms, 1970). Max Planck also emphasized the Academy's link to the Prussian State, for instance, in a speech delivered on Jan. 28, 1937, on the occasion of the anniversary of Friedrich II (1712–1786), king of Prussia from 1740–86, published in the *Sitzungsberichte der Preußischen Akademie der Wissenschaften*, 1937, pp. xxi–xxvii. On the Academy in the Third Reich see also Grau et al. [1979], Walker [1995], chaps. 4–5.

[5]Ernst Heymann (1870–1946) was full professor of German civil law and civil action at Berlin University 1914–38 and dean of the law department there in 1918, 1927 and 1933–34. He became full member of the Prussian Academy of Sciences in 1918 and was secretary for the Philosophical-Historical Section 1926–38. Heymann had drafted the press release in Berlin on a Sunday in the absence of Planck and the presiding secretary H. von Ficker, the latter of whom had only been away for two working days: See Kirsten & Treder [1979], footnote 1, p. 248. On von Ficker and Planck's absence, see footnotes 10–11 of doc. 21. Cf. also M. → von Laue's letter to F. → Paschen, of Apr. 3, 1933 for his vain attempt to convince the Prussian Academy to withdraw its statement, published without the prior consent of its members.

[6]Einstein was just returning to Europe from a trip to the U.S. but dared not re-enter Germany when warned about a serious assassination threat. During their stay in the Villa Savoyard in Le Coq sur Mer the Einsteins were escorted by two body guards assigned by the Belgian government. On Mar. 20, 1933, the *New York Times* reported on "one of the most perfect raids of recent German history' on Einstein's Caputh summer house, allegedly in search of hidden weapons.

[7]Perhaps because of his Zionist contacts Einstein was much more aware than many of his contemporaries of the seriousness of the Nazi rulers' intolerant policy toward their Jewish fellow citizens. Cf., e.g., Hilberg [1961] for the various stages of development in the anti-Semitic

The statements I had given to the press relate to my resignation from my post at the Academy and to my renunciation of Prussian citizenship. I justified this on the fact that I do not wish to live in a state in which individuals are not granted equal rights before the law as well as freedom of speech and instruction. I also described the condition Germany is in today as a psychic disease afflicting the masses and also said a few things about the causes of this condition.[8]

In a document I had given to the International League Against Anti-Semitism for publicity purposes and which was not at all intended for the press, I also called upon rational people and all those who have remained loyal to the ideals of a now threatened civilization to do their utmost to prevent the further spread of this mass-psychosis, which is expressing itself in Germany in such a terrible way.[9] It would have been a simple matter for the Academy to acquire the correct text of my statements before commenting on me in the way it did. The German press has distorted my statements as it is prone to, which can only be expected, considering the extent to which the press is muzzled there today.

I stand behind every word I have published. On the other hand, in view of the fact that the Academy herself has been involved in slandering me before the German public, I expect that she make this statement of mine available to her members as well as to the public, before whom I have been defamed.[10]

Very respectfully yours Albert Einstein

7 Law for the Restoration of the Professional Civil Service [April 7, 1933]

Source: 'Gesetz zur Wiederherstellung des Berufsbeamtentums. Vom 7. April 1933', *Reichsgesetzblatt* (Berlin), Part I, No. 34, published April 8, 1933, pp. 175–177.

---

measures leading eventually to the beginning of the Holocaust in 1940.

[8]For Einstein's statements to the press see, e.g., *Excelsior* [Paris], Mar. 28 and 29, 1933, *Le Soir*, Mar. 29, 1933, *La Dernière Heure*, Mar. 29, 1933, or *Neptune*, Mar. 29, 1933, all referred to explicitly by Ernst Heymann in a memorandum of Apr. 11, 1933 to justify his earlier statement: See the reprint in Kirsten & Treder (Eds.) [1979], Vol. 1, pp. 257–261, esp. p. 260.

[9]For this declaration written for the *Liga für Bekämpfung des Antisemitismus* see, e.g., the *Kölnische Zeitung*, Mar. 30, 1933, or in French in the *Journal des Nations* of Mar. 28, 1933; it is also quoted fully in Kirsten & Treder (Eds.) [1979], Vol. 1, p. 260.

[10]Obviously, Einstein's demand was not met. However, Planck insisted on adding the following statement to the minutes of the Academy's general meeting of May 11, 1933: "I believe I express the opinion shared by my academic colleagues as well as of the overwhelming majority of all German physicists when I say: Mr. Einstein is not only one of many prominent physicists, but Mr. Einstein is *the* physicist from whose papers, published at our Academy, physical knowledge in our century has gained a depth of such significance as can only be measured against the achievements of Johannes Kepler and Isaac Newton. I find it important to express this above all, so that future generations do not come to think that Mr. Einstein's academic colleagues were not yet capable of fully comprehending his importance to science." (Frühauf et al. [1959], p. 9.

The Reich government has passed the following law, which is proclaimed herewith:

§ **1** (1) For the purpose of restoring a national professional civil service[1] and in order to simplify its administration, officials may be dismissed from the service[2] under the terms of the following regulations, even if the conditions required therefor under the relevant law are not met.[...[3]]

§ **2** (1) Officials who have entered into the civil service since the 9th of November, 1918, without the eduational background requisite or usual for their career or who lack other qualifications, are to be dismissed from the service.[4] Their former salaries will be maintained for the duration of three months following their dismissal.

(2) They are not entitled to claims to allowances, pensions or surviving dependents' benefits, nor to the continued use of the official designation, title, uniform, and official insignia.[...[5]]

§ **3** (1) Civil servants who are not of Aryan descent are to be placed in retirement (§§ 8 ff.);[6] in the case of honorary officials, they are to be dismissed from office.[7]

---

[1]The fullest commentary on the background and consequences of this law for the civil service is still provided by Mommsen [1966], esp. pp. 39ff.; cf. also Walk (Ed.) [1981], and the introduction, p. xxvii on its euphemistic title as well as on its reception and impact. On M. → Planck's, M. → von Laue's and W. → Heisenberg's reactions to this law see Wolff [1993], pp. 268f. as well as here docs. 22 and 114.

[2]The first wave of dismissals at universities as a direct result of this law came to a total of 426 scholars in former Prussia and 614 in the rest of Germany, according to official statistics compiled in December 1934 recently published in Gerstengarbe [1994]. By 1938 about 20–25% of all German university teachers had been removed from their offices for political or racially discriminatory reasons (cf. Hartshorne [1937], Beyerchen [1977], pp. 43ff.). In quantitative terms, the impact was even greater for German school teachers; according to Brämer & Kremer (Eds.) [1980] p. 25, in the district of Saxon alone more than 1,000 teachers were dismissed, and more than 60% of all lecturers at Teacher Training Colleges were either dismissed, transferred or reprimanded on the basis of this law. Approximately two thirds of those scholars and scientists dismissed from the service went into exile by 1940; cf. Fischer [1991]a, p. 28, and secs. 3–4 of the Introduction to this anthology.

[3]Nos. 2–4 of § 1, defining who counts as a civil servant in the sense of this law, are omitted; see Noakes & Pridham (Eds.) [1990]b, Vol. 1, Doc. 151, pp. 223–225, 443.

[4]The Weimar Republic was proclaimed on Nov. 9, 1918 in Berlin. This provision gives the false impression that many official appointments made during the Weimar period had been entirely politically motivated. This law served to exclude from professional public service all officials of liberal and leftist conviction. § 4 and the first and third ordinance of this law (cf. docs. 8 and 17) specify further dismissal criteria, i.e., members of the Communist Party and its supporting organizations.

[5]Nos. 3–4 of § 2, providing an exception in extreme cases of need, and applying it to existing pensioners falling under § 1, are omitted here.

[6]According to Hilberg [1961], p. 56, in 1933 the 'non-Aryan' population was about 600,000, or 1 % of Germany's population, whereas approx. 5,000 of these 'non-Aryans', 4,000 being Jewish, were employed in government service (= 0.5% of government personnel).

[7]According to the first ordinance of this law issued four days later (cf. doc. 8), 'non-Aryan

(2) No. 1 does not apply to officials who had already been in the service since the 1st of August, 1914, or who had fought in the World War at the front for the German Reich or for its allies, or whose fathers or sons had been casualties in the World War.[8] Further exceptions may be granted by the Reich Interior Minister with the agreement of the competent departmental minister, or in the case of civil servants abroad, the highest authorities of the states.

§ **4** Civil servants who, based upon their previous political activities, cannot guarantee that they have always unreservedly supported the national state, can be dismissed from service. Their previous salary will be maintained for the duration of three months following their dismissal. From this time on they shall receive three fourths of the pension (§ 8) and the corresponding support for surviving dependents.

§ **5** (1) Every civil servant must accept being transferred to another post—with reimbursement of the prescribed moving expenses—within the same or equivalent career, also to one of lower rank and in a lower budgeted salary bracket, if the needs of the service so require. For transfers to a lower ranking post in a lower budgeted salary bracket, the civil servant retains his previous official designation and the salary of the former post.[9]

(2) The civil servant may request within one month that he be placed in retirement instead of being transferred to a post of lower rank and in a lower salary bracket (No. 1).

§ **6** To simplify administration, civil servants may be placed in retirement,

---

descent' applied to anyone with at least one Jewish grandparent. Because of the evolutionary metaphors used in Nazi ideology, the actual religion practiced by the affected person was essentially irrelevant—only 'racial' heritage was taken into account. On the pseudo-scientific background and motives for the exclusion and discrimination of Jews in the 1933–35 National Socialist legislation, see Stuckart & Globke [1936], esp. pp. 1–30; for critial comments, cf. Harthorne [1937], Hilberg [1961], pp. 43ff., Walk (Ed.) [1981].

[8]This exemption rule was included in the law to appease President Hindenburg: Cf., e.g., Mommsen [1966], p. 48. It did protect some Jewish scientists temporarily from losing their jobs immediately (such as, Richard → Gans and Gustav → Hertz), because they had served as officers in the First World War (cf. Walle [1982] on German Jewish soldiers); but they soon lost this privilege when the Reich citizenship law (*Reichsbürgergesetz*) came into force as part of the infamous Nuremberg Laws in September 1935 (cf., e.g., Noakes & Pridham (Eds.) [1990]b, Vol. 1, pp. 535ff.). James → Franck and Fritz → Haber, who could have stayed in the service based on this exemption, declined to take advantage of it and resigned in April 1933 (cf. docs. 9 and 15). Finally, some scientists lost their positions even though they had fought at the front, sometimes based on § 6 of this law (see the next footnote). One example was Walter Gordan (1893–1939), who lost his position as untenured associate professor at Hamburg University on Sep. 1, 1933, despite his past active military service. Very few 'non-Aryan' scientists stepped down from their civil service positions out of solidarity with their Jewish or Leftist colleagues: Examples are E. → Schrödinger and O. → Hahn.

[9]The purpose of this article was to fill positions more easily with qualified persons from other areas of the country. In 1936/37 the theoretical physicist Richard → Becker, for instance, was forced to abandon the chair he had been occupying since 1927 at the Berlin Polytechnic, in order to fill the vacancy in Göttingen since Max → Born's dismissal in 1933.

even when they are not yet unfit for service. If civil servants are retired for this reason, their posts may not be refilled.[10]

§ **7** (1) Dismissal from office, transfer to another office, and retirement shall be ordered by the highest Reich or state agency, which ultimately decides excluding the right of appeal.

(2) The injunctions per §§ 2 to 6 must be served no later than September 30th, 1933. This deadline can be brought forward with the approval of the Reich Interior Minister, if the relevant highest Reich or state authority declares that the measures authorized by this law have been carried out within its jurisdiction.[11]

§ **8** A pension will not be granted to those civil servants who were retired or dismissed pursuant to §§ 3 and 4, unless they have completed at least a ten-year term of service. This applies also in those cases in which, according to existing regulations of the Reich and state laws, a pension is granted after a shorter tenure. [...[12]]

Berlin, the 7th of April, 1933.
Reich Chancellor Adolf Hitler
Reich Minister of the Interior Frick
Reich Minister of Finance Count Schwerin von Krosigk[13]

---

[10]Forced retirements based on this article needed no justification at all. Many positions in physics were eliminated altogether as a result, including Rudolf Minkowski's chair at Hamburg University in March 1934. Cf. also doc. 35 for a statistical evaluation of the state of affairs in 1934.

[11]The deadline was not kept, and the law stayed in force until the end of the Nazi regime, with several more legal restrictions yet to come; cf. doc. 36.

[12]The remainder of this article on special wartime service benefits and § 9–18, specifying details like the calculation of the pension claim basis, retirement and transfer benefits, etc., have been omitted here. For details on the pension system regulations see Hilberg [1961], p. 57. For other examples of Nazi legislation, see the anthology by Huber et al. (Eds.) [1942].

[13]Hilberg [1961], p. 56, writes: "the sequence of signatures tells us that the decree was drafted by the appropriate experts in the Interior Ministry and that the competent experts in the Finance Ministry were consulted before publication." A. → Hitler was appointed Chancellor Jan. 31, 1933 by the German President Paul von Hindenburg. Minister Wilhelm → Frick was appointed Reich Minister of the Interior on Jan. 30, 1933. Count Johann Ludwig (Lutz) Schwerin von Krosigk (1887–1977) had been Minister of Finance since June 2, 1932. The latter had studied law and political science at Lausanne, Oxford and Halle on the Saale. After passing his state examination in 1909 he filled various official positions at the local level in Strassfurt and Stettin. He volunteered into the Army 1909–10 and was a reserve field officer during the war. After working under the district magistrate in Hindenburg, Upper Silesia, he became adviser to the Reich Finance Ministry in 1920. In 1924 he became ministerial councillor and 1925 chief councillor on the Reich budget, and from 1929 ministerial director and head of the budget division of the Reich Finance Ministry. Contrary to Reich Bank President Hjalmar Schacht (1877-1970), Schwerin von Krosigk openly supported Hitler's antisemitic policy. On the above officials, see, e.g., Führerlexikon [1934], pp. 17–18, and Wistrich (Ed.) [1982].

8 First Ordinance on the Implementation of the Law for the Restoration of the Professional Civil Service [April 11, 1933]

Source: 'Erste Verordnung zur Durchführung des Gesetzes zur Wiederherstellung des Berufsbeamtentums. Vom 11. April 1933.' *Reichsgesetzblatt* (Berlin), published April 11, 1933, Part I, 1933, No. 37, p. 195.

Pursuant to § 17 of the Law for the Restoration of the Professional Civil Service of the 7th of April, 1933 (*Reich Law Gaz[ette].* I, p. 175) the following is ruled:[1]

1. **In re § 2** All civil servants, who belong to the Communist party or to Communist support organizations or substitute organizations, are unqualified. They are therefore to be dismissed.[2]

2. **In re § 3** (1) Anyone descended from non-Aryan, and in particular Jewish, parents or grandparents, is considered non-Aryan. It is sufficient that one parent or one grandparent be non-Aryan. This is to be assumed especially when one parent or one grandparent has practiced the Jewish faith.[3]

(2) If an official did not already enter into the civil service before the 1st of August, 1914, he must prove that he is of Aryan descent or a front-line veteran, or is the son or father of a soldier killed in action during the World War. Proof is furnished by presenting certified documents (birth certificate and marriage certificate of the parents, military papers).

(3) Should Aryan ancestry be questionable, then an opinion must be obtained from the specialist on race research [*Rassenforschung*] appointed by the Reich Ministry of the Interior.[4]

---

[1] Cf. the preceding document. Article 17 empowered the Ministers of the Interior and Finance to draw up the implementation laws jointly.

[2] Cf. Mommsen [1966], pp. 164f., on the debate at the Reich Ministry of Employment (*Reichsarbeitsministerium*) in June 1933 about whether this paragraph also applied to officials who had entered the civil service prior to November 1918 or who had only briefly been affiliated with Communist organizations. It was decided that they should fall under § 4. Thus they could be forced into retirement as 'politically unreliable' and could only claim 3/4 of their pension.

[3] This amendment defines 'non-Aryan' to include so-called 'one-quarter Jews', persons of whom one grandparent was Jewish, thus opting for the more restrictive alternative. Some eugenics experts, including Bernhard Bavink, preferred the 'half-Jew' option affecting the offspring from mixed marriages, see Bavink [1933]. This disagreement was reason enough for the Nazi party to put Bavink's book on the index: Cf. Hentschel [1993]. The Nuremberg citizenship law of Sep. 15, 1935, and the German Civil Service Law (*Deutsches Beamtengesetz*) of Jan. 21, 1937 worsened the situation further by stipulating that civil servants be Reich citizens (*Reichsbürger*) (as distinguished from *Staatsbürger*), which required that not only the civil servants themselves, but also their spouses be of 'Aryan descent'. Cf., e.g., Noakes & Pridham [1990]b, Vol. 1, pp. 530ff., Mußgnug [1988], pp. 95f.

[4] This amendment reflects the increased influence of the controversial disciplines of eugenics and racial hygienics along with other anthropological fields, based on the pseudo-biological foundation of the Nazi ideology: Cf., e.g., Weingart, Kroll & Bayertz [1988], chaps. 4–5 on the institutionalization of racial hygienics and eugenics in Germany before and after 1933.

3. **In re** § **4** (1) In investigating whether the requirements of the first part of § 4 are met, all the political activities of the official, particularly since the 9th of November, 1918, must be taken into consideration.

(2) Every official is obligated to provide information upon request to the highest Reich or state agency (§ 7), on which political parties he has belonged to thus far. Also included as political parties under this regulation are the Black-Red-Gold *Reichsbanner*, the Republican Association of Judges, and the League of Human Rights.[5]

4. All deeds, certified documents and official certificates that become necessary for the implementation of this law, are without charge or seal.

Berlin, the 11th of April, 1933.

Reich Minister of the Interior Frick

Reich Minister of Finance Count Schwerin von Krosigk[6]

9 Voluntary Resignation of Prof. James Franck [April 17, 1933]

Source: 'Freiwilliger Amtsverzicht Prof. James Francks', *Göttinger Zeitung* **71**, No. 23 374, Tuesday, April 18, 1933, 1st supplement;[1] reprinted, e.g., in Becker, Dahms & Wegeler (Eds.) [1987], p. 28a. (Anonymous article).

[5]The *Reichsbanner Schwarz-Rot-Gold* was founded in 1924 as a republican veterans association under the name *Bund deutscher Kriegsteilnehmer und Republikaner* in Magdeburg by the Social Democrat politician, Otto Hörsing. He was succeeded as president in 1931 by the main editor of the association and Social Democrat Karl Höltermann. From 1924 the association published the journal *Das Reichsbanner* as well as the weekly newspaper *Illustrierte Reichsbanner-Zeitung*. In 1932 it united with the Independent Labor Unions (*Freien Gewerkschaften*) and other republican associations and assumed the name 'Iron Front' (*Eiserne Front*) in response to the right (*National*) opposition. In 1932 it had 3.5 million members. The judges association *Republikanische Richterbund* was founded in Leipzig in 1909 with the unification of 17 independent state (*Land*) associations of right-wing judges. Its purpose was to promote legislation and the administration of justice and to advocate the professional and economic rights and status of its members. From 1909 the newspaper *Deutsche Richterzeitung*was its organ. In 1932 membership reached 13,000. The *Deutsche Liga für Menschenrechte* was the German human rights organization formed in 1919 after the French *Ligue pour la Défense des Droits de l'Homme et du Citoyen* on the initiative of the politician Matthias Erzberger (1875–1921) (a U-boat war protester and advocate of the Versailles treaty who was later murdered), in order to promote publically the idea of unity among peoples. It was part of an international organization and issued the publication *Völkerbundsfragen*. For examples of scientists dismissed shortly after April 7, 1933 because of their membership in one or more of these organizations, see Gerstengarbe [1994], pp. 35f.

[6]On the signators, see footnote 13 of the preceding document.

[1]Similar notices about J. → Franck's resignation appeared in most daily newspapers: See, e.g., *Vossische Zeitung* (Berlin), Tuesday, April 18, 1933; *Göttinger Tageblatt*, Wednesday, April 19, 1933, p. 3; *Berliner Tageblatt* **62**, April 19, 1933, no. 180, ed. A, morning ed., ca. p. 9, and other newspaper clippings at the Göttingen University archive, folder J. Franck. For reactions to this resignation see the following two documents.

Fig. 4: 'James Franck.'
'GZ-Klischee, Photo Stanke.'

Voluntary Resignation of Prof. James Franck.

Requested Release from His Official Duties –
Reasons in a Letter to the Rector of *Georgia Augusta*[2]

The director of the Second Physical Institute of the University of Göttingen, Prof. James Franck,[3] has requested from the Prussian Minister of Sciences, Arts, and Culture immediate release from his official duties.[4]

[2]Franck's original letter of resignation to the Prussian Minister of Science, Art, and Education, a carbon copy of which is preserved at the Göttingen University Archive, Kuratorium, folder Franck, p. 94, and to the dean of the university, of which parts are quoted in this article, were dated April 17, 1933. According to Franck's daughter, Mrs. Elisabeth Lisko, Boston, the appearance of the newspaper articles was carefully timed to coincide with the receipt of the letter by the ministry: Cf. Beyerchen [1977], chap. 2, p. 17, esp. footnotes 10–12. *Georgia Augusta* is the official name of the University of Göttingen, referring to King George August of England, Duke of Brunswick and Lüneburg and Elector of Hanover, who granted the university its privilege in 1737. For an account of the University of Göttingen under National Socialism, see Becker, Dahms & Wegeler (Eds.) [1987], especially pp. 27, 349, 377ff., for the background to Franck's resignation.

[3]This opening sentence is in boldface in the original. Franck had been director of the Second Institute of Physics since 1921, which subsequently made major contributions to atomic physics. Cf. the following two documents.

[4]For the full text of this short letter of resignation dated April 17 to Bernhard → Rust see Beyerchen [1977], chap. 2, p. 17. The only mention in this document of the motives behind this step is: "This decision is an inner necessity for me because of the attitude of the government toward German Jewry." In his longer letter to the rector of the university quoted in this newspaper article, he elaborates further. Peter Paul → Ewald drafted a similar letter of resignation to the Minister of Religious Affairs and Education dated April 20, 1933: "Since it is impossi-

This news will cause a major sensation not only in Göttingen, but throughout Germany, and one can even justifiably say, throughout the world. Franck is not just any lecturer of local or preeminent national importance. Franck's international reputation and global fame is unsurpassed by virtually any other German scholar today. When he received the Nobel prize a few years ago,[5] the whole of Germany considered it an exceptional honor—because a German had again spread the fame of German scientific research beyond its borders. When such a man, who is only fifty, voluntarily relinquishes his teaching and research activities, the loss to science is beyond estimation. The reasons for Prof. Franck's decision are the following: Franck is of Jewish descent. But he would scarcely have fallen within the legal standards of dismissal.[6] At the outbreak of the war he immediately volunteered into the service and was at the front line from December 1914 to February 1918, without a break.[7] Severe lameness prevented him from serving during the final months. He was promoted to lieutenant of the infantry already in June 1915 and has earned numerous war decorations in addition to his two Iron Crosses. We are stressing this, since active participation at the front line today determines the fate of Jewish university teachers and civil servants.

At our request Prof. Franck consented to an interview with one of the members of our editorial staff. On this occasion he made available to us an excerpt from a letter to the rector of the university:[8]

---

ble for me to share the opinion of the nationalistic government on the racial issue, I request permission to resign my position as Rector of the Stuttgart Polytechnic with immediate effect and also to be released from my office as Prorector." A carbon copy of Ewald's letter to the Senate of the Stuttgart Polytechnic is included in the Arnold Sommerfeld papers, DMM, No. 1977-20/A,88/2.

[5]Franck received the Nobel Prize in 1926 for the year 1925 jointly with his colleague Gustav → Hertz (1887–1975) for their experimental studies of inelastic collisions of electrons in gas tubes, which they had started in 1911, crowned by their discovery of excitation potentials in mercury vapor in 1914. See their joint papers in *Verhandlungen der Deutschen Physikalischen Gesellschaft* **13** [1911], pp. 967–971, **14** [1912], pp. 167–168, 419–425, 596–604, **15** [1913], pp. 34–44, 373–393, 613–620, 929–934, **16** [1914], pp. 12–19, 457–467, 512–517; cf., e.g., Kuhn [1965], pp. 55f., Lemmerich (Ed.) [1982], pp. 24–34, and Kuczera [1985], pp. 33ff.

[6]This euphemistic term refers to the exemption provision for war veterans in the Law for the Restoration of the Professional Civil Service (see doc. 7, § 3, no. 2), on which Hindenburg had insisted.

[7]Franck actually served in a unit involved in Fritz → Haber's first successful use of chlorine gas as a weapon at the second battle of Ypres on Apr. 22, 1915. Poison gas had been tried initially in battle in the form of gas-shells in a local attack in Poland on Jan. 31, 1915 but had been ineffective due to the extreme cold. At Ypres the gas was discharged from cylinders because the German command was not willing to provide Haber with adequate resources to produce shells as a consequence of the new weapon's performance in Poland; but it still quickly gave the Germans a significant and unexpected advantage in the trench warfare. For more information on this see, e.g., Haber [1924], pp. 75ff., Walle [1982], Lemmerich (Ed.) [1982], pp. 36–37, Becker, Dahms & Wegeler (Eds.) [1987], p. 27, note 61 and p. 378, as well as F. Stern in: Vierhaus & vom Brocke (Eds.) [1990], pp. 526ff.

[8]The German philologist Friedrich Neumann (1889–1978) was rector at Göttingen from

> "I have requested of my superior that I be released from my office. I will try to continue to work in science in Germany.
>
> We Germans of Jewish descent are being treated as foreigners and as enemies of the fatherland. It is expected that our children grow up knowing that they are not permitted ever to prove themselves worthy Germans.
>
> Whoever has been in the war is permitted to continue to serve the state. I decline to take advantage of this privilege, even though I understand the position of those who today see it as their duty to stay resolutely at their posts." [9]

Prof. Franck does not intend, as he explained to us, to accept any possible appointments abroad, as long as he may continue his work in experimental physics, and as long as no serious economic reasons force him to do so. [10]

Apart from the heavy loss that German and international scientific research will sustain from Franck's resignation from his teaching position, it should be mentioned that Göttingen will experience serious economic repercussions as well from this. Numerous students, not only from Germany but primarily also from abroad, came to Göttingen essentially because of Franck, and from now on will seek out other universities. Likewise scholars from all over the world stayed very often for longer periods in Göttingen, in order to take part in scientific conferences together with Franck. Through Franck's resignation these visits and conferences will largely cease. This is a severe loss in activity for our city, economically speaking, since tourism is one of our most important economic factors.

Prof. Franck's decision must be evaluated essentially, indeed even morally. We would like to hope and wish that this step with which Franck shatters his

---

1933–38. He had studied in Marburg, Munich and Göttingen, earning his degree in 1914. He served at the Western front 1914–18 and habilitated in 1921 in Göttingen. Neumann was appointed budgeted associate professor in the same year at Leipzig University and became full professor in 1922. He moved to Göttingen in 1927 as professor of German language and literature and directed the department of German philology. In 1945 he was dismissed primarily because of his activities as rector and his membership in the Party since 1933. He fell under the 'collaborator' category in his denazification proceedings in 1949, which barred him from returning to his position. See Becker, Dahms & Wegeler (Eds.) [1987], pp. 287–288, and the introduction, p. xc.

[9] This letter is preserved neither in the Göttingen University archive nor as a carbon copy in the possession of the Franck family: Cf. Beyerchen [1977], chap. 2, p. 17, footnote 12. There is a much shorter letter there dated April 17, 1933, which only refers to Franck's brief letter of resignation to → *REM* Minister Rust and enclosing as a carbon copy. The *REM* finally responded to Franck's letter of resignation on Dec. 11, 1933 by refusing his "application for release from his official duties in view of the prevailing regulations." However, they offered him the option of applying for dismissal, which entailed the renunciation of further compensation. Franck took that option on Jan. 1, 1934 by applying for dismissal from service to the state (*Staatsdienst*), which was then accepted by the Ministry on Feb. 8, 1934. The fact that Franck had thus been forced into a formal resignation rather than being suspended from his teaching duties proved to be a considerable hurdle in his later legal claim for reparations: Cf. Göttingen University archive, folder J. Franck.

[10] Cf. also Franck's letter to Walther → Gerlach written on May 3, 1933 (doc. 16), in which he comments in more detail on his plans to stay in Germany.

own life's work and substance has the effect that other scientists who are forced to resign as a result of the legal regulations have recourse to our science here. Otherwise, the resulting losses would be irretrievable or only recoupable after a long time has past.

*

James Franck was born on the 26th of August, 1882, in Hamburg and studied physics and chemistry in Heidelberg and Berlin; specifically, he studied physics under professors Warburg and Drude in Berlin.[11] In 1906 he earned his degree at Berlin University, became teaching assistant in Frankfort-on-Main for a short time and then teaching assistant in Berlin under Prof. Rubens.[12] His academic teaching career began in 1911 when he qualified as university lecturer and became unpaid lecturer of physics at Berlin University. In 1918 he became a member of the Kaiser Wilhelm Institute of Physical Chemistry in Dahlem (Berlin) and became the department head. In 1920 he received a call to a regular professorship at Göttingen University; at the same time he became director of the 2nd Physical Institute.

Prof. Franck has thus been in Göttingen for 13 years now and has turned down among others a prestigeous call to Munich to succeed Prof. W. Wien.[13]

Prof. Franck has earned international esteem for his pioneering investigations on the structure of atoms and molecules. His work is of an experimental nature: It deals primarily with the excitation of atoms through collision with electrons. Together with Prof. G. Hertz, Prof. Franck succeeded in confirming the basic concepts of Bohr's spectral theory through electrical analyses as well.[14] In 1926 using a sensational optical method[15] Professor Franck succeeded in determining

[11]Paul Drude (1863–1906) was professor of physics at Berlin from 1905–06 and was known for his textbook published in 1900, which soon also appeared in English: *Theory of Optics*, New York, 1902; his specialties were optics, electrochemistry and electron theory. Emil Warburg (1846–1931) was professor of experimental physics at the University of Berlin from 1895 on. He specialized in photochemistry and cathode rays. He also headed the → *PTR* from 1905 to 1922.

[12]Heinrich Rubens (1865–1922) was director of the physics institute at Berlin University 1906–22, succeeding Drude as professor of experimental physics (see also footnote 13, doc. 77). Ruben's specialized in infrared radiation and developed a spectroscopic method for radiation of wavelengths above 12μm.

[13]Wilhelm → Wien died in 1928 before having retired. On Franck's time in Göttingen see Lemmerich (Ed.) [1982] and Kamp et al. [1983].

[14]Franck and Hertz measured energy changes caused by collisions between atoms and electrons (cf. footnote 5 above for references). They showed that an electron must have a certain minimum energy in order to ionize an atom and that this potential varied for different gases. This result supported one of the fundamental assumptions of Niels Bohr's theory of the atom, namely, that an atom can take up internal energy only in such discrete amounts as transform it from one stationary state to another. Cf., e.g., Mehra & Rechenberg [1982], Vol. 1, pp. 197ff.

[15]See, e.g., J. Franck, 'Elementary processes of photochemical reactions', *Transactions of the Faraday Society* **31** [1925], pp. 536–542.

the temperature at which chemical compounds dissociate.[16] Until that time the dissociation process had only been known on the basis of chemical measurements. Thus Franck could confirm the basic concepts of modern atomic theory in two entirely different ways.

Public recognition of Prof. Franck's epochmaking discoveries in physics did not fall short. The Prussian Academy of Sciences nominated him corresponding member. But the greatest honor conferred on him was the Nobel prize in physics for 1925, which he received together with his friend and collaborator Prof. Gustav Hertz in 1926. The Nobel prize in physics for 1925 was awarded to both scientists equally "for their discovery of the laws on the collision of an electron and an atom".[17]

The funds the Rockefeller Foundation had made available for the construction of the extension of the Physics Institute were another nice sign of recognition of Prof. Franck. Göttingen University received this grant by virtue of Franck's standing.[18]

10 Edith Hahn: Letter to James and Ingrid Franck [April 22, 1933]

Source: Franck Papers, Chicago, excerpt from letter; see the excerpted transcription in Lemmerich [1988], p. 114.

[My dear friends],[1]

I am racking my brains and puzzling over what we could possibly do. If I didn't like you both so much, I could envy you for being Jews—and these are really not just empty words—and thus for having justice completely on your side and we for ever and ever the humiliation and the inextinguishable shame, which never can be made good again. You both did something, this is what is so marvelous;[2] and it has certainly made a big impression. It will please you both to hear perhaps that I have been asked probably twenty times from several quarters: Have you read that thing by Prof. Franck? ... On Wednesday I bought

[16]Dissociation is the reversible resolution or decomposition of a complex substance into simpler constituents caused by variation in physical conditions, as when water gradually decomposes into hydrogen and oxygen under great heat, so that when the temperature is lowered the liberated elements recombine to form water.

[17]Their Nobel prize lectures were given on Dec. 11, 1926 and printed in Stockholm at the *Imprimerie Royale* in 1927. See footnote 5 for references to the original papers.

[18]On the Rockefeller Foundation's promotion of German research even during the Third Reich, see Macrakis [1986], [1993]. See also, doc. 30, footnote 8, and doc. 54.

[1]Ingrid Josephson, a Swedish pianist, was married to James → Franck in Berlin on Dec. 23, 1907. They had two children there, Dagmar and Elisabeth. Franck's second wife was his former assistant in Göttingen Hertha Sponer, whom he married in 1946: See footnote 25 of doc. 20.

[2]This is a reference to J. Franck's open letter of resignation, quoted in the preceding document.

all the remaining issues of the *Voss* at our Ullstein branch[3] and sent them to all the people whom I haven't yet quite given up for lost, because I think your letter should bring them back to their senses; and I hope the whole world will react to it.[4]

Edith[5]

11 Göttingen University Lecturers: Professor Franck's Resignation [April 24, 1933]

Source: 'Der Rücktritt Professor Francks. Francks Rücktrittserklärung ein Sabotageakt der innen- und außenpolitischen Arbeit der nationalen Regierung – Eine Kundgebung Göttinger Dozenten', *Göttinger Tageblatt* **45**, No. 95, Monday, April 24, 1933, p. 3.

Professor Franck's Resignation
Franck's Tender of Resignation an Act of Sabotage of the National Government's Domestic and Foreign Political Activities – A Declaration by the Göttingen University Lecturers

The following was reported to us from among the circle of Göttingen university lecturers:[1]

"Professor James Franck has requested of the Prussian Minister of Culture Rust release from his official duties, and according to the following public statement (*Göttinger*

[3]On April 18, 1933, the *Vossische Zeitung* also reproduced Franck's letter of resignation with the headline: "Prof. Franck legt sein Amt nieder". The *Ullstein A.-G.* in Berlin, one of the largest publishers of books, journals and newspapers in the Reich, was founded as a family business in 1877 by Leopold Ullstein (1826–1899) under the name Ullstein & Co. It grew into a corporation and was finally sold in 1933 to a bank consortium. When the publishing house was reorganized in 1934, the *Vossische Zeitung* was discontinued.

[4]For the only published article known to the editor in support of Franck's action, see doc. 13; for a negative response see the following document. In striking contrast to this weak public support, Franck received many private letters expressing sympathy with his demonstrative action (in a separate folder, among the Franck Papers) of which this is an example. Otto Hahn was one of the few scientists in public service to resign his professorship at Berlin University in protest and solidarity with F. → Haber, but he retained his main position at the *KWIC.* Cf. also footnote 4 of the preceding document. When Otto Hahn proposed that about thirty influential German professors not affected by the Law for the Restoration of the Professional Civil Service prepare a declaration of solidarity with their Jewish colleagues, M. → Planck replied: "If you bring together 30 such men today, then tomorrow 150 will come to denounce them, because they want to take their places". Cf. Heilbron [1986]b, p. 150.

[5]For information on Edith Hahn (1887–1968) see footnote 5 of doc. 5 and, e.g., D. Hahn (Ed.) [1988], pp. 91–101. Her husband Otto, a close friend of James Franck's, was in the USA at the time and only learned about the events many weeks later through American press reports and letters from his family.

[1]On the newspaper's source, cf. footnote 4 below.

Tageblatt, Wednesday, the 19th of April 1933, page 3) had written to the rector of Göttingen University among other things:[2]

"I have requested of my superior that I be released from my office. I will try to continue to work in science in Germany. We Germans of Jewish descent are being treated as foreigners and as enemies of the fatherland. It is expected that our children grow up knowing that they are not permitted ever to prove themselves worthy Germans. Whoever has been in the war is permitted to continue to serve the state. I decline to take advantage of this privilege, even though I understand the position of those who today see it as their duty to stay resolutely at their posts."[3]

We the undersigned Göttingen university lecturers are of the opinion that this step in general, but also in particular the sentence in the declaration:

"We Germans of Jewish descent are being treated as foreigners and as enemies of the fatherland"

can seriously **impede the domestic and foreign political activities of our government** of the national renewal. We are in agreement that the form of the above tender of resignation is tantamount to an **act of sabotage**; and we therefore hope that the government will carry out the **necessary purging measures** expeditiously.

Baß, Behm, Blanck, Blume, Fick, Fuchs, Giesecke, von Grimm, R. Hartmann, Hauberrisser, Hesse, Hilka, Hofmann, Hübner, Jander, Jessen, Kecs, Lauprecht, Fr. Lehmann, Lucke, Manegold, K. Meyer,[4] Nachtigall, Plüchke, Rebel, Rippek, Salfeld, Saller, Schäfer, J. Schmidt, Schole, Schüz, Stadtmüller, Tobeck,

[2]See the preceding documents, esp. doc. 9, paragraphs 3ff., from which these slightly deviating quotes originate.

[3]The next two paragraphs were printed in boldface in the original newspaper article; only the additional emphasis in spaced type is indicated here in boldface.

[4]During his denazification, Konrad Meyer, who later became professor at the University of Göttingen, submitted an official written statement on Oct. 12, 1950: "The declaration signed by these individuals had been intended as a submission to the Reich Interior Minister and was supposed to point out that by no means was Prof. Franck speaking in the name of the Göttingen professoriate. It contained absolutely no attacks against 'Jewish professors' and also refrained from giving any opinion on solving the Jew problem. It had been drafted as an internal memorandum and had been signed as such. Through an indiscretion by the local leadership of the *NSDAP*, this statement was suddenly published in a modified form in the Göttingen press. None of the signers had been informed of this. Upon the complaint of some of the signers, the Party admitted with regret to this indiscretion and to its independent action. I remember, the person in charge at the Göttingen party headquarters who had been responsible was dismissed because of this incident." (Göttingen Univ. Archive, folder James Franck, carbon copy). The context of this declaration rather suggests, though, that the signers had passed their declaration on to the local press in a well-calculated move, just as James Franck had done with his letter to the minister.

Tornau, Hermann Vogel, Voß, Walther, W. Weber, Wehefritz, Zenck, Ziebe."[5]

This declaration is particularly important and justified, since the foreign press has already made the most of Professor J. Franck's tender of resignation in its propaganda against Germany. As far as we are informed, the declaration of the university lecturers reprinted above was also triggered by this commentary in the foreign press.

*

Due to the holidays[6] it was not possible to obtain the signatures of all the professors; but it can be readily assumed that they approve of the above declaration.[7] As we know from a well-informed source, **a majority in the University Lecturer's League disapproves of Professor Franck's declaration**. It has even irritated his fellow Jews at the *Berliner Tageblatt*, which immediately recognized that Professor Franck had made a fatal step that the government cannot overlook idly.[8]

12 Law against the Overcrowding of German Schools and Universities [April 25, 1933]

Source: 'Gesetz gegen die Überfüllung deutscher Schulen und Hochschulen. April 25, 1933', *Reichgesetzblatt*, published April 26, 1933 (Berlin) Part I, 1933, No. 43, p. 225.

The Reich government has passed the following law, which is proclaimed herewith:[1]

[5]Prof. A. Kleinert (University of Halle), is preparing a study on the scholars and scientists who had signed this document, to find out whether and to what extent their signatures had in fact helped to promote their careers (personal communication).

[6]According to the university course catalog, the summer term started officially on April 20, 1933 and ended on July 31, 1933.

[7]This sounds suspiciously like an excuse, since university professors were generally adverse to signing any petition, document or press release of a political nature, regardless of whether they agreed with it or not. These reservations about political activities of any kind were considered a matter of course among university teachers; cf., e.g., Ringer [1969], Freise [1983].

[8]Cf. *Berliner Tageblatt* **62**, Wednesday, April 19, 1933, no. 180, ed. A (morning), ca. p. 9: 'Ein Nobelpreisträger legt sein Amt nieder'. The newspaper *Berliner Tageblatt* was commonly thought to be under the complete control of Jews. Its editors did not agree with Franck's decision to resign: 'From all points of view we deeply regret Franck's resignation. Though we very much respect his motivations, we nevertheless believe that his cause would have been better served if he had stayed at his post as professor." Cf. also *Göttinger Tageblatt* **45**, No. 91, Wednesday, April 19, 1933, p. 3.

[1]According to Götz von Olenhusen [1966], this law was issued on the initiative of the culture policy department at the → *RIM* under W. → Frick, reflecting the fear of a growing 'academic proletariat' since the economic depression of the late 1920's. However, in the early 1930's the number of students was actually already on the decline.

§ **1** The number of pupils and students at all schools, excluding mandatory schools [*Pflichtschulen*[2]] and at universities is to be limited to the point that basic education is assured and the professional demand is satisfied.

§ **2** At the beginning of each school year the state governments determine how many new pupils each school may admit and how many new students each university department may enroll.

§ **3** At schools and university departments where the attendance figure is particularly heavily disproportionate to professional demand,[3] the number of already enrolled pupils and students is to be reduced, without causing undue hardship, during the course of the 1933 school year toward establishing a more proportionate ratio.

§ **4** Acceptance of new enrollees must be carefully proportioned so that out of the total attendance at each school and university department, the proportion of Reich Germans of non-Aryan descent in the sense of the Law for the Restoration of the Professional Civil Service of the 7th of April, 1933 (*Reich Law Gaz[ette].* I p. 175),[4] does not exceed the proportion of non-Aryans in the German Reich's population. The quota will be set uniformly for the entire Reich region.[5]

In reducing the number of pupils and students in accordance with § 3, a suitable ratio must likewise be set between total attendance and the number of non-Aryans. In this event, a differing ratio that is higher than the quota may be taken as a basis.[6]

Paragraphs 1 and 2 do not apply to Reich Germans not of Aryan descent whose fathers had fought at the front during the World War for the German Reich or for its allies;[7] nor do they apply to offspring from marriages that had taken place before this law came into effect where one parent or two grandparents

[2]These cover the first eight years of schooling.

[3]The annual demand for university graduates was later set at 8–10,000. However, see the Introduction, pp. xlviff., on the subsequent shortage of students.

[4]See doc. 7.

[5]The quota later chosen on the basis of a census dated June 1933 was a maximum of 1.5% of new enrollees at each institution. Some of the official statistics published at that time to support the claim that there was a disproportionately large number of Jews in public office had actually been forged, according to Götz von Olenhusen [1966], pp. 179f.; cf. Noakes & Pridham (Eds.) [1990]b, Vol 1, pp. 522f. Golczewski [1988] notes on p. 90 that the quota system was a familiar anti-Semitic tool in modern history, used in csarist Russia, for instance. It reappeared in the programs of almost all right-wing movements. It was considered an efficient measure with the appearance of legality. The United States Immigration Law of 1924 that went into effect in 1929 was a similar quota system aimed at particular ethnic groups; cf. Stuewer [1984], p. 29. The dire situation at American colleges and universities as a result of budget cuts during the Great Depression and indications of anti-Semitic leanings there are described in Rider [1984], pp. 123–129.

[6]A 5 % maximum of non-Aryan's was tolerated: Götz von Olenhusen [1966], p. 179.

[7]This provision is modelled on the exemption rule of the Law for the Restoration of the Professional Civil Service of April 1933 and probably was also adopted on Hindenburg's insistence; see here doc. 7, § 3, No. 2. These restrictions were soon retracted, however, with the Nuremburg Laws of 1935.

are of Aryan origin.[8] They are also excluded from the calculation of the quota and the ratio.

§ **5** Obligations incumbent on Germany from international treaties are not affected by the provisions of this law.

§ **6** The Reich Minister of the Interior issues the implementing regulations.

§ **7** This law comes into force upon pronouncement.

Berlin, the 25th of April, 1933.

Reich Chancellor Adolf Hitler

Reich Minister of the Interior Frick[9]

13 Wolfgang Köhler: Conversations in Germany [April 28, 1933]

Source: Wolfgang Köhler, 'Gespräche in Deutschland', *Deutsche Allgemeine Zeitung*, (Berlin), Friday, April 28, 1933.[1]

The powerful men now governing Germany have inquired on more than one occasion about **the other Germans**, who are now standing on the sidelines and whom it would surely be worth winning over. Anyone wanting to win them over must know why they are distancing themselves. And if they really are worthy, it becomes a patriotic duty to discuss this question in public.[2]

---

[8]Note that this passage initially excludes all 'half-Jews' from the quota system. Universities were finally closed to all persons of Jewish descent as defined in the First Ordinance on the Implementation of the Law for the Restoration of the Professional Civil Service (doc. 8) article 2, No. 2. The primary aim of this law, which was already being drafted in the final days of the Weimar Republic, was to prevent Jews from Eastern Europe, especially Poland and White Russia, from coming to study at German universities. Cf., e.g., Kampe [1985]. It also barred politically 'unreliable' students and women (until 1937) from academic studies. This relatively restrained law, when compared to its successors, also reflects the Nazi's concern about the critical reception abroad of the anti-Semitic tendencies in German lawmaking. The choice of the euphemism 'overcrowding' (*Überfüllung*) as opposed to the original word 'alienation' (*Überfremdung*) is an illustration of this: Götz von Olenhusen [1966], p. 177; Golczewski [1988], p. 90.

[9]On the signatprs see footnote 13 of doc. 7.

[1]As indicated in the signature of the article, Wolfgang Köhler (1887–1967) was then professor at the Friedrich Wilhelms University in Berlin. He was also director of the psychology department. In 1935 Köhler decided to emigrate when he became convinced that it was futile to try to change the course of events from within Germany. He resigned his post voluntarily and took up an appointment as professor at Harvard Univ. in Cambridge, Massachusetts in the same year. In 1936 he was at Swarthmore College, Philadelphia, and in 1958 president of the American Psychological Association. Though not written by a physicist, Köhler's article was apparently the only direct supportive reaction to be published immediately after the appearance of J. → Franck's open letter of resignation (see the references below).

[2]Köhler manipulates the rhetoric of the new Nazi regime in this text without compromising his message. By adopting a martial and moralistic tone he signals to his readers that he shares their patriotic values and norms, with the intent of reaching a much wider audience. Cf. the

It is easy to find out what keeps them away, since they can speak of nothing else right now. Yet, they loathe all gossip, and no political party can outdo the sincerity of their concern for the nation. I have never seen a patriotism more noble than theirs: To them, being German is a [**moral**] **duty**. This duty is answered when the noblest of mentalities and behavior emerge out of the notable and self-critical riches of the German type. They are of the **single** opinion that the recent purge could not have been more necessary; they admire the force of the events which for the first time has made Germany into a lasting Reich within a matter of days; and not least, they are thankful for the abrupt jolt by such uncommon strength of purpose that has torn us all out of the indolence of recent years and has made us so alert.

How can they be unhappy despite all this? Any child could see that they live under great pressure. They are unhappy because they fear for their nation.

This concern has little to do with political events in the narrower sense of the word. The current successes have been won through an unheard of concentration of power, and maintaining this power must be a foregone conclusion for the *Führer*. But—so say these people—where actual politics and power issues meet with the profusion of **practical** issues, Germany itself has not disappeared. The strikingly diverse life of Germans of a hundred professions remains. The new government has specific major goals for the organization of these professions and their social representation, and for the constitution of German work; and in this respect its policies are adequate for all practical issues and for all of German life in general. But the point is that it is adequate for the **practical** issues alone; and with this we have come to the first object of great concern. The nation cannot be indifferent to any of these responsibilities. Therefore, when each and every one of them must be put into the hands of a patriot, then it necessarily follows that this man should be the very best to be found with regard to expertise in the relevant field, personality, and vision. Now, when up and down the land one man after another is thrown out of trade unions, business associations, and even private firms, whose German mentality can be doubted as little as his thorough expertise and suitability of character, then I hear the question posed again and again: Why? Who will be his successor? The powerful movement now in control has shown that it had kept ready at hand enough capable individuals for all essential politically powerful positions. But my friends think it almost impossible that over and above this they are also able to provide a sufficient number of men from outside of the immediate field of politics for every specialized field in German life who are better or best qualified for the administration of universities, vocational associations, and private firms in business, be they for higher up or further down the professional ladder, wherever direction by an expert is needed.

I do not at all think that these people are jealous, for instance. They simply fear for the immediate future of their fatherland when in each aspect of life it is

---

Introduction, esp. sec. 6.

not the plainly most suitable of all German-minded men in the field who assumes the leading role, but rather always the National Socialist. They do not dare hope that the latter will always be identical to the former. And in the meantime, their fatherland is dearer to them than even the Party, which has already achieved so much for the fatherland. Do not misunderstand these people; they are not agitators. On the contrary, they wish nothing more than lasting order in the State; and if they were to ask the powerful men in government for something, then it would be less for a loosening of the government reins than for a just stringency, and a stringency for all Germans equally.

I asked such people why they do not join the Party, since they agree so much on the fundamental points, and in this way make themselves available, each in his own field, for selection to the important positions. The answer was always the same: "Even excellent men had been forced to leave, apparently because they did not belong to the Party. We could not understand this; for, it is Germany that is at stake, not a party, even though it may be the most valuable one. It is hard for us to join a party whose policy we do not yet understand on such a fundamental point."

But when I probe further, it turns out that another concern also exists, which also originates, of course, from the principle of objectivity [*Sachlichkeit*] for the sake of the nation. It regards the racial policy of the National Socialist Party and of the new German State. None of the Germans whom I am talking about denies the existence of a Jewish problem in Germany. The majority of them believes that Germans have the right to control the composition of their national population and to limit through wise regulation the now disproportionately large number of Jews in leadership roles in all essential affairs of the nation. But they can only sincerely condone those measures which do not damage Germany indirectly, which do not suddenly destroy the existence of completely innocent people, and which do not seriously hurt the important, distinguished people among German Jews.[[3]] For, my friends do not want to agree with the thesis that as a Jew, every Jew is a mean, inferior form of humanity. When I asked about this, one of them picked up the Psalms and read:

> The Lord is my shepherd; I shall not want.

---

[3]Cf. Max → Planck's coincident thoughts in doc. 114. Many of those affected by the new laws had never considered themselves Jews before. Cf., for instance, Max → Born's letter to the Austrian physicist Paul Ehrenfest dated June 7, 1933: "I have never felt myself particularly Jewish." Members of his family were "Jews only for today's laws, but have never thought about it before. They (and I also) have absolutely no emotional ties to true Judaism, to its ways and laws. I am—or believe myself to be—basically a liberal Western European with a heavy streak of German culture. [...]. I am as little nationalistically inclined toward *Juda* as I am toward *Germania*, but actually less so, since I know neither Hebrew nor Jewish literature, whereas I have come to love the German language, poetry and art." Ehrenfest scientific correspondence, microfilm No. 535, DMM; cf. also Wolff [1993], pp. 268f.

He maketh me to lie down in green pastures
and He leadeth me beside the still waters. [...]

Yea, though I walk through the valley of the shadow of death,
I will fear no evil; for thou art with me.[4]

He also read the 90th Psalm as well and then said: Hardly any German has ever moved human hearts in distress more deeply and consoled them. And this came to us from the Jews.

Another reminded me that never had a man striven more purely for an elucidation of his world than the Jew Spinoza, whose wisdom Goethe had admired. We should not be afraid to honor where Goethe has honored. Moreover, Lessing's impeccable pen would certainly not have produced *Nathan* if human dignity did not exist among Jews.[5]

Many of the Germans who are now standing on the sidelines are teachers at German universities, and some of them are scientists. It seems that no one can think of the short life and the great work of Heinrich Hertz without an almost tender admiration springing up in his eyes.[6] But Hertz was of Jewish blood.

Someone told me as we were going down the street: "The name of the greatest German experimental physicist today is Franck;[7] some believe he is the greatest experimentalist of this science in the world today. This Franck is a Jew, a kind person if ever there was one in Germany. Until a few days ago he was a professor at Göttingen, a celebrity of Germany, and the envy of the scientific world abroad. A German law has affected this good and great man so strongly, not through its wording but doubtlessly through its intent, that he is leaving us. Have you read the words he used in resigning his post?[8] Had another proof been wanting that Jews can be dignified people, then this man has supplied it." He took his hat off, and it seemed to me that I should do so too.

The gravest reason why all these people stand aside is perhaps apparent here: They feel the foundation of their moral world is affected. To them, a person can only be judged by his true nature; and intellectual importance, noble attitude, and evident merit to German culture remain what they are otherwise, when they are discovered in a Jew. Everything they had learnt since childhood and certainly even more deeply lying forces bind them to this opinion, which they absolutely cannot abandon.—But added to this is their concern for their nation. Germany's current intellectual culture is a mighty edifice built of all sorts of stones. Some

[4]Excerpts from Psalm xxiii.

[5]The last play by the German critic and classic dramatist Gotthold Ephraim Lessing (1729–1781) entitled *Nathan der Weise. Ein dramatisches Gedicht*, written in 1779, is set in Jerusalem, the meeting point between the three religions, Christianity, Islam, and Judaism, and depicts through its hero, the Jewish wiseman Nathan, the ideals of humaneness and religious tolerance.

[6]On Heinrich Hertz, see doc. 31, footnote 5.

[7]On James → Franck, see docs. 9f.

[8]Here, Köhler is alluding to Franck's open letter of resignation; see doc. 9.

stones have proved to be inferior. There is no doubt that these must be removed. However, my friends fear that the stones are not tested for their real properties, but all the stones are rejected only on the basis of origin; they are hastily torn out, and in so doing, parts of the whole construction could inadvertently be threatened. For, a mind blinded by rage never grasps its object clearly. My friends think that there are indispensible pillars in the edifice of German culture which originate from Jews. Each blow now shaking the building makes them tremble for the building and for its future.

I would like to add a word that helps to explain the depressed spirits of these Germans. Hardly any of them wanted me to make their worries known publically. I would only hurt myself by doing so; one should not say these sorts of things right now. I am of another opinion in this, and not least because I have read the Reich Chancellor's book, with all its contempt for pussyfooters and with his constant call for straightforward people.[9] The Germans of whom I speak can be helped immediately at least in this point: by seeing that no one is even thinking of holding our misgivings against us, which arise from a genuine concern for the nation, and that no one will persecute us because of them.

Right after this was written, the broad outlines of the law which deals with the foreign infiltration [*Überfremdung*] of schools and universities was made known.[10] My friends also do not deny that this law was shaped by resolute but at the same time careful hands. Their greatest wish now is probably that only these authorized hands decide upon such things and that all unauthorized reformers defer to the authority of the State.[11]

14 The Spirit at Universities [April 28, 1933]

Source: 'Geist der Hochschule', *Vossische Zeitung*, Friday, April 28, 1933 (Evening edition) No. 202, pp. 1–2. (Anonymous article).

Hardly a day goes by in which a new list of university lecturer suspensions is not issued. Hardly a university remains whose teaching staff has not yet been interfered with.[1] These actions have been taken based on the Civil Service Law, which sets descent and political activity as criteria.[2] Thus in a number of cases the cause of the suspensions is clear. But there are at least as many university

[9] Adolf Hitler's autobiography *Mein Kampf*, written 1925–27, sets forth his political philosophy and his plan for German conquest. Köhler is subtly using it for his purposes by appealing to the same virtues praised in this book, but connecting them to the same people whom Hitler's party targeted and discriminated against.

[10] See doc. 12.

[11] For a personal and warm response to this article, see the letter by the psychologist Kurt Lewin (1890–1947) to Köhler on May 20, 1933, and reprinted in full in Strauss, Buddensieg & Düwell (Eds.) [1987], pp. xi–xvii.

[1] Cf., e.g., N.N. [1933]e for a contemporary chronological overview of dismissals.

[2] See the Law for the Restoration of the Professional Civil Service, doc. 7, § 3–4.

lecturers whose removal from office has no discernible connection to the legal provisions.

Let us set aside the individual's academic standing and personal merit. Controversial men like Dehn had to reckon with disciplinary punishment, along with a few others who had exposed themselves politically.[3] But why did men who taught politically neutral subjects have to resign even though they only represented their specialty and had no visible ties to any party activity or political movement? We call to mind the prominent names on the mathematics and science faculty at Göttingen;[4] we recall Germanists, medical doctors, and chemists, who have never been seen in any political connection whatsoever. These professors had at one time been appointed by their departments on the strength of their academic achievements, not in accordance with the wishes of previous governments. Now their work as researchers and teachers is being declared superfluous, alien or even a resistant element in the fabric of the nation. On what basis was this decision made? Is it perhaps that indiscernible something, that confidence between university lecturer and the student body on which the yes or no verdict rests?

The student league, at any rate, has understood the campaign as such and has set to work on its own initiative.[5] A few days ago in Kiel it demanded the compulsory suspension of a great many professors. The same method is supposed to now be applied at all German universities. The leader of the 'German Student League' [*Deutsche Studentenschaft*] has made an appeal to all the individual student leagues, giving the following directions.[6]

[3]Günther Karl Dehn (1882–1970) was a Protestant theologian with Socialist leanings who was forced to take a one-year leave of absence from the University of Halle due to pressure by the right-wing Student League in October 1932. In November of 1933 he was dismissed completely from public service as 'politically unreliable' as a direct result of the change in government, on the basis of Article 4 of the law referred to above. See Heiber [1991/92], Vol. 1, pp. 82–108.

[4]The 'Law for the Restoration of the Professional Civil Service' had devastating consequences for the Göttingen mathematics and science departments, which lost approximately 30% of their faculties, including among others the internationally renowned professors Max → Born, Richard Courant, James → Franck, Edmund Landau (1877–1938), Emmy Noether (1882–1935), and Hermann Weyl (1885–1955). See Schappacher & Kneser [1990], pp. 26ff.; Becker, Dahms & Wegeler (Eds.) [1987]; *Göttinger Zeitung* **71**, Wesdnesday, April 26, 1933, no. 23381, p. 1: 'Beurlaubte Hochschulprofessoren. Neue Maßnahmen Rusts—Sechs Göttinger Professoren betroffen' and ibid., 1st supplement, p. 1. On Franck's case, see also docs. 9f., on Courant see doc. 22, ft. 4 and Reid [1976].

[5]See, e.g., the declarations "Wider den undeutschen Geist" issued by National Socialist student organizations in German newspapers in April 1933; cf., for example, *Göttinger Tageblatt*, Vol. 45, No. 88, April 13/14, 1933, p. 4. In May 1933 the German Student Leagues (→ *NSDStB*) organized the building of bonfires of 'abominable literature' (*Scheiterhaufen der Schandliteratur*) at all university towns; cf. *Göttinger Tageblatt*, Vol. 45, No. 105, May 5, 1933, p. 3: 'Die Göttinger Studentenschaft im Kampf wider den undeutschen Geist' and No. 110, May 11, 1933, p. 3; also Bleuel & Klinnert [1967] and Götz von Olenhusen [1966] on the activities of right-wing student organizations which supported Nazi ideology at universities most enthusiastically. On this student action in the spring of 1933, see Strätz [1968]. Cf. also, the Introduction, p. xlvi.

[6]In an attempt to woo the majority of students, a law was quickly passed regarding student

"In order to expedite the measures that the Reich government has voted in to purge the civil service, student associations are to provide the leader of the G[erman]. St[udent Association] [*Deutscher Studentenbund*] with the following information as soon as possible:

> a) An enumeration of academic teachers who fall within the law of the 7th of April, 1933, that means teachers who are Jews or who have belonged to Communist organizations or the *Reichsbanner* and the like;[7] likewise, professors who have insulted national leaders, the national renewal movement, or frontline soldiership. For this, specific witness statements (affidavits) are necessary.
>
> b) A listing of all academic teachers whose teaching methods match their liberal, or in particular, their pacifist outlook, and who are thus unsuited to educate German students in the national state. Here also an exact list of references (writings, remarks during course lectures or exercises etc.) is required.
>
> c) Information on professors and assistants (characterization of the size and type of field of study) whose conduct is politically irreproachable in the sense of the National Socialist revolution and whose academic teaching approach matches this conduct. We will disregard reports on professors who despite unobjectionable conduct only show moderate academic talent.
>
> d) At each university a university teacher under paragraph c) is to be named trusted university lecturer by the leader of the local student association and reported to the leader of the German Student League.
>
> e) Student associations are to make all preparations to be able to follow effectively an order by the leader of the German Student League to boycott the lectures and practice sessions of academic teachers whose immediate removal from office by the state is not possible.
>
> f) Reference is made to Exhibit 2, re: Campaign against the un-German spirit."[8]

This appeal clarifies the situation. They are carrying out university reform from the bottom up as well, using standards that are not based on academic considerations. The Berlin student league put it most suggestively recently when

associations (*Reichsgesetzblatt* 1933, I., p. 215), which made membership in the *Deutsche Studentenschaft* compulsory to all German-speaking full-time students, giving these associations the authority to represent all students in fulfilling 'their responsibility toward the nation, the state and the university': Golczewski [1988], p. 88; see also Noakes & Pridham [1990]b, Vol. 1, doc. 317, p. 440, for an excerpt from their new constitution of Feb. 7, 1934.

[7]On the outlawed association *Reichsbanner*, see the footnote to Article 3 of doc. 8.

[8]This campaign "against the un-German spirit" (*Aktion wider den undeutschen Geist*) consisted in the posting of 12 theses at the University of Berlin. The fourth stated: "The Jew is our most dangerous adversary" and the second part of thesis 7: "We will consider Jews aliens and we will take our national identity [*Volkstum*] seriously." See also footnote 5 of doc. 20.

it explained to Rector Kohlrausch that only those professors were acknowledged "who work and teach in the spirit of the *SA*".[9] Will a physicist, an astronomer or a bacteriologist pass muster when he is summoned in such a way before the student tribunal?

Professor Spranger expressed what remains to be said in evaluation of this campaign in his resignation announcement.[10] Even professors who felt politically oppressed under the earlier regime and ideologically united with the oppositional student league no longer see any possibility of agreement today. The *Deutsche Zeitung* voiced this today in the harshest of terms. It described the actions of the student league as a

> "sweeping campaign of braggartliness against teachers—a campaign which on such a scale affects learning itself, but must lead primarily to turning the lecture hall into a tribunal and to totally eradicating authority from universities."

The Culture Ministry is surely also aware that allowing this to continue would seriously hamper all academic work and instruction. It can be inferred from Spranger's statement that Minister Rust has already tried to talk the students out of exaggerating their twelve theses.[11] Indeed, it comes down to reestablishing the *Führer* principle here as well, which this drive flagrantly violates. University management has enough to do in filling the vacancies that have already developed and in carrying on academic activity, which in some departments is seriously jeopardized. But this can only happen with great difficulty and certainly only in a calmed atmosphere.[12]

---

[9]Cf., e.g., *Göttinger Zeitung* **71**, Wednesday, April 26, 1933, no. 23 381, p. 1: 'Schwerer Konflikt an der Berliner Universität' for a report on the conflict between the rector of Berlin University and criminal law expert Eduard Kohlrausch (1874–1948) and the Student League about the posting of the latter's twelve theses (cf. the preceding footnote). Kohlrausch called these theses exaggerations, "that are only suited to discredit the struggle against the un-German spirit but not to promote it" ("die nur geeignet seien, den Kampf gegen den undeutschen Geist zu diskreditieren, nicht aber ihn zu fördern"), whereupon the Student League demanded Kohlrausch's replacement.

[10]The philosopher and pedagogue Eduard Spranger (1882–1963) submitted his resignation as head of the department of pedagogy in May 1933 upon the appointment by Minister → Rust of the Nazi philosopher Alfred Baeumler (1887–1968) as professor of political pedagogy at Spranger's department without the consent of the faculty; but a month later he withdrew his resignation and stayed in office. Cf., e.g., Leske [1990], pp. 221–224, for excerpts of Spranger's own account of the events and his motives.

[11]See also doc. 20 for a contemporary interpretation of the Prussian Minister of Culture Bernhard Rust's approach toward these overly enthusiastic student initiatives.

[12]On the worsening situation for 'non-Aryan' students at German universities see, e.g., Götz von Olenhusen [1966] and Golczewski [1988], chap. 3.

15 Fritz Haber: Letter of Resignation to the Prussian Minister of Sciences, Arts and Culture Rust [April 30, 1933]

Source: Archive of the Max Planck Society, Berlin: I, Rep. 1A, No. 541. Typed transcription of letter. Published in Hahn [1960]; Festschrift [1961], p. 190, Stoltzenberg [1994], p. 581, and elsewhere. The letter is addressed 'An den Herrn Minister für Wissenschaft, Kunst und Volksbildung. Berlin W. 8, Unter den Linden 2–4'.

Esteemed Minister,

I request herewith that as of the 1st of October, 1933 I be retired from my main Prussian post as Director of a Kaiser Wilhelm Institute as well as from my secondary Prussian post as full professor at the local university.[1] According to the provisions of the Reich Professional Civil Service Law of the 7th of April 1933, which as has been stipulated must be applied to the Institutes of the Kaiser Wilhelm Society, I am entitled to remain in office even though I am descended from Jewish grandparents and parents.[2] But I do not wish to make use of this privilege for longer than is necessary to properly dispose of the academic and administrative functions vested in me through my offices.[3]

The context of my request is identical to that of the petitions submitted to the President of the Kaiser Wilhelm Society by Profs. H. Freundlich and M. Polanyi, who are the member scientists and division directors of the Kaiser Wilhelm Institute of Physical Chemistry and Electrochemistry under the employ of the Kaiser Wilhelm Society.[4] I have supported the acceptance of the same.

My decision to request my discharge stems from the contrast between the tradition to which I have adhered up to now concerning scientific research, and the changed attitudes which you, Mr. Minister, and your Ministry represent as

[1]Fritz → Haber had been director of the → *KWIPC* since 1911 and professor of physical chemistry at Berlin University since 1898. On his scientific achievements see Stolzenberg [1994].

[2]See doc. 7, § 3, No. 2, which exempted veterans of World War I.

[3]As a man of principle, Haber refused to accept any privileges from this regime. Consequently, he also rejected any later efforts by Max → Planck and others to build 'golden bridges' for him to remain in office; cf. Haber's letter to Planck, May 9, 1933, AMPG, A1-541, Vierhaus & vom Brocke (Eds.) [1990], p. 363, and Stoltzenberg [1994], chap. 13 for the most extensive account of the last two years of Haber's life.

[4]The colloid chemist Herbert Max Freundlich (1880–1941), the physical chemist and later Nobel Prize laureate Michael Polányi (1891–1976), and also the atomic physicist Hans G. Beutler (1896–1942), being the heads of the section for colloid chemistry, the physical chemistry and reaction kinematics section, and spectroscopy section, were forced to resign during M. → Planck's presidency on April 22, 1933: See, e.g., Vierhaus & vom Brocke (Eds.) [1990], p. 362. Freundlich and Polányi immediately went into exile. Freundlich taught at University College, London, from 1934 on and at the University of Minnesota as professor of colloid chemistry from 1938 to his death. Polányi moved to the University of Manchester as professor of physical chemistry and from there to a personal chair for social studies 1948–58. Beutler stayed initially in Berlin and only emigrated in 1936 to the United States, where he was research associate at the Ryerson Physical Laboratory of the University of Chicago 1937–42.

the vanguards of the great modern nationalist movement.[5] My tradition demands that in my choice of colleagues I take into account the professional and personal attributes of applicants to an academic position without inquiring after their racial characteristics. You will not expect a man of 65 years of age to reject a mentality that has guided him in the past 39 years of his academic life; and you will understand that the pride with which he has served his German native country throughout his life now compels him to make this request for retirement.[6]

Respectfully yours F. Haber[7]

16 James Franck: Letter to Walther Gerlach [May 3, 1933]

Source: Heinrich & Bachmann (Eds.) [1989], p. 71. Illustration No. 43: Facsimile of typed letter signed (No. 90) addressed to 'Herrn Prof. Dr. W. Gerlach, München, Leopoldstr. 6' with sender's address: 'Merkelstr. 4, Göttingen'.

---

[5]According to Goran [1967], p. 159, on April 21, 1933, Haber received a telephone call from Bernhard → Rust as Prussian Minister of the Sciences, Arts and Culture advising him that work at the institute could not continue with the present staff, which included 12 non-Aryan persons (24.5%), of which 3 were division directors and 5 assistants (see Vierhaus & vom Brocke (Eds.) [1990], p. 361). On pp. 579ff., Stoltzenberg [1994] reports that 6 assistants had to be fired along with three others.

[6]Cf. Goran [1967], pp. 160f., and Vierhaus & vom Brocke (Eds.) [1990], pp. 362ff., describing the loss of Haber's two brilliant Jewish department directors, Herbert Freundlich and Michael Polányi, on April 22. On April 27, 1933, the main leadership of the → *KWG* had instructed that all employees be investigated, with directions that dismissals and suspensions could be carried out immediately. This led to Haber's resignation as an act of solidarity. See also Haber's farewell letter to the members of his institute, dated Oct. 1, 1933, quoted in Festschrift [1961], p. 191. Haber's efforts to place former members of his institute at other research and industrial establishments outside of Germany are documented in Stoltzenberg [1994], secs. 13.1 and 13.2.2.

[7]Haber was summoned to the Ministry some days later and was received by one of Rust's assistants; but Haber insisted on speaking to the minister in person. This was refused. According to Goran [1967], pp. 161f., and Stoltzenberg [1994], p. 582, a few days later Rust mentioned in a public address to Berlin studentsthat he had received a letter from a German Jew, without mentioning Haber by name. Rust said, "the man had written that he always chose collaborators solely on the basis of their qualifications. This [said Rust] was a practice absolutely incompatible with the modern development of Germany". When Haber was informed by letter that his resignation had been accepted, he was surprised and "for a time sorely indignant. He had miscalculated its effect upon the Nazis." (Goran [1967], p. 162). An official reply to his letter to Minister Rust has not survived, but in a press conference on May 6, 1933, Rust said, among other things: "Professor Haber may have explained in a letter that he cannot accept being dictated on the choice of his co-workers. The Minister must reply to this, however, that it is the Minister's business to assure that the younger generation is engaged in scientific collaboration. Following these guidelines in the long-run is more important and more beneficial than giving individual scientists the opportunity to make some sort of invention." Less than one year later, on January 29, 1934, Haber died in Basel: Cf. von Laue's obituary, doc. 29, and the materials about the Haber memorial session, which took place on Jan. 29, 1935 in Dahlem (Berlin), quoted, e.g., in Festschrift [1961], pp. 192ff.

My dear Gerlach,

Please accept my heartfelt thanks for your kind letter; and don't be offended that I respond to it only briefly on the typewriter. I am very pleased with your whole attitude towards what affects all of us. But I can say that from what I know of your character it was self-evident to me. You write, I should not despair of Germany and should stay here. This is in fact my firm intention, if I can make a living [here] and have the opportunity to work in science. I have not yet received an answer to my letter from the Ministry up to now; but be it as it may: It is an irreversible process.[1] I plan to stay in Göttingen for the time being and if possible to bring unfinished projects to a close, though not through direct action on my part but only through offering advice. I probably won't be able to examine my students. Apart from that, when everything has become a little clearer I want to look around to see whether I can find employment somewhere. Though naturally, I am not thinking of a professorship but of some sort of scientific laboratory work in industry. I would like most to go to *I. G. [Farben].*[2] But I have no idea whether they can use me and whether I fit there. If I should find nothing in a few months, then I certainly would have to go abroad. But as I said, for the moment I'm not considering it.

My wife and I wish you and Ruth all the best,[3]
Affectionately yours J. Franck.

17 Third Ordinance on the Implementation of the Law for the Restoration of the Professional Civil Service [May 6, 1933]

Source: 'Dritte Verordnung zur Durchführung des Gesetzes zur Wiederherstellung des Berufsbeamtentums. Vom 6. Mai 1933', *Reichgesetzblatt* (Berlin), published May 6, 1933, Part I, 1933, No. 48, pp. 245–252.

Pursuant to § 17 of the Law for the Restoration of the Professional Civil Service of the 7th of April, 1933 (*Reich Law Gaz[ette].* I, p. 175)[1] it is decreed as follows:

[1]See doc. 9 for the article describing J. → Franck's resignation and references there.

[2]On → *I.G. Farbenindustrie* see in particular Borkin [1978] and Hayes [1987]. Since the Law for the Restoration of the Professional Civil Service did not apply to positions in private industry, several dismissed scientists pursued this option; however, especially smaller companies sooner or later made concessions to the new government in acts of 'good will', and fired at least some of their 'non-Aryan' employees.

[3]W. → Gerlach's wife, born Ruth Probst (1905–1994) in Munich, grew up in Stuttgart and studied medicine in Tübingen to become a children's doctor. Her doctoral thesis, submitted to the pharmacology and physics department at Munich University: 'Spektralanalytischer Nachweis von Wismut im Gewebe, quantitativer Nachweis von Quecksilber im Harn' was published in *Archiv für experimentelle Pathologie und Pharmakologie* **169** [1932], pp. 119–129. On Franck's wife Ingrid Franck, see doc. 10; cf. also figure 1.

[1]See doc. 7. On article 17, see footnote 1 of doc. 8.

In re § **1** 1. Regular civil servants fall under this law. [...[2]]

2. Also included as civil servants under this law are judges and teachers in public education, including teachers at academic universities, as well as full and associate professors who have been released from their duties. Furthermore, honorary professors, untenured associate professors and unsalaried lecturers at academic universities are considered civil servants under this law. [...[3]]

In re § **2** 1. § 2 does not apply to a civil servant who had duly attained his office but was later promoted ahead of his time for political reasons. In this case action may be taken against him under § 4.

2. Whoever has engaged in Communist activity is to be dismissed, even if he no longer belongs to the Communist party or to its supporting organizations or substitute organizations. The so called National Communist Movement ('Black Front') also counts as Communist. [...[4]]

In re § **3** 1. Extramarital descent is also included in the definition of descent in § 3. The parent and child relationship in the meaning of § 3 cannot be based upon adoption. [...]

3. (1) Front-line fighters in the meaning of the law includes anyone who has participated in the World War (in the period from August 1, 1914 until December 31, 1918) as a combat soldier in battle, or who has taken part in a skirmish, in trench fighting, or in an occupation force. The entries of the wartime personnel roster or the wartime Army List provide information thereon. It is not sufficient for someone to have stayed in the war zone during the war for official reasons without having confronted the enemy.[[5]]

(2) Front-line fighters include in particular those to whom the Wound Badge has been conferred.

(3) Participation in the battles in the Baltics, in Upper Silesia, against Spartacists[[6]] and separatists, as well as against enemies of the nationalist renewal, is equivalent to participation in battles of the World War.

---

[2]The rest of this paragraph extends this group to supernumerary civil servants, to civil servants on probation and to aspiring civil servants.

[3]On the consequences of the drastic broadening of this law's scope see, e.g., Mommsen [1966], pp. 53ff., Mußgnug [1988], pp. 28f. The remaining amendments to § 1, affecting other categories of civil servants, are omitted here.

[4]The *Schwarze Front* is a general name for small nationalistic and revolutionary groups, particularly the one founded by the former → *NSDAP* member Otto Strasser (1897–1974). Strasser left the Nazi Party on July 4, 1930, and founded an independent activist organisation of 'revolutionary national socialists' (*Kampfgemeinschaft revolutionärer Nationalsozialisten*) six weeks later, representing the anticapitalist left wing of the Nazi movement. It was attacked aggressively by the majority right-wing *völkisch* Nazi ideologues.

[5]On the background to this temporary exemption clause, see footnote 8 of doc. 7, § 3.

[6]The 'Spartacus League' (*Spartakusbund*) was a revolutionary political group co-founded in 1917 by Karl and Rosa Luxemburg advocating extreme Socialistic doctrines. The Communist Party emerged from this organization.

4. "Casualty" also includes those who died from a wound inflicted during front-line fighting. No. 1 above applies accordingly. [...]

6. [...] (2) All non-Aryan civil servants who do not fall under these exemption provisions must therefore be retired. [...]

In re § 4 1. The membership of an official in a political party—except for the Communist party—does not yet alone justify the assumption of nationalist unreliability. This applies even when the civil servant has been a registered member of the party, has paid membership dues to it, and has attended its meetings.

2. The conditions of § 4 Paragraph 1 are met particularly when a civil servant has presented himself orally, in writing or otherwise through his conduct as opposed to the nationalist movement, has insulted its leaders, or has abused his official position to persecute, slight, or otherwise harm nationalist-minded officials. If this is the case, then even his joining a party or an association which supports the government of the nationalist renewal after the 30th of January, 1933, will not suffice to excuse him. No importance should be attached to occasional blunders during an election campaign.

3. A civil servant, who has fought at the front during the World War for the German Reich or for its allies, has proven his national reliability at that time in any case with this deed; a particularly careful examination of his case is thus called for.[7]

4. In examining officials in leading positions, a stricter standard is to be applied than for the rank and file of civil servants, who have often only followed an example set for them or have succumbed to pressure applied on them and are therefore to be regarded as having been led astray.

5. Political parties in the sense of the First Ordinance on the Implementation of the Law for the Restoration of the Professional Civil Service of the 11th of April, 1933 (*Reich Law Gaz[ette].* I, p. 195),[8] also include the Association of Republican Civil Servants and the Iron Front. [...[9]]

In re § 7 [...] 7. Pursuant to this law, professors at academic universities are not to be emerited, but are rather to be retired.[10]

8. In the case of official full and associate professors and unsalaried lecturers at academic universities, the loss of authority to teach is linked to dismissal or retirement on the basis of this law. In the case of honorary professors, and unofficial associate professors and unsalaried lecturers at academic universities,

---

[7]Cf. footnote 8 of doc. 7.

[8]See doc. 8.

[9]This gradual extension of the list of organizations considered incompatible with the civil service illustrates the Nazis' rapid stabilization of power and systematic elimination of all opposing political organizations. The *Eiserne Front* was an activist organisation co-founded by the Social Democratic Party (*SPD*), the General Federation of German Trade Unions (*ADGB*), and the Republican *Reichsbanner* (see footnote 5, doc. 8) in Dec. 16, 1932 in reaction to the growing National Socialist movement.

[10]This group thus automatically lost its authority to teach (*venia legendi*) and were prohibited from being a member of an academic committee.

their authority to teach is to be withdrawn in place of dismissal or retirement. [...[11]]

In re § 18 Whoever remains in office following the implementation of this law, not only again enjoys the full rights of civil servants, but is also entitled to the respect due his office.

Berlin, the 6th of May, 1933.

Reich Minister of the Interior Frick

Reich Minister of Finance Count Schwerin von Krosigk[[12]]

18 Philipp Lenard: A Big Day for Science. Johannes Stark Appointed President of the Reich Physical and Technical Institute [May 13, 1933]

Source: Philipp Lenard, 'Ein großer Tag für die Naturforschung. Johannes Stark zum Präsidenten der Physikalisch-technischen Reichsanstalt in Berlin berufen', *Völkischer Beobachter*, 133rd Edition, second supplement, Edition A (Northern Germany), Vol. 46, May 13, 1933.

This recent appointment is an event for all persons in any way involved in science. It signifies a definite renunciation of the apparently already inescapable predominance of what might briefly be called Einsteinian thinking in physics;[[1]] and it is a move towards reaffirming the scientist's old prerogatives: To think independently, guided only by nature. This kind of thinking had long been unwelcome at such high places in physics; and now just look: Stark, one of the few remaining untouched living examples of this thinking, is himself at the top in such an important post. Many within and outside of the *Reichsanstalt* will have

[11]Max → von Laue reacted to this law as president of the → *DPG* with a letter to all universities on May 10, 1933, in which he requested the names and addresses of all physicists and older students employed in science "who will lose their jobs or whose advancement will be significantly impeded as a result of the Civil Service Law of April of this year". (See von Laue's letter to A. → Sommerfeld, Sommerfeld Papers, DMM) The worsening situation was also observed abroad: On May 19, 1933, *The Manchester Guardian*, for example, published a three-column list of professors from all fields who had been dismissed between April 13 and May 4, filling almost an entire page: N.N. [1933]e. Several emergency committees were formed, such as the Academic Assistance Council on May 24, 1933, and the Emergency Committee in Aid of Displaced Scholars in late May, 1933, in New York. Cf., e.g., Duggan & Drury [1948], Wolff [1993], pp. 269ff., Strauss, Buddensieg & Düwell (Eds.) [1987], and the Introduction, sec. 4.

[12]On the signatories, see footnote 13 of doc. 7.

[1]On P. → Lenard's relation to J. → Stark and in particular on the background to Stark's appointment as director of the *Physikalisch-Technische Reichsanstalt* (*PTR*) (analogous to the American National Bureau of Standards in Washington) against the unanimous advice of his colleagues in the field, see Kleinert [1980]. Both Lenard and Stark had written several anti-Semitic pamphlets and articles against Einstein's theory of relativity, which symbolized to them a totally misleading and dogmatic form of abstract theorizing; see, e.g., Stark [1930]b, Lenard [1918/21] and docs. 32, 56. Stark in particular also protested publically when W. → Nernst was appointed president of the *PTR* in 1922.

already understood and welcomed with relief this decision by Reich Interior Minister Frick.[2] Others, many of whom are already blind, will surely soon be able to see again, because they find before them a visionary leader [*sehenden Führer*].[3]

In physics the limit has been reached, that is, from the top downward. With the massive introduction of **Jews** into influential positions also at universities and academies, the basis of all scientific knowledge, the observation of nature itself, was forgotten and was no longer considered valid. Knowledge of things of the external world was supposed to be based upon the fancies of the human mind. These ideas, immediately called 'theories', had then to be 'proven' by experimenters.[4] The latter usually complied dutifully and promptly with the most superficial research possible. With the repression of candid remarks against such proceedings, 'freedom of research' took on a new flavor. The results can already be felt generally: **Large segments of the population have lost faith in currently accepted science.** Only technology, which is based upon earlier sound research achievements, could continue to earn this trust.

The most prominent example of the damaging influence by Jews on science was provided by Mr. Einstein with his 'theories', which are a combination of sound knowledge that had already existed previously and some arbitrary additions that had been strung together mathematically. These theories are now already falling to pieces, which is the due fate of unnatural products. Even accomplished scientists cannot be spared from reproach, since they had allowed the 'relativity Jew' to gain a firm foothold in Germany in the first place, not seeing or not wanting to see how very mistaken it was, even from an unscientific standpoint, to consider exactly this Jew a 'good German'.

That "developments had to come out this way", as was recently said on the radio, is a bad excuse. Prominently placed theoreticians should have guided these developments better. Now Hitler has guided it. The nightmare has disintegrated; the alien spirit, that wreaked such havoc on the solid foundations of science which had been created as a result of the efforts of the pure Aryan spirit throughout the centuries, is already leaving universities and also the country, even of its own accord. It means nothing that Nobel Prize laureates are also among those who have left.[5] In recent times these prizes have unfortunately become of increasingly

[2]Minister W. → Frick of the → *RIM* appointed the directors of all state institutions, including the *PTR*. After this politically motivated installation Stark quickly established the *Führerprinzip* there and reoriented its research activities; cf. Hoffmann [1993]a, Kern [1994]. In the winter term of 1934/35, the *PTR*'s board of supervisors was formally dissolved on the basis of A. → Hitler's edict: 'Erlaß des Reichskanzlers'.

[3]With this obvious parallel to Hitler, Lenard is attempting to apply the *Führer* principle to the physics community. Cf. also Stark's speech in Würzburg, doc. 28.

[4]As a first step in this quixotic fight against theoreticians, the experimentalist Stark cancelled the *PTR*'s consulting contract with Max → von Laue.

[5]Aside from A. → Einstein, other exiled Nobel laureates included or were soon to include: James → Franck, Fritz → Haber, Erwin → Schrödinger and a number of other future Nobel Prize recipients.

debatable intellectual value; **no one should grieve mistakenly**. Rather, we should rejoice that we have a scientist like Stark now in an influential position, whom leaders in science had thought permanently eliminated.

Not only science may rejoice in this way. Technology also is done a great service in now having Stark at the head of the Reich Physical and Technical Institute. For, he is not only an outstanding scholar and accomplished researcher, but also at the same time a practitioner, who had himself set up and managed technological facilities with success.[6] In all likelihood, never before has such a suitable choice been made for president.

Johannes Stark was born in 1874 in Schickendorf in Bavaria. His father was a landowner.[7] After secondary schooling in Bayreuth and Regensburg, he studied physics, mathematics, chemistry and crystallography at the University in Munich and thereafter was teaching assistent at the Munich physics department. He habilitated in 1900 at the University in Göttingen; and in 1906 he became associate professor at the polytechnic in Hanover. In 1909 he was appointed full professor at the polytechnic in Aachen, then at the University in Greifswald in 1917, and in 1920 in Würzburg. In 1921 he left Würzburg in order to become independently active in industry;[8] his candidness was not appreciated at this university. – His important confirmation of the Doppler effect in canal rays dates back to the year 1905. In 1913 he made the discovery of the influence of electrical fields upon the light emission of atoms; in 1919 he received the Nobel prize.[9] **His published papers have occupied many researchers ever since**. Yet, when he wanted to return from industry to research, Stark was denied renewed access to suitable resources for experimental work. Not a single department, ministry, or research institution appointed him.[10] He had supported Hitler also in public.[11] He thus had to continue to rely on the use of basement rooms in his

---

[6]Stark managed a private porcelain factory in Upper Palatinate from late 1921; see, e.g., Stark [1987], pp. 64ff.

[7]For more biographical details, see Stark's autobiography, Stark [1987], originally written in 1945, edited by Andreas Kleinert.

[8]Stark later claimed that his resignation in Würzburg was the result of a conflict over the habilitation of his assistant Ludwig → Glaser: See, e.g., Stark [1987], p. 61. However, the Würzburg University Archive files show conclusively that Stark had already been planning to leave beforehand: By then he was already spending several days a week on managing his porcelain factory at Weiden, Upper Palatinate.

[9]On Stark's discovery of the influence of electric fields on the splitting of spectral lines, the 'Stark effect', in late 1915 see, e.g., Stark's Nobel Prize acceptance speech, delivered in Stockholm on June 3, 1920: Stark [1920].

[10]Stark had made many enemies, including very influential scientists such as Arnold → Sommerfeld, as a result of his aggressive personality; cf., e.g., Hermann [1967]. Around 1927 Stark continued to conduct experiments at the laboratory of the Jew Rudolf Freiherr von Hirsch at Planegg Castle outside of Munich, later at his own private laboratory: Cf., e.g., Stark [1987], pp. 74ff., or the letter from the dean to the rector dated Sep. 5, 1938, Stark folder, Würzburg Univ. Archive.

[11]See Lenard's and Stark's proclamation of loyalty to Hitler of 1924: doc. 3.

house that had been furnished with modest laboratory equipment; nevertheless, he continued to produce new findings repeatedly. His publications, particularly his books, are very comprehensive. His work on ceramic kaolins stems from his industrial activities.[12]

May luck continue to be with him; and may he be graced with a long life to fulfill all hopes this present appointment may have rightly awakened in him.[13]

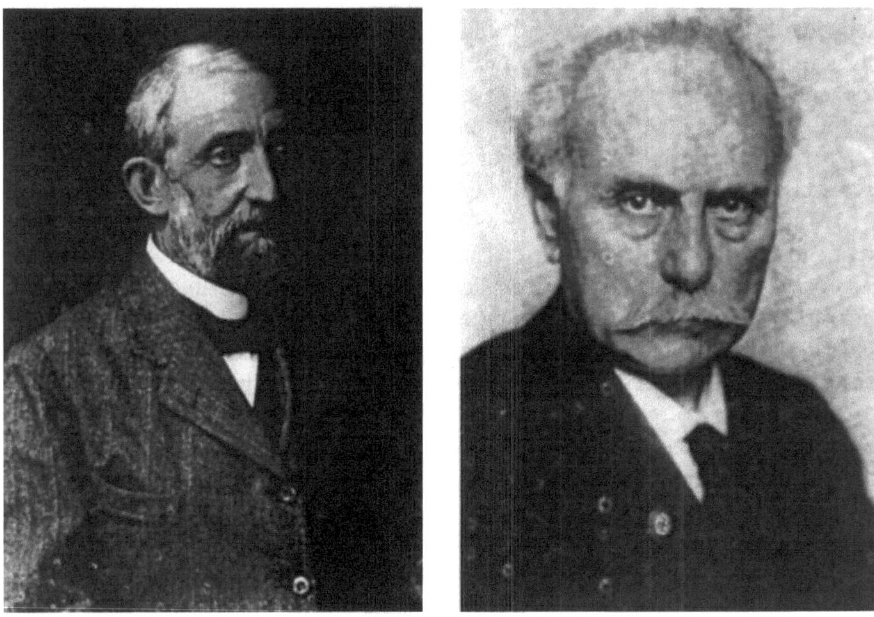

Fig. 5: Philipp Lenard (1936) and Johannes Stark (1933), the latter with a NSDAP party button on his lapel. Reproduced by permission: © *Ullstein Bilderdienst.*

[12]In the years after 1921; cf. footnote 6 and J. Stark: *Die physikalisch-technische Untersuchung keramischer Kaoline*, Leipzig: J. A. Barth, 1922. Other major works published by Stark in this time include: Stark [1927], [1928], [1931].

[13]For Stark's own assessment of his impact after 4 years as head of the *PTR*, cf. Stark [1937]; as well as the end of chapter 6 in Beyerchen [1977], Hoffmann [1993]a and Kern [1994] for overall assessments of the reorientation of research at the *PTR* towards military applications.

19 Max von Laue: The Change in the Reich Physical and Technical Institute Presidency in the Spring of 1933

Source: Archive of the Max Planck Society, Berlin: III, Rep 50, N0. 136. Typed draft of document with 1957 added in pen: 'Der Wechsel in der Präsidentschaft der P[hysikalisch-]T[echnischen] R[eichsanstalt] im Frühjahr 1933.' Published by permission of Prof. Theodore H. Von Laue.

I could go into great length about K. Kuhn's tribute (Nuremberg) to Johannes Stark, which appeared in this journal,*[1] but would risk reawakening old disputes. Therefore I would just like to report a thing or two based on my own recollections, in order to prevent a myth from forming around the issue Kuhn has raised about whether Stark had ousted his predecessor F. Paschen.[2]

Shortly after the '*Machtergreifung*' Paschen invited a few Berlin physicists to see him, Nernst and me among them.[3] He informed us that the new National Socialist Secretary of the Interior had asked him to resign, considering his 68 years of age.[4] He told us, when he had been called to accept the position, he had received a firm promise from the Ministry, though not in writing (legal reasons had precluded this, he said), that he would be able to hold the post beyond the normal age limit, because he had only just assumed it at age [59].[5] He had now referred the Secretary to this promise. But since the success of his appeal was doubtful, Paschen promptly consulted with the physicists he had called together on a suitable successor. At this point Stark's name was mentioned, especially since his ambitions for the presidency were no secret; but serious objections were raised immediately.

In the course of these negotiations Paschen then wrote to at least a half-dozen domestic and foreign physicists for their expert opinions and relayed their answers at a later date to the group mentioned. Although the opinions were completely apolitical, they were completely unanimous with regard to Stark and

* [Original footnote 1):] "*Physikalische Blätter* 1957, p. 370."

[1]The experimental physicist Karl Kuhn, who had submitted his Ph.D. thesis on 'Physical problems of higher strata of the earth's atmosphere' to the University of Erlangen in 1922, wrote an obituary in the journal issued by the → *DPG* in which he did not mention Stark's political activities advocating 'Aryan physics' and confined himself to praising his unquestioned skill as an experimental physicist. M. → von Laue's response was apparently never published.

[2]The spectroscopist Friedrich → Paschen was president of the *Physikalisch-Technische Reichsanstalt* (→ *PTR*) from 1924 to 1933.

[3]Hitler's seizure of power took place on Jan. 30, 1933. The physical chemist Walther → Nernst had been professor at Berlin since 1904 and director of the Institute of Experimental Physics there since 1924. Out of favor with the Nazis, he retired in 1933.

[4]The → *RIM* was in charge of all federal institutions including the *PTR* until 1934. It was headed by W. → Frick, and the relevant secretary of state at the *RIM* may have been *SS-Obersturmführer* Donnevert.

[5]M. von Laue left Paschen's age blank on the typescript. The usual retirement age for civil servants was 68, to which it had been raised from 65 since the economic depression of the late 1920's. It was reduced again to age 65 in the Law on the Retirement and Transfer of Professors in January, 1935; see here doc. 36.

could be summarized in the statement: "Of all the candidates for this post Stark is the least well-suited both scientifically and personally." Obviously, all this correspondence went to the Secretary. He then promptly nominated Johannes Stark, without consulting the board of trustees of the [Reich] P[hysical and] T[echnical Institute].[6]

Now, did Stark oust his predecessor Paschen or did he not?

20 Hans Kopfermann: Letter to Niels Bohr [May 23, 1933]

Source: Niels Bohr Papers, Archive of the History of Quantum Physics (Microfilm no. 22, section 2). Typed signed letter, sent from 'Kopenhagen-Hellerup'.[1]

Dear Professor Bohr,

I have been back from my trip to Germany for 14 days now. If my impressions from 10 days in Berlin, Göttingen and Rostock are not to become outdated and obsolete, I must now finally report them to you.

It is not quite so easy to form an opinion about what is going on in Germany; and it is even more difficult to make any predictions about how things will develop. This is partly because freedom of speech is restricted; but it is mainly because it is hard to judge what the German government itself wants, or how far her supporters will push her. On one point all sensible people agree: The masses are very much more radical than the leaders; and especially students (i.e., the radicals among them, who unfortunately have all the say against a diligent, moderate majority of perhaps 70%) go much further, or at least want to go much further, than the Ministry of Culture.[2] The Berlin student organization's appeal

[6]The original typescript reads '*PTB*', but should be '*PTR*', since it was only renamed a federal institute (*Physikalisch-Technische Bundesanstalt*) after World War II.

[1]The atomic and nuclear physicist Hans → Kopfermann visited Niels Bohr several times at Copenhagen after obtaining his doctorate at the University of Göttingen in 1925. The Danish atomic and nuclear physicist Niels Hendrik David Bohr (1885–1962) took his degree in 1911, then traveled to England to work together with Joseph John Thomson and soon afterwards in 1912 with Ernest Rutherford on atomic structure. Later that year he became teaching assistant at the University of Copenhagen and was promoted to lecturer in 1913. 1914–16 he was invited again to Manchester to work with Rutherford and returned to Copenhagen as professor of theoretical physics, where an institute had been established for him in 1920. He made major contributions to nuclear physics and quantum theory and was awarded the Nobel Prize for 1922. In 1943 Bohr fled his occupied native country to Sweden, from where the British secret service brought him to England. He later participated in the American atomic bomb project, and returned to Copenhagen after the war in 1945. See also footnote 4 of doc. 31.

[2]On the support by radical student organizations for the Nazi movement see in particular, Bleuel & Klinnert [1967]. At some universities, such as at the Berlin Polytechnic in Charlottenburg, the → *NSDStB* already had more than 60% of the votes in the student council in 1931/32. On Apr. 12, 1933, the Prussian Minister of Education Bernhard → Rust issued a new law concerning the student leagues, which introduced racist principles and the *Führerprinzip* into all student organizations. On Apr. 22, 1933, a Reich law concerning the 'formation of student unions' specified that only matriculated students of German origin and language could

illustrates this, for example, which was an attempt to quietly boycott also **those** Jewish professors permitted by the civil service law to stay in office.[3] All students are supposed to be compelled not to attend the courses held by such lecturers. It has actually not been successful up to now; with one exception, the courses have run smoothly.[4] The notorious placard by the students at Berlin University, which states, among other things, that everything that the Jew says in German is a lie, also shows this radical tendency clearly.[5] Neither the Berlin rector nor the Minister of Culture have succeeded in having the placard removed;[6] it is still posted on the bulletin board today. The Culture Ministry's attempt to persuade Haber to stay also seems to indicate that it is trying to attenuate.[7]

At the moment a power struggle is apparently taking place in Germany behind an astonishingly calm façade between the relatively moderate leaders and the radical masses;[8] and the whole problem seems to be whether it is possible to appease the masses, who have been driven so far through privation, the Versailles Treaty and systematic incitement. In order to gain the goodwill of the govern-

be members of the *Deutsche Studentenschaft*; their declaration of 'Aryan' descent had to be certified on oath. On the 'non-Aryan' students at German universities after 1933, see Götz von Olenhusen [1966].

[3]On the student appeal in Berlin see doc. 14, especially footnote 8. The student boycott of Jewish and leftist lecturers was called in Berlin on April 1, 1933.

[4]The exception Kopfermann may have been thinking of is the Institute of Sexology (*Sexualwissenschaft*) directed by Magnus Hirschfeld (1869–1935) in Berlin, which the students had rampaged on May 6, 1933.

[5]A facsimile of a published version of this manifesto can be found, for example, on p. 316 of Eduard Seidler's *Die medizinische Fakultät der Albert-Ludwigs-Universität Freiburg im Breisgau. Grundlagen und Entwicklungen*, Berlin: Springer, 1991. The German Student League posted these '12 theses against Un-German spirit' (*12 Thesen wider den undeutschen Geist*) at Berlin University on Apr. 13, 1933. Thesis 5, for example, stated: "The Jew can only think like a Jew. If he writes in German, he is lying" ("der Jude kann nur jüdisch denken. Schreibt er deutsch, dann lügt er"); thesis 6: "We will root out lies, we will stigmatize treachery"; thesis 7: "We demand of the censors that Jewish works appear in Hebrew. If they appear in German they must be labeled translations". See also Strätz [1968], p. 354, for earlier formulations of this and other theses. On this student action see also doc. 14.

[6]When the rector of Berlin University, the professor of law Eduard Kohlrausch ordered the '12 theses' removed from university bulletin-boards, he was replaced by the Nazi anthropologist and eugenicist Eugen Fischer (1874–1946), who was then instrumental in filling further key positions at the university with Nazi activists. Cf., e.g., Laitko et al. [1987], pp. 507, 513 as well as doc. 14, footnote 9.

[7]On Fritz → Haber's resignation from his position as director of the → *KWIPC*, see doc. 15, footnote 3. Bernhard Rust's and Max → Planck's efforts to convince Haber to withdraw his resignation on the grounds of the war service exemption of the Law for the Restoration of the Professional Civil Service were futile.

[8]Kopfermann was not alone in underestimating the radicalism of the National Socialist movement and the fanaticism of its anti-Semitic and anti-Communistic elements. What they perceived as signs of moderation on the part of the Nazis were only acts of compromise to appease national conservative circles including General Ludendorff and for foreign relations reasons.

ment, serious-minded people are advancing the new education law, according to which only **those** children count as Jewish whose parents are **both** Jews—in contrast to the initially much more radical position.[9] This relaxation in particular caused many parents, of whom one is a non-Jew, to stay in Germany, because they now see opportunities based on this education law for their children which had simply not been there 6 weeks ago.[10]

Naturally, there are also many—and they are usually those most affected by these regulations—of the opinion that radicalism is gaining acceptance; yet the events of recent weeks seem to prove them wrong.

The tactics employed by the individual also vary according to whether or not he believes the one or the other view. Laue, for example, who is certainly the most optimistic, tries to delay all decisions as long as possible.[11] He hopes in this way to be able to save many of those under threat. He bases his optimism on the symptoms I have partially described and which, in his opinion, point to the fact that the Ministry of Culture uses the same delay tactics on the students. He includes all suspensions under this type of action and expects that after a time some of these professors—those who have not laid themselves open politically—will be able to return to their posts.

Haber on the other hand does not believe that this anti-Semitic wave will ebb away so quickly. (The fact that Haber has resigned, and H. Freundlich (the colloid chemist) and Polanyi with him,[12] is due to a large extent to a lack of direction on the part of the leadership of the Kaiser Wilhelm Society and not least to the fact that Planck was in Sicily and could not persuade himself to return to Germany even when urgently pressed.[13] What is to become of the Haber Institute, and thus of me as well, still cannot be estimated at all at the moment.[14]

---

[9] See the Law Against the Overcrowding of German Universities, doc. 12, footnote 8.

[10] Anti-Semitic measures were enforced incrementally, and the full implications only became evident with the infamous Nuremberg Laws of 1935, culminating in the subsequent holocaust during the war years. In the beginning as each of the legal restraints were imposed, many of those affected hoped in vain that it would be the last.

[11] For more on Max von → Laue's tactics during the early years of National Socialism see, e.g., doc. 34.

[12] Haber had resigned because the managing director of the → *KWG*, Friedrich Glum (on Glum see footnote 14 of doc. 30), tried to appease the Nazis with a policy of self-realignment which meant that the Jewish department heads in the → *KWIPC*, Herbert Freundlich and Michael Polányi, lost their jobs. As expressed in his letter of resignation, doc. 15, Haber could not accept the sacrifice of these outstanding scientists for political reasons. See also Stolzenberg [1994], chap. 13.

[13] On Planck's decision not to interrupt his vacation, which also was important in explaining the radical tone of the Berlin Academy of Science's press release concerning Einstein's resignation (cf. doc. 6), see footnote 11 of doc. 21.

[14] After Haber and at least ten other scientists of Jewish origin had left, Haber's successors, Gerhard Jander as acting director (who was installed by the Nazi government without consulting the *KWG*) and Peter → Thiessen (in 1935), transformed the *KWIPC* into a model National Socialist establishment.

As concerns the attitude toward the changed situation in Germany, with the exception of some convinced National Socialists and a small number of people who now hope to obtain a high office, everyone involved in science is against the anti-Semitic measures of the government. But it is very hard to make this known publically. The enclosed article by W. Köhler is certainly the clearest up to now of what has been said in this line.[15] The fact that he was not placed on leave as a result, but rather that careful attempts were made to feel him out, shows that carefully worded criticism (by Aryans, mind you!) is apparently accepted.[16]

There is a tendency among the non-Jewish younger scientists to try to join the movement and to act as much as possible as a moderating element, instead of only standing disapprovingly on the sidelines.[17] The whole issue is, of course, not simply a problem of the Jews alone, although the Jews are affected first and foremost by it, but a problem of universities in general. Culture Minister Rust has said this clearly: University should be not so much a research institution as an educational one, that is, one on an ethically idealistic and nationalistic basis.[18] Subjects in which the students can be influenced in this direction are favored accordingly. Idealistic philosophy, military science(!), racial anthropology, history with a generous nationalistic touch, etc., are rated much more highly than the sciences, for example, which these people presumably suspect instinctively of challenging their view of life.[19] How science is to be conducted on this basis cannot be discerned; this is naturally the main reason why the German intelligentsia is opposed to the new regime. And this opposition naturally endangers an Aryan university lecturer as much as it does a Jewish one. In this case probably one of the most important things to do is to be critical and to explain the facts.

Regarding the persons under threat, at the moment their numbers cannot be estimated roughly yet. Aside from the very few who had been removed from office in the first turbulent days, the younger ones who fall under the civil service law are naturally in the most trouble.[20] Everyone in any kind of official post had to fill out a questionnaire in which he had to specifiy his race going back to his grandfather. Nothing has happened yet as a consequence. It is thought that this

[15]For Wolfgang → Köhler's article see doc. 13.

[16]Nevertheless, in 1935 Köhler also decided to emigrate to the USA because the situation had deteriorated seriously.

[17]This was a common postwar argument used in an attempt to justify involvement in Nazi organizations, even by those who, like P. → Jordan, did not quite act as a 'moderating' factor.

[18]'nationaler' inserted in hand. This probably alludes to Rust's speech during a special conference of the → *RIM* with all provincial university representatives on May 10, 1933, quoted in Laitko et al. [1987], p. 508. See also Rust's reproof to professors in 1937 for having neglected their leadership role, quoted in footnote 5 of doc. 56.

[19]For instance, at Berlin University an institute for 'political pedagogy' was installed as early as May 1933, with the Nazi philosopher Alfred Baeumler as its director; Oskar Ritter von Niedermayer (born 1885) lectured on military science (*Wehrwissenschaften*), and at the medical school there, social hygienics was replaced with 'racial hygienics'; cf. also ft. 10 of doc. 14.

[20]On the civil service law and its implications, see docs. 7, 8, and 17.

information will be processed in one month and that then resignations can be expected.[21]

Mr. Wolfsohn, about whom Rasmussen has written, has been taken on temporarily by Ornstein.[22] The Berlin positivist Reichenbach is in great difficulty.[23] In Franck's institute in Göttingen H. Kuhn's retirement is imminent.[24] Even Miss Sponer will probably have to count on going abroad, even though she is Aryan[25] (the German woman's place is behind the stove!!). Rabinowitsch wrote me today that he will have to leave Germany on the [1?] of August at the very latest for financial reasons; I hope that Professor Brönstäd will settle this matter.[26]

[21]In fact, the first wave of mass expulsions from public office started precisely at the time of this letter (lasting throughout 1933), along with a few voluntary resignations by scientists not actually targeted by the new anti-Semitic and anti-liberal laws including, for instance, Erwin → Schrödinger or Otto → Hahn. For quantitative estimates of the losses, see the Introduction and sources mentioned there.

[22]Günther Wolfsohn (1901–) studied physics with H. Konen at the University of Bonn, submitting his dissertation there entitled: 'Über das Bogenspektrum des Kupfers bei vermin-derter Druck' in 1926 and working on traditional technical applications of spectral analysis. He was a guest at the *PTR* in 1932. Knud Johan Victor Rasmussen (1879–1933) started his studies on eskimoan culture in 1906 in Greenland. In his last year of study 1909 he also founded the missions station Nordstjernen in North Star Bay, which was renamed Thule in 1910. He departed from there on a half dozen expeditions through 1933. Rasmussen received among many awards honorary doctorates from the Universities of Copenhagen in 1924 and St. Andrews, Scotland, in 1927. Leonard Salomon Ornstein (1880–1941) took his doctorate under Hendrik A. Lorentz in Leiden in 1908 and was professor of mathematical physics at the Rijksuniversitaet in Utrecht 1915–41. In 1921 he also became director of the physical institute there and in 1926 professor of experimental physics.

[23]Hans Reichenbach (1891–1953) had been a member of quite radical leftist organizations as a student, which caused problems in his candidacy for a lectureship during the early Weimar period. He emigrated to Istanbul and later to the USA.

[24]James → Franck's student Heinrich Gerhard Kuhn (1904–) started his studies in 1922 at the Universities of Greifswald and Göttingen and took his doctorate at the latter in 1926. He was teaching assistant at the Second Physical Institute there 1927 and became university lecturer 1931–33. In 1934 Kuhn worked in the Clarendon Laboratory at Oxford. He acquired his Master of Arts 1941 and became university demonstrator and lecturer 1945. 1950–54 he was a fellow of Balliol College and from 1955 reader at Oxford University.

[25]Hertha Dorothea Elisabeth Sponer (1895–1986) studied at Tübingen and Göttingen, taking her doctorate at the latter under James Franck in 1920 and working as teaching assistant at the physics department 1921–25. 1925–26 she was a Rockefeller fellow at Berkeley, California. Sponer was university lecturer 1924–32 and supernumerary assistant professor at Göttingen 1932–34 when she became visiting professor at the University of Oslo until 1936. In 1936 she was appointed professor at Duke University in Durham, North Carolina. She married J. Franck in 1946 and was a Guggenheim fellow at the University of Uppsala 1952–53. Sponer worked on electron impact, molecular spectroscopy and their relation to chemical problems, the spectra and structure of polyatomic molecules and their theoretical interpretation, and photochemistry.

[26]Eugene Rabinowitch (born 1904) did stay one year at the University of Copenhagen in 1933, presumably through Brønsted's assistance. Born in St. Petersburg, Russia, Rabinowitch took his degree at the University of Berlin in 1926 and was research associate at the University of Göttingen 1929–32. In 1933 he fled to Denmark to the University of Copenhagen, then to University College, London 1934–38. In 1938 he moved to the United States and was employed

Heitler is going to Bristol for a 1/2 year; and Northeim will probably go to Cambridge.[27]

It is terribly quiet in Copenhagen at the moment; and we look forward to when you are back again. You are hopefully enjoying a pleasant stay in America.

Cordial greetings to you and your wife.[28]

Yours Hans Kopfermann

21 The Scientific Situation in Germany [June 2, 1933]

Source: 'The Scientific Situation in Germany', *Science*, Vol. 77, No. 2005, June 2, 1933, pp. 528–529. (Anonymous article).

A wireless to *The New York Times* from Berlin reports that the Kaiser Wilhelm Society for the Advancement of the Sciences, the foremost scientific organization in Germany, which conducts thirty-two research institutes, held its twenty-second annual meeting here on May 23. It was presided over by Professor Max Planck,[1] who said that nowadays no one in Germany could be permitted to stand aside, "rifle at rest." [2] He declared that there was only one watchword—"the consolidation of all available forces for the reconstruction of the fatherland." He read the following message sent by the society to Chancellor Hitler: "The Kaiser Wilhelm

at the Massachusetts Institute of Technology 1939–46 and was a chemist at the Manhattan Project 1943–46. He was research professor of botany at the University of Illinois from 1947. Rabinowitch wrote, for example, on the periodic system in 1930, a two-volume textbook on photosynthesis 1946–54, and *Uranium Chemistry* in 1951. He also became editor of *Bulletin of the Atomic Scientists* in 1946. The physical chemist Johannes Nicolaus Brønsted (1879–1947) studied at the University of Copenhagen from 1897 to 1902, accepted an appointment there in 1908, and became in addition professor and director of the physical chemistry laboratory at the Copenhagen Polytechnic. He was awarded the Ørsted Medal in 1928 and in 1947 he received an honorary doctorate of science from the University of London.

[27]Walther Heitler (1904–1981) was a research fellow at the University of Bristol 1933–41; see also footnote 46, doc. 77. Lothar Wolfgang Nordheim (1899–1988) studied physics at the Universities of Hamburg, Munich and Göttingen 1919–23. He worked as teaching assistant for the mathematical physics department in 1922 and took his doctorate under the guidance of Max → Born at Göttingen in 1923. 1928 he became unpaid lecturer there and in 1929 was at the institute for theoretical physics. 1930 Nordheim was guest professor at Ohio State University, 1932–33 visiting professor at the University of Moscow. In 1933 he was a member of the scientific staff at the *Institut Henri Poincaré* in Paris, 1934 at the Teylers Stichting, Haarlem, and 1935 at the department of physics at Purdue University. He was professor at Duke University from 1937, and in 1950 he received an honorary doctorate in the sciences from the Karlsruhe Polytechnic. Cf. also doc. 109, footnote 3.

[28]Bohr was married to Margarethe Nørlund (1890–1984), the daughter of a pharmacist, in 1912. She had studied at Frøken Branners Pigeskole, a home economics and lady's finishing school in Slagelse, Denmark. See Pais [1991], pp. 112, 134.

[1]Max → Planck was president of the *Kaiser-Wilhelm-Gesellschaft* (→ *KWG*) between 1930–37. On his role as spokesman for German science see Heilbron [1986]b.

[2]The militaristic tone popular for the time has found its way into the conference rooms of scientific societies.

Society for the Advancement of the Sciences begs leave to tender reverential greetings to the chancellor and its solemn pledge that German science is also ready to cooperate joyously in the reconstruction of the new national state."[3]

Dr. Wilhelm Frick, Minister of the Interior, told the scientists what the state expected from them if it was to look after them. "It is the nature of scientific thought and research so to engross man as to expose him to the danger not only of becoming severed from the greater whole but actually losing his sense of duty—forgetting that he must serve the community," he declared. "With all respect for the freedom of science, let us postulate that service to science must be service to the nation and that scientific achievements are worthless when they can not be utilized for the culture of the people."[4]

No more scientists of Jewish extraction have been eliminated from the Kaiser Wilhelm Society's institutes, and among the elective members of its governing board three persons of the Jewish faith were reelected.[5]

The investiture of Professor Ernst Krieck with the rectorship of the University of Frankfurt, which was renamed Goethe University last year, took place on May 23.[6] The new rector said that in recent years the universities had lost a central idea, had become side-tracked and "never could have struggled from their paralysis but for the folk renascence. The chief characteristic of this rebirth is the replacement of the humanistic ideal by the national and political. Nowadays the task of the universities is not to cultivate objective science but soldierlike, militant science, and their foremost task is to form the will and character of their students."[7]

---

[3]For a description of the Society's move towards self-realignment (*Selbstgleichschaltung*) in 1933 see, e.g., A. Hermann & H. Albrecht in: Vierhaus & vom Brocke (Eds.) [1990], pp. 356–406, in particular, p. 359. Cf. also, docs. 38 and 81. Planck describes his audience with Hitler in May 1933 in doc. 114.

[4]W. → Frick clearly is warning the *KWG* to change its policy, which so far had mainly supported basic research without regard for possible applications and exclusively on the basis of its scientific importance. This 'freedom of science' now came under fire: See also doc. 18.

[5]Because the *KWG* research institutions were governed by the board of the Society as opposed to the Ministry of Culture, the laws affecting the civil service did not apply to these institutions. Nevertheless, the institute directors were presssured to fire staff members of Jewish origin or of leftist or liberal political persuasion: See, e.g., doc. 15 for Haber's reaction to this.

[6]Ernst Krieck (1882–1947) was a philosopher with strong sympathies for the National Socialist movement going back to the Weimar period. He became a Nazi party member in 1932 and lost his job as a result. On May 1, 1933, Krieck accepted a professorship for pedagogy in Frankfort on Main; on July 1, 1933, he was appointed successor to Max Scheeler. During the 1933 summer semester, Krieck was also university president. In 1934 he became spokesman (*Obmann*) of the National Socialist Science Agency (*Amt für Nationalsozialistische Wissenschaft*) at the Reich Association of German Universities (*Reichsverband der Deutschen Hochschulen*), and professor of physics in Heidelberg, where he again became university president in April 1937. On Krieck see, e.g., G. Müller [1978] and Victor Farías, *Heidegger und der Nationalsozialismus*, Frankfurt (Main): Fischer, 1989, pp. 233–235.

[7]In his capacity as university president, Krieck ordered the establishment of a Lecturers Academy to shape university lecturers and their successors and to point them in a common

A correspondent writes from Holland: "The declaration against Einstein, published by the Prussian Academy of Sciences, was not decided upon at a session of the academy.[8] Tactical motives may have caused the subsequent recognition of the declaration by the academy. The first declaration was signed by the presiding secretary Heymann alone. He is, along with the orientalist Lüders, permanent secretary of the Philosophical-Historical Division.[9] The second declaration was signed by Heymann and Von Ficker.[10] Von Ficker is one of the permanent secretaries of the Mathematical-Physical Division. The other permanent secretary is Max Planck, who was in southern Italy at the time of the declarations."[11]

22 Werner Heisenberg: Letter to Max Born [June 2, 1933]

Source: Ehrenfest Papers, Rijksmuseum voor de Geschiedenes de Naturwetenschappen, Leyden, The Netherlands; and on microfilm at the Archive for History of Quantum Theory, Ehrenfest scientific correspondence 2,5 with a copy at the Deutsches Museum, Munich. Typed unsigned transcription of letter in exerpt.[1] Published in Meyenn et al. (Eds.) [1978-93], Vol. 2, p. 167ff.

[...] But I was very startled by the other topic of your letter: I had not reckoned with you also not wanting to return to Göttingen[2] and I still hope that this is not

direction: "die die Dozenten und den Nachwuchs auf Form und gemeinsame Richtung zu bringen vermag". On Kriek's conception of science policy see in particular G. Müller [1978].

[8]See here doc. 6.

[9]On Ernst Heymann see footnote 5 of doc. 6. The Indologist Heinrich Lüders (1869–1943) assumed the full professorship for Indic philology at the Berlin University 1909–35 and was dean of the philosophy department there 1920–21; he was rector of the university 1931–32. Lüders became full member of the Prussian Academy of Sciences in 1909 and was secretary of the Philosophical-Historical Division 1920–38.

[10]Heinrich von Ficker (1881–1957) was full professor of meteorology at the Berlin University 1923–37 and was dean of the philosophy department in 1929. Von Ficker was director of the Prussian Meteorological Institute and of the Potsdam observatory, Lindenberg 1923–34. He was full member of the Prussian Academy of Sciences in 1926, nonresident member in 1937 and again full member in 1949; he was secretary of the Physical-Mathematical Division 1932–37. For the full text of these declarations see Kirsten & Treder (Eds.) [1979], Vol. 1, pp. 241ff.

[11]On the background to this affair concerning the cancellation of Einstein's membership in the Prussian Academy of Science see, e.g., Kirsten & Treder (Eds.) [1979], Vol. 1, chaps. 2 & 6 and the annotation to doc. 6 here. Planck continued his stay in Italy despite being informed about the steps taken in his absence. His partial compliance and ambiguous restraint in this matter is explained by the fact that he was deeply divided between his Prussian loyalty to the state on the one hand and his realization that "in the history of future centuries the name of Einstein will be celebrated as one of the most brilliant stars that have shone in our Academy"; see, e.g., Heilbron [1986]b, p. 157.

[1]Annoyed by Werner → Heisenberg's attitude, Max → Born prepared and sent this transcription with his letter of June 11, 1933, to the Dutch physicist Paul Ehrenfest (1880–1933).

[2]Like James → Franck, Max Born had also decided to abandon his position as professor of theoretical physics at Göttingen in the spring of 1933 and to emigrate (see doc. 9). See also Born [1968], Part 2, and Lemmerich (Ed.) [1982], pp. 111ff. Unlike Franck, however, Born was

a definite decision of yours. For, I have been at Planck's in Berlin and discussed with him the problem of what we could do on behalf of physics. Planck—I believe I may safely pass this on to you—has spoken with the head of government and has received the assurance that the government will not undertake anything that might impede our science beyond the new Civil Service Law.[3] As on the other hand only very few are affected by the law—certainly not you and Franck: surely also not Courant—the political changes could take place without any damage to physics at Göttingen.[4] Naturally, from the human point of view I understand very well your hesitation to return to such thankless surroundings; but you also clearly know how gratefully the students think of you for having been exposed to the scientific atmosphere which pervaded your research. It is obviously not easy for us others either to hold out at our posts: Bloch's departure from my institute[5] has resulted in the partial destruction of a section that had cost me a lot of time and effort. A few nasty things have even been happening within the workings of science itself.[6] Even so, I know that there are leading people also in the new political arena who are well worth our patient endurance. In the course of time the ugly elements will be severed from the fine ones. Therefore I would like to encourage you to the best of my ability not to see only ingratitude at Göttingen. Perhaps in the not too distant future life will become so peaceful here that you will sense how much your work means to a particular group of people. And more we cannot ever attain than that a group of people feels that we are living for them and for our science which has been entrusted upon us. That colleagues know nothing of all this is an indication that we frequently have absolutely no personal relations with those who practice the same profession as

---

informed by the Ministry of Culture by telegraph that he had been suspended on April 25, 1933: See Becker, Dahms & Wegeler (Eds.) [1987], fig. 5-8 preceding p. 29.

[3]See the Law for the Restoration of the Professional Civil Service of Apr. 7, 1933 and subsequent implementation ordinances, here docs. 7–8, 12 and 17. On Max → Planck's audience with Hitler in May 1933 see here doc. 114 and refs. there, particularly Albrecht [1993].

[4]The eminent mathematician Richard Courant (1888–1972) had been professor of mathematics at Göttingen University from 1920 to his dismissal by the Nazis in the spring of 1933. He then emigrated to the UK in 1933 and to the USA in 1934, where he was member of the New York University faculty at the mathematics department 1934–58: See Reid [1976]. This view that the state of a discipline is determined only by the positions taken by the leading figures was shared by many scientists involved in science policy. However, aside from the full professors mentioned, a number of highly qualified subordinate scientists and mid-ranking staff members, whose research contributions met equally high standards, were also affected by the dismissal policy (such as, e.g., Arthur von Hippel, Hertha Sponer, and Emmy Noether), which Heisenberg completely ignores here.

[5]Following his physics studies at Zurich, Felix → Bloch had come to Leipzig to work with Heisenberg on a topic in the new theory of quantum mechanics for his doctoral thesis. Bloch's paper on electron oscillations in crystal grids was a milestone in the history of solid state theory, which has since become the focus of theoretical studies at Leipzig. Bloch went to Stanford University, California in 1933. See Eckert [1993], especially pp. 130–133.

[6]An evasive allusion to the wave of dismissals and emigrations: See the Introduction, section 4, and references there.

us. But often this is not so unfortunate at all.[7]

Therefore I entreat you not to make any decisions now but to wait, to see how our country looks in the autumn.[8] Maybe we will have been able to discuss this personally by then, I would find that splendid.

With best wishes in gratitude yours truly [Werner Heisenberg]

23 Carl Bosch: Letter to Fritz Haber [June 16, 1933]

Source: Archive of the Max Planck Society, Berlin: V, Rep. 13, Haber Collection, No. 944. Typed letter signed, on Bosch's private letterhead: 'Dr. C. Bosch, Heidelberg' and addressed to 'Herrn Geheimrat Prof. Dr. Haber, Berlin-Dahlem'.

Esteemed Privy Councillor,

I heard with great regret in Berlin how very oppressed you feel personally by the present circumstances.[1] You might know that I myself have tried everything possible in order to make the measures against scientists somewhat bearable;[2] and I do not need to assure you that the personal side of the whole movement affects me extremely deeply. It is certainly hard on the nerves;[3] and I can understand that at your advanced age, and above all with your impaired state of health, your inner struggle is not easy. If I can be of any assistance to you somehow, then I am naturally gladly at your disposal.

Assuring you of my sincere esteem and with best regards

Yours very truly C. Bosch.

24 Fritz Haber: Letter to Carl Bosch, June 22, 1933

Source: Archive of the Max Planck Society, Berlin: V, Rep. 13, Haber Collection, No. 944. Unsigned carbon copy of typed letter addressed to 'Herrn Geheimrat Prof. Dr. C. Bosch, Heidelberg'. A copy of Haber's letter of resignation is enclosed. Excerpt in Stoltzenberg [1994], p. 613.

[7]Heisenberg is referring to the rift between the advocates of modern physics (e.g., the Sommerfeld and Bohr schools) on the one hand and the conservative camp led by P. → Lenard and J. → Stark on the other: See the Introduction, sec. 5.3.

[8]This letter reveals that Heisenberg shared the widespread belief that the political situation would stabilize quickly and perhaps even that it was only a matter of months before the National Socialist government would be replaced.

[1]Cf. Fritz → Haber's letter of resignation as director of the → *KWIPC* (doc. 15), which was transformed into one of the most politically streamlined institutes of the → *KWG*. See also Stoltzenberg [1994], chap. 13, for a detailed account of the tragic situation in which the political events had placed Haber in 1933.

[2]Carl → Bosch was head of → *I.G. Farben* and → *BASF* at the time and also member of several commissions of the Kaiser Wilhelm Society.

[3]The word 'hard' is inserted by hand, changing the meaning from 'a trial on the nerves' (eine Aufgabe der Nerven).

Esteemed Privy Councillor,

I have gratefully received your friendly note of the 16th of June.[1] Yours is the only voice to reach me from the *I.G. Farbenindustrie* and Nitrogen Syndicate communities.[2] Presumably from my point of view matters seem to be different to how they appear from your standpoint. Please understand that personally I perceive the appropriate minister's granting of my retirement petition, a transcript of which is enclosed, not as a hardship but rather as a relief.[3] Regarding the general effect of the Civil Service Law[4] on people in science in Germany, I am acquainted with and respect your readiness to help, which you have generously exercised both personally and through your reputation; but I am coming more and more under the impression that all countries abroad are starting to move and are offering care and assistance to those among the affected who had won honor and recognition here in this country,[5] similarly to how the other German states outside of Hanover supported the scholars who had left Göttingen nearly 100 years ago and who remain in academic history under the name of the Göttingen Seven.[6] Our country's cultivation of science will suffer more permanent damage,

[1]This letter is in reply to the preceding document.

[2]The → *I.G. Farben* (Interest Community of the Dye Industry) was a trust of the major chemical and pharmaceutical companies in Germany founded on Dec. 2, 1925 through a merger between the companies *BASF, Bayer, Hoechst, Agfa, Griesheim* and *Weiler-ter Meer*. Between 1925 and 1945 it was Europe's largest private trust; and it made huge profits by cooperating with the Nazi government in creating a national autarky as laid out in the Four-Year Plan (*Vierjahresplan*). It had a major involvement in up-scaling the weapons industry and explosives supplies, etc. After World War II the trust was split again into several independent companies by the Allied forces in the new Federal Republic of Germany, or nationalized in the German Democratic Republic. Hayes [1987] is a history of this trust headed by Carl → Bosch and Karl → Krauch; see also Borkin [1978] on the numerous crimes against humanity committed in labor camps at *I.G. Farben* production sites. The Nitrogen Syndicate (*Stickstoffsyndikat G.m.b.H*) was an industry-wide association of nitrogen producers located in Berlin; see Bruno Waeser, *Stickstoffindustrie*, Dresden & Leipzig, 1924.

[3]On his resignation as director of the → *KWIPC* on April 30, 1933, see doc. 15, footnote 7; cf. Stoltzenberg [1994], sec. 13.1.

[4]See doc. 7.

[5]In 1933 several organizations were founded to help absorb the many thousands of immigrants caused by the systematic expulsion of Jewish, liberal and left-wing scholars in Germany. Among the most important of these were the Emergency Committee in Aid of Displaced German Scholars, founded by Felix M. Warburg and Alfred E. Cohn in the U.S.A., the *Notgemeinschaft Deutscher Wissenschaftler im Ausland*, founded by Philip Schwartz, Fritz Demuth and others in Switzerland, and the Academic Assistance Council (called from 1935 on, Society for the Protection of Science and Learning), founded in England by Lord Rutherford and William Beveridge. Cf., e.g., K. Düwell in: Strauss et al.(Eds.) [1987], pp. viii–x, and here sec. 4.6.

[6]The 'Göttingen Seven' were professor renowned for lodging a protest on Nov. 18, 1837, against the revocation of the liberal constitution of 1833 by King Ernest Augustus of Hanover. The king responded on Dec. 14, 1837, by dismissing them from the university. The seven were: Wilhelm Eduard Albrecht (1880–1876), a jurist who advocated a historically oriented liberalism following the English model, appointed professor of German national law in Göttingen in 1829; (Georg) Heinrich [von] Ewald (1803–1875), a Protestant theologian, orientalist, politician, and

because this intervention has forgotten or at least disregarded the interrelation between the work of the individual and the activity of his field of science as a whole.

To my mind, the regulations that make the present and previous generations of children and grandchildren into second-class citizens, without regard for merit and accomplishment, touches very much more profoundly than the ruling of the Civil Service Law. I will not accept being sentenced the punishment of leaving this inheritance share to my descendants. This ruling has temporarily upset my inner peace; and if I would only have to cross over to the other side of the river like the Göttingen Seven in order to be abroad,[7] then I would scarcely continue living in this country, to which I had believed I belonged inseparably these 64 years.[8] I believe your readiness to come to my aid if I were to ask you.[9] But I do not know what I could ask for, unless it were that you refuse to listen to those people who, subsequent to unfriendly newspaper portrayals, pass me off as a sick old man.[10]

I remain with best regards Yours very truly [Haber]

---

founder of Jewish philology who obtained his professorship in Göttingen in 1827; Georg Gottfried Gervinus (1805–1871), a historian, literary historian and politician who was the first to present German literature in the context of historical development and had been professor at Göttingen since 1835; Jacob Grimm (1785–1863), the founder of German classical studies, and his brother Wilhelm Grimm (1786–1859), both eminent Germanists and collectors of German folk literature who had moved to Göttingen in 1830; and Wilhelm Eduard Weber (1804–1891), a physicist who since his appointment in 1831 worked closely with the famous physicist C. F. Gauß on a telegraph and on measurement instruments. After his dismissal in 1837 he accepted a professorship in Leipzig in 1843, returning to Göttingen in 1849. Weber's electrodynamic measurements later formed the basis of the the international units of measurement. Dahlmann, Jacob Grimm and Gervinus were expelled from the country for actively seeking outside support for their cause. Public opinion and numerous German professors outside of Göttingen sympathized with the seven, whose actions as well as written justifications contributed significantly to the development of German liberalism. In 1948 Albrecht, Dahlmann, Gervinus and J. Grimm were elected to the *Paulskirche* parliament in Frankfurt. Cf. Hans Kück: *Die Göttinger Sieben*, Berlin: Ebering, 1934, reprinted as Vol. 3 of the Gottingensia Series of Edition Herolot: *Die Göttingen Sieben. Ihre Protestation und ihre Entlassung im Jahre 1837*, Aachen: Rader, 1987.

[7]To leave the small Kingdom of Hanover the three expelled leaders of the Göttingen Seven only had to travel a few miles South to the Hessian border near Witzenhausen on the Werra.

[8]Haber remained in Dahlem (Berlin) until late summer 1933, when Gerhard Jander was appointed deputy director by B. → Rust in M. → Planck's and Glum's absence (cf. Goran [1967], p. 163, for Haber's farewell address to his staff and Vierhaus & vom Brocke (Eds.) [1990], p. 375, on Jander's and P. → Thiessen's takeover). Haber then decided to leave Germany and accepted an invitation from Cambridge, England.

[9]The subjunctive leaves open whether Haber really thought that Bosch had supported him sufficiently; cf. Goran [1967], p. 162: "Haber was disturbed about the lack of support from his friends in industry, particularly at the *IG Farben*" See also Stoltzenberg [1994], pp. 613f.

[10]Even Otto → Hahn, who was asked by the president of the → *KWG*, Max → Planck, to replace Haber temporarily as substitute director of the *KWIPC*, later wrote that Haber had been "a sick man for years" (see Hahn [1962]b, p. 91). Haber died during a short holiday in Switzerland on Jan. 29, 1934. Cf. the obituary by Max von Laue, doc. 29.

25 Governor of Hessen Jakob Sprenger: Dismissal of Prof. George Jaffé from the Civil Service [June 26, 1933]

Source: The Bancroft Library, Berkeley, California. For a facsimile see Rider [1984], p. 109, (fig. 1). Typed signed official document, initialled by the preparer: '/Ha.'

In the Name of the Reich

I dismiss from the Hessian Civil Service on the suggestion of the Hessen government the full professor of theoretical physics at the state university

Dr. George Jaffé[1]

on the basis of the Reich Law of the 7th of April 1933 (*R[eich]. L[aw]. G[azette].* I. p. 175), § 4,[2] with effect on the 1st of July 1933.

His current compensation will be continued up to and including the 30th of September 1933.

Another special order will be issued by the Hessen government on the calculation of his pension.[3]

Darmstadt, the 26th of June 1933. Governor of Hessen. Sprenger[4]

[1]George Cecil Jaffé (1880–1965) had been professor of theoretical physics at the University of Giessen since 1926 and specialized in statistical and quantum mechanics. He studied in Munich 1898–99 and then in Leipzig until 1903, taking his doctorate there under the chemist and Nobel laureate Wilhelm Ostwald (1853–1932). 1903–04 he was at Cambridge, then 1904–05 in Paris at the Curie laboratory (see also the following doc. 26, footnote 1). In 1908 Jaffé became university lecturer at Leipzig and in the same year teaching assistant. In 1916 he was appointed associate professor of physics at the University of Leipzig and in 1923 he was promoted to regular associate professor, and receiving a call in 1926 to a full professorship in theoretical physics at the University of Giessen. Jaffé emigrated to the U.S. in June 1933 and became a member of the department of physics at Louisiana State University, Baton Rouge.

[2]See the Law for the Restoration of the Professional Civil Service, doc. 7. Prof. Jaffé was dismissed on the grounds of 'political unreliability' (§ 4).

[3]The fact that most dismissed persons subsequently left Germany gave the government sufficient pretext to cancel their pensions, even where they had been granted initially.

[4]According to Führerlexikon [1934], p. 467, the *Reichsstatthalter* of Hessen Jakob Sprenger (1884–?) volunteered into a Bavarian regiment 1901–02 immediately after completing his schooling and became a postal official in 1902, working his way up to chief inspector (*Oberpostinspektor*) at Frankfort-on-Main. He was a lieutenant in the First World War. Sprenger entered the → *NSDAP* in 1922 and headed the local Frankfort branch and Hessen-Nassau South region of the prohibited *Deutsche Partei* 1923–24. In 1925 he became city councillor and in 1927 regional party leader of Hessen-Nassau. In 1929 he became a member of the Frankfort municipal council and parliamentary party head in the municipal and regional diets as well as substitute member of the Prussian state council (the only Nazi member). He was an expert on civil service matters for the national Nazi leadership. From 1930 also founding head of the National Socialist faction for civil servants at the *Reichstag*. In 1932 he became Reich Party Commissioner for Southwest Germany, becoming the Hessian Governor on May 5, 1933. Sprenger founded and edited a National Socialist newspaper for government officials: *National-Sozialistische Beamtenzeitung*, as well as the dailies *Frankfurter Volksblatt* and *Nassauer Volksblatt* and was named honorary president of the German Civil Service in May 1933.

26 Sir Joseph John Thomson: Letter to George Jaffé [July 18, 1933]

Source: The Bancroft Library, Berkeley, California. For a facsimile see Rider [1984], p. 113, (fig. 2). Handwritten signed letter on J. J. Thomson's letterhead: 'Trinity Lodge, Cambridge.'

Dear Professor Jaffé,

I well remember your stay in Cambridge and your work in the Cavendish Laboratory[1] and I am very sorry to hear that your Professorship has been taken from you:[2] I will certainly take every opportunity I get[3] of supporting proposals for finding you some scientific work in this Country

Believe me Yours most sincerely J. J. Thomson[4]

27 Max von Laue: Opening Address at the Physics Conference in Würzburg [September 18, 1933]

Source: Max von Laue, 'Ansprache bei Eröffnung der Physikertagung in Würzburg am 18. September 1933',[1] *Physikalische Zeitschrift* **34** [1933], pp. 889–890 (reprinted in von Laue [1961], Vol. 3, pp. 61–62).[2]

[1]Georg Jaffé researched at the Cavendish Laboratory immediately after taking his degree 1903–04. In 1904 he went to Paris and served for one year as interpreter for J. J. Thomson and the renowned French physicist and chemist Pierre Curie (1859–1906).

[2]See doc. 25 for his official dismissal from the Civil Service by the Hessen authorities. On emergency placement organizations founded in 1933 to help such victims of the discriminatory German laws, see doc. 24, footnote 5.

[3]The word 'have' is crossed out and changed to 'get'.

[4]The English physicist Sir Joseph John Thomson (1856–1940) was the discoverer of electrons and received the Nobel Prize in physics for 1906 for his work on the conduction of electricity through gases. He held the Cavendish professorship at Cambridge 1884 to 1919 and was master of Trinity College 1918–40, being thus one of the most influential scientists in England. Cf., e.g., Lord Rayleigh, *J. J. Thomson*, Cambridge: Cambridge Univ. Press, 1943, as well as in *Obituary Notices of Fellows of the Royal Society* **3** [1941], pp. 587–609.

[1]For press reports on the meeting see, e.g., *Würzburger General-Anzeiger*, Sep. 18, 1933, no. 214, p. 3; Sep. 19, 1933, no. 215, p. 4; Sep. 21, 1933, p. 3 (obtained by courtesy of archivist Renate Mayer, Stadtarchiv Würzburg). According to Friedrich → Hund, Max → von Laue's speech was added to the program on short notice (personal communication). It directly preceded J. → Stark's speech, thus taking the wind out of his sails: See here doc. 28. See also M. → Planck's letter to von Laue of Sep. 9, 1933, in which he congratulates von Laue for chairing the Würzburg meeting: "Physicists are lucky that you are at the top this time." (DMM, No. 1964-6 138a,b, Sheets 35–36). For other papers given at the conference see *Zeitschrift für technische Physik* **14** [1933], issue no. 11. Laue himself had been implemental in the discovery of X-ray diffraction; see footnote 3 of doc. 76.

[2]The editor has also discovered a carbon copy of the typescript in the von Mises Papers, Harvard University Archives, Cambridge, Mass., Call No. HUG 4574.24, entitled 'Ansprache, gehalten von M. von Laue bei der Eröffnung der Physikertagung am 18. 9. 33 im Platzschen Garten in Würzburg.' The few discrepancies between the published version and the typescript

When we come together tomorrow at the physics department of the local university, we will be standing at a historic site. At the end of 1895 in this building Wilhelm Conrad Röntgen discovered the rays that were named after him.[3] It would be superfluous, not to mention in bad taste, to discuss the importance to physics and the whole gamut of applications of these rays. But we would like to commemorate the great achievements of Röntgen, who was the first to see consciously that which many before him had narrowly missed, and elevate it from shady speculation to transparent and reliable scientific knowledge. We already know from Glasser's Röntgen biography that a genuine photographic X-ray exposure exists on the other side of the ocean dating from the year 1890, which was obviously only recognized for what it was following Röntgen's publications.[4]

Next we commemorate Röntgen's successor at this place: Willy Wien taught and researched in this building for 17 years.[5] His classical theoretical writings on heat radiation, which have become an integral part of our science in the form of Wien's displacement law, originate from his early years, when the young physicist worked under Helmholtz at the Reich Physical and Technical Institute.[6] But nowhere else could he devote so many years in a row unperturbedly to such extraordinarily fruitful experimental research together with so many students and friends. The knowledge on the nature of canal rays that he gained here makes up no small part of the contents of those formidable textbooks now appearing on the subject.[7] I would like to call special attention in particular to two of

affecting the content are indicated.

[3]Roentgen rays, also known as X rays. On Wilhelm Conrad Röntgen (1845–1923) and his discovery of the electromagnetic waves of approximately 1/1000 of the length of visible light see, e.g., Otto Glasser: *Wilhelm Conrad Röntgen and the Early History of the Roentgen Rays*, London: Bale & Danielsson, 1933, and Springfield & Baltimore: Thomas, 1934, especially chaps. 1–4, or Wheaton [1983], chap. 2, and references there. See also: W. Robert Nitske, *The Life of Wilhelm Conrad Röntgen. Discoverer of the X-ray*, Tucson: Univ. of Arizona, 1971; M. von Laue, 'Zum Gedächtnis Wilhelm Conrad Röntgens', *Naturwissenschaften* **33** [1946], pp. 3–7; Jost Lemmerich, *Röntgen Rays Centennial. Exhibition on the Occasion of the Discovery of X-Rays in Würzburg on Nov. 8, 1895*, Würzburg Univ., 1995.

[4]Otto Glasser discussed priority claims by other scientists such as Sir William Crookes, and (in chapter 9) reprinted a shadow photograph taken accidentally by the American A. W. Goodspeed on Feb. 22, 1890 using the rays emitted from a Crookes tube; but he added that nobody took the strange phenomenon seriously, "and the plates were put with the other freaks and forgotten" until after the discovery by Röntgen, who unlike his forerunners, had pursued the new phenomenon systematically.

[5]Wilhelm → Wien worked at the physical lab of the University of Würzburg from 1900 as the immediate successor to Röntgen.

[6]Wien worked at the *Physikalisch-Technische Reichsanstalt* (→ *PTR*) 1890–96 as assistant to its director, Hermann von Helmholtz. On Wien's work on black-body radiation during his Berlin period see, e.g., Kangro [1976], chap. 5, and references there.

[7]These textbooks include Wilhelm Wien's *Kanalstrahlen*, Leipzig, 1927: Vol. 14 of *Handbuch der Experimentalphysik*, Leipzig, 1933; H. Geiger & K. Scheel, *Handbuch der Physik*, Vol. 22, 2nd ed., Berlin, 1932 (therein: E. Rüchardt, 'Durchgang von Kanalstrahlen durch Materie');

Wien's contributions out of this time, since they were fundamental: The first quantum-theoretical determination of the oscillation frequency of X rays, which admittedly produced only an approximate, though essentially correct value;[8] then, the experimental confirmation of the relativity of electromagnetic fields, as required by relativity theory. When examined spectrally, moving canal-ray particles in a magnetic field exhibit the Stark effect, which we are otherwise only familiar with in the case of an electric field with stationary light sources.[9]

But this same building[10] has witnessed quite different goings on as well. During the World War, Wien transformed his department, and with the collaboration of Max Seddig[11] also the neighboring chemistry department, into an enterprise to examine all the instruments essential in modern communications engineering, primarily towards the development of amplifier tubes. The communications system that our Army had installed at the front in 1914 was by no means up-to-date when compared to the rest of its training and equipment; in this the enemy had a great advantage. Any slow yet very significant improvements in this area in the course of 4 years were due to a large extent to the work of the physicists engaged here on these premises.

On the 22nd of June of this year physics could celebrate a particularly special anniversary. 300 years ago on that day Galileo's trial ended before the Inquisition.[12] The grounds for the trial were, as is well-known, Copernicus's

Müller-Pouillet, *Lehrbuch der Physik*, Vol. 4, Part 3, 21st ed., Braunschweig, 1933 (Ch. Gerthsen & W. Kossel, 'Korpuskularstrahlen'); *Hand- und Jahrbuch der chemischen Physik*, Vol. 6, Part 3, Leipzig, 1933 (R. Wierl, J. Hengstenberg & K. Wolf: 'Kanalstrahlen und ihre Wechselwirkung mit Materie'). See also footnote 12 of doc. 18.

[8]See Wilhelm Wien: 'Berechnung der Wellenlänge der Röntgenstrahlen aus dem Planckschen Entropieelement', *Nachrichten der Gesellschaft der Wissenschaften, Göttingen* [1907], pp. 598–601.

[9]See Wilhelm Wien: 'Über die von der elektromagnetischen Theorie geforderte Einwirkung des magnetischen Feldes auf die von H-Kanalstrahlen ausgesandten Spektrallinien', *Sitzungsberichte der Preußischen Akademie der Wissenschaften, Berlin*, math.-phys. Klasse [1914], pp. 49, 70–75; cf. also his contribution 'Kanalstrahlen' in: *Handbuch der Radiologie*, Vol. 4, Leipzig, 1917, pp. 1–210.

[10]The typescript's initial wording of "dies Haus" is corrected in hand to: "das physikalische Institut", but in the published version appears as "dasselbe Haus".

[11]Max Seddig (1877–1963) was a specialist in scientific photography. He obtained his doctorate in 1902 in Marburg and became untenured associate professor of physics in 1915 and president of the department of scientific photography. 1935 50 Seddig was full professor of applied physics at Frankfort-on-Main.

[12]In this section of the talk von Laue, a prominent voice of dissent among German scientists, draws an unambiguous parallel between Galileo and the Jewish theorist Albert → Einstein. Einstein's name is automatically associated with his relativity theories, which the new authorities rejected, just as the Catholic Church had condemned Copernicanism some 300 years earlier. Although von Laue refrains from any outright statement against the Nazi regime, his politically unorthodox message is clear; and it was perceived as an open provocation. See, i.e., Stark [1934]d, p. 9: "Even if Mr. von Laue is a close friend and leading supporter of Einstein, nevertheless he was not allowed to suggest implicitly that the National Socialist government was

theory of the motion of the earth and the other planets around the sun,[13] a theory that caused[14] a similar sensation and stir to that of the relativity theory in our century for being contradictory to the traditional notions of the day. Galileo was not its only advocate, but its most successful one, because he was able to support it so convincingly with his wonderful discoveries of Jupiter's moons, the phases of Venus and the rotation of the sun.[15] The trial ended with a conviction. Galileo had to renounce the Copernican theory and was condemned to lifelong imprisonment.[16] Obviously, this restriction of his freedom was carried out relatively leniently in his case. He was allotted a house to live in that he was not permitted to leave, and he could not receive visitors without permission. In actual fact, during his confinement he could indeed publish his major work in physics, the *Analyses Concerning the Two New Sciences.*[17] Nevertheless he did remain a prisoner for the rest of his life.

There is a well-known myth attached to this condemnation that while giving his oath and signing his renunciation of the theory of motion,[18] Galileo

violating the freedom of scientific research." On Jan. 9, 1934, J. → Stark wrote to the Berlin Ministry of Culture to object to von Laue's succeeding Einstein at the Prussian Academy of Sciences, with specific reference not only to their close friendship, but also to Laue's talk, revealing that its message had not been lost on him nor on the audience, which Stark reported was welcomed with "thundering applause by the present Jews and their friends" (Kirsten & Treder (Eds.) [1979], Document No. 196, p. 269). Twenty years later in his book *The Crime of Galileo*, Chicago: Univ. of Chicago Press, 1955, the Italian historian of science Giorgio de Santillana drew a similar parallel between Galileo and Julius Robert Oppenheimer (1904–1967), who was accused of un-American activities at the height of McCarthyism in the 1950's.

[13]Copernicus's heliocentric conception against Ptolemy's model of the earth as the center of the solar system and the 200-year 'revolution' in planetary astronomy are discussed in, e.g., Thomas S. Kuhn: *The Copernican Revolution. Planetary Astronomy in the Development of Western Thought*, Cambridge, Mass.: Harvard Univ. Press, 1957.

[14]The word 'erregte' (provoked) in the typescript is replaced in the published version with the more neutral word 'hervorrief'.

[15]Galileo's discovery of Jupiter's moons brought into question the uniqueness of the Earth among the planets in the solar system. His discovery of sun spots and of solar rotation damaged further the earlier notion that all astronomical objects are eternal and unchangeable—the cosmos had become dynamic. Von Laue describes Galileo's discoveries in 'Galileo Galilei: Zur 300. Wiederkehr seines Todestages, dem 8. 1. 1642', *Frankfurter Zeitung*, Jan. 8, 1942 (reprinted in Laue [1961], Vol. 3, pp. 182–185). For a Copernican interpretation, see Galileo's *Dialogo* [1632], especially books 2–3, in English translation by Stillman Drake: *Dialogue Concerning the Two Chief World Systems*, Berkeley: Univ. of California Press, rev. ed., 1967.

[16]For a historical account of Galileo's trial see, e.g., Santillana [1955] (cited in footnote no. 12 above, and Pietro Redondi: *Galileo Heretic*, Princeton: Princeton Univ. Press, 1987 (translated from the Italian by R. Rosenthal).

[17]The word 'Unterredungen' (Discussions) in the typescript is changed to 'Untersuchungen' in the published version. See Galileo's *Discorsi e demostrazioni matematiche informe à due scienze attenenti alla mecanica & i movimenti locali*, Leiden: Elsevir, 1638, one of the first scientific treatises published in the vernacular Italian; see the English edition *Two New Sciences*, translated and edited by Stillman Drake, Madison: Univ. of Wisconsin Press, 1974.

[18]In the typescript: "theory of the earth's motion"

was supposed to have said: "And yet, it moves!"[19] This is a myth, historically unverifiable and intrinsically implausible—and yet it is ineradicable in common lore. What is its vitality based on? Surely on the fact that throughout the whole proceedings Galileo must have posed the unspoken question: "What's the point of all this? Whether I or whether any person now believes it or not, whether political or religious forces support or are against it, the **facts** remain unchanged. Might can certainly delay knowledge of the facts for a while, but eventually it will prevail." And this is indeed what happened. The victorious advance of the Copernican theory was irresistible. Even the Church that had condemned Galileo gave up all form of opposition, albeit only 200 years later.[20]

The times were not always favorable to science later either, such as in Prussia under the otherwise so praiseworthy king Friedrich Wilhelm I.[21] But in the face of all the repression, its supporters could stand steadfast in the triumphant certainty that is expressed in the modest phrase: And yet, it moves!

28 Johannes Stark: Organization of Physical Research [September 18, 1933]

Source: Johannes Stark, 'Organization der physikalischen Forschung.' Transcription of a speech at the Würzburg conference in *Zeitschrift für technische Physik*, 14th series, No. 11, 1933, pp. 433–435 (Received on October 31, 1933).[1]

On the 1st of May of this year Reich Minister Frick appointed me head of the Reich Institute of Physics and Technology.[2] Thus a great responsibility has

[19]"Eppur si muove!" According to the *Encyclopædia Britannica* [1960], Vol. 9, p. 978, the earliest appearance of this apocryphal legend is in Abbé Irailh's *Querelles littéraires*, 1761, Vol. 3, p. 49.

[20]All books that advocated the heliocentric doctrine remained on the Church's Index until 1757, and Galileo's *Dialogo* was only struck from the list in 1835.

[21]The rigid economy of Friedrich Wilhelm I (1688–1740) led to his neglect of higher learning and the rejection of foreign cultural influence at the Prussian court, although school became obligatory in Prussia during his reign and he was known for his capable management abilities and his defense of justice. When his enlightened son Frederick the Great (Friedrich der Große) acceded to the throne, however, all trace of frugality disappeared and Voltaire, for example, was welcomed as a frequent guest at the lavish Rokoko palace 'Sans Souci'.

[1]Presented on September 18, 1933: See the newspaper report on the meeting in the *Würzburger General-Anzeiger*, Tuesday, September 19, 1933, Sheet No. 215, p. 4, and footnote 1 of the preceding document. Johannes → Stark apparently made a bid for election as president of the German Physical Society (→ *DPG*) at this meeting. Max → von Laue was reelected, however, and remained president of the society until 1935. Cf., e.g., Max → Planck's letter to Laue of Sep. 9, 1933: "If the majority is really so unreasonable as to vote Stark in as chairman, then it will soon live to regret it; and maybe there is a good side to it as well, if a clarification of the concepts, along with the attached enlightenment, comes very early on" (DMM, No. 1964–6, 138a, b, Sheet 35–36).

[2]J. Stark succeeded the spectroscopist Friedrich → Paschen as president of the *Physikalisch-Technische Reichsanstalt* (→ *PTR*); cf., Stark [1937] for his description of the change in climate

been transferred into my hands, the responsibility of developing and guiding the largest scientific and technical institution of the Reich so that it can properly meet its important tasks in the service of the German people. The institution has responsibilities toward science and responsibilities toward industry. It must promote not only physical science but also the German economy as a whole and thus must act as a mediator and bring about an exchange between two major areas of German culture.

Toward physical science the Reich Institute of Physics and Technology has the following responsibilities: It must promote physical research with the intense research activities of its staff; it must perform particularly difficult analyses and measurements on its own premises; it must provide assistance to physics departments throughout the Reich upon request and must make experimental research equipment available; and it must represent German physics at international conferences on science and technology.

Toward the economy the responsibilities of the Reich Institute of Physics and Technology are the following: It assists all branches of industry by testing or calibrating measuring instruments and machines; it is responsible for counseling as an impartial public office both private firms and government agencies on matters involving physics and technology.

The Reich Physical and Technical Institute is the largest scientific institution in Germany. It already comprises about 45 laboratories with approximately 300 officials and employees. It is subordinate to the Reich Interior Ministry and from this circumstance alone has the importance of a government agency that is centrally active throughout the entire German Reich. In addition, it communicates with all the physics departments and all of industry throughout the German Reich. It is from this central, comprehensive and leading position that its responsibility arises to organize physical research for the benefit of both science and industry.

Some of my listeners may well immediately object to the term 'organization of scientific research'. The question might be raised: Can scientific research be organized at all? Surely scientific progress is always the independent achievement of individuals and cannot be achieved solely and directly through organization.[3] These statements are certainly correct. But they misinterpret the purpose of scientific organization. The latter does not aim to replace the achievements of the individual researcher; on the contrary, through its measures and means it aims to increase the productivity of the individual more than ever. One of the Institute's most important responsibilities is, not to replace the individual researcher, but

at this government institute, which was under the supervision of Reich Minister W. → Frick at the → *RIM*.

[3]Stark alludes here to the familiar idea of the freedom of research which grants to university teachers unhampered choice of research and lecture topics. Nazi efforts to streamline universities and national research institutions ran counter to this conception; but Stark tries to conceal this here in rhetoric, as he does in Stark [1987], pp. 123f.

to promote him.

Following this observation let us turn our attention to the essence, the goals and responsibilities involved in organizing physical research.

Organizing means connecting individuals to a central institution and, by virtue of this centralized link, also to one another. The individual elements of physical research organization include on one hand physicists and physical institutions, on the other hand, industrial enterprises and government agencies. The central institution to unite the individual elements mentioned above, to act as the necessary mediator between them and as the intended source of ideas or support, this central institution is the Reich Physical and Technical Institute.

The goal of organizing physical research through the Reich Physical and Technical Institute is to serve the German national community.[4] As with every fellow German [*Volksgenosse*], it is the duty of the German scientist also to direct his efforts ultimately toward serving the German national community. It is all the more his duty if he is indebted to the national community for his education, his job and his livelihood through governmental institutions and measures. The scientific researcher can serve his national community with three types of achievement: He can widen the sphere of knowledge, the intellectual horizons of his people, with his discoveries; he can develop industrial technology directly or indirectly and thus raise his nation's material standard of living; or he can improve defense technology and thus contribute to securing the survival of his nation.

The central institution must take measures in organizing physics and technology in accordance with these potential forms of service to the common good, in order to exercise its duty to science and the economy. To be able to perform this service and to fulfill its organizatorial duties, the Reich Physical and Technical Institute must expand with needed new buildings, additional technical installations and increased administrative organization. The organization of the *PTR* itself thus forms a part of the organization of physical research.

After taking over the presidency of the Reich Institute of Physics and Technology, I had to examine whether the current state of affairs was adequate for executing its responsibilities. I discovered that not only the existing premises but also the technological installations and administrative organization of the current Reich Institute were insufficient when set against its great responsibilities toward science and the economy. Thus I was faced with the urgent necessity of quickly constructing the new Reich Physical and Technical Institute. I reported this to Minister Frick; and this brilliant, broadminded and farsighted administrator immediately recognized the great importance of the Reich Physical and Technical Institute and the need for its new buildings.[5]

---

[4]From 1923 onwards Stark was a passionate advocate of the National Socialist concept of 'national community' (*Volksgemeinschaft*) and the associated obligation of the individual toward society as a whole; cf., e.g., Stark [1930]c, [1987], pp. 111f.

[5]Stark's flattery was not without purpose, since his ambitious plans to extend the *PTR* were still at a preliminary stage. Unfortunately for Stark, other elements within the polycratic

My listeners may be interested in a brief report on my plans for the Reich Physical and Technical Institute's new building.

First of all a suitable construction site had to be found that is large enough for the projected new buildings and for future extensions. It must have access to a rail connection and must be close to the Reich capital in addition to being located next to a larger town. After a lengthy search I found a construction site that was suitable in every way at the edge of the town of the great Prussian king: Potsdam,[6] the town in which Adolf Hitler ushered in the Third Reich in an official inauguration ceremony.[7] The building site lies between a wide highway and the Templin Lake through which the Havel flows. As you can see (projected image), a railway line runs through it. Its average depth is over 500 m, its length over 2000 m. Potsdam is to the Northeast of it and can be reached by tramway.

I would also like to show a picture of the scale and number of buildings at the Reich Physical and Technical Institute construction site as currently drafted according to my instructions. However, I would like to emphasize that this is only a draft of mine upon which the Reich government has not yet given its opinion. (There follows an explanation of the architect's blueprint which is projected for viewing.)

When the Reich Physical and Technical Institute's new building is completed, it will be able to fulfill its responsibilities properly toward science and industry; and it will be able to complete the organization of physical research and make it fully effective. Unfortunately, several years will pass before the new construction plans are realized. Until then the Reich Institute must try to fulfill all its responsibilities with the available funds and at least start to tackle the organizational duties. I will reveal briefly the current organizational duties as well as those for the future following completion of the new building.

The first organizational responsibility of the Reich Institute of Physics and Technology is to make available to all German physicists experimental apparatus and working space on its own premises. In the future, under my direction the Reich Institute will be available for experimental investigations that require installations that university departments cannot provide adequately. I mention in this regard only the following divisions of the Reich Institute which are already available or are being extended or newly created: The Institutes of Low Temperatures, High Frequency, High Voltage and Heat Engines.[8]

National Socialist system soon outweighed the influence of the *RIM*, such as, for instance, the → *SS* and Hermann → Göring's → *Luftwaffe*, as well as other military organizations.

[6]Potsdam was the residence of Friedrich I (1657–1713), first king of Prussia and elector of Brandenburg as well as of the German emperor king of Prussia Friedrich II, known as 'The Great' (1712–1786). See also footnote 21 of doc. 27.

[7]Lise → Meitner reports on this historical event to Otto → Hahn; see doc. 5.

[8]It is interesting to note that while Stark's plans to expand the *PTR* drastically ultimately failed, in 1937 the Rockefeller Foundation supported the Kaiser Wilhelm Society (→ *KWG*) in building experimental facilities for high voltage and low temperature research at the → *KWIP* in Dahlem (Berlin); see doc. 54.

An additional organizational responsibility toward individual physicists and departments is the loan of equipment for experiments conducted off the Reich Institute's premises. This task has yet to be taken in hand. The ground has already been well prepared by the Emergency Association of German Science.[9] Large storage rooms will be available for the loan equipment in the Reich Institute's new building.

Another of the Reich Institute's organizational responsibilities will involve establishing a Reich Physical Research Agency [*physikalischer Reichsforschungsdienst*] which is to act as an agent between physical research and industry. The purpose of this Agency will be to aid individual industrial firms in locating physicists and physics departments that appear to be especially well suited to execute their particular physical and technological projects.

Furthermore, the Reich Physical and Technical Institute will be involved in the monitoring of physical literature. Since, changes must be made to protect German physics within Germany as well as to maintain its influence abroad.

The Reich Physical and Technical Institute or its president also has other responsibilities with regard to the organization of physical research aside from those already mentioned. These are so dependent on ministerial assistance, however, that it is not yet possible to discuss them as things stand at the present time. But what I have presented should suffice for you to form an idea of the necessity of organizing physical research and the responsibilities involved.

In closing, you will allow me a few more words of a personal nature. As head of the Reich Physical and Technical Institute I am guided in my organizational efforts only by an aspiration to serve German science and the national economy and thus the German national community. In order to accomplish the great tasks before us, I need support from numerous quarters in directing the Reich Physical and Technical Institute. Primarily, we German physicists should hope that Reich Minister Frick will continue to help realize the construction plans of the scientific and technological Reich institutes with the same farsightedness and energy. We should hope that other ministers of the Reich and of the States offer a helping hand in the organization of physical and scientific research in general. We may even hope that the great *Führer* of the German people, Adolf Hitler, will, thanks to his many-facetted genius, be sympathetic to and interested in the planned large-scale building project for the scientific and technological institutes of the Reich, considering that the new Reich institutes will be of great importance to science, to the economy and to national defense.

But I also need your support, gentlemen. You, my colleagues, also can assist directly or indirectly in the organization of the Reich Physical Research Agency and in the reorganization of publication in physics. I ask you for your cooperation

[9] One of the main tasks of the *Notgemeinschaft der Deutschen Wissenschaft* (→ *NG*), founded in 1920, was the distribution of expensive scientific instruments to researchers on loan. On its history see, e.g., Nipperdey & Schmugge [1970] and Zierold [1968].

in the projected organization of physical research for the benefit and honor of the German people.[10]

29 Max von Laue: Fritz Haber † [February 16, 1934]

Source: Max von Laue, 'Fritz Haber †',[1] *Die Naturwissenschaften* **22**, Issue 7, February 16, 1934, p. 97.

On the 9th of December, 1928, Fritz Haber's sixtieth birthday, a small group of friends and co-workers came together in front of the Kaiser Wilhelm Institute of Physical Chemistry and Electrochemistry, **his** institute, and planted a linden tree in his honor, the *Haber-Linde*. He himself stayed in the South.[2] This was the only [official] celebration of his birthday.[3] Like other scientific journals, however, the *Naturwissenschaften* published a jubilee issue composed of contributions by

[10]According to Prof. Friedrich → Hund, Göttingen, who was present at the time, Stark actually ended his speech with the following threat omitted in the published version: "and if you are unwilling, then I will use force!" ("und seid Ihr nicht willig, so brauch' ich Gewalt!") (Personal communication). See also Steenbeck [1977], p. 71, for another eye-witness account of this event.

[1]Otto → Hahn later wrote in his scientific autobiography that Max → von Laue alone had dared to publish an obituary on Haber, "which almost caused his involvement in embarrassing disciplinary proceedings"; see Hahn [1962]b, p. 91. However, other obituaries on Haber were published in German newspapers and physical chemistry journals, as documented in Stoltzenberg [1994], pp. 637–639.

[2]Fritz → Haber had been director of the → *KWIPC* since its foundation in 1911. Following the 'Harnack principle', the → *KWG* built its institutes around first-rate scientists like Haber. For a description of the ceremony see Hahn [1960], Goran [1967], p. 119, or Stoltzenberg [1994], pp. 641ff. Haber himself celebrated the occasion by taking a tour on the Nile river together with close family and friends.

[3]On January 29, 1935, the first anniversary of Haber's death, Max → Planck organized a commemoration in his capacity as president of the *KWG*, in conjunction with the German Chemical Society (*Deutsche Chemische Gesellschaft*) and the *Deutsche Physikalische Gesellschaft* (→ *DPG*). Although professors in the civil service were barred by decree of the Prussian Ministry of Culture from attending this meeting, which the Deputy of the *Führer* accused of being a provocation against the National Socialist state, the meeting was not cancelled. According to Lise Meitner, doc. 120, Max von Laue and Wolfgang Heubner were the only professors who dared to attend in person, but many others were represented by their wives. Leading managers of the → *IG Farben* trust attended as well as personal friends and some members of the *KWIPC, KWIC, and KWIP*. See, e.g., Hahn [1960], p. 6, [1962]b, p. 93; D. Hahn (Ed.) [1988], p. 155, or Vierhaus & vom Brocke (Eds.) [1990], p. 373, for a facsimile of the invitation sheet, and Festschrift [1961], pp. 193f., for the correspondence between the *KWG* and the → *REM* concerning this event. On the intervention of the *REM*, civil servants and members of the → *NSDAP* were forbidden to attend, and the press did not report on the memorial celebration; cf. also, Hahn [1968/86], p. 146: "This open demonstration of opposition to the regime was still possible at the beginning of 1935". It was also widely reported in the foreign press: See, for instance, *The New York Times*, Jan. 12, 1936, **85**, No. 28,477, p. 31: 'Reich Scientists uphold Freedom'.

the most competent experts on Haber's spheres of activity.[4] Whoever is interested in seeing what a loss Haber was, both in war and in peace, to chemistry and physics, agriculture and technology, should take it up. But the reader will also discover there that Haber's greatness is only due in part to his papers: He was maybe even greater as an institute director, in that he allowed his co-workers complete freedom to develop their talents, while still setting the general direction of the institute as a whole. At sixty his activity came by no means to an end. The topics and projects of his own recent investigations included the processes of combustion and explosions and the influence of atoms and chemical radicals on them, autoxidation and reduction in solutions also particularly in biological processes.[5] Other projects from his institute that we will mention include Bonhoeffer's and Harteck's distinction between parahydrogen and orthohydrogen,[6] as well as its application in the analysis of complicated chemical processes and magnetic properties. While it existed, his institute was renowned for its broad range of research in science.

On the 2nd of May, 1933, Haber handed in his resignation.[7]

Themistocles went down in history not as the pariah at the court of the Persian king, but as the victor of Salamis.[8] Haber will go down in history as the genius inventor of the procedure of binding nitrogen with hydrogen, which underlies the synthetic extraction of nitrogen from the atmosphere. He will be remembered as the man who, in the words used at the award of his Nobel Prize, had created in this way "an exceedingly important means towards promoting agriculture and human prosperity". He will be remembered as the man who had made bread out

[4]Cf. *Die Naturwissenschaften* **16** [1928], Issue No. 50 (Dec. 14, 1928) with contributions by Richard Willstätter, Herbert Freundlich, Georg von Hevesy, Otto Stern, E. Terres, Robert le Rossignol, Margarete von Wrangel and James → Franck.

[5]Autoxidation is the slow oxidation of certain substances on exposure to air, or the oxidation induced by the presence of a second substance which is itself undergoing oxidation. On Haber's work on these topics see, e.g., Stolzenberg [1994].

[6]The nuclei of the two hydrogen atoms contained in the parahydrogen molecule have opposite spins, whereas those in the orthohydrogen molecule have spins in the same direction. The existence of these two forms of hydrogen had been predicted in quantum mechanics, in particular by David M. Dennison: 'A note on the specific heat of the hydrogen molecule', *Proceedings of the Royal Society of London* **A 115** [1927], pp. 483–486, as well as by Friedrich → Hund. On the confirming experiments see the papers co-authored by Karl-Friedrich → Bonhoeffer and Paul → Harteck on parahydrogen and orthohydrogen in *Die Naturwissenschaften* **17** [1929], pp. 182, 321f., in *Zeitschrift für physikalische Chemie*, section B, Vols. **4** & **5** [1929], and in *Zeitschrift für Elektrochemie und angewandte physikalische Chemie* **35** [1929], pp. 621–623.

[7]See doc. 15, actually dated April 30, 1933; cf. also, Hahn [1960], pp. 3f., Goran [1967], pp. 161ff., and particularly Stoltzenberg [1994], chap. 13, for a detailed account.

[8]The able Athenian statesman and creator of the Athenian fleet, Themistocles (c. 514–449 B.C.) led the Greeks to victory over the invading Persian fleet at the battle of Salamis on Sep. 28, 480 B.C. by sending Xerxes I (c. 519–465 B.C.) a cunning message that induced him to battle under unfavorable conditions. He fell into disgrace and was ostracized from Athens sometime between 476 and 471 B.C. and settled in Persia where he died.

of thin air and who triumphed "in the service of his country and of the whole of humanity".[9]

And what remains to those who knew him and who now grieve? Reminiscences – – – "His friends and co-workers know his great generosity, loyalty, courteousness and tenderness towards them. They know his indignation at dishonesty and wickedness, and the serenity with which he overlooks trifles. He is never-tiring in his kindness towards his juniors, always intent on encouraging them and pointing out new paths. He has no appreciation for restriction and intolerance. He values the exquisite qualities of other countries and intellects, yet his affections go to Germany—has he not helped this land protect and feed its children in hard times as no one else?" [10]

This is what Margarete von Wrangell[11] wrote in 1928; this is how we remember Haber. For, he was one of our own.[12]

[9]Haber was awarded the Nobel Prize in chemistry for the year 1918. The ceremony took place on June 1–2, 1920, with A. G. Ekstrand, president of the Swedish Academy of Science, giving the opening speech, printed in *Les Prix Nobel en 1919–20*, Stockholm: Norstedt, 1922, pp. 29–38 (see p. 38 for the above quote). Cf. also, Haber's Nobel Prize lecture, delivered at Stockholm on June 2, 1920, reprinted, e.g., in *Die Naturwissenschaften* **10** [1922], pp. 1041–1049, and in Haber [1924], pp. 1–24. See also Stoltzenberg [1994], chap. 10.

[10]Quoted from the last page of M. von Wrangel: 'Fritz Habers Bedeutung für die Landwirtschaft', *Die Naturwissenschaften* **16** [1928], pp. 1071–1075.

[11]Margarete von Wrangel, born Princess Andronikow (1877–1932), was a guest scientist at the *KWIPC* in Berlin in 1922 and was then appointed professor of phytotrophy at the Agricultural College in Hohenheim.

[12]There follows an unsigned description of the circumstances of Haber's death: "Fritz Haber died on the 29th of January, 1934 on a trip from Cambridge to a Swiss health spa. His death was not unexpected to those who knew him. He had long been plagued by a heart condition. Those who witnessed one of his attacks had already wondered years ago that such an attack should pass so quickly and with no apparent after-effects. He was already aware of the incurability of his sickness. He bore it without complaint; only once, in April 1933, the comment slipped out: "This kind of sickness is terrible; it takes so *long* to die." The attacks returned; in August 1933 a particularly serious one caught up with him in the Alps that left quite permanent marks. Then we heard news from Cambridge of his new creative enthusiasm and of the nice progress of begun projects—and now on the 1st of February news that he has died and has already been cremated arrives." Cf. Goran [1967], chap. 14, and Stoltzenberg [1994], sec. 13.3, for a description of Haber's stay in Cambridge at the laboratory of the organic chemist Sir William Jackson Pope (1870–1939), who like Haber himself, had worked on the development of poison gas during World War I.

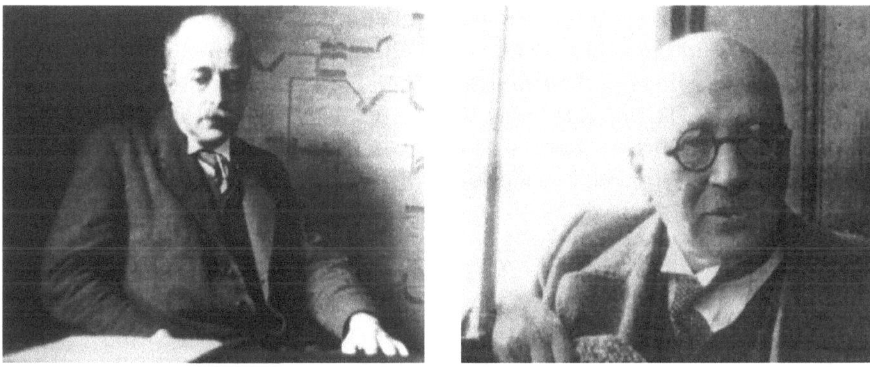

Fig. 6: Max von Laue (April 1936) and Fritz Haber (1931).

30 Kaiser Wilhelm Society: Record of a Conference Regarding the Kaiser Wilhelm Institute of Physics [June 22, 1934]

Source: Archive of the Max Planck Society, Berlin: I, Rep. 1A, No. 1651 (Sheets 126-127). 'Aufzeichnung über eine Besprechung wegen des Kaiser Wilhelm-Instituts für Physik am 22. Juni 1934.' Typewritten document. Distribution instructions and notes are omitted.

Present: the President, Privy Councillor Planck;[1] Ministerial Councillor Donnevert, Reich Culture Ministry;[2] Ministerial Councillor von Manteuffel, Reich Finance Ministry;[3] Ministerial Chief Executive Officer Lebus, Reich Finance

[1] Max → Planck was president of the → *KWG* from 1930 to 1937. Cf. also doc. 38.

[2] Max Donnevert (1872–?) studied law at the Universities of Freiburg, Berlin and Strasbourg and was a lawyer at the regional superior court (*Landgericht*) 1901–18. He obtained his habilitation at the University of Frankfort-on-Main and was a member of the second district of the Alsace-Lorraine diet (*Landtag*) until 1918, and captain of the Bavarian Militia Cavalry. He was adviser to the → *RIM* from 1920 and was officially responsible for affairs of science and the arts. Donnevert was president of an organization promoting the Alsace and Lorraine regions within the Reich and vice-president of the Science Institute for the same regions in Frankfort.

[3] Karl Freiherr von Zoege-Katzdangen Manteuffel (1872–?) (also Zoege or Szöge) was a retired district marschal (since 1904) and lieutenant of the 1st Dragoon Guard of the Reserves. He owned the entailed estate at Katzendangen. Von Manteuffel studied at the Universities of Bonn and Halle and took his doctorate in economics. In 1906 he was a founding member and the first director of a club of Germans in the electorate in the Urals, *Verein der Deutschen im Kurland.* He volunteered into the German Army in 1918 and fought in the Baltics against

Ministry;[4]; as well as the undersigned.

By way of introduction Privy Councillor Planck pointed to the great and as yet incalculable scientific importance of atomic research, which the new institute intends to focus on especially.[5] He reported that Prof. Debye at Leipzig, who is the leading man in the field in question, has been successfully enlisted as director of the institute.[6] He said, the Prussian Cultural Ministry also attaches the greatest importance to the institute being built for Professor Debye, since it intends to call him to the University of Berlin as successor to Privy Councillor Nernst.[7] He hoped that the Rockefeller Foundation, which had already approved 1.5 million reichsmarks some years ago for the construction of the institute, would keep to its promise, even though he could not conceal certain misgivings regarding the current unfavorable sentiment in the U.S.A. towards Germany.[8] The remaining question now is to find out whether the additional financial conditions are met, so that he can submit a new application to the Rockefeller Foundation.

The undersigned reported on the basics of the proposed financing: According to Prof. Debye an annual budget of 150,000.– reichsmarks is considered necessary. There is available for the construction:

---

the Bolsheviks. He earned the Iron Cross, was a member of the Knights of Malta and was a Knight of Honor of the Association of Baltic Veterans. He was first chairman of the Union of Former German Colonists in Russia and a member of the *Nation Club*. In 1935 von Manteuffel was a political writer in Berlin, ministerial director and head of *Amt I*: Reich Budget at the Reich Ministry of Finance (after 1934). See, e.g., Karl Freiherr von Manteuffel: *Meine Siedlungsarbeit im Kurland*, Leipzig: Hirzel, 1941; and his dissertation, *Das Sparen. Sein Wesen und seine volkswirtschaftliche Wirkung*, Halle: Lippert/Naumburg, 1900 (Vol. 26 of *Sammlung national-ökonomischer und statistischer Abhandlungen des Staatswissenschaftlichen Seminars zu Halle a. S.*, Jena: G. Fischer, 1900).

[4]Erich Lebus's title *Ministerialrat* is changed in hand to *Ministerialamtsrat*. Lebus is registered in the Berlin *Addressbücher* of 1929 and 1941 as senior clerk in the civil service (*Minist.-Amtm'*) and chief executive officer (*Amtsrat*).

[5]Plans for the establishment of an institute for pure physical research outside the university system had existed since Philipp → Lenard's petition in 1906, but had not yet been materialized. The Kaiser Wilhelm Institute of Physical Research was founded in 1917 with A. → Einstein as its director; but up till then it had only been an instrument for distributing funds and instrumentation to research groups at universities, since the World War impeded the construction of its physical premises. In 1929 the *KWG* rejected Max von Laue's application to create an institute of theoretical physics. Cf. Vierhaus & vom Brocke (Eds.) [1990], pp. 77, 177f.

[6]Peter → Debye became the director of the *KWIP* in 1935, after the decision for its construction had been made; see docs. 54, 80f.

[7]In 1933 Walther → Nernst retired from his duties as professor of physical chemistry at Berlin University at the age of 69.

[8]Indeed, the Rockefeller Foundation had already committed itself in April 1930 with contributions of 635,000 reichsmarks for the plot of land and 1.5 million reichsmarks for the building. On the Rockefeller Foundation's science policy and the difficult decision by its board on whether to fulfill this promise even though the → *NSDAP* now governed Germany, see Macrakis [1986]; and cf. Düwell in Vierhaus & vom Brocke (Eds.) [1990], pp. 752ff.; on the foundation see Fosdick [1989].

1. the plot that had already been obtained at the time using Rockefeller Foundation funds.[9]

2. capital that has grown meanwhile to about 1 million reichsmarks, so that a capital income of around 50,000.– reichsmarks can be reckoned with.

If the Rockefeller Foundation remains committed to its last grant, then

3. building capital is available that suffices both for the new building and for the institute's furnishings, instruments, etc. A contribution by the Reich for the running costs would be required in the amount of 100,000.– reichsmarks, which only need be made available in the year 1935/36, however; since, construction of the building must be started first, of course.[10]

Ministerial Councillor Donnevert warmly supported the Kaiser Wilhelm Society's requests. Ministerial Councillor von Manteuffel stated that he also would advocate the granting of the requested 100,000.– reichsmarks at the Reich Finance Ministry. Though he had assumed a negative stance earlier, he said, this was due to the fact that he had believed the funds had been requested for a substitute to the Kaiser Wilhelm Institute of Physical Chemistry and Electrochemistry. Aside from the importance of the scientific research, the Reich Finance Administration found it crucial to have a large amount of foreign money come into the country and to have jobs created. He therefore believed he was authorized to face all the consequences. However, he felt obliged to add the condition that no funds be released from other sources. Obviously, his commitment depended on the granting of the budget.

Privy Councillor Planck thanked the representatives of the government for their comments and also communicated that the Society planned to offer Prof. Laue a position at the institute beside Prof. Debye. This was necessary, if not only because the proposal to the Rockefeller Foundation had been granted at that time to Prof. von Laue personally.[11]

The undersigned stated that he would travel to Paris within the next few days in order to negotiate with the Rockefeller Foundation on the release of the previously authorized grant of 1.5 million reichsmarks.

Privy Councillor Planck and the undersigned stated that it was necessary that, with regard to Germany, in the near future nothing happen in German science that might disturb the Americans,[12] because undoubtedly a certain reticence

---

[9]The plot was located in the former Dahlem district, at the outskirts of Berlin near the already existing *KWIC* and *KWIPC* and just North of the Institute of Cell Physiology. Althoff and others looked upon Dahlem as a potential second Oxford. For a description see, e.g., Engel [1984] and Eckart Henning: 'Althoffs Vermächtnis für Dahlem', *Domäne Dahlem. Aus Landgut und Museum*, issue no. 3 [1988].

[10]Cf. doc. 54 for a description of the finished building in 1937.

[11]Although this was not actually the case, Max → von Laue had been instrumental in obtaining the grant as acting director of the *KWIP* since 1921, travelling together with R. → Ladenburg to the U.S. in the autumn of 1930 to visit the physics departments at Schenectady, Baltimore, Pasadena, Chicago and Cambridge. See Macrakis [1986], p. 355.

[12]This part of the text was modified in hand from: "it was necessary that nothing that

can be felt at this time.[13] Glum[14]

31 Max von Laue: On Heisenberg's Uncertainty Relations and their Epistemological Significance [June 29, 1934]

Source: Max von Laue, 'Über Heisenbergs Ungenauigkeitsbeziehungen und ihre erkenntnistheoretische Bedeutung', *Die Naturwissenschaften*, Vol. 22, Issue No. 26, June 29, 1934, pp. 439–441.

Two years ago I indicated in this journal*[1] that the conclusions drawn from the formalism of quantum mechanics, and in particular from the uncertainty relations,[2] do not seem to necessarily indicate a failure of the causality principle, because the conceptual structures on which these conclusions are based stem from Newton's mechanics and, like all of classical mechanics, are of empirical origin. In my opinion this reasoning only refutes applying those [classical] concepts to atomic processes. There is one argument in the discussion of this subject, however, that I had not commented on at the time, which initially sounds quite convincing. I would like to discuss it here. It states: "It is impossible to find a real causal explanation for atomic processes, because there is nothing smaller to measure them with than atomic structures; but their interaction with the atom (or atomic structure) under examination fundamentally disturbs the latter's state. In other branches of physics it has been possible to assume that a test particle can be chosen for measuring purposes that is small enough to exclude noticeable

happens in German science in the near future disturb the Americans, because..."

[13]Here Planck openly links the issue of monetary funds granted by the American foundation with recent actions and laws made by the new government. He hoped to persuade ministerial officials to proceed cautiously, and thus prevent more drastic steps by the government in science policy.

[14]Friedrich Johann Glum (1891–1974) was an official government adviser (*Regierungsreferent*) in 1920 and 1937. He took his doctorate in political science and law and was unsalaried lecturer at Berlin University in 1928 and associate professor 1930–37. Glum was general director and managing member of the administrative section of the *KWG* from 1928 throughout the 1930's. Between 1946 and 1948/1952 he was also ministerial assistant secretary (*Ministerialdirigent*). See also Glum [1964].

* [Footnote 1, on p. 439 of the original:] "*Naturwiss.* **20**, 915 (1932)."

[1]See M. von Laue: 'Zu den Erörterungen über Kausalität', *Die Naturwissenschaften* **20** [1932], pp. 915–916. The main thesis on p. 916 states: "Certain purely physical concepts based on experiment have failed to hold true in more recent experience; at the moment better concepts are lacking. This is not an unusual situation in the sciences; it precedes every major advancement. But these difficulties cannot force anyone to change his epistemological standpoint, whatever it is".

[2]W. → Heisenberg proposed what are known as the 'uncertainty relations' in 1927. See W. Heisenberg: 'Über den anschaulichen Inhalt der quantentheoretischen Kinematik und Mechanik', *Zeitschrift für Physik* **46** [1927], pp. 172–198, and *Die physikalischen Prinzipien der Quantentheorie*, Leipzig: Hirzel, 1930. Cf. also, Cassidy [1992]a, Part II as well as the selection of Heisenberg's semi-popular writings in English translation: F. S. C. Northrop (Ed.), *Physics & Philosophy. The Revolution in Modern Science*, New York: Harper & Row, 1958.

repercussions. This possibility falls away, however, as soon as we examine atoms themselves and their states."[3] In this chain of reasoning replacing the smallest particle by the smallest action quantum h makes no essential difference.

Great credit goes to Bohr and Heisenberg, in my opinion, for pointing out the problem of interaction [between observer and the observed object] and its importance in the exploration of atomic states.[4] Nonetheless, I do not believe that the consideration described points to a completely insurmountable limitation of scientific knowledge, because this conclusion is based upon the tacit assumption that: "New experimental tools are **essential** to opening up new possibilities in measurement." Only when you assume this can you infer further: "Since we have now reached the finest of tools, the atoms themselves, we can never progress any further." Now, is this assumption correct?

Let us have a look at H. Hertz's discovery, which certainly did lead to new understanding.[5] What did Hertz experiment with? With an induction coil along with its appropriate power supply, some conducting wires, metal plates and balls, in short, only with those things others had used thousands of times before him. What did he observe? Small electric sparks that clearly had been known for a long time. So it seems that this scientific advancement did not come from new experimental aids after all, but rather from an ingenious experimental **chain of reasoning**.

One could surely find many an example of this kind in the history of physics as well: from recent times, for example, the Nernst heat theorem.[6] But maybe a story will illustrate what I see as the essence more clearly than any single historical case. Though the fable I would like to tell is completely accurate in physical detail, I will otherwise make some use of an author's license. Well:

"Once upon a time there was a young man who wanted to devote himself to the study of physics, but because he was poor he was forced to earn his bread as a working student. So he went to an electrical engineering company which,

---

[3]This kind of argument forms the basis of the so-called Copenhagen interpretation of quantum mechanics: See the preceding and following footnotes.

[4]In 1927 the Danish physicist Niels Bohr (see footnote 1 of doc. 20) and Werner Heisenberg generalized their understanding of the interdependence of observer and his object, caused by the measuring process, and the latter's influence on the qualities measured. It is known as the Copenhagen interpretation of quantum mechanics; cf., e.g., Jammer [1966] and footnote 50 of doc. 77.

[5]Von Laue is referring to the discovery of electromagnetic waves by Heinrich Rudolf Hertz (1857–1894), who could demonstrate in 1887–88 the transmission of electric oscillations between two physically disconnected electric circuits. Later experiments also confirmed the transmission of signals between electrical dipoles at the speed of light, as predicted by the Maxwell equations and indicated by the identical physical properties of electromagnetic waves and light waves (reflection, refraction, polarization, etc.); cf., e.g., Jed Z. Buchwald, *The Creation of Scientific Effects: Heinrich Hertz and Electric Waves*, Chicago/London: Univ. of Chicago Press, 1994.

[6]In 1905 W. → Nernst formulated what is now known as the third law of thermodynamics, which states that it is impossible to reduce the temperature of a system to absolute zero in a finite number of operations. Cf., e.g., Bartel [1989], pp. 70–82.

because of his poor education in physics, employed him more out of pity than for any other reason. However, they set before him a very simple task, namely, testing galvanic cells for their electromotive force using voltmeters of the standard technical design. He was given a number of such measuring instruments and was instructed to test each galvanic cell with several of these voltmeters, because he could only rely on the accuracy of the measurement when they all showed the same voltage. This the young man did; and for a while everything went quite well as well.

But one day he came across a galvanic cell for which all of his voltmeters showed a different voltage. It was not due to experimental error; no degree of care could eliminate the discrepancies. Then our young man became very thoughtful and arrived finally at the conclusion: [']Obviously, the galvanic cell is affected when a voltmeter is attached. Usually this had no adverse effect. In this case the effect is apparently so strong that it disturbs the state of the cell; and therefore measuring the uninfluenced electromotive force is impossible **with my tools**.[']

He presented this idea to an older friend. He replied: 'You are entirely right about the interaction with the cell; only, it is calculable as soon as the internal resistances of the voltmeters and the cell are known.' And then he gave him a private lecture on Ohm's law.[7]

After this our young man now attached all of his voltmeters at once one behind the other to the same electric circuit, and using this arbitrary unit, immediately determined their internal resistances from their amplitudes. When he now attached two of his voltmeters one after another to the galvanic cell which had given him trouble, he could determine from the two readings by means of simple calculations firstly, the internal resistance of the cell (within the same unit) and secondly, the desired uninfluenced electromotive force. He needed no new experimental aids for this. He did not use a single conducting wire more than before; but he certainly used new **knowledge**.

Are we physicists of today perhaps in a similar situation in atomic theory to the young man in the tale at his **first** stage?

It seems altogether too risky to draw too sweeping epistemological conclusions from the current state of physical knowledge. Quite apart from the fundamental misgivings about giving up on the possibility of fully explaining nature because no one completely understands how to apply it yet, we should surely at least proceed from a basis which is in itself logically secure and contains no inconsistencies. Unfortunately this cannot be said of current physics. I might be permitted here

[7]Ohm's law links current I, tension U, the electrical internal resistance constant R_i and a variable external resistance R_a according to $I = U/(R_i + R_a)$ (in modern notation). It was discovered by Georg Simon Ohm (1789–1854) and published in 1827.

to point to a profound inconsistency in it.†[8]

The time-honored idea of the spacial extension of matter does not agree with the assumption of elementary particles; and yet both stand side by side in current physics. If an elementary particle fills a space in the ordinary sense, no matter how tiny it is, then it, and therefore automatically any corpuscle, should be able to be subdivided. Furthermore, the question necessarily arises: What happens when the particle fragments shift position relative to each other? (Whether this question is thought ripe for final discussion at the present time or whether it would be better left to upcoming generations of physicists to consider cannot be dealt with here.) We are then already speaking of particles smaller than 'elementary' ones. If we still believe in the existence of elementary particles, then in my opinion we can only justify this by the fact that, based on the experimental results on electron and atom diffraction as well as the interpretations of spectra in wave mechanics and the like, the old corpuscular conception of spacial extension is incorrect; since it is inconsistent with this empirical knowledge. And yet in the literature there is still talk of the radius of an electron in the spacial sense, a clear sign of this **unrecognized** contradiction.§

But nothing could be more disturbing to me than having these representations misunderstood as going against modern quantum and atomic theory. There is, it seems to me, a completely objective scale for progress, which comes with it; it has to do with the so sharply criticized lack of intuitiveness. The times determine what is considered intuitive. A theory that forces traditional conceptions of the outside world to change seems always and necessarily to be unintuitive to the contemporaries of its appearance, mostly even to its authors. It was already so with Copernicus, Faraday and Maxwell. With these lines I would only like to prevent physicists from arriving **on principle** at an '*ignorabimus*'[9] on the brilliant mathematical formalism of atomic theory, precisely because it consciously foregoes answering certain questions at the present time. We definitely do admit that those questions could actually be at least partially unanswerable, i.e., they could make no physical sense. We are only arguing against the conclusion that even different formulations of the questions can **never** lead to a full, causal understanding of physical processes. **The uncertainty relations—in my view—place a limit on all corpuscular mechanics but not on all physical knowledge.**

†[Footnote 1 on p. 440 of the original:] "See also M. von Laue, *Scientia* (Milano) **54**, 402 (1933)."

[8]M. von Laue's article on matter and space cited above is entitled, 'Materie und Raumerfüllung', pp. 402–412.

§ [Footnote 2 on p. 440 of the original:] "That the length $\frac{e^2}{mc^2}$ should take on a different meaning for the electron, is naturally entirely possible."

[9]This refers to the proverbial saying: *ignoramus et ignorabimus* ("we do not know and we will never know") by the electrophysiologist Emil Du Bois-Reymond (1818–1896), which appears in his speech entitled: 'On the limits of natural knowledge'.

When exactly should causality be ranked as 'empirically proven'? When the last scientific problem is totally solved, perhaps? This condition may well never come about. There was certainly a time—it lies some decades back—when this point was believed to be already quite close at hand, in physics at least. At that time, the solutions to all physical questions still open were thought simple, and physics was therefore looked upon as essentially closed. Further developments in physics have cured us of this naive optimism. But we are now succumbing instead to a no less uncritical pessimism in thinking that the task of physics is altogether unsolvable. Despite all the physical pretexts offered for it, it appears to me to be only the manifestations in the field of physics of the overall deep cultural pessimism that forms the underlying sentiment of our times. Dealing with it, however, is no longer the business of the scientist; his science is **above** all human sentiment.

32 Johannes Stark: Personal Evaluations of Gustav Hertz and Richard Gans for the German University Lecturers Association [November 8, 1934]

Source: Bundesarchiv, Section Potsdam, 15.19 Physikalisch-technische Reichsanstalt, J. Stark Correspondence, No. 65, Sheet 171. Also published in excerpt in Swinne [1992], p. 93.[1]

There is nothing Jewish in Professor Hertz's outward appearance, behavior, and scientific activity at the Berlin Polytechnic. He is one of our few first-class German physicists, and is also a Nobel laureate.[[2]] He is, in addition, the nephew of the great physicist Heinrich Hertz and is thus the bearer of this famous name.[[3]] It would be a blunder without parallel to deny this man the right to examine

[1]The German University Lecturers Association (→ *NSDDB*)of the party central leadership J. → Stark in a confidential letter for a personal evaluation of the experimental physicists of Jewish extraction Gustav → Hertz and Richard → Gans: "Please send us an evaluation of Dr. Hertz, physicist at the Berlin Polytechnic and Dr. Gans at Königsberg, from the scientific, pedagogic, political and personal point of view. We are especially interested in information on their present irreplaceability". Cf. Swinne [1992], pp. 92f.

[2]Gustav Hertz had been professor of experimental physics at the Berlin Polytechnic since 1928. He was awarded the 1925 Nobel Prize in physics jointly with J. → Franck for work on the laws governing the collision of an electron and an atom. Stark's opening sentence touches upon the anti-Semitic cliché of external appearance. Only then are his other 'qualifications' discussed, the last of which is the Nobel Prize, which was to Nazi ideologues by no means a recommendation, because Jews were also eligible to receive it. Overall, this passage shows that Stark's anti-Semitism was not so much determined by race (according to the official Nazi doctrine) as by personal ideology. From Stark's point of view certain mentalities defined 'Jewishness'. He thus felt legitimized to agitate against the 'Aryans' W. → Heisenberg and A. → Sommerfeld as 'white Jews' (cf. doc. 56), while defending those Jews who happened to please him; cf. also Schröder [1993], pp. 332f., and Hermann [1966] on Stark's initially positive reception of A. → Einstein, whom he later regarded as the personification of Jewishness.

[3]On Heinrich Rudolf Hertz (1854–1894) see, e.g., doc. 31, and footnote 5 there.

students because his grandfather was a Jew.[4] I am convinced that he would not placidly accept such a personal insult but would resign from his office, leave the country, and be welcomed everywhere with open arms.[5]

Professor Gans in Königsberg cannot boast of any accomplishments as important as those of Hertz.[6] Nevertheless, his scientific papers are of value. He has also steered clear of the Einstein circle. According to information from patriotic quarters that I consider reliable, not only did he act irreproachably from the patriotic point of view in Königsberg, but has even participated actively.[7]

33 Gustav Mie: Letter to Max von Laue [November 20, 1934]

Source: Archive of the Max Planck Society, Berlin: III, Rep. 50, Mie Folder No. 1369, sheet 1. Handwritten signed letter, with sender's address: 'Freiburg i[m]. Br[eisgau].' Wintererstr. 67.'

Dear colleague,[1]

Many thanks for your friendly letter of the 9th! I am naturally entirely of your opinion that in this case we must favor J. Stark as the lesser evil.[2] For all that can be said against him, he is nevertheless still a real expert and not a hollow sham.[3] But now I understand why you were so pessimistic already a year ago,

---

[4]This statement shows how arbitrarily the legal definitions of Jewish descent were applied in practice, especially in the case of 'one-quarter Jews', which is also reflected in the laws themselves: See the strict regulations including 'one-quarter Jews' in doc. 8, amendment to § 3 and doc. 12 for a more liberal version.

[5]It is interesting that Stark considered the broader consequences of Gustav Hertz's leaving. The fact that capable scientists forced into exile ended up working for other countries (later, the 'enemy') was usually not even taken into consideration. Hertz resigned nevertheless on political grounds at the end of 1934 and became director of one of the → *Siemens* research laboratories in 1935.

[6]Richard Gans was forced to resign at the end of 1935 on the basis of the Nuremburg Laws, due to his Jewish descent even though he considered himself an atheist. He was employed as technical science consultant to → *AEG* 1936–39. See Swinne [1992], esp. pp. 91ff.

[7]It is not clear in what kind of patriotic activities Gans had been involved in Königsberg. Stark may have meant Gans's participation at meetings, during which 'patriotic songs' were sung; for example, a 'Hindenburg celebration' that had been arranged by the German nationalist wing of that country's Argentinian colony (cf. Swinne [1992], p. 52). This document, written by one of the most influential German physicists in 1934, illustrates how much weight was given to non-scientific criteria in the evaluation of scientists, even by their own colleagues.

[1]Cf. the following doc. for M. → von Laue's reply to G. → Mie.

[2]This refers to J. → Stark's appointment as the new president of the *Notgemeinschaft der Deutschen Wissenschaft* (→ *NG*)on June 23, 1934 and his speech at the meeting of the members of the → *DFG* on Nov. 10, 1934 in Hanover, published as Stark [1934]c, pp. 13–23. To the surprise of many, Stark emphasized continuity and acknowledged the achievements of Friedrich → Schmidt-Ott, his predecessor. See, e.g., Zierold [1968], pp. 173–212, Nipperdey & Schmugge [1970], pp. 53–60; cf. Stark [1987], pp. 123ff., for Stark's own version of his 'fight for the freedom of research' in this period.

[3]The worst-case scenario many scientists had envisioned was that as minister of the →

while I in faraway Freiburg still knew little of how unfavorable the conditions now are for continued progress in our science, which on the other hand is so indispensable to technology and the economy.[4] We are now really quite worried here about the successor to Hevesy.[5] We would like to have Noddack,[6] but as it seems, there are young sycophants around who also have an appetite for the position and are now doing their business in the dark. By the way, we hear that J. Stark has really borne himself out in this case (from which I naturally don't want to draw any general conclusions). I really feel quite ill at ease already when I think about what will happen here when I eventually leave; and that is already coming soon, as you know.[7] Do you know anything about Jena? M. Wien will now also have to resign soon after all.[8] Who will be his successor, I wonder? Do

REM, Bernhard → Rust would choose a politically loyal nonscientist or an incompetent party comrade for the presidency; cf. Stark [1934]c, p. 14: "In this important matter it is crucial that an experienced scientist take over the direction of the German Scientific Research Association [*DFG*]. I would suspect that the consideration that the German Scientific Research Association's organization and internal management can best be performed only by experienced scientists themselves was also decisive for Reich Minister Rust in appointing me. In any case all of German science is greatly indebted to Reich Minister Rust for having appointed a scientist at the head of the German Scientific Research Association."

[4]This letter illustrates the growing unease among German scientists about the status and effectiveness of science under the new regime.

[5]The Hungarian physical chemist Georg von [George de] Hevesy (1885–1966) studied in Budapest, at the Berlin Polytechnic and in Freiburg in Breisgau 1903–08 and took his doctorate there in 1908 as a student of G. Meyer. 1908–10 he was teaching assistant at the Zurich Polytechnic, 1910 assistant to Fritz → Haber at the Karlsruhe Polytechnic, and 1911–13 assistant to Ernest Rutherford, University of Manchester, working on radioactivity. Then he worked in Vienna at the Radium Institute of the Austrian Academy of Sciences, habilitating in 1913. His military service spanned the years 1915–19. In 1918 Hevesy was appointed professor of experimental physics and head of the 2nd Physical Institute at the University of Budapest University, and received a call as professor of theoretical physics in 1920 at Copenhagen University. There he collaborated with Johannes Brønsted on isotope separation and in 1923 he co-discovered with Dirk Coster hafnium 72. From 1926 Hevesy was associate professor of physical chemistry at Freiburg University, resigning on his own initiative in 1934 (he was of Jewish origin) and returning to the University of Copenhagen. In 1943 he moved to Sweden, where he was employed as a scientist at the Research Institute for Organic Chemistry, Stockholm University. Hevesy received the 1943 Nobel Prize in chemistry for developing the tracer method. On Hevesy, see the obituaries by J. D. Cockcroft in: *Biographical Memoirs of the Fellows of the Royal Society* **13** [1967], pp. 125–166 (with photo and bibliography), and Berta Karlik in: *Almanach der österreichischen Akademie der Wissenschaften für 1968*, Vienna, 1969, pp. 261–267, as well as the biography by Hilde Levi, *George de Hevesy. Life & Work*, Bristol/London: Hilger/IOP Publ., 1985.

[6]Walter → Noddack was appointed full professor of physical chemistry at Freiburg University in 1935.

[7]Mie had been full professor of experimental physics at the University of Freiburg since 1917. He retired in 1936, three years past the normal retirement age of 65, which was soon changed by law, effective March 1935; see doc. 36.

[8]Max → Wien in Jena was two years older than Mie, thus the oldest unretired professor in the field. Wien was emerited at the end of March, 1935, but in fact he remained in office

the positions actually still remain unfilled for years on end? I sometimes think about whether we physicists should not rally around a common cause sometime. It is inconceivable what interest leading politicians actually have in tolerating the intrigues that are taking place everywhere, which will surely soon seriously threaten the existence of science. Shouldn't a united effort be made sometime for Gerlach,[9] for example? Of course, I can't quite judge whether it would be of any use; but it is already beginning to become unbearable to watch all this nonsense [*Unfug*] in silence.

With warm greetings Yours Gustav Mie

34 Max von Laue: Letter to Gustav Mie [November 22, 1934]

Source: Archive of the Max Planck Society, Berlin: III, Rep. 50, Mie Folder No. 1369, sheets 2–3. Unsigned carbon copy of typed letter written in Zehlendorf. Published by permission of Theodore H. Von Laue.

Dear and esteemed colleague,

Your friendly letter of Nov. 20th,[1] which voices thoughts that preoccupy me often now, prompts me to ask first of all: Did you know that a few weeks ago Max Wien wrote to the deans of all the responsible departments for information on existing physics professorships and lesser teaching positions and asking how and whether they were filled currently?[2] This information was requested as data for a petition he, as the most senior physical institute director still in office,[3] intended to submit to a high authority (I don't know whether he was thinking of the Reich Culture Ministry).[4] I assume he has since done so. It is maybe doubtful whether it was right of him to take this step on his own. In any case, some time will now have to be allowed to elapse while we wait for the outcome, even if it were only to consist of a letter acknowledging receipt of the petition.[5]

half a year longer than prescribed by the law mentioned above, until Sep. 30, 1935, because no adequate replacement could be found. Cf. Wittig [1988], p. 518.

[9]It is not really clear what Mie means here, because Walther → Gerlach, who had been full professor in Tübingen from 1925 and since 1929 in Munich as successor to Wilhelm → Wien. In 1935 Gerlach became the head of the Appointment Committee for A. → Sommerfeld's chair for theoretical physics, in 1937 member of the → *KWG*'s supervisory board and in 1943 even head of the physics section of the Reich Research Council (→ *RFR*).

[1]Cf. the preceding document.

[2]For M. → Wien's petition on 'Physics at German Universities' see the following doc.

[3]Max Wien was born on Dec. 25, 1866 and was full professor of physics in Jena 1911–35, thus staying in office until the age of 68.

[4]Wien sent his petition to Bernhard → Rust's Ministry of Education, Science and Culture (→ *REM*) on Nov. 25, 1934; for the cover letter see Hoffmann [1989], p. 190.

[5]This is what actually happened—Wien's first memorandum, dated November 1934, was practically ignored by the official authorities; leading industrialists took some interest in it, though: Cf. Hoffmann [1989], pp. 191f.

I know absolutely nothing about the future appointments to physics positions. Also regarding Hevesy's successor, all that I have heard is that apparently the Munich Polytechnic also has its eyes on Noddack.[6] I consider it completely out of the question that leading politicians have an interest in intrigue. They just don't have the time to bother with such minor details as academic appointments. These are just the consequences of the *Führerprinzip*.

Difficult though it is, in my opinion we must now quietly wait and see. Nowadays I console myself about some delayed appointment with the thought: better unfilled than filled with one of those persons whom you quite correctly describe as 'young sycophants'. The occasion may yet arise, maybe very soon, in which something could be done; now it seems hopeless to me. I presume I may save up your letter for such a later occasion.

But perhaps the physics committee of the *Notgemeinschaft* should do something **now** in another very much more minor matter. Mr. von Ficker informs me,[7] he is very often asked by the *Notgemeinschaft* for geophysical or metereological expert opinions, even though he himself had made them aware of the fact that Schmauss was the expert in the field responsible for such opinions.[8] Since in Hanover *Robustus*[9] expressly announced the retention of expert committees also in their present make-up, the members of our committee could surely come together one of these days on a petition to *Robustus* disapproving of such 'deviations' [*'Extraturen'*]. What do you think about this? Should I compose the petition sometime? And should Schmauss be allowed to sign it as well?

Otherwise, *Robustus* must have hardly been recognizable in Hanover. Three times he spoke in the highest terms of the necessity of **free** research, restricted neither by any sort of bureaucracy nor even by the military's commandeering. He also assured us of his special goodwill towards the humanities. What a pity that it was probably his swan song.[10]

---

[6]See the preceding document on Walter → Noddack's candidacy and for biographical information on George Hevesy.

[7]On the professor of metereology Heinrich von Ficker see footnote 10 of doc. 21.

[8]August Schmauss (1877–1954) studied in Munich 1896–1906 and took his doctorate in 1900, including a brief stay in Berlin in 1901. 1904–06 he was the first assistant at the physical department in Munich and 1906–10 first assistant (*Adjunkt*) at the Bavarian Weather Station. In 1910 Schmauss was appointed director at Munich University, and 1917–22 honorary professor. 1922–48 he was full professor, specializing in metereology and aerology.

[9]A sarcastic nickname for Johannes Stark, whose surname translates as 'strong'.

[10]In Stark's autobiography, the relevant section is entitled 'My fight [*Mein Kampf*] for the freedom of research' (Stark [1987], pp. 123ff.). For Stark's speech on the responsibilities of the → *DFG* (which had been called the *Notgemeinschaft* until 1937) in Hanover on Nov. 10, 1934, entitled 'Aufgaben der Deutschen Forschungsgemeinschaft', see Stark [1934]c, Stark [1987], pp. 126f., 147. In it he argued that the *DFG*'s scientific experts would fulfill the task of coordinating research most efficiently. Stark's sudden emphasis on the freedom of research, in direct contradiction to his speech a year earlier (see here doc. 28), is explained by the fact that the *REM* had made drastic cuts to the → *Notgemeinschaft*'s budget to accommodate the new *Reichsakademie*. On Stark's loss of influence in the subsequent years in the Nazi period see, e.g.,

That at least was the general impression in Hanover. It is already evident from his visit to Planck in the middle of October that he is worried about his position—did I write you about that yet? Soon afterwards his opponent Schumann came to Planck.[11] Isn't that amusing? Each accused the other of the blackest atrocities in the past and of the most sinister plans for the future. I fear they may both be right.

With my best regards to your wife[12] and warm wishes

Yours [Max von Laue]

35 Max Wien: Physics at German Universities [Late November 1934]

Source: Bundesarchiv, Section Potsdam, PTR collection, No. 67, document entitled: 'Die Physik an den deutschen Hochschulen', dated late Nov. 1934,[1] published in the appendix to Hoffmann [1989], pp. 198–205.

Recent speeches and writings have pointed out time and again that it is the primary responsibility of physics to bring relief in the current serious economic crisis. Undoubtedly, the great economic upswing at the turn of the century was based above all upon progress in physics; and German universities had played a major part in these advances. I only call to mind spectral analysis, X rays, and electric waves.

Development in physics depends directly or indirectly on the work conducted at university departments. Obviously, other laboratories, such as the Reich Physical and Technical Institute and large industrial plants, also play a major role in this.[2] But those who perform the research and inventive activities there had previously studied at German universities; and thus the latter are everywhere the basis and source of progress.

It is therefore all the more important that instruction at universities and research in their physical laboratories be maintained at the present level. It is

Beyerchen [1977], pp. 103–122.

[11] Max → Planck was still president of the → *KWG*. On Erich → Schumann, at that time just appointed head of the research department of the Army Ordnance Office (→ *HWA*) and on the conflicts between Schumann and various wings of the Nazi science bureaucracy see, e.g., Stark [1945/87], pp. 124ff.

[12] Mie was married to Bertha Heß (1875–1954), the daughter of the surveyor Friedrich Heß, in Heidelberg in 1901.

[1] This dating is based on the cover letter of Nov. 25, 1934 to Bernard → Rust, Minister of Education, Science and Culture (→ *REM*), to whom this petition was directed. On the same day M. → Wien also sent this text to other influential science politicians including Johannes → Stark, the then new president of the *Notgemeinschaft der Deutschen Wissenschaft* (→ *NG*). Cf. Hoffmann [1989], pp. 189f., and here the two preceding documents.

[2] Research at the *Physikalisch-Technische Reichsanstalt* (→ *PTR*) covered the fields of spectroscopy, photometry, electrical engineering and low-temperature physics.

the responsibility of the academic departments, leading members of the physics profession, and the governments, to see to this.

Most universities now have large and appropriately equipped departments. However, there are still too few teaching positions, and this need weighs particularly heavily when existing teaching positions remain unoccupied for a long time. In the following I would like to draw the attention of members in the field as well as other circles to this problem.

The first issue is: Which physical teaching positions at universities and polytechnics should be considered essential. We begin naturally with the **experimental physics representatives** and the physics department head. The latter is kept very busy with the direction of the department, including examinations, meetings, doctoral theses, lectures, etc.; so he will hardly be able to take on more than the broad, general experimental physics [lecture]. Since the general course is not sufficient for proper physicists, but also for mathematicians, chemists, mineralogists, geologists, etc., 'Advanced Experimental Physics' has gained increasing acceptance as a required course. A **second junior** experimental physicist is required for this, who supports the full professor at the same time with the administration of the department and with the guidance of doctoral theses. This person can be an associate professor or an older unsalaried university lecturer,[3] who would have to receive an official, paid teaching assignment, however. These positions are particularly important, first of all because of the additional teaching responsibilities, but then also because, as a rule, department directors will later be selected from all of these second physicists.

Theoretical physics must be represented either by a full professorship, or by an associate professorship. Considering the scope and importance of theory in modern physics, it is not good to have a younger experimental physicist represent this field, however, as was formerly often customary, with the theoretical associate professorship actually only constituting a transitional position to a full professorship in experimental physics.[4] Only in very rare cases does a younger physicist master experiment and theory simultaneously to such a degree that he can represent both completely.

With the close existing relations between physics and technology, and in view of the fact that very many young physicists go into industry or enter the engineering units of the Armed Forces after they have graduated, it is very desirable to expand pure physical instruction by adding a **representative of technical**

[3]An unsalaried university lecturer (*Privatdozent*) had the right to teach at university, but was only paid a small fee by each course participant. This is an intermediary position usually held by 'habilitated' academics aspiring towards a professorship in their field.

[4]On the gradual institutionalization of theoretical physics and the establishment of full professorships at physics departments in the German-speaking states of central Europe, see Christa Jungnickel & Russell McCormmach *Intellectual Mastery of Nature. Theoretical Physics from Ohm to Einstein*, Chicago: Chicago Univ. Press, 1986, especially Vol. 2 for the years 1870–1925.

physics. Considering physics's large breadth, it is no longer possible to fill this gap in the general course through occasional references to technical applications. Instead, students must be introduced in special courses and practical sessions to electrical engineering, high frequency [physics], thermodynamics for engineers and optics, and maybe also to some areas of military science. In other words: There should be a representative specialist in technical physics at every university. This position falls away, of course, at polytechnics, because there engineers can supplement physical instruction.

Thus accordingly at smaller and medium-sized universities adequate instruction in physics would be achieved with a full professor for experimental physics, a second junior experimental physicist, a theorist, and a technical physicist. At polytechnics the last position would be omitted. At large universities the positions need to be filled multiply. Thus, for example, at the Berlin University there were three permanent positions for experimental physics and two for theoretical physics.[5]

The attached table indicates which teaching positions in physics actually exist at universities and polytechnics, and at the same time, how many of these available teaching positions are vacant at present, i.e., at the beginning of the [19]34/35 winter semester. (It was compiled from official data kindly provided by the deans of the departments.) In the table, first, the existing full professorships, personal full professorships [*persönliche Ordinariate*], and associate professorships are listed. In each instance a plus or minus follows the figure, depending on whether the position is filled or not filled. An additional column indicates the missing positions. After that the teaching positions are again added together, first the existing teaching positions, then the unfilled positions, then the missing ones and the sum of the unfilled and the missing positions. I will come later to the last column, the recruit pool. If to begin with we look at the number of missing teaching positions, we find 34, over and above the existing 100 teaching positions. These 34 teaching positions essentially refer to the second [junior] experimental physicist and to positions in technical physics. Considering the current financial difficulties and the drop in university attendance, we naturally cannot demand right now that 34 new teaching positions be created and filled. But we should at least prepare the way for appointments, say, in that teaching assignments are given to older teaching assistants. However, these would have to be paid teaching assignments in order for the lectures in question to be considered official.

But furthermore, out of the 100 existing teaching positions, 17 are not filled at present, while under normal conditions with natural retirements maybe two or

[5]For a systematic comparison of the faculties (as well as funding, average number of students, productivity, and other relevant categories) at European and American physics institutes around 1900 see Paul Forman, John T. Heilbron & Spencer Weart, 'Physics ca. 1900. Personnel, funding, and productivity of the academic establishment', *HSPS* **5** [1975], pp. 1–185. The editor knows of no similar systematic comparison of faculties covering the years immediately before 1933. For a general statistical overview of university teachers, see von Ferber [1956].

three would be vacant. This state of affairs gives cause for the gravest concern, especially in view of the fact that 7 institute director positions out of 37 are also unfilled; and some have been deserted for years. Everyone familiar with the circumstances knows that a department lacking a director quickly goes into decline and even ruin. This is because the teaching assistants and university lecturers charged with its provisional direction cannot, even with the best of intentions, replace an older, experienced director.

Looking for reasons why so many positions are presently not filled, we might blame an absence of suitable recruits. But this is by no means the case; since, as we see from the last column of the table, there are 33 available older unsalaried university lecturers for experimental physics; nine positions must be filled. Among these 33 there are a good many very diligent younger physicists who could also be appointed at any time as department directors. They obviously perceive the continued delay in their appointment as an occupational hardship; and their working morale suffers. In theoretical physics five positions need to be filled, and there are 9 unsalaried lecturers available, so here also there is a large enough recruit pool. In the case of technical physics there are 3 unfilled positions and 6 unsaleried lecturers. Here suitable individuals from engineering could also be drawn upon in addition.

An insufficient number of professors at universities presents a serious danger, because research and student education suffer greatly. If the education of young physicists in Germany is not at least equivalent to that of other nations, however, then as a consequence we will not be able to maintain our present superior position in science and technology, which after a few decades must also have catastrophic repercussions economically as well.[6] Considering the importance both of physics and of the achievements of German physics institutions up to now, which the *Führer* had emphasized especially in his speech on culture,[7] it is our duty to the people and to the fatherland to maintain the level of our science through complete and good staffing of the teaching positions at German universities.[8]

---

[6]The comparison of the German scientific situation against abroad, especially the USA, figured even more prominently in a later petition, M. Wien co-signed with W. → Heisenberg and H. → Geiger in 1936: See here doc. 49.

[7]This was either Hitler's speech on Sep. 22, 1933 on the Law on the Reich Chamber of Culture (*Reichskulturkammer*, which acted as a censor of all of cultural life), or his speech on Oct. 15, 1933 at the cornerstone ceremony of the German Art Building (*Haus der Deutschen Kunst*) in Munich.

[8]Wien draws on the accepted patriotic themes of 'duty', 'people', 'fatherland', etc., to allay any suspicion that this petition to improve the situation in academic physics might be a fundamental critique of Nazi policy on culture.

Universities	Experimental physics					Theoretical physics					Technical physics					Positions				Recruit pool		
	f	p	a	t	m	f	p	a	t	m	f	p	a	t	m	e	v	m	vm	EP	TP	tP
Berlin	1+	1–	1–			1–	1+	1+ 1+							1	7	3	1	4	6		1
Bonn	1–				1		1+							1+		3	1	1	2	1		
Breslau	1+				1		1+					1+				3		1	1	2	3	
Erlangen	1+				1		1+								1	2		2	2	1		
Frankfort/M.	1+				1	1+					1– 1+					4	1	1	2			
Freiburg	1+				1			1+							1	2		2	2	2		
Giessen	1+				1	1+								1+		3		1	1			
Göttingen	1+ 1–					1–					1+					4	2		2	3		
Greifswald	1+				1		1+							1–		3	1	1	2	1		
Halle	1+			1+		1+									1	3		1	1			
Hamburg	1+				1	1+						1+				3		1	1	2		
Heidelberg	1–				1			1+							1	2	1	2	3	1		
Jena	1+				1	1+					1+					3		1	1	1	1	
Kiel	1+			1–		1+									1	3	1	1	2			1
Cologne	1+				1	1+					1–					3	1	1	2	2	1	1
Königsberg	1+				1		1+		1+						1	3		2	2	1		
Leipzig	1+		1+			1+	1+					1+ 1+				6						
Marburg	1+				1		1+								1	2		2	2	1		
Munich	1+		1+			1–			1+						1	4		1	2	1		
Münster	1–				1	1+									1	2	1	2	3	1		
Rostock	1+				1			1+						1+		3		1	1			
Tübingen	1+				1	1–									1	2	1	2	3			
Würzburg	1+				1			1+							1	2		2	2	1		
Polytechnics	Experimental physics					Theoretical physics					Technical physics					Positions				Recruit pool		
	f	p	a	t	m	f	p	a	t	m	f	p	a	t	m	e	v	m	vm	EP	TP	tP
Aachen	1+	–		1+			1+									3						
Berlin	1+		1+			1+										3				1		
Braunschweig	1+			1+						1						2		1	1			
Breslau	1+									1						1		1	1			
Danzig	1+				1	1+										2		1	1	1		1
Darmstadt	1+				1			1–								2	1	1	2		1	
Dresden	1+		1+					1+								3				1		1
Hanover	1+		1+			1–										3	1		1			
Karlsruhe	1–		1–			1+										3	2		2	1		
Munich	1+		1+	1+						1						3		1	1	1		
Stuttgart	1+			1+		1+										3				1	3	1
Totals for Universities and Polytechnics																						
Vacancies	5	1	2	1		4		1			2				1							
Total:			9					5					3									
Missing					19					3					12							

Tab. 1: Max Wien's survey of physics positions at German universities, 1934. The following abbreviations are used here: **f**: Full professorships, **p**: Personal full professorships, **a**: Associate professorships, **t**: Teaching assignments, **m**: Missing positions; **e**: Existing positions, **v**: Vacant positions, **vm**: Vacant plus missing positions, **EP**: Experimental physics, **TP**: Theoretical physics, **tP**: Technical physics.

36 Law on the Retirement and Transfer of Professors as a Result of the Reorganization of the German System of Higher Education [January 21, 1935]

Source: 'Gesetz über die Entpflichtung und Versetzung von Hochschullehrern aus Anlaß des Neuaufbaus des deutschen Hochschulwesens. Vom 21. Januar 1935', *Reichsgesetzblatt* (Berlin), Jan. 25, 1935, Part I, No. 4, pp. 23–24.

The Reich government has passed the following law, which is promulgated herewith:

§ **1** Professors in the Civil Service of the German Reich are released by operation of law from their official duties at the end of the semester in which they have completed their 65th year of age.[1]

§ **2** If preponderant interests of the university require that a particular professor continue to carry out his official teaching duties, the responsible highest individual state authority [*Landesbehörde*] may, with the approval of the Reich Minister of Science, Education and Culture, move the retirement to a later date.[2]

§ **3** Professors in the Civil Service of the German Reich may be transferred to a chair appropriate to their field at another German university if it is required in the Reich's interest in the context of the reorganization of the German system of higher education.[3]

§ **4** If due to the reorganization of the system a chair is eliminated or is added to another field of study, then the current occupant may be released from his official duties.[4]

§ **5** Retired professors continue to receive their lawful salaries; however, they are no longer promoted. Special remunerations and additional salaries terminate upon retirement.

[1]The purpose of this article was to prevent university professors from teaching past the age of 65, which had become general practice after 1933—cf., e.g., the correspondence between G. → Mie (who taught until age 68, the official retirement age in the Weimar Republic) and M. → von Laue, discussing appointments in 1934, docs. 33–34. On Sommerfeld's case see also doc. 83, footnote 18.

[2]By demanding the explicit approval of the → *REM* for extending academic teaching activity beyond the age limit, the Nazis hoped to gain control of the academic sector more quickly, by allowing only politically 'reliable' individuals to continue teaching past the official retirement age or to assume new positions.

[3]This article was another serious encroachment on academic autonomy, where such obligatory transfers had been unthinkable. The *REM* also frequently used this article to refill crucial academic positions that had become vacant as a result of the blunt and categorical policy of racial and political discrimination. Few scientists refused to comply out of solidarity when designated to replace those who had gone into exile or who had been dismissed. An example of such shuffling of chairs is the cancellation of Richard → Becker's position as professor of theoretical physics at the Berlin Polytechnic in 1936, which was called "superfluous at a polytechnic", and he was ordered to transfer to Göttingen to assume Max → Born's former chair.

[4]This article opened the way to a complete restructuring of universities in accordance with the needs of the new legislators, who envisioned a quite different orientation for 'science'.

Transferred professors continue to receive their lawful salaries. Special remunerations and additional salaries for their teaching activities are redetermined.

§ **6** The Reich Minister of Science, Education and Culture is to take the measures based on §§ 3 to 5 of this law directly.

He is to decree the implementation regulations necessary for carrying out this law and is to specify the remaining rights of professors following their retirement.

§ **7** In the case of forestry colleges, the Reich Forestry Master takes the place of the Reich Minister of Science, Education and Culture.

§ **8** This law becomes valid on the day of its promulgation and expires on the 31st of December, 1937.

Berlin, the 21st of January, 1935.

The *Führer* and Reich Chancellor Adolf Hitler[5]

Reich Minister of Science, Education and Culture Rust[6]

37 Martin Nordmeyer: Letter to Lise Meitner [May 21, 1935]

Source: Meitner papers, Churchill College, Cambridge. Typed signed letter with sender's address: 'Dr. Martin Nordmeyer, 'B[er]l[in].–Lankwitz, Franzstr. 9 b'. Published by permission of Dr. J. Peter Nordmeyer in Hagen.[1]

Dear Professor,

I have chosen after all to wait here for the decision and to leave only on Saturday morning.

I would like to emphasize once more that I naturally subordinate my personal fate to the requirements of the situation as a whole. But since this actually does involve a decision on my future, I do want to present the facts of the matter

[5]In contrast to the laws in the *Reich Gazette* of 1933 (cf., e.g., docs. 7, 8, 12 and 17), these laws were not signed by the Minister of the Interior or another minister, but by Hitler himself, who had gradually focussed the legal and executive powers on to his person, creating a dictatorship (*Führerdiktatur*): On Aug. 1, 1934, for instance, the presidency and the chancellery of the Reich were merged in the 'Law on the Head of State of the German Reich' (*Gesetz über das Oberhaupt des deutschen Reiches*), which became valid on the day Hindenburg died, Aug. 2, 1934.

[6]Bernhard → Rust held this office from 1934–45, but he was one of the less influential actors in the Nazi polycracy: cf., e.g., Seier [1964] on the *REM*'s science policy.

[1]Martin Nordmeyer (1904–1981) commenced his physics studies at Darmstadt in 1923. He later studied under A. → Sommerfeld at Munich, and in 1926 under Rutherford at Cambridge. He submitted his thesis to the Munich Polytechnic with H. Kulenkampff as his adviser. A shortened version of his thesis was later published: 'Untersuchungen zur Stoßwirkung langsamer positiver Ionen in Edelgasen', *Annalen der Physik* **16** [1933], pp. 697–719. 1936–37 Nordmeyer worked at the *Marine Observatorium* in Wilhelmshaven and was married in 1938. 1952–69 he was appointed director of the Spectrometry and Optics Department of the Materials Testing Office in Dortmund. According to Nordmeyer's son Dr. J. Peter Nordmeyer, the sender's address was the residence of Nordmeyer's cousin, the lawyer Dr. Ludwig Bock.

briefly to Privy Councillor Planck, about which Prof. Bavink could not have been accurately informed.[2] Basically, I cannot imagine that a person of the purest German blood, according to the National Socialist world view, should be excluded forever from scientific research work because he felt obliged to follow his conscience in expressing his reservations about compulsory enrollment in the *SA*; and I cannot believe that such a person is not even allowed to volunteer at an institute that has nothing to do with the education of young people.[3]

On the other hand, employment at a Kaiser Wilhelm Institute seems to me to be the only solution, precisely because it has nothing to do with academic instruction[4]—and I have been advised to that effect from all sides. Now it is crucial for me to be reintegrated in some way in science.

With sincere thanks for your attentive interest

I am very truly yours Martin Nordmeyer.

38 Max Planck: Progress Report by the Kaiser Wilhelm Society for the Advancement of the Sciences (April 1934 to the End of March 1935) [June 28, 1935]

Source: Max Planck, 'Tätigkeitsbericht der Kaiser Wilhelm-Gesellschaft zur Förderung der Wissenschaften (April 1934 bis Ende März 1935). Allgemeiner Bericht.' *Die Naturwissenschaften*, **23**, Issue 26/28 dated June 28, 1935, p. 411.

Once again the Society looks back on a year of scientific work, the 24th since its foundation.[1] She is especially grateful to the Reich government, which has

[2]Max → Planck was president of the → *KWG* at the time; the secondary school teacher and natural philosopher Bernhard Bavink (1879–1947) was a personal friend of Planck's and had apparently acted as an intermediary for Nordmeyer, who hoped to be employed in Meitner's physical section at the → *KWIC*. On Bavink's activities during the Nazi period see Hentschel [1993]. For an example of Planck's and the *KWG*'s timid politics of moderate political alignment (*Selbstgleichschaltung*) see, e.g., the following doc. 38 and references there.

[3]This letter shows that sometimes even small courageous acts of civil disobedience could have severe consequences even outside of institutions directly controlled by the State. Enrollment in the → *SA* had become obligatory when the organization in which Nordmeyer was a member was integrated into it in 1935. According to his son, political pressure compelled Nordmeyer to join the National Socialist Party, and he volunteered into the Armed Forces. In 1936 or 1937, he got a position at the Naval Observatory in Wilhelmshaven.

[4]See the preceding doc. 36 for the recent National Socialist legislative efforts to gain full control over the sector of academic teaching; cf. also the Introduction, sec. 3.2.

[1]For a history of the *Kaiser-Wilhelm Gesellschaft* (→ *KWG*) founded in 1911 see, e.g., Burchardt [1975], Vierhaus & vom Brocke (Eds.) [1990]. In May 1933, at the Society's 22nd annual meeting, Planck reportedly said that Germans could not be permitted to stand aside "rifle at rest" and demanded the "consolidation of all available forces for the reconstruction of the fatherland"; see also Planck's retrospective account of his audience with Hitler in the same month (doc. 114), and his foreword in *Forschung tut Not*, Issue 6: *Der Geist meistert den Stoff*, Berlin: VDI-Verlag, May 1933, pp. 1–2, where he emphasized the significant contribution of the research conducted by the Kaiser Wilhelm Institutes to the nation's economy, health and

shown full appreciation of the important problems to be solved and has rated its financial support accordingly.[2] Notably the Kaiser Wilhelm Institute of Physics should be mentioned here, the construction of which can now be begun with the support of the Rockefeller Foundation,[3] once the Reich government has most generously removed the obstacles that still stand in the way by guaranteeing the annual budget.[4] The Institute will be erected on a site in Dahlem (Berlin), which had already been acquired for this purpose some years ago.[5] Prof. Peter Debye, presently still at Leipzig,[6] has been chosen to assume the directorship. With the beginning of this construction a wish that the Society had been fostering for years will be fulfilled.[7]

The Society's foreign relations were also carefully cultivated during the year under review.[8] Aside from the fact that these relations are of great importance to German scientific research, through them many a misleading view abroad about the new Germany can be rectified.[9] Even though the Kaiser Wilhelm Society is not in the position to accept all invitations to conferences of foreign scientific societies, she nevertheless does make an effort to maintain these neighborly contacts. [...[10]]

---

well-being (*Volksgesundheit*; *Volkswohl*); cf. Goran [1967], p. 162.

[2]Cf. the similar paragraph in Planck's progress reports for the periods April 1932–March 33 and April 1933–March 1934, which all demonstrate Planck's active involvement in the Society's policy of self-realignment, in *Naturwissenschaften* **21** [1933], p. 417, and **22** [1934], p. 339.

[3]The Rockefeller Foundation granted 635,000 reichsmarks for the grounds and 1.5 million reichsmarks for the building construction in April 1930. Cf. Macrakis [1986] and doc. 30.

[4]On the *KWG*'s budgets from 1932 to 1944, see Vierhaus & vom Brocke (Eds.) [1990], pp. 377–387. From 1936 'special allocations' (*Sondermittel*) from the Ministries of Aviation and Nutritian covered a growing portion of the Society's budget.

[5]See footnote 9 of doc. 30.

[6]The Dutch physicist P. → Debye was full professor of experimental physics there from 1927.

[7]Cf. footnote 5 of doc. 30

[8]On the contrary, statistics on the visitors staying at the Society's 'Harnack House' in Dahlem (Berlin) illustrate the breakdown of international links with the *KWG*: From April 1933 to March 1934 there were 287 visitors there, 221 of these being German, only 16 American, and 9 English. Between April 1934 and March 1935 there were only 60 visitors, of which 11 were Americans and 20 Englishmen. The fact that Planck found it especially noteworthy to mention that the house was used several times to accommodate courses on genetic biology for physicians and civil servants in medicine and law is characteristic of the *KWG*'s close adherence to the new rules. Furthermore, the building was used for the regular meetings of the National Socialist public welfare organization *Volkswohlfahrt* and the local Dahlem contingents of the → *NSDAP*. See *Die Naturwissenschaften* **22** [1934], p. 349, and **23** [1935], p. 424.

[9]This statement alludes to the negative image the science policy of National Socialist Germany had earned, reflected in foreign scientific journals like *Nature* or *Science*: Cf., e.g., doc. 41, or N.N.[1933]e-f, [1936]b, [1938]c-d.

[10]This sentence is a typical example of Planck's careful and rather indirect criticism of the official science policy, which consisted in attacking the traditional 'internationality of science', favoring instead an autarchical concept of 'German science'. For other examples of Planck's between-the-lines cautions see here doc. 62, footnotes 8f.

39 Philipp Lenard: Foreword to 'German Physics' [August 1935]

Source: Philipp Lenard, Foreword to: *Deutsche Physik in vier Bänden. Erster Band: Einleitung und Mechanik.* Munich: J. F. Lehmann, 1936, pp. IX–XV.[1]

The foreword stems from today's conflict,
The work seeks values infinite.

"**German** physics?" you will ask.[[2]]—I could also have said Aryan physics or physics of the Nordic type of peoples, physics of the probers of reality, of truth seekers, the physics of those who have founded scientific research. —"Science is international and will always remain so!", you will want to protest. But this is inevitably based upon a fallacy. In reality, as with everything that man creates, science is determined by race or by blood. It can seem to be international when universally valid scientific **results** are wrongly traced to a common **origin**[[3]] or when it is not acknowledged that science supplied by peoples of different countries is identical or similar to German science, and that their science could only have been produced because and to the extent that other peoples are or were likewise of a predominantly Nordic racial mix. Nations of different racial mixes practice science differently.

Yet, no people has ever embarked on scientific research without basing themselves on the fertile ground of already existing Aryan achievements.[[4]] Initially, foreigners had always only kept pace with or imitated these. The racial characteristics of these foreign forms only become recognizable after they have developed over a longer period. Based upon the available literature, one could perhaps already talk about Japanese physics.[[5]] Arabian physics existed in the past.

---

[1]The following appears on the title page under the author's name: "Written to delight all who seek intellectual satisfaction [*geistige Ruhe*] in well-founded scientific knowledge." The translated titles of the four volumes are: Vol. 1: *Introduction and Mechanics*; Vol. 2: *Acoustics and Heat Theory*; Vol. 3: *Optics and Electricity Theory, Part 1*; Vol. 4: *Electricity Theory, Part 2*. The dedication on a page of its own reads: "Respectfully dedicated by the author to the Reich and Prussian Minister of the Interior Dr. Frick, promoter of large-scale research in the Third Reich".

An advertisement for Lenard's book containing excerpts from its foreword is reproduced in Sugimoto [1989], p. 125. The heading of the notice adds: "The Nobel prizewinner of physics shows ways to a race-specific [*arteigenen*] understanding of natural science in his new book". See also doc. 45 for a review. Discrepancies with later editions are indicated.

[2]As is already clear from this first sentence, Lenard is attacking primarily the already widely held view that science is an international endeavor independent of the personal characteristics, such as race or background, of the individual researcher. For more on this Nazi concept of 'Aryan science' versus 'Jewish science', see the Introduction, sec. 5.3.

[3]This word is not emphasized in later editions.

[4]Lenard provides his own definition of these 'existing Aryan achievements' in his work cited in footnotes §§ and 30 below.

[5]On the rapid developments in Japan after 1900 see, e.g., James R. Bartholomew, *The*

Nothing has yet emerged about Negro physics. In contrast, a characteristically Jewish physics has developed and become prevalent, which has only rarely been recognized until now, because literature is usually categorized according to the language in which it is written. Jews are everywhere; and whoever still contends that science is international today clearly means unconsciously Jewish science, which is, of course, similar to the Jews everywhere and is everywhere the same.[6]

It is important to examine the 'physics' of the Jewish people a bit here, because it stands as a conspicuous counterpart to German physics, and because for many, the latter will only be brought into the right light by identifying its opposite. As with everything Jewish, Jewish physics also only recently came under some unbiased public scrutiny.[7] It had remained hidden for a long time and developed haltingly. At the end of the war[8] when Jews in Germany began to dominate and to set the tone, the full force of its characteristics suddenly burst forth like a flood. It then promptly found avid supporters even among many authors of non-Jewish or of not really pure Jewish blood. To characterize it briefly, let me best refer you simply to the activities of its undoubtedly most prominent representative, to the unquestionably pure-blooded Jew A. Einstein. His 'relativity theories' attempted to transform and dominate the whole of physics; but they have now already completely played themselves out against reality.* [9] Apparently they never were even intended to be **true**. The Jew conspicuously lacks any understanding of **truth** beyond a merely superficial **agreement with reality, which is independent of human thought**. This is in contrast to the Aryan scientist's drive, which is as obstinate as it is serious in its quest for truth.[10]

The Jew has no noticeable capacity to grasp reality in any form other than as it appears in human activity and in the weaknesses of his host nation [*Wirtsvolk*].[11]

Formation of Science in Japan, New Haven: Yale Univ. Press, 1989, or Erwin Müller-Hartmann (Ed.) *Die Geschichte der Physik in Japan*, Stuttgart: Steiner, 1984.

[6]This is a typical prejudicial generalization held by Nazi ideologues and many contemporary ethmologists, falsely assuming that Jews share unique racial characteristics throughout the world and that their culture is completely independent of their local sociocultural environment.

[7]The suspicion that Jewish press agencies and newspaper staff were conducting a deliberate propaganda campaign in an attempt to manipulate the masses was frequently raised by right-wing ideologues: See, e.g., Gehrcke [1924]a, footnote †† of Glaser's anti-Semitic tirade (doc. 77), or Stark's accusations in doc. 56, p. 160.

[8]In the third edition, the year 1918 is specified here in a footnote.

* [Footnote 1), p. x of the original:] "It goes without saying that a discussion of this misguided theoretical system has no place in this work. It will become evident that this does not leave a gap in our comprehensive and interrelating presentation of general scientific knowledge. The same also applies to other unrealistic or unreliable 'theories', which will remain unmentioned. The insignificance of their absence is best supported by their own unauthoritativeness."

[9]On Lenard's agitation against A. → Einstein's theories of relativity see, e.g., Lenard [1918/21]; cf. Hentschel [1990], secs. 2.3 and 3.2 and references there.

[10]The following paragraph is printed in smaller type in the original.

[11]The term *Wirtsvolk* also evokes the common anti-Semitic image of Jews as 'parasites'.

Astonishingly, truth and reality do not appear to be anything at all special or different from untruth to Jews, but are equivalent to any one of the many different theoretical options available.†[12] It is obvious that this attitude is thus totally inappropriate for science; yet this fact was concealed through computational tricks. The characteristic audacity of the uninhibited Jew, together with the deft collaboration of his fellow Jews, enabled the construction of Jewish physics, which already fills libraries. The characteristic haste of the Jewish mentality to come up with untested ideas was actually contagious; though it provides personal advantages (Jewish applause, priority), it has a negative effect on the whole. The great Aryan scientists were reluctant to come forward with contestable conclusions. On the contrary, they calmly turned primarily to testing their new ideas against reality, in order to arrive at acknowledged facts, not assumptions. Thus publications of rich and novel factual content appeared, each itself a milestone in the progress of scientific knowledge. In Jewish physics every assumption that later proves not to be completely false is already considered a milestone. But with such an evaluation the Aryan approach is abandoned, and this has had very striking consequences. The alien spirit has a paralysing effect. Everything that is racially foreign is damaging to the German nation.

Jewish 'physics' is therefore only an illusion and a degenerate manifestation of fundamental Aryan physics. It was necessary to bring this up explicitly here, since we can only recover full appreciation of the latter by clarifying the antithesis between Aryan and Jewish physics.

This revival is necessary, however, regardless of whether you expect a realistic **conception of the world** in conformance with science—as opposed to illusory verbiage—or, say, **technological advancement**, which is so important for Germany's existence. The greatest technological achievements have arisen without exception from investigations and findings made by no means for their practical use, but solely out of the Aryan scientist's drive towards an ever better, tested understanding of nature. It is thus important not just to improve technology[13] but primarily to keep the Aryan researching spirit alive that constantly comes to

† [Footnote 2), p. x of the original:] "Obviously there are Aryans who also lack a feel for truth, its value and its tremendous importance, as well as for the possibility of investigating it unambiguously. But in the case of Aryans this is doubtlessly based chiefly on ignorance, which can be adequately explained by the bad state of scientific instruction at schools and which is apparent in the many 'new worldviews' that are continually being thought up without regard for the existing reliable knowledge."

[12]In footnote † above of the second edition, the words "bad state of scientific instruction" in general is emphasized (*Unterrichtswesen* replacing *Schulwesen*) with the addition "(see next page)". An example of such worldviews is provided with the parenthetical: "(E.g., 'Cosmic ice theory')". In the third edition, the following sentence is added: "This widespread ignorance originating in schools applies just as much to facts of scientific knowledge as to the mentality of successful Aryan scientists, who should be role models in preventing thoughts from flying to flippant starry-eyed fantasies."

[13]"to improve technology" is emphasized in subsequent editions.

our rescue with new fundamental knowledge. The great investigation of nature itself, which had always produced unexpectedly valuable results, has already gone far too much astray in Germany, due to the cultivation of the oppositely disposed Jewish mentality that only looks for superficial 'verification' of predictions.[14]

Now that Germany is taking the lead in making some attempt to rescue the Aryan spirit on earth, we can surely call this mentality the 'German' spirit and speak of German physics with respect to its effect on scientific research. **The German nation has a right to resolutely foster its own nature in science as well.** And this is so not only for the sake of the Fatherland, but because in this way we are cultivating the best that mankind has to offer. Helping this along is the aim of this work. What remains to be mentioned in advance about the book's organization and about the deficiencies it hopes to redress, is based almost entirely on the contrary racial natures just described and on the already very extensive, yet far too little noticed influence of the Jewish spirit on science, which has also already installed itself securely in public instruction.

The unspoiled German national spirit [*Volksgeist*] seeks depth; it seeks **theoretical foundations consistent with nature**, and irrefutable knowledge of the cosmos. Surely it should be able to refer to the results of scientific research in this context. In general, however, we can hardly get any of this at school and university.[15]

Physics teachers at schools are completely glutted with too much new mathematics to be able to think along with nature in simple terms; and increasingly rapidly physics professors are becoming exclusively specialists, who have no personal connection with the general fundamentals that they should be presenting in the main.[16]

The same applies to books. They are almost without exception not at all equipped to relay simply information on the fundamentals and on their reliability. And yet everywhere doubts are being expressed about the validity of the foundations of physics. It is being assumed that all knowledge is uncertain and needs to be revised, as though it were a matter of course to consider what we have inherited from the great scientists as useless. Anyone wanting to see unquestionably verified knowledge separated from the dubious usually searches in vain in books. There knowledge of nature is often presented, seemingly intentionally, as an almost inpenetrable jumble of ideas that can only be understood by a true specialist. This knowledge thus appears to be a heap of patchy work, theories that are usually only hypotheses and that resemble oracles that give prophecies of dark, mathematical origin. In addition, usually mostly the newest and most

[14] "The great investigation" and "in Germany, due to the cultivation of the oppositely disposed Jewish mentality" are emphasized in later editions.

[15] "school and university" is emphasized in subsequent editions. On physics instruction at schools see the Introduction at the end of sec. 3.2.

[16] The following two paragraphs appear in smaller type in the original.

uncertain but superficially pleasing ideas are brought to the forefront; and calculations almost always conceal the essential points. Ordinary small books are completely inadequate, because they only string together the minimum knowledge generally needed for examinations. Larger physics textbooks mostly present everything that some author or other had propounded or thought, as if paying tribute to authors, especially living ones, were its main purpose. Other books cater more to technicians and indulge in rough illustrations but remain unsatisfactory because they neglect the deeper issues.

The present book has a different aim: to provide precisely **these missing things** as described, but for which I have received a lot of requests. Otherwise I would not have taken the trouble to write this complete revision of my former lectures.

All the well-verified knowledge of inorganic nature can be found here in a uniform and totally coherent context; and it is carefully supported. Occasional glances at living nature do occur; but they must be kept within modest limits, inasmuch as there is no adequately precise measurement-based (exact quantitative) control in the case of living organisms. It was just this aid that permitted inorganic nature to become understood so well. Yet it should become apparent with this publication that it is **possible** to make such a **uniform presentation that is properly backed up throughout**; for, the current dominant opinion is to the contrary. Here we again find nature as **an interconnected whole**—as far as only thorough, mature investigation is already available. This is precisely in keeping with how the great scientists had long thought and could prove with their own work to clarify obscure areas.

We attach great importance everywhere to **convincing proof of the validity of findings**. This proof is always best provided by presenting the method by which the finding was made through the observation of nature. That is why we often follow the chains of reasoning of the time at which each new finding was made. Boldly advanced assumptions have often made more of an impression, sometimes lasting long periods of time, than they have been fruitful. At all times this was a sign of intellectual weakness and was damaging to science. Since this is just as applicable still today, we always emphasize scientific observations and the conclusions based upon them, each of which has led to a determination. We test these determinations further by searching out the relations between the new discoveries and other already well-verified knowns. We can only concede as fact those findings which have such a basis and which withstand such verification. We do not waste our time with contradictions or even with the obviously unrealistic.§[17] We do not proffer here a handbook for researchers, who of course

§ [Footnote 1), p. xii of the original:] "Newton said, '*Hypotheses non fingo*' (I do not allow a simple assumption—a derivation not related to the phenomena—to pass off as science)."

[17]In footnote § Lenard quotes Newton's 'Scholium Generale' in *Philosophiæ naturalis principia mathematica* [1687]. For a commentary on this statement, which referred to the lack of a fully developed mechanical model for gravity and indicates Newton's disdain for hypotheti-

have to constantly consider all kinds of assumptions; nor is this a book for curious 'know-alls'. Rather we want to present the solid foundations and the already broad scope of secure scientific knowledge.

Scientific knowledge is gained without exception on the slow path of step-by-step refinement whose reliability improves gradually. Thus only old and **therefore sufficiently tested** material is at all worth widespread attention. This has been completely overlooked recently—not to mention, how this all came about. Indeed, the exact opposite has been happening. 'German physics' does not concede such a distorted need for the latest uncertainties—which all newspapers are full of;[18] this need is also not at all German. We, however, work our way steadily through all the more recent solidly based findings—starting from facts that were verified long ago—to the limits of the coherently related secure knowledge of today. Aspects of these always clearly defined limits might well become uncertain—yet, the distant barely visible horizon is quite different to fog in the foreground, which robs the eye of all certainty.[19]

The currently very common division of physics into two sections, 'classical' and 'modern' physics, also seems to be such a misty area: An artifice presupposing historical ignorance. It is a part of Jewish physics,‡[20] since, the Jew wants to create contradictions everywhere and to separate relations, so that preferably, the poor naive German can no longer make any sense of it whatsoever. 'Modern' physics usually consists in an exaggerated exposition of as yet unclear, unfinished findings that obviously do not have to agree in every way with already well-founded knowledge; but the latter is then called 'classical' with an aftertaste of having outlived its time or of having become obsolete.

Many of the newer findings are related to the previously inaccessible interior of the atom. It is obvious that these results must be novel; since, even the knowledge we had before about the impenetrability of matter, which is subdivided into atoms, already demonstrated the peculiarity of the interior of the atom in comparison to the surrounding space. The space outside of the atom is easily penetrated by other atoms; the interior of atoms is not.[21]

---

cal explanations, see, e.g., I. Bernard Cohen's *Introduction to Newton's Principia*, Cambridge: Harvard University Press, 1978, pp. 240–245.

[18]The 'latest uncertainties' (*neuesten Unsicherheiten*) might well also be an allusion to Heisenberg's uncertainty relations proposed in 1927 and widely discussed ever since.

[19]The remaining text, aside from the closing paragraph preceding the acknowledgments, is printed in small type in the original.

‡ [Footnote 2), p. xii in the original text:] "In some applications the term 'modern physics' seems altogether synonymous with Jewish physics, which as mentioned above certainly has a quite recent, 'modern' origin."

[20]The following extra paragraph is added at the end of footnote ‡ in the third edition: "The division of physics into 'experimental' and 'theoretical' physics is also one of these Jewish smoke screens (see the following footnote and the Introduction 12.)" (Subsection 12 there is entitled: 'Theoretical or mathematical physics').

[21]Lenard and Stark attempted to construct classical models of atoms and their interaction

We will acknowledge other, in part only preliminary, newer observations as well at the relevant place,[22] but always only as far as specific facts and usable relations are supported by already secure knowledge. Three most universally important things will be mentioned here in this regard, which we base individually and in detail on experience:[23] 1. The view that all masses are energy masses [*Energiemassen*] and that gravitation is an effect of energy;[24] 2. The assumption that each energy mass has its own ether; 3. The theory that all energy involves electromagnetic force fields.[25] The overall content of our description will show that these three things are linked sufficiently reliably not only to individual observations but also to all previously held knowledge.¶[26]

The **use of mathematics** in the present work requires special comment.

Only elementary mathematics, simple (Euclidian) geometry and trigonometry and some basic knowledge of conic sections will be assumed. The gist of infinitesimal calculus cannot be dispensed with. We explain it—which is, by the way, very simple—in just those cases that at one time gave rise to the development of this form of calculation with infinitely small values, the origin of which lies completely within science. The pertinent calculation rules are only necessary if a given calculation with infinitely small values is required using the simplest method. But this execution is mathematical craftsmanship and has absolutely nothing to do with the purpose of the calculation—which is the essential point. For this reason we only apply knowledge of these calculation rules in the **appendices**, which taken together, offer a basic summary of all characteristic essential

with the surrounding ether. Cf., e.g., Lenard [1910], [1920/21], and Stark's textbooks cited in doc. 18, footnote 12.

[22]In the third edition 'acknowledge' (*würdigen*) is replaced with the more neutral 'mention' (*erwähnen*).

[23]This sentence along with the rest of the paragraph and incorporating the footnote is made into a separate paragraph in larger type in the later editions.

[24]This is Lenard's reformulation of Einstein's famous energy-mass equivalency.

[25]These points sum up Lenard's own ether theory which is laid out in Lenard [1910], [1918/21], and [1920/21].

¶ [Footnote 1), p. xiii of the original text:] "By introducing these three assumptions thoroughly we go beyond what has been offered up to now in the summarizing literature."

[26]The clause: "which illustrates that they actually do belong within German Physics." ("wonach sie tatsächlich in die Deutsche Physik gehören.") is added in the third edition.

points of 'theoretical' or 'mathematical' physics.**[27] It thus becomes apparent how very modest the scientist's complete mathematical equipment is. The larger amounts of calculations that usually appear in more detailed works offer nothing new with regard to scientific knowledge. If the observational basis of any conclusion is incontestable and if the means by which the conclusion is arrived at is explained in words, then the correct presentation of the appropriate formula can be a matter of course. The detailed execution of a particular calculation teaches nothing more than calculational skills which is better placed in books on mathematics.

Thinking along with nature—following its processes systematically—**is very seldomly done correctly**; usually you are confronted with formulas instead. It is peculiar to see physics texts filled with mathematical derivations that offer absolutely nothing about the origin, value and significance of the topic under discussion and only prompt you to go over the figures again, which is superfluous, of course, while certainly making an easy job of it for the author. To me, authors like that seem similar to a composer, for example, who discusses the technique of his musical instrument instead of appealing to the mood his artistic work attempts to convey.[...[28]]

Only the basics are given in engineering as well as metrology. We do not concern ourselves with all sorts of expressions and terms that technicians and specialists have introduced and use in their papers. They are not necessary to understand natural science. This arbitrary vocabulary only distracts. Their absence in the alphabetical index (Volume 4) distinguishes them as such.††[29]

** [Footnote 2), p. xiii of the original:] "The only scientific knowledge lost to those who prefer to disregard the appendices is a more detailed examination of how the quantitative relations, from the simplest to the most comprehensive ones, are generated. The purpose of all these relations will have already been explained in the main text without many formulas. The opinion that 'mathematical' or 'theoretical' physics is the more advanced or main aspect of physics is propagated by mathematicians who like to see their art applied. How very superficial mathematical application often is, however, is illustrated by the expressions used to indicate calculation results, such as: 'mathematically' it is like this, but 'physically' like this; 'theoretically' like this, and 'practically' it is different again."

[27]The last sentence in footnote ** is removed and the preceding one is changed in the third edition to read: "The opinion that 'mathematical' or 'theoretical' physics is the more advanced or main aspect of physics or is even just a special part of it, demonstrates a limited understanding of scientific knowledge (see Introduction 12)."

[28]The following two paragraphs agaom describing the contents of his book in general terms against other textbooks and arguing against too much detail are omitted.

††[Footnote 1), p. xiv, of the original:] "It should also be noted that we do not adhere to the extreme attempts to Germanize [*Verdeutschung*]. It is just too trivial when set against the lack of substance. Modern German scientific papers stand out more for their un-German mentality than for un-German vocabulary. However, I have completely avoided terms that would have to be read with a foreign pronunciation (such as, 'niveau', for example)."

[29]This sentence is removed in the subsequent editions, with footnote †† moved to the preceding sentence. The omission of 'modern' technical terms not only reflects Lenard's support for

Historical material is only touched upon as far as it comes up of its own accord in closer analysis of the fundamentals.§§[30] Authors are only rarely named, only in the case of great, fundamental advances.[31] References to more contemporary individuals are out of the question, because it is impossible to come to a final decision on the still living and on the only recently deceased.‡‡ Where this is done, frequently the wrong names are actually mentioned the most often.

Although we have thus omitted a lot, this *German Physics* tome has nevertheless become more voluminous than originally intended. Had we only wanted to summarize what has been generally attained on the whole toward an understanding of nature and a knowledge of the macrocosm, a modest volume would indeed have been sufficient. But without the detailed supporting explanations offered here, such a book could not hope to stand out and be understood today; since, it is clearly already generally assumed in the Jewish muddle that no secure intellectual foundation toward understanding natural science is available at all.

In our times of overabundant technological benefits from science along with the universal enjoyment of them, it is truly a sign of great cultural degeneration [*Unkultur*] that so little is understood—or even known—about the intellectual achievements in science, as the writings of today reveal.[32] For already 30 years now the German nation has been fed with the achievements of a racial and cultural alien and with that of his supporters and successors;[33] and this is still going on. But a people that has produced the likes of Copernicus, Kepler, Guericke, Leibniz, Fraunhofer, Rob[ert] Mayer, Mendel, Bunsen and Kirchhoff, will know how to find itself again, just as it has again found a *Führer* of its own blood in politics as heir to Frederick the Great and Bismarck, who saved it from the chaos of Marxism, which is equally alien racially.[34] It was with this confidence in our people that I had written this work; and it is with a special confidence in the

Nazi efforts to reform the German language, but also is an attempt to conceal the fact that his textbook was simply not up-to-date. See in this respect → von Laue's critical review, doc. 45.

§§ [Footnote 2), p. xiv, of the original:] "For a thorough historical account *Große Naturforscher* is available with detailed alphabetical indices at the back to facilitate quick consultation (J. F. Lehmann Publishers, 2nd edition, 1930)."

30 Lenard's hagiographical work on 'Great Scientists' cited in footnote §§ attempted to demonstrate that all truly 'great' scientists were Nordic or 'Aryan'. The 3rd edition (1937) and the 4th edition (1941) are referenced in the second and third editions, respectively.

31 This and the following sentence along with its footnote are removed in the third edition.

‡‡ [Footnote 3), p. xiv, of the original:] "We place approximately the same time limit for name references as in *Große Naturforscher* (see the Foreword there, p. 7)."

32 "about the intellectual achievements in science" is emphasized in the subsequent editions.

33 Albert Einstein was this so-called 'racial and cultural alien', whom he considered the incarnation of 'Jewish physics'—Einstein's *annus mirabilis* was three decades earlier, in 1905.

34 The fear of Communism was just as alive in Europe at this time as it was 20 years later in the U.S. Lenard sees himself as a *Führer* leading the way out of the 'chaos' within physics caused by 'racially alien' scientists; Stark's speech at the inauguration of the 'Philipp Lenard Institute' in Heidelberg (doc. 40) illustrates that he was accepted as such by his readers.

German nation's leadership in the Third Reich that I publish it. [...[35]]
Heidelberg, August 1935.

40 Johannes Stark: Philipp Lenard: An Aryan Scientist. Speech at the Inauguration of the Philipp Lenard Institute in Heidelberg [December 13, 1935]

Source: Johannes Stark,[1] 'Philipp Lenard als deutscher Naturforscher. Rede zur Einweihung des Philipp-Lenard-Instituts in Heidelberg am 13. Dezember 1935', *Nationalsozialistische Monatshefte*, Vol. 7, Issue No. 71, February 1936, pp. 106–112.[2]

From now on the physical institute in which we have gathered together today in celebration will bear the name 'Lenard Institute'. This is a distinction for the institute and a caution to all its visitors, indeed to all German scientists. It is an honor for me personally to be given the opportunity to point out what the name Lenard Institute should alert German scientists to in the future.

Certainly, this institute's new name is also meant to pay tribute to Lenard himself, since he had built the institute and had worked there successfully for many years as a researcher and teacher.[[3]] But the name Lenard Institute should be more than just in honor of and in memory of the man himself. It should place Lenard the scientist before all German scientists as a great model to be emulated. From Lenard's example they should learn how the scientist can acquire knowledge and make discoveries. From Lenard's example they should be encouraged to fight steadfastly for truth and against falsehood.

Certainly all scientists concur in the desire to acquire new knowledge or indeed to make great discoveries. But they differ greatly in their choice of the ways to arrive at this goal; and these days they usually go astray. A large group of people, primarily in physics, believe that to be able to arrive at results, or at least to come up with impressive articles or even with sensational formulations, they must produce a mathematically lavishly dressed theory or base their work on the formulas of such theories. This type of approach is consistent with the Jewish peculiarity of making their own opinion, their own desires and advantage into the measure of all things and thus of scientific knowledge as well. Jewish physics, which had

[35]Lenard's acknowledgments regarding the preparation of the manuscript are omitted here.

[1]The author's name is followed by J. → Stark's official title as president of the *Physikalisch-Technische Reichsanstalt* (→ *PTR*) and of the German Scientific Research Society (→ *DFG*), formerly called the *Notgemeinschaft* (→ *NG*).

[2]This is the published version of a speech about Stark's colleague and friend P. → Lenard. On the inauguration see also 'Einweihung des Philipp-Lenard-Instituts' in *Völkischer Beobachter*, No. 349, Dec. 15, 1935, p. 5, cols. 2–3, and N.N. [1936]a.

[3]Lenard taught at the University of Heidelberg between 1896 and 1898, and then again continuously from 1907 as successor to Quincke and director of the local physics and radiology laboratories. The labs now being renamed 'Philipp Lenard Institute' had been built in 1914 under his supervision.

come about in this way in the last three decades and which has been practiced and propagated by Jews as well as by their non-Jewish students and emulators, has logically also found its high priest in a Jew, in Einstein.[4] Jewish propaganda has tried to portray him as the greatest scientist of all time.[5] However, Einstein's relativity theories were basically nothing more than an accumulation of artificial formulas based upon arbitrary definitions and transformations of the space and time coordinates. The sensation and propaganda of the Einsteinian relativity theory was followed by Heisenberg's matrix theory and Schrödinger's so-called wave mechanics, one as impenetrable and formalistic as the next.[6] Despite the accumulation of mountains of this kind of theoretical literature, however, it has contributed no important new knowledge of actual fact in physics.[7] This could not have been otherwise; since its point of departure, formalistic human opinion, was false.

Jewish formalism in science must be rejected under all circumstances. The overemphasis on technology [*Technizismus*] in science is not as damaging, but is also basically unfruitful or is at least unable to lead to great discoveries and new knowledge. In the last four decades technology has made extraordinary progress in the construction of measuring instruments and machines as well as in the development of new methods. For physics this development is illustrated by comparing the stout volume of the latest edition of Kohlrausch's *Praktische Physik* against its slim first edition.[8] The development of technology in physics has become so

[4]Stark had not always been opposed to Albert → Einstein: On his previous support of Einstein's light quantum hypothesis and the reasons for the later drastic change in their relationship. which deteriorated into downright hostility, see Hermann [1966].

[5]Such Jewish propaganda allegations and the singling out of Einstein as the paradigm of the style of physics adherents to *Deutsche Physik* abhorred are recurrent themes in such texts.

[6]Extending antirelativistic sentiment to include W. → Heisenberg's and E. → Schrödinger's new versions of quantum mechanics, introduced in 1925, was a common tactic. This is surprising insofar as Schrödinger's approach to physics in particular was by no means as abstract and formalistic as Stark claims here, since Schrödinger's original aim had been to reduce the new quantum mechanics to well-known eigenvalue problems of continuum mechanics, illustrated by a whole series of very concrete visual images: Cf., e.g, Moore [1989], chap. 6. For historical accounts see, e.g., Edward MacKinnon 'Heisenberg, models, and the rise of matrix mechanics', *Historical Studies in the Physical Sciences* **8** [1977], pp. 137–188; cf. Jammer [1966], chap. 5, and Mehra & Rechenberg [1982], Vols. 3 and 5.

[7]Any textbook on the history of quantum mechanics will contradict this charge. Quantum mechanics did in fact help in understanding many previously unexplained empirical facts (for instance, hydrogen and helium spectra, ferromagnetism, and specific heats), and it predicted new phenomena (e.g., tunneling): Cf. references in the preceding footnote.

[8]Friedrich Kohlrausch's textbook, first published in Leipzig by Teubner in 1870 under the title *Leitfaden der praktischen Physik. Zunächst für das physikalische Prakticum in Göttingen*, consisted of 123 pages in addition to the front matter to p. xxxi. The 17th edition, reworked by Fritz Henning and published in 1935 by the same publishing house under the title *Praktische Physik. Zum Gebrauch für Unterricht, Forschung und Technik*, had grown to the imposing size of 958 pages of text, including front matter to p. x and 512 illustrations; and it continued to expand: The 19th edition of 1943 appeared in 2 volumes of 544 and 578 pages each.

diffuse and detailed that a scientist prepared for simplicity and clarity could almost become revolted by it. Lenard made a striking comment to me once in this regard. It was in the autumn of 1921. We had gone together to see the exhibition on physical instruments and machines which was linked to the physics conference at Jena. Lenard went along silently beside me through the exhibition. When we finally made our way to the exit and were climbing down the stairs of the exhibition building, he broke his silence with the exclamation: [“]This highly developed technology is madness![”] We certainly understand what Lenard meant with his exclamation, namely, that should technology go too far, the advancement of scientific knowledge could be threatened. Indeed, quite a few experts in all sections of science assume that being at the cutting edge of technology is the precondition to acquiring new knowledge or even to making new discoveries. I in particular, as President of the German Scientific Research Association,[9] often have the occasion to observe this overestimation of technology in grant applications for the most expensive of instruments. I regret such applications not simply because they make heavy demands on the financial means of the Research Association, but because I detect in them the danger of rating technology too highly. Advances in the development of measuring instrument technology certainly alleviate observation and permit an increase in precision. Even experts who do not possess the talents of a scientist can produce and amass observational data with their aid. But this alone does not guarantee an advancement in research. Indeed, sometimes it even causes a step backwards, in that the performance of a task is guided by the available technology, instead of the technology being newly developed to fit the special features of the task. And it is instructive that the great scientists themselves used simple technical aids in making their discoveries.[10]

Against the misguidedness of mathematical formalism and the one-sided overemphasis on technology [*Technizismus*], Lenard's example teaches us the right way to conduct scientific research. He justifiably sees reality as the point of departure and also the final goal of scientific research. The scientist should not keep to himself; he should not look at processes and states in nature through the distorting or blurring mirror of preconceived opinions or formulas. He should rather get to the bottom of real nature and should observe and measure. He should solve questions that arise from observation through suitably conducted experiments under properly defined conditions. His final goal is to understand the conformity of observed phenomena with natural laws, and where applicable, to discover and identify reliably a process or body that had been previously unknown. Whether and how the new knowledge can be formulated mathematically or exploited technically and commercialized is of subordinate importance to him.

Lenard's great discoveries arose from this mentality of the true scientist. I

[9]Stark was president of the *Deutsche Forschungs-Gemeinschaft* (→ *DFG*) June 1934–1936.

[10]Consequently, during his time as president of the *PTR* and head of the *DFG*, Stark supported several experimental projects of questionable scientific value. His unsuccessful project to extract gold from German swampland is the most famous; cf. e.g., Zierold [1968], pp. 188–190.

would like to illustrate this with only two examples.

Cathode rays had already been observed and described by Hittorf as a luminescent phenomenon of glow discharge; but he offered no explanation of its character. Crookes's almost playfully conducted manifold experiments left its nature equally unexplained.[11] Then Lenard, the observer and scientist, came along. He posed the experimental problem of the nature of cathode rays. He recognized that to obtain a reliable solution to this question it was necessary to direct the cathode rays away from the vicinity of the source electric current. He guided it into a special testing chamber where its characteristic behavior towards matter and in electric and magnetic fields could be analyzed. And as a scrupulous scientist should do, he carried out his observations and measurements under well-defined experimental conditions. In order to eliminate possible electrical influences on the part of the cathode ray tubes, he encased the tubes in a housing of sheet zinc with an aluminium window through which the cathode rays emerged. By using such care and such clean experimental conditions, Lenard laid a secure basis for the discovery of the novel nature of cathode rays.[12]

Lenard proceeded similarly in explaining a phenomenon already familiar to him of the tendency of ultraviolet light to induce the release of negative electricity from the surface of certain metals. Under well-defined and appropriately chosen experimental conditions he investigated the nature of the negative electricity carrier detected in this phenomenon, which metal surfaces emit in reaction to light; and he identified it as slow cathode rays. Furthermore, he set the experimental problem of the relationship between the velocity of photoelectric cathode rays and the intensity and color or wavelength of actinic light and thus made the fundamental discovery that the velocity of photoelectric cathode rays is independent of the light's intensity and is determined solely by its frequency.[13] It is typical that Einstein later formulated this discovery of Lenard's mathematically,[14] and

[11]See Julius Plücker & W. Hittorf: 'On the spectra of ignited gases and vapours', *Philosophical Transactions of the Royal Society London* **155** [1865], pp. 1–30; William Crookes: 'On radiant matter', *Nature* **20** [1879], p. 419. For a survey of early cathode-ray physics see, e.g., J. J. Thomson: *The Corpuscular Theory of Matter*, London: Constable, 1907, chap. 1.

[12]Cf., e.g., Lenard's papers: 'Über Kathodenstrahlen in Gasen von atmosphärischem Druck und im äußersten Vakuum', *Annalen der Physik* (3) **51** [1894], pp. 225–267; 'Über Kathodenstrahlen', Nobel lecture, Leipzig, 1906; *Über Kathodenstrahlen*, Berlin & Leipzig, 1920. However, Stark carefully talks around the fact that it was not Lenard but J. J. Thomson at the Cavendish Laboratory in Cambridge, England, who demonstrated that cathode rays consist of electrons as elementary particles with a specific amount of electric charge e and a specific charge-to-mass ratio e/m. See, e.g., Isabel Falconer, 'Corpuscles, electrons and cathode rays: J. J. Thomson and the 'discovery' of the electron', *British Journal for the History of Science* **20** [1987], pp. 241–276.

[13]Cf. P. Lenard's contributions in the *Annalen der Physik* (Series 4), Vol. 1 [1900], pp. 486–507, **2** [1900], pp. 359–375, **3** [1900], pp. 298–319, **8** [1902], pp. 149–198, and **12** [1903], pp. 449–490. Cf. also Bruce R. Wheaton: 'Philipp Lenard and the photoelectric effect, 1889–1911', *Historical Studies in the Physical Sciences* **9** [1978], pp. 299–322.

[14]See Albert Einstein: 'On a heuristic point of view concerning the production and transfor-

received a Nobel Prize especially for this formulation.[15]

Thus Lenard should serve as a model scientist for us and for coming generations because of his approach to discerning reality in nature. But he should also be our model because of his convictions; for, the true scientist should acknowledge openly and courageously the insights he has gained. He should identify as erroneous and combat with equal frankness and courage the error of theoreticians in interpreting or presenting experience. Lenard satisfied this requirement of a scientist amply as well. When 15 years ago relativity theory was made out to be the supreme goddess of science and everyone was obliged to adore it, Lenard rose undauntedly against the general relativity mania and declared relativity theory nonsense.[16] His brave appearance in opposition to it at the conference of scientists in Nauheim in 1920 will always do him as much credit as it will be to Planck's discredit for having served as Einstein's second and for having promoted Einstein and his line of thought in every way previously and subsequently.[17] Well, today Einstein has vanished from Germany; and no serious physicist considers his relativity theories as an unassailable revelation anymore. But unfortunately his German friends and supporters still have the opportunity to continue acting in his spirit. His main supporter Planck is still at the head of the Kaiser Wilhelm Society; his expounder and friend Mr. von Laue is still permitted to act as referee in physics at the Berlin Academy of Sciences;[18] and the theoretical formalist Heisenberg, the essence of Einstein's spirit [*Geist vom Geiste Einsteins*], is supposed to even be honored with a call to a professorship.[19] In view of

mation of light', translated by Anna Beck in *The Collected Papers of Albert Einstein. The Swiss Years: Writings, 1900–1909*, Princeton: Princeton Univ. Press, 1989, Vol. 2, pp. 86–103 ('Über einen die Erzeugung und Verwandlung des Lichtes betreffenden heuristischen Gesichtspunkt', *Annalen der Physik* (4th series), **17** [1905], pp. 132–148, esp. pp. 145ff.); cf. also the editorial headnote in John Stachel et al. (Eds.) *The Collected Papers of Albert Einstein*, Princeton: Princeton Univ. Press, Vol. 2, 1989, pp. 134ff.

[15]In late 1922 Einstein did indeed receive the 1921 Nobel Prize for his theory of the photoelectric effect: Cf., e.g., Pais [1982], chap. 30. Stark's implicit plagiarism charge in this paragraph is completely unfounded, however, since the physical interpretation of the photoelectric effect was very unclear until the publication of Einstein's paper (cf. Wheaton's paper in footnote 13 above).

[16]Cf., e.g., the supplement to Lenard [1918/21] regarding the Bad Nauheim discussions. On Lenard's earlier relations with Einstein see Kleinert & Schönbeck [1978].

[17]During the annual meeting of the Society of German Scientists and Medical Doctors at Bad Nauheim in September 1920, Einstein and Lenard had a fierce debate on Lenard's common sense arguments at the session on relativity theory. Aside from Lenard's account of the event (referred to in the previous footnote) see, e.g., Weyl [1920], and Weyl's dispute with Gehrcke, *Umschau* **25** [1921], pp. 99, 123f., 227. For recent studies on these debates, cf., e.g., Beyerchen [1977], sec. 5.3., Hentschel [1990]a, pp. 82f.

[18]Stark's particular bitterness toward Max → von Laue is explained by the latter's recent success in preventing Stark's entry into the Prussian Academy of Sciences; cf. also doc. 19 for von Laue's retrospection on these events.

[19]Stark attempts to conceal the problem that although Planck's and Heisenberg's physics were similar to Einstein's, absolutely no Jewish blood coursed through their veins. In 1937,

these regrettable circumstances which contradict the National Socialist spirit, Lenard's struggle against Einsteinianism should be a warning. The responsible official advisers at the Culture Ministry should preferably consult with Lenard regarding the filling of professorial chairs in physics, including those in theoretical physics.[20]

Observation, measurement, and interpretation is a matter of the mentality held by a scientist; acknowledgment of truth and the struggle against falsehood is a matter of his character. In both respects Lenard is a shining example of the model scientist for us. But the German scientist should not merely be a narrow specialist. He should rather also feel and act as a German national comrade [*Volksgenosse*]. He should not lock himself up in his laboratory and say: "What happens out there in politics is all the same to me. I'll obey a red or black minister just as well as a National Socialist one;[21] surely it is adequate that I work in my field and accomplish something."[22]—Certainly, the German scientist does not need to worry about the political and economic conflicts of interested parties; but he should take part in the great destiny of his people. He should contribute to the struggle for the freedom and security of his nation. In this regard also Lenard is a great example to German scientists. In the years following the Jewish-Marxist revolt when the Jewry dominated in government and on the street, Lenard saw the overpowering influence of Jewry as the crucial cause of the German people's misfortune; and he did not keep this knowledge to himself, but spoke out.[23] When all state institutions in Germany were supposed to observe the funeral day of the assassinated Jew Rathenau, he instinctively ignored this order and exposed himself to the danger of violence by a Marxist band under Jewish leadership; and he barely escaped being thrown into the Neckar at that time.[24] But he did

Stark solved this dilemma by inventing the term 'White Jew', denoting all people who displayed elements of formalistic 'Jewish thinking', even though they were not of Jewish descent: Cf. doc. 55. For Heisenberg's reaction to this vicious attack see, e.g., his letter to A. → Sommerfeld, Feb. 14, 1936, DMM, 1977-28, A-136, p. 1: "An official of the district leadership of the *NSDAP* had advised me to submit an official protest at the very highest level to Stark's speech in Heidelberg, which has since been published. I intend to consult with my ministry [the → *REM*] about this option in the next few days."

[20]On the successor to Arnold Sommerfeld in Munich see Cassidy [1992]a, chapter 18.

[21]Red and black were the party colors of the Socialists or Communists, and the Conservatives, respectively. Together they represent the Weimar system, with the 'brown' Nazis in opposition.

[22]Stark describes here the dominant attitude shared by German scientists and scholars, who preferred not to take any public political stance: Cf., e.g., Ringer [1969].

[23]Beyerchen [1977] points out on p. 93 that Lenard interpreted the word 'professor' to mean 'one who professes his convictions', which was quite in contradiction to the prevailing view of the apolitical scholar.

[24]The Neckar is a river in SW Germany flowing northwards through Heidelberg. Stark alludes to when Lenard as the acting head of the physics department adamently refused to do honor to the German foreign minister Walther Rathenau (1867–1922), who had been assassinated by right-wing extremists on June 22, 1922. He prevented the flag on the roof of the building from being dropped to half-mast and did not close the institute to honor Rathenau's funeral day

not let himself be intimidated by this violence and later again raised his voice against Jewish influence, most notably at Heidelberg University.[25] He was firmly convinced that Jewish insolence would have to collapse eventually. I remember with pleasure to this day a conversation I had with Lenard in 1928 or 1929. With some consternation I presented to him how powerful Jewish influence was in commerce, in the economy, in politics, in the press, and in science. At this Lenard exclaimed with eyes ablaze and pumping his arm up and down:

"And that is why, more than ever, the Jews must be sunk right down to the center of the earth!"[26]

But Lenard's contribution toward the reestablishment of the German nation did not merely consist of a fight against Jewish influence. His prophetic eye also made him one of the first German scientists to recognize Adolf Hitler as the coming savior and *Führer* of the German people and to fearlessly proclaim his loyalty to him. When Hitler was sentenced at the beginning of 1924 by the Munich people's court [*Volksgericht*] following the collapse of his liberation attempt,[27] and when in the eyes of most Germans his activity seemed to have been finally disposed of, in May 1924 Lenard set before me a proclamation of loyalty to Hitler. This proclamation is of such fundamental importance that I may repeat the text verbatim: [...[28]]

Every word of this proclamation was authored by Lenard alone. He showed it to me for possible modification; but it had been written so much from my own heart that I did not change a word and signed it together with Lenard for publication in the *Großdeutsche Zeitung*. National Socialists have frequently reproached German professors justifiably for having failed nationally concerning Adolf Hitler.[29] Nevertheless, it should not be forgotten that at the early date of May 1924 two German physicists of international esteem openly declared their allegiance to Hitler as the future *Führer* of the German people.

My German compatriots! We have the great fortune that Lenard is still among

---

on June 27 in compliance with a suggestion by the leader of the student body. Students and the labor unions demonstrated in protest. Lenard was taken into protective custody and was temporarily banned from entering his institute to appease the demonstrators, which Lenard found degrading; but this incident came to his credit when the Nazis came to power: Cf. Beyerchen [1977], p. 94.

[25]For an account of Lenard's activities at the university of Heidelberg see, e.g., Neumann & Putlitz [1985].

[26]On p. 94, Beyerchen [1977] mentions two likely causes for Lenard's rabid anti-Semitism: (i) In February 1922, he lost his son Werner, partially as a result of malnutrition suffered during the wartime blockade, and (ii) soon afterwards, his money (including his Nobel Prize savings), which had been invested in government bonds suddenly rendered worthless by inflation. Lenard blamed what he saw as the swindling Jewish Weimar government.

[27]Hitler's failed *putsch* of Nov. 8, 1923; for more information on this attempted coup and the subsequent trial, see the footnotes in doc. 3 as well as, for instance, Noakes & Pridham (Eds.) [1990], Vol. 1, part I, chap. 1 (iv).

[28]Lenard's and Stark's declaration of solidarity is reproduced here as doc. 3.

[29]See, e.g., the last section of doc. 3, and doc. 56.

us. Whoever has come to know him personally will have felt the radiant strength of character of this luminary. But Lenard's example as a German scientist and as a German compatriot should and will act as inspiration and encouragement beyond the past and present, for generations to come.[30]

41 Nazi-Socialism and International Science [December 14, 1935]

Source: 'Nazi-Socialism and International Science', *Nature*, Vol. 136, No. 3450, Saturday, December 14, 1935, pp. 927–928. (Anonymous article). Reprinted by permission from *Nature*; copyright © 1935 Macmillan Magazines Ltd.

Events of the past few years have accustomed us to the knowledge that the position of 'non-Aryan' scientific colleagues in official positions in the German Reich has become intolerable, even in the few cases where they have not been expelled from office.[1] Now a new situation is developing which may be fraught with considerable danger to the continued co-operation of workers in other countries with their German colleagues. This new development carried to the limit can only result in the isolation of German men of science in all matters of international organisation.

The unwelcome move, to which attention has been directed in a recent issue of the *Basler National Zeitung*, is the use that the Nazi authorities intend to make of international congresses for their own purposes. The *Reichsminister* of Propaganda[2] has created a 'Science Congress Centre' (*Wissenschaftliche*

[30]This speech marked the beginning of about two years of heavy campaigning for 'Aryan physics' and against 'Jewish physics': See also docs. 42, 55, 56. Cf., e.g., Beyerchen [1977], sec. 8.1–2, for Alfred → Rosenberg's reception of Stark's speech. For reactions abroad to the highly ideological speeches delivered at the ceremonial opening dedication of the Philipp Lenard Institute at Heidelberg see, e.g., N.N. [1936]a and P.F.F. [1936].

[1]In its May 19, 1933 issue *The Manchester Guardian Weekly* published on p. 399 a list of scholars dismissed from German universities and colleges between April 13 and May 4, 1933: N.N. [1933]e. These included the experimental physicist James → Franck (see doc. 9) and the theoretical physicist Prof. → Born, the latter dismissed on April 26, 1933 from Göttingen. The untenured associate professor of theoretical physics at the University of Berlin Alfred Byk (1878–?), also lost his position on May 2, 1933, which he had been holding since 1921. Byk took his doctorate there in 1902 and habilitated at the Charlottenburg (Berlin) Polytechnic in 1905. The physical chemists Fritz Haber and Prof. Polányi from Berlin were dismissed on May 3, 1933 (see doc. 15). Of course, later much more detailed lists were published of people mostly forced into exile: See, e.g., the *List of Displaced German Scholars* of 1936, and the *Supplementary List* of 1937, both republished in Strauss, Buddensieg & Düwell (Eds.) [1987]. This briefer list of about 200 names shows that foreign countries were well informed early on about what was going on inside Germany. In the summer of 1933, the Academic Council in London had already counted about 700 displaced scholars, and the above-mentioned *List of Displaced Scholars* lists 1628 scientists and scholars. At its height, dismissals were announced almost daily. See also the Introduction, sec. 4, for further references to recent studies on the exile of physicists.

[2]Joseph → Goebbels became the head of the newly created Public Information and Pro-

Kongres[s]-Zentrale), one of the objects of which is to use scientific congresses as opportunities to influence public opinion through the medium of the important persons attending. No longer is this to be left to the chance of personal and friendly relations between the foreigners and the Germans present, but, where possible, steps are to be taken to introduce ideas of a Nazi-socialistic stamp into the discussions of matters of world moment.

For meetings held in Germany, the new centre would act as an intermediary between the Government and party officials on one hand, and the president and officers of the congress on the other. The programmes of the meeting and the lists of speakers would be subject to the approval of the centre: thus international conferences held in Germany are to be sifted of undesirable elements by the authorities beforehand and to survive as meetings approved by the Nazi party. For meetings held outside Germany the control must necessarily be much less, but here too it is possible to influence matters by sending German members to a meeting as a delegation under an appointed leader chosen for his reliability as a member of the Nazi party.[3]

The meeting of the *Astronomische Gesellschaft* held at Berne last summer[4] may be cited as an example of the working of the new machine. The *Astronomische Gesellschaft* is a long-established body with headquarters in Germany but with strong international affiliations, which have been particularly stressed by the Germans in recent years.[5] One of the valued secretaries of the *Astronomische Gesellschaft* came up for re-election at the meeting and in the ordinary course of events would have been re-elected without question: on purely scientific grounds there was every reason for his continuance in office.[6] This was, however, against

paganda (*Volksaufklärung und Propaganda*) Ministry on March 13, 1933, which executed the Nazi's tactic of attempting to influence the masses in all aspects of cultural life through the media.

[3]For the National Socialist regulations concerning scientists wanting to travel abroad, see Walker [1992], pp. 344ff.

[4]For the official report on this meeting in Berne of July 24–27, 1935, see: *Vierteljahresschrift der astronomischen Gesellschaft* **70** [1935], pp. 286–380, esp. p. 280, for a list of the 129 participants from 19 countries.

[5]The *Astronomische Gesellschaft* was founded in 1863 mainly by German astronomers and with German as the official language for its proceedings, but early on it also accepted members from Scandinavia, Russia, the Benelux countries, France, England, USA, etc. On the Society's history, see *Vierteljahresschrift der astronomischen Gesellschaft* **70** [1935], p. 289; on the long-running tradition of close international cooperation in astronomy and astrophysics see, ibid., pp. 288–290; cf., e.g., D. B. Herrmann & J. Hamel: 'Zur Frühentwicklung der Astrophysik. Das internationale Forscherkollektiv 1865–1898', *NTM* **12** [1975], pp. 25–30, or Karl Hufbauer: *Solar Physics since Galileo*, Baltimore: Johns Hopkins Univ. Press, 1991, pp. 76ff., 82ff.

[6]The secretary in question was Richard Prager (1883–1945), who started his studies in astronomy at the Universities of Marburg, Göttingen and Berlin in 1901, taking his doctorate in 1908 and becoming teaching assistant in Berlin. He subsequently assisted on the History of Fixed Stars project at the Academy of Sciences in Berlin and was engaged as department head at the Observatory in Santiago in Chile 1909. In 1913 Prager became assistant and then observer at the Babelsberg Observatory near Berlin 1916–38, and accepted a professorship in

the views of the German delegation, whose members were instructed not to vote for a Jew: alternative names were put forward and it became clear in the course of the meeting that something more vital was at stake than the question as to the holder of a particular office; in fact, nothing less was at issue than the question whether an organisation of international standing and repute should come under the control of an anti-Semitic clique.[7] When the vote was ultimately taken by secret ballot, it was found that the Nazi element had failed in its attempt, and the threatened secession of non-Germans from the *Astronomische Gesellschaft* was averted for the time. But the danger of subsequent development remains a source of anxiety to those concerned in international scientific relations.

It should be added in fairness to the Germans that internal persecution of a minority is not limited to their country, nor even the use of international gatherings to spread nationalist political propaganda. Both these undesirable practices are to be found in more than one country: we have no sympathy for either. But when these tendencies spread and a nationalistic movement tries to secure the control of international scientific work, it is time to call a halt.[8]

Berlin in 1924. In 1939 he emigrated to the U.S. and was research associate in astronomy at Harvard University until 1945. Prager was considered an outstanding authority, noted especially for his bibliography on variable stars.

[7]According to the *Vierteljahresschrift* **70** [1935], pp. 292 and 311, the following persons were to be confirmed as members of the council of the *Astronomische Gesellschaft* or replaced: Knut Lundmark (1889–1958), professor of astronomy and director of the observatory in Lund (Sweden) since 1929; Anton Michael Tass (1896–?), director of the observatory at Budapest since 1923; Franz-Josef Hopmann (1890–1975), full professor and director of the observatory at the University of Leipzig since 1930; and R. Prager. The names proposed to replace Prager were: William Otto Brunner (1878–1958), full professor of astronomy at the *Eidgenössische Technische Hochschule* Zurich since 1926 and associate professor and director of the observatory of the University of Zurich; Julius Dick (1891–?), assistant at the observatory Berlin-Babelsberg since 1922; Arnold Kohlschütter (1883–1969), full professor and director of the observatory at Bonn since 1925; [Hans] Rolf Müller (1898–1981), professor of astrophysics at Potsdam since 1930 and observer at the Potsdam observatory; and Paul ten Bruggencate (1901–1961), main observer at the *Astrophysikalisches Observatorium Potsdam* since 1935. However, from the 93 ballots, 62 were for Prager, 2 were not filled out, and the other candidates only got one vote each. The political overtones of the debate were not mentioned in the official proceedings.

[8]Statements of this type, published in many of the most widely-read scholarly journals, were also easily accessible to Germans and used for quite different purposes: in support of efforts to uphold good relations with foreign countries, as in the case of Planck (cf. doc. 38), or as evidence of alleged anti-German sentiment, as in the case of Stark [1934]a,b.

Part III
The Ideological Schism: 1936–1939

42 Willi Menzel: German Physics and Jewish Physics [January 29, 1936]

Source: Willi Menzel,[1] 'Deutsche Physik und Jüdische Physik', *Völkischer Beobachter*, Vol. 49, No. 29, Wednesday, January 29, 1936, Northern Ed./Edition A, p. 7.

There is no such thing as National Socialist physics or mathematics. Likewise, machines and motors are also not considered National Socialist. But every technician in the world prizes **German** machine building, German electrical engineering, **German** medicine.

"What should all of this have to do with physics? Physics is international and will always remain so."[2] And yet I contend there is a German way of practicing science, and particularly, physics. German physics is the study of reality in inanimate nature through experiment and arises out of the joy of observing its structure and processes.

As an illustration, we shall take a brief look at the history of the development of physics. Each small advance in the physical conception of the world [*Weltbild*] costs the lifework of a great, genius mind.[3] Copernicus and Kepler collided with a worldview that was completely dictated by the Church. Slowly, very slowly, the truth prevailed; slowly we began to become aware of the rules of physics; slowly people's conception of the world became that of physics, and the nation became receptive to new ideas. In the wake of the revolution in physics came theoreticians like Einstein who then strove to turn physics into a purely mathematical system of concepts. They propagated their ideas in the manner characteristic of Jews and forced them upon physicists. They tried to ridicule men who criticized this new type of 'science' with the argument that their intellect just could not aspire to the lofty spheres of the Einsteinian intellect—an intellect which, says Lenard, does not consciously seek after the truth.

---

[1]For the context of this newspaper article at the beginning of the campaign for 'Aryan' physics launched by Stark and others in late 1935, cf. doc. 40. The author was a chemistry student and protégé of Stark, at that time leader of the professional branch (*Fachschaftsleiter*) of the → *NSDStB* at the Berlin Polytechnic's physics department; see Beyerchen [1977], p. 142.

[2]This text closely follows the structure of Philipp → Lenard's foreword to his textbook on 'Aryan physics', *Deutsche Physik* (cf. doc. 39). Both texts open with an effort to refute that physics is a supranational science, because this deep-rooted belief prevented their colleagues from accepting Nazi claims that the quality of physics depended on the race of its scientists.

[3]These hagiographic leanings are also typical of Lenard: Cf. Lenard [1929].

I will now briefly contrast the views and practices of 'Einstein physics', which is most prevalent in Germany, against what one might call Nordic physics, which is despised as heretical.[...[4]]

The German way of doing research comes from a completely different point of departure. The desire of the German scientist is to explore truth, to genuinely investigate nature. He does not attempt to harness a natural phenomenon within an arbitrary and abstract thought structure, but rather he probes nature by posing carefully defined questions through his experiments and tests, which nature answers. The great Aryan scientists perform clean experiments. They present nature with a question, to which they can expect one definite and clear answer. Therefore, observation is the starting point of research for Nordic man; and conclusions about the conception of the world are drawn from it and generalizations are made only if they are in agreement with experience.

"The unspoiled national spirit [*Volksgeist*] seeks depth; it seeks theoretical foundations consistent with nature, and irrefutable knowledge of the cosmos." ... "The great Aryan scientists were reluctant to come forward with contestable conclusions. On the contrary, they calmly turned primarily to testing their new ideas against reality, in order to arrive at acknowledged facts, not assumptions. Thus publications of rich and novel factual content appeared, each itself a milestone in the progress of scientific knowledge. **In Jewish physics every assumption that later proves not to be completely false is already considered a milestone**. But with such an evaluation the Aryan approach is abandoned, and this has had very striking consequences." (Lenard: *Deutsche Physik*, Munich 1936, from the Foreword.)

The German, Lenard, recognized early on how things stand in research, but only recently was he heard.

"**In reality, as with everything that man creates, science is determined by race or by blood**.[[5]] It can seem to be international when universally valid scientific **results** are wrongly traced to a common **origin** or when it is not acknowledged that the science supplied by peoples of different countries is identical or similar to German science, and that their science could only have been produced because and to the extent that these other peoples are or were likewise of a predominantly Nordic racial mix. Nations of different racial mixes practice science differently."

During the German nation's time of decline only a very few men saw the danger that threatens German science and fought against it effectively. A well-deserved honor has now been bestowed on one of these, only after he had stepped

[4]Emphasis in the original article is in spaced type. Menzel reproduces here the third paragraph of J. → Stark's speech (doc. 40), adding the following: "(From the speech by the President of the *Physikalisch-Technische Reichsanstalt* and the *Deutsche Forschungsgemeinschaft*, P[arty] M[ember] J. Stark, held at the dedication of the *Philipp-Lenard-Institut* in Heidelberg on the 13th of December 1935.)"

[5]Emphasis in these quotes from Lenard was added by Menzel.

down bitterly from the podium in 1931: Philipp Lenard. The Physical Institute in Heidelberg was named in his honor. This is National Socialist Germany's gratitude to the scientist who single-handedly, with only the support of a few students, held the proper name 'German' above the adjective 'Jewish'.

We, the younger generation, want to continue the fight today for German physics; and we will succeed in elevating its name to the same heights that German technology and science has already been enjoying for a long time.[6]

43 Werner Heisenberg: On the Article: 'German and Jewish Physics' [February 28, 1936]

Source: W. Heisenberg, 'Zum Artikel: Deutsche und jüdische Physik. Entgegnung von Prof. Dr. Heisenberg.' *Völkischer Beobachter*, February 28, 1936, Northern edition (A), No. 59, p. 6.

We present the following reply by Prof. Heisenberg to our article 'German and Jewish Physics', published in issue No. 29 of the *V[ölkischer]. B[eobachter].* by Willi Menzel.[1]

Since we can by no means agree with the views expressed in this reply, we have turned to Prof. Stark as an authority in the field of physics asking him for his opinion, which is printed subsequently.[2]

We herewith consider the discussion of this topic concluded.

Reply by Prof. Dr. Heisenberg

On the authority of Ph. Lenard and J. Stark, two of the most senior and meritorious German physicists, W. Menzel offers arguments against theoretical physics in his article 'German Physics and Jewish Physics' (*Völkischer Beobachter* of January 29th, 1936) that appear erroneous and misleading to the majority of younger scientists. In the following lines I will attempt to portray with what hopes and

[6]For W. → Heisenberg's reply to this article see the following document.

[1]Cf. doc. 42. The first three sentences introducing the following article appear in bold in the original. All other emphasis appears in spaced type. German press law prescribes that when newspapers and journals publish an allegedly libelous story, they are obliged to print the injured party's opposing view upon request. Although legal proceedings were usually necessary to enforce this right, Werner → Heisenberg had enough influence to get his commentary published in the daily newspaper of the → *NSDAP*. Cf. his letter to Arnold → Sommerfeld, Feb. 14, 1936, DMM, 1977-28, A-136, p. 1: "In the meantime, I have tried to get an article against Menzel's published in the *V[ölkischer]. B[eobachter].*, probably without success." See also docs. 56, 62 and 83.

[2]See the following document. It was and still is quite common in fact that such protests are followed by commentary reflecting the publication's stand on the matter; but it cut short further debate, preventing Heisenberg from responding again. For additional comments on Heisenberg's reply cf., e.g., Rechenberg (Ed.) [1992], p. 72, and pp. 78–80 for a reprint of Heisenberg's unabridged reply.

aspirations young Germans of today approach this science.[3]

The overall goal of scientific research is not only to observe nature, but ultimately to understand correlations identified in it that have the character of laws. Because the precise prediction of a particular phenomenon in theory is often the most reliable touchstone for a resulting insight, the mathematical formulation of natural laws is regarded in physics as the clearest form in which to represent our understanding of nature. But the actual substance of our scientific findings is quite unaffected by this mathematical form. What do determine this content, however, are the systems of concepts, which are the final result of many observations and which can be comfortably applied in nature.

The discovery on which quantum theory was based in the year 1900 revealed new types of correlations in radiation phenomena that could not be explained with the physical concepts of the time.[4] Thus the birth of a new system of concepts was heralded in; and we have learned since that it forms the basis of the entire field of atomic physics. Only in the last decade did quantum theory acquire its final form—and thus could be expressed in mathematical terms.[5] This form makes it possible to predict the behavior of atoms in every experiment quantitatively.[6] Thus, with its aid the position of 'spectral lines' (that is, the color of light emitted from matter under suitable conditions) can be calculated—and in the simplest of cases with extraordinary accuracy.[7] Thus quantitative analysis is now possible in the century-old field of spectroscopy. One of the nicest examples of this is that theory can offer a complete explanation for the splitting of spectral lines induced by an electrical field.[8]

Furthermore, this theory makes the principle facts in chemistry comprehensible; and in some cases it delivers us important chemical data more precisely than observation can. Significant chemical facts were predicted and discovered

---

[3]Note that, like Menzel, Heisenberg also presumes to speak on behalf of the 'young Germans of today'.

[4]Heisenberg alludes here to early quantum theory introduced in Max → Planck's work on the energy distribution of black-body radiation, that is, radiation in a thermal equilibrium within a totally black cavity; see Planck's contributions in *Verhandlungen der Deutschen Physikalischen Gesellschaft* **2** [1900], pp. 202–204, 237–245, and cf., e.g., docs. 46f. and Kangro [1976].

[5]For a detailed historical account and references concerning developments connected to the rise of quantum mechanics from 1925 on, described here briefly see, e.g., Jammer [1966], chaps. 4–6, and Mehra & Rechenberg [1982], Vols. 2ff.

[6]With the following survey of some of the virtues of the new quantum mechanics, Heisenberg tried to counter the charge so often but unjustifiably made by Lenard, Stark and their allies that his theory was empirically unfruitful; cf., e.g., doc. 40.

[7]According to Bohr's quantum theory of 1913, spectral lines of frequency ν are formed when electrons 'jump' discontinuously from one definite quantum state of energy E_1 to another of energy E_2, and $\nu = (E_1 - E_2)/h$, being absorption lines if $E_2 > E_1$, and emission lines if $E_2 < E_1$.

[8]This is a deliberate reference to the Stark effect, discovered in 1913 by Johannes → Stark, one of Heisenberg's main antagonists. The new quantum mechanics explained this effect theoretically, despite Stark's objections.

on the basis of quantum theory, like the fact that hydrogen can decompose into two forms.[9] Through modern atomic physics the elastic, thermal and magnetic properties of solid bodies can be explained in great detail. A look into the current physical literature reveals that young scientists throughout the world are engaged in experimental and theoretical work to expand upon and deepen the new knowledge about atoms which is laid out in quantum theory.

In the same way, the theory of relativity is also the obvious basis for further research, because every detail in it that can possibly be verified by experiment has stood the test. For example, there is a treasure trove of data from the results of spectroscopic experiments that could only be understood and used once it had been brought within the context of quantum theory and relativity theory (multiplet structure of spectral lines).[10] The existence of an important elementary building block, the 'positron', had already been predicted on the basis of the relativity and quantum theories even before its discovery in experiment.[11]

But aside from the practical successes, there is a completely different side to this development, which makes theoretical physics especially important to us Germans. The system of concepts in modern physics is so different from that of earlier exact science that our acceptance of it significantly alters our position on the ultimate questions of scientific knowledge. A serious analysis of this changed situation leads the exact sciences away from the naive materialistic conception of the world that assumes that it is possible to simply translate all experience into material occurrences in space and time. That this view of physics's principle importance did not date back to the popular postwar overestimation of physical theory is evidenced by a few lines from a public speech that Max Planck gave in 1909.[12]

"Only now in our times does a final decision seem to be in the making as the end result of a profound movement gripping theoretical physics—a movement so radical and revolutionary that its waves beat well beyond the shores of physics proper into the neighboring fields of chemistry, astronomy, and even into epis-

[9]On C.-F. → Bonhoeffer's and P. → Harteck's work on orthohydrogen and parahydrogen, see doc. 29, footnote 6.

[10]The multiplet structure of spectral lines was discovered by J. Stark, Miguel Catalan and others, through close observation of the splitting patterns in electric and magnetic fields. It could also be accounted for in quantum theory, which postulates that multiplicity m is related to the azimuthal quantum number κ via $m = 2\kappa + 1$.

[11]This particle, which is the anti-particle to the electron in that it has precisely the same mass but an opposite charge, had been predicted in Dirac's relativistic generalization of quantum theory and was confirmed experimentally by Carl Anderson in 1932 in cosmic ray showers. Cf., e.g., Norwood Russell Hansen: 'Discovering the positron', *British Journal of Philosophy of Science* **12** [1961], pp. 194–214, **12** [1962], pp. 299–313.

[12]Planck's speech about the position of modern physics on the mechanical view of nature was delivered at the 82nd meeting of the *Deutsche Naturforscher und Ärzte* in Königsberg, on September 23, 1910(!): 'Die Stellung der neueren Physik zur mechanischen Naturanschauung', published separately by Hirzel in 1910. Reprinted in *Physikalische Rundblicke*, Leipzig: Hirzel, 1922, pp. 38–63, see p. 40.

temology; and in its wake battles are being waged in science that can only be compared with the struggle over the Copernican worldview."[13]

In the meantime, experimental research in the last decades has helped the new system of concepts to victory. But atomic physics, now the open research area, will certainly permit an even deeper penetration into the epistemological foundations of our science. If it is up to Germans in particular to develop these correlations, this will be in continuation of the great philosophical tradition that Kant introduced with his epistemological analyses of the foundations of natural science. The continuation of this research, which may well exert the greatest influence on the structure of our intellectual life as a whole, is one of the noblest missions of German youth in science.[14]

44 Johannes Stark: Comment on W. Heisenberg's Reply [February 28, 1936]

Source: Johannes Stark, 'Zum Artikel: Deutsche und jüdische Physik. Stellungnahme von Prof. Dr. J. Stark', *Völkischer Beobachter*, North German ed. (A), No. 59, February 28, 1936, p. 6.[1]

Comment by Prof. J. Stark

For clarity's sake it is essential that the preceding article by Heisenberg be rectified immediately.[2]

It is designed to give the impression to readers who are not physics experts that the great discoveries in physics of recent decades were an achievement of theory, and wherever possible, even of Jewish theory. Actually, theory was only involved to a very limited extent. These discoveries followed almost exclusively from the careful observations and measurements of experimental physicists.[3]

[13]On the intellectual upheaval caused by Nicolaus Copernicus's (1473–1543) heliocentric model of the solar system, put forward in his book *De revolutionibus orbium cœlestium* (1543), and its slow assimilation see, e.g., Thomas S. Kuhn, *The Copernican Revolution*, Cambridge, Mass.: Harvard Univ. Press, 1957.

[14]The epistemological discussion on the interpretation of the new quantum mechanics was still in its early stages at the time of this article and is still continuing today: Cf., e.g., doc. 31 and Max Jammer: *The Philosophy of Quantum Mechanics*, New York: Wiley, 1974.

[1]This commentary immediately follows the preceding document under the same main headline: 'On the article 'German and Jewish physics' ' of this daily newspaper of the → *NSDAP*.

[2]The opening sentence appears in bold in the original text. All other emphasis appears in spaced type. The animosity between Johannes → Stark and Werner → Heisenberg is partly explained by the fact that Heisenberg was a pupil of A. → Sommerfeld, who had been a personal enemy of Stark's since their controversy over the light quantum hypothesis and X-ray physics in 1909; cf., e.g., Wheaton [1983], part III.

[3]The issue of the relative importance of experimental and theoretical physics was the subject of intense debate at the time: Cf. also, the Introduction, sec. 5.3.

Thus **theory was not involved** in the discovery of the electron.[4] What theory came up with on its dynamics following its discovery is today no more than a pile of useless literature.[5] Theory was **furthermore not involved** in the discovery of the X ray and of radioactive phenomena. It was **not involved** in the discovery of the influence of magnetic fields on spectral lines (Zeemann effect).[6] Following this discovery Lorentz did indeed develop a theory on the subject and even received half of a Nobel prize for it; but today this theory is recognized as inapplicable.[7] Theory was **not involved** in the discovery of the influence of an electrical field on spectral lines. What is more, the discovery was even made contrary to a theoretical prediction;[8] and if Heisenberg makes it appear in his article as though the manifestations of this effect have been established and described exhaustively by theory, then this is **also incorrect**. Not even in the simplest case of the hydrogen atom is theory in full agreement with experiment, not to mention the remaining chemical elements.[9]

Not even the discovery of Planck's constant, which is of itself a commendable theoretical achievement, was made without the crucial spade work of experimental physics. When Planck attempted to apply to thermal radiation a mathematical representation of the thermal condition of gases that had been developed by the great physicist Boltzmann, he was only successful when he became aware of measurements on thermal radiation that had been carried out at that time at the

---

[4]Cf. doc. 40 and the references in footnote 12 there.

[5]For an historical account of the electron theories by Lorentz, Drude and others around 1900, which were soon superseded by the early quantum theories by Planck, Einstein, Ehrenfest and others see, e.g., Whittaker [1951/53], Vol. 1, sec. xiii; Vol. 2, sec. iii–iv.

[6]Pieter Zeeman (1865–1943) discovered this effect in 1896; see, e.g., Zeeman's 'On the influence of magnetism on the nature of light emitted by a substance', *Philosophical Magazine* (5) **43** [1896], pp. 226–239. Cf., e.g., Whittaker [1951/53], pp. 410ff., and Theodore Arabatzis: 'The discovery of the Zeeman effect', *Studies in History and Philosophy of Science* **23** [1992], pp. 365–388.

[7]On the phenomena listed in the foregoing, cf. Whittaker [1951/53], Vol. 1, sec. xiii, pp. 410ff., and references there to primary literature.

[8]J. Stark detected this long sought-after effect in late 1913, in which intense spectral lines of hydrogen split when exposed to a strong electric field. The electrodynamical theory by Woldemar Voigt (1850–1919) predicted any such effect to be about 1,000 times smaller than it turned out to be. Cf., e.g., W. Voigt: 'Zur theoretischer Einwirkung eines elektrostatischen Feldes auf die optischen Eigenschaften der Körper', *Annalen der Physik* (3) **69** [1895], pp. 297–318.

[9]This depends on how you interpret 'in full agreement'. Epstein's and Schrödinger's 1926 theory of the Stark effect certainly fully agreed with the observed intensities of the various components of spectral lines split in an electric field; and Sommerfeld's semiclassical theory of 1921 as well as the full quantum mechanical theories published since 1925 could even explain qualitatively the so-called quadratic Stark effect with asymmetric shifts of spectral lines proportional to the square of the electric field. But small unexplained differences between experiment and theory remained, especially for elements with higher atomic numbers. See, e.g., the survey of the literature in Stark's contribution to Wilhelm Wien (Ed.), *Handbuch der Experimentalphysik*, Vol. 21: J. Stark, *Starkeffekt*, Leipzig: Akademische Verlagsgesellschaft, 1927, pp. 339–548.

Physikalisch-Technische Reichsanstalt.[10] No productive experimental physicist uses Einstein's relativity theories as a point of departure for research, and all of its premises are controversial from the experimental point of view to this day; or they have even been shown to be unfounded.[11]

In his article Heisenberg continues to advocate the fundamental attitude of Jewish physics even today. Indeed, he even expects that young Germans adopt this basic attitude and take Einstein and his comrades as their models in science. He sees 'systems of concepts' [*Begriffssysteme*] as the source of physical knowledge as well as its goal and speaks of a victory of the new conceptual systems.[12]

In the face of this, members of the branch of physics to which we are truly indebted for progress and great discoveries must emphatically assert that the source of knowledge in physics is experiment based upon observation and measurement. **Theory must be brought back within its bounds from its so presumptuous position; and it must be acknowledged that in physics, theory is but a method, an arithmetical and representational aid.** We must insist furthermore that the type of physics advocated by Heisenberg no longer be allowed to assume a leading role, as it has up to now in the appointment of professorships in physics.[13]

Theoretical physics under Einstein, Planck, and Sommerfeld has not only almost completely dominated chairs in theoretical physics in recent decades, but no representative of experimental physics could receive a professorship against their objections, not even when he could demonstrate recognized experimental accomplishments.[14]

The article by the student Menzel is a welcome sign that young Germans are shunning the influence of Jewish physics and that they want to study physics in the same spirit that pervades Lenard's recently published text-

[10]See Stephen G. Brush (Ed.), *Ludwig Boltzmann, Lectures on Gas Theory*, Berkeley: Univ. California, 1964. See also Max Planck's papers: 'Ueber irreversible Strahlungsvorgänge', *Annalen der Physik* **1** [1900], pp. 69–122; 'Ueber eine Verbesserung der Wienschen Spectralgleichung', and 'Zur Theorie des Gesetzes der Energieverteilung im Normalspectrum', *VDPG* **2** [1900], pp. 202–204, 237–245; 'Ueber das Gesetz der Energieverteilung im Normalspectrum', *Annalen der Physik* **4** [1901], pp. 553–563. On the experiments conducted at the → *PTR*, see Kangro [1976]. See also doc. 46, where Gehrcke makes the same point on Planck, as well as Debye's reply, doc. 47.

[11]Clear-cut high-precision tests of the general theory of relativity only became feasible in the 1960's; cf., e.g., Jean Eisenstaedt, 'The low water mark of general relativity', in: Don Howard & John Stachel (Eds.), *Einstein and the History of General Relativity*, Boston et al.: Birkhäuser, 1989, pp. 277–292.

[12]In using this holistic terminology Heisenberg was influenced mainly by the Danish physicist Niels Bohr (see footnote 4 in doc. 31), as opposed to Albert → Einstein.

[13]On contemporary new appointments to chairs in physics see doc. 33.

[14]The three theorists, Sommerfeld in particular, were indeed very influential in appointments to chairs in physics all over Germany. Stark experienced this influence personally when his attempts to return to academia in the mid-1920's failed: See, e.g., Beyerchen [1977], sec. 6.3.

book *Deutsche Physik*, which reflects physical reality without the 'new systems of concepts'.[15]

45 Max von Laue: Review of Philipp Lenard's 'German Physics' [February 29, 1936]

Source: Max von Laue, 'Deutsche Physik', *Frankfurter Zeitung*, Saturday, February 29, 1936, Vol. 80, No. 110/111, p. 10.

This work by Philipp Lenard* clearly originated from the lectures on experimental physics the author had held over a number of decades as professor in Heidelberg.[1] They were famous, and justifiably so. Whoever heard the author lecture, even if it were only once, could not avoid coming away with the impression that he is a great personality gifted with a deep understanding of diverse aspects of physics and instilled with a sincere love for nature. This book seeks to preserve as much of this as possible for posterity. Thus naturally it cannot exactly be a modern textbook of physics, which fact might become more apparent in the volumes to follow the present one under discussion.

In any event, Lenard presumably did not present the foreword in his lectures.[2] I fear it may well gain him some enemies. No doubt he also felt that an apology was called for, since he offered the following motto to it:

> "The foreword stems from today's conflict,
> The work seeks values infinite."[3]

Let us therefore turn to the work itself. The volume in question covers mechanics and generally keeps within the standard program: Statics, general kinetics of solids, gravitation, planetary motion, molecular energy, elasticity, and hydrostatics including surface tension, one of Lenard's favorite subjects; mechanics of gases and hydrodynamics then follow. It even adheres to the standard in an omission: From physics textbooks the reader learns virtually nothing on the crystalline state, which is fundamental to understanding the state of aggregation in solids. Historical developments have relegated this responsibility entirely to the minerologist, much to the detriment of physics. At best, the only way this book

---

[15]It is not clear whether Willi Menzel really wrote this article himself (presented here as doc. 42) or whether he only served as a strawman for Stark or Lenard, as is suggested by the similar vocabulary and style.

* [Footnote *) in the original:] "Philipp Lenard: *Deutsche Physik.* 4 Volumes. J. F. Lehmann Publishers, Munich, 1936. First volume: *Introduction and Mechanics.*"

[1]Philipp → Lenard was full professor at Heidelberg University from 1907 to his retirement in 1931.

[2]See doc. 39 for this foreword.

[3]"Der Vorwort entstammt der Kampf der Zeit, Das Werk sucht Werte der Ewigkeit."

distinguishes itself is in the few lines that it does devote to the molecular structure of crystals, which are quite wrong.[[4]]

An itemization like the one above does not reveal much in itself. What count are the individual presentations, and here they are mainly delightfully clear and simple; and the way they affectionately emerse the reader in natural processes will hopefully leave a very lasting impression on the budding disciple of physics.

Surely of more interest for a review in a daily paper is the 13-page general introduction dealing with the position of physics among other sciences and outlining its research principles. The author certainly does not want to explain philosophical foundations of physics. He rather portrays the mood in which he pursued his research work; and in many essential points we will thoroughly agree with him. But a few things are astounding.

There is the ether question, for example. No one will be surprised that Lenard's book avidly supports the preservation of ether, since he already has emphasized this position frequently and poignantly elsewhere.[[5]] But then he writes on Page 12: "Ether is apparently harder to understand than matter; it already seems to reveal the limits of comprehensibility." From there the step to the theory of relativity is really not so faraway anymore. Presumably only its mathematical dress prevents the author from taking this step, since he also certainly ought to appreciate the magnificent union of various branches of physics in the theory of relativity.

But Lenard does not have a single good thing to say about the mathematics and theoretical physics that the theory makes full use of. "The meaning of all well-tested natural laws is simple; they can thus be formulated not only using equations, but also using normal words. All undisputed scientific knowledge can be presented without a significant amount of mathematics." (Page 8) "Theoretical or mathematical physics sets out to make all calculations that can be connected to natural processes." [(Page 9)] This is a bit malicious. I would say, it selects the interesting problems out of all the possibilities, excepting exercises, of course, which are a part of theoretical physics only inasmuch as glass-blowing is a part of experimental work. Then again a few lines later there appears: "Some great, solitary conceptual structures stand out that can be presented fully only in the language of mathematics, even though simple words again suffice in summarizing their meaning."

I would think in science every means that leads to the end is justified. Should a prominent scientist be capable of achieving something without actually employing mathematical concepts, fine! But he should not belittle others who use mathematics to attain their goals; and in addition he should not consider that teachability is inherently attached to the concept of science. And given the way

[4]Laue reveals here his special interest in this topic since his involvement in the discovery of X-ray diffraction in crystals (see footnote 3, doc. 76).

[5]Cf., e.g., P. Lenard's pamphlets: Lenard [1910], [1918/21] and [1920/21]. The last two of these also appeared in Stark's *Jahrbuch der Radioaktivität und Elektronik.*

humans are normally gifted, mathematical formulation is anyway quite often unavoidable in physical instruction.

Experiment, and in particular the experiment free of all interference, or what Lenard calls the 'pure' experiment (Page 5), is and remains, of course, the basis of physics. The book names Galileo's experiment as an example of one of these, in which he dropped quite heavy objects, in order to reduce the influence of air resistence to the point of negligibility, which was impossible before the invention of the air pump.[6] This is, of course, a good example; and today, 300 years later, we are quite easily convinced on the basis of more advanced knowledge that Galileo actually did achieve this purity in his experiment. But what guarantee did **he** have that he was not mistaken, that no other yet unrecognized, and consequently not yet eliminated, sources of error were involved? Could the 'genius inspiration' to which Lenard objects on Page 9 not have played a role here after all? Here is a more recent example to illustrate how hard it is to judge the accuracy of an experiment in time. For many years famous experimental scientists worked on the laws governing the passage of electrons through matter. They were convinced that they had achieved the purest conditions—Mr. Lenard will confirm this. And as a matter of fact, they did arrive at valuable results, but they overlooked the fundamental, purest and actually epoch-making phenomenon of the **diffraction** of electrons. It was discovered only much later in the course of research on quantum theory, which really does describe it in extremely mathematical terms.[7]

But we do not want to close on a polemical note. On Page 9/10 he writes: "Whoever wants to come closer to God in nature through science surely needs to have an untiring diligence, a reckless spirit of initiative, and the imagination of ten poets, aside from an innate logic and an inherited delight in emersing oneself in observable truths together with the intractible impulse to understand the same; and much more is required, but with an unbounded modesty toward the great unknown one [*dem großen Unbekannten*], whom he wants to serve exclusively and devotedly, selflessly and loyally." If **all** scientists **truly** take **this** stand, then it should not be hard to achieve a peaceful unity among them, over and above all differences of opinion. M. von Laue, Zehlendorf (Berlin).[8]

[6]Otto von Guericke (1602–1686) built the first airpump around 1650 with which he was able to create a vacuum in an air-tight container and demonstrate the influence of air pressure. On the history of the airpump see Edward Neville da Costa Andrade, 'Zur Geschichte der Luftpumpe', *Endeavour* **16** [1957], pp. 29–35, and Steven Shapin & Simon Schaffer, *Leviathan and the Air-Pump. Hobbes, Boyle, and the Experimental Life*, Princeton: Princeton Univ. Press, 1985.

[7]On electron diffraction see Clinton Joseph Davisson & Lester Halbert Germer: 'The scattering of electrons by a single crystal of nickel', *Nature* **119** [1927], pp. 558–560, and 'Diffraction of electrons by a crystal of nickel', *Physical Review*, Series 2, **30** [1927], pp. 705–740. Cf. also, Arturo Russo: 'The discovery of electron diffraction, *Historical Studies in the Physical Sciences* **12** [1981], pp. 117–160.

[8]Max von → Laue was full professor of theoretical physics at Berlin University 1919–43.

46 Ernst Gehrcke: How the Energy Distribution of Black-Body Radiation Was Really Found [April 3, 1936]

Source: Ernst J. L. Gehrcke, 'Wie die Energieverteilung der schwarzen Strahlung in Wirklichkeit gefunden wurde. Mitteilung aus der Physikalisch-Technischen Reichsanstalt'. Received April 3, 1936. *Physikalische Zeitschrift* **37** [1936], pp. 439–440.[1]

Physics has assumed a course for some time now that makes our science appear so convoluted and in many respects so unclear that one would think physics were only a hotbed of complicated abstractions, where the invented abstract things rather than scientific empirical knowledge were what mattered. In other words: The distinction between the descriptive **form** and the **object** being described has become blurred.

On the other hand, a few people, particularly Lenard,* have upheld the view that all valuable physical knowledge can be explained simply, because it is simple.[[2]] What seems convoluted is only so, because a clear and simple explanation is lacking or because one must first be worked out. Often paradoxical formulations of astonishingly simple facts contribute to the confusion. This struck me for the first time in 1901 when I saw an instrument in Lummer's laboratory at the *Physikalisch-Technische Reichsanstalt.*[[3]] It consisted entirely of pure white pipes, both inside and out, and I was informed: "This is a black body!"[[4]] There's something wrong here! I said to myself. This psychological and as yet unresolved problem in science is that findings that can just as well be expressed simply, intelligibly, and clearly, are discussed in paradoxical and artificial terms. The tendency towards convoluted and paradoxical forms of expression becomes especially evident in the area of black-body radiation energy distribution. Now that this controversy has come to some conclusion and things have calmed down, we can trace objectively the simple course of development of this field and thus contribute toward understanding it.

The energy distribution of the black body has been the object of an immense amount of research work. I gave a brief description of this in Volume 2 of my *Handbuch der physikalischen Optik*, p. 846 ff. (see also p. 1033).[[5]] I will not

---

[1]This undated text by Ernst J. L. → Gehrcke carries the subheading 'Communication from the *Physikalisch-Technische Reichsanstalt*' (→ *PTR*), where he was employed from 1921, becoming the director of its optical laboratories in 1926.

∗ [Original footnote 1) (of column 1, p. 439):] "Ph. Lenard, *Deutsche Physik* **1**, 12, 1936."

[2]For the foreword to Philipp → Lenard's textbook referenced in footnote ∗ see doc. 39.

[3]Otto Richard Lummer (1860–1925) took his doctorate in Berlin in 1884 was professor at the *PTR* from 1894. In 1902 he became university lecturer of physics at Berlin University and in 1905 full professor of physics at Breslau. In 1903, Gehrcke and Lummer developed the so-called Gehrcke-Lummer plate for high-precision interferometry.

[4]A black body in physics is a body which completely absorbs any heat or light radiation falling upon it, remaining in equilibrium with the radiation reaching and leaving it. Black-body radiation is the radiation emitted from an ideal black body.

[5]E. Gehrcke edited his 'Handbook of Physical Optics' together with Felix Auerbach, Azeglio

repeat this description here but will expand upon it. The energy distribution function held to be correct, i.e., the energy distribution function that is valid within the current limits of observational error, had not been predicted theoretically but was an extremely surprising experimental finding, namely, at the *Physikalisch-Technische Reichsanstalt*, where the relevant experimental research had been performed primarily by Kurlbaum, Lummer, and Pringsheim.[6] Lummer was the first to conceive of, and advance the idea publicly that Wien's energy distribution function, which had been generally accepted as valid until then, does not accurately reflect the established facts. The larger the wavelength is, the more inaccurate is Wien's formula, while conversely, Rayleigh's formula fails in the case of short waves.[7] Lummer, who executed the conclusive series of measurements together with Pringsheim, had to have a lot of courage to make this observation.[8] He found little support among his colleagues; and Planck, favoring the the prevailing view, also adopted the position against the experimental results

Bemporad, Otto Berg, et. al. It was published in Leipzig by Barth: Vol. 1, 1927; Vol. 2, 1928. The two articles mentioned deal with black-body radiation (pp. 835–856); and Gehrcke's brief addendum (*Nachtrag*) on p. 1033 describes Boltzmann's introduction of energy quanta as 'fictions' in his famous paper on the interconnections between the second law of thermodynamics and probability theory, which appeared in the *Sitzungsberichte der Wiener Akademie der Wissenschaften*, sec. II, **76** [1877], pp. 376–435. Gehrcke refers to this paper in an attempt to remove Planck's priority claim: "These words, which Lenard kindly drew to my attention, indicate that Boltzmann already had thought of the idea of allowing energy to change in discrete amounts [*sprunghafte Beträge*] which are whole-number multiples of elementary charge [*Elementarquantum*] ϵ." This is actually not correct, since Boltzmann always performed the limes of $\epsilon \longrightarrow 0$ in his calculations after calculating the equipartition function on a combinatorial basis with a finite ϵ. Thus he never assumed discontinuous energy in the physical sense. On Ludwig Boltzmann see also footnote 10 of doc. 44.

[6]Ferdinand Kurlbaum (1857–1927) studied in Berlin and Heidelberg 1880–86, took his doctorate in Berlin in 1887, and became full professor at the Polytechnic in Charlottenburg (Berlin) in 1899, chairing the physics department. He became a member of the *PTR* in 1901. Ernst Pringsheim (1859–1917) studied at Heidelberg, Breslau and Berlin University 1877–82, taking his doctorate as a student of Helmholtz. 1886 he was unsalaried lecturer at Berlin University and in 1896 titular professor. In 1905 Pringsheim was appointed full professor of theoretical physics at the University of Breslau. (See, e.g., Kangro [1976] chap. 7 for a detailed historical account of experimental work at the *PTR* on the spectral density of black bodies, and Kuhn [1978] for the theoretical background.

[7]Willy $\longrightarrow$ Wien's formula for the electromagnetic energy distribution per unit of volume u as a function of the frequency ν and temperature T was $u = \frac{8\pi\nu^3 b}{c^3} \cdot e^{-a\nu/T}$, with c as the velocity of light in vacuo, and a and b constants to be determined through experiment. The formula by Lord [John William Strutt] Rayleigh (1842–1919), $u = \frac{8\pi\nu^2}{c^3} \cdot kT$, which he first suggested in 1900, was only valid for low frequencies in the red and infra-red sections of the spectrum (all formulas are converted to modern units and conventions). For contemporary measurements of this function in the high-frequency and low-frequency regimes, see again Kangro [1976], chap. 4.

[8]Lummer's and Pringsheim's papers were: 'Die Vertheilung der Energie im Spectrum des schwarzen Körpers', *Verhandlungen der Deutschen Physikalischen Gesellschaft*, Series 2, Vol. 1, [1899], pp. 23–41, 215–235; and 'Über die Strahlung des schwarzen Körpers für lange Wellen', ibid., Series 2, Vol. 2, [1900], pp. 163–180.

and declared, for example, that the validity of Wien's formula was just as far-reaching as the validity of the second law of thermodynamics.† [9] Only when the observational evidence gradually became overwhelming did people start to doubt Wien's equation. Planck sent for Lummer's latest measurement results, in order to see how they fit in an interpolation formula. Lummer, who has shown me his correspondence with Planck on this matter on more than one occasion, was accustomed to representing the observations **graphically**, together with Pringsheim; and thus in this way they were the first to find the correct energy distribution. The physical discovery of the distribution of energy must therefore be credited to Lummer and Pringsheim; and that is why one should rather have referred to a Lummer-Pringsheim energy distribution, since clothing the curve in an analytic expression is only a mathematical accessory. It is characteristic of the physics of the past epoch that Planck's elementary mathematical trimmings were esteemed more highly than the original, fundamental physical discovery by Lummer and Pringsheim. This manifests itself, among other things, in the term 'Planck's spectral equation' rather than 'Lummer-Pringsheim energy distribution'. In a zealous effort to find an analytic expression for the distribution curve, formulas were drawn up by the mathematician Jahnke, by Max Thiesen of the *Physikalisch-Technische Reichsanstalt*,[10] and finally also by Planck. The latter undertook to correct Wien's equation by extending Wien's expression: $\frac{c_1}{\lambda^5} \cdot e^{-\frac{c_2}{\lambda\vartheta}}$ with an expansion series made up entirely of Wien's expressions with rising power exponents, namely: $\frac{c_1}{\lambda^5} \cdot \left(e^{-\frac{c_2}{\lambda\vartheta}} + e^{-\frac{2c_2}{\lambda\vartheta}} + e^{-\frac{3c_2}{\lambda\vartheta}} + \cdots\right)$.

This geometric series can easily be summated, thus leading to the familiar expression: $\frac{c_1}{\lambda^5} \cdot \frac{1}{e^{\frac{c_2}{\lambda\vartheta}} - 1}$.

Only when this analytic expression, which reflects the observations more accu-

† [Original footnote 1) (of column 2, p. 439):] "*Ann[alen]. d[er]. Phys[ik]*. **1**, [p.] 118, 1900."

[9]Max → Planck's paper on irreversible radiation processes appeared on pp. 69–122 of the preceding reference and was in fact based on earlier papers dated 1897–99. On p. 118 Planck wrote: "I believe I must conclude from this that the definition of radiation entropy given in § 17, and thus also Wien's energy distribution law, is a necessary result of applying the principle of the increase of entropy to the electromagnetic theory of radiation, and therefore that the limits of validity of this law, should such limits exist at all, coincide with those of the second principle of heat theory." But in his next paper, also appearing in 1900, Planck retracted this thesis and proposed a new formula for the energy distribution of black-body radiation, soon called the 'Planck distribution': $u = \frac{8\pi\nu^3 b}{c^3} \cdot \frac{h\nu}{e^{-a\nu/T} - 1}$. See, e.g., Kangro [1976] as well as Armin Hermann's introduction to the reprinted version of Planck's paper in *Dokumente der Naturwissenschaft*, issue no. 11 [1969], pp. 7–16.

[10]Eugen Jahnke (1863–1921) studied in Berlin 1881–85 and took his doctorate at Halle in 1889. In 1892 he became senior teacher at the *VIII. Realschule* secondary school and from 1900, teacher at the *Friedrich Werder Oberrealschule* upper school in Berlin. He was privy councillor and mining adviser at the Berlin Mining Academy (*Bergakademie*). Max Ferdinand Thiesen (1849–?) studied at Königsberg from 1867–72 and then at Berlin from 1876–78, when he took his doctorate and became unsalaried university lecturer there in 1881. In 1883 Thiesen became *Adjunkt* at the *Bureau International des Poids & Mesures*. He was appointed professor of physics 1890 at the *PTR* and retired in 1910.

rately than the earlier ones, had been found was the search begun for its theoretical derivation and foundation. The details on this can be found in the previously referenced source, p. 851, etc.[11] It is interesting that at a conference the French mathematician Poincaré dressed the inner inconsistencies of Planck's effort at a theoretical derivation in the piquant argument: If, as in Planck's derivation, two contradictory assumptions are incorporated into one theory, then **everything** can be derived from it.[12] Debye§[13] also pointed out this 'weak spot' in Planck's derivation.[14]

47 Peter Debye: Comment on the Preceding Article by E. Gehrcke: 'How the Energy Distribution of Black-Body Radiation Was Really Found' [May 4, 1936]

Source: Peter Debye, 'Bemerkung zu dem vorstehenden Artikel von E. Gehrcke: "Wie die Energieverteilung der schwarzen Strahlung in Wirklichkeit gefunden wurde".' Received May 4, 1936, *Physikalische Zeitschrift*, **37** [1936], p. 440.[1]

My opinion of the importance of M. Planck's work does not in any way concur

[11]See footnote 5.

[12]Gehrcke is paraphrasing a comment by Henri Poincaré (1854–1912) at the first Solvay congress at Brussels in 1911: "On the other hand, it struck me in the discussion we have just heard to see the same theory sometimes applying to the principles of the old mechanics, sometimes to the new hypotheses which negate them; we should not forget that any proposition can be easily proved if you allow two contradictory premises to enter into the proof as well." See Paul Langevin & Louis de Broglie (Eds.): *La Théorie du Rayonnement et les Quanta. Rapports et Discussions de la Réunion tenue à Bruxelles, du 30 octobre au 3 novembre 1911. Sous les Auspices de M. E. Solvay.* Paris: Gauthier-Villars, 1912, p. 451.

§ [Original footnote 1) (of column 1, p. 440):] "*Ann[alen]. d[er]. Phys[ik]*. **33**, 1427, 1910."

[13]In P. → Debye's article cited in footnote § above on the concept of probability in the theory of radiation, 'Der Wahrscheinlichkeitsbegriff in der Theorie der Strahlung', pp. 1427–1434, he derived Planck's formula from the energy quantum hypothesis alone, dispensing with Planck's resonator model.

[14]Gehrcke alludes to the following introductory passage in Debye's paper (see the preceding footnote): "Planck's Law is verified entirely by experience; and yet the derivation has a *weak spot*, to the extent that the basic assumptions of both parts on which the proof of the law of radiation is constructed differ from one another. Namely, on the one hand, as is generally known, a connection is established between the average energy of the resonator, using a formulation that is completely determined by its dependence on its momentum and its rate of change, and the average energy density of the radiation in the ether. But then for the second part of the proof, the pioneering assumption of the existence of elementary energy quanta is made, which nevertheless is in no way connected to the energy description of the resonator which is assumed in the first part; indeed, it virtually contradicts the latter." See the following document for Debye's reply to this sly attempt at claiming the support of one theoretician at the expense of another, when they are actually both in one and the same camp.

[1]See the preceding document for Ernst → Gehrcke's article and the context. Gehrcke had quoted the Dutch physicist Peter → Debye's article of 1910 in his critique of Max → Planck's theory of black-body radiation.

with that of the foregoing article. I am cited because I have expressed the opinion that when seen from a logical standpoint and when supported by the then current knowledge, the derivation of Planck's radiation formula is not flawless. This is correct. But it is precisely in this fact that I see great merit. A lot of courage and an intimate feel for nature was needed to overcome all constraints and to give us the idea of the energy quantum and the quantum of action. This was true theoretical physics, which I also readily rank above mathematical physics. How magnificently Planck has been borne out by the developments is most evident, incidentally, when you try to imagine how deficient today's physics would be without Planck's quantum of action.

Dahlem (Berlin), Kaiser Wilhelm Institute of Physics.[2]

48 Peter A. Thiessen: Physical Chemistry in the National Socialist State [May 9, 1936]

Source: Peter A. Thiessen, 'Die physikalische Chemie im nationalsozialistischen Staat', *Der deutsche Chemiker. Mitteilungen aus Stand / Beruf und Wissenschaft* (supplement to *Angewandte Chemie. Zeitschrift des Vereins Deutsche Chemiker*, No. 19.) Vol. 2, No. 5, May 9, 1936, pp. 19–20.[1]

In his article: 'Intellectualistic Science'* Dr. Christian J. Hansen makes observations about the position of chemistry and the mission of physical chemistry, which cannot be disagreed with more earnestly.[2] There are inaccuracies and grave errors in every sentence of his representation. It is stated that "so-called physical chemistry **dominates** in chemistry". Where exactly is it supposed to be dominating? In professorship appointments in chemistry? In the distribution of research establishments? In the expenditure of funds? The real situation looks

[2]Debye had just been appointed director of the new → *KWIP* (cf. also docs. 30 and 54), which was still under construction.

[1]The author is identified as "P[arty] C[omrade]. [(*Pg.*)] Prof. Peter A. Thiessen, Director of the Kaiser Wilhelm Institute of Physical Chemistry and Electrochemistry, Dahlem (Berlin)". P. → Thiessen had been a member of the → *NSDAP* since 1925 and became director of the → *KWIPC* in 1935, transforming it into a model Nazi institution. See Karlson [1990], p. 98.

∗ [Footnote 1) in the original:] "*V[ölkischer]. B[eobachter].* of March 14, 1936 under *Volkstum, Kunst und Wissenschaft.*"

[2]Christian Johannes Hansen (1886–1958) studied in Zurich, Berlin, Leipzig, and Heidelberg from 1905–08, then taking his doctorate in science in Heidelberg. After two years as laboratory assistant in Leipzig he was employed in 1911 in industry as a chemist at → *IG Farben, AG Leverkusen.* In 1927 he was head chemist and authorized company officer (*Prokurist*), *Heinrich Koppers AG* in Essen. In 1931 he became an independent consulting chemist and in 1938 authorized officer and technical director of the chemical factory, *Byk-Gulden Werke chem. Fabrik AG* in Oranienburg (Berlin). In 1945 he again took up consulting. His article cited in footnote ∗ entitled 'Intellektualistische Wissenschaft', appeared in the 'culture, arts and science' section of the Northern edition, No. 74 of the Nazi Party propaganda newspaper → *Völkischer Beobachter.*

entirely different. At all universities the branches of 'classical' chemistry are much better represented in numbers than physical chemistry. In the field of chemistry organic chemistry takes the lion's share of regular professorships and especially of full professorships. Then inorganic chemistry follows, and only then, physical chemistry. At some universities the latter cannot claim any regular professorships of its own at all; and at a number of universities, it does not even have any research establishments to call its own that are independent of the main chemistry chair. The university policy of the National Socialist state is moving perceptibly towards strengthening the influence of physical chemistry, which is still widely underrated, and towards giving this field a greater role in the chemist's training, as befits its importance.[3] Physical chemistry is **not** the 'semi-science' [*Halbwissenschaft*] that Hansen seems to think it is. It is **not** a hybrid of physics and chemistry unfit for the rigors of life and actually only belonging in a curiosity cabinet. On the contrary, it is an experimental science of fundamental importance to theoretical science and praxis. Accordingly it has had the greatest influence on research and industry, contrary to the effect Dr. Hansen wishes to attribute to it.

The term 'Physical Chemistry' has historical origins. Its substance and scope would be much more appropriately described by the label 'General Chemistry', since, it covers the general foundations of each specialized branch of chemistry, and none of the latter's specialties can dispense with it without suffering injury. Where 'Classical Chemistry', or preparative and analytical chemistry, initially poses the question: '**How**' do bodies react to each other, and which new materials result from their interrelations?, 'General Chemistry', read 'Physical Chemistry', asks: '**Why**' do bodies react in a specific manner to each other? How much energy must be applied to bring about transformations, or which forces must be overcome in the process? Furthermore, it reveals information about the anticipated yield of chemical processes, about the speed of transformations, and even about the expected properties of new substances yet to be created. Its results, which are based upon an abundance of carefully screened material compiled through experiment, thus make it possible to predict through which interrelations between materials and energy a reasonable yield of certain scientifically, technically, and economically important substances can be produced profitably.

Dr. Hansen declares that the returns either for science or for technology brought in by physico-chemical 'semi-science' do not even approach what one

[3]This article exemplifies the struggle for recognition and for improved public standing by many fields within physics as well as those bordering other sciences. Scientists felt the need to protect the existing position of their fields during the unstable Nazi period. Physical chemistry had a curious position right between the much more powerful fields of general chemistry and general physics. Thiessen thus thought it necessary to reply immediately and unequivocally to Hansen's inflammatory article, to prevent any influence it might have on future science policy. Rather than emphasizing the importance and relevance of his specialty in science, he draws on the popular and politically correct norms of the time, such as, not only rejecting the stigma of being too abstract and theoretical, but also pointing to his field's crucial role economically and in industry, as well as to Germany's standing in the world.

should have fairly expected from the amount of funds and personnel invested in it. What does the real situation look like? Very large commercially successful branches of our German chemical industry are based on so-called 'Physical Chemistry'. The manufacture of light metals in particular of aluminium, alkalis, and chlorine, takes place today via electrochemical processes almost without exception with the greatest commercial success. Yet, electrochemistry is a branch of physical chemistry. Our chemical engineering's major syntheses—whether ammonia based on nitrogen in air, organic solvents of inorganic base, or whether the high-pressure syntheses or products of catalytic hydrogenation, known as 'coal hydrogenation', which leads to fuels—have developed directly on the basis of 'Physical Chemistry'. Ceramics as a whole, glass technology, and the appropriate utilization of cement and hydraulic mortars, are no longer conceivable without the principles of colloid science, a branch of physical chemistry. Finally, metallurgy in general, the theory of alloys, their production and working, their functional shaping and hardening, are simply the steady application of physical chemistry.

That is why it is clear that Hansen's statement that supposedly almost without exception none of the most important minds, particularly in theoretical and practical chemistry, were physico-chemists, is also completely incorrect.[4] Van t'Hoff, who had proven the theory of solutions, and whose findings on the structure of molecules are among the self-evident and indispensable tools of organic chemistry, and whose investigations of the equilibriums of salt mixtures also are still today the uncontested basis of the production and preparation of our potassium salts, received the Nobel Prize as a physico-chemist.[5] Wilhelm Ostwald, who recognized and articulated clearly the influence of catalysts on the acceleration of chemical reactions, received the Nobel Prize as well. Without his findings in the field of physical chemistry, many processes of the chemical industry would be unthinkable.[6] Through the well-directed and consistent application of thermodynamics and kinetics on the basis of experimental physical chemistry, Tammann, the 'father of metallurgy', created the essential foundations for the current state of theoretical and applied science of alloys.[7] Zsigmondy, the origi-

[4]For a history of physical chemistry see, e.g., James R. Partington: *A Short History of Chemistry*, New York: Dover, 3rd ed., 1989, chapter XIV, or John W. Servos, *Physical Chemistry from Ostwald to Pauling: The Making of a Science in America*, Princeton: Princeton Univ. Press, 1990.

[5]The founder of stereochemistry, Jacobus Henricus Van't Hoff (1852–1911), was chemistry professor 1878–96 in Amsterdam and subsequently at Berlin University. He was awarded the first Nobel Prize in chemistry for 1901 for his research on stereochemistry.

[6]The monist [Friedrich] Wilhelm Ostwald (1853–1932) was professor of physical chemistry at Leipzig from 1887–1906 and received the Nobel Prize in chemistry for 1909 for his work on catalysts and electrochemistry as well as on the thermodynamics and kinetics of solutions. Cf., e.g., Wilhelm Ostwald, *Elektrochemie, ihre Geschichte und Lehre*, Leipzig: Veit & Co., 1896.

[7]The Russian-born Gustav [Heinrich Johann Appollon] Tammann (1861–1938) studied at the University of Dorpat 1879–82, then becoming teaching assistant at the chemistry department. In 1887 he became unsalaried university lecturer, in 1891 lecturer and a year later pro-

nal founder and most avid promoter of exact colloid chemistry, received the Nobel Prize as a physico-chemist.[8] Clearly a large number of influential leaders in the chemical industry are physico-chemists. This range of the finest names in this field can only be overlooked out of complete ignorance.[9]

It is a grotesque mistake to believe that by using physical **methods** the chemist can adequately substitute the destruction or limitation of general chemistry. Pure methodology can never replace a **specific** way of thinking and working. It is doubtlessly true that inadequate people have also been active in the field of physical chemistry. But, where in science or in life is this not the case? That it is often particularly the unqualified persons who proclaim their opinion the loudest is a misfortune that physical chemistry shares with many other fields; however, the blame for this should not be allowed to fall on this vigorous and realistic [*lebensnah*] branch of science.

Today when competent and responsible promoters of science in the National Socialist state energetically support physical chemistry, which is, of course, none other than 'General Chemistry', this has its reasons. We cannot deny support for the general foundations just at a time when chemistry—'Physical Chemistry' is a branch of chemistry!—has major tasks to fulfill within the framework of the German raw materials issue.[10] Dr. Hansen's attitude is therefore completely incomprehensible; it is explained by a very obvious ignorance of the true context.

'Physical Chemistry' is practiced in Germany almost without exception as an experimental science by fully trained chemists. No branch of chemical research and application can dispense with it without causing serious damage to themselves. Impairing it would be an inexcusable step backwards that would surely cost Germany its lead in chemistry.

49 Werner Heisenberg, Hans Geiger & Max Wien. Petition [Spring 1936]

Source: Werner Heisenberg, Hans Geiger & Max Wien, 'Denkschrift 1936. (Heisenberg-Geiger-Wien Memorandum, 1936)', main text published without signatures as appendix to: Hoffmann [1989], pp. 206–208, and in: Rechenberg (Ed.) [1992]a, pp. 81–83. Addressed: 'An den Herrn Reichsminister für Erziehung, Wissenschaft und Volksbildung.'[1]

---

fessor of chemistry there. 1903–30 he was full professor of physical chemistry at Göttingen Univ.

[8]The Austrian Richard Adolf Zsigmondy (1865–1929) was professor of inorganic chemistry in Göttingen 1908–29. In 1925 he received the chemistry Nobel Prize for his colloid research.

[9]For additional historical analyses of the field of physical chemistry in Berlin in particular, see Harteck [1960].

[10]For background on the Four-Year Plan, see footnote 1 of doc. 52 as well as doc. 58.

[1]The petition, co-authored by W. → Heisenberg, H. → Geiger and M. → Wien was circulated among physics professors at German universities and polytechnics in the spring of 1936. The names of the 75 signers are provided on the basis of a typed carbon copy of the petition in the private papers of Friedrich → Hund, Göttingen.

To the Reich Minister of Education, Science, and Culture.[2]

Reich Minister:

Considering the grave concern within the profession arising out of recent attacks against theoretical physics, the undersigned spokesmen of experimental, theoretical, and technical physics in Germany are particularly thankful to you, Reich Minister, for giving them the opportunity to present their own views in the following on the mutual relationship between experimental and theoretical physics.

Before the undersigned examine this matter in detail, they would like to explain in general terms why these attacks appear particularly unsettling. As Reich Minister knows, at the moment physics in Germany is in serious crisis. A shortage of suitable recruits is imminent, considering the high demand for physicists in technology and in the Army. Great difficulties are often encountered in filling vacated professorships; and the number of physics students in the beginning semesters is far too low.[3] These dangers, determined in part by external circumstances, are being exacerbated by the attacks mentioned above, which in the opinion of the undersigned appear to be unfounded and harmful. They generally scare students away from the study of physics and they scare physics students in particular away from the study of theoretical physics, even though the necessity of a good theoretical education has surely been emphasized often enough not only by representatives of pure science, but also by leaders in technology.[4] After all, these attacks considerably damage the reputation of German science abroad, as is evidenced by several articles in the foreign scientific and daily press.[5] However, a comparison of the duties of these two courses of research shows most plainly that the attacks are in fact unjustified.

The aim of science is to gain knowledge about natural laws. Analysing phenomena experimentally is the primary and categorical condition to acquiring this

---

[2]Bernhard → Rust became minister at the → *REM* in 1934, which replaced the former Prussian Ministry of Science, Art and Education and supervised universities, polytechnics, and all state research institutions. For the ministry's reaction in October 1936 see doc. 50.

[3]See also doc. 35 and the Introduction, sec. 3.4. for statistics on the number of students and faculty members in physics in the years following 1933. Max Wien continued to compile data on student enrollment in physics, which was published posthumously in Wien [1944]. By systematically comparing the numbers of participants in courses and practical sessions for beginning and more advanced students as well as the numbers of doctoral students and qualifying examinations (*Promotionsprüfungen*), Wien was able to show that the "the proportion of mathematicians and physicists studying in the period 1930–39 dropped from 20% to 4%, thus 5 times more than the average [... Thus we can expect] in the future at most as many finished physicists per year as are immediately required to fill the demand in research and university departments for degree-holding assistants; and accordingly the other positions [in industry, official agencies, the military, etc.] will come away empty-handed." (p. 56).

[4]See, e.g., Max Wien's memorandum of 1934 on physics at German universities, doc. 35.

[5]See, for instance, doc. 41.

knowledge. On the one hand, by revealing previously unknown features of nature, it leads to new correlations that have the character of laws. On the other hand, experimental analysis makes the definitive decision on the validity of a conjectured law. The aim of research is not, however, to enumerate many experimental results, but to establish conclusively the natural laws governing the relevant field. It is the task of theoretical physics to make a more accurate determination of these laws. Since Kepler and Newton it has been considered necessary to formulate natural laws in mathematical terms, because this can guarantee the absence of inconsistencies and a clear application of the formulated laws. Thus theory has always been an unavoidable and necessary complement to experiment.[6] Every experimental physicist who has observed something new will feel a sense of fulfillment when his new results are incorporated into theory and are thus related to previous findings; and acknowledgment of this relationship [between theory and experimental results] is crucial for the advancement of science.

In recent decades as well physics has developed as a result of a close collaboration between experimental and theoretical research. The abundance of new and singular experimental results in atomic physics and in radiation physics have led to the novel conceptions of natural processes frequently labeled with the catchwords, relativity theory and quantum theory. These theories are highly abstract. One group of physicists today sees them as a transitional phase to be surmounted; another group believes that this unintuitiveness [*Unanschaulichkeit*] is based upon the nature of the object at hand and is therefore unavoidable. In any case, this unintuitiveness is completely compensated by the fact that on the one hand modern theory forms a comprehensive basis for understanding new experimental results.* On the other hand, however, it also is a fruitful stimulus for further experimental research.†[7] Quantum theory in particular has been an extraordinarily productive stimulus to physics and especially also to experimental physics; and we Germans can be proud of the fact that the foundation stone of this theory was laid by a German.[8]

Thus in the future both directions in physics, experiment and theory, must work together. Through their interaction they are equally involved in the advancement of science. The undersigned would welcome it, however, if in the future it

[6]For another text on the relationship between theoretical and experimental physics see, e.g., doc. 21; cf. also the Introduction, sec. 5.3.

* [Footnote 1) in the original.] "E.g., through mastering the spectra of the elements at the theoretical level."

† [Footnote 2) in the original:] "E.g., diffraction of matter waves, separation of hydrogen in two modifications, annihilation of positrons, etc."

[7]Footnote † alludes to de Broglie waves, the separation of orthohydrogen and parahydrogen by Bonhoeffer and Harteck (cf. footnote 6 of doc. 29) and Dirac's theory of electrons, which predicted anti-electrons with the same mass and opposite charge, and processes of pair-creation and annihilation of positron-electron pairs. On Dirac, see also footnote 16 of doc. 55.

[8]This is an allusion to Max → Planck, who initiated quantum theory with important papers in 1900; cf. doc. 46 and references there.

were possible to cease all public discussion in the daily press that attempts to belittle one line of research to the advantage of the other.

[Signed:[9]]

Aeckerlein, Dir. Phys. Dept., Mining Academy, Freiberg. **Back**, Dir. Phys. Dept., Univ. Tübingen. **von Bayer**, Prof. Univ. Berlin. **Barkhausen**, Prof. Polytechnic, Dresden. **Bechert**, Prof. Univ. Giessen. **Bothe**, Prof. K.W.I. of med. Research, Heidelberg. **Braunbek**, Prof. Univ. Tübingen. **Buchwald**, Prof. Polytechnic, Danzig. **Busch**, Prof. Polytechnic, Darmstadt. **Cermak**, Prof. Univ. Giessen. **Czerny**, Dir. Phys. Dept., Univ. Frankfort. **Debye**, Dir. K.W.I. of Physics, Dahlem (Berlin). **Diesselhorst**, Prof. Polytechnic, Braunschweig. **Esau**, Councillor of State, Rector of Univ. Jena. **Füchtbauer**, Dir. Phys. Dept. of Univ. Bonn. **Fues**, Prof. Univ. Breslau. **Geiger**, Dir. of Phys. Dept. of Polytechnic, Charlottenburg (Berlin). **Gerlach**, Dir. Phys. Dept., Univ. Munich. **Gerthsen**, Dir. Phys. Dept., Univ. Giessen. **Grüneisen**, Dir. Phys. Dept., Univ. Marburg. **Gudden**, Dir. Phys. Dept., Univ. Erlangen. **Hahn**, Dir. K.W.I. of Chemistry, Dahlem (Berlin). **Harms**, Prof. Univ. Würzburg. **Heisenberg**, Dir. Inst. of theor. Phys., Univ. Leipzig. **Hettner**, Prof. Univ. Jena. **Hoffmann**, Dir. Phys. Dept., Univ. Leipzig. **Hückel**, Prof. Polytechnic, Stuttgart. **Hund**, Prof. Univ. Leipzig. **Joos**, Prof. Univ. Göttingen. **Jordan**, Prof. Univ. Rostock. **Kalähne**, Prof. Polytechnic, Danzig. **Karolus**, Prof. Univ. Leipzig. **Koch**, Dir. Phys. Dept., Univ. Hamburg. **Kühl**, Prof. Univ. Jena. **Kratzer**, Prof. Univ. Münster. **Kulenkampff**, Dir. Phys. Dept., Univ. Jena. **Kunze**, Dir. Phys. Dept., Univ. Rostock. **Leithäuser**, Prof. Polytechnic, Charlottenburg (Berlin). **Lenz**, Prof. Univ. Hamburg. **Madelung**, Prof. Univ. Frankfort. **Mie**, Dir. Phys. Dept., Univ. Freiburg/Br. **Möller**, Prof. Univ. Hamburg. **Nernst**, Prof. Univ. Berlin. **Ott**, Prof. Univ. Würzburg. **Paschen**, emer. Prof. Univ. Berlin. **Picht**, Prof. Polytechnic, Charlottenburg (Berlin). **Planck**, emer. Prof. Univ. Berlin. **Pohl**, Dir. Phys. Dept., Univ. Göttingen. **Prandtl**, Prof. Univ. Göttingen. **Rajewsky**, Prof. Univ. Frankfort. **Rau**, Dir. Phys. Dept., Polytechnic, Darmstadt. **Rausch von Traubenberg**, Prof. Univ. Kiel. **Regener**, Dir. Phys. Dept., Polytechnic, Stuttgart. **Reich**, Prof. Univ. Göttingen. **Reiger**, Prof. Univ. Erlangen. **Rogowsky**, Prof. Polytechnic, Aachen. **Rüchardt**, Prof. Univ. Munich. **Schaefer**, Dir. Phys. Dept., Univ. Breslau. **Schiller**, Prof. Univ. Leipzig. **Schmidt**, Prof. Univ. Münster. **Schulze**, Prof. Univ. Marburg. **Schumann**, Prof. Polytechnic, Munich. **Seeliger**, Prof. Univ. Greifswald. **Seitz**, Prof. Polytechnic, Aachen. **Smekal**, Prof. Univ. Halle/S. **Sommerfeld**, Dir. Dept. of theor. Phys., Munich. **Starke**, Prof. Polytechnic, Aachen. **Straubel**, Prof. Univ. Jena. **Unsöld**, Prof. Univ. Kiel. **Valentiner**, Prof. Mining Academy, Clausthal. **Wachsmuth**, Prof. Univ. Frankfort. **Waetzmann**, Prof. Univ. Berlin. **Wehnelt**, Dir. Phys. Dept., Univ. Berlin. **Wien**, Prof. emer. Univ. Jena. **Zenneck**, Dir. Phys. Dept., Polytechnic, Munich.

50 Reich Education Ministry: Memorandum to Reich Education Minister Bernhard Rust [October 2, 1936]

Source: Bundesarchiv Koblenz: R21 (Rep 76.)/203, sheet 29. Typed document headed with the administrative notation: 'Dem Herrn Minister weitergeleitet.' and initialed: 'Z/Sch.'[1]

Seen from the point of view of the Ministry, the matter appears to me to be as

[9] In order of appearance on the typed transcription. The petition was signed by a total of 75 physics professors from all specialties and of various political persuasions: both known critics of the Nazi regime and loyal party members. The more outspoken critics, like Max von → Laue, were not invited to sign, however, because it may have damped the effect of the petition.

[1] This internal memorandum was probably by Werner Zschintzsch, who was undersecretary at the → *REM*: See footnote 7 of doc. 53.

follows:

In a controversy among professors that has been taken up by political newspapers,[2] one side has sought the support of the Reich Minister of Science. Its representative Mr. Heisenberg has visited the responsible official and has been advised to find a common position for all German physicists on the scientific problem supposedly under debate: 'Theoretical Physics/Experimental Physics'.[3] Following this advice, a circular letter announces that the Reich Science Minister, represented by Prof. Mentzel, wishes to settle the dispute. It thus may appear as though the Science Minister were standing on Heisenberg's side in the dispute and were using him.

In the case of a purely scientific dispute, in my opinion, the Minister should keep himself out of it. In the case of a political dispute, there would be even less reason to intervene in the matter. In all events the impression has been made that the Minister had placed himself on the side of those being attacked for promoting the Jewish spirit in physics. It would certainly have been more correct had the official consultant not attached his name, and that of the Science Minister's, to interventions of this kind.

Berlin, the 2nd of October 1936.

51 Germany and the Nobel Prizes [February 12, 1937]

Source: 'Germany and the Nobel Prizes', *Science*, **85**, February 12, 1937, 'Scientific Events' section, p. 171. (Anonymous article).

Following the award of the Nobel Peace Prize to Carl von Ossietzky, the German pacifist who was confined at the time in a concentration camp,[1] Chancellor Adolf Hitler issued on January 30 a decree which reads:[2]

---

[2]See Willi Menzel's 1936 article, Werner → Heisenberg's response, and Johannes → Stark's comment, docs. 42–44.

[3]See also the Himmler correspondence in 1938, docs. 62–64, and the character evaluation by the → *SS*, doc. 73, on W. Heisenberg's drawnout case.

[1]Carl von Ossietzky (1889–1938) was a German pacifist leader and editor of *Die Weltbühne*, a left-wing political weekly, in which he exposed secret rearmament preparations by *Reichswehr* leaders. He was sentenced to 18 months imprisonment for treason in November, 1931, but was amnestied in December 1932. When Hitler came to power von Ossietzky was sent to the Papenburg-Esterwegen concentration camp in February 1933 where he contracted tuberculosis and died while still in custody in a Berlin hospital on May 4, 1938.

[2]When Ossietzky was awarded the Nobel Peace Prize in 1935, A. → Hitler was so enraged that he prevented the internee from accepting it and forbade all Germans from accepting any Nobel Prizes, to prevent similar affairs in the future. The decree entitled 'Führer's and Reich Chancellor's Decree on the Foundation of a German National Prize for Art and Science.' ('Erlaß des Führers und Reichskanzlers über die Stiftung eines Deutschen Nationalpreises für Kunst und Wissenschaft'), was published in *Reichsgesetzblatt*, Part I, No. 33, dated Jan. 30, 1937, issued in Berlin on March 16, 1937, p. 305.

> In order to avert such shameful occurrences for all future time, I decree with this day the foundation of a German national prize for art and science.
>
> This national prize shall be divided annually among three worthy Germans to the amount of 100,000 marks each.
>
> Acceptance of a Nobel Prize is herewith forbidden to all Germans for all future times.
>
> Executive orders will be issued by the Reich Minister for Popular Enlightenment and Propaganda.

At the Propaganda Ministry it was emphasized that the prohibition applied not merely to the peace prize, but to all Nobel awards.[3]

The New York Times prints in full the statement made to the *Reichstag* by its president, Colonel General Hermann Goering, which reads:[4]

> Ridiculous insults which proceed partly from rage and partly from the bad taste of others can neither offend nor shame Germany. They merely fall back on those from whom they come, and especially on those who always pride themselves on their special good breeding.
>
> When we see attempts to insult Germany before the world by awarding a peace prize to a traitor, to a person punished with penal servitude, then such action does not shame Germany but merely makes those ridiculous who are responsible for it.
>
> But because Germany will not tolerate such shameful things in the future and does not want any dispute about them at all, the *Fuehrer* has created this day a national prize for art and science.
>
> May the world realize from this that everything which it may undertake to insult the German people will always fall back on the other. It is, as the *Fuehrer* has repeatedly emphasized, a singular characteristic of present-day democracies to ignore facts and realities. But one must learn in time that the once torn and impotent Germany has been transformed into a proud, strong, honest, honor-loving, freedom-loving people—a people that has a right to be proud of its achievements before all nations, before history, before the future.

An Associated Press dispatch from Stockholm of the same date reports that Germany's ban on acceptance of Nobel Prizes by Germans will have no effect on the granting of awards by the Nobel committee. Professor Karl Manne Siegbahn, a committee member who won the prize for physics for 1924, asserted that the

[3]The Party rulers ignored the fact that the two prizes are granted by entirely different institutes. The Nobel Peace Prize award is granted by the Norwegian parliament while the scientific prizes are granted by the Swedish Academy of Science.

[4]H. → Göring's statement to parliament appeared in the Sunday, Jan. 31, 1937 issue of the *New York Times*, Vol. 86, No. 28,862, pp. 1 and 26.

awards would be without regard for German laws, on the ground that no distinctions were possible between German and other scientists.[5] However, whether payment is possible is a matter between the winner and his government.[6]

52 Bernhard Rust: Creation of a Research Council [March 16, 1937]

Source: Bernard Rust, 'Bildung eines Reichsforschungsrats', *Deutsche Wissenschaft, Erziehung und Volksbildung* **3** [1937], Amtlicher Teil, section 'Wissenschaft, a) Für das Reich', No. 172, pp. 151–152.

The Four-Year Plan's major goals for German science necessitate that all efforts towards these goals in the field of scientific research be uniformly concentrated and systematically applied.[1] The principle of independent research will not be affected by such a coordination of certain branches of science to meet the Four-Year Plan target; nor will a systematic assignment of projects and a regulated distribution of resources affect it. This is because the freedom of research is based not upon an arbitrary choice of projects, but rather upon its independent operation.[2] In this historical moment as science takes on goals of immense scope that are vital to the whole nation, no special justification is needed for scientific

[5][Karl] Manne [Georg] Siegbahn (1886–1978) studied in Lund, Sweden 1908–09 as well as in Göttingen and Munich and took his doctorate at Lund University in 1911. He was appointed professor of physics at the University of Lund and then at the University of Upsala in 1923. He was awarded the physics Nobel Prize for 1924 for his contributions to X-ray spectroscopy. In 1937 Siegbahn was made the first director of the Nobel Institute of Physics in Stockholm, founded by the Swedish Royal Academy of Sciences; he held this position until 1964 in conjunction with a professorship at Stockholm University, his research primarily covering atomic structure.

[6]For instance, the Nobel Committee of the Swedish Academy of Science awarded the 1939 Nobel Prize in chemistry jointly to Adolf Butenandt (1903–) and Leopold Ružička (also Ruička) (1887–1976), and retroactively for the year 1938 to Richard Kuhn (1900–1967); Gerhard Domagk (1895–1964) received the Nobel Prize in physiology and medicine. They were all forced to decline the award, however. For Butenandt's vivid description of the pressure he was subjected to by R. → Mentzel and his colleagues, see Karlson [1990], pp. 105–106. See also *Time*, Nov. 6, 1939: "Since Adolf Hitler forbade Germans to accept Nobel Prizes, Domagk has already politely refused to take the prize money. [...]. Kuhn and Butenandt will probably do the same, unless they want to perform the scientific experiment of living in a concentration camp." The prizes were handed out to them retroactively in 1947 when they were able to explain the circumstances of their initial involuntary refusals.

[1]The crisis in foreign trade and foreign exchange since 1935 prompted the announcement of a 'Four-Year Plan' on Sep. 9, 1936, enacted on Oct. 18, 1936, placing H. → Göring in charge of its implementation. The plan encouraged the production of substitute and raw materials to make Germany economically autonomous and intended to make the Army operational within four years. Cf., e.g., Noakes & Pridham (Eds.) [1990], Vol. 1, part II, chap. 13.

[2]This is a redefinition of the concept of 'freedom of research'. It is also an attack on the earlier evaluation process in which committees consisted solely of experts in the relevant scientific discipline, which from the Nazi point of view was seen as 'arbitrary'.

research to work with particular emphasis precisely in these target areas and thus be obliged where necessary to postpone less important and less urgent tasks, even in the case where the performance of the latter more adequately matches the researcher's previous field of work or better coincides with the customary practices of resource distribution.[3] In order to guarantee the systematic and consistent implementation of the required measures as quickly as possible, I create herewith a Research Council to cover the agencies engaged in research that are subordinate to me.[4]

The Research Council's establishment does not eliminate the agencies within my purview heretofore extant in this area; and it also encroaches upon their jurisdiction and administration only insofar as this is naturally required in fulfillment of the Council's goals. It is not set up, however, for the purpose of calling into existence an additional parallel organization to take over, supervise or otherwise assign the current responsbilities of existing agencies.

The specific responsibilities of the Research Council arise out of the general objectives laid down in Par. 1 and 2.[5]

The Research Council is subject to my supervision. It consists of the following bodies:

1. a president,
2. a vice-president,
3. the administrative manager,
4. scientists to be appointed especially as directors of the future specialty divisions within the Research Council,
5. a legal advisory board.

The members of the Research Council are appointed to and removed from office by me.[6]

I have appointed as president of the Research Council the dean and full professor at the Berlin Polytechnic's Defense Engineering Department, Artillery General Karl Becker, and as his vice-president, the head of the Science Office, Minister of State Dr. Otto Wacker.[7]

---

[3]This is also a critique of the evaluation process conducted by the *Notgemeinschaft* (→ *NG*, now called German Scientific Research Association (→ *DFG*)) and other research foundations.

[4]See also doc. 75 for E. → Schumann's account of the establishment of this council.

[5]The fact that the new research agency's powers were not clearly defined with respect to the existing establishments, such as the *DFG*, which were neither eliminated nor explicitly reduced in scope, illustrates the typical counterproductive duplication of authority in the polycratic Nazi system.

[6]The top-downward hierarchy followed the '*Führer* principle', replacing the democratic structures of elected representatives.

[7]Both the weapons engineer at the *Wehrtechnische Fakultät* of the Berlin *Technische Hochschule*, General Karl → Becker (see doc. 75, footnote 6), and the *SS* senior troops leader (*Oberscharführer*) and ministerial director at the *Amt Wissenschaft* of the → *REM*, Otto Wacker (1899–1940), were typical Nazi bosses, outsiders to the existing scientific establishment. They were responsible for, and controlled the political orientation of research policy.

The remaining members of the Research Council will be appointed shortly at the president's recommendation. [...[8]]

Berlin, the 16th of March, 1937.

Reich and Prussian Minister of Science, Education and Culture Rust.[[9]]

53 Reich Ministry of Science, Education and Culture: The Taking of a Doctorate by Jews of German Nationality [April 15, 1937]

Source: 'Erwerb der Doktorwürde durch Juden deutscher Staatsangehörigkeit', *Deutsche Wissenschaft, Erziehung und Volksbildung* **3** [1937], Issue 9, May 5, 1937, Amtlicher Teil, section 'Wissenschaft, a) Für das Reich', No. 229, pp. 224–225.

In agreement with the *Führer*'s deputy, with the Reich and Prussian Interior Minister[[1]] and with the Foreign Ministry, I rule with immediate effect that Jews (§ 5 of the First Ordinance of the Reich Citizens Law of the 14th of November, 1935, *R[eich] L[aw] G[azette]* I, p. 1333),[[2]] who are of German nationality, are no longer to be admitted to examination for a doctor's degree. In addition, renewal of doctor's diplomas held by the same is to be discontinued.

The taking of a degree [*Promotion*] by part-Jews [*jüdische Mischlinge*] (§ 2 Par. 2 of the First Ordinance of the Reich Citizens Law of the 14th of November, 1935)[[3]] remains admissible. In cases of doubt my decision must be obtained.

German citizens must enclose with the application for admittance to examination for a doctor's degree a filled out questionnaire according to the attached sample, together with the candidate's birth certificate, the birth and marriage certificates of the parents and grandparents on both sides, or in place of the certificates, the certified geneological tree [*Ahnenpaß*].[[4]] There is no objection to

[8]The omitted three paragraphs specify the organization of the Research Council and its various commissions and committees, as well as its relation to the *DFG*.

[9]Bernard → Rust held this position at the *REM* from June 1, 1934 until 1945.

[1]Hermann → Göring was appointed Prussian Interior Minister on Jan. 30, 1933 and Prussian Minister-President in April, 1933. He was named Commander-in-Chief of the Air Force and → Hitler's deputy on May 1, 1935. Wilhelm → Frick became Reich Interior Minister on Jan. 30, 1933 (until August 1943) and in addition head of the Prussian Interior Ministry in May 1934.

[2]This article of the infamous Nuremberg Laws provided the official definition of a Jew—that is, those descended from at least three grandparents who were fully Jewish by race, or from two such grandparents and is married to, or a child of, a Jew when this law came into force. See Noakes & Pridham [1990], Vol. 1, p. 539, for a translation of this article, and pp. 530–538 for the background to these discriminatory laws.

[3]Article 2 defines Jews of mixed blood as those descended from one or two fully Jewish grandparents, and specifies further that any grandparent who is a member of the Jewish religious community at the time of the issuance of the law is also considered a 'full' Jew. See ibid., p. 538.

[4]On the required geneological information, cf. *Deutsche Wissenschaft, Erziehung und Volksbildung* [1937], Amtlicher Teil, section 'Amtliche Erlasse, Allgemeine Verwaltungssachen a) Für das Reich', No. 139: 'Verwendung von Ahnenpässen zum Nachweis der Abstammung', issued

the return of the certificates or the *Ahnenpaß* after inspection. In this case an appropriate notation is to be made in the degree conferral files [...[5]].

There is no objection to the delivery of doctor's diplomas to Jews who have already completely fulfilled the graduation requirements upon this edict's coming into force, that is, who have also provided the prescribed obligatory copies of their doctoral thesis to the academic department. The same applies to those cases in which I have already exceptionally given my approval of admittance to examination prior to this edict, provided registration for the examination follows at latest within three months after this edict comes into effect. This deadline must be brought explicitly to the attention of the candidates concerned. Yet the following regulations must be noted regarding the delivery of a doctor's diploma to students of medicine and dentistry who are citizens of Jewish blood. [...[6]]

These regulations apply accordingly to the qualification of university lecturer [*Dr. habil.*]. Applicants for habilitation, who have taken their degree prior to the publication of this edict are required to supply in addition to the aforementioned certificates the questionnaire provided in § 4 Num. 2 of the R[eich]. Habil[itation]. Reg[ulation]. following the sample attached [thereto] regarding citizenship in addition to the aforementioned certificates, unless the certified genealogical tree is submitted.

For foreigners the prevailing regulations remain in force.

I request that the academic departments be informed accordingly.

Berlin, the 15th of April, 1937.

The Reich and Prussian Minister of Science, Education and Culture.

By proxy: Zschintzsch.[[7]]

54 Peter Debye: The Kaiser Wilhelm Institute of Physics [April 23, 1937]

Source: Peter Debye, 'Das Kaiser-Wilhelm Institut für Physik', *Die Naturwissenschaften*, Vol. 25, Issue 17, April 23, 1937, pp. 257–260.

The new building of the Kaiser Wilhelm Institute of Physics was built in the course of the last year in Dahlem, in close proximity to the other scientific in-

---

on Feb. 16, 1937, pp. 115–116.

[5]A related citation and the following paragraph on the repeal of conflicting regulations is omitted.

[6]The omitted two paragraphs of text lay out the conditions persons with one Jewish parent must meet in order to obtain their diplomas; the stateless are treated as foreign students.

[7]Werner Zschintzsch (1888–?) was engaged at the Prussian Interior Ministry in January 1925. He became a senior executive adviser (*Oberregierungsrat*) there in April 1925 and cabinet minister (*Ministerialrat*) in August 1926. In February 1933 he was named district president in Wiesbaden and later served as Minister Bernhard → Rust's undersecretary at the → *REM*. On Zschintzsch, cf. Führerlexikon [1934], p. 548; Heiber [1990], Vol. 2.1, pp. 60, 195f., 229f., 371, 382.

stitutes of the Kaiser Wilhelm Society.[1] With the exception of certain interior installations the new institute is now complete enough for scientific work to begin already in a few months. The Kaiser Wilhelm Institute of Physics had initially been established as a scientific fund as early as towards the end of the war.[2] This fund mainly supported physicists in loaning instruments to conduct their experiments. It only became possible for the Institute to procure its own building with the Rockefeller Foundation's grant of building funds to the Kaiser Wilhelm Society,[3] as well as through the Reich government's active support of the project, both in its conception and materially by making available the most important part of the fixed budget for the period, thus guaranteeing the feasibility of the scientific enterprise.

Professor Sattler demonstrated great insight in drawing up the plans to accommodate various requirements.[4] (They were realized under the very capable direction of the architect, Schrank, director of the building office of the Kaiser Wilhelm Society.) For one, the laboratories had to be designed as generally as possible to suit experiments in all fields of physics. At the same time, however, two special fields of work were planned from the outset requiring especially constructed installations; namely, firstly, experiments in the field of nuclear physics employing very high electrical voltages, and secondly, research at extremely low temperatures very close to absolute zero. In addition, the laboratories had to be separated as far as possible from the workshops and engine rooms to avoid vibrations and noise. In fulfilling these requirements the basic arrangement of the building resulted as shown in the attached diagram. The main wing extending roughly from E[ast] to W[est] contains all the general laboratories, while the engine rooms and workshops are housed in a lower side wing perpendicular to it. A special tower for the extra-high voltage facilities is annexed on the West side of the main wing, in which the nuclear physics experiments are carried out.[5] All installations and working areas connected to the production and analysis of low temperatures are brought together within a separate single-story building, the low temperature laboratory situated to the North of the main building.

---

[1]For background on the construction of the → *KWIP* and its funding by the Rockefeller Foundation, see doc. 30. For a description of Dahlem, the site of most of the Kaiser Wilhelm Institutes, see footnote 9 of the same document.

[2]On the foundation of the Institute as merely a funding agency in 1917 see Kirsten & Treder (Eds.) [1979], Vol. 1, pp. 146–157.

[3]On the early history of the *KWIP*, cf. Macrakis [1986] and Kant [1987], [1992].

[4]Carl Sattler was director of the state School of Commercial Arts in Munich and main architect of the Kaiser Wilhelm Society. He designed the *Harnack-Haus* (1927–29) the *KWIP* (1936–37) and all the Kaiser Wilhelm Institutes in Dahlem on the outskirts of Berlin between 1918 and 1939. See Vierhaus & vom Brocke (Eds.) [1990], pp. 238, 323f., 880.

[5]Cf. doc. 74 for a photograph of these facilities.

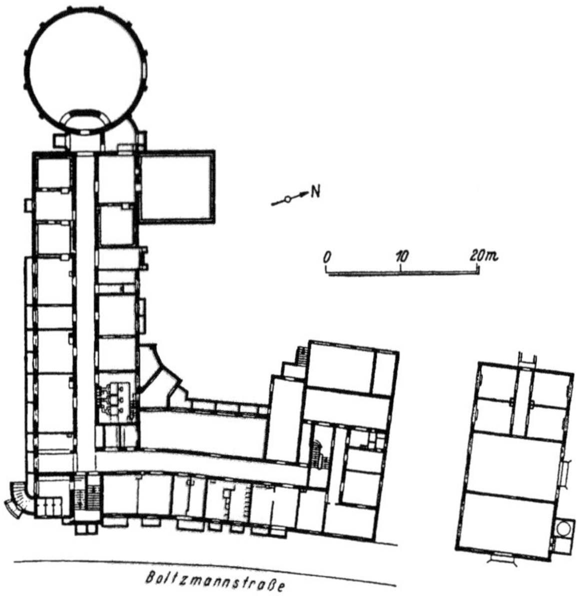

Fig. 7: Floor plan of the KWI of Physics facilities in Dahlem [on p. 257 of the original; the big circular structure is the so-called Max Planck Tower – cf. here also fig. 16, p. 202].

The diagram shows the dimensions and layout of the buildings. The main wing of the institute has three floors. It consists of basement (which lies halfway below street level), ground floor and upper story. In the attic above it, directly under the roof, there is still room, however, for later extension if necessary. The laboratories, altogether about 20, are located chiefly in the basement and upper story in this wing. The ground floor mainly contains, aside from a few laboratories, rooms for the director and the administration, book storage room, reading room, and a colloquium room that is suitable for smaller-scale lectures and conferences. The basement laboratories are all provided with vibration-free foundations that are embedded in concrete in the earth separately from the remaining ground. In addition, some rooms here were built without windows and with walls lined with insulating tiles in order to enable maintaining particularly constant temperatures. One of these rooms, 9 × 8 meters in size, is annexed to the main wing and is intended for experiments with a concave grating of a large radius of curvature. Some of the laboratories on the upper story also have special installations, namely, for researches using X rays and cathode rays. A small room to accommodate high-voltage instruments (transformers, vents) is located exactly between each pair of laboratories it supplies with high voltage; and also next to these laboratories are the observation rooms from which the observer can monitor his instruments through a lead glass window to avoid having to stay in the rooms, which are always a bit contaminated with radiation, when taking longer X-ray photographs. But even those laboratories planned primarily for specific purposes have been provided besides with the same equipment as all the other laboratories, in order to also be available for other kinds of experiments if the need arises. [...[6]]

In designing and furnishing the laboratories, we have, of course, also drawn on all the experience from other new large physics and chemistry buildings. [...[7]]

We have now become generally acquainted with the facilities and the most important installations of the main building and its laboratories. Now the **extra-high voltage tower** and the **low temperature laboratory** must be considered in a little more detail. The round windowless tower, which is to accommodate the extra-high voltage system, is 15 m[eters] in diameter and has a total height including the tower spire of about 20 m[eters].[[8]] A single cylindrical room fills the whole interior, which is completely open from its floor, which is level with the basement of the institute, up to a height of $14\frac{1}{2}$ m[eters]. At this height a gallery runs along the inside wall of the tower from which a trolley can be operated which is suspended from the tower's roof frame and which can sweep all points situated not too close to the tower wall through movement in two directions perpendicular to each other. The actual extra-high voltage system is being built by the company Siemens & Halske.[[9]] The funds for this are being made available by the Reich government. The system consists of two columns of about 7 meters height erected symmetrically from the center of the tower, which are constructed following the principle of a Greinacher voltage duplicating circuit of condensers and rectifying valves.[[10]] Each column is supposed to produce at a total current consumption (corona + usable current) of 3 mA a maximum voltage of about 1,400 kV against earth (no-load voltage about 1,600 kV). The voltage can be adjusted continuously from zero up to this maximum voltage. Therefore, there is a voltage available between the highest conductors of both columns of 2.8 million Volts at 3 mA when one column is switched to positive and the other to negative against earth. The voltage constancy is so efficient as a result of using 500 periodic current for the input transformer, that the total ripple for one column will amount to only 7 kV per milliampere of current consumed. The fact that each column can be run separately and its voltage lies already significantly above 1 million Volts is particularly important. Thus experiments on [electron and positron] pair generation from radiation requiring at least 1 million Volts can already be carried out with **one** column with probably quite good efficiency.

---

[6]Two paragraphs follow with a detailed description of the wing containing maintenance, electrical, and storage rooms, as well as workshops, heating and employee facilities.

[7]A more detailed description of architectural installations, permanent fixtures and utilities of the laboratories follow in this and the next two omitted paragraphs.

[8]See doc. 74 for a photo.

[9]The → *Siemens* Company was a major supplier of a variety of specialized research equipment.

[10]In a Greinacher voltage duplicating circuit, two condensers are charged alternately. They then decharge in a series ciruit whose voltage is twice the amplitude of the transformer voltage. See Heisenberg [1943]b (3rd. ed.) pp. 144f., or [1989], C series, Vol. IV, p. 430, for a full description and circuit diagram of the Greinacher voltage duplicating circuit and the *KWIP*'s high tension generator, which could produce up to 1.2 million Volts.

It goes without saying that the radii of curvature of all structural parts in the tower were made as large as possible to avoid as much corona loss as possible. But special protective measures against radiation also had to be implemented, however, due to the hardness and intensity of the γ rays formed during full operation (corresponding to several kilograms of radium). All exterior walls are therefore of quite massive construction. In addition, a special radiation-proof 80 cm thick protective wall was built in front of the doors leading from the separate floors of the institute into the tower, lined one above the other with balconies extending in front, so that no in any way hazardous amounts of radiation can escape from the tower into the institute. An observation room separated from the tower by an especially thick wall was also installed from which the entire high-voltage system can be operated. Since absorption of very hard rays is only proportional to density, the shielding effect of masonry is very much more efficient relative to that of lead at high voltages than is the case with lower voltages.

Whereas the high-voltage tower is annexed directly to the institute's laboratory wing, the **low-temperature laboratory** is isolated from the main building as a separate building. Here experiments are to be conducted on the behavior of matter at very low temperatures, which can be produced according to the adiabatic demagnetization process specified by Debye and carried out practically by de Haas.[11] For this process to be fully effective it is necessary to start from the lowest possible temperatures; and it was thus necessary to erect a large enough system for the successive liquefaction of nitrogen, hydrogen, and helium. Since working with liquid hydrogen always requires special ventilation and other safety measures due to the hazard of explosion, separating the low-temperature laboratory from the rest of the institute emerged as the most convenient solution. The one-story building contains, apart from 2 engine rooms, a small workshop and three smaller laboratory rooms. In addition, there is also an extension for a hydrogen gasometer. Raising additional floors in the future as needed in the West section containing the laboratories is easily feasible. One of the engine rooms serves to accommodate the compressor and the liquefaction system with an ex-

[11]Peter → Debye and W. F. Giauque (1895–1982) proposed independently in 1926 a new cooling technique, adiabatic demagnetization, which made it possible to arrive at temperatures below 1 K (see footnote 13 below). See, e.g., W. F. Giauque & D. P. MacDougal, 'Attainment of temperatures below 10 Absolute by demagnetization', *Physical Review* **43** [1933], p. 768. Wander Johannes de Haas (1878–1960) studied physics at Leiden 1905–11, becoming teaching assistant there and obtaining his doctorate in 1912. Between 1911–13 he was assistant at the Bosscha Labs, Berlin, then 1913–15 at the → *PTR*. After a one-year teaching position at each of two schools in Deventes and Haarlem, de Haas took on a professorship in physics 1917 at the Delft Polytechnic. 1922 he accepted a call to Groningen, moving finally to Leiden in 1924 (until 1948) also to direct the Kamerlingh-Onnes lab of physical chemistry. Aside from his work on adiabatic demagnetization, in 1915/16 he worked on what later became known as the Einstein-de Haas effect, and on paramagnetism and superconductivity. For a list of his publications in the *Proceedings of the Kgl. Akademie van Wetenschappen Amsterdam* between 1923–31 see *Poggendorff*, Vol. 6 (1937), pp. 992–993.

pansion engine that functions on the Linde-Heylandt process for the production of liquid nitrogen.[12] The system will produce 20 l[iters] of liquid nitrogen per hour. In addition, the large magnet which is needed to apply the magnetic process is also housed in this room.[13] The liquefaction systems for hydrogen and helium are in the other engine room. The engine rooms are particularly well ventilated by means of a so-called lantern attached to the gable with windows to the exterior. All the major switches for the motor drive, etc., are mounted in separate switchboxes outside of the building on its exterior wall, the switches affixed inside for lighting and to control the outside switches are specially protected to rule out an explosion triggered by sparking. The general set-up of the 3 laboratories is entirely the same as that of the other laboratories of the institute, only here special arrangements have also been made to avoid switch sparking.

The research program automatically emerges from the description of the institute's installations. This institute testifies to the extraordinary foresight of the sponsors and the governmental authorities responsible for science: the Kaiser Wilhelm Society for the Advancement of the Sciences, which drew up the plans for the building of the institute; the Rockefeller Foundation, which gave the final impetus through a grant; and the Reich government, which gave its support in every way to the venture, both conceptually and materially. They have created here a **unique opportunity** to approach **fundamental problems** of physics on a large scale, thus supporting a project whose practical fruits will perhaps only appear in decades to come. Here again, in a time of extreme economic strain, which justifiably causes many agencies to build upon the familiar and to fix their eyes on the nearest at hand, the highest authorities of the government leadership demonstrate wise and far-sighted interest in independent, unappropriated, and unrestricted research, which the latter indeed can and must claim for the future.

[12]Linde-Heylandt liquefiers differ from the usual expansion cooling machines in their use of higher pressure and higher initial temperatures. In 1922 *Linde Eismaschinen A.G.* bought up the Heylandt Company, including its manufacturing unit in Britz (Berlin) for air-separation machines based on the Heylandt system; in 1944 the Linde Company was destroyed in an air raid, partly because it was involved in the production of heavy water (deuterium) for the *Uranverein.* Cf. *75 Jahre Carl v. Linde Gesellschaft für Eismaschinen*, Frankfurt: Brönner, 1954, pp. 123–125, and Walker [1989]a, pp. 144f.

[13]Adiabatic demagnetization is a method of obtaining very low temperatures. A paramagnetic salt is cooled to 1 K by liquid helium. The salt is magnetized under isothermal conditions and then magnetized under adiabatic conditions (avoiding heat exchange with surroundings). As a result, the temperature falls.

55 'White Jews' in Science [July 15, 1937]

Source: ' "Weiße Juden" in der Wissenschaft', *Das Schwarze Korps*,[1] issue no. 28, July 15, 1937, p. 6; reprinted, e.g., in Sugimoto [1989], p. 126. (Anonymous article).[2]

There is a primitive type of anti-Semitism that limits itself to fighting against Jews alone. Its proponents are content to believe that there is a clear dividing line between Germans and Jews. They believe the problem would be solved, if a ban were set on intermarriage and Jews were no longer allowed to participate in the political, cultural, and economic life of the nation.[[3]] They have a vague notion that a Jewish exodus to Palestine or elsewhere was the ultimate solution.[[4]] It follows from this that the moment Germany were rid of the Jews [*judenrein*] to this extent, anti-Semitism should also come to an end.

Attractive though this view may be for its simplicity, it contains a fallacy. If we were to combat the Jews following the old and even obvious characteristics of hooked noses and frizzy hair, this fight would be a fight against windmills. The fact that we had to fight Jewish influence in politics and in cultural life, however, and that we have to continue to fight against Jewish influence in German business already proves that we are not dealing with Jews 'per se', but rather with the mentality or rather bad mentality [*Ungeist*] they spread; we are actually dealing with what we call influence.[[5]]

---

[1]This weekly journal appearing on Thursdays since March 6, 1935, was a propaganda organ of the → *SS* and → Himmler's mouthpiece, carrying the subtitle: 'Newspaper of the Defense Squadrons of the German National Socialist Party, Organ of the Defense Squadron Reich Leadership' ('Zeitung der Schutzstaffeln der NSDAP, Organ der Reichsführung–SS'). It was published by Franz Eher Verlag, which was owned by the → *NSDAP*, and was edited by Himmler's confidant Gunter d'Alquen (born in 1910). Around 1937 approximately 500,000 copies were printed weekly, growing to more than 750,000 copies in World War II. It represented the views of radical, revolutionary National Socialism, sometimes even in opposition to the official party line. Its issues were full of attacks against intellectuals, clerics and many other social groups Himmler was displeased with for one reason or another. Cf., e.g., Höhne [1984/90]b, pp. 205–207 on the close collaboration of the *Das Schwarze Korps*'s editors with Heydrich's security service (*SD*) since 1937, and William L. Combs, *The Voice of the SS. A History of the SS Journal 'Das Schwarze Korps'*, New York, 1986.

[2]This article was followed by a comment by Johannes → Stark; see doc. 56 and the first two footnotes there on authorship. Doc. 57 presents the reaction of one of Heisenberg's colleagues.

[3]Such a ban was imposed by the Nuremberg 'Law for the Protection of German Blood and German Honor' of Sep. 15, 1935, in which marriages between "Jews and citizens of German or kindred blood are forbidden". Cf., e.g., Noakes & Pridham (Eds.) [1990]b, Vol. 1, part II, sec. 23 (ii), pp. 535ff.

[4]During the early years of the Nazi regime, various vague and mostly unrealistic plans emerged even into 1940, such as the large-scale resettlement of the Jewish population and mass emigrations to Madagascar and Palestine, followed by the accelerated deportation of Jews to the Eastern territories acquired at the beginning of the war. Only gradually did the 'final solution' or *Endlösung* emerge in the form of a systematic holocaust, officially resolved at the Wannsee Conference in January 1942: Cf., e.g., Hilberg [1961], Mommsen [1991], pp. 184ff.

[5]With this turn in the debate and the invention of the term 'White Jew' (see below) Nazi

Unfortunately, in the aftermath of the ideal solution of a Jewish emigration, we will still have to struggle against Jewish influences; and a wide field of action will still remain for aggressive anti-Semitism, even if not a single hooked nose were left in the entire German Reich. It is unfortunately the case that the terrible threat of the Jewification [*Verjudung*] of our public life and the power of Jewish influence that National Socialism was forced to curb was not propagated by the Jewry alone, which was numerically weak. Rather, people of Aryan blood who proved to be disposed toward the Jewish mentality and who became enslaved by it also helped them to no small degree.

For this reason any victory of racist anti-Semitism can only be rated as a **partial victory**. We should not content ourselves with insisting upon a thorough implementation of the Nuremburg Laws and with finding a solution to the continuing problem of '**Jews in business**'.[6] We must also exterminate the **Jewish spirit** that can flourish today undisturbed as never before, while its carriers plant its seed among the best examples of Aryans.

This is because the **ethnic Jew himself** threatened us less than the **mentality** he was spreading. When the carrier of this mentality is not a Jew but a German, it is twice as important to fight him than it is to fight an ethnic Jew, who cannot conceal the source of his mentality.

Jews by Mentality ['Gesinnungsjuden']

The vernacular has stamped such germ carriers with the label, 'White Jew', which is extremely suitable, because it extends the concept of the Jew beyond race.[7] In the same sense we could also speak of Jews in spirit [*Geistesjuden*], of Jews by mentality [*Gesinnungsjuden*], or of Jew types [*Charakterjuden*]. They complaisantly adopted the Jewish mentality, because they lack one of their own. They are admirers of a sophistical intellect, because they lack natural instincts and those moral values that force men to develop their own abilities, and to limit themselves to them, if necessary.

One field above all in which the Jewish spirit of 'White Jews' is evident in pure culture [*Reinkultur*], and in which the intellectual ties of 'White Jews' to Jewish role models and masters can be irrefutably demonstrated, is **science**. Cleansing it of the Jewish spirit is of utmost urgency; since, we can surely catch up with the 'White Jew', whom we meet in everyday life, through the police and through

ideologues attempted to gain control of the academic sector that had limited itself to following strictly racial criteria. Now scientists who refused to adapt to the 'spirit' of the new regime but so far had been protected from legal measures because of their purely German descent, came under increased fire.

[6]On the so-called Nuremburg Laws arising out of Nuremburg meeting of the *NSDAP* see footnote 1 of this doc. The public discrimination of Jewish businessmen had already started with the call for a systematic official boycott of Jewish shops in April 1933.

[7]The origin of this term 'White Jew' is not clear, but it apparently was also used as an abusive term against the authors Thomas and Heinrich Mann in Nazi propaganda.

more stringent laws; but **Jewish contaminated science** is the key position from which intellectual Jewry can repeatedly win leading influence on all aspects of life of the nation.

Thus, it is characteristic, for example, that in our times when the German medical community is faced with entirely new challenges and when decisive performance is expected from medical research in the fields of genetics, racial hygiene [*Rassenhygiene*], and public health,[8] out of a total of 2,138 papers produced in medical journals within a half-year, 1,085 papers are from foreign authors, with 116 Soviet Russians among these, which hardly cover problems that would appear to be of particular importance to us today. Hiding behind the screen of an 'exchange of knowledge' is the thesis that science is international, which the Jewish mentality always introduced and propagated because it created the pre-condition for its uncontrollable self-glorification.

The Jewish spirit is certainly most clearly recognizable in the field of **physics**, where Einstein rose up as its most 'prominent' representative.[9] All great scientific discoveries and findings are due to the extraordinary abilities of the Germanic researcher to make constructive scientific observations through patience and diligence. The Germanic scientist always regards so-called theory only as an aid, which simplifies scientific observation where necessary, but which can never become an end in itself. The knowledge of true reality is the only goal of his research, to which he even sacrifices his own working hypotheses, should they turn out to be incorrect or inadequate. In contrast, in the last decades, the advancing Jewish spirit managed to push **theory, which was taught dogmatically and was far removed from reality**, into the forefront. By sophistically generalizing available knowledge, cleverly juggling with mathematical formulas, and using obscure ambiguities, it established the autocracy of such theories. They corresponded so well to the Jewish spirit and Jewish 'research methods' because they made diligent, patient and constructive scientific observation appear superfluous.

The Dictatorship of Gray Theory

The Jewish professor in Munich Leo Græetz made the significant statement that in the course of time experimental physicists would fall below theoreticians to the level of a good mechanic, to whom the theoretician would assign tasks for experimental analyses.[10] Einstein also proclaimed in a lecture in 1922: "We

[8]For a history of these disciplines during the National Socialist period see, e.g., Weingart, Kroll & Bayertz [1988] or Proctor [1988].

[9]A. → Einstein had become the figurehead of radical change in theoretical physics in the public eye since the 1920's. See, e.g., von Brunn's review of a book authored by non-physicists directed against the revolutionary theory of relativity, doc. 4; cf. also Hentschel [1990]a, sec. 3, on the popular debate about the physical theory.

[10]Leo Graetz (1856–1941) studied at the University of Breslau 1877–81 and took his doctorate in 1880. He became unsalaried lecturer at Munich University in 1883 and then associate professor in 1893. In 1908 he was appointed full professor there until 1926 and worked on electricity

can expect that theory will soon be in a position to calculate the properties of chemical atoms and their reactions in advance, so that the chemist's tedious, time-consuming experimental work will become superfluous." [11]

This aim of driving the scientists who are in close contact with reality into the background, was put very quickly into practice.

The Jews Einstein, Haber, and their mind-mates Sommerfeld and Planck were in almost total control of the issue of German professorship appointments.[12] Sommerfeld alone could boast of having provided ten of his students with full professorships. The publications of Jewish theoreticians and their propagandists swelled in just under fifteen years to 50,000 printed pages. Young students were educated almost exclusively in their 'spirit' [*'Geist'*]. Had they been allowed to have their way, in a few decades the type of scientist that is productive and close to reality would have died out. Ruminating, unfruitful theoreticians would have taken their place.

National Socialism's seizure of power has staved off this danger; but it has not eliminated it yet. The Jewish mentality attempts to hold on to its positions with all its might in those branches of science at least, which like **physics**, do not have an obviously ideological character; and these positions could one day become the point of departure for a renewed advance.

Einstein as a Cornerstone ['Eckstein']

How secure 'White Jews' feel in their positions, is evidenced by the actions of the professor for theoretical physics in Leipzig, Professor Werner **Heisenberg**, who succeeded in 1936 in smuggling an essay into an official party organ, in which he declared Einstein's relativity theory to be "the obvious basis for further research" and saw "one of the noblest missions of German youth in science as the continued development of theoretical systems of concepts".[13] At the same time

and thermal conductivity.

[11]The editor was unable to locate this quote in Einstein's lectures of this period: in Kyoto, Japan, on Dec. 14, 1922; in Paris, April 6, 1922; and in Gothenburg, July 22, 1923. Cf., however, Stark [1987], p. 62, for another version of this 'quote'.

[12]Fritz → Haber, Arnold → Sommerfeld and Max → Planck were also favorite targets in the campaign against theoretical physics.

[13]These are modified quotes from W. → Heisenberg's paper, which appeared in the *Völkischer Beobachter* on Feb. 28, 1936 (cf. here doc. 43). Heisenberg decided to respond to what he perceived as a severe personal attack. He submitted an official letter to the → *REM* in which he requested a thorough examination of his case: "Either the ministry regards the standpoint of *Das Schwarze Korps* as correct, in which case I request my dismissal. Or, on the other hand, the ministry disapproves of such attacks, in which case I believe that I have the same rights to protection that, say, the *Wehrmacht* itself would render its youngest lieutenant in such a case." (quoted in Cassidy [1992]a, p. 384). Cf. also ibid., chap. 20, here doc. 64, or Heisenberg's correspondence with Sommerfeld in 1938, DMM, call no. 1977-28, A-136 for the later developments in this affair, which led on one hand to the exclusion of Heisenberg as Sommerfeld's successor at the University of Munich, but on the other hand did not lead to

he attempted to impress influential persons and to silence critics of his work with a poll of German physicists on the theory's merit.

This puppet of the Einsteinian 'spirit' in new [Weimar] Germany and **this model pupil of Sommerfeld's** became professor in Leipzig at the age of 26, thus at an age that hardly afforded him time to pursue thorough research.[14] He began his vocation by letting the German teaching assistants at his institute go and hiring instead first the Viennese Jew Beck, then the Jew Bloch from Zurich.[15] Up to 1933 his seminars were mainly attended by Jews; and even today the core of his students still consists of Jews and foreigners.

The 'Ossietzky' of Physics

In 1933 Heisenberg received the Nobel Prize along with the Einstein disciples Schrödinger and Dirac—a demonstration of the Jewish influenced opposition by the Nobel Committee to National Socialist Germany, which is comparable to their 'distinction' of Ossietzky.[16] Heisenberg returned his thanks in August 1934 by refusing to sign a proclamation by the German Nobel laureates in support of the *Führer* and chancellor. His response then was: "Although I personally vote 'yes', political declarations by scientists seem to me improper, since [this] was never a

---

his forced resignation at Leipzig and exile, as he had feared in his letter to Sommerfeld of April 14, 1938: "Now I actually don't see any other option than requesting my dismissal, if the protection of my honor is denied here. But I would like to ask your advice beforehand. You know that it would be very painful for me to leave Germany; I do not want to do it, if it is not absolutely necessary. But I also don't feel like living here as a second-class citizen. At the moment, therefore, it is completely unclear to me what should happen next."

[14]On Heisenberg's Leipzig period, which turned out to be very successful, both in terms of research and in particular in terms of teaching, see Kleint & Wiemers (Eds.) [1993], pp. 136ff., for a listing of his students, scientific guests, lectures and seminars.

[15]Guido Beck (1903–1988) took his doctorate at Vienna University under Hans Thirring in 1925. Two years later he derived the photoelectric effect from Schrödinger's quantum mechanics. In 1928 he worked in nuclear physics, becomin teaching assistant of theoretical physics in Leipzig in the same year under Heisenberg. In 1932 Beck was visiting lecturer at the German University in Prague, in 1934 professor of mathematical physics at the University of Kansas, and a year later for theoretical physics at the University of Odessa in the USSR. In 1938 he joined a research project at the University of Lyon. In 1942 Beck was visiting professor at the Universities of Coimbra and Oporto, Portugal, for a year, then from 1943 until 1951 astrophysicist at the National Observatory in Cordoba, Argentina. In 1951/52 he was appointed titular Prof. Centro Brasileiro de Pesquisar Fisicas, Rio de Janeiro, Brazil, returning to Argentina at the Physics Institute in Barilocha in 1963. On Beck see *Physics Today* **43** [1990], pp. 89–90. Felix → Bloch started his studies at the Zurich Polytechnic 1924–27, before coming to Leipzig.

[16]On Carl von Ossietzky and his Nobel Peace Prize, see footnote 1 of doc. 51, Hitler's decree of Jan. 30, 1937, forbidding Germans from accepting the Nobel Prize. In 1933, Heisenberg was awarded the reserved 1932 Nobel Prize "for the creation of quantum mechanics, the application of which has led, among other things, to the discovery of the allotropic forms of hydrogen". Erwin → Schrödinger and Paul Adrien Maurice Dirac (1902–1984) received the 1933 Nobel Prize 'for the discovery of new and fruitful forms of atomic theory"; cf., e.g., Cassidy [1992]a, pp. 324f.

normal practice even formerly. Therefore I do not sign."[17]

This response exemplifies the Jewish mentality of its author, by regarding the national solidarity and responsibility of scientists as 'improper'.

Heisenberg is only one example among several others. All of them are puppets of Jewry in German intellectual life and must disappear just as the Jews themselves.

The urgency of this demand and the importance of the present problem confronting the future of German scientific research induced *Das Schwarze Korps* to solicit the opinion of Professor Johannes Stark, president of the *Physikalisch-Technische Reichsanstalt.*[18] The eminence of this expert, who was of course not only an old champion of National Socialism, but is also a recipient of the Nobel Prize—which he received at a time when its distribution was still unaffected by hateful and vindictive political motives[19]—should open everyone's eyes to the fact that German science and its responsible caretakers face a decision that can no longer be avoided.

56 Johannes Stark: 'Science' is Politically Bankrupt [July 15, 1937]

Source: Johannes Stark, 'Die "Wissenschaft" versagte politisch', *Das Schwarze Korps*, July 15, 1937, issue 28, p. 6; reprinted, e.g., in Sugimoto [1989], p. 127.

Professor Stark wrote us in response:[1]

The preceding article is basically so appropriate and complete that further additions would really be superfluous. But on the invitation of the editorial staff I would still like to add the following remarks.[2]

It is generally known that during National Socialism's time of struggle [*Kampfzeit*] the great majority of professors at German universities and colleges **failed disgracefully, nationally speaking**. They were unsympathetic toward Hitler and his movement; and some even repudiated him. At many universities serious

[17]This telegram is also quoted in the *SS* report on Heisenberg, doc. 73.

[18]See the next document for Johannes → Stark's comment on this article. Stark was elected president of the → *PTR* in 1933 (cf. doc. 18).

[19]J. Stark received the 1919 Nobel Prize in physics for his experimental studies on the influence of electrical fields on spectral lines; Stark [1920]; [1987], pp. 53–57.

[1]Johannes → Stark's article is a supporting comment on the foregoing anonymous article; both appeared in the weekly journal of the → *SS*. For an immediate reaction by a close colleague of Heisenberg see doc. 57.

[2]Although this sentence suggests that Stark was not the author of the preceding document, the latter's content and style point clearly to the contrary. If a second author had been involved, he would probably have been alluded to in docs. 62 and 64. It is not inconceivable that Stark chose this form specifically not only to strengthen his argument by presenting the appearance of two independent opinions, but also to disguise his very personal attack on Heisenberg. Cf. Beyerchen [1977], chap. 8, Cassidy [1992]a, chap. 20, and Schröder [1993] on Stark's ideological anti-Semitism.

conflicts arose between the National Socialist student body and the faculty linked to the black-red political system.[3] In 1933 Reich Minister Rust[4] justifiably had bitter words in this regard for the professoriate in Berlin.[5] The dominant Jewish influence at German universities was the decisive cause of the political failure of the majority of German professors in the National Socialist struggle for German liberty.[6]

It's strength lay not only in the fact that in numerous departments 10 to 30 percent of all lecturers were Jewish or closely related to Jews,[7] but it was also primarily because Jews had the support of their Aryan comrades [*Judengenossen*] and pupils [*Judenzöglinge*].[8]

The Jewish mentality's political influence at universities was well-known. Its influence scientifically speaking was less well-known but equally damaging, in that it crippled Germanic research, which is oriented toward reality, by instilling Jewish intellectualism, dogmatic formalism, and propagandistic commercialism. Furthermore, this Jewish influence attempted to educate students in the Jewish

[3]Black and red are the colors adopted by the German Socialist and Conservative parties, respectively. The Weimar governments formed by these parties, led by the Social Democrat Hermann Müller in 1924 and by the Centrist politician Heinrich Brüning from 1930, generally shunned the National Socialists in domestic politics until 1933.

[4]Bernhard → Rust only became *Reichsminister* of the → *REM* in 1934; in 1933, he was still Prussian Minister of Culture.

[5]In the *Göttinger Tageblatt* of Monday, May 8, 1933, Vol. 45, No. 107, p. 1, it was reported that during his official announcement to the Berlin student body (*Studentenschaft*) about their revised legal rights on May 6, 1933, the Minister of Culture supposedly said, while addressing himself to the Berlin professors: "In these years [of the Weimar Republic] when this un-German government blocked the German youth's way, in your professorial isolation and in your dedication to your research work, you failed to see that the youth was looking to you for leadership into the German nation's future. The youth took to the streets, but gentlemen, you were not at the head! That's the point."

[6]Cf., e.g., docs. 77 and 83 for other incidents of this complaint, which indirectly show the failure of Nazi efforts to gain complete control over the academic sector. Since ethnic Jews had been forced into exile or to resign due to the National Socialist legislation, Stark was again clearly thinking in terms of his own ideological definition of the Jew.

[7]A census completed on June 16, 1933 listed the place of residence and profession of the 503,000 Jews still in Germany at that time, comprising 0.76% of the population; but in cities like Frankfort-on-Main, Berlin or Breslau they came to more than 3% (cf., e.g., Noakes & Pridham (Eds.) [1990], Vol. 1, pp. 522f.). A retrospective examination of the situation at specific universities shows that in some universities, such as for instance, the physics and mathematics section of the University of Göttingen, the proportion of Jews in the faculty was indeed above the population average; cf., e.g., Volkov [1987] on possible causes for the strong representation of Jews particularly in the natural sciences, and Nachmansohn [1979] on their scientific achievements.

[8]Stark's use of these terms as well as 'white Jews' in the preceding article was aimed at extending the racially based attack to his 'Aryan' opponents. His ideological conception of 'Jewishness' allowed Stark to occasionally defend ethnic Jews like Gustav → Hertz (see here doc. 32) while still feeling free to attack any 'Aryan' colleagues for supposedly having been ideologically brainwashed: Cf. Schröder [1993], pp. 332f. See also footnote 5 of doc. 55.

mentality, including primarily the upcoming academic generation.

Tactics Changed

Though ethnically Jewish lecturers and teaching aids were forced to resign their positions in 1933 and though Aryan professors married to Jewesses are presently also being discharged, the majority of Aryan comrades and pupils of Jews, who had previously supported Jewish power in German science openly or covertly, have retained their positions and keep the influence of the Jewish spirit alive at German universities.

In their unworldliness they counted on a swift end to the National Socialist government right up to the election of the *Führer*, and therefore withheld public announcements in support of him.[9] Two years ago, however, their tactics changed. Now they behave outwardly nationalistically; former pacifists throng to military service; pupils of Jews, who had published numerous scientific papers together with domestic and foreign Jews and who had participated besides in congresses held by Soviet Jews, now seek connections within party and government agencies.

Aside from their nationalistic or even National Socialist activities, they also try to exert their influence at high places with the following arguments: They say that as scientific experts they and their candidates are indispensable to the implementation of the Four-Year Plan.[10] Moreover, they say, they are recognized abroad as great German scientists and must have a leading influence on German science in the interest of its reputation. With this bluff they believe they can count on the fact that the authorities are not informed that their 'fame' abroad is an over-exaggeration of the result of their collaboration with foreign Jews and comrades of Jews.

The following is indicative of continued Jewish influence in German academic circles: Not long ago an influential doctor said to me: "I cannot at all imagine medical science without Jews."

New Flood of Jews

The department of natural sciences at a major university recently nominated three pupils of Jews for a professorship, two of whom had published numerous scientific papers **together with domestic and foreign Jews**. The scientific book market in Germany is being deluged lately, primarily in physics, with books penned by domestic and foreign Jews and pupils of Jews, particularly through

[9]This may allude to Heisenberg's refusal to sign a petition in support of Hitler in 1933/34, quoted in the previous document.

[10]On the Four-Year Plan see footnote 1 of doc. 52. On its repercussions for scientific education see doc. 58.

the Julius Springer publishing company in Berlin and Vienna, which was formerly completely Jewish and is today supposedly 50 percent Aryan.[11]

While the Jewish spirit's influence on the German press, literature and art, as well as on the German legal system has been eliminated, it has found its defenders and proliferators in German science among the Aryan comrades and pupils of Jews at universities. Behind the screen of scientific impartiality and referring to its international recognition, the Jewish spirit continues on unweakened and even seeks to dominate by securing and consolidating its tactical influence on important positions.[12]

Under these circumstances *Das Schwarze Korps* does a great service with its courageous and fundamentally important observations by turning public attention to the damage inflicted by 'White Jews', which threatens a segment of German intellectual life as well as the education of the academic youth.

57 Friedrich Hund: Letter to Reich Minister Rust [July 20, 1937]

Source: Provided and published by permission of Prof. Emer. Friedrich Hund, Göttingen. Carbon copy of typed letter unsigned. Addressed: 'An den Herrn Reichsminister für Wissenschaft, Erziehung und Volksbildung. Berlin.'

Complaint about the President of the Reich Physical and Technical Institute, Prof. Johannes Stark.[1]

Reich Minister,

The newspaper *Das Schwarze Korps* of the 15th of July prints two articles under the headline 'White Jews in Science', the first without naming the author and the second bearing Mr. Stark's signature.[2] In addition to containing insults to Planck and Sommerfeld, the first article makes very serious abusive statements about my colleague W. Heisenberg,[3] specifically by drawing a parallel between

[11]The publishing house was founded by the Berlin bookseller Julius Springer (1817–1877) in 1842. Springer's grandchilden directed the company in the 1920's: Ferdinand, who stressed publishing in medicine, the natural sciences and art; and his cousin Julius, who stressed the engineering sciences. The company also published the journals → *Die Naturwissenschaften* and from 1918 *Mathematische Zeitschrift*, which was initially edited by Leon Lichtenstein. From 1946 the publishing house's main place of business was Heidelberg; cf. Holl [1992].

[12]Cf. doc. 83 for another assessment of the fight over the filling of vacant chairs in physics, written three years later. Stark had in mind the continued influence in particular of A. → Sommerfeld, who was frequently asked for an expert opinion on candidates to professorship appointments in theoretical physics.

[1]J. → Stark was president of the → *PTR* from May 1, 1933 until 1939.

[2]See docs. 55, 56 and 64 here.

[3]Werner → Heisenberg was professor of mathematical physics at the University of Leipzig from 1927. Max → Planck became professor at Berlin University in 1889 and was a renowned quantum theorist. Arnold → Sommerfeld started his career as professor of mathematics in 1897

him and the traitor Ossietzky.[4] In the second article President Stark supports these representations. He thus attacks what he considers a misguided line of physical research which is pursued exceptionally well by Planck and Heisenberg and which the rest of us physicists esteem very highly, by calling it Jewish and by supporting insults to Heisenberg that exceed all bounds of decency. I regret that a German physicist has stooped to this.

As a close colleague of Heisenberg and as a member of the disparaged line of research, I think it fitting to inform you of this. I have confidence that you, Mr. Reich Minister, will prevent the President of the *Physikalisch-Technische Reichsanstalt* from injuring the honor of our science any further in this manner.

[Friedrich Hund]

Prof. of Mathem[atical] Physics at the University of Leipzig.[5]

58 Carl Krauch: Youth to the Front Line. New Blood in Science and Technology [August 1937]

Source: Carl Krauch, 'Jugend an die Front. Die Nachwuchsfrage in Wissenschaft und Technik', *Der Vierjahresplan*, Vol. 1, 8th Series, August 1937, pp. 456–459.[1]

The outcry against overcrowding, cautions against academic studies, and complaints about an academic proletariat are still ringing in our ears.[2] Then unnoticed to the wider public a significant change suddenly took place. Today there is already a shortage of around 5,000 engineers in industry. From 1930 to 1935 the total number of students at polytechnics dropped to a half. Enrollments decreased to a third of its former level.[3] Labor service and military service cause current students to complete their career training $2\frac{1}{2}$ years later.[4] With a total

and was full professor of theoretical physics at Munich 1906–38.

[4]On Carl von Ossietzky see footnote 1 of doc. 51, a report on the German government's response to the award of his Nobel Peace Prize.

[5]Friedrich → Hund was appointed to this position in 1929.

[1]As head of the Research and Development Department of the German Raw Materials Office *Abteilung Forschung und Entwicklung im Amt für deutsche Roh- und Werkstoffe*), and member of the board of the → *IG Farben* concern, Carl → Krauch had an immediate interest in a sufficient supply of university graduates.

[2]Cf. doc. 12, the 1933 law on the 'overcrowding' of schools, and footnote 1 there on the feared development of an academic proletariat.

[3]For data on student enrollment and passed academic examinations during the National Socialist period, see the 10-year statistics prepared by the → *REM*: Lorenz [1943], and the Introduction, sec. 3.4.

[4]During the rearmament period, 1935–45, unmarried German students were required to spend their holidays working as a kind of obligatory community service (*Arbeitsdienst*). Aside from bed and board, the only compensation received for such heavy agricultural or factory labor was weekly pocket money of the nominal amount of 3 marks (in 1938). For eyewitness accounts see, e.g., Sopade (Ed.) [1980], 1934 vol.: pp. 126f., 221f. 224, 419–424, 640–646; and 1936 vol.: p. 1341 on students. Compulsory military service for all Germans 20–25 years of age

supply of 250,000 engineers by 1942, we will be short of 30,000–35,000 engineers. Chemists, the second large group within the scientific profession, are in a similar position.

Today already, universities are in need of instructors and primarily of teaching assistents. This fact has already led to a substantial drop in the examination performance of graduates at universities and technical schools. Before I go into the reasons for these developments and make practical suggestions, I think a few words are necessary on the importance of these trends to the fate of our nation in the future.

The German nation is forced to live within a much too restricted space. Barred from owning the world's sources of raw materials, it is forced to produce the raw materials it needs for its national security out of its own deposits—from coal, salts, lime and other substances, as well as from air and water through chemical means. This is the point of the Four-Year Plan, as described by the *Führer* in these words: "I am establishing this today as the new Four-Year Program. In four years Germany must be completely independent of foreign countries with regard to all materials that German talent, our chemistry and machine industry, as well as our mining industry can itself produce in some way." [5]

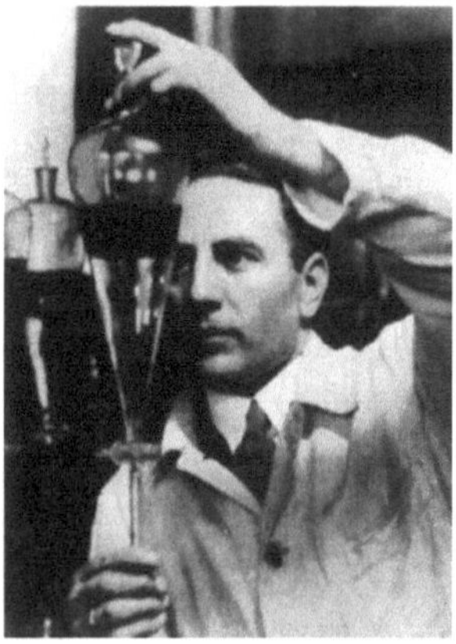

Fig. 8: 'The chemist', 'Phot[ographer]. Dr. Paul Wolff'

Alongside of this great overall mission, the German people's second vital charge involves producing export goods and constantly improving their quality. This allows the German nation to acquire goods and services from abroad

(lasting 18 months in peacetime) was reintroduced on May 21, 1935 with the passage of the new defense law (*Wehrgesetz*). See Sopade (Ed.) [1980], 1935 vol.: pp. 275–282, 387–392, 409–412. Altogether, three years were normally spent in such work camps and barracks; students attempted to fulfill these obligations during their holidays.

[5]Hitler introduced the Four-Year Plan in 1936 to attain national economic self-sufficiency (autarky) and to promote industrial production especially for rearmament purposes; see, e.g., Noakes & Pridham (Eds.) [1990], Vol. 1, chap. 13, and footnote 1 of doc. 52.

through international commerce which is necessary for the nation's security and for raising its living standard.

Every person should do his best in his own position towards fulfilling these great obligations. Ultimately, we can only be successful, however, if no one fails or slackens his performance. Despite the best of intentions and the greatest effort, the factory worker will not be able to reach the set goal if engineers or chemists fail or are not available in the necessary numbers. The scientist and technician is indispensable to the farmer because he provides him with the means of maintaining and increasing his productivity and because he makes available to him machinery and fertilizers essential to his enterprise. The soldier, too, needs the collaboration of science and technology. As armaments are being constantly improved throughout the world, it provides him with the weapons to successfully counter also on a technical level all attacks that could ever threaten our nation.

It is apparent, therefore, that the scientist and technician has a great and important mission to fulfill for the life destiny of our nation, right beside the worker, the farmer and the soldier. Scientists and technicians only rarely make an appearance in public: Their work only becomes known in the case of great and revolutionary inventions. And yet, it is they who can and must make crucial contributions in the German nation's current struggle for survival [*Lebensbehauptung*] to assure its security and continued development in the future, which is a goal that we all wish together with the *Führer*.

We thus see that maintaining and developing German science as well as the next generation of German engineers and chemists is of utmost importance to the whole, and that lagging behind in this section of the front line could in the long-run endanger the success of the German nation and its leadership in its struggle for renewal.

Fig. 9: 'In the laboratory', '[Photographer Dr. Paul Wolff]'

What are the reasons for the developments presented above?

The economic recovery accomplished by the National Socialist leadership and rearmament have engaged all of the available young talent practically in the areas of technology and chemistry. In estimating future developments, we will base our considerations upon how the young person confronted with the prospect of entering professional life tends to see his future today. Just as in school, the student in the last grade of highschool acquires a sincere feeling of responsibility as a member of the H[itler] Y[outh], the *SS* or the *SA*;[6] he sees it as his duty to put his life and his work into building and advancing his fatherland. He should acquire extensive skills in his youth to this end, in order to be able to apply them soon for the good of his country at the same time as starting a family. Above all, he will be given repeatedly to understand the value of the 'practical' professions of the craftsman, the businessman and above all, of the [military] officer and political leader. In comparison, the scientist and scholar, and in particular the natural scientist dealing with 'mathematical' matters, is unfortunately often seen as an unworldly stay-at-home, to whom the stream of life passes by unnoticed. That is why an alert and vivacious boy who is considering his future profession and exhibits an inherent talent for science tends more towards becoming an officer, or under suitable preconditions at home, towards becoming a merchant or possibly also a doctor, as the man who can offer practical help to suffering humanity. He can much less easily decide to engage in such a 'theoretical' matter as science. Added to this the young engineering and chemistry student initially sees ahead of him a course of study and training that is possibly too long. Naturally even now he cannot yet survey adequately the chances that undoubtedly await him in trade and industry if he is diligent and has the appropriate skills. It is by no means an exclusively American privilege anymore for young technicians to arrive at the highest peak of success.[7] In modern Germany trade and industry also offer diverse opportunities to develop one's knowledge and abilities in a responsible position.

Now, how can the developments leading necessarily to a neglect of scientific and engineering careers be checked? An attempt will be made in the following to provide possible solutions to this question. The situation is such that things can be changed through the unanimous cooperation of all agencies and organizations involved in these matters.

---

[6]The Defense Squadron (→ *SS*) and Storm Detachments (→ *SA*) actively recruited young members and there was considerable social pressure to join the boy-scout type Nazi youth organization *Hitler Jugend*. Cf., e.g., Peter D. Stachura in Bracher et al. (Eds.) [1986], pp. 224–244, and refs. there.

[7]In the 19th Century, the "scientist-engineer hybrid" dominated during the developing years of American science. On H. Draper, L.M. Rutherfurd, H.A. Rowland and A.A. Michelson, for example, cf. Nathan Reingold (Ed.) *Science in Nineteenth-Century America: A Documentary History*, New York: Hill & Wang, 1964. On the subsequent rise of physics see, e.g., Kevles [1987], and here doc. 91.

The following steps appear to be important:

I. Informing the public of the importance of science and technology to our nation, namely, the following points:

1. The implementation of high-quality scientific and technological achievements is indispensable to realizing our political aims.
2. Scientific and technical work at university, in the laboratory, in the technical design room, and in the factory directly contributes to safe-guarding our nation's present and future.
3. Supreme effort is required of every scientist and engineer, as an unconditional duty to the national community during their period of training just as later in practicing their trade.

II. Specially instructing the youth on the aims and prospects in science and technology.

Presentations should be made before highschool seniors by suitable gentlemen from universities and industry which include slides (and perhaps tours) in which the importance of technology and chemistry to the German national economy is demonstrated along with the chemist's vital role and mission.

III. Improving the future prospects of the student in the natural sciences.

The training and future prospects of the would-be technician and chemist must be adjusted so that the highschool graduate can be told that as a chemistry student he can take his doctor's degree at 25 and start his family at 26 years of age.

Fig. 10: 'From the plant laboratory of a large concern', 'Photogr. Dr. Paul Wolff'

This is to be achieved by

1. shortening the studies preceding the first professional examination [*Verbandsexamen*] to seven semesters;[8]

2. furthermore, seeking ways and means of giving the young soldier the opportunity to attend introductory courses in chemistry, physics, and mathematics during the course of his period of service.

By taking these points into consideration it will thus be possible for the high-school graduate to leave school at 18; he must still serve for another year at $19\frac{1}{2}$; but he can attend the fundamental science lectures and take his doctoral examination at 25 years of age.

3. Through recognizing publically and honoring good scientific and technical accomplishments, the engineer's professional pride can be raised, the regard of the whole nation for technical achievements can be heightened, and the overall understanding of technology deepened. It must become generally perceived that science and technology in particular give the young person an opportunity to develop and apply his intellectual abilities directly for the benefit of his nation.

IV. Improving scientific research at universities. Scientific research at university must be enriched by new problems relevant to life [*lebensnah*]. Students working practically in university laboratories should often have problems presented to them, in addition to abstract theoretical analyses, that are important to the concrete development of technology and chemistry. Stimuli must arise out of the collaboration between practice and academia in this field. A great torrent of new knowledge must flow again from Germany's higher educational institutions into industrial application. However, an increase in the number of university teachers and teaching assistants in Germany is required for this.[9]

To this end it will be necessary to create an endowment that makes available two to three million [reichsmarks] annually. The purpose of this endowment is to enable diligent young academics in chemical and engineering science to prepare for a career as professor after having obtained a doctorate from university [*Promotion*] by granting scholarships for the duration of the necessary training period.

Translating all these suggestions into reality naturally depends upon the energetic determination and commitment of all persons concerned. The nation has a right to the most extensive development of its sons' intellectual gifts. A young person who has the energy and feels able to do this should join the proud front line of those men who repeatedly give the German people new scientific and technical creations. Like everyone in our nation, he must give his best. The tradition and work of his fathers obligate him. Just as everywhere today younger generations

[8]Shortening the curriculum also in physics was a typical measure in the second period of the Nazi regime. The *Diplomphysiker* degree was introduced, reducing scientific studies by about 2 years. Cf. Ramsauer [1938], Orthmann [1939], Stuart [1939], and the Introduction, sec. 3.4.

[9]For statistics on the shortage of academic staff see, e.g., Max → Wien's petition to Education Minister Rust of 1934, doc. 35.

storm past the goals the foregoing ones have reached, a new impetus must likewise come from the youth in science and technology. The achievements of Runge,[10] Liebig,[11] Kekulé,[12] the Krupps,[13] Siemens[14] and others have brought Germany international standing. Nothing should prevent our youth from constantly winning this international recognition with new major accomplishments produced by the best minds among them. Our young engineers and chemists must be shock troops in the struggle for our nation's economic future, fighting with maximum commitment for the highest of aims. Thus, a menacing gap at the German nation's front will close; and thus, nothing will prevent the marching advance into a great future.

[10]Friedrich Ferdinand Runge (1795–1867) was professor of chemistry at Breslau before moving to industry in 1830, specializing in the chemistry of dyes. Carl David Tolmé Runge (1856–1927) was professor of applied mathematics at the Polytechnic in Hanover 1886–1904 and in Göttingen 1904–23. On Runge's multifacetted scientific work ranging from Fourier series approximations to the establishment of series of formulas for band spectra, see Iris Runge: *Carl Runge und sein wissenschaftliches Werk*, Göttingen: Vandenhoek & Ruprecht, 1949; Gottfried Richenhagen, *Carl Runge: von der reinen Mathematik zur Numerik*, Göttingen: Vanderhoek & Ruprecht, 1965; or Paul Forman's article in the *Dictionary of Scientific Biography* **11** [1975], pp. 610–615.

[11]Justus von Liebig (1803–1873) was professor of chemistry at the University of Giessen from 1825 to 1852, when he moved to the University of Munich. Liebig was very active in establishing chemistry as a discipline and was editor of the first scholarly journal in chemistry, *Annalen der Chemie*. He did pioneering work in many branches of the field, including agricultural chemistry, biochemistry and nutrition. Among his famous contributions are the discovery of chloroform, isomerism, the identification of a process involving alcohol that opened the way to a method of silvering mirrors without the use of toxic mercury, and his investigations on protein and carbohydrates that resulted in his inventions of meat extract and baby cereal.

[12]Friedrich August Kekulé von Stradonitz (1829–1896) was professor of chemistry in Ghent from 1858 and in Bonn from 1867. His main work was on the structure of the carbon atom and its components, which revolutionized organic chemistry.

[13]The steel magnate Friedrich Krupp (1787–1826), his son Alfred Krupp (1812–1887), and grandson Friedrich Alfred Krupp (1854–1902), built up their Essen-based → *Krupp* steelworks to become the dominant manufacturer of cast steel in Germany.

[14]Ernst Werner von Siemens (1816–1892), artillery officer 1838–49, founded together with the mechanic Johann Georg Halske (1814–1890) a telegraph manufacturing company in 1846 (→ *Siemens*) that soon gained international importance. He introduced scientific methods of industrial production and made some important inventions like the self-correcting dynamo and electrical pyrometers. Cf. 'Dem Andenken an Werner v. Siemens', *Naturwissenschaften* **4** [1916], pp. 759–827.

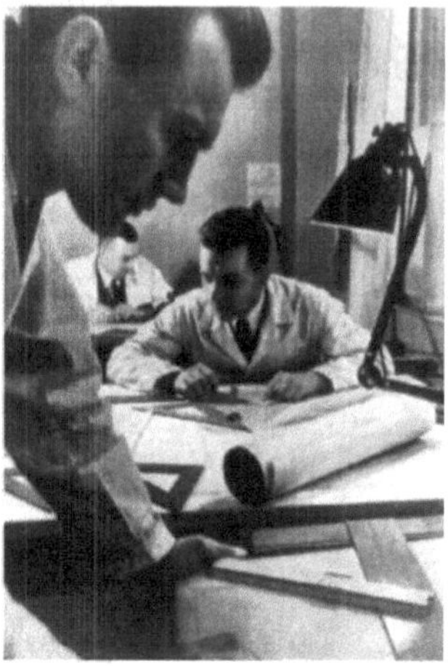

Fig. 11: 'Inside an engineering office', 'Phot[ographer]. Dr. Paul Wolff'

59 Fritz Heyer: Review of Erich Günther's *Military Physics. A Handbook for Teachers* [April 18, 1938]

Source: Dr. Heyer, review of Erich Günther (Ed.): *Wehrphysik. Ein Handbuch für Lehrer*, in *Der Deutsche Erzieher. Reichszeitung des Nationalsozialistischen Lehrerbundes*,[1] Issue No. 2, April 18, 1938, p. 46; reprinted in Brämer & Kremer (Eds.) [1980], p. 199.

Günther, Dr. Erich:[[2]] *Military Physics. A Handbook for Teachers.* 212 text illustrations, a color cloud table, two weather maps. Frankfort-on-Main, M. Diesterweg, 1936. 189 pp., 5.80 R[eich]M[arks].

Military physics is neither supposed to be a novel form of physics, nor even ought it to force its way into the already overloaded curriculum at our high-schools. Rather, through general physics the pupil should also be able to form an opinion on defense technology matters, as the editor himself also says so correctly elsewhere: "A solid instruction in physics still remains the best prepa-

[1]Since the coming into force of the law on civil servant professional organizations of May 27, 1937, the Nazi Teachers League (→ *NSLB*) was the only organization for teachers from all kinds of schools, and the membership rate was 97% of all teachers (according to its own reports). According to Eilers [1963], pp. 11f., the publication *Reichszeitung Deutscher Erzieher* had a total circulation of around 314,000, and its successor, the journal *Deutscher Erzieher*, no longer gratis for *NSLB* members, had about 240,000 subscribers.

[2]The physics teacher (Otto) Erich Günther (1886–1951), who had studied under Otto Wiener, was senior master 1928–45 in Dresden, and coauthored several physics textbooks. He was also the first chairman of the professional association *Deutscher Verein zur Förderung des mathematisch-naturwissenschaftlichen Unterrichts* until 1936, as well as regional physics expert for Saxony; cf. in particular W. Lietzmann in *Der mathematische und naturwissenschaftliche Unterricht* **4** [1951/52], pp. 193–194, Brämer & Kremer (Eds.) [1980], p. 48.

ration also for military purposes."[3] From this outlook this book provides us with a selection of interesting problems as they relate to military preparedness and where they are particularly suited as practical examples or illustrations in general physics instruction for all age levels. Thus the exercises and problems given under 'Seeing, Measuring, and Aiming' fit will in introductory geometry or physics classes and offer welcome material for practice and review (e.g., prism telescope, range-finder, prismatic compass). In the area of acoustics, problems are posed on directional listening; and sound measurement procedure and head wave formation are also addressed. But in the general physics courses at high-schools there will scarcely be time for a discussion of Behm's echo ranging.[4] Weapons technology and ballistics, which can least dispense with a knowledge of physical concepts, are represented in a number of important problems; of these it will be possible to determine projectile velocity in the classroom using a ballistic pendulum,[5] while other problems like the pressure build-up in the barrel or the output of gunpowder gases exceed the needs and means at schools. Under 'Means of Communication', exercises and experiments are assembled on ways of communicating acoustically, optically and electrically. What is offered on wireless telegraphy [or] on the properties of thermionic tubes can probably be found in any physics textbook, however. Aeronautics[6] is mainly dealt with from the angle of gadget construction; the choice and division of the material into specific problems is very successful. In another section an informed selection of the essentials of meteorology is presented, accompanied by good figures. A variety of applications from the field of pioneer engineering, gas-mask use, the photoelectric cell, and photographing in fog conditions bring the volume to a close.[...] Dr. Heyer.[7]

---

[3]On Günther's strategy employed also by teachers in other subjects of underscoring the importance of physics in the Third Reich so as not to be overshadowed by other rival subjects like biology, see in particular Brämer & Kremer (Eds.) [1980], pp. 43ff.

[4]The relevant section in Günther (Ed.) [1936], pp. 49–52, was in part taken directly from a publication by the range-finder manufacturer *Behm-Echolot-Fabrik* in Kiel.

[5]This experiment is described in more detail in Günther [1933], p. 233: "A little cigar case filled with sand is hung on four strings like a parallelogram pendulum. You now shoot horizontally along the pendulum's oscillation level at the front side of the little case. The bullet gets lodged in the sand in the case and the extent of its movement is transferred on to the pendulum, which swings in a certain manner. According to the law of momentum, you obtain $mv = (m + m_1) \cdot v_1$, whereby m is the bullet's mass, m_1 the pendulum's mass, v the bullet's velocity, and v_1 is the velocity of the oscillating object when the bullet has come to a halt it in. [...]. Some figures are provided on the experimental conditions: German precision carbine (22 short, 5.6 mm calibre) with a rifle barrel; $m = 1$ gram, $m_1 = 1000$ grams, pendulum length 100 cm, sideways swing of the oscillating object $a = 6$ cm yields v at approximately 200 m/sec."

[6]On Nov. 17, 1934 the Education Ministry (→ *REM*) had issued an edict 'On the fostering of the aviation spirit at schools' (Zur Pflege des Luftfahrtgedankens in den Schulen) which prescribed the addition of a comprehensive course on aeronautics in the school curriculum.

[7]Presumably Fritz Heyer (1871–?).

60 Carl Bosch: Letter to Reich Minister of the Interior Wilhelm Frick [May 20, 1938]

Source: Facsimilie in D. Hahn (Ed.) [1979] p. 147; also reprinted in Krafft [1981], p. 173.

Esteemed Reich Minister,[1]

In my capacity as President of the Kaiser Wilhelm Society I must consider a petition that in our opinion only you can decide.[2] It concerns Professor Meitner, who conducts scientific research within the framework of the Kaiser Wilhelm Institute of Chemistry.[3] Mrs. Meitner is a non-Aryan but was retained with the approval of the official authorities (Ministry of Culture), since she has a great amount of scientific experience; to be more specific, she ranks as a leading physicist. She has solved various major problems in long-term scientific collaboration with Professor Hahn, who is a chemist in this field.[4] Mrs. Meitner is of Austrian citizenship. With the reintegration of Austria,[5] however, she has become a national; and it can be assumed that sooner or later the issue of her leaving will become pressing.[6] Since Mrs. Meitner is highly esteemed in the scientific world, I consider it very desirable in the interest of the Institute that a way be found for a smooth resolution of this case.[7] Mrs. Meitner is willing of her own accord to resign at any time and to assume another scientific position abroad.[8] She has also received suitable offers. The question now is whether Mrs. Meitner, who holds an Austrian passport, is to receive an endorsement that she may re-enter

[1]W. → Frick was Interior Minister 1933–43 and as such supervised state institutions, including the employment of civil servants at government-funded institutions and universities.

[2]Carl → Bosch was elected president of the → *KWG* as successor to Max → Planck in 1937. Aside from his scientific reputation (Nobel Prize in chemistry for 1931), Bosch had probably been chosen because, as an important industrialist (chairman of the board at → *IG Farben*), he was much more independent politically than any civil servant (*Beamter*).

[3]Lise → Meitner worked at the → *KWIC* from 1913; cf., e.g., Krafft [1981], chaps. 2–3.

[4]On Meitner's earlier work and correspondence with Otto → Hahn see Ernst [1992].

[5]After mounting political pressure in the spring of 1938, the German Army marched into Austria on March 12, 1938 to forestall an Austrian plebiscite on the Nazi's role in their land; only then did Hitler finally decide to annex the land of his boyhood instead of allowing it satellite status. Its new Nazi Interior Minister declared its annexation (*Anschluß*) the following day. See also footnote 4 of doc. 62.

[6]According to Krafft [1981], footnote 29 on p. 173, Max von Laue reported that he had heard of an order by Heinrich → Himmler that no university scholars were allowed to leave Germany, regardless of whether or not they were Jewish.

[7]Both Bosch and Hahn were very concerned about the fate of the whole institute, because several 'problematic' co-workers including Lise Meitner, as a Jew, and the liberal scientists, Otto Hahn, Max Delbrück and Fritz Straßmann, lent a politically unfavorable reputation to the institute. This was aggravated by actions by employees loyal to the party line, such as in particular the attempt by the head of the institute's visitors division (*Gastabteilung*), Kurt Heß, to take over leadership of the institute.

[8]Bosch had consulted with Meitner shortly before writing this letter. She was nevertheless very disappointed that her friends had not tried to arrange for her to remain in Berlin; cf. Hahn [1962]b, pp. 126–127, Krafft [1981], p. 172.

Germany, since otherwise a trip abroad for interviewing purposes is impossible, or whether Mrs. Meitner is to be issued a German passport. I would be very grateful to you, esteemed Reich Minister, if you could put me into the position of settling this case in the interest of the Institute and its scientific relations.[9]

Heil Hitler![10] Bosch.

61 Reich Interior Ministry: Letter to Carl Bosch [June 16, 1938]

Source: Facsimilie of first page and excerpted transcription in D. Hahn (Ed.) [1988], p. 165. Also reprinted in full in Krafft [1981], p. 174. From Meitner's stenographic notes on hotel letterhead: 'Hotel Adlon, Berlin W., Unter den Linden 77, am Pariser Platz 9.'[1]

Esteemed Privy Councillor,

On behalf of Reich Minister Dr. Fr[ick] I am authorized to respectfully inform you with regard to your letter of the 20th of the past month[2] that political objections exist to issuing a foreign passport to Prof. M[eitner]. It is considered undesirable that renowned **Jews** travel from Germany abroad to act as representatives of German science or even, using their name and their experience, to act in accordance with their inherent attitude against Germany.[3] The K[aiser] W[ilhelm] S[ociety] could surely find a way for Prof. M[eitner] to continue to remain in Germany following her resignation as well and, as the case may be, to also work privately in the interest of the Society.[4] This opinion was held in particular by the *SS Reichsführer* and Chief of the German Police at the Reich Ministry of the Interior.[5]

Very respectfully yours[6]

---

[9]See the next document for the Ministry's reply.

[10]This closing became a required formality in all official correspondence during the Nazi regime and does not necessarily reflect the signer's private political convictions.

[1]Lise → Meitner was staying at Hotel Adlon for her personal safety while she awaited the official response to C. → Bosch's letter, the previous document. Her stenographic notes of the above text which she had received by telephone are preserved (see also Berninger [1974]).

[2]Cf. the preceding doc.

[3]'Jews' is underlined in the original. See footnote 6 of the preceding document on Himmler's order regarding the travel rights of university scholars.

[4]More Bosch was unable to achieve. Meitner's Dutch colleagues, the professors Dirk Coster, Fokker and de Haas managed to convince their government to admit her into the country without a visa, and she escaped to Holland in July 1938. Cf. Hahn [1962]b, pp. 126–127; Sime [1990].

[5]Heinrich → Himmler's involvement as head of the → *SS* and police chief in Meitner's case indicates that it was discussed at the highest levels of the → *RIM*.

[6]The letter was signed by an assistant to Reich Minister Wilhelm → Frick.

62 Ludwig Prandtl: Letter to Heinrich Himmler [July 12, 1938]

Source: Prandtl Papers, Göttingen. Initialed carbon copy of typed letter from: 'Professor Dr. L. Prandl, Göttingen, Calsowstr. 15' addressed: 'An den Reichsführer SS Herrn Himmler, Berlin SW' with file reference: 'Pp.502/38'. Partially quoted in Vogel-Prandtl [1993], pp. 140–141.

Esteemed Mr. Himmler,

When I was your neighbor at table on the occasion of the ceremonial meeting of the German Academy of Aeronautic Research[1] on the 1st of March of this year,[2] I turned the conversation to certain difficulties German representatives of the field of 'Theoretical Physics' have been facing due to unjustified attacks on the side of a group of experimental physicists, and I mentioned in particular the personal plight of Mr. Heisenberg.[3] You then criticized the behavior of Mr. Heisenberg vis-a-vis Mr. Einstein and requested that I advise Mr. Heisenberg that he dissociate himself personally from Einstein, should he discuss the Einsteinian doctrines in his lectures, which, as I would like to remark here again explicitly, are acknowledged as correct, of course, by the majority of physicists and which have already become a firm component of the conceptual framework of theoretical physics.

I then did so right away, and Mr. Heisenberg also answered me immediately in a thoroughly satisfactory manner. I immediately had transcriptions of this letter made at that time and forward to you as enclosures two of these for your discretionary use. That this should only happen now has to do with the fact that at that time the annexation of Austria was taking place, then the propaganda for the plebiscite, the latter of which was followed by the *Führer*'s birthday and the Rome trip, in which, as you know, you also took part.[4] I said to myself that during such animated times it was better to leave you in peace regarding this matter. I now take up this affair again because I hear that new steps have just been taken in the Heisenberg affair and that just now my report will possibly

[1]See footnote 16 of doc. 75 for more on the *Deutsche Akademie der Luftfahrtforschung*.

[2]L. → Prandtl himself had already been cleared by the party authorities. His character evaluation dated May 28, 1937 read as follows: "Prof. Prandtl is the ivory-tower scientist type. His entire interest lies in his scientific work, for which he enjoys world renown. Prof. Prandtl has absolutely no interest in politics [...]. In summary, Prandtl can be characterized as the old-style respectable, conscientious scholar concerned about propriety and an unstained reputation, whom we cannot, nor wish to, dispense with in view of his extraordinarily valuable scientific achievements in promoting the *Luftwaffe*." See Trischler [1992]a, pp. 144–145.

[3]On the 'Aryan Physics' campaign against Werner → Heisenberg led by Johannes → Stark and others see, e.g., Beyerchen [1977], chap. 8 and docs. 42, 44, 55, 56, 77 and 83.

[4]The annexation of Austria took place on March 13, 1938 after German troops had marched in on the previous day. A plebiscite of April 10, 1938 retroactively legalized the *Anschluß* with 99.08% of the votes in favor in Germany and 99.73% of the votes in favor in Austria according to official information. A. → Hitler's 49th birthday was on April 20, 1938; and on May 2, 1938 Hitler and several top Party functionaries arrived in Mussolini's fascist Italy on a state visit.

come at the right moment.

I would also like to add in connection with this a few words about Theoretical Physics.[5] The difficulties encountered by this field are brought about primarily by the fact that a relatively small group of experimental physicists, who were unable to keep pace with the theorists' research, has vehemently resisted the newer developments in theoretical physics, basing themselves chiefly on the fact that significant parts included in the conceptual framework of current theoretical physics stem from non-Aryan scientists. It must be conceded that among these non-Aryan scientists were also those of inferior quality, who called out their sham wares [*Talmiwaren*] with the industry characteristic of their race. It is only just and good that such articles disappear; but there among the non-Aryans are also prime scientists who strive ardently to advance science and who have actually promoted it in the past. I call to mind here as only one example the prematurely deceased Heinrich Hertz, who with laboriously and cleverly arranged experiments first proved the existence of electric waves, the very same waves that have gained great technical importance today in wireless telegraphy and in radio broadcasting.[6] In the case of Einstein a distinction must be made between the man and the physicist. The physicist is first-class through and through; but his early fame seems to have gone considerably to his head, so that he has become intolerable as a person.[7] However, science cannot be concerned with these human qualities. It simply acknowledges the fact that laws had been discovered that in their turn had become the motive for further discoveries; and these laws cannot be omitted without destroying the subsequent conceptual framework upon which they had been based.

Theoretical physicists will always have to work alongside of experimental physicists. Their strength lies in conducting experiments; they leave the drafting of the underlying theory to the theorists. It is outrageous, however, when such experimentally minded people now simply declare theories pernicious or reprehensible, because they are incapable of following them and therefore believe they can sling mud at the theory's advocates. In order to present the opinion of the well-known theoretical physicist Max Planck, I would like to submit three transcriptions of an excerpt from an address by this scholar.[8] It is astounding how

---

[5]The bad state of theoretical physics, with only 24 out of a total of 35 existing chairs occupied by theoretical physicists is also documented in an unsigned report of the summer of 1938. Cf., e.g., Beyerchen [1977], p. 172, Wolff [1993], p. 271, and here docs. 83 and 92.

[6]On Heinrich Hertz (1857–1894) see doc. 31, footnote 5.

[7]It is not clear whether this was Prandtl's real opinion; but scientists used this distinction between the personal characteristics of their Jewish colleagues and their scientific achievements as a way to circumvent anti-Semitic policy while defending the validity of their theories. Himmler takes up this approach in the postscript to his reply to Prandtl of July 21, 1938, doc. 65.

[8]Presumably, the address referred to is M. Planck: 'Die Physik im Kampf um die Weltanschauung', delivered at the → *KWG*'s *Harnack-Haus* in Dahlem (Berlin) on Mar. 6, 1935, published by Barth in Leipzig, 1935. In this address, Planck first tries to demonstrate that physics also can provide a sharp weapon in the fight over the appropriate worldview (p. 4).

this man, who has also been seriously insulted by Mr. Stark, explains his position on the tasks of his field in the most distinguished manner, without addressing the ugly attacks with even a single word.[9] This short text would deserve the widest dissemination.

Now, as for Mr. Heisenberg himself, he has certainly been insulted even more seriously by Stark;[10] and it is extremely urgent that he be removed from disgrace in some effective way. It should also be made clear[11] that the Party's central authorities and top agencies are not of the same opinion as Mr. Stark. This is necessary simply because if his bad image persists, Mr. Heisenberg would be severely hampered as an academic teacher. Students would be taught the opinion, to the detriment of their later professional effectiveness, that what they could learn from him was worthless and maybe even harmful to them.

One option would be to inform the student leadership of the affair. I consider Planck's comment particularly suited for this.[12] I would also consider it very beneficial if it were possible to allow Mr. Heisenberg to express himself personally by means of some appropriate contribution out of his field of expertise in the *Zeitschrift für die gesamte Naturwissenschaft*, which is the official publication of the Reich Science Section [*Reichsfachgruppe Naturwissenschaft*] of the Reich Student Leadership.[13] I would be very willing to get in touch with Mr. Heisenberg on the appropriate selection of this field of expertise. I close with the

---

Later on, he gives a brief account of the development of physics up to relativity theory and quantum theory (pp. 8–25), in an attempt to give a balanced evaluation of contributions from both the experimental and theoretical sides. He thereby implicitly refutes Stark's one-sided antitheoretical attitude. Planck also addressed science policy matters in the guise of a historical description in his speech on Jan. 28, 1937 at the official ceremony on the 150th anniversary of the death of Friedrich II, published in the *Sitzungsberichte der Berliner Akademie der Wissenschaften*, math.-phys. division, pp. xxi–xxvii. Planck was concerned that the government realize the importance of a flourishing science: "But here we should also not forget here the part in these successes that must be attributed to the national government, which has appreciatively and readily fostered the true interests of the Academy throughout the changing times by not only making available the necessary funds, but also by granting the equally essential freedom of action. This applies to today and allows us to look with confidence to the near future."

[9]For further examples of Planck's indirect style, avoiding open confrontation during the National Socialist period in an attempt to prevent the worst excesses—though at the cost of moderate political self-alignment—see also doc. 38 or Planck [1933]a, where he objected to the pseudoscientific tendencies that had begun to find widespread support among Party officials.

[10]Prandtl alludes to Stark's labeling Heisenberg a 'White Jew' in doc. 56.

[11]'made clear' (*deutlich gemacht wird*) was inserted by hand to replace 'clearly acknowledged' (*deutlich erkannt wird*), probably by Prandtl.

[12]See footnote 8 above.

[13]Himmler was convinced to allow Heisenberg the opportunity to publish a rebuttal in the → *NSDStB*'s journal, as indicated in his letters of July 21, 1938 to Heydrich and Prandtl (cf. the next docs.). Heisenberg's contribution on the evaluation of 'modern physics' was written in 1940 but only appeared after some delay in 1943; see doc. 95; cf. also Rechenberg (Ed.) [1992]a, pp. 90–106.

observation that I consider this whole matter has less to do with individuals than with preventing the devastation of this field, which is extremely important for the continued development of our scientific knowledge.

With respectful greetings and Heil Hitler!

Yours very sincerely L. P[randtl].

63 Heinrich Himmler: Letter to Reinhard Heydrich [July 21, 1938]

Source: Prandtl Papers, Göttingen. Typed transcript of letter from: 'Der Reichsführer-SS, Berlin'; file reference: 'Tgb.Nr.AR/453,RF/Pt.' to 'SS-Gruppenführer Heydrich, Berlin.'

Dear Heydrich,[1]

I have received the very factual and positive report on Prof. Werner Heisenberg, Leipzig.[2] I am enclosing a very sound letter by Prof. Prandtl, Göttingen, which I must very much endorse.[3] In addition, I am enclosing my letter to Heisenberg in transcript for your information.[4]

Please do advise the Reich Student Leader to follow Dr. Prandtl's suggestion that Heisenberg be able to publish something in the *Zeitschrift für die gesamte Naturwissenschaft.*[5]

I ask you furthermore to settle the whole case through Six,[6] not only with the Student League, but also with the Reich Student Leadership, since I likewise believe that Heisenberg is decent; and we cannot afford to lose this man or have him killed, since he is relatively young and can bring up the next generation.

[1]Reinhard [Tristan Eugen] Heydrich (1904–1942) became head of the Secret Police (→ *Gestapo*) and the Security Service (*SD*) on June 17, 1936—see also footnote 1 of doc. 73. He was H.→ Himmler's closest associate. Upon election as president of the International Criminal Police Commission in 1940 he sought to establish a German espionage system outside of Germany. He was promoted to *Obergruppenführer* of the → *SS* (general) in 1941, and was chosen to administer the 'Final Solution' of the Jewish question in January 1942 Shortly after his appointment as Reich Deputy Protector of Bohemia and Moravia on Sept. 27, 1941, he was murdered by Czech partisans in May 1942.

[2]A copy of this *SS* report was enclosed with a letter to R. → Mentzel as well as to the University Lecturers League (→*NSDDB*) in the following year, see doc. 73.

[3]See the preceding doc. 62.

[4]See the following doc. 64.

[5]Cf. the preceding document and Heisenberg's text on 'modern physics' written in 1940, which finally appeared in the → *NSDStB*'s publication in 1943, doc. 95.

[6]Franz Alfred Six (1909–?) was a foreign policy expert. In 1935 he was lecturer at the University of Königsberg in 1935, untenured associate professor in Berlin in 1939, and full professor in 1940. He later became dean of the department of foreign studies and president of the German Institute of Foreign Studies (*Deutsches Auslandswissenschaftliches Institut*) in Dahlem (Berlin). In 1936 Six was also head of the press department of the main division 'against opponents of differing worldviews' (*Weltanschauliche Gegenerbekämpfung*) within the 'Worldview Research Office' (*Amt II: Weltanschauliche Forschung*) of the Security Service (*SD*). Six held the title of senior *SS* battalion commander (*Obersturmbannführer*) in 1938.

Beyond this, I would consider it beneficial if Six would bring Prof. Heisenberg together with Prof. Wüst someday.[7] Please conduct this entire correspondence on to Wüst with the request that he send the personal staff back again upon cognizance. Wüst should then try to probe Heisenberg, since we might need him for the ancestral heritage program [*Ahnenerbe*], once it should become a complete academy, and we may be able to get this man, who is a good scientist, to cooperate with our people on the cosmic-ice theory.[8]

Heil Hitler! Yours H. Himmler.

64 Heinrich Himmler: Letter to Werner Heisenberg [July 21, 1938]

Source: Prandtl Papers, Göttingen. Typed letter signed on Himmler's official letterhead: 'Der Reichsführer *SS*, Berlin SW 11, Prinz-Albrecht-Straße 8', file reference no: 'Tgb.Nr. AR/453, RF/Pt.' Addressed to 'Herrn Prof. Heisenberg, Leipzig 0 27, Bozener Weg 14.'

Esteemed Professor Heisenberg,

Only today do I come to finally responding to your letter of the 21st of July 1937, in which you turned to me regarding Prof. Stark's article in the *Schwarzes Korps*.[1]

I have had your case examined with particular care and scrutiny, since you were recommended to me by my family.[2]

[7]Walther Wüst (1901–), an Indo-Europeanist, was head of the Defense Squadron's → *Ahnenerbe* division, the ancestral heritage organization which became increasingly involved in scientific research. Cf., e.g., Kater [1974]. Wüst was unsalaried lecturer at Munich University in 1926, unbudgeted associate professor in 1932, becoming full professor in 1935. He was later rector, vice-president of the German Academy, and curator of the *Ahnenerbe* research and educational association, and regular member of the Bavarian Academy of Sciences. Wüst was also battalion commander (*Sturmbannfüher*) and main company commander (*Hauptsturmführer*) of the *SS*. He wrote a number of works promoting 'Aryan' culture; see, e.g., *Das Reich: Gedanken und Wirklichkeit bei den alten Ariern. Festrede aus Anlaß der von den Münchener Hochschulen veranstalteten Reichsgründungsfeiern*. Munich: Gässler, 1937, which includes a list of his publications on p. 9.

[8]On the pseudo-scientific cosmic-ice theory (*Welteislehre*) invented by Hanns Hörbiger (1860–1931) see, e.g., Nagel [1991]; on the background to Himmler's intention to link Heisenberg with the *Ahnenerbe*, see Schröder [1991].

[1]Cf. doc. 56 and the preceding docs. Heisenberg had reacted to this intentional vilification by writing a letter to the head of the → *SS*, H. → Himmler, dated July 21, 1937, in which he demanded either approval of Stark's attacks (in which case he would resign), or objection (in which case he expected protection against such attacks and the restoration of his honor (*Ehre*)). On July 28, 1937, he applied to the → *REM* to initiate disciplinary proceedings against himself so that the accusations raised in the *Schwarzes Korps* could be investigated officially (copies of both letters are at the Heisenberg Archive, Max Planck Institute of Physics, Munich): Cf., e.g., E. Heisenberg [1984]b, pp. 66f., Cassidy [1992]a, pp. 385f., [1992]b, p. 657, and doc. 55, footnote 13.

[2]Heisenberg's mother (née Anna [Annie] Wecklein) (1871–1945) had forwarded the letter

I am happy to be able to inform you today that I do not approve of the offensive article by the *Schwarzes Korps* and that I have put a stop to any further attack on you.[3]

I hope that I can see you at my office in Berlin someday in the autumn—though only very late, in November or December—so that we can have a man-to-man talk about this.[4]

With friendly greetings and Heil Hitler!

Yours H. Himmler

PS. I do find it appropriate, though, that in the future you separate clearly for your students acknowledgment of scientific research results from the scientist's personal and political views.

65 Heinrich Himmler: Letter to Ludwig Prandtl [July 21, 1938]

Source: Prandtl Papers, Göttingen. Typed letter signed on Himmler's official letterhead: 'Der Reichsführer *SS*, Berlin SW 11, Prinz-Albrecht-Straße 8' ref. no: 'Tgb.Nr. AR/453, RF/Pt.'

Esteemed Dr. Prandtl,

I have received your letter of the 12th of July 1938.[1] I have examined the Professor Heisenberg matter thoroughly in the course of the last few weeks and have also been thoroughly briefed.[2]

I have likewise become convinced that Prof. Heisenberg has a sincere personality and is a man of integrity. I have seen to the matter and have informed

mentioned in the preceding footnote to Himmler's mother (née Anna Koch) (1866–1941), whom she knew through her father Nikolaus Wecklein (1843–1926) who was a member of the same hiking club for Bavarian secondary school rectors as Himmler's father, [Joseph] Gebhardt Himmler (1865–1936). H. Himmler had his personal staff conduct a special investigation of Heisenberg's case. Heisenberg was interrogated several times at *Gestapo* headquarters in Berlin, microphones were secretly installed in his home, and spies attended his lectures and seminars in Leipzig. Cf., e.g., E. Heisenberg [1984]b, pp. 69f., and Cassidy [1992]a, pp. 385f., [1992]b, p. 657.

3Cf. the *SS* report on Heisenberg issued on May 26, 1939 (doc. 73), after which the *SS* looked upon Heisenberg more positively.

4According to Elisabeth Heisenberg [1984]b, p. 73, and Cassidy [1992]a, pp. 393f., such a meeting between Himmler and Heisenberg never took place. In a letter to Sommerfeld of July 23, 1938 (DMM, 1977-28, A-136), written directly after receiving this letter, Heisenberg mentions that he had asked Himmler to distribute copies of his letter to several officials including B. → Rust at the *REM* and the dean of the University of Munich, who was involved in the pending appointment of Sommerfeld's successor. But this official action did not help Heisenberg's candidacy to the chair in theoretical physics because of strong resistance by the Munich-based Lecturers League (→ *NSDDB*), radical Nazi student activists (→ *NSDStB*) such as B. → Thüring and Kubach (on Kubach see footnote 16 of doc. 85), and Lenard supporters, like R. → Tomaschek in the faculty.

1Cf. doc. 62.

2Cf. the evaluation on W. → Heisenberg, a copy of which was sent to the University Lecturers League (→ *NSDDB*), doc. 73.

Prof. Heisenberg in a personal letter that I do not approve of the attacks by the *Schwarzes Korps* and have put an end to further attacks.[3]

I think your suggestion that Dr. Heisenberg be given a chance to speak in the *Zeitschrift für die gesamte Naturwissenschaft* is very worthy of consideration and will convey it on to the Reich Student Leadership.[4]

I consider it important that in his written and oral lectures and remarks Heisenberg clearly distinguish for outsiders as well between facts of scientific knowledge and political stance.[5]

Heil Hitler!

Yours H. Himmler.

66 Reich Ministry of Science, Education and Culture: Memorandum to State Secretary Otto Wacker [October 3, 1938]

Source: Geheimes Staatsarchiv, Dahlem (Berlin): I/76 No. 1398. Carbon copy of typed document, unsigned: '1. Vermerk: Betr.: Deutsche Physikalische Gesellschaft (Vors. Prof. Dr. Debye) und Deutsche Gesellschaft für Technische Physik (Vors. Dr. Mey). 2. Herrn Amtschef W Staatsminister Dr. Wacker'.

1. Memorandum: Re: German Physical Society (President, Prof. Debye) and the German Society of Technical Physics (President, Dr. Mey).[1]

Both societies' scientific and organizational activities can be rated as good. However, despite the past 5 years of National Socialist government leadership, both societies have made only slight progress in their general National Socialist conduct.[2] Even after the election of new boards in September 1937, such progress was not particularly encouraged.

At the last meeting of the societies in Baden-Baden the welcoming speech of the President of the German Society of Technical Physics, Dr. Mey, conspicuously lacked substantial National Socialist references. Moreover, in the commemorative address—which must be described as having miscarried—on the 50th anniversary of the discovery of electromagnetic waves, remarks were made about the half-Jew

[3]See doc. 64.

[4]Heisenberg's article was written in 1940 and appeared after some delay in the → *NSDStB*'s journal in 1943: See doc. 95; cf. Rechenberg (Ed.) [1992]a, pp. 90ff.

[5]H. → Himmler thus adopts Prandtl's separation of the scientist from the person (cf. doc. 62).

[1]Peter → Debye was president of the *Deutsche Physikalische Gesellschaft* (→ *DPG*) 1937–39. Karl → Mey was president of the *Deutsche Gesellschaft für technische Physik* (→ *DGtP*) 1931–45.

[2]Both of these private societies were reluctant to align themselves politically. See the Introduction, sec. 5.2. Cf., however, footnote 7 below.

Hertz,[3] which in view of the numerous foreigners present must be judged at the very least as extraordinarily gauche. This is all the more astonishing since Dr. Mey has participated in the meetings that took place here on the premises under *W E*[4] and was thoroughly informed about the position and wishes of the involved ministries, from the discussions as well as from the decrees that were forwarded to him following the conference.

Added to this, Dr. May gave a poor speech at the foreigners' evening that took place on the occasion of the Baden-Baden meeting; and at the official festive night he also gave an impossible incoherent talk before 800 mathematicians and physicists about the most insignificant matters—he was doubtlessly already tipsy—at the end of which he did not propose the usual toast to the well-being of the *Führer* and the *Reich*, but drank the first glass to the societies instead. This blunder [*Entgleisung*] was even more striking since the first historical dialog between the *Führer* and Chamberlain was just taking place on that same evening (15th of September) and the mood was extraordinarily depressed because of the events in Sudeten Germany.[5] Dr. Mey's blunder became particularly conspicuous when Prof. Esau,[6] with the best of intentions, made up for the toast to the *Führer* shortly after Dr. Mey's speech.

A large number of physicists and mathematicians present did not accept Dr. Mey's slip up without objection and directed their complaints to me. Numerous professors in particular stressed that it would be impossible for them to continue to participate in the societies if conditions there were not changed along with the Jew question regarding membership and contributions to the societies' publications.[7]

I discussed this matter with Prof. Esau, the director of the physics section at the Reich Research Council,[8] and won his support for the proposed plans:

---

[3]On Heinrich Rudolf Hertz (1857–1894) and the discovery of electromagnatic waves, see doc. 31, footnote 5.

[4]Internal file references probably abbreviating the departments of science (*Wissenschaft*) and education (*Erziehung*) at the → *REM*. An ellipsis of 8 points follows the initials in the original text.

[5]The English prime minister in the years 1937–40 [Arthur] Neville Chamberlain (1869–1940) was mediating in the escalating dispute over the autonomy of Sudetenland, an area in Bohemia adjoining Germany with a large German-speaking minority. He met Hitler outside of Munich on the 15th for a private discussion and in the resulting agreement signed on Sept. 29, 1938 Britain accepted the removal of Czech domination over the region and allowed German troops to march in to enforce a plebiscite there. For an analysis of contemporary nationalistic propaganda in German newspapers concerning the Sudeten crisis, see Engelbert Schwarzenbeck: *NS-Pressepolitik und die Sudetenkrise 1938*, Munich, 1938.

[6]Abraham → Esau was then professor of technical physics at the University of Jena. In 1939, he accepted a call to the Berlin Polytechnic and became president of the → *PTR*.

[7]See doc. 67 for the subsequent compulsory exclusion of Jews from membership in the *DPG*, carried out in December 1938.

[8]Esau was director of the physics section (*Fachsparte Physik*) of the → *RFR* from its foundation in 1937 to 1943.

On the 16th of September and the 17th of September I conducted lengthy negotiations with Dr. Mey, drew his attention to the above-described conditions, and attempted to induce him to support their being changed as soon as possible, by setting before him the choice of having the same taken up—depending on the wishes of the societies—either by the Reich Education Minister or by Dr. Todt, the head of the *NSBDT*.[9] Dr. Mey declared his support for our Ministry's supervision, if only due to the large number of its university members. I promised the most inconspicuous solution to the affair possible, namely, that this Ministry would suggest to both societies the appropriateness of a merger of both societies into one, since their separation is only historically based and they pursue practically the same goal and have held their main meetings jointly for years. The articles of the societies could then be conformed in the process to current requirements. The new articles of the Kaiser Wilhelm Society of the 22nd of June 1937 could serve as a model of a sort here.[10] In particular, the new regulations would also have to provide that only citizens of the *Reich* can apply for regular membership and foreigners would be admitted as special members, if necessary.[11] Furthermore, it would also have to be pointed out that Jews may not be involved in the societies' journals, either as editors or as contributors. The acceptance of contributions from members of the Jewish race will be restricted to exceptions (and only if they are very exceptionally valuable); and in the review section reviews of papers by Jews should also be avoided.

The appointment of a new president, which would have to be subject to the approval of the Reich Education Ministry, would have to necessarily follow such a regulation.

Berlin, the 3rd of October 1938.

2. To Science Department Head, State Secretary Dr. Wacker[12]

[9]Fritz → Todt had been vice-president of the National Socialist League of German Technicians (*NS Bund Deutscher Technik*) since its foundation in November 1933, advancing to president on Nov. 26, 1934.

[10]According to the new statutes passed by the → *KWG*'s council (*Senat*) on May 29, 1937 and ratified at the general meeting on June 22, 1937, the *KWG* was 'subordinate to the Reich Minister of Science, Education and Culture' (§ 1). The minister was now authorized to name the president of the society 'on the advice of the council' (§ 10) and had the right to confirm the election of the 28 council members by the general assembly (§ 15). The *Führerprinzip* was adopted by transferring many significant powers of the general assembly, the council and the management to the president, who was to be the society's 'responsible leader' (§ 10); and the general assembly could only makes changes to the statutes with the explicit approval of the minister (§ 23). See Albrecht and Hermann in: Vierhaus & vom Brocke (Eds.) [1990], pp. 385f.

[11]§ 3 of the *KWG*'s new statutes limited membership in the society to 'natural persons with the full civil rights of the Reich', that is, those of Aryan descent in the sense of the Nuremburg Laws of 1935. Foreigners could become members on a 'case-by-case basis'; ibid., p. 386.

[12]Otto Wacker (1899–1940) was Culture Minister of Baden and successor to → Vahlen as head of the science office at the *REM* (abbreviated here as *W* (*Wissenschaft*) and *SS* senior squad leader (*Oberscharführer*). He also was vice-president of the *KWG* 1937–40 and vice-president of the *RFR*.

respectfully submitted through official channels, requesting acknowledgment and, if applicable, statement of approval. I would invite both presidents to a conference and only then send the official 'proposal' out from here.

Berlin, the 3rd of October 1938.

67 Peter Debye: Letter to the board of the German Physical Society and enclosed draft letter to the German members [December 2 & 8, 1938]

Source: Archives of the Physical Society, Berlin; also in the Debye Papers, Archive of the Max Planck Society, Berlin: Reprinted in Rechenberg [1988]a, p. 418; see also Swinne [1988]b, pp. 95–96. Typed signed letter on the society's letterhead: 'Deutsche Physikalische Gesellschaft E.V., Charlottenburg 5, Wundtstr. 46'. Salutation: 'Sehr verehrter Herr Kollege!'

Dear Colleague,

I intend to send out on Wednesday the 8th of December the enclosed letter to all the German members of the German Physical Society.[1] Unless I receive notice from you before the 6th of December, I will presume your consent.[2]

P. Debye,[3] President.

To the German Members of the German Physical Society.

Under the compelling prevailing circumstances I am forced to view as no longer acceptable the continued membership in the German Physical Society of German Reich Jews in the sense of the Nuremburg Laws.[4]

---

[1]See the preceding doc. 66 for the background on the increased pressure to expel Jewish members from the *Deutsche Physikalische Gesellschaft* (→ *DPG*).

[2]M. von → Laue responded promptly on Dec. 12, 1938 with the notation "Agreed". The society's treasurer, Walter Schottky (see footnote 6, doc. 68), had objections, however: "the members in Germany affected by this circular form only a relatively small group, [...] there is the fear that not only non-Aryan foreign members would withdraw their memberships [...]. If we were to lose a large number of these members, this would mean a loss which in my opinion would be regrettable also from the political standpoint of the State; as Treasurer I must also point out that we register quite considerable foreign currency receipts (from membership dues and membership-based book and journal purchases) on the part of foreign members." He even advised, "before taking a step with such grave consequences, ought the Culture Ministry not be contacted once again to put forward the above-mentioned aspects?" (All quotes from Rechenberg [1988]a, p. 418).

[3]The *DPG* was in the awkward position from 1935 on of having a foreigner as its president, Peter → Debye. When in 1939 Debye was forced to decide between retaining his Dutch citizenship and his directorship at the → *KWIP*, he relinquished his post and was suspended from office by the → *KWG*.

[4]The *Reichsbürgergesetze* of Sep. 15, 1935 specified that only those 'of German or related blood' could be a Reich citizen with full civil rights, and intermarriage between Jews and those of 'national German or related blood' was forbidden, thus stripping the Jewish citizens of their

I therefore request, in agreement with the Board of Trustees, that all members who fall within this provision inform me of their withdrawal from the Society.[5]
Heil Hitler sig[ned] P. Debye, President

68 German Physical Society: Minutes of the Meeting of December 14, 1938 [December 14, 1938]

Source: Archive of the German Physical Society, Berlin. Excerpt from the original document (item no. 6 of the agenda); a draft of the minutes of the Physikalische Gesellschaft zu Berlin meeting on December 14, 1938: Published in Swinne [1988]b, p. 96.

The Chairman[1] reports on the developments of the non-Aryan issue to date in the German Physical Society. Following the initiative expressed in a letter by Messrs. Stuart and Orthmann[2] and after privately consultating with some of the members of the Board of the German Physical Society, and asking all the members of the Board in writing, the President of the German Physical

basic rights. For an English translation of excerpts from these Nuremberg Laws, issued between September and November 1935 to define the legal status of Jews in Germany see, e.g., Noakes & Pridham (Eds.) [1990], Vol. 1, chap 23 (ii), pp. 535ff.

[5]The final version dated December 9th was modified as follows: "Under the prevailing compelling circumstances the membership of German Reich Jews in the sense of the Nuremburg Laws in the German Physical Society can no longer be upheld. In agreement with the Board of Trustees I therefore summon all members who fall within this provision to inform me of their withdrawal from the Society." The initial polite request in the second sentence was also given stronger emphasis—though both verbs used have double meanings: 'bitte' (request/invite) to 'auffordere' (summon/bid). Cf. the following doc. 68 about objections to this wording at the meeting of the *DPG* on Dec. 14, 1938. Note that in both versions Jewish members were requested to submit their own resignations instead of being expelled by the society. Jewish members outside Germany had generally already cancelled their memberships or had at least considered this step: See, for instance, Samuel Goudsmit's letter to Walther Gerlach, June 24, 1936, printed in full in Heinrich & Bachmann (Eds.) [1989], p. 75, where he describes his mixed feelings towards the society. According to Rechenberg [1988]a, p. 418, about 350 of a total of 1350 members were affected by this action - cf. also the introduction, sect. 5.2. The following were among the membership withdrawals: Richard → Gans (on Dec. 10); E. Cohn, G. Jaffé and L. Graetz (on Dec. 13); W. Kaufmann and Dr. Boas (on Dec. 17).

[1]The chairman of this → *DPG* meeting was Carl → Ramsauer, director of the → *AEG* research labs. It was by virtue of his politically immune position in private business that he was elected president of the society in 1941, in an effort to maintain its independence from the → REM.

[2]Herbert → Stuart had been lecturing at Berlin University since 1936 as the appointed substitute for the theoretical physics chair vacated by E. → Schrödinger; Wilhelm → Orthmann had become unsalaried lecturer of physics at Berlin University in 1931. Thus both these avid supporters of the Party line were in lower-level temporary positions—this was soon to change as a result. See also Stuart's letters in the Stuart file among the Goudsmidt Papers at the American Institute of Physics. For Orthmann's and Stuart's initiative to reform physics education, see their reports at the first German mathematics camp (*Mathematikerlager*) organized by the *NS Dozentenbund* (→ *NSDDB*) on July 1-3, 1938: Stuart [1939] and Orthmann [1939].

Society, Mr. Debye, sent a letter addressed to all the German members of the German Physical Society, calling upon Jews in the sense of the Nuremburg Laws to withdraw membership in the Society.[3] The second step to be taken is a change in the Society's statutes. The same are to be conformed to the wishes of the Ministry and are to be presented for resolution at a special general meeting.[4]

In the discussion Mr. Orthmann points out that the first sentence of the letter addressed to the German members of the Society is formulated in such a way that the letter could be misunderstood.[5] Mr. Debye asks that this sentence be taken just as it is meant and assumes responsibility for the wording selected. Messrs. Trendelenburg, Picht, Westphal, Stuart, and Schottky participate in particular in the subsequent discussion.[6]

---

[3]Cf. the preceding doc. 67 for Debye's letter to the members of the board of the → *DPG* on Dec. 2, 1938 and the attached draft letter sent a week later to all the German members of the society on Dec. 9, 1938.

[4]This demonstrates the Education Ministry's increased influence in the internal affairs of private organizations that had avoided political realignment up to that point.

[5]The wording of the opening sentence in the draft version in particular, but also in the revised text clearly suggests criticism of the measures and implies their temporariness and lack of justification. Subtle formulations became one outlet for passive resistance, which did not go unnoticed. A sarcastic comment in the files of the information service of the Reich University Lecturers Leadership dated February 1939 demonstrates this: "They seem to be very far behind the times in the Ger. Physical Society, and still very devoted to their dear Jews. It is indeed remarkable that membership of Jews can only no longer be maintained intact 'under the compelling prevailing circumstances'!" Quoted in Beyerchen [1977], p. 230. See the modified text in footnote 5 of doc. 67.

[6]Ferdinand Carl Trendelenburg (1896–1973) studied at Edinburgh, Berlin, Tübingen and Göttingen, where he took his doctorate under Max Reich in 1922. His career started as a physicist at → *Siemens* in the same year. In 1929 he became unpaid lecturer at Berlin University and in 1935 supernumerary associate professor of physics there, becoming in 1940 honorary professor in Berlin, and transferring to Freiburg University in Breisgau in 1949 and to the Munich Polytechnic in 1957. He was also head of the *Siemens-Schuckert* research labs at Erlangen from 1951. Trendelenburg was an acoustics specialist and also worked on electron diffraction and phonetics. [Karl Wilhelm] Johannes Picht (1897–1973) studied at Berlin University 1919–25, taking his doctorate under Max → von Laue in 1925. In the same year he was employed at the Geodesic Institute in Potsdam and 1927–30 at the Einstein Institute, Potsdam. In 1930 he worked at the Askania Works in Friedenau (Berlin). Picht became unpaid university lecturer at the Berlin Polytechnic in 1931 and then tenured associate professor of theoretical optics there 1933–45 at the Defense Technology Department (*Wehrtechnische Fakultät*). After 1945 he headed his own research institute of mathematics, physics and optics. 1946–49 he worked at the research department of the Academy of Sciences in the Soviet Union (Akad. Nauk SSSR). Picht took on a lectureship 1949–50 and became professor in Brandenburg in 1950 until 1952 when he accepted a chair and department directorship at the Pedagogical College in Potsdam. His specialties included theory of optics and astrophysics. Picht may well have supported his former teacher von Laue, who was also present at the meeting—his work at the Einstein Tower had involved verifying the theory of relativity through astronomical observations under Erwin F. Freundlich. Wilhelm → Westphal was associate professor at both Berlin universities from 1934. Walter Hans Schottky (1886–1976) studied at Berlin University and took his degree there under M. → Planck in 1912. In 1922 he became associate professor and in 1926 full professor of

69 Percy W. Bridgman: 'Manifesto' by a Physicist [February 24, 1939]

Source: Percy W. Bridgman,[1] ' 'Manifesto' by a Physicist' in *Science*, Vol. 89, February 24, 1939, p. 179. Copyright 1939 by the AAAS.

Many scientists must have been profoundly disturbed by the revelations of recent events as to what the implications of the totalitarian philosophy of the state really are.[2] There would seem not to be room on the same planet for totalitarian states and states in which the freedom of the individual is recognized. Many scientists must have been moved to try to find something to do about it. In my own case this urge to find something to do has resulted in the decision to close my laboratory to visits from citizens of totalitarian states.[3] I have had the following statement

---

theoretical physics at the University of Rostock. 1927–45 he was scientific consultant to *Siemens* in Berlin and became a contractual employee there in 1946. 1953–57 he additionally chaired the → *DPG*'s semi-conductor committee. Schottky's specialties included thermionic cathodes, electricity theory, quantum theory and solid state physics.

[1]Percy Williams Bridgman (1882–1961) studied physics at Harvard, graduated in 1904 and got his Ph.D. there in 1908. He was a Harvard fellow 1908–10, instructor of physics 1910–13 and assistant professor 1913–19 when he became professor of physics there until 1927. 1926–50 he was professor of mathematics and natural philosophy, working at the Harvard Research Lab and 1950–54 Higgins Professor. Bridgman introduced the concept of 'operationalism' to the philosophy of science, which states that operational definitions, which specify how physical properties are to be measured, provide the meaning of scientific terms denoting physical properties. He received the Nobel Prize in physics for 1946 for his research on extremely high pressures of the order of up to 100,000 kg/cm^2, which he achieved using his ingenious packing method described, e.g., in his Nobel lecture: 'General survey of certain results in the field of high pressure physics', *Les Prix Nobel en 1946*, Stockholm: Norstedt, 1948, pp. 149–166; see also ibid., pp. 20–23.

[2]Official German emigration statistics published at the end of 1938 indicated that 16,561 Jews had left Germany in that year—a sharp rise from 7,155 in 1937; 10,173 of these immigrated to the USA. On Jan. 11, 1939, the US government reminded the German government that the American Constitution guaranteed the civil rights of all their citizens regardless of race or religion, including the American Jews in Germany; and on Jan. 12, 1939, the fate of about 3,000 Jewish and 2,000 political German and Austrian emigrants living in the the rump Czech state again came under discussion because Germany demanded their forced return.

[3]Contrasting opinions existed among his fellow countrymen on his action. In a letter to Bridgman dated April 7, 1939, the Harvard alumnus and psychologist [Abram] Ellis Freeman (1896–) applauded his actions and described the astronomer Harlow Shapley's (1885–1972) repudiation of the manifesto, which Freeman had first mistakenly attributed to him: "I was deeply gratified to read of your position toward Nazi visitors [...]. [Dr. Shapley responded] taking pains to disavow not only his deserts but also the attitude which I imputed to him. While I regret the error, I cannot help feeling provoked into observing that Dr. Shapley is living according to a tradition of scientific intercommunication and internationalism which we all naturally prefer but which, in the face of what Nazism has done to science, is now a sheer fantasy. One need only recall the words of Professor Lenard, a Nobel laureate in physics at that, to grasp this. [Freeman then quotes the first few lines of doc. 39]. Does one need more than this to realize that if the internationalism of science, free inquiry, and intercommunication between scientists are desirable, this cruel and solemn farce should be repudiated by scientists

printed, which I hand to any prospective visitor who may present himself.

Statement

I have decided from now on not to show my apparatus or discuss my experiments with the citizens of any totalitarian state. A citizen of such a state is no longer a free individual, but he may be compelled to engage in any activity whatever to advance the purposes of that state. The purposes of the totalitarian states have shown themselves to be in irreconcilable conflict with the purposes of free states. In particular, the totalitarian states do not recognize that the free cultivation of scientific knowledge for its own sake is a worthy end of human endeavor, but have commandeered the scientific activities of their citizens to serve their own purposes. These states have thus annulled the grounds which formerly justified and made a pleasure of the free sharing of scientific knowledge between individuals of different countries. A self-respecting recognition of this altered situation demands that this practice be stopped. Cessation of scientific intercourse with the totalitarian states serves the double purpose of making more difficult the misuse of scientific information by these states, and of giving the individual opportunity to express his abhorrence of their practices.

This statement is made entirely in my individual capacity and has no connection whatever with any policy of the university.

Science has been rightly recognized as probably the one human activity which knows no nationalisms; for this reason it has been a potent factor making for universal civilization. Action such as this is therefore to be deeply deplored and to be untertaken only after the gravest consideration. But it seems to me that the possibility of an idealistic conception of the present function of science has been already destroyed, and the stark issues of self-survival are being forced upon us. Perhaps the only hope in the present situation is to make the citizens of the totalitarian states realize as vividly and as speedily as possible how the philosophy

themselves? / It would seem the beginning of wisdom for scientific men to recognize what can help or hurt science. Certainly the Nazis, even to the scientists who are articulate, understand how science can help or hurt Fascism. And what is more, were a German scientist entirely innocent of all extraneous motives in visiting a laboratory in this country, the German publicity machine would see in the visit an opportunity of coupling the name of the American scientist with German science, and by that means gaining a kind of oblique endorsement for Nazism. This much is elementary. If an American does not realize this, it is because he is uninformed of contemporary methods of publicity. It may be taken as a rule that to Fascism there is no scientific detachment or neutrality; there is only overt support or hostility, and neutrality becomes in their view endorsement by default." In his response dated April 14, 1939 Bridgman adds: "Although I know that a great many of my colleagues on the Harvard Faculty do not approve of my action, it so happens that Shapley was the only one who took the trouble to write me and express his **violent** disapproval." (Harvard University Archive, call no. HUG 4234.10)

of their states impresses and affects the rest of the world. Such a realization can be brought about by the spontaneous action of the individual citizens of the non-totalitarian states perhaps even more effectively than by their governments. Here I think is one of the few conceivable situations in which the popular conception of the social 'responsibility' of 'science' can touch at all closely the individual scientist.

70 Ludwig Prandtl: Mechanics [c. April 1939]

Source: Ludwig Prandtl,[1] 'Mechanik' in *Deutsche Wissenschaft. Arbeit und Aufgabe*,[2] Leipzig: Hirzel, pp. 155–157, and Plate 1 (facing p. 156).

Mechanics, which is the theory of the balance of forces and the motion of mass systems, rigid bodies, elastic bodies and liquid and gaseous media is, as far as the fundamental laws are concerned, already an old and thoroughly established science. Yet it is still filled with pulsating life even today; because advancing technology ensures a constant and abundant supply of new problems. In our times advances in higher arithmetic alone have now opened many possibilities toward the successful completion of research that had been considered too hard to tackle before. What is more, teamwork is increasingly replacing the individual scholar.[3] An individual's research work is fed by an inner drive and is dependent only to a very slight degree upon the political and economic circumstances of the outside world. The contrary is true in the case of teamwork, since it can only come about when important goals are set and the government leadership is willing to grant it funding. It can thus be clearly demonstrated that in the wake of the seizure of power six years ago, and especially following the return of the German nation to military fitness, very great progress became possible also in the various branches of research in mechanics. This is closely related to the fact that skilled research groups were now able to investigate the problems thoroughly. A most obvious example is the work connected to the development of our new German Air Force.[4] This area will be given special attention in the

[1]Ludwig → Prandtl was full professor of applied mechanics at the University of Göttingen from 1907. In the same year he also headed the newly founded aerodynamics testing site (*Modellversuchsanstalt für Aerodynamik* later called *Aerodynamische Versuchsanstalt: AVA*) and from 1925 on the Kaiser Wilhelm Institute of Fluid Dynamics Research (→ *KWIS*). Cf., e.g., Rotta [1990].

[2]This collection of brief summaries of disciplines in the sciences—in the broader sense of the word—was dedicated to Adolf → Hitler in celebration of his 50th birthday on Apr. 20, 1939.

[3]On the emphasis of teamwork as the research counterpart to the new 'national community' (*Volksgemeinschaft*), see also Jordan, doc. 87.

[4]On the 'new upswing' in aerodynamics research applications after the National Socialist seizure of power in 1933 and a ten-fold increase in research personnel by 1938, see in particular Tollmien in Becker et al. (Eds.) [1987], pp. 466ff. Under the *Reichskommissariat für Luftfahrt*, founded in January 1933 and transformed into the Reich Aviation Ministry (→ *RLM*) on May 8, 1933, H. → Göring and his substitute General Milch rearmed the German Air Force

examples presented in the following. In addition, applications to ship building can also be mentioned.[5] Naturally many other sciences apply the laws of mechanics to other fields of technology, such as engine building, automobile construction, metallurgical engineering, the engineering of chemical apparatus, bridge building and surface engineering, road construction and hydraulic engineering, etc. But the purpose of these exposés seems rather to call for brevity over a fair attempt at all-inclusive breadth.

Among the applications of the **theory of mass systems**, resonance problems are uppermost today. An important chapter in this field, the study of vibrations in the mechanism of rapidly running aircraft engines, has recently succeeded in making very noteworthy advances. In particular, through a more precise understanding of all the individual processes, we can determine the revolutions per minute that are hazardous to the engine very much more reliably. Our knowledge about ship-hull vibrations has also been substantially improved through experimental and theoretical analyses.

The physics of gyroscopes is primarily in the forefront in the field of the **mechanics of rigid bodies**. Gyroscopes used in airplane measuring instruments provide an artificial horizon, making the airplane's turns visible in fog. On the other hand, they also provide the basis for automatic pilot systems. These relieve the aircraft pilot of the routine work during actual flight, thus making unmanned remote-controlled airplanes feasible. The aircraft industry has just recently brought out very finished automatic pilot designs of this type.

Also in the area of **elasticity theory** very important investigations have been conducted in recent years towards solving new problems posed particularly in 'lightweight construction', which have resulted in great progress in exploiting the stress characteristics of component parts. Here the theories put forward and experiments conducted to elucidate the stress properties of so-called shell construction must be mentioned. This construction allows the metallic fuselage and wing fabrics to contribute substantially in withstanding forces, with the natural result that weight can be saved at other component parts. The tension distribu-

---

dramatically, which was only officially reestablished in 1935. Prandtl's *AVA*, the *KWIS* in Göttingen and other research institutions for applied aerodynamics also made significant gains as a result. Prandtl, who presided over the fluid dynamics committee and became one of the aerodynamics and hydrodynamics experts (*Fachgruppe für Aerodynamik und Hydrodynamik*) of the newly founded aviation research association *Vereinigung für Luftfahrtforschung* in 1933, strictly opposed the intrusion of Nazi ideology into science. Nevertheless, he collaborated a lot in practical terms with the military branches of the National Socialist polycracy, especially with Göring's *Luftwaffe*, to promote the importance of his field of research and win funding for it. On Prandtl's defense of Hitler's policy in correspondence with the English aerodynamicist G. I. Taylor in 1932 and 1939, see Tollmien in Becker et al. (Eds.) [1987], pp. 476f.; cf. also Prandtl's own retrospective account from 1947 in Vogel-Prandtl [1993], pp. 176–180.

[5]At the Göttingen *AVA*, for instance, in addition to a high-speed wind tunnel (see Fig. 12), three water channels were constructed between 1938–40 for hydrodynamic research with possible applications for the German Navy in mind, but aerodynamics was the main focus.

tion and the deformation of such shells is very involved; but today teamwork on a large scale has resulted in an excellent command of this subject.

With regard to the **mechanics of liquid and gaseous media** we should point first of all to the advances made in the theory of ship resistance and propulsion. Let me draw special attention to the theoretical mastery of the formation of waves produced by the ship itself, which significantly determines the ship's resistance. A related problem is the behavior of aircraft floats and gliding boats on water. This also has been largely clarified in recent years with the aid of experiment and theory. In aircraft engineering the distribution of air currents around the airplane wing is a central problem. Through the work of domestic and foreign theoreticians and research institutes in the last two decades, an extensive clarification of the processes here has already been achieved.[6] But it has become evident precisely here, how much deeper we can go into detail when sizeable teams devote themselves to a problem, applying all the advances both in applied mathematics and in research technology. Today we have already gained welcome knowledge of the finer details of many problems; and designing engineers are now in the position to continually improve the aerodynamic characteristics of their airplanes. Technological progress makes sure that there is no shortage of problems to solve. The huge increase in travel speed of airplanes in recent times, for instance, which is a fruit of the research mentioned previously, has caused completely new problems to become urgent related to flight speeds approaching the speed of sound. Research has already refocussed itself on these new questions, since theory has already prepared the way for it well; and it will not be long in bringing practical results.

The above descriptions should not and could not be more than a narrow cross-section of a very expansive field; but they should bear witness to the vitality of the **science of mechanics** today.

[6]After the outbreak of World War II, the *KWIS* continued its defense technology and military research unimpeded, thanks to its 'indispensable' priority rating: *uk* (*unabkömmlich*). Cf., e.g., C. Tollmien in Becker et al. (Eds.) [1987], pp. 477–480, and Trischler [1992], pp. 241ff., on the German mobilization of aeronautical research in World War II.

Fig. 12: "High pressure wind tunnel at the Aerodynamics Research Institute in Göttingen. The picture shows both alternately usable jets that blow currents of air at test objects. The housing surrounding the jets can be closed by means of a drum-shaped sealing cap. Increased air pressure within the housing makes it possible to imitate the conditions caused by currents on objects that are larger than the related test models. Low pressure permits attaining very high speeds. The testing installation presented in the foreground of the picture serves to measure the speed distribution in the air current."

71 Hans Geiger: Experimental Physics [c. April 1939]

Source: Hans Geiger, 'Experimentelle Physik' in *Deutsche Wissenschaft. Arbeit und Aufgabe,*[1] Leipzig: Hirzel, pp. 165–167, and Plates 6 and 7 (facing pp. 166, 170).

Modern experimental physics concentrates to a substantial degree on problems in the atomic and subatomic fields.[2] The reason for this is twofold: Firstly, there is a lot of promise of success in implementing the already highly developed experimental technology. Secondly, this direction of research leads right to the heart of the major question of the essence of matter and energy. The importance of these questions for a deeper understanding of nature and for technological progress is generally recognized. German research stations have also been actively involved in these areas and in recent years were able to score several fine victories over the competing institutes abroad. The three following examples leading us

[1]See footnote 2 of doc. 70 on the dating of this collection.

[2]H. → Geiger's own specialty was radioactivity. He was full professor of physics at the Polytechnic in Charlottenburg (Berlin) from 1936, succeeding Gustav → Hertz. For more on Geiger, see in particular Swinne [1988]b.

into the range of the smallest dimensions offer a glimpse into the areas of study German physics has adopted.

First of all to the internal structure of atoms. Earlier studies have shown that atoms are made up of electrical charges and are held together by electrical forces. It is also known that every atom is controlled by a center of force, the so-called nucleus, which is itself infinitely small in comparison to the atom. Today, it is primarily necessary to explore the nature of these nuclei. The structure of these nuclei can be exposed by bombarding them with particles of matter with the aid of high electrical tensions, because the nuclei react in different ways, depending upon the type and force of the impact.[3] Often an atomic transformation takes place in the process, so that, for example, a nickel atom turns into a copper atom. In recent years German institutes have traced many kinds of such reactions and have examined more closely how these reactions unwind. Two of the latest discoveries are particularly significant: By means of the one, it becomes clear that atomic transformations are possible not only through collision with particles of matter but also with light rays of very short wavelength;[4] an extremely peculiar interaction between light and matter has thus been revealed. The second new discovery is that when appropriately bombarded, heavy atoms can break apart into two almost equally large parts.[5] This contradicts all previously known nuclear transformations, which without exception result in atoms that do not differ in weight from the parent atoms, or if so only a little. Thus new light has been shed on the complicated forces that hold the individual elements of the nucleus together.

The analysis of the remarkable fact that some metals and alloys lose their electrical resistance at extremely low temperatures is also included among the fields of atomic research.[6] At the Reich Physical and Technical Institute an important discovery was made that in such cases the electrical current does not penetrate into the interior, but in complete contradiction to its usual behavior, flows in thin layers at the surface.[7] Simultaneously, the technology developed to produce very low temperatures has progressed further.[8]

---

[3]See here Fig. 13 (Plate 6, facing p. 166 in the original text).

[4]For a discussion of the atomic transformations induced by neutron bombardment, see doc. 74 and references there; similar results can be obtained by exposure of matter to intense gamma radiation.

[5]This fission process was discovered by Hahn & Straßmann [1939] and was correctly interpreted theoretically by Meitner & Frisch [1939]; cf. also doc. 74.

[6]Heike Kamerlingh Onnes (1853–1926) discovered this effect called superconductivity in Leiden in 1911 and was awarded the 1913 Nobel Prize.

[7]This is called the Meissner-Ochsenfeld effect, discovered by [Fritz] Walther → Meissner and Robert Ochsenfeld (born in 1901) at the → *PTR*'s low-temperature laboratory in 1933. See W. Meissner & R. Ochsenfeld: 'Ein neuer Effekt bei Eintritt der Supraleitfähigkeit', *Die Naturwissenschaften* **21** [1933], pp. 787–788.

[8]For a historical overview of low-temperature physics see, e.g., Kostas Gavroglu & Y. Goudaroulis: *Methodological Aspects of the Development of Low Temperature Physics 1881–*

The second example concerns an interaction between electricity and light which also is only observable within the range of the smallest dimensions. Each moving electron is linked to a wave field,[9] thus it is also possible to conduct experiments with fast electrons that are completely analogous to those performed in optics. The practical repercussions of this have already manifested themselves in the microscope. As is generally known, microscopical magnification is limited by the fact that the structure of light is not fine enough for objects smaller than one ten-thousandth of a millimeter. Though X rays have a much finer structure, they cannot be used, however, because they cannot be refracted through lenses. We Germans have now succeeded in building an 'electron microscope', in which light rays are replaced by much more finely undulating electron beams. The lenses are replaced by fields of magnetic force.[10] Thus objects as small as a hundred-thousandth of a millimeter can be studied.[11]

Cosmic rays from outer space will be mentioned as the third example, which offer us an abundance of still unsolved problems. This radiation consists in part of electrons of immense penetrating force, in part of photons (light particles) which exceed the energy of even the most potent X rays a thousandfold. Without doubt we will gain new insights into the structural make-up of the cosmos once we fully understand the origin and nature of this extremely peculiar radiation.[12] Thus these problems are being worked on intensely in all civilized countries; and Germany also has made valuable contributions to clarifying the unsettled problems. We will mention primarily the numerous systematic investigations aimed at understanding these beams, which can consist of hundreds of energy-rich electrons emitted by individual cosmic rays, by releasing them within a narrow space confined by lead or another material. Such electron beams caused by cosmic rays have also even been detected at depths of several hundred meters under the earth's surface.

1956, Dordrecht: Kluwer, 1989.

[9]Geiger's choice of the term 'wave field' (*Wellenfeld*) reflects his lack of familiarity with quantum mechanics, since according to Schrödinger, the commonly used term is 'wave function' (*Wellenfunktion*).

[10]Electron microscopes were primarily developed at the high-voltage lab of the Berlin Polytechnic in Charlottenburg and at → *Siemens*, where the first commercial models were produced in 1939, with a resolution of 250–500 Angstrom. In 1986, Ernst August Friedrich Ruska (1906–1988) shared the Nobel Prize in physics with Gerd Binning (1947–) and Heinrich Rohrer (1933–) for his work on the electron microscope. On the contemporary technology in electron microscopy see, e.g., M. Knoll & E. Ruska, 'Das Elektronenmikroskop', *Zeitschrift für Physik* **78** [1932], pp. 318–339; M. Knoll: 'Elektronenmikroskop' in Goubau & Zenneck (Eds.) [1948], pp. 57–78. See also the richly illustrated and extensive retrospective account by Ernst Ruska, *Die frühe Entwicklung der Elekronenlinsen und der Elektronenmikroskopie*, Halle: Saale, 1979 (= Acta Historica Leopoldina, **12**).

[11]Cf. also fig. 14 (Plate 7 facing p. 170 in the original text).

[12]For contemporary overviews of cosmic ray physics, the precursor of today's elementary particle physics see, e.g., Euler & Heisenberg [1938]; cf. Bothe & Flügge (Eds.) [1948], chap. 2 by E. Bagge, G. Molière, K. H. Höcker & A. Ehrmert on cosmic radiation.

It would be incorrect to rate the discoveries presented here as examples of lucky strikes. Rather, they emerged from a broad groundwork as the final results of assiduous and careful experimentation. The importance of such persistent research cannot be valued highly enough. It offers the secure basis for technological progress and at the same time, it prepares the ground upon which the ultimate reward of a far-reaching discovery can develop. X rays and broadcast waves also came to us as the final and almost inevitable consequence of tedious and most scrupulously detailed work.

Added to the successes that the Third Reich can register in physics is the heightened interest more and more areas of physics encounter today. Even talks on specialized questions of atomic structure or of the structure of light find a large audience extending far beyond the academic profession. This growing intellectual activity offers the best guarantee that the next generation will never be wanting for the very indispensable work in physical research.

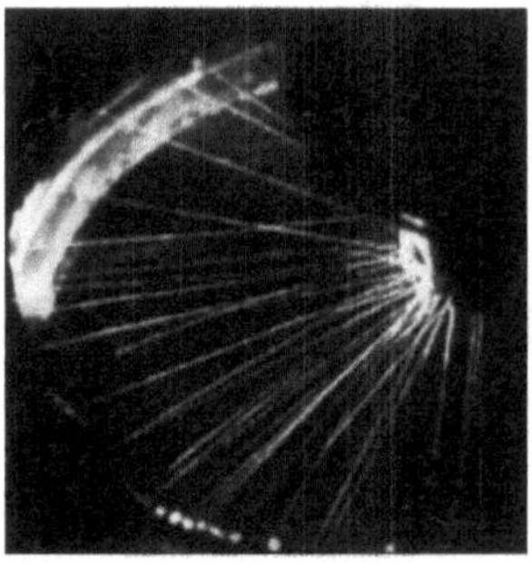

Fig. 13: "At the right edge of the picture matter particles coming from above, but not visible here, strike lithium atoms at high speed, which explode apart under the impact of the collision. The paths of the explosion fragments can be seen in the illustration as light lines."

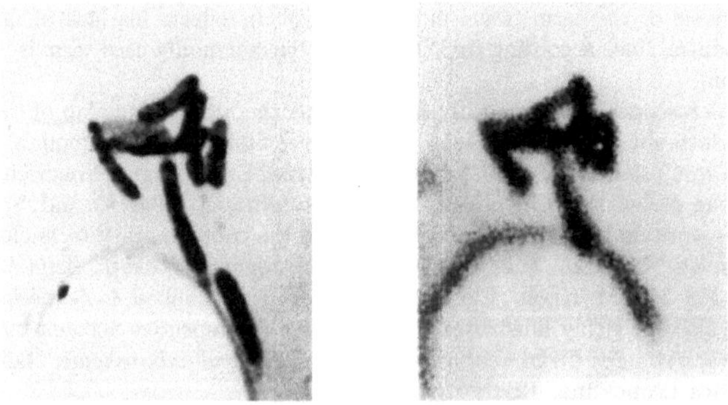

Fig. 14: [Left:] "a) Electron-optical exposure, [and right:] b) photographic exposure of coli bacteria at the same magnification (by 10,000). Details inside the bacteria are clearly visible on the electron-optical exposure."

72 Abraham Esau: Technical Physics [c. April 1939]

Source: Abraham Esau, 'Technische Physik' in *Deutsche Wissenschaft. Arbeit und Aufgabe,*[1] Leipzig: Hirzel, pp. 171–172.

As in all fields of technology, technical physics has been facing a large number of new challenges and goals since 1933 and in particular since the Four-Year Plan,[[2]] Thanks to the funds provided and the working morale of German scientists, they have already been solved partially; and the remainder are well on the way to a solution.

The large number of new materials produced out of native raw materials have posed new and comprehensive problems in materials testing. This regards materials used in machine building, therefore, those subjected to mechanical stress, and also regards those that increasingly find applications as insulating materials in electrical engineering. The results are not only of importance to the designing engineer and to the operational safety of installations; they also facilitate using these materials in exceptionally economical ways. The time required to execute such types of tests, which a few years ago still lasted months and weeks but today can be shortened to days and hours, is of economic importance. In recent years new procedures also had to be developed to test new materials produced by the ceramics industry for electrical applications; and these procedures have already proven their worth in practice. The results of the tests performed on these new materials reveal that materials manufactured from domestic raw materials are not only equivalent in quality to those used previously, but in very many cases they are even superior.

Particularly great progress has been achieved recently in the field of high-frequency engineering, which has always been a particularly successful branch of study in German science.[[3]] This pertains not only to the production of high energies but also to the generation of ever shorter wave[length]s, which hold in store great application possibilities in engineering in general and in high-frequency engineering in particular. Thus it became possible to produce an output of 1 kilowatt from wavelengths of about 1 meter, and to attain the still very considerably high yield of 150 watt from a wavelength of only 25 centimeters.[[4]]

[1]See doc. 70, footnote 2 on the dating of this collection.

[2]The Four-Year Plan, introduced in 1936, was aimed at attaining economic self-sufficiency; see, e.g., Noakes & Pridham (Eds.) [1990], Vol. 1, chap. 13, and doc. 58.

[3]High-frequency engineering was Abraham → Esau's specialty. Esau was full professor of technical physics at the University of Jena from 1927; but in 1939 he changed to a full professorship at the Polytechnic in Charlottenburg (Berlin). He also became president of the → *PTR* at this time and was director of the physics division of the → *RFR*. In the winter of 1943/44 Esau became H. → Göring's plenipotentiary of high-frequency and radar technology. Because he had been appointed councillor of the state of Thuringia (*Staatsrat*) in October 1933, Esau had direct access to A. → Hitler; thus he was politically one of the most influential physicists in Germany.

[4]Cf. C. Schmelzer et al.: 'High frequency measuring technique' in Goubau & Zenneck (Eds.)

After overcoming quite significant difficulties, German science has conquered a new wave range in this field. As a result it is possible today to produce electrical waves of a centimeter and a millimeter's length and to test successfully their possible practical usefulness. Suitable fields of application include communications systems requiring particularly accurately developed signal aiming, which today already approaches the quality of optical means of communication. On the other hand, new possibilities are opening up in solving the most important and urgent problems in air navigation.

Regarding the use of electrical waves in medicine, which had originally been conceived and begun by German scientists, further progress has also been made in recent years which will affect treatment in the near future. The use of shorter waves (1 meter as opposed to previously 6–10 meters) has resulted in improving its efficacy four times over on internal organs of the body.[[5]] More favorable results can still be expected from a further reduction of the length of the wave, and this is already being looked into.

Theoretical and experimental studies on the reflection and absorption of sound have substantially improved our knowledge of the difficult problems of room acoustics and soundproofing.[[6]] The production of inaudible sound waves (ultrasound) has also been developed further with good result.[[7]] They are of particular importance in underwater acoustics and ship navigation; and similar to electrical waves, they are also already beginning to find applications in medicine.

Precision instrumentation, which has always been particularly cultivated in Germany, has also made important advances in recent years, not only in the electrical, but also in the magnetic, optical, and mechanical branches. These advances have had a fruitful impact on all areas of engineering and will continue to do so in the future to an even greater degree.

[1948], Part II, pp. 211–278. On the crucial importance of radar research, see also the Introduction, sec. 5.4, footnote 285.

[5]On applications in therapy of high-frequency electric fields, see J. Pätzold et al. in Rajewski & Schön (Eds.) [1948], pp. 1–62, and Trendelenburg [1975], pp. 250f.

[6]On industrial research in this area see, e.g., Trendelenburg [1975], pp. 173–204; In 1928 Max Wien pushed for the creation of an acoustics lab at the *PTR*, which was finally built with → *DFG* funds on Franklin Street under Stark's presidency. Martin Grützmacher (1901–) was its director from 1934, and by 1938 there were 22 staff members; see, e.g., Stark [1937], pp. 97f., 120ff.; Kern [1994], pp. 197, 230ff.

[7]For an overview of the state of the art in these areas, see Ludwig Bergmann: *Der Ultraschall*, Berlin: VDI-Verlag, 1944 (4th ed.); or L. Bergmann: 'Schallausbreitung in Flüssigkeiten' and 'Schallausbreitung in Gasen', in Kappler (Ed.) [1948], pp. 121ff., 206ff. The *PTR*'s ultrasound laboratory was established in 1936.

73 SS Head of the Central Office of Public Safety: Letter to Rudolf Mentzel enclosing Report on Heisenberg [May 26, 1939]

Source: Bundesarchiv, Abteilungen Potsdam, Zentrales Staatsarchiv: 49.01, 'Reichsministerium für Wissenschaft, Erziehung und Volksbildung', 2943: sheets 370–372. Typed transcript of enclosure letter with the sender's address: 'Der Reichsführer SS, Der Chef des Sicherheitshauptamtes, *II 2111 -5*,[1] PA 7897/36 Js./Me., – WP Heisenberg b –' and addressed: 'An das Reichsministerium für Wissenschaft, Erziehung und Volksbildung z. Hd. v. SS-Standartenführer Prof. Dr. Mentzel in Berlin W 8, Unter den Linden 69.'[2] with the heading: 'Abschrift zu WP Heisenberg c' on both the letter and the attached two-page report, and: 'Previous correspondence: None. Encl.: 1' and no salutation.

Re Prof. Werner Heisenberg, Leipzig.

Enclosed you will find an evaluation of Prof. Heisenberg for your information. An identical evaluation went to the leader of the Reich University Lecturers Association.

Following a severe attack on Heisenberg in an article in the *Schwarzes Korps* in 1937,[[3]] thorough investigations into Prof. Heisenberg's personality and working approach were conducted by this office, which are now more or less concluded.[[4]] After checking the situation as a whole, it seems advisable to transfer Heisenberg to the Theoretical Physics Chair in Vienna, which is presently vacant, because a call to Munich would have to be interpreted as a victory over the local Party authorities there, who had expressed their opposition to his transfer to Munich some years earlier.[[5]] On the other hand, most of the professionally competent physics professors in Vienna had already belonged to the N[ational] S[ocialist]

[1]The main office of the Security Service (*Sicherheitsdienst-Hauptamt*) was the second secret system of public surveillance aside from the → *Gestapo*, the secret police. The *Sicherheitsdienst* (*SD*) was set up by → Himmler's → *SS*, with R. Heydrich at its head. The Central Office *Zentralabteilung II-2* of the *SD-Hauptamt* dealt with the surveillance of German territory (*Deutsche Lebensgebiete, SD-Inland*). From 1936 *SS-Gruppenführer* Otto Ohlendorf (1908–1951), a trained economist, was its head. Four months after this document was written (on Sep. 27, 1939), the *SD* and the central units of the secret police, which often pursued the same groups of people (e.g., Marxists, leftists, fugitive Jews, saboteurs, deserters, etc.) were merged into the *Reichssicherheitshauptamt* (*RSHA*), again with Ohlendorf directing the domestic office (*RSHA Amt III* = *SD-Inland*) and with Dr. Franz Six heading the new division II covering ideology (*Weltanschauliche Forschung*). Following another reorganization in 1940, this unit was renamed division VII. On the *SD* and the *RSHA* see, e.g., Hilberg [1961], p. 185, Höhne [1984], chap. 10, esp., pp. 237ff., Broszat & Frei (Eds.) [1989], p. 207.

[2]Rudolf → Mentzel had been appointed Undersecretary at the → *REM*'s Science Office in that year.

[3]Cf. docs. 55 and 56.

[4]Heisenberg himself had initiated the investigation with his letter to Himmler of July 21, 1937; see also docs. 57 and 64, footnote 2. Cf. Beyerchen [1977], p. 254, Cassidy [1992]a, p. 385, [1992]b, p. 657.

[5]On the difficulties involved in filling Sommerfeld's chair for theoretical physics see, e.g., Cassidy [1992]a, chap. 18, and further references there, as well as here doc. 85, No. 4.

Party during the time when it was illegal [*illegalen Zeit*[6]] and are as a group politically and ideologically reliable. So this would seem to ensure that Heisenberg will be influenced into developing an interest in political events and in the National Socialist worldview.

Head of the Central Office of Public Safety, J. A.
Director of Central Division II 2, J. V.
Signature,[7] *SS* Battalion Leader and Chief of Staff

Re Prof. Werner Heisenberg, Leipzig.

Heisenberg is a man of great standing in science. His strength lies in training the next generation, which includes among others von Weizsäcker and Flüg[g]e.[8]

Regarding the current dispute on the foundations of physics, Heisenberg is of the opinion that a conflict of a difference in mentality cannot really arise between experimental and theoretical physics, because every theoretical physicist considers experimental physics absolutely necessary to his work, with the converse equally applying.

Heisenberg makes a clear distinction between 'good' and 'bad' scientists in the field of theoretical physics; and he defines as bad physicists those who are faraway from intuitive experience. His definition of the bad physicist corresponds completely to the concept of the alien theoretician [*artfremder Denker*] in physics. Heisenberg includes in particular in this definition some Jews and Aryan scientists from the Jewish school against whom Lenard and Stark (his opponents) are fighting.

Heisenberg himself comes from a school of physics that teaches the Jewish approach and methodology.[9] According to this school the only result of any value is formalism, which is built upon theory regardless of how far it is confirmed in nature or how it relates to natural occurrences. Therefore, Heisenberg's first big efforts also carry this alien stamp. In later years, however, the intuitive Aryan [*artgemäß*] way of thinking becomes more and more prevalent in Heisenberg's working method.[10]

[6]After the failure of the attempted coup and the murder of the Austrian Chancellor Engelbert Dollfuss on July 25, 1934, the National Socialist Party (→ *NSDAP*) was an underground organization in Austria until its annexation (*Anschluß*) to the German Reich on March 13, 1938.

[7]Proxies signed and initialed this document for Rudolf Heydrich and Otto Ohlendorf (on Heydrich, see the first footnote above as well as doc. 63).

[8]Carl Friedrich → von Weizsäcker wrote his Ph.D. thesis under the supervision of Heisenberg in Leipzig in 1933. Siegfried → Flügge had worked together with Heisenberg at the University of Leipzig 1936–37. On Heisenberg's activities and academic influence during his Leipzig years see, e.g., Geyer et al. (Eds.) [1993], parts 2–3, and Kleint & Wiemers (Eds.) [1993], part 2.

[9]Heisenberg had studied physics under A. → Sommerfeld and had specialized early on in problems of quantum theory, then considered a typically Jewish field.

[10]This strange evaluation may have been based on Heisenberg's interest in ferromagnetism and nuclear physics, two fields in which there was a lot of interaction between experiment and

To Heisenberg theory is merely a working hypothesis with which the experimenter investigates nature by means of suitable experiments. It is only through experiment that theory can be evaluated. Theory that is confirmed by experiment thus clearly describes observations made in nature using the precise tools of mathematics.

As a person Heisenberg is decent, both in character and attitude. Heisenberg is a typical apolitical scholar.[11] He is ready at any time to defend Germany whole-heartedly; however, he is of the opinion that you are either "born a good German or not".

Heisenberg was a member of the Lützow volunteer corps. He served in the Armed Forces of his own free will and enlisted voluntarily for military service during the September Days of 1938.[12] But he declined to participate in an election rally in 1933 with the reason that this action could be misunderstood by foreign colleagues with whom he maintained good relations. Heisenberg declined in the following telegram to sign a proclamation of loyalty to the *Führer* that had been initiated by Prof. Stark in the summer of 1934: "Although I personally vote 'yes', political declarations by scientists seem to me improper, since [this] never was a normal practice even formerly. Therefore I do not sign".[13] However, through the years Heisenberg became increasingly convinced by the successes of National Socialism; and he now views it positively. However, he is of the opinion that it is not the place of a professor to be actively involved in politics, aside from the occasional participation in camps and the like.[14] Today Heisenberg also rejects in principle the infiltration [*Überfremdung*] of foreign Jewish influence into the German living space [*Lebensraum*].

74 Siegfried Flügge: Exploiting Atomic Energy. From the Laboratory Experiment to the Uranium Machine – Research Results in Dahlem [August 15, 1939]

Source: Siegfried Flügge,[1] 'Die Ausnutzung der Atomenergie. Vom Laboratoriumsversuch zur Uranmaschine – Forschungsergebnisse in Dahlem', *Deutsche Allgemeine Zeitung*, August 15, 1939, No. 387, Supplement.

---

theory in these years.

[11]On the apolitical attitude of professors in Germany see, for example, Döring [1974], Ringer [1969], and the Introduction, sec. 3.1.

[12]The *Septembertage* refer to the period of intense and volatile negotiation between Germany and the later Allies on the fate of the Sudetenland in Czechoslovakia in 1938.

[13]For the proclamation, see doc. 3. Another description of the above events is in the polemical article that started the whole affair against Heisenberg, doc. 55; cf. Cassidy [1992]a, pp. 323f.

[14]On Heisenberg's stance in these matters and his occasional participation in political indoctrination retreats and Army sports camps "in order to acquaint [himself] a little more with this politics", see Cassidy [1992]a, pp. 215f.

[1]Siegfried → Flügge also wrote another influential article in the same year on the same topic: Flügge [1939]. Cf. also footnote 16 below for other popular texts on this topic.

The world of atoms is the great new scientific territory conquered by physics in the first thirty years of this century. It is widely known today also among nonprofessionals that at the center of each atom there is a tiny atomic nucleus, the diameter of which is about a one-hundred thousandth of the atom's diameter and which, similar to a planetary system, is surrounded by an orbiting cloud of electrons.

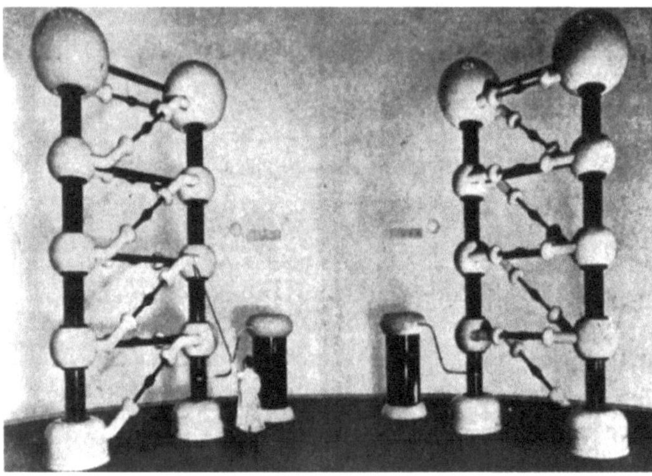

Fig. 15: 'Two high-voltage generators that generate 3 million volts of constant direct current. The equipment is used in Dahlem (Berlin) for experiments in the field of atomic physics. Plant photo'[2]

Protons and Neutrons

The mass of these electrons is so small that the nucleus makes up almost the entire mass of the atom. As Heisenberg recognized clearly for the first time in 1932, this nucleus consists of **protons** and **neutrons**.[3] The proton is the nucleus of the lightest of all atoms, the ordinary hydrogen atom. Its mass and its positive

[2]The → *KWIP*'s high-voltage apparatus was not only a powerful scientific instrument but also an object of prestige. The fact that a member of the rival → *KWIC* was using a photograph of this equipment without specifying its location became a sensitive issue. Flügge then attempted to extricate himself from this "embarrassing position vis-à-vis the men from the Kaiser Wilhelm Institute of Physics" by requesting a correction in the newspaper and assuring its president, P. → Debye, that swindling himself "into higher circles using the equipment of your institute" had been the least of his intentions: See Weiss [1994], p. 274; see also doc. 54 for a description of the *KWIP*'s installations and instruments in Dahlem (Berlin).

[3]See Werner → Heisenberg's three-part paper: 'Über den Bau der Atomkerne', *Zeitschrift für Physik* **77** [1932], pp. 1–11; **78** [1932], pp. 156–164; **80** [1933], pp. 587–596.

electric charge is usually chosen as a unit, which thus means it has the mass 1 and the charge 1. The other nuclear building block, the neutron, has likewise almost exactly mass 1 but no charge. The overall charge of an atomic nucleus is therefore equal to the number of protons bound within it. The electron's charge is exactly opposite (thus negative) to the proton's charge. If the atom contains just as many electrons in its shell as there are protons in its nucleus, then the whole structure is electrically neutral, which is the normal state of matter. The nuclear charge therefore determines the structure of the electron shell and thereby also the chemical properties of the atom. Each chemical element is furnished with a distinct nuclear charge: hydrogen with charge 1, helium with 2, lithium with 3, and so on up to the heaviest element, uranium with nuclear charge 92 and with just as many electrons around its nucleus. The nucleus of uranium thus contains 92 protons. In addition, 146 neutrons are also bound within it, so that its mass is $92 + 146 = 238$. This is the chemical atomic weight of the chemical element uranium. The heaviest nuclei of all, for example, those of radium, thorium, as well as of uranium, are not stable but decay spontaneously in the course of time either by chipping off so-called alpha particles, that is, helium nuclei which are made up of two protons and two neutrons each or, on the other hand, by emitting so-called beta rays, that is, high-velocity electrons. This is because in some nuclei a neutron can split into a proton and an electron; and the electron then shoots out of the nucleus. These are the well-known phenomena of natural radioactivity, which have been known and studied in detail for almost forty years.

A New Chapter in the History of Physics

Then in 1919 Lord Rutherford used such alpha particles, that is, fast helium nuclei, for the first time as projectiles to bombard other less heavy nuclei and in such a way to artificially stimulate transmutations.[4] For example, he allowed the alpha particles of a radioactive preparation to react with nitrogen. Upon collision with a nitrogen nucleus an alpha particle then happens to combine with it to form an 'intermediate nucleus', which after an inconceivably short time splits apart again, not into its original components, but into nitrogen (with nuclear charge 7) and helium with charge 8 and a proton with charge 1. Thus for the first time the transmutation of one chemical element into another had been achieved; and along with it at least in principle, the alchemist's ancient dream was realized. For, nothing prevents us from writing down this process as a chemical equation: Nitrogen + helium = oxygen + hydrogen. Thus a new chapter had been opened in the history of physics.

[4]The New Zealand-born English physicist Ernest Rutherford (1871–1937) received the 1908 Nobel Prize in chemistry for his experimental research on radioactive decay and the chemistry of radioactive elements. See his paper 'Collision of alpha particles with light atoms', *Philosophical Magazine* (6) **37** [1919], pp. 537–587, and the summary in *Nature* **103** [1919], pp. 415–418. Cf. also A.F.: 'Atomic Transmutation', *Nature* **132** [1933], pp. 432–433.

The Energies of the Atomic Nucleus

Power generation appeared to have an even greater potential than the artificial transformation of elements. Here also such processes can be compared with chemical reactions: The burning of coal is amongst the most frequently used technical energy sources, that is, the chemical reaction of combining carbon and oxygen into the gas carbon dioxide. Energy is thereby released; and this heats in the same way both the boilers of super-power-stations and the heaters of our homes. In many nuclear transformations free energy is generated exactly in this way. But—and this is what is so marvellous here—the amount of energy released per atom is a million times greater than in the chemical reactions upon which our technology currently still depends.

Two serious obstacles made the realization of such technological dreams of the future appear completely hopeless until recently. Firstly, expensive radioactive preparations were needed as projectile sources, i.e., radium and similar elements, which cost in the tens of thousands; and secondly, even then only isolated atoms could be transmuted successfully, while for technical purposes we would have to transmute trillions upon trillions of atoms.

Significant advances have already been made some time ago regarding the first obstacle. Today, for example, protons are used as projectiles in place of the alpha particles of radioactive preparations, which can reach high velocities with the aid of a high-voltage system as a result of its charge in an electric field; and thus they not only completely replace alpha particles, but in many cases are even very much more effective. The first installation of this kind was built in 1931 by Cockcroft and Walton, two of Lord Rutherford's colleagues at the Cavendish Laboratory in Cambridge. They split lithium nuclei with it, for example, which is a reaction in which for each proton causing a transmutation, a multiple of its own kinetic energy is released.[5] In England the magnificence of this pioneering achievement was quickly recognized. The first now historic installation, which was soon replaced by a bigger one, has already been standing in the Science Museum in London for a number of years now, next to the models of the oldest locomotives and the first airplanes.

Thus a powerful trend had been launched in nuclear research. Notably in America, where large financial resources had been made available to science, gigantic installations were built; and just recently Professor Lawrence reported in a scientific journal that he had even produced in his laboratory in Berkeley (California) protons of such high velocity as if they had traveled through an

[5]The English physicist Sir John Douglas Cockcroft (1897–1967) won the Nobel Prize in 1951 with the Irishman Ernest Thomas Sinton Walton (born 1903) for their achievement of 1932. See, e.g., John Cockcroft & E. T. S. Walton, 'Experiments with high velocity positive ions', *Proceedings of the Royal Society of London* **A 129** [1930], pp. 477–489; cf. *Nature* **129** [1932], pp. 242, 649.

electric field of 18 million volts![6] Research was performed also at other sites with fewer means but with equal enthusiasm. Of the German research institutions I will name in particular the Kaiser Wilhelm Institute of Chemistry in Dahlem (Berlin) and the younger Institute of Physics at the Kaiser Wilhelm Institute of Medical Research in Heidelberg.[7] The results of this research period were: The construction of an entire field of 'nuclear chemistry', in which innumerable reactions of all sorts of nuclei were analyzed with a great variety of projectiles.[8] Many hundreds of new radioactive kinds of atoms were produced. The neighboring sciences of chemistry and physiology also profited from the new findings. Using the American installations[9] it was even possible to produce such artificial radioactive materials in very large quantities and at a much lower cost than, for example, that of naturally produced radium and similar materials, upon which radiotherapy had previously been exclusively dependent.

But nothing had changed regarding the technical problem. Not in a single case had the transmutation of atomic nuclei in weighable quantities succeeded. Thus the medieval alchemist's dream of making gold had not yet been realized practically; it had neither been drawn at least within reach, nor had the release of mentionable amounts of energy succeeded. About a million projectiles had to be shot in order to transform a single atom. As a result of this tiny yield, one gram of artificially produced gold, for example, which would incidentally have to be produced through the transmutation of precious platinum, would cost many trillion marks. With the available installations only a few hundred atoms are transformed for each second of bombardment; thus the shooting of the platinum would have had to already have been started a billion years ago! Accordingly, the reservations of physicists seem truly comprehensible.

---

[6]On Ernest Orlando Lawrence (1901–1958), the inventor of the cyclotron which accelerates charged particles in powerful magnetic fields along a spiral path, see the Introduction, footnote 320, p. lxxxix.

[7]The research conducted at the *KWIC* is described, e.g., in Hahn [1943], and Bothe [1943] provides an overview of the research conducted at the facilities for nuclear physics in Heidelberg until 1943 as well as of the contemporary accelerator technology. See also the contributions by W. → Gentner and others in Bothe & Flügge (Eds.) [1948], part II, pp. 28ff.

[8]For an overview of the achievements within this field, see Hahn [1943], as well as the contributions by Otto → Hahn and others in the FIAT review, edited by Bothe & Flügge [1948], part I, pp. 171f.

[9]Flügge is referring to the cyclotrons developed in the USA (see the previous footnotes). Only a very few machines existed in Europe by 1945: See Heilbron [1986]a.

Fig. 16: 'The Max Planck Tower in Dahlem (Berlin). Plant photo'

Nevertheless, the situation has changed fundamentally since about half a year. The trigger for this was the discovery of a completely new sort of nuclear reaction by Professor Otto Hahn and Dr. Straßmann at the Kaiser Wilhelm Institute of Chemistry in Dahlem (Berlin).[10]

In 1934 the Italian physicist Fermi was the first to have bombarded uranium with neutrons.[11] Some new radioactive materials emerged in the process which he believed to be elements with nuclear charges 93 and 94 to be designated beyond uranium and therefore lying outside of the periodic system of the elements of the time.[12] These so-called 'transuranic elements' were analyzed further by Prof. Hahn together with Lise Meitner and Straßmann throughout several years of careful work.[13] In the process a series of physical inconsistencies arose; and some chemical questions remained that particularly Mrs. Curie in Paris also stressed

[10]For a detailed description of O. → Hahn's and Fritz → Straßmann's researches, which led to the discovery of nuclear fission see, e.g., Krafft [1981].

[11]Enrico Fermi (1901–1954) showed that by bombarding virtually all chemical elements with neutrons new radioactive isotopes would be formed; see, e.g., E. Fermi: 'Radioactivity induced by neutron bombardment', *Nature* **133** [1934], p. 757; 'Possible production of elements of atomic number higher than 92', ibid., p. 898; and E. Fermi, E. Amaldi, O. D'Agostino, F. Raseti, E. Segrè: 'Artificial radioactivity produced by neutron bombardment', *Proceedings of the Royal Society of London* **A 146** [1934], pp. 483-500, and **149** [1935], pp. 522–558; cf., e.g., E. Segrè: 'Fermi and neutron physics', *Reviews of Modern Physics* **27** [1955], pp. 257–263; Pierre de Latil: *Enrico Fermi. The Man and His Theories*, London, 1965, and Samuel K. Allison: 'Enrico Fermi', *Biographical Memoirs of the National Academy of Sciences* (Washington) **30** [1957], pp. 125–155 (with a complete bibliography of his writings).

[12]For this discovery of elements purportedly heavier than uranium, Fermi won the Nobel Prize in physics for 1938. It turned out soon afterwards that his interpretation was erroneous. Cf., e.g., Menke & Herrmann [1971].

[13]See, e.g., their retrospective accounts: Meitner [1963] and Hahn [1962]a. For the collaboration between Hahn and Lise → Meitner on transuranic elements, see Krafft [1981], secs. 2.2, 3.2; cf. also Noddack [1939] and Meitner [1954].

repeatedly.[14] Towards the end of last year, Hahn and Straßmann re-examined one of these questions.[15] In doing so they came to the completely surprising result that the chemical element barium emerged from uranium which had been irradiated with neutrons. The nucleus of uranium had split as a result of the impact with the neutron. Out of a single nucleus of charge 92 two nuclei had been formed: one barium of charge 56, and one krypton nucleus of charge 36!

Here for the first time was a process that really deserves the name 'atom-smashing'.[16] News of the discovery spread quickly throughout the whole world. In France, the Netherlands, Denmark and Sweden, in England and above all, in resource-rich America specialists threw themselves upon this new phenomenon; in a few months dozens of papers appeared about it, and the flood of publications is still continuing.[17] Out of the profusion of experimental results the following picture gradually came together: As the nucleus splits apart, a great many chemical elements form which fall somewhere in the middle of the periodic system between the nuclear charges 34 and 58. So uranium can split into barium and krypton or into strontium and xenon and probably also into a series of other pairs. The elements that originally had to be considered transuranic are surely in part, and probably throughout, mixtures of such medium-heavy elements. As uranium splits, the two 'fragments' fly apart with an energy that is around 50 times larger than what we are used to seeing in other nuclear reactions. The fact that during fission some more neutrons chip off is critical for our statement of the problem. This was first proved by Prof. Joliot and his colleagues in Paris.[18] They amount to about 2 or 3 neutrons for each splitting uranium nucleus; but this is not yet known exactly.

Fantastic Energies

At the turn of the year when the splitting of the uranium nucleus was verified on the basis of Prof. Hahn's chemical results, we immediately posed ourselves

[14]The French nuclear physicist Irène Joliot-Curie (1897–1956) received jointly with her husband [Jean] Frédéric Joliot (1900–1958) the Nobel Prize in chemistry for 1935 for their research on the nucleus. See, e.g., F. Joliot & I. Curie, 'Artificial production of a new kind of radio-element', *Nature* **133** [1934], pp. 201–202, esp. p. 202: "These experiments give the first chemical proof of artificial transmutation"; cf. Krafft [1981], sec. 3.2.3.2.

[15]See in particular their joint papers: Hahn & Straßmann [1939]a,b; cf. again Krafft [1981], secs. 1.3, 3.2.

[16]The German word *Atomzertrümmerung* was sometimes used in the semi-popular literature, cf., e.g., Flechtner [1942] or P.K. [1940]. Meitner and O. → Frisch coined the term 'fission' (*Spaltung*), which was predominantly used in technical papers in analogy to the biological process of cell division: See Meitner & Frisch [1939]. On the terminology see also Wohlfarth (Ed.) [1979], p. 100.

[17]For a chronology of the main events, see Rechenberg [1988]b. This newspaper article is itself an example of the wide attention these discoveries received.

[18]See, e.g., Hans von Halban, Frédéric Joliot & Lew Kowarski: 'Liberation of neutrons in the nuclear explosion of uranium', *Nature* **143** [1939], pp. 470–471.

the question: When some neutrons are released through the impact of a neutron during fission, what happens to these neutrons from then on? They obviously have the opportunity of splitting other uranium nuclei; in doing so, such a neutron will again produce new neutrons and so on, so long as fissionable uranium is still present. Thus a rapidly swelling avalanche of neutrons must disintegrate the entire available uranium. This is precisely what is called a chain reaction in chemistry. Thus that which had never succeeded before had been achieved: Through a single neutron which 'ignites', a weighable and even arbitrarily large amount of uranium is transformed, and nuclear energy is thereby released. How much energy can be produced in such a way can be specified quite accurately. Uranium occurs naturally in the compound uranium oxide; this is the purified form of the ore pitchblende, which is found in the St. Joachimsthal mines in the Sudeten district. One cubic meter of this oxide weighs 4.2 [metric] tons and contains 9,000 trillion trillions of uranium atoms. During the fission of one uranium atom, about a 3 trillionth of a meter-kilogram of energy is released; the transformation of the whole amount, therefore, releases 27,000 trillion meter-kilograms.[19] Since a cubic kilometer of water weighs a trillion kilograms, this energy can lift a cubic kilometer of water 27 kilometers high. In other words, it is powerful enough to hurl approximately the water content of the Wannsee upwards into the stratosphere![20]

Taming Atomic Energy

Two principle obstacles in creating such energy sources are the following: Firstly, large amounts of uranium are always needed so that the reaction chain does not immediately break off. This is because the neutrons produced travel a few centimeters within the uranium oxide before they transmute an atom again; therefore, if the entire amount of uranium used is only a few centimeters in diameter, then the neutrons will leave the uranium for the most part before they have another chance to split an atomic nucleus. We can calculate that the amount of uranium required would have to have a diameter of several meters; exact fig-

[19]The meter kilogram (force) is the metric analog of the foot-pound measure of energy or work, abbreviated as either m kgf or kgf m. Also a completely earthbound unit, it specifies the amount of energy or work that lifts 1 kilogram mass a distance of 1 meter, where gravitational acceleration is just g_o. See Herbert Arthur Klein: *The Science of Measurement. A Historical Survey*, New York: Dover, 1974, p. 227.

[20]A lake in the Western outskirts of Berlin, formed by the river Havel. Such an impressive description of the possible implications of releasing nuclear energy must have made an impact on the military as well. On April 29, 1939, German physicists formed the 'Uranium Club' (*Uranverein*) and researched possible applications of nuclear chain reactions. See, e.g., Rechenberg [1988]b, pp. 456f.; Walker [1989]a, pp. 30ff. This article shows that up to its publication in August 1939 there was evidently little secrecy on this issue. This was to change, however, immediately after the onset of World War II on Sep. 1, 1939, when the → *HWA* took over the reins. On the later research on nuclear energy in Germany see, e.g., Walker [1989]a or Irving [1967].

ures cannot yet be given at the present time.[21] The second obstacle is that the whole neutron avalanche develops within less than a hundredth of a second; thus, the entire huge amount of energy is released in such a very short time. We are therefore dealing with an explosion of an intensity approximately equivalent to a volcanic eruption. In order for the process to be made technically useful, it is essential to be able to slow it down at will.[22] It is of great credit to French physicists that a means to do this has been found.[23] The mechanism generating the avalanche must obviously be dampened somehow by adding substances that absorb the majority of the neutrons that are continually being produced. One element that does this just in the right way even when added only in small quantities is the metal cadmium.[24] When adding small amounts of cadmium, a steady burning would have to be possible at a calculable temperature, which is higher, the smaller the amount of added cadmium is. This element only captures slow neutrons though. Therefore the emerging neutrons must be slowed down quickly. One tested means for this is the addition of water. Then the emerging neutrons pass their energy on very soon as they collide with the light hydrogen atoms. Working with these slow neutrons simultaneously has the advantage that the neutrons also cover shorter distances in the substance, so that the required amount of uranium only has to have a diameter of around one meter to prevent the chain reaction from breaking off.

The 'Uranium Machine'

Such a 'uranium machine', which I have described in detail elsewhere,[25] would look sort of like this: 4.2 [metric] tons of uranium oxide are thoroughly mixed with 56 grams of cadmium; and the whole thing is made into a paste with 280 liters of water. Such an arrangement should burn slowly—subject to the still great uncertainty with which all these figures are laden at the present time—at a temperature of 350 degrees Celsius. It is easily imagined that at this high temperature water evaporates away constantly. It is replaced continuously by

[21] The precise amount depends very much on the magnitude of the neutron absorption coefficient of uranium, and this in turn depends on the neutron's energy range and the degree of purity of the uranium sample. For later measurements of these quantities which also proved to be decisive in the development of the atomic bomb see, e.g., G. → von Droste's contributions in Bothe & Flügge (Eds.) [1948], part I, pp. 197ff.

[22] Here Flügge was already thinking of the future application of nuclear energy in a power plant, an 'uranium machine', as was later developed in the German uranium project. No doubt many of his readers were equally attracted to its impressive explosive potential described above.

[23] Flügge [1939] refers specifically to a paper by F. Adler & Hans von Halban: 'Control of the chain reaction involved in fission of the uranium nucleus', *Nature* **143** [1939], p. 793; see also, ibid., pp. 470, 680.

[24] According to contemporary measurements, its absorption cross section was estimated to be 2800×10^{-24} cm^2 for energies of up to 0.4eV, whereas cadmium was practically ineffective for neutrons at higher energies; see Flügge [1939], sec. 5.

[25] See sections 4–5 of the expanded version of this article, Flügge [1939].

means of a spray, and the escaping steam is used to feed a steam engine. The total energy, which is released bit by bit in this set up, is about sufficient to cover the entire demand for electricity of the German Reich in one year.[26] More specific details naturally cannot be touched upon within the scope of this article. It also should not be forgotten that such a 'machine' has not been built till now; on the contrary, for the time being all of these considerations only exist on paper. It is thoroughly conceivable, even probable, that unexpected obstacles still stand in the way of practical implementation. For example, the fact that trial experiments with small amounts of uranium cannot be done poses a serious problem for work in the laboratory; many cubic meters of uranium must be processed at the start, whereby to the quite considerable cost is added the hazard of explosion. But even after these difficulties have been overcome, which will certainly be soon, the fact that the uranium has to be cleaned of the products of fission from time to time remains particularily disturbing for practical implementation, since these products also absorb neutrons themselves without producing new ones.

Yet even if upon closer analysis it should turn out that the practical exploitation of uranium fission is not possible (which I consider unlikely), enormous gains have still been made by the fact that for the first time a way has been spied which makes it basically possible, regardless of whether in the near or more remote future, to put the vast energy reserves of atomic nuclei in the service of humanity. Once again it is demonstrated that while going its own way quietly and steadfastly for decades, not in search of material advantage, but in the quest of pure knowledge, scientific research suddenly and unexpectedly bears fruit; practical implementation can be left confidently in the hands of the engineer and be of use to all of mankind.[27]

[26]Such comparisons obviously were written to attract government support for the research. The National Socialist government had been striving to achieve economic self-sufficiency at least since the adoption of the Four-Year Plan in 1937. The author is nevertheless being diplomatically cautious with his projections, quite in contrast to the speculative popular science article on the 'fuel of the future': P.K. [1940].

[27]This sentence defends the policy of scientific institutions like the → *KWG* for supporting pure research, regardless of whether or not promise of immediate applications existed. In contrast to this, many National Socialist science policy-makers preferred only to fund projects of immediate use to industry and society (*Volksgemeinschaft*).

Part IV
Physics at War: 1939–1945

75 Erich Schumann: Armed Forces and Research [1939]

Source: Erich Schumann,[1] 'Wehrmacht und Forschung', in: Richard Donnevert (Ed.):[2] *Wehrmacht und Partei*, 2nd expanded edition, Leipzig: Barth, 1939, pp. 133–151.

The same discipline schools
Both scholar and soldier. Nietzsche.

Wherever defense research [*Wehrforschung*] is mentioned, the term 'military sciences' [*Wehrwissenschaften*] automatically appears, which is in any case a vague concept to the competent scholar. At universities primitive military science talks, defense politics courses, lectures discussing the war of the future, special seminars on questions of defense and the economy, and evening-long events organized by clever self-advertisers, all sailed under the flag of 'military sciences'. The success of general staff officers, appointed to hold defense politics lectures for a short time at universities a few years before military sovereignty had been regained, could not compensate for the **astonishingly** negative influence of these so-called 'military scientists'.

From the diffuse definitions originating from this time in the military and war sciences, it is not so remote a step to extract such peculiar analyses as 'the mental state of soldiers before captivity' or the concept of 'magic in a soldier's life' (see the journal "*Soldatentum*", 1938, No. 5 and 3).[[3]]

From the outset the Armed Forces' University Central Office [*Hochschulzen-*

---

[1]The author's functions are described in the original document as "Ministerial Assistant Secretary, Full professor of physics [since 1933] at Berlin University, Head of the Army High Command Research Department, Head of the Armed Forces High Command Science Department [since 1934]". Especially following his appointment as ministerial assistant secretary at the Science Section of the Reich Ministry of Defense/War (→ *RWM*) and as a member of the Reich Research Council (*RFR*) in 1939, Erich → Schumann was one of the most influential physicists in Germany.

[2]On the titlepage of the work the editor's title appears as follows: "Ministerial Councillor in the Armed Forces High Command, Berlin. Reich Administrative Head in the Staff of the *Führer*'s Deputy." Richard Donnevert studied medicine in Göttingen, submitting his thesis 'Combination eines Odontoms mit einer Verwachsung' in 1902.

[3]See Gustav Naß: 'Das Gefühl des Magischen', pp. 102–105, Martin Lezius: 'Seltsame Vorgänge im Soldatenleben', pp. 115–122, and Leonhard Fritzsching: 'Der seelische Zustand von Soldaten vor der Gefangennahme', pp. 221–226, all in: *Soldatentum. Zeitschrift für Wehrpsychologie, Wehrerziehung, Menschenauslese* **5** [1938]; cf. ibid., p. 2, for a contemporary definition of 'military psychology' (*Wehrpsychologie*). During the National Socialist period the professionalization of psychology was justified precisely on its military usefulness, such as in selecting officers on the basis of character: See Geuter [1984/92], chaps. 3–4.

tralstelle der Wehrmacht] imposed the requirement of appointing only specially schooled staff (university lecturers or active officers) to lecture at universities on military science and defense policy. Furthermore, it allowed only habilitated university lecturers to substitute in subjects of research and instruction that **might** be considered military science. In these cases it granted special teaching assignments only in connection with the main discipline to which the relevant special field belongs—e.g.: in geography (with especial emphasis on military geography), history (... war history), philosophy (... military philosophy), economics (... war economics), physics (... ballistics and physics of explosives), chemistry (... military chemistry), hygiene (... military hygiene), surgery (... war surgery). [...[4]]

Though the efforts of the postwar years **to encourage students to serve in the military** arose mainly out of a healthy resistance to the Dictate of Versailles, it was especially the collaborating scientific and technical experts of the Armed Forces who attracted the next generation to the problems of defense technology, in spite of all obstacles. Today through the victory of National Socialism, a defense mentality has been securely anchored again in the entire nation; and it goes without saying that all German scientists take this fact into account in their proper teaching and research fields. This attitute was by no means a matter of course before. As a result, the sciences and technology suffered particularly severely under the constraints of the humiliating peace [*Schmachfrieden*].[[5]] **But thinking could not be banned!**

Thus branches of science which could be engaged in **expanding practically the potency of weapons** were promoted systematically, at first the theoretical bases of these branches, and then to an increasing degree also their experimental investigation.

The Armed Forces had both a soldier and a scientist in the current General of the Artillery and prof. of engineering Becker.[[6]] Knowing that **highly developed weapons technology**—especially when it leads to the unexpected deployment of new kinds of weapons—can be crucial to a victorious war in addition to excellent leadership, military discipline and good morale among the troops, he constantly strove for a **close relationship between the Armed Forces and science.** As early as in 1924 he recommended the establishment of a central laboratory of physics and chemistry, based on the fact that results in physics and chemistry have a considerable influence on the future development of weapons technology.

[4] The following paragraph on war history is omitted.

[5] The Treaty of Versailles of 1919 included the famous war-guilt clause, Article 231, which held Germany morally responsible for all loss and damage to the Allies as a result of the war. The German population's refusal to accept this, along with the tough reparations demands, formed the basis for Germany's alienation from the Allied countries in the following decades.

[6] The general of the artillery Karl → Becker was appointed full professor of applied physics in 1933, and also in the autumn of 1935 of ballistics and military technology at the Berlin Polytechnic's Defense Technology Department in Charlottenburg, which he expanded significantly in 1937; cf. Ludwig [1974], pp. 220f.

But only as director of the 'Ballistics and Ammunition Department' of the then Weapons and Equipment Inspection [*Inspektion für Waffen und Gerät*], now the Army Ordnance Office [*Heereswaffenamt*], was he put in the position to implement his idea. The 'Central Office of Army Physics and Army Chemistry' [*Zentralstelle für Heeresphysik und Heereschemie*] was formed in his competency in 1926, which was only recognized officially as an Armed Forces department in 1929, however. It was given the responsibility of **taking advantage of German university facilities for Army purposes and of producing accurate groundwork through its own research in Army technology.**

Although any link between the Armed Forces and universities was prohibited in the Dictate of Versailles, and even the official German agencies were mindful of the strictest adherence to this compulsory regulation, the Central Office succeeded in enlisting a substantial number of personnel willing to go into the service who seemed suited to its special research projects. It is indicative that a large portion of the acquired staff were members of the *NSDAP**[7] They were instilled with

* [Footnote 1), p. 136 of the original text:] "As editor of the book *Wehrmacht und Partei*, for which the present contribution was written, I think it important to point out the following: "Long before the seizure of power, the Party was well-acquainted with the Central Office, which then already belonged within the sphere of responsibility of the current general, Becker. For example, circuit heads (Dr. Mentzel), parliamentary representatives of the *NSDAP* (Dr. Hunke), *SA* leaders (Dr. Dinse) and *SS* leaders (Dr. Bauer, Dr. Winkel) were among the scientists working within or in close connection with it. Dr. Donnevert."

[7]Referring to footnote * above, Rudolf → Mentzel had been a member of the → *NSDAP* and the → *SS* since the early 1920's, then serving as the circuit head (*Kreisleiter*) in Göttingen. Heinrich Hunke (1902–) studied national economics, geography, physics and mathematics at the Universities of Münster, Berlin, Göttingen and Halle. 1923 he became engaged in the National Socialist movement; 1924–25 he was circuit leader in Lippe and Westfalia. Hunke took his degree in 1927 and habilitated in 1929. In 1929 he was *Organisations-Leiter II* of the greater Berlin district (*Gau*) and Economic Commissioner of the Reich Leadership for greater Berlin. In 1930 he was teaching assistant and passed his teaching certificate examination in the same year. Hunke acted as consultant for the *RWM* 1927–33 and was a member of the *Reichtag* from 1932. He founded and edited the journal *Die Deutsche Volkswirtschaft* and also acted as managing director and permanent deputy to the president of the economic advisory committee *Werberat der Deutschen Wirtschaft* in Dahlem (Berlin) and was district leader of the National Socialist business organization *Handels- und Gewerbe-Organisation* (*NS-Hago.*). In 1935 he was appointed honorary professor of national economics at the Berlin Polytechnic. See, e.g., H. Hunke: *Luftgefahr und Luftschutz mit besonderer Berücksichtigung des deutschen Luftschutzes*, 1932; *Grundzüge der deutschen Volks- und Wehrwirtschaft*, 1938. Oswald Adolph Friedrich Theodor Bauer (1876–1936) is possibly meant in footnote * above, who was full professor of ferrous metallurgy in Breslau and director of the State Materials Testing Office (*Staatliches Material-Prüfungsamt*) in Berlin in 1925; Bauer became deputy director of the Kaiser Wilhelm Institute of Metallurgy in 1923 and its head in 1931. It is also unclear whether the hydraulic engineer Richard (Adalbert Eduard Heinrich Ludwig) Winkel (1883–1951) is referred to here, who took his doctorate in 1924 at the Berlin Polytechnic and became full professor there two years later, then in the same year at the Polytechnic in Danzig, 1926–45. After the war he was lecturer of earth pressure theory, soil and hydromechanics at the Polytechnic in Braunschweig. On the editor, Richard Donnevert, see footnote 2 above.

the spirit of self-sacrifice, which was a precondition to the often quite hazardous job.[...[8]]

What immense difficulties General Becker, then the department head, and his assistants [*Referenten*] had to overcome in finding financing and in carrying out the scientific investigations are **not** mentioned in the files. But it is evident that since the foundation of the Central Office of Army Physics and Army Chemistry until 1933 not a single minister made his way into or even looked for its laboratories and research establishments. But a lot would have been concealed from them anyway, so that they would not have had to have a bad conscience in this regard at the regular inquiries of the parliamentary commission.[[9]]

Only through Adolf Hitler were these working methods changed decisively. Immediately in the summer of 1933 the *Führer* had the results of the detailed work that had been carried out at the testing sites and in the laboratories since 1926 presented to him; and he inspected every last corner of the existing installations. He decided on the spot—far beyond our hopes—to continue on the same course at an accelerated pace, in order to drive the development and testing of new weapons forward, and **to place future research on the broadest foundation.** For General Becker and his co-workers this was the finest acknowledgment of the judiciousness of their previous actions.

Soon after the *Führer's* visit, the Culture Ministers of the States, who were by this time National Socialists, and the remaining responsible leaders of the German cultural authorities were invited to presentations at the main test site and in the laboratories of the research establishments. Thus those authorities which the research branch of the Armed Forces had to reckon with were also to be convinced of the importance of the scientific work and and their cooperation and support secured. After gaining an insight into the nature of the work there, in a concluding talk they were convinced **in the interest of the country's defense to make available the scientific establishments under their authority while avoiding serious impediments to the general practice of science.**

With the sudden boom in defense research, a large number of professors rushed forward to collaborate who until then had fancied themselves 'above any form of applied research'. They did not want to miss the boat and hoped to be able to provide proof of their political reliability through a connection with the Armed Forces.

The Central Office, which had been elevated to the 'Research Department of the Army Ordnance Office', was nevertheless able to make a quite fair selection, because since around 1927 it had been well aware of all professors and other scientists who could possibly come into consideration as collaborators.

The Research Department established for itself—with the Reich's generous fi-

[8]The following paragraph on staff and working conditions is omitted.

[9]Officially, Germany was then still controlled by the Allied forces under the Treaty of Versailles, which forebade Germany from rearming itself. Any such steps had to be taken covertly: Cf., e.g., Ludwig [1974], p. 219. Large-scale rearmament started in 1935.

nancial support—**top-quality research institutions and laboratories equipped with the most modern scientific facilities.**[10] Its highest priority was to only recruit first-class researchers—regardless of whether they came from higher education, from other scientific institutions, or from industry—who had proven themselves in science, and who were also able to ensure a congenial collaboration from the political standpoint. The information obtained today from the leader of the University Lecturers Association via the Reich Education Ministry is decisive in this.

Once their special project is concluded, in general, the collaborating consultant scientists return again **to their original sphere of activity**—where working within a military institute was essential to maintaining secrecy. There in their civilian occupation they will continue to keep the problems of defense research in mind and, as the case may be, offer of their own accord new suggestions or ideas to the Armed Forces. The problems of military science extend in this way beyond the narrow scope of those most immediately involved, and the continually rejuvenated German university staff is influenced and utilized in a specific direction.

Any bureaucratic regimentation is tantamount to confining free initiative and to paralysing the German scientist's working morale. The science administration in National Socialist Germany must be constantly aware—and this also applies to a full extent to the research establishments of the Armed Forces—**that its task is confined to exerting its influence on organizing scientific activities** and to establishing research sites, to facilitate the execution of major research projects crucial to the State and to the nation. But full responsibility for how research is conducted cannot be taken away from the scientist. **It is precisely the most energetic and original scientists who will always resist being harnessed to a bureaucratic millwheel.**

But if absolutely no permanent scientists were employed at Army research institutions, then the constant flux in the research leadership would result in a loss of experience and knowledge of the experiments underway and of their apparatus. That is why **science officials** who are firmly based at the laboratories and the testing grounds are required. They remain **indispensable collaborators alongside of the consultant scientists.**

In the specifically **military** fields (e.g., ballistics, the study of explosives and chemical weaponry) **scientists who have made significant contributions in research** and who on the basis of their successes are sure to continue to prove themselves as leaders in this area **are employed permanently** in particularly responsible positions. Some of them are simultaneously employed in addition to their regular duties as professors at the Polytechnic or at Berlin University.[11]

---

[10]One such research site is the Army Ordnance Office's (→ *HWA*) *Heeresversuchsanstalt* in Gottow established in 1940, where Kurt → Diebner conducted research on the 'uranium machine'. Cf., e.g., Walker [1989]a, pp. 94–104.

[11]General Becker is an exceptional example, being employed simultaneously by both the

Obviously, **industry** also attempts to recruit scientists (including teaching assistants and research assistants) suited for the projects at the Armed Forces laboratories. That is why there is a noticeable exodus of the best employees, since industry pays salaries for exceptional performance far exceeding the rates paid by the State. **Fundamental change must be made here, if the State wants to get real peak performance in responsible and important posts.**†

So that the Armed Forces research offices could be informed about all facilities, results and current experimental work in German science, it was necessary to create an **organization** capable of **providing appropriate surveys at any given time**. Out of the 26 existing German science organizations, which are hardly in contact with one another, **none** was capable of fulfilling this task even only approximately—even the German Scientific Research Association [*Deutsche Forschungsgemeinschaft*] was unable to do so; when making decisions it was completely uncertain about whether similar scientific projects were already being carried out at another location.

For this reason in 1933 at the suggestion of the Armed Forces a proposal was made to create a 'Reich Research Academy' [*Reichsakademie der Forschung*] that was to be responsible for applying appropriately the available research funds in the interest of German science as a whole;§ since, it is obvious that **an establishment that can draw on the combined experience of all the governmental and numerous private research departments within the institutions of the Reich and the individual States, as well as of university departments and Kaiser Wilhelm Institutes, is best suited** to advise the State and the scientific community.[12]

It was clear that freedom of research must be maintained on principle and even that **scientific freedom, which had been gagged** in the Weimar period [*Zwischenreich*], should be **restored completely**.[13]

Berlin Polytechnic and the University as well as by the → *HWA*.

† [Footnote 1), p. 139 of the original text:] "Not only to the outside observer does it remain incomprehensible why, for example, in government institutes it should be impossible to employ only **one** first-class staff member only asking for 400.– RM instead of two average engineers with together an 'as per wage scale' monthly salary of 550.– RM."

§ [Footnote 1), p. 140 of the original text:] "In 1934 in a contentious essay, 'Nationalsozialismus und Wissenschaft', which counters the untruths about the National Socialist government's attitude toward science, Joh[annes]. Stark refers to the 'Research Council' created by Mussolini. Stark then proposed in a **Memorandum** of the 13th of August, 1934, the creation of a 'Reich Academy of Scientific and Technological Research' and underscored the need for this organization 'in the interest of scientific and technological research, in the interest of national defense, in the interest of the entire German economy, and in the interest of German prestige in the world'."

[12] J. → Stark's attempts to centralize control of research is also evident in his speech as the new head of the → *PTR* on Sep. 18, 1933, doc. 28.

[13] On the science policy during the Weimar period see Forman [1973], [1974], and Richter [1973], [1977]. On later efforts to reorganize research activities see, e.g., Ludwig [1974], chaps. 6–7; Macrakis [1993], chaps. 4–5; and here docs. 52 and 98.

Science cannot be ordered around. But 'freedom' is not synonymous with licentiousness. Politics has primacy over science, and this primacy must be upheld fundamentally in every nationalistic state without restricting the autonomy of scientific research. Today the German people can no longer be indifferent to whether the considerable funds being allocated for scientific purposes be used, for example, for mollusk research in the South Seas—the special value of which is not at all being denied—or for the development of new materials of vital importance. For fourteen years any vagrant—and often, non-German—university professor had the opportunity to squander State funds on completely alien, pseudoscientific projects. The State **must** lay emphasis on things that **concern the entire nation.**

There simply is a **hierarchy** of projects **which in science, as in all other aspects of life of the nation, is set according to its urgency to the whole.** It may well seem painful to the individual, but looked at on a large scale, it is a national necessity.

The importance of an organization that provides a **survey of the current state, course and goal of all of German science**; that **may intervene on behalf of** a scientific project, depending on its importance to the nation; that **acts as a balance** with regard to the most economical use of the available funds; and that **advises** all the establishments involved in applying the research funds; the importance of such an organization is obvious.

The following budgeted material costs were spent on research in 1934, for example (excluding private research establishments and those owned by the Armed Forces):

Technology	5 773 511	RM
Science	8 007 740	"
Agriculture and forestry	11 642 295	"
Medicine	36 033 812	"
Humanities	6 678 886	"
Scientific academies and general scientific research facilities	5 078 372	"
Total	73 214 616	RM

Tab. 2: Overview of research expenditures in 1934 (in reichsmarks)

One should consider here that the budgeted material costs make up only a portion of the total expenditures for research. In 1934, for example, at German universities they amounted to only 22.5 % of the total.

Though the objections made against any form of centralization tended initially to repel the plan for a Reich Academy, they could not kill the idea as such, however. Aside from its supporters in the Armed Forces, men of science also advocated it unremittingly, such as Professors Bachér, Mentzel and Thießen.[14] In the first two years in which National Socialism was being set up, the **need for a comprehensive research organization** proved already to be so **urgent** that individual special ministries took steps to find solutions in their own areas. Thus suitable establishments were created: For example, the Reich Transport Ministry set up its 'Transport Science and Technological Motor Vehicle Research Council'; the Reich Forestry Bureau, its 'Reich Association for Forestry and Timber Research'; the Reich Postal Ministry, its 'Research Institute of the German Reich Post'; and the Reich Agricultural Association [*Reichsnährstand*] formed its 'Research Service'.[15] Finally, Field Marshal Göring united within the 'German Academy of Aviation Research' which he had founded, "noteworthy scholars and engineers devoted to aviation research, employed in the fields of engineering and science, in order to extend the scientific foundations of aviation technology through teamwork, and to promote the practical application of the knowledge gained".[16]

All these establishments pointed clearly to the urgent need for a **unified orientation of research throughout the Reich**. On the 16th of March, 1937, this was attained, after innumerable obstacles had been overcome, with the **founding of the 'Reich Research Council'** in the presence of the *Führer*. General of the Artillery Prof. Becker was summoned as its first President by Reich Minister Rust.[17] The fact that an active general had been appointed to head this

---

[14]The organic chemist Franz Herman Theodor Bachér (1894–) took his degree in 1921 in Rostock. 1934–45 he was full professor of organic chemistry at the Berlin Polytechnic and in 1939 also worked for the → *REM* Science Office (*W*-I) as deputy chief in Berlin; in 1954 he was appointed full professor at the Berlin Technical University. The *SS* officer and chemist Rudolf → Mentzel headed the '*W*-II' Science Office for military research at the same ministry; and the physical chemist Peter → Thiessen was Mentzel's top scientific adviser and head of the Chemistry and Organic Materials Division of the Reich Research Council (→ *RFR*) after its foundation in 1937. On the roles of Bachér and Mentzel in the affair over the theoretical physics chair appointment at Munich University to replace Sommerfeld in these years see, e.g., Cassidy [1992]a, pp. 347f.

[15]This multiplication of research institutions illustrates the polycratic fragmentation of competencies within scientific research during the Nazi regime.

[16]Hermann → Göring founded the *Deutsche Akademie der Luftfahrtforschung* only weeks before the *RFR* was installed, with the clear intention of creating a rival institution in the broad area of scientific research to support the needs of the German Air Force. See: *Jahrbuch 1937 der deutschen Luftfahrtforschung; Ausgabe Flugwerk*, Munich & Berlin: Oldenbourg, 1937, in particular, the front matter, p. vii, on the Research Department at the Reich Aviation Ministry (→ *RLM*), the above-mentioned aviation academy, and for the above quote; cf., e.g., Zierold [1968], pp. 222f., Ludwig [1974], pp. 219f., Trischler [1992], pp. 236ff., and (Ed.) [1992], pp. 127ff., 161ff.

[17]See the official announcement of the creation of the *RFR*, signed by the Minister, here doc. 52; on the complex motives behind its foundation, cf. Zierold [1968], pp. 214ff., Nipperdey & Schmugge [1970], pp. 60ff., Ludwig [1974], pp. 217ff.

organization **is a marked expression of the great political struggle of our times, where "statesmen and soldiers stand in the forefront of the action"** (Rosenberg).[18]

The goals of the Reich Research Council were defined to the effect that "through systematically assigning projects and distributing funds", the Council would point out the way and make long-range research possible—not least in conformance with the great aims of the Four-Year Plan.[19] **Thus for the first time German science appeared publicly as an integrated whole.** It proclaimed its firm desire to bring the diverging researchers together in a joint effort in the service of the German **nation** and to call them over to where the battle still needed to be won. Far from, let's say, introducing new bases of scientific practice or even eliminating the already existing and tried organizations of the individual branches of science, the Reich Research Council saw as its **main mission, guiding the course of research according to the national** [*völkischen*] **and political needs acknowledged by the new government leadership.**[20] Its administrative apparatus is thus relatively limited.[21] The **principle of self-government** is its guiding principle, which has long since proven itself in Germany in scientific life, especially in the science departments and the academies of the sciences; and this principle has been imitated in many other countries with more or less success. Tested men from individual fields are the responsible leaders of **specialty divisions** and direct the course of their narrower field of study through the distribution of funds, the assignment of projects, and through keeping track of critical research results as they appear. Academic instructors and independent scholars are engaged and government-agency and industrial laboratories are mobilized according to their efficiency, so as not to miss any possibility of searching out suited staff and employing them for the benefit of the German people. **Importance is attached to close cooperation with the responsible leaders of large business enterprises in the technical exploitation of results.** The Reich Research Council's method revives the idea of professional committees, long since proven viable in the German Scientific Research Association, which had emerged from the former Emergency Association of German Science and which is, **incidentally, closely linked organizationally to the Reich Research Council.**[22]

---

[18]This is refers to the Nazi ideologue Alfred → Rosenberg.

[19]See doc. 52; cf. Zierold [1968], pp. 215ff., Nipperdey & Schmugge [1970], pp. 62f.

[20]The *RFR* never fulfilled the task it had been designed for because of the continued existence of rival organizations supported either by the ministries, the military, or private industry. Cf., e.g., Simon [1948], chap. VI, on the ineffectiveness of the *RFR*; Ludwig [1974], p. 222: 'despite declarations to the contrary, during the 'Third Reich' neither cooperation nor any real coordination of all the research agencies was achieved'; cf. also Zierold [1968], p. 217, Nipperdey & Schmugge [1970], pp. 66f.

[21]The *RFR*'s limited number of bureaucrats resulted in its being treated administratively as a part of the *DFG*: See Ludwig [1974], p. 218.

[22]On the link between the Emergency Association of German Science (→ *NG*, founded in

Time and again, the Reich Research Council points out issues and results of particular importance **also to the Armed Forces** from out of the abundance of material in its extensive sphere of activity. Research projects in progress or results of scientific research that look promising for Armed Forces applications can thus **be brought most quickly to the attention of the Armed Forces departments** and can be implemented most rapidly and practically. The days are past when, like the English Major General Fuller,[23] one could say—if only in jest—that new inventions in the area of bacteriological warfare or of lethal radiation do not need to be banned at Geneva, since the war ministries would already take care that such innovations stay unused, gathering dust in their drawers.

The **close ties between the Reich Research Council and the Armed Forces**, which is particularly emphasized by the appointment of an active general as its president, do not mean, however—as it might appear—that the Reich Research Council and therefore German science had gotten caught in a possibly unnecessary, dependent relationship with the Armed Forces. In anticipation of such misunderstandings, upon taking his office the president immediately **ordered that absolutely none of the Reich Research Council's or the German Scientific Research Association's funds be used for Armed Forces research purposes. The total funds should be fully available to 'civilian' science**, where the problems of race, prehistory, medicine and general technology had to be brought to the foreground.[24] This order also finally silenced those who previously could not be convinced that the men responsible for the management of science in the Armed Forces **really were only concerned** with creating **a single** organization outside of the Armed Forces **that could provide at any time exhaustive information on all investigations performed at all Ger-**

1920), which was named German Scientific Research Association (*DFG*) under Johannes → Stark's presidency, and the *Reichsforschungsrat* see, e.g., Zierold [1968], sec. II, pp. 173ff; Nipperdey & Schmugge [1970], pp. 53–60. According to Ludwig [1974], p. 218, Rust used the outbreak of World War II to incorporate the *DFG* and thus also the Council within the competency of his ministry.

[23]John Frederick Charles Fuller (1878–1966) was a British soldier, military analyst and war historian. After an education at Malvern College and Sandhurst 1897–98, he fought in the South African war and served in India. He fought in France 1915–16 and was posted as chief general staff officer of the tank corps in 1916; his successful surprise attack at Cambraie in 1917 demonstrated the potentials of armored warfare. 1913 Fuller enrolled at Staff College, 1923 he was senior instructor there, was promoted to major general in 1930 and retired in 1933. Fuller wrote articles advocating a new model army based on tanks and other modern equipment. His sometimes provocative books on military strategy and the potential effectiveness of the armored offensive to subdue an unprepared opponent were rejected by contemporary British military circles; however, they played an influential role in forming German and Soviet military tactics. His ten books include: *The Foundations of the Science of War*, published 1925, *Memoirs of an Unconventional Soldier*, 1936, and *The Decisive Battles of the Western World, and their Influence upon History*, in 3 volumes, 1954–56.

[24]For an overview of the kinds of projects funded by the *DFG* during the National Socialist period see, e.g., Zierold [1968], pp. 184ff., 211ff.

man scientific locations and that was thereby also in the position to occasion the necessary mutual **exchange of experience** for the best possible economical use of the available funds.

[...[25]] **The national** [*völkisch*] **unity created by the National Socialist movement and the penetration of the spirit of military preparedness in the German nation provided a firm backing for generously enlarging defense technology.** Professorial chairs, assistant professorships, laboratory personnel, apparatus, funds, etc., immediately became adequately available, so that very quickly research in defense technology could be set on a broader basis.

Sponsorships of officers assigned to study at the Polytechnic and for whom General Becker had been responsible even before 1933 could now be continued with far better funding. Only now was it possible for the Military Technical Academy to achieve the goal set in 1903 of training officers "not only to understand **how to exploit technological achievements for military purposes**, but also to be capable of **presenting technical problems** and **participating** in solving them **with a proper grasp of the matter**". These 'academic officers' [*Studienoffiziere*] thus became **bearers of defense technological thinking at headquarters and among the troops** and are in the position to augment advances in military technology with new technical ideas from the field.

In 1933 Reich Minister Rust transferred the permanent office of dean of the newly founded 'Department of General Technology' [*Fakultät für allgemeine Technologie*] to General Becker, then full professor of defense technology at the Polytechnic in Charlottenburg (Berlin). Ministerial Department Head Prof. Gerullis was instrumental in this.[[26]]

After military sovereignty had been reestablished in 1935 the academic department was able to discard its name chosen for reasons of disguise and to call itself 'Defense Technology Department' in accordance with its actual purpose. Two and a half years later **Adolf Hitler laid the foundation stone** for the new buildings of the department and in so doing simultaneously took the first step in reshaping the capital city of the Reich.[[27]] The new buildings presently under

[25]The omitted five paragraphs and subsequent lead-in sentence discuss English science policy, the German military's supposed neglect of technology before the World War, and the fate of German military science institutes after the Versailles Treaty.

[26]The Baltic philologist Georg Gerullis (1888–?) was unsalaried university lecturer in Königsberg in 1919 and associate professor in Leipzig in 1922. In 1933 he declined an appointment at the Ministry of Culture in Saxony and became undersecretary at the Prussian Ministry of Culture. 1933 he was appointed full professor at the University of Königsberg.

[27]The planned expansion of the *Wehrtechnische Fakultät* through the establishment of the institutes described in the next seven paragraphs were looked upon as exemplary for the new 'Universal University' (*Universale Hochschule*) of the Third Reich, which merged technical colleges and academic universities, incorporating military technology within the curriculum. The cornerstone ceremony in late 1937 was an official event that also signalled the beginning of Albert → Speer's bombastic building plans to transform the city of Berlin into the Reich's new capital city: See Ludwig [1974], pp. 220f.

construction are in keeping with the scale and design of the great goals of this educational and research center, which, together with numerous institutes, bears witness to the great importance of defense technology within German science.

The Institute of Weapons Construction is a research center for the field of weapons technology as a whole. The most basic research areas involve weapons mechanics, the interrelations of forces and the material stress in a weapon, kinematics and oscillation theory. Also covered are dynamic strength of materials, specialized metrics, materials technology, and forms of ammunition and firing aids.

In the Institute of Explosives Chemistry researches are being carried out in the area of explosives analysis (now with particular attention to German raw materials and to the best possible exploitation of all raw materials currently in use). The institute is also studying perfecting production processes with a view to mass production, as well as producing ignition devices. Finally, investigations are being conducted in the area of interior ballistics (gunpowder) and the physiochemical transformation of explosives.

The Institute of Technical Physics and Ballistics studies the application of physical problems to defense technology through theory and experiment, particularly the creation of ballistic measurement methods, taking into account the newest findings. It also studies the application of these methods to other areas of technology and physics.

Research areas of the Institute of Military Means of Reconnaissance include: Optical (including electro-opt[ical].) reconnaissance and measurement devices (e.g., comparison of various observation and range-finding methods); issues of acoustic position finding (the sound of artillery, aircraft and tanks), signaling, warning and camouflage. The scientific bases for the construction of mechanical and electromechanical instruments (transmission devices, anti-aircraft command sets, artillery computing machines, etc.) are being worked on in conjunction with the Professorial Chair of Instrument Technology.

The subjects of research of the Institute of Military Constructional Engineering lie in the area of civil engineering, insofar as they are suited to military application: Foundation and hydraulic engineering, reinforced concrete construction, bridge, rail and fortification building, field fortification, road building (especially expressways, bridging of destroyed roads) and building materials testing; in addition, also geology as a complementary science to military construction engineering.

The Institute of Military Motor Vehicle Engineering conducts basic research on the construction of caterpillar-tread vehicles (e.g., the analysis of the static and dynamic conditions of a vehicle) and motor vehicle engines (motor engineering and internal combustion technology), furthermore, marginal issues touching military motor vehicle engineering.

The responsibilities of the Institute of Military Communications Technology include the treatment of the scientific and technical questions concerning communications devices; moreover, perfecting the construction of the communications device, particularly the transferal of common telecommunications engineering methods to uses in defense technology (examples: communication device with light as the form of transmission, angle measurements, remote controls, encoding techniques); in addition, telecommunications engineering aids for other fields of instrumentation.

The magnificent facilities of the Defense Technology Department are **not reserved exclusively** for the benefit of **the academic officers**. For example, they also cater to the **Armed Forces officials serving in advanced engineering**, who can be employed among other things as unit engineers. Moreover, the department is **also open to all civilian students** through its close connection to the rest of the Polytechnic's departments. It gives those particularly interested in problems of defense technology the opportunity of thorough professional training and thus provides the German armaments industry with the next generation of scientists.

The academic officers, who are also occupied as department heads or consultants in the High Command of the Army or Armed Forces, constantly change and return to the ranks, just like the general staff officers. Likewise, constant change takes place among the scientists brought into the Armed Forces to complete special projects. Therefore, scientists in addition to the official scientists at the Armed Forces testing fields (laboratories) must be available **for research in the long term and for continuity in defense technology in general**, who simultaneously serve permanently as advisers. This responsibility is assumed by the professors of the Defense Technology Department.

It was left to the new Germany to acknowledge the full importance of Armed Forces research. The fact that General Becker was appointed as member of the time-honored Prussian Academy of Sciences, an honor bestowed for the first time on an active general, can only be looked upon as an expression of this high regard for military research.

With the elevation of the Central Office for Army Physics and Army Chemistry to the Research Department in the Army High Command and with the installation of numerous subordinate scientific institutes and testing grounds of the Army; with the founding of the Research Department in the Reich Air Force Ministry, the expansion of the 'Research Institute for Aviation' [*Versuchsanstalt der Luftfahrt*] and the creation of various other scientific aviation facilities;[28] with the establishment and enlargement of many research locations of the Navy; and finally, with the Science Division [*Abteilung Wissenschaft*] of the Armed Forces High Command, decreed by the Chief of the Armed Forces High Command on the 1st of October 1938, an endpoint of a sort has been reached in the

[28]For more on these organizations under Reich Marshal → Göring's supervision in his capacity as commander in chief of the *Luftwaffe* see, e.g., Simon [1948], Trischler (Ed.) [1992], pp. 132ff., as well as the following document; see also doc. 80.

organization of research within the Armed Forces.

The Defense Technology Department and (for the Reich Air Force Ministry) the German Academy of Aviation Research are necessary additions. The Reich Research Council—appointed to provide German science as a whole with a goal and a direction—is thus also an indispensable adviser to German Armed Forces research.

76 Max Planck: Max von Laue. For the 9th of October 1939. [October 6, 1939]

Source: Max Planck, 'Max von Laue. Zum 9. Oktober 1939', *Die Naturwissenschaften*, Vol. 27, Issue No. 40, October 6, 1939, pp. 665–666.

Ten years have elapsed since this journal has noted the important day on which some 60 years ago Max von Laue came into this world and followed his calling in science.[1] We could take a number of different courses to illustrate briefly what he has contributed to science. But since we have occasion today to turn our attention once more to his personality and to his lifework, it is all the more important to express a sentiment we share with all physicists from every nation: a feeling of joy and gratitude that even now at the full prime of life this celebrated person continues his work untiringly and is at the same time a blessing to his family, to his fatherland, and to humanity.

The exceptional fruitfulness of his labors is demonstrated by the sheer number of his publications.[2] No year has gone by in the last decennium without von Laue's having risen to speak at various places and for various reasons. His primary interest, quite naturally, was the continued development, both physically and mathematically, of his theory of X-ray interference in crystals,[3] specifically through the introduction of electron beams and their phenomena of diffraction on the one hand, and through the refinement and simplification of the calculation methods governed by wave mechanics on the other.[4] But aside from this, he remained open to all the other current issues in physics, even those relatively

---

[1]Max → Planck alludes to the article celebrating Max → von Laue's 50th birthday in *Die Naturwissenschaften* **17** [1929], pp. 787–788.

[2]For a full bibliography of von Laue's writings, see the obituary Ewald [1960], pp. 149–156. Cf. von Laue [1961], for only a selection of his papers filling 3 volumes.

[3]In 1912, von Laue (at that time unsalaried lecturer of theoretical physics at Munich University) discovered together with the experimentalists Walther Friedrich (1883–1968) and Paul Knipping interference patterns (also called 'von Laue diagrams') caused by the diffraction of X rays by crystals, which was instrumental for further investigations of crystallographic structures. Von Laue was awarded the Nobel Prize for 1914 for this work. (On Knipping, see footnote 27 of doc. 77.) See, e.g., Peter Paul Ewald (Ed.): *Fifty Years of X-Ray Diffraction*, Utrecht: Oosthoek, 1962; cf. Paul Forman: 'The discovery of the diffraction of x-rays by crystals', and Ewald's remarks, both in *Archive for History of Exact Sciences* **6** [1969], pp. 38–71, 72–81; and Hildebrandt [1993].

[4]For a summarizing textbook on X-ray interferences, see M. von Laue: *Röntgenstrahl-*

faraway from his original field of work. A whole series of individual, uniquely conceived, and carefully executed studies show this, such as those on the secrets of superconductivity, on the propagation of light in temporally variable curved space, on the formation of chemical elements, and on cosmic rays.[5]

On the whole, as commonly tends to occur with aging, a growing interest in more general, fundamental observations and in historical context is becoming apparent also in the case of von Laue. His comments on the problems that gave Galileo, the patriarch of mechanics, such trouble, or on the chain of reasoning that guided Newton in constructing his optics, read magnificently.[6] The difficulties to which the conception of matter-filled space leads, as an outgrowth of classical physics, have occupied von Laue repeatedly.[7] So it was inevitable for him to give his opinion also on the deep-seated question of causality, which at the present time seems to split physicists into two opposing camps, depending upon whether they view the duality between corpuscular and wave mechanics in its current formulation as a finally closed principle that disregards any causal significance of individual processes in atomic events, or on the contrary, whether they feel the current state of the theory is a provisional arrangement that needs to be reduced to a higher unity. Von Laue never left any doubt that he ranked himself among the second group, yet always without failing to emphasize clearly at the same time that this position should not be perceived as an argument against the importance of research into statistical conformities with natural law, which are the only ones within our grasp with current methods and which also in many cases are only of interest to the experimenter.[8] But surely the epistemological postulate of causality does not have to be given up because of this. Its proof some day, once the time will have become ripe for it, promises to round off the last corners of quantum theory.[9]

Interferenzen, Leipzig, 1941. On electron-beam diffraction, see also his *Materiewellen und ihre Interferenzen*, Leipzig: Akademische Verlagsgesellschaft, 1944.

[5]Cf. von Laue [1961], particularly Vol. 2, pp. 115–123, for his work on light propagation according to the general theory of relativity; Vol. 2, pp. 156ff., 259ff., 313ff., 383ff., for his work on superconductivity, performed in collaboration with the experimental physicist Walther → Meissner, head of the → *PTR*'s low-temperature physics laboratory. For the last two topics, see von Laue: 'Entstehung der Elemente und kosmische Strahlung', *Naturwissenschaften* **19** [1931], pp. 530–531, 641.

[6]Cf., e.g., von Laue's *Geschichte der Physik*, 1st ed., Frankfurt/Main: Ullstein, 1947, 4th ed., 1958, especially chapters 2–4, and several articles on the history of physics, reprinted in von Laue [1961], Vol. 3, part II.

[7]See, e.g., von Laue: 'Materie und Raumerfüllung', *Forschungen und Fortschritte* **9** [1933], pp. 283–284, and his article 'Materie und Raum in der neueren Physik', *Vossische Zeitung*, No. 21, Jan. 25, 1934, both reprinted in von Laue [1961], Vol. 3, pp. 63–77.

[8]See von Laue's article on Heisenberg's uncertainty relation, doc. 31.

[9]This seems to be Planck's own view in the debate about causality and the new quantum mechanics. From among his pupils, von Laue was the one with whom Planck was on the closest terms; and they often agreed on scientific matters. However, von Laue was a much more outspoken critic of the Nazis.

Von Laue's activity as a teacher and adviser to the younger generation is inseparable from his scientific lifework. Through his effectiveness at the university and at the Kaiser Wilhelm Institute of Physics, he has been for a number of years the foremost and most often called representative of the gloriously expanded field of theoretical physics in Berlin, which had been founded by Helmholtz and Kirchhoff.[10] He directs research with a watchful eye and indefatigable patience in seminars and colloquia, where a community of students, teaching assistants, and university lecturers gather around him, brought together solely by a common interest in science, for weekly open discussion of more recent publications.[11] Without him a thorough cultivation of theoretical physics as a specialty at Berlin University would be unthinkable for the time being. What has now yet again come to the foreground of public interest has been his special concern all along: the guidance of the rising generation of young scientists, whom he supports persistently and warmly in word and deed.

He stayed at the helm through hard times, inspired by a love of science and by a deep-rooted feeling of responsibility; and in so doing he won the gratitude and the hearts of all those whom he could somehow help along the way, either officially or personally. His fair-mindedness, his courageous readiness to speak for what is right, his sterling loyalty to friends, and not least, his bubbling humor, which has surely helped him through some dark moments of annoyance and disappointment from which no one is spared, these qualities have served him as an inexhaustible source of energy. May he preserve the same vigor and creativity well beyond sixty and for many more years to come, for the sake of the scientific world, his fatherland, and his friends!

[10]Hermann von Helmholtz (1821–1894) and Gustav Robert Kirchhoff (1824–1887) are considered the fathers of German theoretical physics. For an account of their importance to the institutionalization of theoretical physics in Heidelberg and Berlin see, e.g., Christa Jungnickel & Russell McCormmach: *Intellectual Mastery of Nature*, Chicago: Univ. of Chicago Press, 1986, especially chaps. 12 and 14.

[11]After Heinrich Rubens's death in 1922, von Laue was the main organizer of the regular Wednesday physics colloquium at Berlin University, which provided ample opportunity to all interested physicists in the area for lively discussions on new developments in physics; cf., e.g., Hubert Laitko et al. (Eds.) [1987], pp. 481ff., esp. p. 484.

77 Ludwig Glaser: Jews in Physics: Jewish Physics [November 1939]

Source: Ludwig Glaser, 'Juden in der Physik: Jüdische Physik',*[1] *Zeitschrift für die gesamte Naturwissenschaften*, Vol. **5**, Issue No. 8, November 1939, pp. 272–275. Square brackets in the original have here been replaced with parentheses.

I. Jews among Helmholtz's Students

In almost no other science did Nordic man in Germany take a more active part than in physics. But also none has the same sorry distinction of producing a Jew like Einstein.† This German citizen, expatriated in 1933, was revealed to be not simply the docile scholar his numerous followers passed him off as, but a Jewish politician, who in gratitude to the host country that had taken him in, denounced, scorned and abused it throughout the world, in true Hebraic style.[2]

One would think that the Third Reich's distancing itself from a parasite[3] was enough cause to lead to some soul-searching. However, Jewish propaganda had already confused our minds too thoroughly. This incident arose on ground that had been prepared long ago. Who is to blame for this? Well, it was a person of no lesser importance than Helmholtz.[4] From the patriotic [*völkisch*] point of view, the life of this unquestionably great scholar is marred by certain facts which have pernicious consequences that only now are apparent: The refusal to

---

∗ [Footnote 1) in the original text:] "P. Lenard, Zur Festlegung des Begriffs 'Jüdische Physik', *Deutsche Physik*, Vol. 1—Foreword, also, Talks and Lectures at the Opening Ceremony of the Philipp Lenard Institute at Heidelberg University."

[1]In footnote ∗ above the engineer L. → Glaser refers to Philipp → Lenard's essay on 'laying down the concept 'Jewish Physics' in his textbook entitled 'German Physics' of 1936, the foreword of which is translated here as doc. 39; see also Stark's inaugural eulogy, doc. 40.

† [Footnote 2) in the original text:] "According to *The International Who's Who* 1938, p. 307, despite being born in Ulm [Germany] on the 14th of March 1879, now a 'Swiss physicist'. An emigrant in Princeton N[ew]. [Jersey]., USA; permanent member of the Institute for Advanced Study, Princeton University from 1933; and as member of the Prussian Academy of Sciences (since the 12th of November 1913) wrote *About Zionism* 1930, *Why War* (with Prof. Sigmund Freud) 1933, *My Philosophy* 1934, *The World as I See It* 1935; excluded from the Prussian Academy since the 28th of March 1938."

[2]For Albert → Einstein's withdrawal from the Prussian Academy of Science, see doc. 6; for the Nazi criticism of Einstein as the incarnation of 'Jewish physics', see also docs. 55, 42.

[3]The use of terms like 'parasite' (*Schädling*), or elsewhere 'inferior being' (*Untermenschen*), was a typical Nazi tactic to dehumanize their opponents.

[4]Glaser's incrimination of one of the 'fathers' Berlin theoretical physics, the physiologist and theoretical physicist Hermann [Ludwig Ferdinand von] Helmholtz (1821–1894), is a striking demonstration of the degree of absurdity such accusations reached. Helmholtz taught at Königsberg from 1849, Bonn from 1855, Heidelberg from 1858, and from 1871 in Berlin, where the biggest physics institute in the new *Kaiserreich* was established for him. He became world renowned particularly for his formulation of the law of conservation of energy.

acknowledge the merits of Robert Mayer;§[5] Helmholtz's weak stance as the representative of German science at international congresses, resulting in Gauß's and Weber's‡ names being eliminated in the naming of the conventional international electrical units;[6] and the complete lack of appreciation for one of his most important German contemporaries, the physicist and mathematician Hermann Graßmann.¶[7] That the blame must be laid on Helmholtz for having granted the Jews admission to physics is proven by the case of the physicist and anti-Jewry pioneer Eugen Carl Dühring.[8] His removal from the academic faculty

§ [Footnote 3) in the original text:] "Cf. in this matter Eugen Carl Dühring: *Julius Robert Mayer, der Galilei des 19. Jahrhunderts.* [*Eine Einführung in seine Leistungen und Schicksale*], Chemnitz, 1879."

[5]Julius Robert Mayer (1814–1878) independently discovered the first law of thermodynamics on the conservation of energy in 1842 which appeared in book form in 1867: *Die Mechanik der Wärme*; however, there was much controversy over the priority of his ideas. See footnote 8 below on the author of the work cited in footnote § on Mayer as 'the Galileo of the 19th Century', published by Schmeitzner in 1880, .

‡ [Footnote 4) in the original text:] "P. Lenard, 'Über Kathodenstrahlen', Nobel Prize acceptance speech, 2nd ed., as well as *Deutsche Physik*, Vol. 3, 289–290, 1937. *E.T.Z.* 1919, Issue 10.—Friedrich Zöllner, *Erklärung der universellen Gravitation* [*aus der statischen Wirkungen der Elektricität und die allgemeine Bedeutung des Weber'schen Gesetzes*]"

[6]The mathematician Carl Friedrich Gauß (1777–1855) built together with Wilhelm Eduard Weber (1804–1891) the first working telegraph in Göttingen in 1833. On the establishment of the system of electromagnetic units, closely connected to Weber's work see, e.g., Wilhelm Weber & Rudolf Kohlrausch: *Fünf Abhandlungen über absolute elektrische Strom- und Widerstandsmessung*, Leipzig: W. Engelmann, 1904 (= Ostwalds Klassiker der exakten Wissenschaften, no. 142, with a slightly altered selection in the new series, no. 5). The units finally adopted were the Ampère (for currents), Ohm (for resistance) and Volt (for tension). Cf. Herbert Arthur Klein: *The Science of Measurement. A Historical Survey*, New York: Dover, 1984, chap. V. Lenard's works cited in footnote ‡ include his speech on cathode rays and his textbook on 'Aryan' physics; and the book by [Johann Karl] Friedrich Zöllner (1834–1882) on an 'explanation of universal gravitation', first published by Fock in 1882. Zöllner is best known for his work in spectral photometry and his invention of the polarization photometer in 1859.

¶ [Footnote 5) in the original text:] "Hermann Graßmann's *Gesammelte mathematische und physikalische Werke*, Vol. III, Part II. 1911, pp. 164ff., 333ff."

[7]Hermann Günther Graßmann (1809–1877) was a mathematics teacher in Stettin. He published on mechanics, non-Euclidean geometry, geometrical analysis and algebraic curves. The edition cited in footnote ¶ on Graßmann's collected mathematical and physical works was published by the mathematical and physical sections of the *Königlich Sächsische Gesellschaft der Wissenschaften* through B. G. Teubner publishers, with contributions by J. Lüroth, Eduard Study, J. Graßmann, Hermann Graßmann, Jr., Georg Scheffers and was edited by Friedrich Engel. The second part to the third volume is a biography of Graßmann by Friedrich Engel entitled *Graßmanns Leben*, including a list of his published writings and an overview of his unpublished handwritten papers.

[8]Eugen Karl Dühring (1833–1921) studied law in Berlin 1853–56, then upon losing his sight, philosophy, in which he took his doctorate in 1861 in Berlin, also habilitating in economics and becoming unsalaried lecturer at Berlin University in 1864. In 1877 he was struck from the list of university lecturers for criticizing Berlin professors and was thus excluded from an academic career; but he continued to publish in his field, including his well-received critical history of mechanics *Kritische Geschichte der Principien der allgemeinen Mechanik*, Berlin,

at Berlin University in 1877 is indicative and marks the beginning of the Jews' unscrupulous invasion of German professorships in physics.

The part-Jew [*jüdische Mischling*] Heinrich Hertz, the Jew Wertheimer, the Jew E. Goldstein, the Jew Grunmach and many others are students of Helmholtz.[9] How welcome Helmholtz's influence was to Jewry is demonstrated by the fact that the Eastern Jewish mathematician Leo Koenigsberger** became his biographer.[10] The Jew Emil Gabriel (!) Warburg[11]—offspring of the race of second-hand dealers and money-changers in Hamburg—was brought in from Freiburg to succeed Helmholtz at Berlin University. He worked there from 1895 to 1905 and finally occupied the highest position the German Reich can offer in physics: This very Jew assumed Helmholtz's place as President of the Reich Physical and Technical Institute from the 1st of April 1905 until the 31st of March 1922.[12] Warburg made a thorough job of it for Jewry. With him in such a position and since then, the Jew came to feel at home in academic teaching positions in Berlin. In 1905 the Jew Heinrich Rubens became his successor after he had already been appointed professor at the Polytechnic in Berlin in 1900 and

---

1873; Leipzig: Fues Verlag, 1878. Dühring later became a well-known German positivist and political economist who also wrote on topics in mathematics, physics and literature, and was severely attacked by Friedrich Engels for his anti-Semitic and anti-Marxist stance. See, e.g., his autobiography *Sache, Leben und Feinde*, Karlsruhe: Reutter, 1882.

[9]On Heinrich Rudolf Hertz (1857–1894), see doc. 31, footnote 5. The physicist from Bielefeld Eduard Wertheimer published a book on natural processes in the earth's atmosphere in 1933 and another on electromagnetic and kinetic equations for vapors and gases in 1937. Eugen Goldstein (1850–1930) was an experimental physicist specializing in spectroscopy, who taught in Berlin. Leo Grunmach (1851–1923) took his Ph.D. in Berlin in 1881, was unsalaried university lecturer the following year at the Polytechnic in Charlottenburg (Berlin), and then employed at the Prussian bureau of standards *Normal-Aichungskommission*. Between 1893–1921 he was professor of physics at the Berlin Polytechnic in Charlottenburg.

** [Footnote 6) in the original text:] "This Eastern Jew was salaried instructor of physics (!) and mathematics at the Berlin Cadet Corps (!) from 1861 to 1864. In 1868 the discoverer of vector and tensor calculus Hermann Graßmann was supposed to be called to a professorship at Greifswald. The friends of the Jew Koenigsberger prevented this; see L. Koenigsberger: *Mein Leben*. 1919, pp. 79–80."

[10]See Leo Königsberger: *Hermann von Helmholtz*, Braunschweig: Vieweg, 1901/03 (3 Vols.). Leo Königsberger (1837–1921) took his doctorate at Berlin University in 1860 and was full professor of mathematics at Heidelberg University 1884–1914. He was a close friend of Helmholtz. Königsberger's autobiography cited in footnote ** was published by Carl Winter in Heidelberg

[11]Warburg was born in 1846; after studying science at Heidelberg (under Bunsen, Helmholtz and Kirchhoff) and experimental physics under Hans Magnus in Berlin, he became associate professor at the German *Reichsuniversität* in Strasbourg, where he worked with August Kundt on effects of the kinetic theory of gases. In 1876 Warburg became full professor at the University of Freiburg/Breisgau. In 1895 he was called to the Berlin University as successor to Kundt, not Helmholtz, as erroneously mentioned in the text.

[12]Warburg immediately succeeded Friedrich Kohlrausch (1840–1910) to this position at the → *Physikalisch-Technische Reichsanstalt*; under his directorship, the *PTR* expanded rapidly: See, e.g., Cahan [1989].

at the Technical Military Academy (!) in 1903.[13] Then the Jews R. Börnstein, James Franck, Peter Pringsheim, Max Born, F. Reiche, A. Korn, F. Grünbaum and many others followed later.[14]

At the same time, the Jewish-minded influence of the councillors of the former Kaiser Wilhelm II was devastating, who was anyway repaid this confidence by the Jewry world-wide in true Jewish style, of course. The founding of the Kaiser Wilhelm Institutes in Dahlem was the prelude to a new flood of Jews into phys-

[13]Heinrich [Henri Leopold] Rubens (1865–1922) studied at the Darmstadt Polytechnic and at the Polytechnic in Charlottenburg (Berlin), then in 1885 at the Universities of Berlin and Strasbourg (under August Kundt and Franz Stenger). In 1889 Rubens took his doctorate at Berlin University and became assistant the following year at the *Königlich Physikalisches Institut* there. In 1896 he was appointed professor at the Polytechnic in Charlottenburg (Berlin) and director of the physics laboratory, becoming full professor in 1900 and director of the *Physikalische Sammlung* in 1904; he also worked at the *PTR*. 1906 he succeeded Drude as professor of experimental physics at Berlin University as well as at the Royal Physical Institute. 1908 he became a member of the Prussian Academy of Sciences.

[14]Richard Börnstein (1852–1913), was appointed professor at the Agricultural College in Berlin in 1881. James → Franck was teaching assistant at Berlin University upon taking his degree in 1906. 1911 he habilitated and 1915 became associate professor there. In 1933 he emigrated to the United States. See doc. 9 on his resignation. Peter Pringsheim (1881–1963) took his Ph.D. in Munich, was in Göttingen 1906–07 and in Cambridge 1907–08. He was teaching assistant in 1909 at the physics department at Berlin University, then unsalaried lecturer in 1920. In 1921 Pringsheim was untenured then 1923 regular associate professor of physics, becoming full professor in 1929. In 1933 he emigrated to Brussels and accepted a position as associate professor at the University of Brussels. In 1941 Pringsheim was visiting professor at the University of California, Berkeley, 1942–44 he worked together with James Franck at the University of Chicago, 1944–46 was director of research at the Penz Control Company, Pasadena, 1946–47 visiting professor at Chicago University, and 1947–55 senior scientist at the Argonne National Laboratory, Chicago. He took up residence in Antwerp in 1955. Pringsheim researched on the photoelectric effect, spectroscopy, fluorescence, phosphorescence and optical properties of crystals. Max → Born habilitated in mathematical physics in 1909. 1915–19 he was associate professor at Berlin University prior to his appointment to Frankfort-on-Main. He was dismissed in 1933 and emigrated to England. Fritz Reiche (1883–1969) took his Ph.D. in physics at the University of Berlin in 1907 and was teaching assistant to Max Planck in the field of quantum physics 1920. 1921–33 he was appointed full professor of theoretical physics at the University of Breslau. Reiche was an emigré in Prague 1934–35, moving to the United States in 1941. Arthur Korn (1870–1945) was full professor of technical physics at the Berlin Polytechnic. In 1935 he was dismissed and emigrated to the United States in 1939. A list of scientific staff members and visitors compiled by J. → Stark in 1937 does not mention F. Grünbaum. This may refer to Eduard August Grüneisen (1877–1949) who was engaged as a staff scientist at the *PTR* 1899–1927 in the area of electrical metrology and while employed there was promoted to department director. He was full professor of experimental physics at the University of Marburg 1927–47 and returned to Berlin as a guest 1932–33 to work on the characteristic values of elasticity.

ical science.[15] The Jew F. Haber,††[16] nephew of the Jewish bigtime operator Koppel,[17] secured the directorship of the Kaiser Wilhelm Institute of Physical Chemistry and Electrochemistry. Work there was reserved almost exclusively for Jews: O. Sackur (killed in an accident on the 17th of December 1914, was involved in a strange investigation for high treason);[18] F. Kerschbaum (following his resignation from the K[aiser] W[ilhelm] I[nstitute], worked in the USA and was finally director of the then still Jewish Frankfort o[n the]. M[ain]. Gold and Silver Refinery); Leyser; [19] the Jewess Liesa Meitner;[20] James Franck (now an emigrant, member of the Prussian Academy of Sciences, went in 1933 first to Soviet Russia, then to Niels David (!) Bohr,[21] where he stayed from 1933–1935, since 1935 in USA, prof[essor]. of physics at Johns Hopkins University in Baltimore); R. Ladenburg (son of the Jewish chemist Ladenburg, related to the Jewish American bankers, since 1933 an emigrant, prof[essor]. of physics at Palmer Physical Laboratory, Princeton University);[22] H. Mark (later at *I. G. Farben*, then prof[essor].

[15]Institutes like the → *KWIC* were built shortly after the → *KWG*'s foundation in 1911; the → *KWIP* was only built in 1937; cf., e.g., Vierhaus & vom Brocke (Eds.) [1990], and docs. 30, 38, 54.

†† [Footnote 7) in the original text:] "Cf. the tasteless obituary of the Jew, deceased on the 30th of January 1934 in Basel, by the member of the Prussian Academy of Sciences M. von Laue, in *Die Naturwissenschaften* 1934, at that time edited by the Jew A. Berliner."

[16]For Max → von Laue's obituary on Fritz → Haber, see doc. 29; for Haber's resignation of his directorship of the → *KWIPC*, see doc. 15. Arnold Berliner (1862–1942) was a prominent editor of the scientific journal *Naturwissenschaften* 1913–35.

[17]The banker and owner of the → *Auer* Company Leopold Koppel (1854–1933) provided the necessary funds for the foundation of the *KWIPC* in 1911. Einstein's appointment as a fully paid member of the Prussian Academy of Sciences in 1914 was also made possible through a donation by Koppel.

[18]The physical chemist Otto Sackur (1880–1914) studied in Heidelberg, Berlin and Breslau between 1898 and 1901 when he took his doctorate at the University of Breslau and became subsequently teaching assistant at the university chemistry laboratory. 1902–03 he was assistant (*Hilfsarbeiter*) at the Reich Department of Health (*Reichsgesundheitsamt*). He completed his habilitation and became unsalaried lecturer in 1905 and was teaching assistant at the chemistry department 1905–09. In 1911 Sackur was appointed professor in Breslau, and from 1914 was department head of the *KWIPC* in Dahlem (Berlin). He was killed in an explosion in his lab while testing poison gas during World War I. It is not clear what investigations Glaser is referring to. See, e.g., W. Herz's obituary in *Physikalische Zeitschrift* **16** [1915], pp. 114–115.

[19]The physical chemist Richard Leiser (1876–1955) took his degree at Vienna University in 1901. Between 1899–1907 he was teaching assistant at the Vienna Polytechnic, becoming lecturer at the Karlsruhe Polytechnic in 1909. Subsequently, Leiser was department head at the *KWIPC* at Dahlem (Berlin), returning to Vienna in 1914. After the war he managed his own physical engineering research lab. Leiser was also an inventor and patented an audio warning device used in mining to indicate the presence of explosive gas; cf. Vierhaus & vom Brocke (Eds). [1990], p. 63.

[20]On Lise → Meitner, see docs. 5, 37, 60, and 61.

[21]On Niels Bohr, see footnote 1 of doc. 20.

[22]Rudolf → Ladenburg's father, Albert Ladenburg (1842–1911), was full professor of chemistry at the University of Kiel 1874–99, then at the University of Breslau 1899–1909.

in Vienna, dismissed during the seizure of power in Austria);[23] M. von Polanyi (Hungarian Jew, emigrant, now prof[essor]. at Manchester University)[24] and H. Kallmann (F. Haber's nephew).[25] Through F. Haber and Finlay Freundlich (a Jew) as Acting Director in 1921,[26] the well-known discoverer of X-ray interference, the deceased Professor Paul Knipping,[27] was driven from the Institute under the most degrading circumstances and had to flee to the National Prize recipient P. Lenard.§§

The K[aiser] W[ilhelm] I[nstitute] of Physics, today's Planck Institute, had been set up for the Jew A. Einstein in 1914. His assistant, the Jew Erwin Freundlich, was his traveling preacher. (Emigrated first to Ankara, then to Prague.)[28]

---

[23]Hermann Francis Mark (1895–?) taught physics and physical chemistry 1919–21 in Vienna, where he earned his degree in 1921. 1921–22 he taught organic chemistry in Berlin, advancing between 1922–26 from resident fellow to group leader of the Kaiser Wilhelm Institute of Fibrous Materials Chemistry at Dahlem. 1927–28 he became resident chemist at → *IG Farbenindustrie*, group leader from 1928 and assistant resident director from 1930–32. Between 1927–32 Mark was also associate professor at the Karlsruhe Technical University and was appointed full professor of chemistry 1932 at the University of Vienna. 1938 Mark emigrated, in 1940 became assistant professor, and in 1942 full professor of organic chemistry at the Polytechnical Institute of Brooklyn. 1945 he was named director of the Polymer Research Institute, Brooklyn. 1956 he acquired the title *Dr. rer. nat.* in Berlin. Mark was a specialist in the use of X rays and electrons in synthesis, characterization, reactions and properties of natural and synthetic macro-molecules (polymer chemistry).

[24]The Nobel laureate Michael von Polányi (1891–1976) contributed a formula in calculating adsorption. On his resignation from the *KWIPC* in 1933, see doc. 15. He was professor of physical chemistry between 1933–48 at Manchester, professor of social studies between 1948–58, and in 1949 and 1954 visiting professor at the University of Chicago, emeriting at the Victoria University of Manchester in 1958. Polányi was senior resident fellow at Merton College, Oxford 1959–61.

[25]Hartmut Paul Kallmann (1896–?) was group leader at the *KWIPC* in Dahlem (Berlin) between 1920–33 and also unsalaried lecturer at the University of Berlin 1927–33. Then he was employed as a scientist by → *IG Farben* 1933–45 and was full professor and director of the Institute of Theoretical Physics at the Technical University in Berlin 1945–48. 1948–49 he was research fellow at the US Army Signal Corps Laboratories in Belmar, New Jersey. 1949–68 he was professor at the Physics Department and director of the Radiation and Solid State Laboratory at New York University. He was emerited there in 1968 and in the same year was guest professor at the Technical University in Munich.

[26]Herbert Max Freundlich (1880–1941) was an important colloid chemist and the department head at the *KWIC* in Berlin; cf. doc. 15 on his resignation in 1933. He emigrated to England.

[27]Paul C. M. Knipping (1883–1935) was associate professor at the Darmstadt Polytechnic from 1928–35. According to correspondence between von Laue and Knipping in the Lise Meitner Papers at Churchill College, Cambridge, Knipping attempted to habilitate under Lenard at Heidelberg University, but was rejected because of his positive attitude toward A. Einstein. He died in a motorcycle accident. On his discovery, see also doc. 76, footnote 3.

§§ [Footnote 8) in the original text:] "The supporting material in this matter consists of private reports by my friend and *SA* comrade."

[28]The official foundation of the *KWIP* was delayed by the outbreak of World War I until 1917. It was made contingent on Albert → Einstein's moving to Berlin to become its first director. See, e.g., Kirsten & Treder (Eds.) [1979], Vols. 1 & 2, chap III. On Erwin Finlay Freundlich (1885–1964), who was director of the Einstein Tower in Potsdam 1920–33, emigrating to Istanbul—not

Einstein's friend M. von Laue‡‡[29] became the second director.[30]

II. Munich: A crystallization point for Jewish physicists.

Munich became the last, but therefore all the more sturdy stronghold for Jewry. Here the Jew Arthur Korn[31] operated beside the Jew Leo Graetz,¶¶ the son of the notorious rabbi and historian of Jewry Heinrich Graetz.[32] A new crystallization point formed in this city with A. Sommerfeld as its patron to the very end,[33] similar to Göttingen with its Jews, Max Born (son of the Eastern Jew and prof[essor]. of anatomy Born (Breslau),[34] following the seizure of power emigrant in Belgrade, then 1935 until 1938 Stokes Lecturer of Mathematics at Cambridge University, since 1938 prof[essor]. at Edinburg[h] University) and James Franck (1920 to 1934 prof[essor]. of physics and director of the 2nd Physical

---

Ankara—and then to Prague in 1937 and St. Andrews, Scotland, in 1939, see Hentschel [1996].

‡‡ [Footnote 9) in the original text:] "The kinds of peculiar methods M. von Laue employed to suppress troublesome opinions on relativity theory are revealed in *Zehnders Briefe an Röntgen.* Pp. 157–158, Zurich 1935."

[29]The correspondence edition described in footnote ‡‡ is correctly cited as Wilhelm Conrad Röntgen: *Briefe an L. Zehnder* with contributions by Ludwig Albert Zehnder, published by Rascher.

[30]Von Laue, who had long served as the acting director of the *KWIP*, officially became director around 1922.

[31]On Korn, see footnote 14 above.

¶¶ [Footnote 10) in the original text:] "It is probably little known that the *Kurzer Abriß der Elektrizität*, which experienced its 13th edition in German in 1921, appeared in Polish in 1921 in Berlin and in Hebrew in 1922 in Vilnius."

[32]On Leo Graetz, see doc. 55, footnote 10. His father Heinrich Graetz (1817–1891) was a German-Jewish historian and author of an 11-volume history of Jews: *Geschichte der Juden: Aus den ältesten Zeiten bis auf die Gegenwart*, Berlin: Veit and Leipzig: Leiner, 1853–1875, (English translation 5 vols, 1889–1895; *History of the Jews*, 1949). He matriculated in Breslau 1842 and became professor there in 1869. Graetz's 'brief survey of electricity' cited in footnote ¶¶ was first published by J. Engelhorn in Stuttgart in 1897.

[33]Arnold → Sommerfeld was professor of theoretical physics at the University of Munich from 1906. On his great influence as a teacher see, e.g., Eckert et al. (Eds.) [1984], part II. A thorough investigation into Sommerfeld's ancestry had the following outcome: In a letter to the leader of the University Lecturers League of Apr. 18, 1939, the rector reports: "The expert on race research at the Reich Interior Ministry has reported in a letter of the 21st of Sep, 1933 F 434, that Privy Councillor Prof. Sommerfeld's lineage has been verified up to his great grandparents, it was thereby ascertained that Prof. Sommerfeld's ancestors are of Aryan origin." (Munich University Archive): Eckert [1993], pp. 149f.

[34]Max Born's father Gustav Jakob Born (1851–1900), an ethnic German born in Posen, Poland (then a part of Prussia), studied medicine at Breslau, Bonn, Strasbourg and Berlin and obtained his doctorate in 1873. After working briefly in Heidelberg he became teaching assistant at the anatomy department in Breslau where he was made reponsible for performing dissections (*Prosektor*) in 1876 and habilitated in the same year. In 1886 he became associate professor and in 1898 honorary full professor. He published on histology and embryology, one paper on the influence of gravity on frogs' eggs. His most important paper of 1897 discussed deformity experiments on amphibian larvae.

Institute).***[35] During the World War the following two Eastern Jews operated in Munich: Paul Sophus Epstein (born 20th of March 1883 in Warsaw, from 1909 to 1913 unsalaried university lecturer in Moscow, since 1921 prof[essor]. of theoretical physics at the California Institute for Technology in Pasadena)[36] and A. Rabinowicz (born the 22nd of February 1889 in Sadagora, 1919 to 1920 unsalaried university lecturer in Czernowitz, 1920 to 1922 prof[essor]. at Ljubljana, from 1922 prof[essor]. at Lodz).[37]

The Eastern Jew Kasimir Fajan (born 27th of May 1887 in Warsaw),[38] whom the Jew Haber had 'discovered' in Karlsruhe and who could never speak a proper German even up to his retirement from Munich University, was called to Munich during the World War. After the artificial ties with the Rockefeller Foundation[39] no longer yielded anything, he emigrated to Poland where he

*** [Footnote 11) in the original text:] "See *Große jüdische National-Biographie, mit mehr als 12000 Lebensbeschreibungen namhafter jüdischer Männer und Frauen aller Zeiten und Länder* Vol. V, p. 570 by S. Vinniger Czernowitz, 'Aria' [sic] Printing Press and *Jüdisches Lexicon [Ein enzyklopädisches Handbuch des jüdischen Wissens*], Vol. V, 1930, p. 493."

[35]The titles of the reference works cited in the preceding original footnote are translated as: 'Great Jewish National Biography with More Than 10,000 Biographical Descriptions of Noteworthy Jewish Men and Women of All Time and from All Countries' (published by Arta and edited by Salomon Wininger (1879–1968), who survived the Hitler era in Czernowitz and moved to Palestine in 1945), and 'Jewish Lexicon. An Encyclopedic Manual of Jewish Knowledge' (initially edited by Georg Herlitz and Bruno Kirschner and published by Jüdischer Verlag in Berlin). On Franck and Born in Göttingen see, e.g., Born [1968], part I, sections 18–19, and part II, sections 1–2, Lemmerich (Ed.) [1982], and Kamp et al. [1983].

[36]The Pole Paul Sophus Epstein (1883–1966) studied in Moscow 1901–06 under Peter N. Lebedev, becoming teaching assistant there 1906 and university lecturer 1909. 1910–11 he was assistant professor at the Agricultural Institute in Moscow. 1911–13 Epstein studied under A. Sommerfeld, obtaining his doctorate in Munich in 1914 on a topic in optics. 1919 he habilitated in Zurich and was private lecturer until 1921 when he emigrated to the U.S. He was professor of physics at Caltech 1921–53. Epstein worked on applications of quantum theory and thermodynamics.

[37][Wojciech] Adalbert Rubinowicz (1889–1974) submitted his dissertation in 1914 under M. Radakovič and was full professor of theoretical physics 1922–37 at the Technical University of Lwów (previously a part of Poland, now within the Ukraine) and 1937–46 also director of the Institute of Theoretical Physics there. He was full professor at Warsaw University 1946–60, and president of the Polish Physical Society 1961–73.

[38]The Polish physical chemist Kasimir Fajans (1887–1975) discovered simultaneously with Frederick Soddy (1877–1956) the displacement rule governing radioactive transformations resulting from alpha and beta-ray emissions. After acquiring his doctorate in Leipzig 1909, he studied in Heidelberg; in 1910 he was resident fellow at Zurich. Fajans went to Manchester in 1911 and was assistant and unsalaried lecturer of physical chemistry at the Polytechnic in Karlsruhe, 1911–17. At Munich 1917–35 he advanced from associate professor to the directorship of the department of physical chemistry. In 1936 he was professor of chemistry at the University of Michigan, Ann Arbor and was emerited in 1957. Fajan specialized in thermochemistry, the theory of chemical forces, photochemistry, and light absorption. He co-discovered with O. H. Gohring uranium X_2.

[39]One of the Rockefeller Foundation's main activities involved funding the education of scientists abroad. For references on this American institution, see also footnote 8, doc. 30.

was soundly thrashed by the students. Then he went to the United States. He was followed there by the Jew K. F. Herzfeld,[40] who had submitted his habilitation thesis at Munich University in 1920. In 1926 Herzfeld emigrated to the United States, where he has been prof[essor]. of physics and director of the physics department at the Catholic university (!) in Washington since 1936. In addition, other active Jews in Munich were A. Landé (now an emigrant at Ohio State University, Columbia, USA),[41] W. Pauli, Jr. (now professor of theoretical physics at the Zurich Polytechnical Institute),[42] F. and H. London (formerly Clar[e]ndon Laboratory, Oxford, now Institut Henri Poincaré, Paris, Duke University, Durham, North Carolina),[43] K. Lanczos (Department of Mathematics, Perd[u]e University),[44] H. A. Bethe (son of the Jewish physiologist Prof. Albrecht Bethe, 1930 until 1933 unsalaried university lecturer in Munich, emigrated firstly to England and worked at Manchester and Bristol, now at Cornell Uni-

[40]Karl Ferdinand Herzfeld (1892–1978) took his doctorate at Vienna in 1914. From 1920 he was unsalaried lecturer and assistant at the University of Munich. In 1923 he became associate professor of theoretical physics and chemistry and was appointed full professor of physics at the Johns Hopkins University in 1926.

[41]Alfred Landé (1888–1975) studied at Marburg and Göttingen and took his doctorate in Munich under the guidance of Sommerfeld in 1914. 1919 he was unsalaried university lecturer at Frankfort-on-Main, and 1922 associate professor at Tübingen University. 1931–60 Landé was professor of theoretical physics at Ohio State University, in Columbia. He was a specialist in atomic structure and quantum theory, spectral lines, Zeeman's effect and multiplet theory.

[42]Wolfgang → Pauli wrote his doctorate under Sommerfeld's guidance at Munich, and taught at the *Eidgenössisch-Technische Hochschule* in Zurich from 1928 but was at the Institute for Advanced Study in Princeton 1940–46.

[43]Fritz London (1900–1954) took his Ph.D. at the University of Munich in 1921. 1928–33 he was unsalaried university lecturer in theoretical physics and assistant to Erwin Schrödinger, Berlin University. In 1933 he emigrated to the United Kingdom. 1933–36 he received a grant from the Imp[erial]. Chem[ical]. Industries, teaching at Oxford; 1936 he moved to France. 1936–39 London was *maître de recherche* then *directeur de recherche* in physics at the *Institut Poincaré*, Sorbonne. In 1939 he moved to the United States and he was a member of the faculty at Duke University 1939–54. Heinz London (1907–1970) took his doctorate in 1933 in Breslau. 1934 he emigrated to the U.K. and 1934–36 was at Clarenden Lab., Oxford. 1936–41 he researched at Will lab., Bristol; and 1941–46 worked on isotope separation through gaseous diffusion. Both men made important contributions on superconductivity, low-temperature physics and atomic physics; cf., e.g., F. & H. London: 'The electromagnetic equations of the superconductor', *Proceedings of the Royal Society* **A 149** [1935], pp. 71–88.

[44]Cornelius Lánczos (1893–1973) studied at Budapest University and took his doctorate in mathematical physics at the University of Szeged in 1921. 1927 he was unsalaried lecturer at the University of Frankfort-on-Main. From 1931–46 he was professor of mathematics and mathematical physics at Purdue University, Lafayette, Indiana, 1943–44 staff mathematician at the National Bureau of Standards, 1946–49 senior resident engineer at Boeing Airplane Company, Los Angeles, and 1947–48 Walker-Ames Lecturer at Seattle, Washington. 1949–52 Länczos was staff mathematician and visiting lecturer at the Institute for Numerical Analysis, at the University of California, Los Angeles; 1952–53, visiting professor, 1954–68 senior professor; and 1968 emeritus at the school for theoretical physics. He subsequently consulted as a computing specialist at various centers in the USA and Ireland.

versity, Ithaca, New York),[45] W. Heitler (working in Bristol in England at the university),[46] S. Bochner,[47] and R. Peierls (an emigrant in England at Manchester University).[48]

The notorious Jew Theodor Lasarus Lessing exposed what these persons were up to in his book *Der jüdische Selbsthaß*,[49] in which he celebrated the destruction of physics as the accomplishment of the Jewish intelligentsia. W. Heisenberg, who took his degree on the 23rd of September 1923 in Munich, also once lifted the veil on these matters when he said in the foreword to his *Physikalische Prinzipien der Quantentheorie* (Leipzig 1930) that the purpose of his book would be fulfilled if it serves a bit in spreading that 'Copenhagen spirit of quantum theory'.[50] But it was precisely there in Copenhagen in the presence of the half-Jew Niels

[45]Hans Albrecht → Bethe's father, Albrecht Th. Julius Bethe (1872–1954), was full professor of physiology at the University of Frankfort-on-Main and was emerited there in 1937. Glaser was incorrect in taking the physiologist for a Jew—he evidently had assumed this, knowing that Hans A. Bethe was a 'part-Jew'. Bethe himself referred ironically to his 'birth defect' (his mother was Jewish) in a letter to Sommerfeld dated Apr. 11, 1933. See Eckert [1993], pp. 152f.

[46]Walther Heinrich Heitler (1904–1981) studied physics at the Karlsruhe Polytechnic and at the Universities of Berlin and Munich. 1926 he took his doctorate under K. F. Herzfeld. In 1927 he habilitated and was unsalaried lecturer at the University of Göttingen 1929–33. 1933 he was dismissed and emigrated to the U. K. and was research fellow 1933–41 at the University of Bristol. 1940 he was interned for several months, moved to Ireland in 1941, was appointed professor at the Dublin Institute for Advanced Study, and director there 1946–49. In 1949 he became full professor at the University of Zurich. His specialty was on the quantum mechanical explanation of chemical valence forces (together with Fritz London).

[47]Salomon Bochner (1899–), born in Podgorze close to Cracow, Poland, studied mathematics at the University of Berlin from 1918 and wrote his Ph.D. under the guidance of Erhard Schmidt in 1921. In 1927 he became assistant at the University of Munich, emigrating in 1933 first to the U.K., then to the USA where he was assistant 1933–34, assistant professor 1934–39, associate professor 1939/45, and finally full professor of mathematics at Princeton University.

[48]Rudolf Ernst Peierls (1907–1995), took his doctorate at Leipzig in 1929. 1932–33 he received a Rockefeller fellowship in Rome and Cambridge and did not return to Germany. 1933–35 he was honorary research fellow at the University of Manchester, and 1935–37 research associate at the Royal Society Mond Lab, Oxford. Peierls worked on quantum electrodynamics, solid state physics and nuclear physics, and soon became professor at Birmingham 1937–63, and Oxford. He participated in the Manhattan project at Los Alamos between 1943–46.

[49]The writer, cultural critic and philosopher Theodor Lessing (1872–1933) became known for his polemical articles against all forms of discrimination and was vehemently opposed to Hindenburg's nomination as president. His radical outspokenness led to public outcry and the Prussian Ministry of Culture pressured him to relinquish his lecturer's position at the Hanover Polytechnic in exchange for a research grant abroad in 1926. He was murdered by Nazi Security Service (*SD*) agents while in exile on Aug. 31, 1933 in Marienbad, Bohemia. The title of his book, which translates as 'Jewish selfhatred', was published in 1930 by Jüdischer Verlag, Berlin.

[50]On Werner → Heisenberg, Niels Bohr and their collaborators as well as on their role in the 'Copenhagen interpretation of quantum theory', see, e.g, John Heilbron: 'The earliest missionaries of the Copenhagen spirit', *Revue Internationale d'Histoire des Sciences* **38** [1985], pp. 195–228. See also doc. 31, footnote 4. Heisenberg's textbook was originally published by S. Hirzel. See also the translation by Carl Eckart & Frank C. Hoyt: *The Physical Principles of the Quantum Theory*, New York: Dover, 1930 (reprinted in 1950).

David Bohr—at a scientific institute—that the American R. A. Millikan from Pasadena made a nasty attack on the Third Reich, which was duly dismissed in the *Metallarbeiter Zeitung* of the G[erman] W[orkers] F[ront].[51] The National Prize recipient Phil[ipp]. Lenard characterized this physics for what it is: "These schematic calculations with spectra can be called 'stockjobber optics' [*Börsianer-Optik*], because they consider the processes within the atom as scantily as the people dealing merely with a factory's 'stock certificates' consider its physical operations." ...

As early as 1925 (the 17th of October) the doyen [Lenard] wrote me:

> ... "But this can only improve when reasonable circles realize that though Einstein ...... and all the likes of him may well be quick at figures, they cannot understand at all what scientific research is. They have been supplied with so much material to compute along with other fundamental things from previous scientific research that they do not notice their own deficiency at all. But if they do not permit genuine scientific researchers to get back on their feet again—which is impossible in their system—then the prospects are dismal indeed."

But at that time the few who saw daylight and took up the fight to save real scientific research—Lenard, Gehrcke, Stark, Dingler, Uller, and I—were unable to win recognition.[52]

But then in 1938 a November storm raged throughout the land—it swept away the withered foliage.[53] The remaining Jews, part-Jews and those related to Jews through marriage have disappeared from the academies and libraries, from the lecture halls and from wherever else they had clung as supposedly indispensable persons. Many of them have taken to their heels; and the Ahasuerus,[54] the

[51]Robert Andrews Millikan (1868–1953) was professor of experimental physics at the University of Chicago from 1896; around 1909 he developed the oil-drop method for measuring the unit of electric charge; and in 1912 he confirmed the validity of Einstein's theory of the photo effect. He received the 1923 Nobel Prize for these precision measurements. Millikan taught at Caltech in Pasadena from 1921, which became a world famous scientific center in that time.

[52]Philipp → Lenard, Ernst → Gehrcke, Johannes → Stark, Hugo Dingler, Karl Uller and Ludwig Glaser were amongst the most avid anti-relativists even during the Weimar period: On the objections by the Munich philosopher Dingler to relativity theory, see in particular Hentschel [1990]a, sec. 4.5.4. See also doc. 83, footnotes 19 and 20; on Uller see footnote 12 there.

[53]An allusion to November 9, 1938, the 'Night of Broken Glass' (*Kristallnacht*), when concerted terror attacks were made on Jewish synagogues and stores. Reinhard Heydrich, leader of the Security Service (on Heydrich, see the first footnote to doc. 63) ordered the destruction of all Jewish places of worship in Germany and Austria on the pretext of the assassination of a German diplomat in Paris two days earlier by a Polish Jew. 101 synagogues were burnt down and 76 demolished; and 7,500 Jewish-owned stores were systematically destroyed. Pillaging and looting continued throughout the night and the streets were littered with broken glass. For eyewitness accounts, documentation and historical analysis see, e.g., Kurt Pätzold & Irene Runge, *Pogromnacht 1933*, Berlin: Dietz Verlag, 1988.

[54]The king of ancient Persia, known to the Greeks as Xerxes (486–465 B.C.) and the husband

Wandering Jew,[[55]] roves again restlessly from country to country in order to rob Nordic people of their jobs, just like in Germany.

We thank our *Führer*, Adolf Hitler, for having freed us from the plague of Jews. Yet the chaff must now be sifted from the wheat. The Jew in Germany has stopped working in German teaching positions and in science; but now his traces must be removed, and the Jewish mentality, which has become lodged in minds here and there throughout the land, must be eradicated. The fight goes on for an untainted youth.

78 Philipp Lenard: Foreword to 'Ideological Continental Blockade' [February 1, 1940]

Source: Philipp Lenard, Foreword to: *Ideelle Kontinentalsperre*, Issue 2 of *Kriegsschriften der Reichsstudentenführung*, edited by Reich Student Leader Scheel, Munich: Franz Eher & Sons, 1940, pp. 5–6.[1]

Comments 25 Years Later (December 1939):

When this text was written, nothing was known generally about the important consequences of racial mixing among peoples or about the nature of Jewry. Only

of Biblical Esther (also called Assuerus). According to the Book of Esther he saved the Jews from their enemies. At the Jewish festival Purim commemorating this event, theater pieces were performed in Europe, and medieval Christian spectators mistook the travelling actor of Ahasuerus for the legendary 'Wandering Jew' figure.

[55]'The Wandering Jew' (*Ewige Jude*) is a legendary character condemned to roam until the second coming of Christ for having taunted Jesus, or in a later version, for having struck him on the day he was crucified. This Protestant myth first appeared in print in Leyden in 1602. The demagogic passage above reflects the anti-Semitic prejudices of contemporary German propaganda writings, such as the pictoral study: Hans Diebow (Ed.), *Der ewige Jude*, Munich & Berlin: Zentralverlag der NSDAP, 1937, containing 265 photographs and illustrations, published in a series on Jews in Germany by the 'Jewish Studies' Institute (*Institut zum Studium der Judenfrage*) through the main Party publishers; or the film of the same title.

[1]P. → Lenard's original pamphlet was written in August 1914. A list of the issues appearing in the volume reveals the belligerent nature of the series (the titles are translated here from the German original): "Ernst Krieck, *England, Ideology and Reality*; Philipp Lenard, *Ideological Continental Blockade*; *Documents on the British Plutocracy*; Karl Alexander von Müller, *England and the Continent*; Heinrich Ritter von Srbik, *Westphalian Peace and German National Unity*; Johannes Paul, *Germany, Scandinavia and England*; Friedrich Wagner, *Napoleon and England*, other issues to follow. The subsequent general description read: "Noteworthy German scientists and new talent at universities contribute the results of their scientific research to the cause of our National struggle for survival [*Völkischer Lebenskampf*], publishing in the Reich Student Leadership's War Writings [series]. The War Writings' documentary publications in particular will offer a thorough explanation of the form, essence and goal of subversive plutocratic forces. The Reich Student Leadership's War Writings should prove that the sword of German science can also slash its way to victory and is thus in its own way a sharp and irreplaceable weapon in our nation's struggle for existence." The publishers *Franz Eher Nachf. GmbH* appear on the title page as the main publishing house of the → *NSDAP*.

after the war did the effects of this become evident. Earlier studies and warnings by great minds were kept concealed and covered up purposefully.

What I accuse the English of in the pamphlet applies especially to Jews, who are basically warmongers and greedy, in keeping with their mentality. But actually, England's ruling mentality was to a large part Jewish (and is very much more so today), so my pamphlet still rightly hits the mark.

My knowledge of England was acquired mostly from the scientific point of view. To me J. J. Thomson in particular is the archetypal modern Englishman with his scientific grabbiness [*wissenschaftlich raffsüchtig*] and filthy impertinence.[2] I came to know him quite well enough.

I have already described Germany's shameful position in the international scientific literature before the World War. The Jewish advance in science has led to complete capitulation in England. See, for example, the lecture schedule of the once distinguished Aryan 'Royal Institution' (London)—now predominated by Jews. It is high time that there be an end to the favoritism of English authors over Germans (which I have encountered throughout my life) in German publications. The international mentality at German physical societies was really the creeping Jewish mentality [*geistige Verjudung*]. Proof of this is the praise that the Einstein mentality [*Einsteingeist*] constantly receives.[3]

Now we have a second major war by England upon us!

Heidelberg, the 1st of February, 1940.

79 Contract between the Army Ordnance Office and the Kaiser Wilhelm Society for the Advancement of the Sciences [March 6, 1940]

Source: Geheimes Staatsarchiv, Dahlem (Berlin): 33 Dept. I/Rep. 151, Folder No. 1256, Sheet Nos. 119–120. Typed transcript of document with the heading: 'Abschrift von Abschrift zu WN Nr. 218/40 g.' and stamped 'Secret'. Administrative notations and signatures are omitted.[1]

---

[2]On the face of it Lenard's anti-English diatribes must be seen in the context of World War I (on the boycott of German science after the war see Grundmann [1965]). However, Lenard's singling out of this eminent English physicist had its origin in a bitter priority conflict over the discovery of the electron. Twenty years earlier their relationship had been a quite different one: See Lenard's friendly letter dated Nov. 20, 1896, in which he thanked Thomson profusely for his invitation to Cambridge on the occasion of his award of the Rumford Medal, reprinted in Lord Rayleigh, *The Life of Sir J. J. Thomson*, Cambridge: Cambridge Univ. Press, 1942, p. 55. On Joseph John Thomson see footnote 4 of doc. 26.

[3]See, e.g., doc. 4 for A. → von Brunn's review of a collection of essays protesting against this supposed infatuation with A. → Einstein's relativity theories.

[1]Final certification of the contract was dated March 6, 1940. The term of this contract between the → *KWG* and the → *HWA* began January 1, 1940. However, this transcript was enclosed with the following letter dated March 15, 1940 (doc. 80), in which Minister B. → Rust of the → *REM* Rust attached new conditions before giving his approval to the contract.

Contract

between the German Reich Treasury—Army—Army High Command—,
represented by the Army Ordnance Office [*Heereswaffenamt*],
and
the Kaiser Wilhelm Society for the Advancement of the Sciences,
registered society, in Berlin C2, the Castle,
hereinafter named in short 'the Society'.

§ **1.** (1) The Society entrusts to the Army Ordnance Office the building, furnishings, etc., of the Kaiser Wilhelm Institute of Physics in Dahlem (Berlin), 20 Boltzmann St., for use by the Armed Forces for its own purposes. Both parties will agree on a case-by-case basis upon the continuation of current research projects, as well as upon the release of sections of the Institute that are not needed by the Army Ordnance Office at present or during the term of the contract.[2]

(2) The powers of the present Director and Co-Director of the Institute are suspended during the period in which the Army Ordnance Office assumes control.

The Army Ordnance Office will appoint an authorized agent vested with all the Director's rights and duties to replace the present Director.[3]

(3) Furthermore, the Society places at the disposal of the Army Ordnance Office the existing staff employed at the Institute at the signing of the contract.[4] Absolutely no reductions may be made in salaries to be paid by the Army Ordnance Office to these connected staff members. The personnel do not enter into any kind of employee relationship with the Army Ordnance Office through this settlement.

[2]This contract ended with the return of the → *KWIP* to the Society (under supervision of the Reich Research Council (→ *RFR*) at the beginning of July 1942 after the *HWA*'s interest in the 'uranium project' had waned.

[3]This replacement had become inevitable, because the *KWIP*'s Dutch director, Peter → Debye, declined to apply for German citizenship, which had been placed as a condition to his leading the secret German uranium research project. Debye applied instead for an indefinite leave of absence to Cornell University in Ithaca, New York. He was succeeded by K. → Diebner, who became acting director at the institute in January 1940. After July 1942, W. → Heisenberg became acting director at the institute (as Debye was then still officially on leave). After World War II, Debye was invited to resume his directorship, but he decided to stay at Caltech in Pasadena, California.

[4]The *KWIP*'s staff included: F. → Bopp, E. → Fischer, S. → Flügge, C. F. von → Weizsäcker, K. → Wirtz and Alexander Lindley Deubner (born 1905). Deubner studied physics at the University of Freiburg (Breisgau) in 1922 and at Munich University in 1924. He took his doctorate under G. → Mie at Freiburg in 1927 and was teaching assistant to W. → Nernst and A. Wehnelt at the University of Berlin 1932–36. He worked at the *KWIP* 1937–41, and habilitated in 1938. Deubner worked under C. Gerthsen at the University of Berlin 1945–48 and became university lecturer in 1946. In 1947 he became associate professor and in 1950 department director at the first physics institute at Humboldt University in Berlin, and in 1952 was appointed full professor of physics at the agricultural and horticultural department of the University of Berlin.

(4) Dismissals from among the acquired personnel may only take place in agreement with the Society.[5] All other staff changes (engagement of additional temporary staff for the purposes of the Army Ordnance Office and their dismissal) may be performed independently by the appointed authorized agent regardless of how they affect the Army Ordnance Office.

(5) The Army Ordnance Office is not liable for accidents that occur during the performance of the work. [...[6]]

§ **2.** (1) In addition, the Society will make available to the Institute the sums it receives for the Institute's research purposes from the Reich Ministry of Science, Education, and Culture. Should cuts be made in the public funds granted to the Society, then the contribution amount will also be lowered in proportion to the reduced amount allotted to the Institute. Presently, the sum totals *RM 85,000.–* (in words: Eighty-Five Thousand Reichsmarks) annually. The annual sum will be paid in monthly installments of 1/12 of the relevant annual total.

(2) The Institute's obligations existing on the 1st of January 1940 will be satisfied by the Society out of the current annual total.

(3) The rendering of accounts of the Institute's available funds will be performed as before by the Institute's administration; the review following the close of each fiscal year, by the Army Ordnance Office's auditors.

§ **3.** (1) The Army Ordnance Office will pay the Society monthly compensation to be paid in advance in the amount of *RM 1,000.–* (in words: One Thousand Reichsmarks) in restitution for expenses incurred to the Society through the normal maintenance of the Society's loaned building and furnishings.

(2) Since this compensation by no means covers the Society's real existing expenses covered in Numeral (1), all types of cost increases as well as new charges incurred after the closing of this contract will be assumed by the Army Ordnance Office.[...[7]]

§ **5.** (4) At the termination of the contract the Army Ordnance Office must restore the space to its original condition at the request of the Society and must perform all necessary restoration work at its own cost. The Society has the right, however, to accept the space in the condition it is in at the end of the contract including the alterations and extensions made by the Army Ordnance Office. Contingent compensation for the Army Ordnance Office's built-in equipment and instruments will be arranged by mutual agreement.[...[8]]

[5]This ensured that the *HWA* could not simply replace staff members for political reasons.

[6]The remainder of the paragraph concerning accident liability is omitted here.

[7]§ 4 in which the Army Ordnance Office assumed the Institute's running and physical maintenance costs, as well as the first 3 paragraphs of § 5 involving the physical maintenance and alteration of the premises are omitted here.

[8]§ 6 regarding the existing installed equipment, their condition upon return, joint ownership conditions of new equipment acquired out of the Institute's budget, and the Army Ordnance Office's own equipment acquired with its own funds, is omitted here.

§ **7.** (1) The contract is valid from the 1st of January 1940 for the duration of the war, on the understanding, however, that the Army Ordnance Office has the right at any time to revoke it at the end of any month with 3 months notice.[9]

(2) The Army Ordnance Office must return the Institute to the Society no later than 3 months after the end of the war. The Army Ordnance Office must fulfill any outstanding obligations of the Institute existing at the time of its return.[...[10]]

Berlin, the 5th of January 1940.

For the Kaiser Wilhelm Society for the Advancement of the Sciences, registered society signed, Dr. Telschow.[11]

Berlin, the 17th of January 1940.

For the Army High Command, Army Ordnance Office: signed, Becker.[12]

80 Bernhard Rust: Letter to the Army High Command [March 15, 1940]

Source: Geheimes Staatsarchiv, Dahlem (Berlin): 33 Dept. I/Rep. 151, Folder No. 1256, Sheet No. 117. Unsigned typed certified transcript of letter on official stationery: 'Der Reichsminister für Wissenschaft, Erziehung und Volksbildung, Berlin W 8, Unter den Linden 69' with file reference 'W N Nr. 200 g' and stamped 'Secret'. Addressed: 'An das Oberkommando des Heeres, z. Hd. von Herrn Regierungsrat Nasner, o.V.i.A., in Berlin W 35, Tirpitz Ufer 72/6'. Other administrative recipients and notations omitted.[1]

Re: Kaiser Wilhelm Institute of Physics, contract between the Army Ordnance Office and the Kaiser Wilhelm Society.

I approve the contract submitted with the above-mentioned letter[2] to be concluded between the German Reich—Army—and the Kaiser Wilhelm Society, which establishes the temporary entrustment of the Kaiser Wilhelm Institute of Physics to the Army Ordnance Office, under the condition that within the scope of this contract the Institute's low-temperature laboratory together with its trained staff be placed at the disposal of the Reich Minister of Aviation for his own purposes. The grants that the Kaiser Wilhelm Society receives from my Ministry for the Institute will continue to be made available, taking into account any special budgetary management resolutions made during the war.

---

[9]In fact the contract was terminated prematurely in the summer of 1942, because the Army Ordnance Office lost interest as soon as it realized that the uranium project could not yield practical results of immediate military relevance in the near future—see Heisenberg's retrospective account, doc. 115, footnote 53, and doc. 111, footnote 14.

[10]§ 8 referring to the relevant Reich laws governing contracts and § 9 covering the incidental costs incurred in the writing of the contract, are omitted here.

[11]On the secretary of the society Ernst Telschow, see doc. 81, footnote 6.

[12]General Karl → Becker became head of the *HWA* on Feb. 4, 1938; see also doc. 75.

[1]This is the cover letter to the preceding document, doc. 79, by Minister Bernhard → Rust.

[2]The previous correspondence referenced is dated December 20, 1939, file reference: '-91a 2220 Wa F/Ia (Ch H Rüst u. BdE)-'.

I request that the Army Ordnance Office's authorized agent to be appointed to the Institute be made known to me.[3]

Signature [B. Rust].

81 Ernst Telschow: Foreword to Yearbook of the Kaiser Wilhelm Society for the Advancement of the Sciences [August 1940]

Source: Ernst Telschow, Foreword to *Jahrbuch für 1940 der Kaiser Wilhelm Gesellschaft zur Förderung der Wissenschaften*, Berlin, 1940, p. 7. Signed published document.

With the onset of the great fight over Germany's freedom and future,[1] the year 1939 has posed new and important challenges also for the Kaiser Wilhelm Society. Almost all of the Society's 36 Institutes could become engaged in solving the problems vital to the war and to the economy without having to neglect basic research in the process, which is in keeping with the tradition of the Society.[2] Although a large number of the Institutes' results cannot yet be discussed today in detail, the following extracts from the Society's activities will prove, nevertheless, that German science sees as its principle duty being committed to the people and to the fatherland both in war and in peace.[3] It can be said furthermore that with its combination of independence and close contact with the various Reich Ministries and offices of the Party and the State, the organization of the Kaiser Wilhelm Society makes it outstandingly well suited to prompt and multifarious action.[4] It is thus demonstrated anew precisely in our times how correct the underlying conceptions is upon which the Society was built, and which had already found its first champion in Alexander von Humboldt.[5]

Berlin C 2, in August, 1940. Dr. Telschow.[6]

---

[3]This refers to § 1 No. (2) of the foregoing contract, which specifies the nomination of a new director at the → *KWIP*, since its former director Peter → Debye had applied for indefinite leave of absence. Kurt → Diebner, the *HWA*'s expert in nuclear physics, became his substitute.

[1]This is a euphemistic reference to the outbreak of World War II triggered by the German invasion of Poland on Sep. 1, 1939.

[2]See H. Albrecht & A. Hermann in Vierhaus & vom Brocke (Eds.) [1990] for a history of the → *KWG* during the 'Third Reich'; in particular, pp. 359ff., for the society's early self-imposed political alignment (*Selbstgleichschaltung*) and pp. 385ff. for its later political realignment resulting in the new articles of the society.

[3]This issue of the yearbook covers the period Apr. 1, 1939 to Mar. 31, 1940.

[4]See doc. 80 for an example of how close this relationship with governmental authorities was during the war period.

[5]Baron Friedrich Heinrich Alexander von Humboldt (1769–1859) was an eminent naturalist, writer and statesman.

[6]Ernst Telschow (1889–1988) was assistant administrator at the society from 1931, then second manager of the general administration from October 1933, and finally general secretary of the *KWG* from July 15, 1937 to May 18, 1960(!). He was a student of Otto → Hahn and became a member of the → *NSDAP* in May 1933. Cf. Vierhaus & v.Brocke (Eds.) [1990], p. 52.

82 Hans-Joachim Flechtner: Where does German Physics Stand Today? A Wartime Conference in Berlin [September 8, 1940]

Source: Hans-Joachim Flechtner, 'Wo steht die deutsche Physik? Eine Kriegstagung in Berlin', *Das Reich*, Berlin, Issue No. 16, September 8, 1940, pp. 17–18.

On the 1st and 2nd of September a joint conference of the German Physical Society and the German Society of Technical Physics took place at the Polytechnic in Charlottenburg, Berlin.[1] The fact that such a scientific conference can take place in Germany in 1940, a war year, while it had to be cancelled in the previous year at the outbreak of the war[2] is, as the chairman of the Physical Society J. Zenneck emphasized in his opening speech,[3] a noble and fine indication of what the German leadership and the Armed Forces of the German nation have achieved in the past year: Within the country scientific work can continue on its course. The wartime conference of German physicists stems from the indispensability of the work itself, which is at the service of the people and the Reich today as seldom before. But now that new physical discoveries deeply affect the development of individual branches of physics, it is particularly essential that a researcher in a specialty sustain links to other branches, in order to preserve the possibility of constantly establishing new contacts with scientists in other specialties and benefiting from their working methods. It is the product of the times and a necessity today that we confine ourselves strictly to what is important and focus ourselves on these tasks. The absence of public discussion of the talks is a manifestation of this attitude. It was left to the scientists themselves to clarify points of controversy and to draw intellectual conclusions from the lectures in private conversations.

---

[1]For some of the papers delivered at this joint conference, which took place at the *Technische Hochschule* in the Charlottenburg suburb of Berlin, see the proceedings of the *Deutsche Physikalische Gesellschaft* (→ *DPG*) and the *Deutsche Gesellschaft für technische Physik* (→ *DGtP*), in *Physikalische Zeitschrift* **41** [1940], pp. 514–579; cf. the *Verhandlungen der deutschen Physikalischen Gesellschaft* (3) **21** [1940], pp. 31–50 for abstracts.

[2]A physics conference had been planned in Marienbad which was to have been the first joint meeting of the *DPG* and the *DGtP*. Because of the strained situation in the succeeding years during and after World War II—in the postwar years the *DPG* was split initially into many regional subunits—no major physics conferences were held before 1950 in Bad Nauheim: Cf. Heinicke [1985], pp. 5, 9, 11.

[3]In his opening speech Jonathan → Zenneck emphasized the many applications of physics for the German Armed Forces, such as observation and information technology to extend acoustical and optical range: The development of prismatic binoculars, shear-jointed telescopes, sighting instruments, range finders, sound and flash ranging techniques, radio bearing, sonic depth finders, sonar, aerial photography, etc. Zenneck's speech was printed in full in the *Verhandlungen der Deutschen Physikalischen Gesellschaft* (3) **21** [1940], pp. 31–34, and in a shortened version in the *Physikalische Zeitschrift* **41** [1940], p. 514.

The Polytechnic's large physics auditorium was tightly packed at the opening—and yet, upon closer examination the scene was different to earlier conferences: Many familiar faces among German physicists were missing here in Berlin;[4] uniforms were interspersed among the rows of participants; and the more or less official air that otherwise pervaded German physicist conference inaugurations through the presence of representatives of government authorities and branches of the armed services, was also lacking. The German physicists' wartime conference was not a festive congress, but a practical symposium.

Sunday was devoted to pure physics. The modern developments of nuclear physics formed the prelude to the conference, in keeping with the significance and importance of the subject. These developments were summarized in two longer talks and supplemented by shorter detailed presentations. G. Hoffmann (Leipzig) provided a survey on 'Methods and Results of New Research in Nuclear Physics'.[5] Nuclear research can be called the third stage in atomic theory—if the philosophical atomic theories are considered a preliminary stage. In the first stage with the help of atomic theory it was possible to explain physical, and above all, chemical processes. The theory withstood testing and could be considered secure. Then with the discovery of radioactivity and related phenomena the second stage began: Research on atoms themselves. Atoms proved to be complex—complicated structures made up of a nucleus and arranged around this core, electron shells. How the various atoms, of which there are now 93 sorts, are exactly structured, and how the various numbers of electrons are arranged in the separate types of atoms, on the one hand, and in the shells, on the other, the answers to these questions were discovered and confirmed in this second stage. The third stage at which we are today and which far exceeds the other two in difficulty and in its theoretical level and possibilities thus emerged: It appears that nuclei are complex as well; and the question is how they are put together. Essentially only the proton, neutron, the light and 'heavy' electrons, or positrons, and the neutrino are suitable building blocks. Earlier theory supposed that the strong bonds within nuclei could be explained by forces arising between the various nuclear components. Today, in order to be able to explain the make-up of nuclei, we must show that exchange forces also exist between these same nuclear building blocks. In recent years the 'field theory of exchange forces' by the Japanese physicist Yukawa was able to solve a large number of problems at issue here.[6] This theory had the two

---

[4]Zenneck mentioned on p. 33 (see the preceding footnote) that many physicists were not able to attend "either because they are enlisted into the Services or cannot be spared from urgent work at Army offices or laboratories in industry."

[5]Gerhard → Hoffmann was full professor of experimental physics at the University of Leipzig 1937–45. See G. Hoffmann [1940]; cf. the abstract in *Verhandlungen der Deutschen Physikalischen Gesellschaft* (3) **21** [1940], p. 35.

[6]Hideki Yukawa (1907–1981) postulated the existence of a particle he dubbed a mesotron, which has a medium-size mass like the π-meson, and is thus heavier than an electron, but is lighter than a proton or neutron. The exchange of these mesotrons (referred to since as mesons) explained the existence of a nuclear force between protons and neutrons.

crucial characteristics that determine a useful physical theory: firstly, it correctly considered the quantitative relations upon first approximation; secondly, however, it predicted as a result of this theory a building block of matter not yet known at the time, the mesotron (also called meson), which has since been detected. Yet, enough difficulties still remain also in this theory, to which E. C. G. Stückelberg v. Breidenbach (Geneva) referred in an interesting talk.[7]

The difficulties of overcoming problems of nuclear physics at the theoretical level notwithstanding, the mastery of current questions at the observational level is no less difficult. The downright inconceivable minuteness of the nucleus—the nucleus's diameter is in the order of trillionths of a centimeter!—requires the inquiring mind to invent completely novel experimental tools. On the other hand, it is necessary to work with enormous energies and electrical tensions and to invent instruments that are very different from those employed when 'atom-smashing' was first carried out, tapping the natural energy of radioactive radiation. From Professor Hoffmann's description, the cyclotron created by Lawrence in America is pivotal today. With it the tiny elementary particles of matter artificially obtain tremendous speeds before being used to radiate the nuclei under analysis.[8] A survey of the development of the size of the cyclotron specially mentioned the American plan for a new cyclotron with a 'dish' of one and a half meters in diameter, which provides matter particles with a velocity equivalent to a total tension of 100 (a hundred!) million volt. It concluded with the promising statement that essentially all was ready for the construction of a medium-sized cyclotron in Germany; so that at some future date German science will also be able to make greater use of this important research tool.[9] Among the advances in nuclear physics, the discovery of uranium decay and the related elimination of transuranic elements were mentioned primarily by the speaker, as well as the discovery of two kinds of atomic nuclei of mass three: a hydrogen and a helium atom.[10] (We have already reported on both discoveries in *Das Reich*, so that this reference will suffice here.)

---

[7]Ernest Carl Gerlach Stueckelberg von Breidenbach (1905–) was full professor at the University of Geneva from 1939. In 1942 he was lecturer at the University of Lausanne, 1950 honorary professor at the University of Geneva, and in 1957 associate professor (*extraordinaire*) at the University of Lausanne. See E. C. G. Stueckelberg von Breidenbach: 'Schwierigkeiten in der Feldtheorie der Austauschkräfte', *Physikalische Zeitschrift* **41** [1940], pp. 523–524, and the abstract in *Verhandlungen der Deutschen Physikalischen Gesellschaft* (3) **21** [1940], pp. 35–36.

[8]On Ernest Orlando Lawrence's development of cyclotrons, in which charged elementary particles are accelerated on spiral paths, see the Introduction, p. lxxxix, and footnote 320; for a contemporary description cf. G. Hoffmann [1940], pp. 518f., and Watzlawek [1942]a. According to Osietzki [1988], Hoffmann and W. → Bothe competed to get the first → Siemens cyclotron, which was only operational in autumn 1943, however.

[9]Cf. Heilbron [1986]a.

[10]The nucleus of deuteron (or heavy hydrogen) consists of one proton and two neutrons, while helium's nucleus consists of two protons and one neutron; see again G. Hoffmann [1940], pp. 519f.

It is difficult to discuss here in more detail the wealth of stimulating and important expositions presented on the first day of the conference. The problems themselves as well as the methods to overcome them and the results found are of so specialized a nature, that explaining them requires more interest than can be presumed here—and more space than is available. We will mention Pohl's fine interference experiment (Göttingen),[11] G. Hertz's experiments (Berlin) on sound pressure on steel,[12] as well as E. Lau's presentation (Berlin) on the interference fringes of light passing through screens connected in a series, which will be of technical importance in determining the spacing of light signals.[13] Gehr[c]ke's experiments (Berlin) on shape and spacial vision with polarization spectacles were also interesting.[14] Unsöld (Kiel) spoke about the abundance of light chemical elements in the universe, based today on precise spectral analyses of stars and nuclear physics.[15] Studies on the electrical conductivity of metals and semi-conductors, on the distribution of conduction electrons in metal under the influence of gravity, and other matters, brought this first day of the conference to a close.

Contrary to the past, current technological development is marked not so much by the revelation of new energy sources and by the constantly improved refinement in exploiting existing energy options. Rather, it is determined chiefly by what can almost be called a sensational broadening of choice of materials.

[11] Robert Wichard Pohl (1884–1976) was professor of experimental physics at the University of Göttingen from 1919; he worked mostly on the properties of alkaline halide crystals and other problems in solid-state physics. At the conference Pohl demonstrated an experiment on the interference of diverging rays of light with the intention of replacing Fresnel's mirror experiment: Cf. *Verhandlungen der Deutschen Physikalischen Gesellschaft* (3) **21** [1940], p. 37. On Pohl's research see, e.g., Jürgen Teichmann: *Zur Geschichte der Festkörperphysik. Farbzentrenforschung bis 1940.* Stuttgart: Steiner, 1988.

[12] See Gustav → Hertz, 'Der Schallstrahlungsdruck in Flüssigkeiten und Gasen im Zusammenhang mit der Zustandsgleichung', *Physikalische Zeitschrift* **41** [1940], pp. 546–549; cf. *Verhandlungen der Deutschen Physikalischen Gesellschaft* (3) **21** [1940], p. 38.

[13] Ernst Gustav Lau (1893–1978) was Ernst → Gehrcke's collaborator at the → *PTR* from 1920 and finally became senior executive officer (*Oberregierungsrat*) there. In 1929 he was awarded the Prussian Ministry of Commerce Medal in Berlin. 1948 he became director of the Institute of Optics and Precision Engineering and developed his double microscope there. In 1951 he was named Inventor of Merit of the German Democratic Republic. Lau was professor at the German Academy of Sciences in Berlin from 1953. See E. Lau, 'Abstandssignale', *Physikalische Zeitschrift* **41** [1940], pp. 544–546; cf. *Verhandlungen der Deutschen Physikalischen Gesellschaft* (3) **21** [1940], pp. 37–38.

[14] See E. Gehrcke, 'Neue Erscheinungen der physiologischen Optik', *Physikalische Zeitschrift* **41** [1940], pp. 540–544; cf. *Verhandlungen der Deutschen Physikalischen Gesellschaft* (3) **21** [1940], p. 37.

[15] Albrecht Unsöld (1905–1995) was a student of Sommerfeld and took his doctorate in 1929 at the University of Munich. He became full professor at the University of Kiel in 1932 and was one of the foremost astrophysicists in Germany of the time. See A. Unsöld, 'Die kosmische Häufigkeit der leichten Elemente', *Physikalische Zeitschrift* **41** [1940], pp. 549–552; cf. the abstract in *Verhandlungen der Deutschen Physikalischen Gesellschaft* (3) **21** [1940], pp. 38–39.

Today's technology is characterized by: the progressive improvement of known and commonly used materials, the discovery of new possibilities in using and refining existing natural substances, and finally, the advance into the new territory of purely synthetic materials which are superseding traditional raw materials extensively. But this development is only possible on the strength of a thorough analysis of materials and their properties, based on scientific experiments and results, which reveal to us the materials' properties in general, and along with it also their technologically important characteristics. The new developments in physics—above all, the knowledge gained from atomic and molecular physics—have provided us with the necessary foundations: perceiving the properties of materials in terms of their recognized atomic and molecular structure.

At the conference of the German Society of Technical Physics, this central role of substances—materials, raw materials, and synthetics—in current technology was taken into account. In its presentations due emphasis was laid on the treatment of a new group of materials developed in Germany in the last decades: hard metals.[16] Dr. E. Ammann (Essen) first gave an overview of the historical development of hard metals, which is virtually typical of technology's new approach to materials.[17] Hard metals are chiefly tool material; thus they serve in preparing other materials—and this presupposes that they are not only very hard but also very tough and resistant to wear. The historical route of producing such materials was set when it was discovered that iron could be transformed into steel through binding it appropriately with carbon, and that this steel could be substantially refined technically by adding certain amounts of other metals, such as chrome, nickel, tungsten, etc.—adding only one or also several of them. Modern high-speed steel was more or less the last step on this path.

---

[16]The *DGtP* met on Monday, Sep. 2, 1940; Flechtner chose to report in more detail on the one session that was obviously related to the needs of the military, but other topics covered had no such use. See, e.g., W. Weizel & F. Faßbender's paper on the cause of hissing in a solid carbon arc: 'Über die Ursache des Zischens beim Homogenkohlebogen', abstract in *Verhandlungen der Deutschen Physikalischen Gesellschaft* (3) **21** [1940], p. 47. On the metallurgical research conducted at the Kaiser Wilhelm Institute of Metal Research in Dahlem (Berlin) from its foundation in 1920 to its dissolution on Sep. 30, 1933, and on the research conducted at the Kaiser Wilhelm Institute of Metallurgy founded in Stuttgart in 1934, see the contributions by O. Bauer & E. Schmid, and W. Köster in Hartmann (Ed.) [1936], pp. 178–191. See also Hansen (Ed.) [1948] on contemporary metallurgical research.

[17]Ernst Ammann (1898–1949) studied engineering at the Polytechnic in Munich and ferrous metallurgy at the Polytechnic in Aachen. He graduated in 1925 and while working as assistant in the Ferrous Metallurgical Institute (*Eisenhüttenmännisches Institut*) in Aachen worked toward his doctorate which he obtained in 1927. He was then employed at the → *Krupp* Company and advanced into its management in 1932. After successive promotions he gained recognition for his capable direction and knowledge of the manufacture and application of hard metals and powder metallurgy and was noted for introducing hard metals into the mining and drilling industries. See E. Ammann's abstract: 'Über die Entwicklung und die wirtschaftliche und technische Bedeutung der Hartmetalle', *Verhandlungen der Deutschen Physikalischen Gesellschaft* (3) **21** [1940], p. 42.

Yet, the results won were by no means satisfactory. An abundance of experiments investigated the field in all conceivable directions—the science of iron, of steel and of its alloys has already become a major field of research today—but all improvements were not able to bring about the critical step. This step only came with the brave relinquishment of iron as the base metal. The degree of hardness of the steel types turned out to be predominently dependent on sintered carbides—bonds between metal and carbon—which develop within steel. Steels with admixtures of tungsten and titanium carbides became especially important. A lot of courage was needed to take the crucial step forward: namely, to choose tungsten and titanium carbides themselves as the base material for the new materials—hard metals. Iron, which had been the critical material for thousands of years and which had dominated technology in the form of steel wherever a combination of hardness, strength, and durability was needed, had been unthroned: The heavy metals, tungsten and titanium, began to replace it in special applications.

Then it soon became evident that, though very hard, materials out of pure tungsten carbide or alloys of tungsten and titanium carbide were also quite brittle and therefore were not very useful in engineering. Efforts were now begun, conversely, to alloy these hard materials with metals from the iron group, to make them more tensile, more elastic. Cobalt has proved to be the best metal. The speaker presented comparative overviews of the properties of high-speed steel and modern hard metals. Tungsten carbide with 5 percent cobalt emerged as the best material. Its hardness when cold, for example, was about 50 percent better than the best quality steel; but when heated—that is, after being heated up to about 500 degrees [Celsius]—it is seven to eight times harder than steel under the same conditions.

The very informative presentations by Dr. A[m]mann were augmented by Dr. W. Dawihl's more detailed studies (Berlin) on the processes involved in the wear of hard metals.[[18]] Here also the enormous superiority of hard metals over the best steels became apparent beyond doubt. If we take the data and research results of the last decades presented here, together with the current aggressive advance of light metals in technology, the prospect of modern materials science largely 'dethroning' iron among metals becomes clearly apparent. This image emerges not out of necessity but through a penetrating knowledge of the nature of previously 'second-rate' metals and of the technical possibilities they afford.

[18]Walther Dawihl (1904–) studied at the University of Berlin between 1923–27; he took his doctorate in 1928 under M. Bodenstein's guidance and habilitated in 1942, becoming university lecturer of metallurgy at the Berlin Polytechnic. In 1944 he was at the Stuttgart Polytechnic, and in 1953 accepted a lectureship at the University of the Saar province. His specialties included powder metallurgy and corrosion prevention. See the abstract of W. Dawihl's paper on the wear of hard-metal tools: 'Untersuchung über die Vorgänge bei der Abnutzung von Hartmetallwerkzeugen', *Verhandlungen der Deutschen Physikalischen Gesellschaft* (3) **21** [1940], p. 42.

Here again we must forego entering into the numerous interesting individual talks of the second day. Subjects had been chosen out of all fields of technical physics in acknowledgment of the smart progress in this branch of physics. Attendance at the lectures on this day also proved that there was no less of an interest in the questions of applied physics than for purely theoretical problems. The wealth of ideas that every individual participant takes away with himself will be the best proof to each of them of the success of this meeting of German physicists in the war year 1940.

83 Wilhelm Müller: The State of Theoretical Physics at Universities [November 1940]

Source: Wilhelm Müller, 'Die Lage der theoretischen Physik an den Universitäten', *Zeitschrift für die gesamte Naturwissenschaft*, Vol. 6, Issue 11–12, November/December 1940, pp. 281–298.

Looking at scientific activity mainly in the exact sciences at universities, we cannot help gain the impression that the German movement has hardly affected these fields at all, in the revolutionary sense at least. Moreover, work seems to go on as if these fields could not be directly affected or influenced by the patriotic political [*völkisch-politisch*] events of the last seven years. With few exceptions, the exact sciences still survive on the dogmas that had been hatched and disseminated during the episode of German disgrace [*deutsche Schmach*[1]] and that in a certain sense illustrate the awkward and deceptive retreat to a falsely interpreted internationality and fraternization among peoples. These dogmas are a considerable disturbance to science still today, blinding it to the compelling and vital reality of current world events and shutting out the struggle of new values against the forces of a dying epoch in history. Science is thus deprived of its liberating confrontation with today's tough demands. It must be emphasized repeatedly that more things take place now "between heaven and earth" than our professors can ever dream up—with their chronically retrospective bookish learning.[[2]]

The case of theoretical physics is particularly instructive in this regard. It has gradually separated itself from real or experimental physics, partly with the cooperation of mathematics and so-called mathematical physics, and has barely kept in touch with the needs of practical science. Just the fact that most theoretical advances—precisely the far-reaching progressive ones—date from the period after the World War must sound suspicious; since, they arose not from a predominating will to shake off the defeat but from a general submission to alien

[1]After the German defeat in 1918 the Treaty of Versailles, often called the 'peace of shame' (*Schmachfriede*), imposed severe conditions on Germany, including a ban on participation by German scientists at international conferences.

[2]W. → Müller is alluding to Shakespeare's *Hamlet*: "There are more things in heaven and earth, Horatio, Than are dreamt of in your philosophy." (Act I, scene v, lines 183–184).

and anti-nationalist slogans that were designed to extend the defeat even further into the intellectual realm. Through this international influence, this discipline has grown and expanded so rapidly, mainly by integrating the special fields of structure analysis,[3] atomic theory and physical chemistry, but has also become so complicated in the process because of increasingly difficult mathematical formulations, that it is utterly impossible for an individual scientist to master all the parts equally, to be in the position to verify critically the foundations and basis of his knowledge. Overhastily produced detailed researches are beginning to appear in relentless rapid succession, only rarely allowing leisure for contemplation and for getting a general overview. Many not exactly beneficial side effects go hand in hand with this clearly unhealthy development. First of all, science has burdened itself with an apparently increasingly home-grown dogmatism that almost appears to be a defensive reaction to the constant threat of collapse, like a building that has long since outgrown the bounds of its own foundations. It is true that a certain dogmatism develops wherever academic theoretical science is conducted and knowledge is gathered and laid down within a system—though the representatives of the scholarly world will not admit this. But in theoretical physics in particular, this dogmatism has borne such odd fruits, which will be described later in more detail, that we can take this dogmatism directly as typical of this cast of mind; and thus this field has least of all to do with scientific freedom. Naturally, the student is in the most serious predicament, because he cannot even find the time and leisure to cope with the foundations of classical theory, let alone with the Babelism of the latest atomic and nuclear physics.

I. University Policy Tactics and Methods

On the whole, this science is dominated by an almost incomprehensible dependence on dogmatic doctrines that are dictated and propagated by a limited circle of experts. Most of the textbooks we have (Planck, Cl[emens]. Schaefer, Haas, Joos etc.),[4] all sound the same, as if they were all based upon the same, naturally not immediately apparent, source. The same ambiguities can be found in all of them, even down to the very last detail and at exactly the same places.

[3]The term *strukturphysikalisch* was not commonly used at the time; Müller may have meant solid state physics.

[4]These textbooks on theoretical physics are: Max → Planck's lectures: *Einführung in die theoretische Physik* (5 vols.) Leipzig: S. Hirzel, 1916/30; Clemens Schaefer, *Einführung in die theoretische Physik* (3 vols.) Leipzig: B. G. Teubner, first ed., 1929/33; Arthur Erich Haas, *Einführung in die theoretische Physik mit besonderer Berücksichtigung der modernen Probleme* (2 vols.) Leipzig: Veit, 1919. Probably the most important textbook was Georg Joos's *Lehrbuch der theoretischen Physik*, Leipzig: Akademische Verlagsgesellschaft, 1932 (2nd ed., 1934; 3rd ed., 1939; 4th ed., 1941; 5th ed., 1942; and 6th ed. still in preparation in Dec. 1944/published in 1945); Arnold Sommerfeld's very influential lectures on theoretical physics in 6 volumes: *Vorlesungen über theoretische Physik* were only published in 1943 (Leipzig: Akademische Verlagsgesellschaft).

For example, in the transition from the mechanics of continua to etherless field theory [*Feldphysik*] or in the introduction to wave theory, fundamental difficulties and historical origins are omitted, side-stepped and concealed with always the same feints and distorting tricks. In another example, Joos's book, which is given to most students, is aimed from the outset at suppressing any independent thought, because it gives the student the false impression that after he has taken in the mostly dogmatically formulated material, he will have mastered the field of theoretical physics; whereas, in actuality he is not guided to the real problems at all, which almost always only start where the text of the book stops. In contrast to that we must show our delight with a book like Ph[ilipp]. Lenard's *Deutsche Physik*, which is clearly written by a great personality.[5]

Scientific practice in theoretical physics does in fact give the impression that the system had been arranged and fixed by some sort of esoterical group of authorities to be binding upon scientists as a whole, somewhat like how the interest rates and stock values are fixed at the stock exchange.[6] It is basically exactly the method used earlier in liberalist Germany by Jewish newspaper publishers and those organs catering to Jews, to sedate the soul of the people so that they comply with foreign aims.

Dogmatic coersion is particularly clearly evident in the very successful physics worldview literature (of the likes of Eddington or Zimmer).[7] In their full-blown arrogance these authors favor presenting the results of physical science in the context of humanity's general cultural advancement. By placing absolutely trivial ideas by mathematical physicists on the same level with a 'Copernican' discovery, the student is actually prevented from the start from making any unbiased critical re-evaluations, because if he rejects anything, he is subjected to the accusation of being culturally and intellectually backward. But these publications, as well as increasingly the daily press, really rate theorists like Planck, Heisenberg, and Schrödinger far too highly when they present them as the founders of a new natural philosophy or of a new physical worldview.[8] These theorists understand

[5]Cf. doc. 39 for the foreword to Philipp → Lenard's textbook on 'Aryan' physics.

[6]Here Müller takes up Lenard's 'stockjobber' metaphor (see doc. 77, p. 233), which incorporates the general distrust of the free market since the disastrously inflationary and volatile interwar period.

[7]The eminent English astronomer and physicist Arthur Stanley Eddington (1892–1944) published popular expositions about relativity theory and cosmology. The school teacher Ernst Heinrich Zimmer (1887–1965) also wrote didactical and popular books on physics. Zimmer had studied at the Universities of Freiburg (Breisgau) and Berlin 1906–12. 1914 he obtained a position as *Studienrat*, a secondary school teacher with civil service status, in Stettin, and 1917–41 was in a similar position in Lübeck at the time Müller's article appeared. In 1941 Zimmer was promoted to senior master (*Oberstudienrat*) in Lübeck and in 1947 to headmaster. 1946–52 he directed the *Staatliches Studienseminar* in Lübeck and was named honorary professor in 1949, retiring in 1952. See, e.g., Zimmer's book *Umsturz im Weltbild der Physik* with an introduction by Max Planck, published in Munich: Knorr & Hirth, 1st ed., 1934; 6th ed., 1942; 12th ed., 1961.

[8]Max → Planck was—and still is—considered the founder of quantum theory because of his

virtually nothing of the true limits of knowledge and of the concepts of natural philosophy—for example, the concept of causality, which they frivolously believe they may do without; and they only have a very cloudy notion of all the Kantian difficulties. Certainly a different calibre would be needed to master Kantian philosophy![9] The ideological and exaggerated literature of academicians (who are in any case not very talented at criticism) caused altogether the worst distortions and confusion. In it they use slogans to circumvent the difficulties, and research is degraded to an arena for empty chatterers. The younger generation cannot be warned earnestly enough against this sensation-hungry worldview prattle, which is neither physics, philosophy nor even ideology. All these books and brochures still adhere all too noticeably to the aftermath of the nasty tradition of relativistic propaganda literature, which we can only look back upon with the same contempt as for the whole nightmare of Jewish management after the war.

In these papers you immediately meet that almost mystical commitment to a specific program that permeates the entire system of theoretical physics and threatens anyone who dares to express his own opinion independently from the acknowledged clique. On the other hand, the fact that theoretical physicists are so surprisingly in agreement with each other on the basic issues is simply explained in that all contrary opinions were opposed and blighted systematically; every transgression of the prescribed laws resulted in total exclusion and financial ruin; and all influential university professorships were filled by obedient emissaries of a small group of academic high priests. It can be proved as a matter of fact that these few high priests of physics (Munich was previously a center of this kind)[10] were not only considerably involved in the appointment of all full professorships in physics, but they placed in all professorial chairs of this type almost exclusively men who could guarantee the strictest adherence to the laws of the invisible priesthood, just as reliably as parish priests and Jesuit students could to secure the creed of the 'holy' Roman [Catholic] church and papal authority. For the other side of the story just remember scientists like Uller, Zehnder, Glaser, Gehrcke, and Dingler, among others.[11] Some had to content themselves with

introduction of the energy quantum in 1900; Werner → Heisenberg and Erwin → Schrödinger created two different versions of quantum mechanics (matrix mechanics and wave mechanics) in 1926. For historical commentary see, e.g., Jammer [1966].

[9]According to Kant, causality is one of the categories of human reasoning (*Verstand*) which cannot be avoided because it is projected into nature by the very act of understanding it. This Kantian creed was challenged by quantum theory and quantum mechanics, which describe atomic processes such as radioactive decay as intrinsically probabilistic.

[10]Arnold → Sommerfeld, who made important contributions to Niels Bohr's quantum theory, held the chair for theoretical physics in Munich. Many of the later representatives of quantum mechanics were his pupils. On Sommerfeld's influence on professorship appointments in theoretical physics elsewhere, and on the next generation of founders of quantum mechanics, see Benz [1975], pp. 62ff., Eckert et al. (Eds.) [1984], pp. 111ff., Eckert [1993].

[11]On Uller and Zehnder see the following footnotes. Ludwig → Glaser, Ernst → Gehrcke, and Hugo Dingler (see below) were all ardent anti-relativists.

completely subordinate positions; others even went hungry and were subjected to the fate of being considered condemned and ostracized as not quite mentally sound, or as pitiable misfits or fantasts, or as somehow hereditarily tainted. And yet, in the midst of utter confusion and an unprecedented process of intellectual subjugation, precisely these men have retained that which is the precondition to every scientific achievement: The ability to think and make judgments independently and unburdened by prejudice.

To stay with these examples: K. Uller,[12] who has proved the absolute conceptual vagueness and lack of foundation of theoretical physics as a whole, and particularly also of Maxwell's electrodynamics, using the example of wave theory and through a brilliant command of the material, has remained incomprehensibly as good as unnoticed to this day.[13] It is true that his papers are hard to read; but whoever makes the effort to penetrate deeper into Uller's truely original chains of reasoning will doubtlessly benefit greatly from his papers and will maybe also get an idea of the incredible thoughtlessness of our scientists in the civil service, who actually only understand what they want to understand and what suits their purposes. Uller is unquestionably one of the most important creative minds in conceptual physics, which are few and far between. He absolutely should have been secured a full professorship in the interest of real productive scientific criticism. It is a disgrace to our leaders, to put it mildly, that no one had the nerve to give a professorship to such a man against the dictate of the high priests.

In his day the aging Zehnder,[14] a contemporary and former co-worker of Röntgen, was ruined by the Munich clique, because he also had dared to support a mechanically based physical theory of the universe, which was in conflict with the dogmas laid down by the group of esoterics and with non-Euclidian relativistic physics.[15] Regardless of whether one can agree with him or not on all points, it

---

[12]Karl Uller (1872–1959) studied at the Polytechnic in Charlottenburg (Berlin) and at the Universities of Berlin and Rostock, worked as an electrical engineer in Berlin and Nuremberg 1900–03 and submitted his Ph.D. thesis in Rostock in 1903. He was subsequently assistant at the physical institute of the University of Giessen 1905–26, specializing in wave physics (including acoustics) and electrodynamics, and was associate professor of physics there 1916–37.

[13]See, e.g., Karl Uller, 'Eine Kritik der Elektrodynamik und Relativistik', *Sitzungsberichte der Akademie der Wissenschaften, Heidelberg*, No. 10 [1919]; *Das Grundgesetz der Wellenfortpflanzung aus bewegter Quelle in bewegten Medien. Der Michelson-Versuch und die Raum-Zeit-Lehre von Einstein*, Munich & Berlin: Oldenbourgh, 1935; 'Der Sturz der relativistischen Feldphysik durch die Wellenkinematik', *Zeitschrift für die gesamte Naturwissenschaft* **3** [1937/38] 399–414.

[14]Ludwig Albert Zehnder (1854–1949) studied mechanical engineering in Zurich 1873–76 and founded his own electrotechnical factory in Basel in 1880. Later he also studied physics in Berlin 1885–87 and at Giessen 1887–88 (with Röntgen), submitting his thesis there in 1887. He later became assistant at the physical laboratory, moving to Würzburg 1888–89. Zehnder submitted his habilitation thesis in 1890 at Basel and was first assistant at the University of Munich in 1900, later transferring his permission to teach to Freiburg University in 1891. In 1904 he became chief scientist of practical training in physics at the post office (*Reichspostamt*) in Berlin; and in 1919 he was appointed associate professor of physics at the University of Basel.

[15]See Zehnder's *Die Entwicklung des Weltalls aus mechanischen Grundlagen*, Tübingen:

is stirring to read how Zehnder relentlessly seeks substantiation and development of this theory (which incidentally deals with modern scientific results), regardless of whether or not he is acknowledged. In any case this also is the mark of a true scientist, that what matters to him ultimately is not personal advantage but an inner commitment to truth, which can only be classical because it has to be forever.

L. Glaser, one of the best authorities on precision measurement in physics—in particular spectro-optical measurement—fearlessly confronted Einstein and his cohorts together with E. Gehrcke as early as in 1920;[16] and in this way sacrificed his prospects to a teaching position in the civil service during the Weimar period [*Systemzeit*[17]]. Until recently, that is, for seven years under the new government, he received only very minimal grants and was thus forced to set up his high-quality experimental apparatus that had been built out of his own funds in a room in his private residence. It is nevertheless surprising that no one has taken the opportunity up to now to assist this meritorious physicist in getting an important position. At the same time Privy Councillor Arnold Sommerfeld was offered the opportunity of keeping his venerable seat (although it has since been robbed of its former glory) as high priest of theoretical physics, despite his international connections and his pledged support of Einstein, and although he was five years beyond the age limit.[18] This happened in the middle of the making of the new Germany and in the middle of the racial decontamination [*Entjudung*] of our cultural life.

The case of Dingler is similar.[19] Hugo Dingler is the only philosophical thinker today of greater stature to really understand mathematics and physics;

Laupp, 1928; *Der Aether im Lichte der klassischen Zeit und der Neuzeit*, ibid., 1933; cf. also W. Müller [1940]a.

[16]On the Bad Nauheim incident see footnote 17 of doc. 40. See, for instance, Ludwig → Glaser's attack against Einstein's collaborator Erwin Finlay Freundlich in: *Sitzungsberichte des Vereins zur Beförderung des Gewerbefleisses* **99** [1920], pp. 105–106; and against L. Grebe in *Physikalische Zeitschrift* **23** [1922], pp. 100–102; cf. Glaser [1920] and his article on attempts to verify relativity, 'Versuche zur Bestätigung der Relativitätstheorie', *Glasers Annalen für Gewerbe und Bauwesen* **87** [1920], pp. 29–33; **88** pp. 30–33, 42–43. Ernst Gehrcke wrote numerous articles against relativity from 1911 on including the inflammatory anti-Semitic popular pamphlets suggesting subversive manipulation of the masses: Gehrcke [1920]b, [1924]a. Glaser's virulent anti-Semitism in later articles is reflected in doc. 77.

[17]Müller uses the derogatory term *Systemzeit* for the Weimar Republic, which connotes obligatory and submissive conformity within the system.

[18]Arnold Sommerfeld had just reached the age limit of 65 when the Nazis came to power in 1933; the dean could postpone his retirement until early 1935 (see the new retirement law, doc. 36), but he was able to stay in office as substitute until his successorship was settled, which took until the end of 1939 (see, e.g., Benz [1975], pp. 177–184, Cassidy [1992]a, chaps. 18–21).

[19]Hugo Dingler (1881–1954) was habilitated in Munich in 1912 and was associate professor there 1920–32. In 1932 he answered a call as full professor at Darmstadt until his dismissal in 1934. He subsequently became lecturer at Munich. On his opportunistic efforts to create a Nazi philosophy of nature, see, e.g., Dingler [1934], [1937/38], [1938], [1940], [1942], [1943]; cf. Wolters [1992].

and he seems to be the only one destined to create a guiding German method of physics that will prove to be valuable and fruitful also for the experimental physicist. Due to his early opposition to Einstein—which is evidenced by that high priest's efforts against him—he also was practically excluded from the urgently needed instruction and reorientation of students.[20] They thus have hardly any chance of being shown the most essential points from the fundamental and epistemological point of view and of learning to distinguish the real from the unreal, and true knowledge from apparent knowledge.

Many other examples could be named in this regard. We might even have had to lament Joh[annes]. Stark, the Nobel Prize recipient and physicist of the highest order, as a victim of this boycott by the theorists, if a far-sighted National Socialist minister had not appointed him later as President of the Reich Physical and Technical Institute in Berlin.[...[21]]

II. Science and Dogmatism in Theoretical Physics

[...[22]] Most of the concepts currently dominating theoretical physics set themselves apart by revealing inevitable attempts at comprehending physics mechanistically. Yet at the same time they are continually inconsistent with these mechanical elements, and thus with themselves. Finally, in desperation they rush for the safety of the mathematical formula, where everything is presented nice and cleanly and precisely; and the whole confusion of physical concepts appears to be concealed from the eyes of the uninitiated: Consider for instance the incompatible notions of the wave-like nature of light, the corpuscular and emission theories of cathode and other rays, the concept of matter waves, the energy-quantum-like conception of radiation and the comparison of matter to 'wave packets'.[[23]] A lot, even utter nonsense, can be proven by means of language. This applies to a greater degree to equations, which after all are capable of giving a mathematical meaning to any absurdity. The hypnotic effect of mathematical formulas had already crippled critical assessment of Maxwell's theory in its day and is the reason for the spread of a dangerous inflation of theory today as well.

[20] In 1929 Dingler wrote a surprisingly sympathetic article in celebration of Einstein's 50th birthday: 'Albert Einstein. Zu seinem 50. Geburtstag.' *Münchener Neueste Nachrichten* **82**, No. 72, Thursday, March 14, 1929, pp. 1–2. After 1933 he transformed into a vehement critic; see, e.g., Dingler [1938/39]. Dingler lost his professorship at Darmstadt in 1934 as a result of intrigues that placed into question his political 'reliability'. On the Munich philosopher's opposition to relativity theory see also Hentschel [1990]a, sec. 4.5.4; cf. Wolters [1992].

[21] On this appointment at the *Physikalisch-Technische Reichsanstalt* (→ *PTR*), cf. doc 18. The following three paragraphs on the supposed silencing of, or disregard for, the experts Julius Robert Mayer, Hermann Graßmann, Gustav Jaumann and Philipp Lenard is omitted here.

[22] The opening polemical discussion of Maxwell's mechanics, and the role of mathematics in physics (six paragraphs) are omitted here.

[23] These notions form a part of quantum mechanics which incorporated earlier insights on the wave-matter duality; cf. Wheaton [1983].

[...I]t follows furthermore that mathematical representation cannot be seen as an adequate rendition of physical reality and thus also never can express the 'essence' [*das 'An sich'*] of phenomena. Mathematical and physical reality are ontologically two entirely separate spheres, even though their relation to each other can be illustrated graphically. It is absolutely necessary to reassign mathematical tools their naturally given limitations as an aid and instrument of calculation and as an arithmetical means to penetrate reality; and we must realize that it is completely unacceptable to misuse these implements virtually to **create** reality and to construct physics purely through mathematics or from axioms that were laid down arbitrarily, as is the case in non-Euclidian physics, for example. The formulations of quantum theory also involve at best only mathematical analogies without a trace of physical evidence. For example, just as the eigenvalues of the Schrödinger equation, which naturally had long been known in mathematics, only result in multiples of energy quanta because the Planck constant had already been factored in, in the same way, no real physical statement can ever be won through a formula that had not already been anticipated somehow. Heisenberg's matrix model also only provides an analogy to the discrete sequence of the series of lines in atomic spectra, which is incidentally related to the theory of eigenvalues; but it certainly does not provide more physical statements than had been put into this model at the outset; and above all, it provides no physically based theory of line spectra that had not already been proposed by then.[24]

The extremest form of this improper disregard of the limits and of the reversal of status of the two fields of physics and mathematics can be found in theoretical physics, which has ravished reality in favor of mathematical structure, having given up all consideration of the possibility of having a normal empirical basis. The origin of the dogmatic fixing of certain laws and theoretical opinions in science can also be found here at the same time. An example is the totally unacceptable principle, from the point of view of rationally unbiased scientific thought, of the absolute constancy of the propagation velocity of light waves, uninfluenced by any factors, which at the same time is supposed to be the upper limit of all occurring speeds. Indeed, this is the origin of the dogmatically intolerant paralysis of theoretical science as a whole, which ultimately retards all progress. The substantial promotion of this kind of border-blurring and dogmatic and violating physics [*dogmatisch-vergewaltigender Physik*], which contaminates our academies and has completely confused our youth, stems in any case from the **Jews**,* who had already started to creep into universities under Helmholtz, partly with his approval and support.[25] Later they produced

[24]Cf. footnote 8 above.

* [Footnote 7) in the original:] "See Wilh. Müller, *Judentum und Wissenschaft.* Leipzig 1936."

[25]Wilhelm Müller's 61-page pamphlet cited in footnote * above on 'Jewry and science' was published by Fritsch. For another defamatory view of Helmholtz's influence, see doc. 77. On the historical roots of anti-Semitism see, e.g., the Introduction, p. lxxii.

through their great emissary Albert Einstein the great masterpiece of Talmudic inflationary physics and the most unscrupulous distortion of reality: the theory of relativity,†[26]which gave its scientific seal to the time of inner disintegration and national impotence. Advocates of the subsequent atomistic theories arising out of this atmosphere and bound to the Einstein program consciously believed they could offer up mechanically based intuition [*mechanische Anschauung*], which is basically the only creative principle, in favor of mathematical form, which was supposed to be the real 'thing-in-itself'[27] and which could be carried as far as they wished. Ultimately these theories are only the final outcome of intense Jewish influence on the masses [*Massensuggestion*], which has already largely been overcome in political and cultural life, but which unfortunately still continues to be felt in science to the present day, without the great majority of academics realizing what is happening and what is at risk.[28] Relativistic field theory [*Feldphysik*] assumes a four-dimensional vacuum, void of any medium, in which all processes are supposed to take place and which was constructed through artificially coupling space and time. As soon as this original dogma is relinquished and abandoned as a cumbersome alien element to science, all the dogmatic and ultimately aphysical fictions drawn from it will also have to lose their meaning—that is, the independence of the velocity of light from its source and its medium, the derivation of the equivalence of energy and mass, the relativistic implications of fine structure[29] in atomic physics, and the contradictory and questionable concept of matter waves.[30]

Most of these dogmas, which are being promoted propagandistically from this country, originate primarily from the plutocratic states out of a 'consensus' of scientific societies and cliques that had been grafted together. In connection with this fact, it is fitting to say a word about our science's degrading dependence on foreign countries—in particular on English-speaking countries—and on their

† [Footnote 8) in the original:] "See Br. Thüring, 'Albert Einsteins Umsturzversuch der Physik und seine Inneren Möglichkeiten und Ursachen.' *Forschungen zur Judenfrage*, Vol. 4, pp. 134–162. Hamburg 1940."

[26]The translated title of Thüring [1941/43] cited in footnote † is 'Albert Einstein's attempt to subvert physics and its inherent potentials and causes.' Bruno → Thüring was one of the most active Nazi ideologists; see, e.g., Litten [1992], pp. 148–151, 256.

[27]An allusion to Immanuel Kant's distinction between phenomena and *Ding an sich* in his *Kritik der reinen Vernunft* (1st ed., 1781 and 2nd ed., 1787), translated by J. M. D. Meiklejohn as *Critique of Pure Reason*, London: J. M. Dent, etc., 1991 (first published 1934).

[28]See also footnote 16 above for Gehrcke's pamphlet based on this suspicion.

[29]Sommerfeld's theoretical calculation of the relativistic corrections for hydrogen's fine structure and subsequent experimental confirmation by F. → Paschen count as a decisive verification of relativity theory and quantum theory; cf., however, Helge Kragh: 'The fine structure of hydrogen and the gross structure of the physics community', *Historical Studies in the Physical Sciences* **15** [1984/85], pp. 67–125.

[30]Also known as de Broglie waves, in which every homogenous stream of particles with momentum p can also be attributed a wave-length $\lambda = h/p$. Cf. Wheaton [1983], pp. 286ff., and doc. 45, footnote 7, for the experimental verification of this concept.

approval which is naturally also responsible for the questionable international 'fame' of some of our relativity and quantum physicists.§[31] There is no doubt that wherever English examples are quoted, put forward and copied, German scientists who have said the same and better in our own equally worthy language are being stifled or passed over in silence; and if in this country verification of foreign accomplishments had only been postponed, there is no doubt that in all fields German science would not have made minor progress, but in many fields substantially more progress, because they would have been less hampered. This does not mean that we are now justified to take away the scientific standing of the great classical English or Western scientists of Nordic origin like Newton, Faraday, Lagrange and others, only because they spoke a different language; nor can they now be replaced with arbitrarily chosen German representatives. Rather, a scrupulous historical investigation must be conducted to find out where the real priority of discoveries lies and whether a foreign active and greedy propaganda had only been attempting to belittle, conceal or push into the background German scientific accomplishments to the unwarranted advantage of foreign scientists.

III. New Reponsibilities for Research in Theoretical Physics

When looking at the situation described here and applying German standards, it seems that an inevitable reorientation is absolutely necessary in theoretical physics, especially after the end of this war. We have now arrived at a turning point; the conviction must take the upper hand that theoretical physics's former danse macabre into nebulous, empty formalism cannot continue. The narrow-minded and conceited view of progress [*spießerhafter Fortschrittsdünkel*] of the slant of popular science worldview chatter is least appropriate. What we need is scientific modesty and critical self-examination. In the first place universities should focus on guiding students to the fundamental questions and on analyzing more closely the epistemologically related border issues in mathematics and physics, while taking into account the historical course of scientific development. They should not halt scientific progress by encouraging students to work on irrelevant isolated projects associated with the doctrines based on faith [*Glaubenslehre*], merely as a means of acquiring an academic title. Research of recent decades in physics in any case is purposefully ahistorical, because it must be afraid that it cannot survive historical scrutiny. In all events it is striking that applications for teaching positions in the history of physics were mostly rejected precisely by the dogmatic leaders. Historical clarification is more important for productive work than simply muddling along on courses set by fashion. Above all, reading the **original versions** of the works by Newton, Maxwell, Euler, Gauß and oth-

§ [Footnote 9) in the original:] "See Ph. Lenard, *Ideelle Kontinentalsperre, Schriften der Reichstudentenführung* Munich 1940."

31 Müller changes the subtitle of the pamphlet cited in footnote § from 'war writings' (*Kriegsschriften*) simply to 'writings' (*Schriften*). See P. Lenard's foreword to this pamphlet, doc. 78.

ers should be a requirement instead of using exclusively the standard superficial textbook extracts, which only detract from the real problems.[32]

Another major area of responsibility involves supplementing experimental findings. Theoretical physics departments that own their own experimental installations and laboratories should aim their research primarily at a completely unbiased and undogmatic analysis of those facts and phenomena that had previously been ravished for the benefit of theoretical prejudice—namely, verification using the most stringent precision methods that modern technology has at its disposal, primarily in experimental optics. One urgent task would have to be, for example, measuring scrupulously under differing experimental conditions the constancy of physical values that had been assumed up to now but had not yet been proven, such as light velocity, the relation of elementary charges to the mass of the electron, and other structural elements of atomistic doctrine [*Glaubenslehre*]. The question of **longitudinal light waves**, usually rejected today as groundless, must and will also play a big role in the future.[33] But in the process science must have enough self-control to delay working out and fixing theory until all the empirical data has come in; and before the theory is laid down science must make unambiguous theoretical explanation a priority. [...[34]]

Theoretical physics is constrained by the stuffy and oppressive atmosphere of mathematical formalism; and it definitely needs fresh air. This reinvigoration can only come from the rallying spirit [*Frontgeist*] in **engineering**, in which incidentally, classical and graphic scientific method is still very much alive today and will also certainly remain so in the future. I myself have been at university and know enough to thoroughly appreciate the science that sprang up from its soil; but I also have worked at polytechnical institutes for over fifty years, at three different ones, that is.[35] I was very much impressed by the mentality in the technical sciences and by the determined seriousness of the working method there. It

[32]History of science, particularly of physics, was taught at very few German universities including Leipzig and Hamburg. Nazi ideologues like the *NSDStB-Führer* Fritz Kubach who represented typical National Socialist historiography did not hold regular university positions in history; see, e.g., Meinel & Voswinckel (Eds.) [1994], pp. 265ff. On Kubach see also footnote 16 in 85.

[33]Modern electrodynamics has excluded the existence of longitudinal light waves (analogous to sound waves) on basic physical grounds since Maxwell's time, only allowing transversal oscillation modes.

[34]The following three paragraphs suggesting expanding the responsibilities of theoretical professorships, increasing their number and linking them closer with applied physics are omitted here. Essentially, Müller was pleading for a complete redefinition of theoretical physics, abolishing most of its current research agenda, which only alienated him further from the physics community.

[35]Müller was associate professor at the Polytechnic in Hanover in 1928 and became associate professor at the Prague Polytechnic in the same year. In 1934 he was appointed full professor at the Polytechnic in Aachen, then at Munich in 1939 to assume the Sommerfeld chair for theoretical physics at the University of Munich in 1940 under very controversial circumstances. Cf. doc. 84, footnote 9, and doc. 85, p. 265.

is subject to the powerful imperative of practicality and technical requirements and thus to the vital needs of the nation; and it is already for this latter reason alone that their method must avoid any superfluous side-tracking. It must be possible to bring into universities even only a trace of this mentality, especially when we note the outstanding successes of our efforts in technical science and defense technology in the German people's present struggle for survival. I am thinking primarily of those branches of science connected to **aircraft engineering**, which certainly belong within the course of study of future instructors and teachers, who are supposed to introduce the rudiments of aviation physics to the impressionable youth. Since aeronautical problems have been imposed as a school requirement, even by decree of the [Reich Education] Ministry,[36] university lecturers in theoretical physics really ought to finally pull themselves together as well and take this situation into account in their lectures. I made an attempt of this type at Munich University; and I was able to conclude with satisfaction from this experience that a lecture on mechanics and flight aerodynamics fills a real need among students. The motion or the static and dynamic stability of an aircraft certainly is an extraordinarily attractive problem, which is related so variously to almost all branches of theoretical physics: to the dynamics of rigid bodies, the mechanics of continua, hydrodynamics, aerodynamics, the dynamics of gases, and thermodynamics. It can thus be included in physics with the same justification as any other field of the official curriculum, such as spectroscopy or electron optics. In my opinion this problem is in any case more vital than the issue of the motion of an electron, which has neither ever been seen nor ever will be seen by anyone and which can only be imagined very indirectly and vaguely.[37] Other special fields of importance to technology and defense technology could be described similarly, for example, ballistics and artillery measurement methods, which have just as versatile connections to physics as aircraft engineering has.

Beyond this, institutes of theoretical physics should strive to extend their current working methods by searching out contacts with groups working on special projects in industrial physical research. In this way they would not only get new ideas for instruction and research, but also would have the opportunity to serve the general good productively. As a result those theoretical speculations of no use whatsoever to front-line duty, in the wider sense of the word, would then be eliminated or corrected of their own accord.[38]

---

[36]The Education Ministry (→ *REM*) issued this decree on Nov. 17, 1934: 'Zur Pflege der Luftfahrtforschung an den Schulen' followed by a series of implementation regulations on the creation of aeronautical work groups and teacher training courses. Cf. Brämer & Kremer [1980], pp. 62f., 169, 184.

[37]See, however, Hans → Geiger's contribution on experimental physics, in which he considers atomic analysis very much within its scope, doc. 71.

[38]Note the martial vocabulary, as here *Frontdienst*, which was then very much in vogue, imparting a feeling of power but also of the individual's duty and self-sacrifice toward the State.

These are just a few general guidelines for now in reforming instruction in theoretical physics. A more thorough discussion of them within a wider context must be reserved for a later time and for a more extensive analysis. Whoever was actively involved in the developments of the last decades, but more specifically whoever consciously informed himself of the latest efforts of the German renewal of recent years, will **not be able to ignore the necessity of a new orientation of our scientific conscience.**

IV. Science and the Turn of an Era

Today we are at a turning point of an era, at the dividing line between two epochs that have compared strengths once again in this war, surely for the last time in Europe. The world now under Western Jewish influence is in the process of breaking apart to make room for a new European order, a new nationalistically and socially based attitude to the world [*völkisch-sozialen Weltgefühl*]. This is obviously not purely a political conflict nor only the succession of two political systems, but a sweeping global change with at the same time the deepest cultural consequences that penetrate into all branches of intellectual life and can make no exception of science. This is because science is not an isolated cell of human effort or human will, even though the outward appearance of scholarly life could often support this impression. Science is rather the conceptual expression of a people's will to live and is thus somehow based on and oriented toward ideology. It is therefore also inevitable that the victory over Western Jewish plutocracy and democracy finds its expression in the new scientific orientation; and thus it must succeed. Unlike National Socialism's domestic political victory over German defeat and over the party state, science may under no circumstances stand back and let itself be taken by surprise and caught off its guard, only to become aware retrospectively of its responsibility, for which it was prepared neither by instinct nor through practice. Previous events must serve as a warning. We live in the middle of a new **time of struggle**; and we scientists also should not allow it to slip by wasted. Rosenberg says on one occasion: "What happens after fifty years is not decisive for the victory of any ideology, the first ten steps made on the way to the goal are what are critical".‡[39] Rosenberg also speaks of the entrance of the German Army in Paris in June of 1940 as marking an end to the racial defilement of Europe [*Rassenschande*]. This is to be understood not only literally

‡ [Footnote 10) in the original:] "From Alfred Rosenberg's speech, 7th November 1934, at the opening of the w[inter] s[emester] at Munich University."

[39]The summarized newspaper version of this speech on scientific freedom quotes Alfred → Rosenberg as follows: "Our ideological fight will also be of service to all other nations; though spurned today, after only 10 years it will be admired." 'Freiheit der Wissenschaft. Eine Rede von Alfred Rosenberg', *Münchner Neueste Nachrichten*, No. 306, Friday, Nov. 9, 1934, p. 2. For a report on the opening ceremony of the 1934/35 academic year organized by the Munich University Student League, see also 'Student der Bewegung – Freiheit der Wissenschaft. Feierliche Semestereröffnung des NSDSt.', *Münchener Zeitung*, No. 312, Thursday, Nov. 8, 1934, p. 3.

but also symbolically; at the same time it outlines a complete program for science that the responsible leaders at our universities are entrusted to carry out. It is intellectual sabotage of this duty to demand exclusive rights for science that stem from the closing epoch of European values. It is necessary to drop the last habits and methods of this sentimentalistic and liberalistic era in which no racial law existed; and instead of evading the new fate along old paths, we must concentrate on the ultimately decisive **imperative of blood**, which lies deeper and extends further than the reach of belated formal lucidity.[40] The exact sciences must also finally join those racially aware individuals striving to root out all racially alien elements; and they must also be committed ultimately to reassigning everything its destined place and to reinstating the laws that had been pillaged by a racially alien dialectic which are anchored in the order of the universe.

84 Ludwig Prandtl: Letter to Reich Marshal Hermann Göring [April 28, 1941]

Source: Prandtl Papers, Göttingen. Initialed carbon copy of typed letter from: 'Professor Dr. L. Prandl, Göttingen, Calsowstr. 15' addressed: 'An den Beauftragten des Vierjahresplanes, Berlin'; marked personal and sent by registered mail with file reference: 'Pp.314/41', and enclosure. Salutation: 'Hochzuverehrender Herr Reichsmarschall!'

[Re:] Averting a serious threat to the rising generation of German physicists.

Esteemed Reich Marshal,

At German universities recent developments in physical instruction have led to a situation that harbors very serious dangers for the rising generation of leaders in this field. If it were to persist, it would inevitably lead to the inferiority of Germany in this militarily and economically important subject, and it would place us in a completely hopeless position especially against the American competition. The fact that you have so kindly remembered my birthday already many times now gives me the courage to present to you in more detail the above circumstance, particularly since clearly only your personal intervention could alleviate this situation.

To put it briefly, the matter centers around the fact that a group of physicists, who unfortunately have the *Führer*'s ear,[1] is denigrating **theoretical physics** and disparaging the most deserving theoreticians. Moreover, it is capable of carrying through completely intolerable appointments to professorships, etc., and

[40]As the spiritual leader of the National Socialist ideology, Rosenberg adapted Arthur Comte de Gobineau's and Houston Stewart Chamberlain's racial doctrine to the anti-Christian neoromantic 'myth of blood'. On the intellectual roots of this pseudoanthropological racism see, e.g., Weingart, Kroll & Bayertz [1988], part II, or Bäumer-Schleinkofer [1990].

[1]This alludes to Philipp → Lenard's personal connections to A. → Hitler, who saw him as one of the very few scientists to openly support the National Socialist movement in the early years (see Lenard's and Stark's joint statement of loyalty, doc. 3); cf. also the Introduction, sec. 5.3.

does this under the pretext that modern theoretical physics was a Jewish pretence that could not be eradicated soon enough, with the intent of replacing it with 'German physics'. I have elaborated on what is of consequence in this regard in an attachment enclosed herewith.[2] In any case it is indisputable that theoretical physics is an indispensable subject precisely in the education of the next generation of physics leaders. Its function is, of course, to arrange logically all physical data and to develop out of this the conformities with natural law. In turn, the **technical** physicist can use these laws as an aid in designing his reconstructions systematically to predict their effectiveness. A physicist's education that omits theoretical physics can generate good handymen [*Handlanger*] but never leaders with a proper command and overview of the whole field.

The members of the aforementioned group of physicists led by Professor Lenard share the characteristic of lacking the ability to assess mathematically based material acutely and critically.[3] Because of their inability to judge papers critically due to the requirement in modern theoretical physics of considerable mathematical proficiency, they are also unclear as a result about the indispensability of this field. In contrast to these individuals, all really authoritative physicists who possess the required mathematical skills unanimously recognize that theoretical physics is essential. Honorable Reich Marshal, I would like to request that, based on my representations, you arrange an interview with two well-known physicists employed in industry: Professor Ramsauer, Director of the research laboratory of *AEG*[4] and Professor Joos, Chief Physicist at the Zeiss Works.[5] Their positions assure that they are both sufficiently unaffected by the terroristic behavior of the Lenard group.[6] I myself would prefer to remain in the background in this matter, since it can be argued that I am not a physicist by profession (although I took the examination for a doctor's degree in physics in my youth and have always followed attentively the subsequent developments in physics, so that I do not lack expertise!).

When clamoring that modern theoretical physics is a Jewish pretence, the group of 'German physicists' referred to have the ear not only of the Science

[2]The following doc. 85 was enclosed with this letter.

[3]For a list of Lenard's followers, see Hentschel [1989].

[4]Carl → Ramsauer was director of the research division of → *Allgemeine Elektrizitäts-Gesellschaft*, the major German electric company from 1931 to 1945.

[5]Georg → Joos acquired this position in 1941 and was in the management of the → *Zeiss-werke*, a leading optical research company in Jena, until 1945.

[6]Civil servants, which include professors and teaching staff at universities, were vulnerable to pressure and actions by the Ministry of Science and Education (→ *REM*).

Ministry,*[7] but also of the student leadership.[8] As already indicated, this group launches vile attacks on the most worthy and reputable advocates of theoretical physics and disaffects students from this field as well. Recently it has also succeeded in pushing through a downright scandalous new appointment, which can only be described as completely senseless—unless perhaps the 'sense' is to destroy.[9] The attached description gives more specific information on these matters as well.[10] The struggle to put an end to the trouble is hopeless without winning over the *Führer* in this matter; and that is why I ask you for your personal intervention.[11]

Heil Hitler!

Always faithfully yours, L. P[randtl].

85 Ludwig Prandtl: Attachment to the Letter to Reich Marschal Hermann Göring [April 28, 1941]

Source: Prandtl Papers, Göttingen. Signed carbon copy of typed document with handwritten emendations: 'Anlage zum Brief vom 28.IV.1941 - Pp.314/41 - an Herrn Reichsmarschall Hermann Göring'.[1]

I. On Theoretical Physics

The purpose of theoretical physics is to erect a logically irrefutable theoretical framework upon which a classification of observational data can be established that offers the most accurate single explanation possible of even distantly related factual findings. The laws resulting from this work can then also serve in their turn to predict new experimental data. A theory should contain no logical contradictions on its own; and its implications should do justice to the facts. Depending on the hypotheses underlying the theory, several admissible theories can also exist for one and the same group of factual findings. Now, if a new fact is observed that is in conformance with one of these theories but not with another, however, then this other theory must be abandoned. This was the case at the turn

* [Original handwritten footnote *):] "*Postscript* End of May 1941. This comment was erroneous, and is corrected in my letter of the 27th of May 1941 to [the Second?] Secretary of State at the office of the Commissioner of the Four-Year Plan."

[7]At the *REM* not only the minister, Bernhard → Rust, but also many of the lower-ranking functionaries like Theodor → Vahlen and Rudolf → Mentzel were National Socialist hardliners.

[8]The → *NSDStB* was firmly set on an anti-Semitic and pro-'Aryan Physics' course through Lenard's followers, which included the Nazi officials Fritz Kubach (on Kubach see footnote 16 of the following document) and Bruno → Thüring.

[9]This is a reference to Wilhelm → Müller's appointment as successor to Arnold → Sommerfeld; cf., e.g., Cassidy [1992]a, chaps. 18 and 20f., as well as Beyerchen [1979], chap. 8.

[10]See the following document.

[11]However, Prandtl had still not heard from H. → Göring over a month later (see doc. 86).

[1]Enclosed with the preceding doc. 84. A modified excerpt of this attachment was also sent by Carl → Ramsauer to Minister → Rust, see doc. 90.

of the century with the observational results supplied by the so-called Michelson interference experiment. This experiment should have proven a statement on the relative speed of the earth to cosmic ether,[2] on the basis of the Newtonian space-time conception that had been considered irrefutably correct until then. But despite numerous sophisticated repetitions, the results were always negative (instead of a shift of the interference bands, no shift). Thus the Newtonian space-time conception was proven fundamentally incorrect. However, the deviations are so slight for all terrestrial processes that this old space-time conception could still continue to be used practically. At that time H. A. Lorentz in Leiden was able to show that the Maxwell electrodynamic equations do not lead to any contradiction with the new experimental findings, therefore, that no contradictions arise if the physical world is understood as being composed of electrodynamic processes.[3] From a scientific standpoint a clearly formulated space-time conception was now needed that would take the new facts fully and accurately into account. A. Einstein was the first to supply one such conception.[4] His system is without inner contradiction; yet in so saying, it is naturally not the only possible solution, but probably the simplest.

In the face of this situation, today the Lenard group is hiding its head like an ostrich in the sand and is clinging to the Newtonian space-time conception, even though it has been shattered long ago by the Michelson experiment.[5] Though the Einsteinian space-time conception is recognized by all clear-thinking physicists throughout the world as the best solution for the time being and has long since become a secure foundation of the later developments in physics, in the view of the Lenard group it ought not be true, simply because its author was a Jew. Therefore, in the eyes of the Lenard people, all the very extensive investigations that have been carried out since then by Aryans using new instruments belong within 'Jewish physics'.[6]

The situation is a bit different in the case of quantum theory, which also originated from the beginning of this century. Human knowledge of physics has

---

[2]In the typed document, 'was supposed to prove' (*erweisen sollen*) was corrected by hand to read 'should have proven' (*erweisen müssen*). On Albert Abraham Michelson's (1852–1931) interference experiment, first performed in 1881 and then repeated with increased precision in 1887 together with Edward Williams Morley (1838–1923) see, e.g., Whittaker [1951/53], chap. 13, and Lloyd Swenson, *The Ethereal Aether*, Austin: Univ. of Texas Press, 1972.

[3]Cf. Hendrik Antoon Lorentz's (1853–1928) book *Versuch einer Theorie der elektrischen und optischen Erscheinungen in bewegten Körpern*, Leiden: Brill, 1895, and his paper 'Electromagnetic phenomena in a system moving with any velocity smaller than that of light', *Proceedings of the Academy of Sciences, Amsterdam* **6** [1904], pp. 809–831.

[4]Cf. Albert → Einstein's paper 'On the electrodynamics of moving bodies', the English translation by Anna Beck, *The Collected Papers of Albert Einstein*, Vol. 2, Princeton: Princton Univ. Press, 1989, doc. 23, pp. 140–171, of 'Zur Elektrodynamik bewegter Körper' *Annalen der Physik* **17** [1905], pp. 891–921.

[5]On P. → Lenard and his supporters see the annotations in the preceding doc. 84.

[6]On so-called 'Jewish physics' see, e.g., W. → Menzel's article and W. → Heisenberg's and J. → Stark's responses, docs. 42–44.

developed from the visible world around us. According to current proven knowledge this world is made up of small particles (atoms), which are themselves also built of even smaller components. All visible processes involve incredibly large numbers of such particles, even within the smallest of spaces;[7] and it was thought that a strict regularity existed between cause and effect, from which it was possible to derive a 'principle of causality'. All earlier theoretical physics is built upon such causal interconnecting links between separate values. Science has long since penetrated into the interior of atoms and is dealing with the ultimate constituents, so to speak, of matter (through certain methods these separate objects become individually perceptible); and it has become evident today that here the concept of causality fails and must be replaced with a statistical concept of probability.[8] (With decaying radium atoms, for example, there is no way of specifying when a particular radium atom will explode apart. But the average number that explode in one hour or in one year within a given amount of radium can be specified. Such matters are not restricted to radium but play a part in all atomic processes.) Thus old-fashioned causality proves simply to be the 'law of large numbers', which appears in all kinds of problems on statistical ensembles. The larger the number, the smaller are the random deviations from those expected from the 'causal' standpoint. So in addition to not being able to know in each individual case exactly when and where energy conversion takes place in atoms, atoms also have the characteristic, depending on the kind of conversion, of always converting very specific quanta of energy at once (hence the name quantum theory). With such considerations it was possible to collect an enormous abundance of old and new observational data within a logically uniform system. This also counts as 'Jewish physics' to the Lenard people, perhaps following the motto that I have occasionally heard mentioned: "Whatever's beyond my comprehension, I regard as Jewish pretension".[9] Naturally Jews have also contributed to this part of scientific development, but the bulk of it had been created by Aryans, starting with Planck and Sommerfeld through to Heisenberg and Schrödinger.[10]

[7]'of such particles' (*solcher Teilchen*) was inserted by hand. The order of magnitude of this number is given by the Avogadro number $N_A = 6.022 \times 10^{23}$ atoms per mole (one mole of the O_2 molecule, for instance, is only 32 grams).

[8]This shift from the description of physical processes in causal terms to an intrinsically statistical description was completed in what is known as the statistical interpretation of quantum mechanics. It was advanced by Max → Born, Werner Heisenberg and others after 1925. See, e.g., M. Born, 'Quantenmechanik der Stoßvorgänge', *Zeitschrift für Physik* **38** [1926], pp. 803–827; W. Heisenberg, *Die physikalischen Prinzipien der Quantentheorie*, Leipzig: Hirzel, 1930, chap. 4. See also the contributions in L. Krüger et al. (Eds.) *The Probabilistic Revolution*, Cambridge: MIT Press, 1990, Vol. 1; and Max Born, 'Statistical interpretation of quantum mechanics', *Science* **122** [1955], pp. 675–679.

[9]"Was ich nicht verstehen kann, sehe ich als jüdisch an".

[10]For a brief history of quantum theory and quantum mechanics and the roles played by Max → Planck, Arnold → Sommerfeld, Werner Heisenberg and Erwin → Schrödinger see, e.g., Jammer [1966].

II. Particulars about the Lenard Circle

1. Lenard himself has deservedly made a name for himself with experimental investigations that would probably have also led him to the discovery of X rays at that time, had Röntgen not beat him to it.[11] Mathematical theory never suited him, though, and all would have been fine, had he not started to quarrel about mathematical matters. In his later years he wrote a textbook of several volumes, which he entitled: *Deutsche Physik*.[12] This textbook lacks the necessary clarity and rigor in describing even the simple laws of ordinary mechanics, but might pass as a textbook for the lower level; on no account is it adequate for the upper level.

2. Another exponent of the group is Johannes Stark, the ex-president of the Reich Physical and Technical Institute. I know Mr. Stark from his time as unpaid university lecturer in Göttingen to be a very impulsive and peculiar man.[13] Yet he is a completely brilliant experimenter and made two first-rate discoveries, one of which immortalized his name by being known to the whole profession as the 'Stark effect'.[14] As far as I know, he never concerned himself with theory; but this did not prevent him from extravagantly insulting Messrs. Planck and Heisenberg in unbelievably venomous articles, one of which appeared in the *Schwarzes Korps* a number of years ago.[15]

3. There are a number of sympathizers in this group who generally have experimental leanings and who do creditable work in their own way. But a fair number of people among them also consider themselves unappreciated geniuses, who could not improve their position in earlier times due to their insufficient achievements but who have naturally seen a glimmer of hope following the National Socialist revolution. Some of these participate very extensively in the reformation of the philosophical foundations of physics, especially regarding the

[11]Applying Hertz's discovery that thin metal sheets transmit cathode rays, Philipp Lenard constructed a gas-discharge tube with a 'Lenard window', as it was later referred to, which directed the cathode rays outside of the tube. Lenard had provided Wilhelm Conrad Röntgen (1845–1923) with the fine sheet metal necessary for the latter's discovery of X rays. Lenard felt that his contribution should have been better acknowledged in Röntgen's articles: 'Ueber eine neue Art von Strahlen', *Sitzungsberichte der Physikalisch-Medizinischen Gesellschaft in Würzburg* **137** [1895], pp. 132–141, and **138** [1896], pp. 11–17, and insisted on using the term 'high-frequency radiation' for X rays instead of the generally accepted term 'Röntgen rays'.

[12]See Lenard's forword to his physics textbook on German (i.e., 'Aryan') physics, doc. 39.

[13]J. → Stark habilitated in Göttingen in 1900 and worked at the physics department until 1906. He was president of the *Physikalisch-Technische Reichsanstalt* (→ *PTR*) from 1933–39. See Lenard's celebration of his nomination to that office, doc. 18, and von Laue's retrospective account, doc. 19.

[14]This phenomenon is described in the title of Stark's paper announcing the discovery: 'Observation of the separation of spectral lines by an electric field', *Nature* **92** [1913], p. 401. See also his book *Elektrische Spektralanalyse chemischer Atome*, Leipzig: Hirzel, 1914.

[15]See here docs. 55–56 for the anonymous article followed by Stark's supporting comment. On Heisenberg's response to this aggressive attack see doc. 55, footnote 13; doc. 40, footnote 19.

space-time conception. However, the Michelson experiment plays absolutely no part for all these people—in other words, they still adhere to the Newtonian space-time conception.*[16]

4. The matter of the successorship to the world-renowned theorist Sommerfeld of Munich, to whom we are indebted for organizing the laws of spectral lines clearly, let alone many other things, is extraordinary. This successor by the name of Wilhelm Müller was formerly teaching assistant and later professor of technical mechanics, the former in Hanover, the latter in Prague and Aachen; and so far, he has mainly dealt with fluid dynamics and aerodynamics and has also written textbooks on branches of these fields.[17] But he contributes nothing, absolutely nothing, to theoretical physics. Instead of lecturing on what physicists would need for their education, that is, on electrodynamics, electron theory, optics and radiation theory, thermodynamics, mathematical statistics and partial differential equations in physics, according to his own statements he lectures on aerodynamics and other mechanical engineering topics. It cannot be denied that it is useful for university students to learn something about these things as well; but if a really essential part of physics is therefore withheld from them, then the resulting situation can only be described as a sabotage of necessary instruction. Mr. Müller

* [Footnote *) on p. 6 of the original:] "See various articles, some venomous ones among them, by Kubach, Thüring, Dingler, among others in *Zeitschrift für die gesamte Naturwissenschaft*, the subtitle of which identifies itself as the 'Organ of the Reich Science Specialty Division of the Reich Student Leadership'. (On the verso of the title page, [see] the listing of the editorial committee!)"

16 Fritz Kubach (1912–1945), mentioned in footnote * above was a historian of science and → *NSDStB* official. In 1943, he published a biography in which he claimed the famous astronomer Copernicus to be a 'Great German', disregarding that he came from Thorn (Eastern Europe) and had studied in Cracow and Italy: *Nikolaus Kopernikus - Bildnis eines Großen Deutschen*, Munich: Oldenbourg. He was subsequently engaged in preparing the Copernicus edition. 1931–35 Kubach studied mathematics at Heidelberg. He became a member of the → *NSDAP* and the → *SA* in 1933. 1934–36 he was the main official science director for the Heidelberg Student League and deputy regional League leader in Baden. 1935 he was promoted to assistant at the regional observatory in Heidelberg and was made honorary member of the → *Gestapo* in the same year. In 1936 he transferred from the *SA* to the → *SS* and became until 1937 head of the Science and Professional Training Office. From May 1941 Kubach served in the *Wehrmacht* where he was promoted to lieutenant. 1944–45 he was chief of staff of the Lecturers League (→ *NSDDB*) and deputy director of the Trainee Office in the → *RFR*. He was missing in June 1945 and declared dead in 1947. On Kubach see Grüttner [1995], p. 509. In 1937 Bruno → Thüring became one of the editors of the → *Zeitschrift für die gesamte Naturwissenschaft*, which was published by the *NSDStB*; and Hugo → Dingler was one of its key authors, contributing to the anti-Semitic and antitheoretical impact of this ideological journal.

17 See also doc. 73 on W. → Heisenberg's candidacy to Arnold → Sommerfeld's chair in theoretical physics at the University of Munich in 1939. W. → Müller's textbooks published at this time include: *Dynamik*, Leipzig: de Gruyter, 1925 (2nd ed., 1952); *Mathematische Strömungslehre*, Berlin: Springer, 1928; *Einführung in die Theorie der zähen Flüssigkeiten*, Leipzig: Akademische Verlagsanstalt, 1932; and *Einführung in die Mechanik des Fluges*, Leipzig: Jahnecke, 1936 (2nd ed., 1942, 3rd ed., 1953), which reproduce his lectures held at the German Polytechnic in Prague between 1929–34.

discussed his program in detail in the November-December, 1940 issue of the *Zeitschrift für die gesamte Naturwissenschaft*, pp. 281–298.[18] In this article, which has the title 'The State of Theoretical Physics at Universities', he did not miss the opportunity to say a lot of unfriendly things about the leading German theoretical physicists. Otherwise, the physics credo [*Glaubensbekenntnis*] of the Lenard group was given a generous hearing. One can read there that "a mysterious group of authorities" .....[19] on the side of theorists forces "an almost mystical commitment to a specific program", and "threatens anyone who dares to express his own opinion independently from the acknowledged clique." (I must remark on this that the "mysterious authorities" are nothing less than the experimental facts, and that regarding the threat to those of a different opinion, we are only dealing with the justified repudiation of cranks who are not willing to face the facts.) Incidentally, at the moment the Heling publishing house, Leipzig, is advertising a pamphlet, *Jüdische und Deutsche Physik* by Messrs. Stark and Müller,[20] in which a most pointed challenge is made "against a dogmatism still dominated by the Jewish spirit that does not serve truth and knowledge but strives to do violence [*vergewaltigen*] to nature and to reduce it [*erniedrigen*] to a mere servant of formulas". It also supposedly presents the "fundamental difference between Jewish or Jewish-influenced theoretical caprice and German pragmatic theory" through examples "that do not tamper with [*vergreift*] the given order of phenomena but that endeavor to explain this order with the simplest possible, concretely comprehensible system of causal laws".†[21]

I think that these text samples suffice to characterize adequately the attitude of the Lenard circle.[22]

---

[18]See Müller's article, here doc. 83.

[19]Müller refers specifically to a 'limited circle of experts' (*engeren Kreis von Autoritäten*) and an 'esoterical group of authorities' (*esoterische Gruppe von Autoritäten*).

[20]The pamphlet on 'Jewish and German physics', published in 1941 was edited by Müller alone, containing opening speeches by Müller and Stark at the colloquium of theoretical physics at the University of Munich. The *Helingsche Verlagsanstalt* in Leipzig (like Lehmann's in Munich) was one of the key publishing houses for Nazi literature, particularly literature on 'racial hygiene'; see also Müller's *Kampf in der Physik*, ibid., 1944.

† [Footnote ∗) on p. 8 of the original inserted by hand:] "*Note* (November 1941). The book has since appeared. It contains absolutely no positive contributions of any relevance to physical knowledge. P[randtl]."

[21]These are not direct quotes from the pamphlet but close paraphrasing. Their vocabulary carry a violent parallel message: 'vergewaltigen, erniedrigen, vergreift' can also mean 'rape, humiliate, and assault', illustrating the extreme language frequently used in this time also by Stark.

[22]The term 'Lenard circle' mimics Lenard and his followers' own allegations of the existence of Sommerfeld and Einstein cliques, as does the term 'credo' (*Glaubensbekenntnis*) in the preceding paragraph.

Fig. 17: Ludwig Prandtl and Carl Ramsauer.

86 Carl Ramsauer: Letter to Ludwig Prandtl [June 4, 1941]

Source: Prandtl Papers, Göttingen. Signed typed letter on printed company stationery: 'AEG Forschungs-Institut, Berlin-Reinickendorf/Ost 1, Holländerstraße 31–34'. Addressed to: Herrn Professor Dr. L. Prandtl, Göttingen, Calsowstr. 15.' Penciled file reference: 'P.g401 R1'.

Dear Colleague,

I did not answer your friendly letter of the 28th of April, 1941 earlier, because I wanted to maintain my complete impartiality toward the inquiry by the Reich Marshal, which I expected very shortly.[1] But now the matter seems to be dragging out longer after all.

I would like to express my sincere thanks to you and am very pleased that you have taken the initiative in this important question. I myself am **very** willing to oppose the menace to theoretical physics in principle and on behalf of industry.[2] My relations with my old teacher Lenard will certainly be completely broken off in the process; but they have already suffered considerably anyway, since I have always maintained my own opinion independently of his.[3]

As the newly elected and confirmed president of the German Physical Society, which according to the new regulations is vested in very far-reaching powers, I

[1]L. → Prandtl had also written to Hermann → Göring on the same day: See doc. 84.

[2]'in principle and' (*grundsätzlich und*) was inserted in Ramsauer's hand.

[3]Carl → Ramsauer, the director of the research division of → *AEG* from 1928 to 1945 had been Philipp → Lenard's teaching assistant at Heidelberg in the years 1907–09.

have now selected Prof. Finkelnburg from Darmstadt (or Strasbourg) as vice-president, primarily because he personally advocated theoretical physics in a very commendable manner.[4] Besides, I considered it necessary to have a moderate party member on the board of trustees, in order to be able to cope better with the extreme-minded colleagues.

With best regards and greetings,

Yours very sincerely, C. Ramsauer.

87 Pascual Jordan: Revolution in Science [September 1941]

Source: Pascual Jordan, 'Naturwissenschaft im Umbruch', *Deutschlands Erneuerung*, Gustav Borger, Harold Steinacker, Otto Streck, Robert Wetzel (Eds.), Munich and Berlin: J. F. Lehmanns Verlag. Vol. 25, Issue 9, September, 1941, pp. 452–458.[1]

1. The task of espousing the correct position within the framework of the new Reich was partly more difficult for the sciences than for the humanities. The latter received directly applicable criteria and productive guiding ideas—in the presentation of history, for example—from the National Socialist world view, which at once set sharply defined historiographical objectives oriented towards new values. In contrast, no direct reference points existed for ideologically reevaluating electron physics, the mathematical problems of higher algebra, or enzyme chemistry. The question of how these and the numerous similarly placed fields of research are to be pursued in the future within the new spirit has prompted long-standing diverging debates, proposals of the most diverse and contradictory solutions, and endless misunderstandings. But the reality of National Socialist development strode unwaveringly past all theoretical debate. The dominating events in the young Greater-German Reich's struggle—Four-Year Plan, armament, and war[2]—have shown scientific research its secure place in the life of the nation; and German science—German chemistry, German physics, German geology, etc.—filled this place well: 'German Science' lies behind the amazing victories of the National Socialist Armed Forces as one of the shaping preconditions of the predominance of German arms, shoulder-to-shoulder with the German worker.

In any attempt to outline the position and importance of German science this historical fact must be acknowledged as the actual pivotal point at which the relationship between National Socialism and science is fixed—though not

[4]Ramsauer was appointed chairman of the *Deutsche Physikalische Gesellschaft* (→ *DPG*) in 1940. Wolfgang → Finkelnburg became associate professor at the Polytechnic in Darmstadt in 1936 and at the University of Strasbourg in 1942.

[1]P. → Jordan, who had made important contributions to quantum mechanics and quantum field theory in the 1920's, is identified here as member of the National Socialist German University Lecturers League: '*NSDDB*, Professor of Theoretical Physics, Rostock University; currently Armed Forces'. On Jordan's activities and political position during and after the National Socialist period, see N. Wise in: Renneberg & Walker (Eds.) [1994] and Beyler [1994].

[2]Cf. footnote 1 of doc. 52 for references on the Four-Year Plan.

only in an 'external', merely 'practical', or other secondary sense. Rather, it must be understood in the most serious, deepest sense possible—since, there can be no more important topic, even for our philosophizing, than that over which our science also is fighting full force today: German survival. With this view—as with every aspect of the National Socialist world view—we are obviously in sharp contradiction to Liberalistic thought.[3]

Although the magnificent technological developments which are based upon the sciences had attained the most sweeping proportions already in the period of Liberalism, nevertheless, the Liberalistic standpoint included soundly rejecting the idea that the value of science is based upon its technical 'application'. A certain moral exasperation was behind the rejection of this idea, which concealed Liberalism's bad conscience about having made possible and having sanctioned the improper use of technology: The awe-inspiring magnitude of the technical achievements themselves was set against a plutocratic power system, which not only allowed the fruits of this technological development to go solely to the benefit of a narrow class of capitalists, but furthermore, also applied the technology in the service of unrestrained plutocratic profiteering, which is a destructive force throughout the whole world. In this situation it was inevitable that placing technology and 'pure' science on the same level or even seeing science at the service of technological development would appear to be disreputable 'utilitarianism'. While fearfully suppressing thought of technological repercussions, the true value of scientific research was sought in the 'purely intellectual', whose dignity supposedly could only be preserved through this disregard of the consequences. By virtue of its seeking to separate the mere notion of technological utility from its scientific ideal, the pacifist-Liberalistic ideology only gained more resolve in pushing aside any thought of war technology development arising out of the results of science. This happened as such a matter of course and with such persistence, that today we have hardly even become aware of the glaring contradiction that exists between the Liberalistic harmless image of science and its actual politico-military impact.

In the era of National Socialism neither technology as such nor its military utilization are a matter of bad conscience for us. We now see matters as they really stand: Our outlook's newly won impartiality lets everything fall automatically into its natural place. That this has not only 'practical' bearings, as already stressed, which must be accepted to some extent as a necessary evil but which can lay no claim to consideration in a philosophical evaluation of science, could be discussed in more detail in connection with the epistemological state of modern physics—indeed a stimulating and significant assignment!—but can only be mentioned here in passing. Physics's intrepid and successful advances into the realm

[3]Jordan's style is characterized by militaristic metaphors. 'Liberalism', or excessive attention to conflicting and minority points of view, was a favorite target of the National Socialists, which they blamed for the vacillation and inherent weakness of the previous Weimar government.

of the unknown throughout the last few decades in its campaign all the way to the atomic nucleus have placed us physicists before such astonishing, totally novel circumstances, that most thorough thought had to be given to the potential, responsibility, objectives, purpose and limits of physical knowledge, in order to be able to master the problems. Yet one epistemological interpretation has stood the test against precisely the most difficult, initially apparently virtually hopeless cases. This radical interpretation sees the sense and purpose of forming physical concepts as creating orientation guides to experimentally observable reality. But this interpretation (developed primarily by German physicists) immediately leads logically to evaluating scientific research as an expression of the 'drive for power', which presses for technological domination over nature, and lately, for the display of power through weapons technology.[4]

2. In the statements of the last few years, it has been made sufficiently clear and hardly needs particular emphasis anymore that the new trend toward treating 'applied research' [*Zweckforschung*] on an equal level with 'pure' research cannot lead under any circumstances to henceforth considering pure research as comparatively unimportant. Even if, in the extreme case, the evaluation of science and its individual subfields were to be adjusted exclusively to its potential for technological exploitation, only the most short-sighted approach could fail to recognize that pure science must precede applied research and that the most crucial technological successes have resulted from the evaluation of precisely those findings which had to be acquired without any direct consideration for their practical usefulness, since potential technological applications could only be discovered afterwards but could not have been anticipated in advance.

On the contrary, the 'militarization' of science, which in the age of total war enlists more and more scientific fields of research, requires that science reorientate itself and make trenchant changes in the way it functions and in the prevailing working conditions. Although the subject to be broached here eludes by its very nature exhaustive and concretely detailed public comment, a few basic comments must nevertheless be devoted to it, since nowhere else is the break with the Liberalistic ideology of science expressed so bluntly and unyieldingly as here.[5]

Unrestricted publicity was an obvious vital element of Liberalistic science. Though the individual researcher tended to envelop his current project with some secretiveness (usually to the amusement of his colleagues), it was in any case taken for granted that the final result belonged to the public. The work had reached its culmination and completion upon publication; and the results presented in a professional journal (or in a book), available to everyone, gave other scientists

[4]Here Jordan touches on the pseudo-Darwinian concept of the 'struggle for survival' which was also taken as an argument for the genocide and other crimes against humanity committed by the Nazis in their effort to 'exterminate' what they chose to consider as 'less well adapted' and 'inferior' members of the human race.

[5]On the secrecy of defense technology see also, e.g., doc. 75. On Allied scientific research on similar issues see, e.g., Baxter [1947], as well as the Introduction, sec. 5.4.

fresh impetus and links to further research of their own. Though capitalistic manufacturing interests required keeping individual findings in applied science secret, these were nevertheless too remote from the affairs of pure research—in the strict separation of pure and applied science—to be able to cause a perceptible restriction on the principle of publicity in science.

Surely many a scientist longs still today for the return of the 'good old days' of this unrestricted public access to scientific findings; it is often believed or hoped that the end of the war will bring it back. In reality, however, the liquidation of this former state of affairs, which already has largely taken place during the war, will certainly be an essential and irrevocable step in the total elimination of the Liberalistic era. No matter how the future may develop, there is no going back in global historical development. The totality of war, which is progressively including all areas of the economy and of technology, cannot be diminished in the future but can only become more keenly pronounced; and although the post-war period will, of course, surely be a time of peace, it will nevertheless be a time of constant preparedness. This same development, which causes scientific research laboratories to become increasingly the nerve centers and brain cells of the national fighting force, must prohibit with growing firmness that the results of their work be revealed to the enemy through unrestricted publication. Thus it is to be expected that there will be no halt to the trend in which not only applied research, but also increasingly comprehensive fields of pure research, will be removed from public view.

This development causes—in the post-war period to a fundamentally even more widespread extent—objectives of great national and political importance to emerge, which are at the same time missions of the most delicate nature and of the greatest consequence. These missions will surely find their solutions through experiment, not on the basis of theoretical programs; but it can never be too soon to consider their basic direction, which requires the most thorough deliberation possible.

In the competition in science, which after the war (as it does already now during the war) will constitute an essential component of the struggle between states, that nation which best understands how to organize the exchange of up-to-date scientific results will gain a considerable advantage. A hermetically isolated research autarky would most damage that nation which imposes this restriction upon itself. The new goal cannot be a halt to international scientific exchange, but rather, the replacement of the current indiscriminate give-away of scientific results with a systematically regulated mutual exchange. In its execution, scientific, technological, national economic, cultural propagandistic, politico-diplomatic and military aspects claim equally well thought-out, weighed consideration.

In tackling these problems on a large scale, where techno-scientific, politico-diplomatic and military interests become increasingly tightly intertwined, a serious difficulty will have to be addressed: the traditionally held apolitical attitude of most German scientists, which tends to be especially prominent among the creative researchers in science in particular.[6] This attitude is also an inheritance of times past and will have to give way to the demands of the present. On the other hand, the future developments mentioned above will place the authorities and individuals outside of science collaborating with science before the difficult and increasingly responsible task of being able to distinguish, despite their lack of professional competence in science, the real, creative researcher from egoistical careerists, who seek to conceal the deficiency in their own scientific achievements with superficial self-important and officious zeal [*Gschaftelhuberei*], and who possibly know how to present themselves impressively through reciprocal personal recommendations.[7]

3. But National Socialism will not be content with harnessing German scientific research; on the contrary, in appreciation of the fundamental importance of this research work for the life of the nation, it will provide German science with impetuses from its own energy and conception, out of which the dawn of a new, great epoch of development can be expected. Yet by no means is it only a matter of allocating substantially increased amounts of material support for research purposes. It can be considered a foregone conclusion that after the end of the war, the lead apparently irretrievably lost particularly to American science, as far as available funding is concerned, will be regained.[8] The injection of energy that the essence and spirit of National Socialism will contribute to scientific work will be of much greater importance and significance. This current situation must be explained very briefly, in order to illuminate how much scientific research with its present objectives is virtually expecting and depending upon the intervention of National Socialism.

The most obvious characteristic is the multiple confluence of previously distinctly separate fields, which through the natural course of their development are drawn to close interaction with one another or are dissolved point-blank within a higher unit. Thus out of physics, which deciphered the mysterious world of atoms, arose a superior science, which describes the laws of matter so uniformly from the chemical and the physical point of view that the traditional distinction between 'chemistry' and 'physics' gradually only retains historical importance. But today it is everywhere precisely in the 'border areas', in the zones of contact between different branches, that the most active work is taking place and the most fruitful results are appearing—similarly to how 'physical chemistry' had

[6]See, for example, Werner → Heisenberg's refusal to sign the proclamation of allegiance to the National Socialist cause in the early years, described in the *SS* personal evaluation on Heisenberg, doc. 73. The traditional apolitical self-image of German academicians is analyzed in depth in Ringer [1969].

[7]This passage hints at resentment against the success of his colleague Heisenberg, who was known for his ambitiousness. Heisenberg had previously co-authored papers on the foundations of quantum mechanics with M. → Born and Jordan: See, e.g., 'Zur Quantenmechanik II', *Zeitschrift der Physik* **35** [1926], pp. 557–615. Jordan, on the contrary, received only a fraction of the prestige, which may have been due in part to his speech defect.

[8]America's lead in physics is focussed on in C. Ramsauer's petition to the → *REM*, doc. 91.

prepared the way for the complete union of the two fields it had linked together. Thus in analyzing the structure of protein molecules, for example, chemical and the most sophisticated of physical methods are combined with biological ones: Protoplasm and enzyme research, serology and other disciplines of biology have a place and interest in the treatment of this topic. Simultaneously, a powerful trend is running through the most up-to-date and active biological fields of research today—genetics, mutation research, virology—towards merging with the physics-chemistry of molecules and atoms: The analysis of the heredity carrier the 'gene' and its mutations is being continued today with the most modern tools offered by radiation physics (X rays, neutron beams, etc.).[9] Thus, the river of genetic research gushes into the waters of genetic molecular physics.[10]

The most promising perspectives lie, incidentally, in this effectuating fusion of biology and atomic physics. The inflexibility of the 'mechanistic' representation of nature, which had been cultivated by materialistic philosophy, has recently roused biological scientists and thinkers to raise fundamental objections. But the conviction that the wondrous natural phenomenon of organic life cannot be understood through the conceptional means of the mechanistic world view seemed to tear organic and inorganic nature apart. The surprising findings of modern atomic physics touched upon above point here to completely new paths; since, they have shown that electrons, atoms and molecules also already no longer agree with the basic hypotheses of the mechanistic world view.

But even taking aside these connections, which extend into the philosophical, with the intertwining of biology and the most recent physics, an inevitable progress is manifesting itself today, whose vigor is underscored by the fact that even the newest findings of nuclear physics of only a few years ago (neutron beams, artificial radioactivity, etc.) are already being employed as unexpectedly effective instruments of biological research and may also prove to be valuable aids in medicine (cancer therapy).[11]

4. This situation—which could only be outlined most cursorily here—places fundamental and completely novel demands on scientific work and on those working in science. Not long ago it was feared that the extreme specialization of individ-

[9]For contemporary research on the effect of radiation on genes see, e.g., Hans Banes's contribution in: Rajewski & Schön [1948], part I, pp. 66–88.

[10]Jordan himself was very interested in these interconnections, which were then still very speculative—Watson and Crick's DNA model was only published in 1953; see, e.g., Jordan's 'Zur Frage einer spezifischen Anziehung zwischen Genmolekülen', *Physikalische Zeitschrift* **39** [1938], pp. 711–714; his review article on contemporary theories about the duplication of genetic information and reproduction, 'Zum Problem der Eiweiß-Autokatalysen', *Naturwissenschaften* **32** [1944], pp. 20–26; or his more general overview 'Die Stellung der Quantenphysik zu den aktuellen Problemen der Biologie', *Archiv für die gesamte Virusforschung* **1** [1940], pp. 1–20; and 'Zur Quanten-Biologie', *Biologisches Zentralblatt* **59** [1939], pp. 1–39; cf. also Deichmann [1992], pp. 136ff., 148f.

[11]Cf., e.g., Rajewski & Schön (Eds.) [1948], part I, p. 132 on the carcinogenic effects of radiation.

uals, which resulted unavoidably from the inflation of our body of knowledge, would lead to a disintegration of the sciences into miniscule incoherent branches. Developments in the sciences, which led us everywhere virtually by force to a new synthesis of scientific knowledge, have proven this fear groundless. But the objectives now resulting from the—often most unexpected—contact and interaction between the remotest of branches exceed to a growing degree the energy and potentials of an individual, isolated researcher. Since the division of labor and specialization must stand as preconditions to the highest quality work, the 'border problems' in the different specialties' zones of contact and interaction are turning more and more towards teamlike collaboration among specialists of different subject fields. In actual fact nowadays the best accomplishments are made increasingly possible by such collaboration.

This development in which the individual's efforts along with the individual himself give way to the achievement of the community, heralds to the keen eye the coming hour of National Socialism also in the huge battlefield of scientific research. It is, of course, not the case that advancing and probing penetration into nature's secrets gave us the prospect of arriving at an endpoint or a finish line. New, larger problems emerge on the heels of the solved ones; and today we have come so far that the immense tasks confronting us require an unprecedented deployment of means to attack them. Beyond the necessity just touched upon of pooling together personnel from the most diverse subject fields in a comradely, teamlike fashion—a necessity which has already begun to be carried out in practice—the state of affairs demands comprehensive, large-scale research planning. The execution of this research will include a radical increase not only with respect to the material and apparative means, but also regarding highly qualified staff. These necessities arise from the almost incalculable abundance of urgent tasks lying before us today, especially in the overlapping zones of research in biology, chemistry and atomic physics; and carrying them out in the former style of individualistic, isolated projects would involve such an intolerable waste of the most valuable manpower that in doing so, a suitable rate of progress could no longer be maintained.

According to the ideal of Liberalistic 'free' science, in principle every individual researcher chose the problems that attracted him and that in his own judgment seemed promising. Though junior colleagues at larger institutes took their places within the confines of a research program pursued by the institute head, this was usually looked upon as only a transitional stage from which the younger staff members (inasmuch as they stayed in research) moved on as soon as possible to complete independence. But successful analysis of the current problems of today and tomorrow will demand to a growing degree carrying out examinations of equivalent quality on a very much larger scale. Accomplishing these tasks requires appropriate organization of series of researches.

As an example (one out of many) let me mention that in analyzing an enzyme it is important to measure its inactivation[12] so comprehensively by means of monochromatic ultraviolet light, as to obtain an entire 'inactivation spectrum', which provides valuable information when compared against the ultraviolet absorption spectrum of the same enzyme. Only in the case of a single enzyme (urease)[13] is this comparison available today in desirable full detail. Carrying it out for as many of the numerous known enzymes as possible is a task of great urgency. But due to the quantitative scope of the researches required, it lies totally beyond the reach of what can be attained with the previous organizational forms of scientific research.

The unorganizable performance of ingenious individuals will pioneer, bear fruit, and lead in scientific progress also in the future. It should not and must not be stifled by organizational constraints. On the contrary, the highest possible efficiency in their performance should be obtained by organizing scientific research, to make it as easy as possible for leading individuals to set about addressing most energetically and in the greatest breadth those questions they deem of most importance.

If we execute resolutely what is becoming imperative because of the current state of science, and if we materialize what is becoming possible through National Socialism, the result will be not a restriction of ingenious individual achievements, but on the contrary, greater freedom. The collaboration of members of the most disparate fields of study; an efficient approach to the tasks on a large scale, as required in the above; and ambitious planning of far-reaching major thrusts forward into the realm of the unexplored: These are the emerging demands of today, which National Socialism guarantees to meet. It is simply a question of turning **from individualistic to socialistic research**, to be accomplished with the most rational impartiality but also with the boldest resolution—and in full awareness of its immense importance.

88 Carl Ramsauer: Letter to Ludwig Prandtl [October 31, 1941]

Source: Prandtl Papers, Göttingen. Signed typed letter on printed stationery: 'Deutsche Physikalische Gesellschaft E. V., Der Vorsitzende: Prof. Dr. C. Ramsauer Berlin-Reinickendorf-Ost, Holländerstr. 31./34'. Addressed to: 'Herrn Professor Dr. L. Prandtl, Göttingen, Calsowstr. 15.' Marked 'special delivery', penciled file reference: 'P p 734/41'.

Esteemed Colleague,

In my efforts as President of the German Physical Society to act on behalf of German physics and in particular on behalf of theoretical physics, I have succeeded today after a longer discussion with Colonel-General Fromm, Commander of the Reserve Army and Chief of Defense, in convincing him of the serious menace

[12]Denoting a compound which does not rotate the plane of vibration of polarized light.

[13]An enzyme that changes urea into ammonium carbonate, found in bacteria, fungi, etc.

to German physics.[1] He promised to plead for these issues from the standpoint of the Army and is of the opinion that his resources of power are large enough to carry his intentions through in the [Reich] Culture Ministry. In addition, he will assign a suitable high-ranking military man to summarize the Army's interests in physics and to work together with the German Physical Society.

I now have the following request: Will you allow me in this connection to forward to Colonel-General Fromm your letter to the Reich Marshal along with the enclosures (possibly also the enclosures by themselves)?[2] This would be of great help to me. Formally, we could then take the view that you had placed this material at my disposal as the President of the German Physical Society.

I would be grateful for a decision as soon as possible, since the new physics curriculum will be under consideration in the next few days at the Ministry, at which time the role of theoretical physics will also be under discussion. I have succeeded at present in being included in this meeting as the President of the German Physical Society.

With best compliments and greetings, Yours very sincerely, C Ramsauer

89 Walter Weizel: Review of 'Jewish and German Physics' [January 1942]

Source: Walter Weizel, review of W. Müller (Ed.) [1941]a: *Jüdische und deutsche Physik* in: *Zeitschrift für Technische Physik*, Vol. 23, Issue No. 1, January, 1942, p. 25.

Jewish and German Physics. Opening talks at the colloquium of theoretical physics held at the University of Munich, edited by Dr. Wilhelm Müller, 56 p[ages]. Heling'sche Verlagsanstalt, Leipzig 1941. Broch[ure], 1.80 Reich Marks.[1]

Because of the fundamental importance of physics to the sciences as a whole, the Faculty of Science and Mathematics at the University of Bonn examined this work and agreed unanimously with the opinion expressed in the following review.[2]

---

[1]C. → Ramsauer was president of the → *DPG* from 1940 to the end of the war. *Generaloberst* Friedrich (Fritz) Fromm (1888–1945) was Commander in Chief of the Reserve Army and Chief of Armament from Sep. 1, 1939 to July 20, 1944. Fromm was later implicated in the attempted *putsch* against Hitler on July 20, 1944, and executed on March 12, 1945.

[2]For Ramsauer's previous correspondence with L. → Prandtl, see doc. 86. For the latter's correspondence with *Reichsmarschall* Hermann → Göring, see docs. 84–85.

[1]The typewritten draft of Walter → Weizel's review of Wilhelm → Müller's pamphlet *Jüdische und deutsche Physik*, among the Prandtl Papers, Göttingen, was originally titled: 'Jewish and non-Jewish Physics' ('Jüdische und nicht jüdische Physik') and indicates the price 1.50 RM. Other minor discrepancies are omitted. See also doc. 96 which encloses the draft version. Ludwig → Prandtl's description of the pamphlet is at the end of his submission to Hermann → Göring, doc. 85. See also Ramsauer's attachment, doc. 92.

[2]It is rather unusual for a book review to represent the opinion of anyone other than the

In this polemical work the authors attempt to use the National Socialist worldview as a springboard for their opinions on physics which have not been well received in science. The first talk by Müller uses common phrases in objecting to not only the theory of relativity, but also the quantum theory, with which the author is evidently not very familiar. In the second talk Stark first clarifies the distinction between what he considers dogmatic Jewish physics and pragmatic Germanic physics. Right at the outset he presents the Aryans Planck, Heisenberg, and Sommerfeld, and later Schrödinger and Jordan, as the main representatives of the 'Jewish' spirit [*'jüdischer' Geist*], to whom he also associates Bohr and Born.[3] Born was evidently the only purely Jewish representative of the 'Jewish' spirit he could find in quantum theory. The fact that even Stark must list 5 Germans out of 7 founders of quantum theory should convince the open-minded reader of this work more than anything else that the development of this most modern branch of theoretical physics is an achievement precisely of German science. Following this unsuccessful introduction, Stark discusses his own analyses of canal rays. The brief overview he offers reveals the mark of a prominent physicist.[4] In this section, to the reader's surprise, Planck comes back into favor, Bohr in part also, and even Einstein, no less.

It is completely unscientific to suspect quantum theory of being Jewish. This theory will prove to be right; and all attacks will not harm it in the least. Works like this one that attribute the merit of a great achievement of German science to the Jews are, however, of lasting detriment to the German nation. They will only lead to the propagation of pro-Jewish culture, a result that the authors certainly do not intend.[5]

---

reviewer himself. It shows that Weizel had solicited the support of his colleagues in severely criticizing the editor W. Müller, the inappropriate successor to A. → Sommerfeld as professor of theoretical physics at the University of Munich since 1940.

[3]On J. → Stark's attacks on Albert → Einstein's and Niels Bohr's contributions to quantum theory and Werner → Heisenberg's and Max → Born's contributions to quantum mechanics see, e.g., doc. 31, footnote 4, and docs. 55 and 56 where he dubs Arnold Sommerfeld and Heisenberg 'White Jews'. Müller wrote in the same vein about Max → Planck, Heisenberg, Sommerfeld, and Erwin → Schrödinger in his article on theoretical physics, doc. 83. The attack against Pascual → Jordan is particularly revealing since Jordan, who had contributed to the algebraic version of quantum mechanics in papers co-authored with Born and Heisenberg, was himself a fairly active Nazi propagandist.

[4]For Stark's experimental research on canal rays, including the first verification of the Doppler principle with moving particles in 1905, which brought him fame as an outstanding experimentalist and the Nobel Prize in 1919 see, e.g., J. Stark's papers in *Physikalische Zeitschrift* **6** [1905], pp. 892–897; **7** [1906], pp. 353–355; and **8** [1907], pp. 79–81 as well as his detailed paper 'Über die Lichtemission der Kanalstrahlen in Wasserstoff', *Annalen der Physik* (4) **21** [1906], pp. 401, 456.

[5]Not only does Weizel question the argument that quantum theory was exclusively a Jewish creation, but he also demonstrates the counterproductiveness of Stark's and Müller's polemics, thus adeptly turning the tables on the assailants.

90 Carl Ramsauer: Letter to Bernhard Rust [January 20, 1942]

Source: Prandtl Papers, Göttingen. Typed transcript of letter addressed to: 'Herrn Reichsminister Rust, Berlin W.8, Unter den Linden 69. Reichsministerium für Wissenschaft, Erziehung und Volksbildung.' 'Registered!' added in pencil. An appended list of the attachments is omitted. Published in Ramsauer [1947]b.

Esteemed Reich Minister,

As President of the German Physical Society,[1] I regard it my duty to lay before you my anxiety about the future of German physics as a science and as a factor of power. May I note in this connection that I am fully acquainted with the state of German physics through my position as director of a large industrial research institute and through my many years of university experience.[2] However, I have absolutely no personal interest in any improvement in position of German academic physicists.

German physics has lost its former supremacy to American physics and is in danger of continuing to lag behind. The accuracy of this view, which is shared by many of Germany's leading physicists, is also clearly reflected in the statistical data in the enclosed Attachment I.[3] Individual details in it might possibly be disputable; but the conclusiveness of the data on the whole speaks for itself all too clearly.

The Americans have made very great progress indeed. This is not solely due to the fact that the Americans employ far larger amounts of material resources than we do. It is due at least as much to the fact that they have succeeded in attracting a large new generation of carefree and motivated young scientists, the individual achievements of whom are equivalent to those of our own at the best of times with the advantage that they are able to work in teams.

In contrast, on our side the category of persons to whom we are indebted for our predominance in physics before the World War is being eliminated at an ever increasing pace; yet it is this group, based entirely on a now historic structure, that should be best suited today as well to take up the competition with American physics and to lead the way to victory. This group consists of the professors, university lecturers, teaching assistants, and doctoral students at German universities and polytechnics. The career opportunities and working

---

[1]C. → Ramsauer was appointed president of the → *DPG* in 1941. This letter to Minister B. → Rust at the → *REM*, is part of his and W. → Finkelnburg's initiative to fight the 'Aryan Physics' movement's influence on German science policy. See also the following texts, as well as docs. 84 and 85 for an earlier initiative by Ludwig → Prandtl.

[2]Ramsauer became head of the → *AEG*'s research laboratory in 1928 and was appointed honorary professor at the Polytechnic in Charlottenburg (Berlin) in 1931.

[3]For Attachment I: 'American Physics Outdoes German Physics', see doc. 91. See also Heisenberg's at that time unpublished memorandum on the progress of American physics, recently published in Rechenberg (Ed.) [1992]a, pp. 54–56.

morale of this category of persons has suffered very severely, both ideologically and materially, which the foundation of large governmental research institutes or research activity in industry could not adequately compensate for. Germany cannot afford to neglect this nature-given asset either in its struggle for survival.

The causes of this are the following:

1) The normal tangible budgets of physics departments at universities and polytechnics receive only a fraction of the funds that are absolutely necessary in our times of advanced technology for physical research, teaching, and education. The same applies to an even greater degree to indispensable staff assistants, such as engineers, glass blowers and laboratory technicians. In this connection it should be readily acknowledged though that the needs of capable physicists are being provided for exemplarily by the German Scientific Research Association and the Helmholtz Society, and that the necessary resources are being made generously available for projects of military importance.[4] However, demonstration lecture courses and laboratory sessions all have to continue to make do with completely outdated technology; and the necessary equipment installed at the institutes, such as electrical power supplies, gas liquifaction machines, compressed air, etc., as well as engineering and glass blowing workshops usually remain completely inadequate. The many projects that cannot be or are not yet incorporated into the above-mentioned organizations' programs also continue to have insufficient options at their disposal. **It is a fact that a large number of diligent university physicists are constantly under the paralysing impression that they are unable to compete against German industrial laboratories and foreign countries.**

2) In our country one main branch of physics, theoretical physics, is being pushed more and more into the background. The legitimate struggle against the Jew Einstein and against the excrescences of his speculative physics has spread to the whole of modern theoretical physics and has brought it largely into disrepute as a product of the Jewish spirit (see Attachment II).[5] The working morale of our theorists is being crippled, and the rising generation is being scared away from taking up theoretical research.

In contrast to this, it must be observed that it is impossible for physics as a whole to thrive unless theoretical physics thrives. Modern theoretical physics in particular has a whole series of the greatest positive achievements to offer that could also be of vital importance to the economy and the armed forces; and the very general accusations made against the advocates of modern theoretical

[4]This may allude to the → *DFG* and the *Helmholtz-Gesellschaft* having funded research on nuclear physics on behalf of the → *HWA* since October, 1939. Its secret status prevented Ramsauer from being more explicit in this context.

[5]Attachment II, a list entitled 'Publications against Modern Theoretical Physics', is omitted here. Ramsauer incorporates the politically correct anti-Semitic point of view (to which renunciating Albert → Einstein is an essential part) in lobbying for his cause—Minister Rust had been a Party member since 1922.

physics of being pioneers of the Jewish spirit, are as unsubstantiated as they are unjustified (see Attachments III and IV).*[6]

All this is even more regrettable since the potential of the German spirit in theoretical physics is in itself very large. There would be a chance precisely in this field in the race against America, which in the beginning was very far behind and took, and still is taking, the greatest pains in order to keep pace with us.

3) Professorship appointments in physics do not always follow the principle of preference according to ability which has proven successful both in the past and the present. I do not want to go into the details of well-known and obviously inappropriate appointments, as this would not change the circumstances and would only greatly annoy individuals. But I am prepared to support my opinion more specifically upon request. Nevertheless, in order to elucidate the absolute seriousness of the state of affairs, I would like to make an exception, all the more so since the case in point is of symptomatic significance precisely to the state of theoretical physics in Germany. I enclose an opinion on the successor to Sommerfeld, Prof. W. Müller (Munich), by Prof. L. Prandtl (Göttingen), which the latter had submitted to me in another connection but had placed at my disposal (Attachment V).[7]

4) An academic career in physics is rapidly losing its former appeal among our best students. This is supported among other things by the fact that transfers into this career from positions in industry are hardly attempted anymore, quite in contrast to the past. The following causes should be mentioned: The general lowering of status of physics professorships, as well as of the next generation, both socially and materially; limited technical aids, and the too heavy workload from additional duties—as a result of the shortage of technical and administrative staff; various impediments involved in professorial appointments.

The following requests arise from these 4 points:

Re 1: The tangible budgets of physics chairs should be raised substantially in accordance with modern requirements. I believe I can justifiably assert that in the whole of Germany hardly any project could make a greater impression on the

---

* [Original footnote +):] "With this defense of modern theoretical physics I will surely be accused of propagandizing for the Jews. The contrary is true! I am interceding precisely for **German** physics, demonstrating that our modern theory has made great advances and that it is indebted not to the Jewish, but to the German spirit for these advances. Moreover, one could justifiably blame the opposing side for—unintentionally—strengthening the already far too exaggerated claims of the Jewry by ascribing German achievements to the Jewish spirit."

[6]Attachment III: 'The Crucial Importance of Theoretical Physics and Particularly of Modern Theoretical Physics' is not included in this anthology; cf., however, doc. 85, for a similar text; for Attachment IV: 'Refuting Allegations that Modern Theoretical Physics is a Product of the Jewish Spirit', see doc. 92.

[7]This Attachment V: 'Excerpt from a Submission by Professor Prandtl' was taken from Prandtl's memorandum of 1941, here doc. 85. A modified two-page excerpt entitled 'Threat to the Rising Generation of Physicists' ('Gefährdung des Physikernachwuches') is among the Prandtl Papers.

population as a whole than in this case, were a million granted it.

Re 2: The internal conflicts within German physics **must** be settled, if we want to bring about a recovery. I suggest that a debate between some prominent representatives of each of the two sides take place again, similar to the one in Attachment VI[8] under the neutral chairmanship of an acknowledged physicist with no immediate theoretical interests (such as Gerthsen or Gerlach).[9] My own formulation of the question can serve as the point of departure here, while my reasoning would have to be verified carefully. However, this matter should not peter out again but should be brought to a close, if need be through pressure by the [Reich Education] Ministry.[10]

Re 3 & 4: The deficiencies described here should be eliminated in the future or mitigated as far as possible.

Most honorable Minister, I request that you give me the opportunity to also present my concerns to you in person. I would like to mention in this regard that I have been exchanging thoughts on these questions with Colonel-General Fromm already for a while now, and that furthermore the authorized persons at the Reich Aviation Ministry have approached me regarding these issues.[11] I have therefore forwarded a transcript of this submission to these two agencies upon their request. I am convinced that the whole Armed Forces would readily apply their influence in order to help obtain approval of the required funds from the Finance Ministry according to Point 1 of this letter. In addition, the Reich University Lecturers Leadership, which has been very interested in these questions for some time now, has received a transcript of this submission.

Heil Hitler !

Yours very sincerely, signed C. Ramsauer.

91 Carl Ramsauer: American Physics Outdoes German Physics [January 20, 1942]

Source: Prandtl Papers, Göttingen. Typed document: 'Eingabe der Deutschen Physikalischen Gesellschaft. Anlage I. Die Überflügelung der deutschen durch die amerikanische Physik.' Tran-

[8]Ramsauer's Attachment VI: 'The Munich Conciliation and Pacification Attempt' is reproduced here as doc. 93.

[9]Christian Gerthsen (1894–1956) studied physics at Heidelberg and Munich until 1914, and 1919–22 at Göttingen and Kiel, where he submitted his Ph.D. thesis with the theoretical physicist W. Kossel as his supervisor in 1922; but he subsequently specialized in experimental work on canal rays. In 1928 he became lecturer in Tübingen, 1932–39 full professor of experimental physics in Giessen, 1939–48 at the University of Berlin, and from 1948 at the Polytechnic in Karlsruhe. Walther → Gerlach was appointed full professor of physics at the University of Munich in 1929.

[10]See footnote 3 of doc. 94.

[11]On Fritz Fromm see doc. 88, footnote 1.

scription of Attachment I to letter to Reich Minister Rust:[1]

1) An American statistical survey that was made for the use of libraries and therefore has no bias whatsoever in our direction, offers a broad overview of physics as a whole. ('A Study of Scientific Periodicals', Ruth H. Hooker, *Rev[iew of]. Sci[entific]. Instr[uments].* **6**, 1935, p. 333).

In a study of the physics literature world-wide for 1934, each of the 5 countries Germany, England, France, USA, and Russia, is represented by a major journal. All the citations appearing in these journals are ordered by country of origin and year of publication. Out of the very extensive material totaling 11,400 citations a brief excerpt is presented here, converted into percentages of the total figure:

Countries of Origin		
Year of Publication	Germany	USA
1897	64 %	3 %
1912	54 %	7 %
1933	36 %	33 %

Tab. 3: Citation percentages by country of origin, according to Hooker [1935].

This table is fairly objective, since national differences are compensated as best as possible by the equal treatment of the five leading countries in physics. The percentages of the totals can thus be taken as a measure of the share of the total world assets in physics that each nation had in the relevant year.

Germany's originally outstanding share of these total assets has thus dropped by almost a half from 1897 to 1933; the USA's originally almost infinitesimal share has risen more than tenfold.[[2]] The percentages of both nations for the year 1933 have become almost equal. However, the trend of the figures clearly shows that a similar statistic for 1938, the last year before the war, let alone for 1941, would give a substantially bleaker picture.

My own observation is in the same direction, as follows: American citations in the 1913 volume through to the 1938 volume of the *Annalen der Physik* have risen from 2.9% to 14.9%. In the same period German citations in the *Physical Review* have fallen from 29.9% to 16.0%.

[1] For the enclosure letter see the preceding doc. 90. Cf. C. → Ramsauer's much more detailed report, Ramsauer [1943], which he presented at the second official meeting of the Aviation Research Academy (*Deutsche Akademie der Luftfahrtforschung*) on Apr. 2, 1943, and which was one of the secret *Kernphysikalische Forschungsberichte* reports of very limited distribution.

[2] On the rise of physics in the USA, see footnote 7 of doc. 58 and the literature described in Albert E. Moyer's contribution 'History of physics' in: Sally Gregory Kohlstedt & Margaret W. Rossiter (Eds.), *Historical Writings on American Science. Perspectives and Prospects*, Baltimore: Johns Hopkins, 1985, pp. 163–182.

2) A more ad valorem survey of physics as a whole is provided by the distribution of Nobel Prizes to Germany and USA. The comparison is naturally only carried out up to the year in which for compelling reasons Germany had to refuse continuing to accept Nobel Prizes, i.e., up to the year 1935.[[3]]

For the sake of clarity the numbers have been converted into percentages, despite their actually being too low for this purpose.

	For Discoveries in Physics from the Years	
Number of Prizes	1890 - 1920	1921 - 1935
Total	33 (100 %)	9 (100 %)
To Germany	12+) (36%)	1++) (11%)
To USA	2 (6%)	2.5 (27.5 %)

Tab. 4: Percentage of Nobel Prize awards by country to 1935. ' +) of these, two Nobel Prizes to Jews; ++) in addition a half Nobel Prize to an emigrant.'

These involve individual top achievements; but just as in sports, they can also be regarded as a good measure of the total performance.

3) The role that the American journal *Physical Review* plays in physical literature provides a similar picture of physics as a whole. The journal has moved from a virtually ignored position in 1905 to become the internationally **acknowledged leading** physics journal.

4) We will look at nuclear physics as a single example, but one which reaches the outermost limits of research and has the greatest future prospects.

The following small table gives a summary of the number of papers written in German and English (mostly American) in this field according to the *Physikalische Berichte*.

The number of German papers on this most modern and promising field has thus risen 3.5-fold in this time, whereas the number of papers written in English have risen 13.5-fold. Yet, as every nuclear physicist will confirm, the quality of American papers is at the very least equivalent to that of German papers.

5) A synopsis of cyclotrons around the world gives an even more dismal picture. This most important experimental tool of nuclear physics was signficantly first invented and built in USA. This is because, beyond the cleverness of the in-

[3]On Jan. 30, 1937, Hitler issued a decree according to which the acceptance of Nobel Prizes was categorically "forbidden to all Germans for all future times" (*Reichsgesetzblatt* 1937, part I, p. 305, issued March 16, 1937). For an English translation of the decree and international commentary, see doc. 51. On the nationalistic aspects of Nobel Prize awards, see also Günter Küppers, Peter Weingart, Norbert Ulitzka (Eds.) *Die Nobelpreise in Physik und Chemie 1901–1929, Materialien zum Nominierungsprozess*, Bielefeld: Report Wissenschaftsforschung, 1982, Part 2; and Elizabeth Crawford, *Nationalism and Internationalism in Science, 1886–1939. Four Studies of the Nobel Population*, Cambridge: Cambridge Univ. Press, 1992, chap. 6.

	1927	1931	1935	1939
Germany	47	77	129	166
USA and England	35	77	329	471

Tab. 5: Comparison of numbers of articles published in German and English, according to the *Physikalische Berichte*.

vention idea itself, the technical and pecuniary means required for its realization and testing would not have been remotely available in this scale in Germany. The current hopes of experimental nuclear physics rest primarily on such cyclotrons which existed in 1941 in following numbers:[4]

USA	about 30	Germany	1 each
England	about 4	Russia	"
Japan	at least 1	France	"
		Denmark	"

Tab. 6: Numbers of cyclotrons throughout the world by January, 1942.

6) We would have a match for this American superiority in our theoretical physics achievements, which has long been part of the legacy of the German spirit and which obviously only requires as external aids a pencil and a piece of paper. But instead of promoting a man like Heisenberg in every way, to whom we owe vital contributions to the theoretical development of nuclear physics, he is subjected to stinging attacks because of his theoretical research. This must weaken his own productivity and prevent the next generation from studying under him. In this way the creation of a great German school of theoretical nuclear physics is being blocked, and we are thus surrendering one of our chances in the competition with USA (see also the Attachments II, III, IV).[5]

[4]On the early development of cyclotrons, an early form of accelerator used in nuclear physics in which charged elementary particles are guided by powerful magnetic fields along spiral paths to collide with targets outside the instrument at high speeds see, e.g., Heilbron [1986]a. The USA led in the development of cyclotrons, with Ernest Orlando Lawrence the inventor of the instrument and M. Stanley Livingston (1905–) at the University of California at Berkeley. In England the first cyclotrons were installed at Cambridge, Liverpool and Birmingham. The first working cyclotron in Germany was built at the physics section of the Kaiser Wilhelm Institute in Heidelberg for W. → Bothe and W. → Gentner's team; but the Paris cyclotron of the French team headed by (Jean) Frédéric Joliot and the Copenhagen cyclotron in Niels Bohr's institute were also available to German researchers during the German occupation. Cf. Watzlawek [1942]a, p. 325, where the totals are 35 in the USA, 3 each in England and Japan, 2 in Russia, and 1 in Zurich.

[5]Ramsauer's Attachments II: 'Publications against Modern Theoretical Physics' and III:

92 Carl Ramsauer: Refuting Allegations that Modern Theoretical Physics is a Product of the Jewish Spirit [January 20, 1942]

Source: Prandtl Papers, Göttingen. Typed document: 'Eingabe der Deutschen Physikalischen Gesellschaft. Anlage IV. Die Widerlegung der Vorwürfe gegen die moderne theoretische Physik als ein angebliches Erzeugnis jüdischen Geistes.' Transcription of Attachment IV to letter to Reich Minister Rust:[1]

Germany's prospects are good, considering its people's talent for theoretical physics, and it should therefore do its best within its ability to promote this potential. Instead of this, Germany has allowed a numerically small group of extreme-minded physicists, astronomers, and philosophers to harshly attack German theoretical physics and its most notable representatives, as the publications listed in Attachment II show.[[2]] No objection is made to impartial criticism nor to the fight directed against Einstein's presumptuous Jewish speculations.[[3]] But beyond this, as a cursory perusal of the list shows, Germany's entire modern theoretical physics is portrayed as contaminated by the Jewish spirit; and in this connection deserving German theorists like Sommerfeld, Planck, and Heisenberg are severely attacked. Thus in article No. 3 Sommerfeld and Planck are characterized as mind-mates [*Gesinnungsgenossen*] of the Jews, while a bit further on following lengthy remarks about Heisenberg that can be proven contradictory to the truth there appears verbatim: "Heisenberg is only one example among several others. All of them are puppets of Jewry in German intellectual life and must disappear just as the Jews themselves."[[4]]

These articles exclude the possibility of objective discussion by drawing in political factors. They deliberately insinuate that working with modern physical theories is not consistent with National Socialist conduct. The trend of these

'The Crucial Importance of Theoretical and Particularly of Modern Theoretical Physics' are not included in this anthology; see, however, doc. 85 for a related text and Attachment IV: 'Refuting Allegations that Modern Theoretical Physics is a Product of the Jewish Spirit', doc. 92. Comparing this text with an earlier memorandum by W. → Heisenberg on the progress of physics in the USA (dated 1937, but only recently published in Rechenberg (Ed.) [1992]a, pp. 54–56), it is evident that Ramsauer made much more effort in supporting his claims with astonishingly up-to-date statistics.

[1]For this letter see doc. 90.

[2]C. → Ramsauer's Attachment II: 'Publications against Modern Theoretical Physics' listed publications by some of the most ardent advocates of 'Aryan physics' such as Philipp → Lenard, Johannes → Stark, Ludwig → Glaser, Wilhelm → Müller, Bruno → Thüring, and Hugo Dingler (on Dingler see doc. 83, footnotes 19–20).

[3]Ramsauer may have had personal reasons in opposing A. → Einstein. The scientific community only acknowledged an experimental effect named after Ramsauer (the Ramsauer or Ramsauer-Townsend effect of the energy dependence of the scattering cross-section of electrons in atoms and molecules as they travel through a gas) after Einstein proposed a theoretical interpretation for it.

[4]See the anonymous article entitled ' "White Jews" in Science', doc. 55, which attacks Arnold → Sommerfeld, Max → Planck and Werner → Heisenberg.

attacks reveals further that for the most part these attacks were not made in the scientific press for the purpose of discussion, but that the authors frequently turn to the public, which is by no means competent to judge here, with the intention of influencing its opinion; and they have also partly succeeded, as the commentaries of the press show time and again.

Besides, it is extremely unfortunate that the authors of these articles have succeeded in gaining access to publication in official student organs (*Bewegung* and *Zeitschrift für die gesamte Naturwissenschaften*) to these ends.[5] A completely wrong idea of theoretical physics is generated in this way, particularly among young students who are not yet competent to judge; and thereby **serious damage** is done to **professional education**. If no public response was made on the part of the affronted professional physicists, then this was only to avoid damaging the reputation of German physics abroad even more through public discussions of this kind.

In analyzing these accusations we will select out of the numerous publications on Attachment II's list the book *Jüdische und Deutsche Physik* by J. Stark and W. Müller, because the title of this work promises to thoroughly clarify this whole question, and because Stark—in contrast to the majority of the authors on this list—is a really important physicist.[6]

Stark sets himself the task "of identifying clearly the Jewish dogmatic spirit in physics and the German pragmatic approach to physics and of comparing their real successes as reflected in lasting scientific advances". He thus arrived at the following characterizations: The dogmatist "seeks to extract scientific findings through the human intellect". He believes, "he can arrive at new conclusions through mathematical operations at his desk". He "builds up his theory for its own sake and is only interested in the results of experience insofar as they seem to confirm his theory." ["]The pragmatist gathers his findings from careful observation and from properly performed experiments." "For the pragmatist, theory is a fiction [*Erscheinung*], a precise and brief description using mathematically described equations." Then follow in conclusion the tenets: "The dogmatic approach is characteristic of the Jewish spirit." "The pragmatic approach to physics is primarily characteristic of the Germanic race."

These characterizations of the pragmatic and the dogmatic spirit are completely arbitrary and are one-sided to the extent that they already contain at the outset the conclusions that are supposed to be drawn from them later. The assignment of the two trends of Jewish and Germanic mentality are expressed as if generally recognized truths were involved, while they are, of course, merely unproven assertions. The following catch phrases could be equally justly or unjustly coined: "The extraction of new knowledge from one's own intellect is typically

[5]The latter journal, issued by the → *NSDStB*, was edited by the Nazi activist B. Thüring and Lenard's pupil F. Kubach (on Kubach see footnote 16 in doc. 85).

[6]For a review of this publication, Müller (Ed.) [1941]a, which reproduces speeches by Müller and Stark, see doc. 89.

Germanic", as true as Kant and Hegel were Germans. "Fearfully clinging to the results of the immediate experiment, i.e., to the tangibly material, is typically Jewish".

From this basis, which is equally arbitrary as it is shaky, then follows an unrestrained attack on the intellectual achievements of German scientists and university teachers. Three objections in particular are raised against modern theory:
a) It supposedly is not based upon experimental results.
b) It supposedly is not subject to examination against experimental results.
c) It supposedly overrates mathematical theoretical work.

The objections under a) and b) simply do not correspond to the facts. This applies even to the most extreme case, the general theory of relativity,*[7] the founder of which can, by the way, really be accused of Jewish spirit and Jewish arrogance in the widest sense. Its point of departure is the Michelson experiment. Its goal, which it had itself always advocated, is its acknowledged verification through checking its astronomical consequences.[8] Were a repetition of the Michelson experiment at high altitudes to lead to other results than on level ground, or were the measurement of the deflection of light rays near the sun to lead irrefutably to other numerical values than those the theory demands, then it would not occur to any modern theorist to continue to support this theory any further. What applies to a real Jewish product, applies all the more to our German theorists; since every theory would become absurd if it abandoned its basis on observations or if it hoped to reject experimental verification. Thus Heisenberg deliberately constructs his theory, quantum mechanics, in such a way that it only contains observables [*prinzipiell meßbare Bestimmungsstücke*]. In the introduction to his fundamental paper 'Über quanten-theoretische Umdeutung kinematischer und mechanischer Beziehungen' (*Z[eitschrift] f[ür] Phys[ik]*, p. 879, 1925) after characterizing the failure of classical mechanics for atoms he says in this regard:[9]

"Under the circumstances it appears more advisable to give up completely that hope of observing as yet unobservable quantities (such as position and rotation period of the electron). Thus we should admit at the same time that the

* [Original footnote +):] "Yet, I am scarcely one of Einstein's followers. At the height of the Einstein stir, I prevented Einstein from holding one of his pro-Jewish lectures in German Danzig, despite strong opposition by the other side."

[7]On the sensation referred to as the *Einstein-Rummel* in the early 1920's, when the results of Sir Arthur S. Eddington's expedition to test Einstein's light deflection prediction became generally known, see Elton [1986] and Hentschel [1990]a, chap. 2. C. Ramsauer, himself a pupil of Lenard, had been appointed full professor of physics at the Polytechnic in Danzig in 1921.

[8]On the famous experiment that sought to verify the existence of ether using a sensitive optical interferometer, conducted by Albert A. Michelson in Potsdam in 1881, and then with increased precision together with Edward Morley in 1887, see footnote 2 of doc. 85.

[9]Quoted from Heisenberg's paper on 'a quantum theoretical reinterpretation of kinematical and mechanical relations'. See also Jammer [1966], p. 197.

partial agreement of the quantum rules mentioned is more or less coincidental with experience; and we should try to formulate a mechanical theory for quanta analogous to classical mechanics in which only relations between observables occur."

The fact that a new theory gropes its way gradually towards the truth and must leave some points to later experimental or theoretical explanation is based on the nature of the human mind, and in particular of the German mind. This is not reproachable as long as experimental results are recognized in principle as the final authority.

The objection under c) redounds on Stark himself. The modern theorist does not **over**rate the value of theoretical mathematical work; but Stark evidently **under**rates it when he says: "The pragmatic approach takes its findings from careful observation and from properly performed experiments. Individual interpretations are then used solely as an aid in thinking up experiments. If these interpretations are not confirmed by the latter, it is immediately replaced by another one that corresponds more closely to reality." This view is appropriate when penetrating step by step into basically familiar areas; but it proves to be far too narrow for totally novel problems. Could thermodynamics, for example, have developed in this manner? Certainly not! In this case pure theoretical work, to a large part German theoretical work, has led "through mathematical deskwork" to "new insights" into a theoretical physical system, which despite its "dogmatic" construction, is one of the most important bases of our physics and technology. This is true of all great, new bodies of experience [*Erfahrungskomplexen*]. It is also true of atomic structure and nuclear physics. Initially, a physical system must always be created through mathematical operations on a broad foundation, naturally starting out from experimental results and always open to examination against additional experimental findings. Such comprehensive theoretical work is **German** in the best sense, as true as Robert Mayer, Clausius, Boltzmann, and Helmholtz were Germans.[10] Characterizing such theoretical work as Jewish can only serve to recognize and elevate the claims of Jewish mentality, which is really more than unnecessary.

Stark does not even satisfy the minimum requirement that a system built on an arbitrary foundation might be required to meet, namely that of being able to put into practice consistently and to be applied systematically. Suddenly and surprisingly the knowledge of Planck's radiation law is supposed to have been "won through pragmatic research", while this discovery is surely a typical example of the dogmatic view, since these new insights had been the product of mathematical deskwork.

[10]On Robert Mayer and Hermann Helmholtz see footnotes 5 and 4 of doc. 77. The Prussian mathematical physicist Rudolf Julius Emanuel Clausius (1822–1888) was a pioneer in the field of thermodynamics, who formulated the second law of thermodynamics and introduced the term 'entropy'. The Austrian theoretical physicist Ludwig Eduard Boltzmann (1844–1906) made important contributions to the kinetic theory of gases, energy distribution and thermodynamics.

Stark's view of atomic theory is similarly contradictory. He rejects Bohr's theory as dogmatic, because he considers electron rotation without any radiation of energy unacceptable.[11] He opposes—and this justifiably—the attempts to offer a concrete interpretation of Schrödinger's equation through "blurring" of the electron charge or through "irratic motion" [*Wimmelbewegung*] of the electron.[12] But he forgets that his own assumption of an electron staying in a particular position at a specific distance from the nucleus without the centrifugal force of an orbit is physically just as unacceptable.

Besides, at the bottom of his heart, Stark is after all too important a physicist not to place the real achievements of science in the forefront ahead of his definitions of pragmatic and dogmatic physics. According to his own definition, "precise and brief description through mathematically formulated equations" is what counts. Therefore, the way that had led to these equations or the meaning that could be given to these equations are of lesser importance. He recognizes in this sense, for example, the Epstein formula on the Stark effect of hydrogen lines as a real achievement of dogmatism.[13] Yet, it is incomprehensible that he should stop at this isolated acknowledgment, since he himself had set out "to compare" the "real achievements" of pragmatic and dogmatic physics "in lasting scientific advances".

These actual successes of modern theory are evidenced in Attachment III, with examples that no physicist can pass over.[14] Thus there is really no reason for us to attack these achievements as products of the Jewish spirit, but every reason to be proud as Germans of German achievements.

---

[11]In 1913 Niels Bohr postulated the existence of stable Kepler-like orbits of electrons around a positively charged atomic nucleus. Because electrons are negatively charged particles and should lose energy on these accelerated orbits according to J. Larmor's theorem, there was an irresolvable contradiction with classical electrodynamics built into Bohr's atomic theory. On Bohr, see also footnote 1 of doc. 20.

[12]Erwin → Schrödinger's wave mechanics of 1926 substituted Bohr's theory for most practical applications of quantum mechanics. In 1926 Max → Born and others interpreted Schrödinger's wave equation as describing the amplitude of a probability distribution of particles as a function of space and time. Though Heisenberg's uncertainty relation prohibits a simultaneous definition of momentum and position or of energy and time, and thus classical representation of a particle is not possible, Schrödinger's spatially blurred charge distribution was consistent with Heisenberg's result. Cf. Jammer [1966], section 6.1; Mehra & Rechenberg [1982], Vol. 5.

[13]See Paul Epstein: 'Zur Theorie des Stark-Effektes', *Physikalische Zeitschrift* **17** [1916], pp. 148–150 and in *Annalen der Physik* (4) **50** [1916], pp. 489–520. On Epstein see footnote 36 of doc. 77.

[14]Ramsauer's Attachment III: 'The Crucial Importance of Theoretical and Particularly of Modern Theoretical Physics' is omitted here; see, however, doc. 85 for a similar text.

93 Carl Ramsauer: The Munich Conciliation and Pacification Attempt [January 20, 1942]

Source: Prandtl Papers, Göttingen. Typed document: 'Eingabe der Deutschen Physikalischen Gesellschaft. Anlage VI. Der Münchener Einigungs- und Befriedungsversuch.' Transcription of Attachment VI to letter to Reich Minister Rust:[1]

In an attempt to avoid the continuation of the well-known attacks on German theoretical physics and in an effort to clarify the situation, the Reich leadership of the National Socialist University Lecturers League organized a debate at the 'Physicians House' [*Ärztehaus*] in Munich on the 15th of November, 1940. Members of the group of assailants met with representatives of the remaining German physics community, 14 participants in all.[[2]] During the debate the assailants had to withdraw practically all of their factual objections, so that in the end consensus could be reached on the following five points:*[[3]]

1. Theoretical physics together with all its mathematical aids is a necessary part of physics as a whole.

2. The observed facts summarized in the special theory of relativity are an established part of physics. The applicability of the special theory of relativity[[4]] to cosmic relationships is not so certain, however, as to eliminate the necessity of further verification.[[5]]

[1]For C. → Ramsauer's letter to Bernhard → Rust see doc. 90.

[2]W. → Finkelnburg, leader of the Lecturers League (→ *NSDDB*) in Darmstadt since the summer of 1940, organized the event later known as the 'Munich synod' (*Münchner Religionsgespräche*) with the intention to combat the Aryan Physics movement. He later described in detail this debate as well as the representatives of both sides in an article that he intended to publish anonymously; see doc. 110; see also Beyerchen [1977], pp. 176ff.

* [Original footnote +):] "Present at the **closing** session: Messrs.: [Rudolf] Tomaschek, [Alfons] Bühl, [Johannes] Malsch, [Harald] Volkmann, [Ludwig] Wesch, [Herbert] Stuart, [Wolfgang] Finkelnburg, [Georg] Joos, [Otto] Scherzer, [Hans] Kopfermann, [Otto] Heckmann, [Carl Friedrich] v. Weizsäcker. Not present: [Bruno] Thüring, W[ilhelm]. Müller."

[3]For an explanation for the absence of the most hardened opponents of 'mathematical' theoretical physics at the closing session, B. → Thüring and W. → Müller, see footnote 25 of doc. 110. Johannes Malsch (1902–1956) started his physics studies in Jena in 1920 and acquired his doctorate under Max → Wien in 1924. He became teaching assistant in that year at the Institute of Theoretical Physics and university lecturer in 1927. In 1934 Malsch was appointed associate professor at the University of Cologne and became a civil servant in 1938. 1951 he took up a position as staff scientist at the *Telefunken* company in Ulm. Malsch worked on conductivity measurements, especially of dielectric materials and in the ionosphere. On the other participants see the annotation to doc. 110.

[4]Here the general theory is meant, which is an error in the original text. For a contemporary comparison of relativistic cosmology with neo-Newtonian alternatives, see Heckmann [1942].

[5]Particularly the deflection of light in a gravitational field and gravitational redshift had not yet been measured conclusively due to many disturbing influences in the sun's outer atmosphere: See, e.g., Harald von Klüber, 'The determination of Einstein's light-deflection in the gravitational field of the sun', *Vistas in Astronomy* **3** [1960], pp. 47–77; Eric Gray Forbes, 'A history of the solar redshift problem', *Annals of Science* **17** [1961], pp. 129–164.

3. The four-dimensional description of natural processes is a useful mathematical tool; but it does not imply the adoption of a new conception of space and time.[6]

4. Any link between the relativity theory and a general relativism is rejected.[7]

5. Quantum and wave mechanics are the only known ways at present to describe atomic processes quantitatively. It is desirable to go beyond formalism and its correspondence rules [*Deutungsvorschriften*] to come to a deeper understanding of atoms.[8]

A basic clarification seemed to have thus been reached. The debate closed with an appeal by the Reich University Lecturers leadership to all physicists to refrain from publishing any more polemical papers in the future. In spite of this, as the list of publications, Attachment II shows, the assailants continued the fight with a growing number of polemical articles![9]

The appropriate Ministry did not take steps against this situation, which is just as damaging to the reputation of German science as it is to research and education. The general conclusion drawn from this, that even the Ministry itself is unfavorably disposed to modern theoretical physics and supports the group of attackers, seems to be supported by the following actions:

a) Just recently several of the authors mentioned in Attachment II were appointed to responsible positions by the [Reich Education] Ministry or were proposed for such positions.

b) Sommerfeld's chair for theoretical physics at Munich University was conferred upon Prof. W. Müller, which amounts to the destruction of the Munich theoretical physics tradition, which is highly acclaimed abroad as well.[10]

c) The full professorship for theoretical physics at the Berlin Polytechnic was

[6]This pragmatic attitude was a renouncement of Minkowski's interpretation of Einstein's special theory of relativity in which he had postulated in 1908 that "space and time per se will sink into oblivion; only a union of both shall have some remaining value"; cf., e.g., Peter Galison, 'Minkowski's space-time: From visual thinking to the absolute world', *Historical Studies in the Physical Sciences* **10** [1979], pp. 85–121.

[7]This link had often been made in bad popularizations of Einstein's theory and as well as by some of its strongest opponents who wanted to discredit it. Therefore, the rejection of this link was in the interest of both camps.

[8]This point clearly reflects the 'Aryan Physics' side's desire for clearly identifiable, concrete definitions as well as their claim that quantum mechanics was purely formalism. However, the first sentence shows that they could offer no better alternative to quantum mechanics, which had proved very successful empirically in a broad range of applications: See, e.g., Jammer [1966], chap. 8.

[9]In Attachment II: 'Publications against Modern Theoretical Physics', Ramsauer lists publications of the Aryan Physics movement appearing since, particularly by Thüring and Müller. On the other hand, both R. → Tomaschek and A. → Bühl did respect the truce; cf. Beyerchen [1977], p. 179.

[10]On the controversial conferral of Arnold → Sommerfeld's chair on Wilhelm → Müller, see here the second part (no. 4) of doc. 85 as well as Cassidy [1992]a, chaps. 18 and 21.

eliminated.[11]

On the other hand it must be emphasized once again that any injury to German theoretical physics amounts to harming the whole of German physics; and quite apart from the consequences on science, unjustifiable harm is also done to the German economy and German defense technology.

94 Ludwig Prandtl: Letter to Carl Ramsauer [January 28, 1942]

Source: Prandtl Papers, Göttingen. Unsigned typed carbon copy of letter with handwritten emendations, dated January 28, 1942. Addressed to: 'Herrn Professor Dr. C. Ramsauer, Berlin-Reinickendorf-Ost, Holländerstr. 31/34.' referring to a letter dated January 23, 1942, and File Reference No. Pp.62/42. 'eigene Mappe' added in hand at the top of page 1.

Esteemed Colleague,

I have read with the greatest interest the transcript of the submission to Reich Minister Rust that you kindly forwarded to me[1] regarding the plight of theoretical physics in Germany; and I find it exceptionally impressive and convincing precisely because of its impartial style.[2] The military agencies will doubtlessly make sure that the matter does not get stuck in the sand, which from previous experience with the [Reich] Education Ministry would otherwise certainly have been irrecoverably the case, since various lower-ranking positions seem to pull together with the Munich clique.[3]

First and foremost I think that the military agencies should enforce through the official channels a ban on any publishing activity against theoretical physics by the sabotaging Munich group;[4] and then the essential new appointments would just have to be made, where necessary by creating new positions. This internal war [*innere Krieg*] is naturally complicated by the fact that it is also a war against the 'Brown House';[5] and the military agencies would have to be

[11]This was the chair formerly occupied by the leading theoreticians Max → Planck and Erwin → Schrödinger, the latter of whom left Germany after 1933. In 1936 H. A. → Stuart was commissioned to teach theoretical physics at the University of Berlin; after 1939 W. → Westphal continued to teach this subject in a lower ranking position.

[1]See C. → Ramsauer's letter to the head of the Reich Education Ministry (→ *REM*), B. → Rust, dated Jan. 20, 1942, doc. 90.

[2]This is an allusion to the use of Nazi jargon in Ramsauer's letter to Rust; cf. doc. 90, footnote 5.

[3]The word 'also' (auch) is replaced with 'seem to' (zu scheinen) in pen. Max → Wien submitted similar petitions in 1934 and, together with Werner → Heisenberg and Hans → Geiger,, in 1936 to the *REM* but had also received no response; see docs. 35 and 49. For this reason in 1941 L. → Prandtl also sought the support of the head of the Air Force, Hermann → Göring; see doc. 84.

[4]'Munich' is inserted in pen. The 'Munich group' refers to Sommerfeld's successor Wilhelm → Müller and his clique, including B. → Thüring and Wilhelm → Führer.

[5]*Das Braune Haus* refers to the → *NSDAP* headquarters on Brienner Street in Munich,

made particularly aware of this as well.

You also ask me what I have to report on my campaign. To this I can answer that I myself have undertaken nothing new recently. I might add here, though, that our former Göttingen mineralogist Professor Drescher-Kaden,[6] now in Strasbourg, 18 Herder Street, has requested a transcription of my submission and has forwarded one of these to Secretary of State Stuckart of the Reich Interior Ministry.[7] You might ask Professor Drescher-Kaden yourself, with reference to me, what he has to report on how Secretary of State Stuckart received my text and possibly on further steps.

Furthermore, the famous aircraft producer Professor Willy Messerschmitt[8] (Augsburg) has received from me upon request a transcription of my submission; but he will hardly proceed otherwise than in conjunction with the Aviation Ministry.

With best regards and Heil Hitler!

Yours very sincerely [L. Prandtl]

which had previously been called 'Palais Barlow', but was acquired by the Party in 1930 and modified by the architect Paul Ludwig Troost and was the seat of the Nazis until 1933, when it took over the national government. The building became the central point of a system of secret underground tunnels and bunkers under Königsplatz, planned by the Nazis as early as 1933, to connect a dozen or so existing or planned governmental and public buildings.

[6]Friedrich Karl Drescher-Kaden (born 1894) studied at the University of Breslau in 1915 and 1919–22, taking his doctorate in 1922 and serving as teaching assistant to L. Milch and H. Cloos 1921–23; 1923–29 he was at the Darmstadt Polytechnic as teaching assistant and then later unsalaried lecturer under A. Steuer. 1929 Drescher-Kaden was appointed full professor at the Mining Academy in Clausthal, moving to the Berlin Polytechnic in 1933 and to Göttingen in 1936, where he was professor of mineralogy. In 1942 he was at the University of Strasbourg and was named honorary professor of the University of Munich in 1949. In 1952 Drescher-Kaden accepted a chair for mineralogy and geological deposits at the University of Hamburg.

[7]Wilhelm Stuckart (1902–1953) entered the → *NSDAP* in 1922 and became the party's legal adviser in 1926. After the National Socialist seizure of power he became Secretary of State in the Prussian Ministry of the Interior; from 1935 he was Secretary of State in the → *RIM*, where he worked notably on the Nuremberg Laws against the Jewish population. This loyal *SS* member of Jews, was awarded Hitler's 'Distinguished was appointed Interior Minister on May 5, 1945, and was captured soon after the war but was only convicted to four years imprisonment.

[8]'Augsburg' is inserted here in pen. Wilhelm [Willy] Messerschmitt (1898–1978) founded the famous airplane construction company that carried his name in 1923. In 1927 the Messerschmidt Company merged with the Bavarian Aircraft Works (*Bayerische Flugzeugwerke*) in Augsburg, and in the following year Willy Messerschmidt became a member of the board of directors of the corporation, plant manager and chief engineer in 1928. In 1930 he accepted a lectureship on aeronautical engineering as honorary professor at the Polytechnic in Munich and was awarded the German National Prize (*Deutsche Nationalpreis*) in 1938 for his contributions to the field.

95 Werner Heisenberg. The Theoretical Basis for the Generation of Energy from Uranium Fission [February 26, 1942]

Source: Werner Heisenberg, 'Die theoretischen Grundlagen für die Energiegewinnung aus der Uranspaltung'. Manuscript to a talk held on February 26, 1942 at the *Haus der Deutschen Forschung* in Dahlem (Berlin), published in: Heisenberg [1989], Ser. A, Part II, pp. 517–521, and Figs. 1–3. Originally published in *Zeitschrift für die gesamte Naturwissenschaft*, Vol. 9, 1943, pp. 201–212.[1]

The following experimental findings were known when work was begun on the uranium problem within the framework of the Army Ordnance Office study group:[2]

1) Common uranium is a mixture of three isotopes: ${}^{238}_{92}$U, ${}^{235}_{92}$U, and ${}^{234}_{92}$U, which occur in natural minerals approximately in the proportion 1 : 1/140 : 1/17000.

2) Uranium nuclei can be split, according to Hahn and Straßmann, by neutron radiation.[3] Specifically, the ${}^{235}_{92}$U nucleus can be split by neutrons at **all** (including low) speeds (Bohr);[4] and ${}^{238}_{92}$U and ${}^{234}_{92}$U nuclei can only be split by high-energy neutrons.

3) An energy of about 150 to 200 million electron-volts per atomic nucleus is released during fission. This energy is approximately 100 million times greater than the energy normally released per atom during chemical reactions. Moreover, every time an atomic nucleus is split, some neutrons are ejected from it.

One can draw from these established facts: If it were possible to convert all the atomic nuclei of 1 t[o]n of uranium, for example, through fission, then the enor-

[1]Also summarized in Walker [1992], pp. 369ff. For the figures, see W. → Heisenberg's talk presented on May 3, 1943: 'Die Energiegewinnung aus der Atomspaltung', the text of which is reprinted in Heisenberg [1989], Ser. A, part II, pp. 570–575. For the full program of this meeting also featuring E. → Schumann, W. → Bothe, H. → Geiger, K. → Clusius, P. → Harteck and A. → Esau as speakers, see D. Hahn [1988], p. 192. In his retrospective summary of German nuclear research Heisenberg mentions that the talk was presented before the Minister of Education B. → Rust and several war research directors: See also the references to the primary literature and more detailed annotation to doc. 115.

[2]Research conducted by the *Heereswaffenamt* (→ HWA) on nuclear research started immediately after the outbreak of World War II in September 1939, with Kurt → Diebner in charge of the recently formed second 'Uranium Association' (*Uranverein*). On the development of the theory of atomic nuclei by Heisenberg and his colleagues in Leipzig, see H. Rechenberg in Kleint & Wiemers (Eds.) [1993].

[3]See Otto → Hahn's and Fritz → Straßmann's papers, Hahn & Straßmann [1939]a,b. See also Hahn's Nobel Prize speech delivered on Dec. 13, 1946, Hahn [1946/47] for his own account of the discovery of uranium fission; cf., e.g., Krafft [1981], secs. 1.3. and 3.2.

[4]See Niels Bohr's paper written together with J. A. Wheeler: 'Mechanism of nuclear fission', *Physical Review* **55** [1939], p. 1124; **56** [1939], pp. 426–450 (on Bohr, see footnote 1 of doc. 20).

mous amount of energy of approximately 15 trillion kilocalories would thereby be released. It has long been known that such large amounts of energy can be converted through nuclear reactions. Before the discovery of fission, however, there was no chance of carrying out nuclear reactions on larger quantities of substances. This is because the energy expenditure in reactions that are artificially induced in high-voltage sites, cyclotrons, etc., is always much greater than the resulting energy yield.[5]

However, the fact that several neutrons are emitted during the fission process opens the possibility of forcing the transformation of large amounts of substances by means of a chain reaction: The neutrons emitted during fission should also in their turn split other uranium nuclei; new neutrons thereby again emerge, etc. Through numerous repetitions of this process the number of neutrons that multiply begins to accelerate and only stops when a large portion of the substance is converted.

Prior to examining the question of whether this plan could be implemented, the various processes that can draw a neutron out in uranium had to be studied in more detail. Fig. [18] is a diagram of these processes.[6] Supposing a neutron is liberated by fission, after travelling a short distance it can either collide with a uranium nucleus, split it, and thereby produce new neutrons, provided it has a sufficient amount of energy; or in such a collision it can simply release energy to the nucleus without destroying it, which is unfortunately much more likely, whereupon the neutron flies on with reduced energy. In the latter case after a few collisions the energy of the neutron will have reduced so much that its fate can then only be one of the two following possibilities: At one point it can become lodged in the uranium nucleus it collides with. Then any further 'multiplication' is impossible. Or it can collide with a $^{235}_{92}$U nucleus—which unfortunately is relatively unlikely—and split it. Then new neutrons are again created in this process, and the described events can begin anew. A portion of the neutrons can escape at the surface of the uranium and are thereby lost to further multiplication.

A more thorough analysis of the probabilities of the occurrence of these various processes was an important point in the team's program; Mr. Bothe will report on its results.[7]

For the following it suffices to state that the process of neutron absorption in common uranium (the capture of a neutron in $^{238}_{92}$U in forming a new isotope, $^{239}_{92}$U)

[5]On contemporary German high-voltage sites and the few cyclotrons at the time, see F. Kirchner and W. Gentner in: Bothe & Flügge (Eds.) [1948], Vol. 2, pp. 24ff., 28ff, and Bothe [1943]. The high-voltage instrumentation at the → *KWIP* in Dahlem (Berlin) is described in doc. 54. On the first European cyclotrons outside Germany, see also Heilbron [1986]a.

[6]Figure 1 of the original text, here Fig. 18, illustrates the path of neutrons reacting in common uranium and a layer of slow-down material (left), and in uranium 235 (right).

[7]Walther Bothe's unpublished report on this issue was entitled 'Ergebnisse der bisher untersuchten Anordnungen zur Energiegewinnung', according to the program of the meeting reproduced in D. Hahn (Ed.) [1988], p. 192. See, however, the contributions by Bothe and Flammersfeld in *Kernphysikalische Forschungsberichte* quoted in doc. 115, footnote 58.

occurs much more frequently than that of fission and multiplication. Therefore the desired chain reaction cannot be triggered in normal uranium; and new ways and means of inducing the onset of the chain reaction must be devised.

The neutron behavior in uranium can, of course, be compared to population density, where the fission process is analogous to marriage and the absorption process, to death. In common uranium the death rate by far outbalances the birth rate, so that a given population must invariably die out after a short time.

This situation can obviously only be improved if it is possible either: (1) to raise the number of births per marriage; or (2) to increase the number of marriages; or (3) to reduce the probability of death.

Possibility (1) does not apply to the neutron population, since the average number of neutrons per fission is a constant determined by natural laws that cannot be influenced at all. (For the determination of these important constants, refer to Mr. Bothe's presentation.)[8]

Therefore, only avenues (2) and (3) remain. An increase in the number of fissions (2) can be obtained, if isotope $^{235}_{92}U$ is enriched, which though rarer can also be fissioned at lower energies. If it were somehow possible even to isolate isotope $^{235}_{92}U$, the relationships represented on the right side of Fig. [18] would then apply. After one or more collisions, each neutron would split another, provided it does not escape at the surface beforehand, for example. In this case the probability of death through absorption, as compared to the probability of multiplication, is infinitely low. Thus if just enough of $^{235}_{92}U$ is accumulated for the neutron loss at the surface to remain small against the multiplication in the interior, then the neutron count will increase tremendously in a very short time, and the entire fission energy of 15 trillion [15×10^{12}] calories per t[o]n will be released in a small fraction of a second. The pure isotope $^{235}_{92}U$ is thus undoubtedly an explosive of completely unimaginable power. However, this explosive is very hard to produce.

A large portion of the Army Ordnance Office team's work is devoted to the problem of enriching or isolating the $^{235}_{92}U$ isotope. American research also seems to be following this course particularly energetically. This meeting includes a report by Mr. Clusius[9] on the status of this issue; therefore I need not go into it any further.

[8]See the previous footnote.

[9]Klaus Clusius's paper on the enrichment and isolation of this isotope was entitled 'Anreicherung der Uranisotope'. A published account of the Clusius-Dickel separation tube appeared in *Ergebnisse der exakten Wissenschaften* **20** [1942], pp. 121–182. For a later report on the same topic, see also his survey for the *Deutsche Akademie der Luftfahrtforschung*, Clusius [1943].

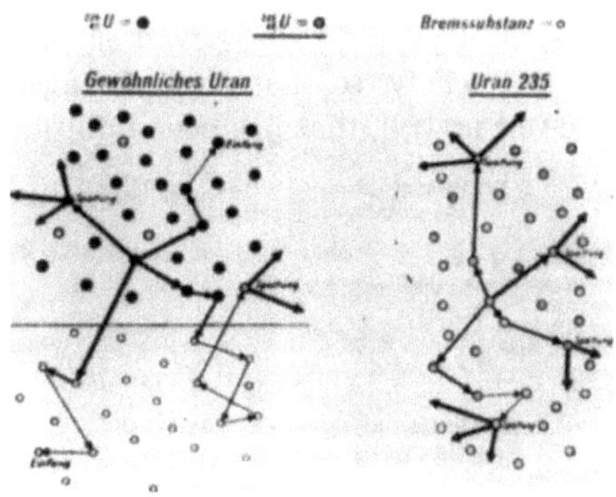

Fig. 18: 'Diagram of the processes involved when neutrons travel in uranium.'

Now only the third possibility in inducing the chain reaction remains to be mentioned: dropping the death rate, i.e., the probability of the neutrons being absorbed. It could be assumed from general observations in nuclear physics that the probability of absorption becomes greater only when the neutrons are at very specific energies. (The analyses of the last year have contributed important new information specifically on this point.)[10] Therefore, if it is possible to bring the neutrons quickly within the range of the lowest feasible energies (i.e., within energies produced by heat motion), without their colliding too often with uranium nuclei, in this way the death rate can be reduced substantially. In practice, a rapid reduction in the speed of the neutrons can be achieved by the addition of suitable slow-down materials; that is, substances that remove some of the neutron's energy when hit by a neutron. Only when enough of the moderator is added can the neutrons be brought safely within the range of the lowest energies. But unfortunately most slow-down materials have the characteristic of occasionally capturing neutrons as well, so that too large an amount of the moderator raises the probability of absorption, i.e., the death rate, again. These proportions are illustrated in the diagram on the [left] side of Fig. [18].

It is therefore a matter of finding a slow-down material that removes energy from the neutrons quickly, but that absorbs as few of them as possible.

The only substance that does not absorb at all, helium, unfortunately can hardly come into consideration practically, because of its low density. Deuterium must then be considered the most suitable substance, which is also available in a dense enough form in its simplest compound, heavy water. However, it is also not easy to produce heavy water in large quantities. The team analyzed thoroughly the suitability of heavy water and other feasible substances (beryllium and carbon).[11]

[10]For a more technical description of the progress in determining neutron diffusion through absorption in various materials see, e.g., Heisenberg's 1947 account, doc. 115, and Heisenberg & Wirtz [1948], sec. 7.1, and references there.

[11]Paul Harteck delivered a paper on the production of heavy water: 'Die Gewinnung von

Following Harteck's idea, it proved to be advantageous to separate the uranium and the moderator spacially, so that then arrangements result like the layered globe, illustrated in Fig. [19], that has been built for a model experiment at the Kaiser Wilhelm Institute.[12]

Fig. 19: [Left:] 'Cross-section of a layered arrangement of uranium metal and paraffin which had been constructed at the Kaiser Wilhelm Institute of Physics in Dahlem to analyze neutron multiplication.' [Right:] 'Exterior view of model fission generator'.

For the time being it had to remain an open question whether such a layering of common uranium and slow-down material can lead to a chain reaction and along with it to the release of large amounts of energy, i.e., whether the 'death rate' can be lowered far enough for the 'birth rate' to prevail and for an increase in the 'population' to result, since the properties of the few substances at all usable in slowing down the reaction are given, of course, and cannot be changed.

Clarifying this point was another one of the study group's most important objectives.

Let us assume for a moment this question were resolved in the positive sense. It must then be investigated how this chosen design performs as the neutron population continues to multiply. It became evident that the process of multiplication does not just stop when a large portion of the uranium is converted, but

schwerem Wasser'. See also Irving (Ed.) [undated], nos. 29-625ff., for original documents relating to Harteck's research on the above.

[12]For a description of these experiments at the Berlin *KWIP* see, e.g., Heisenberg & Wirtz [1948], pp. 152ff., and Heisenberg [1989], Ser. A, part II, pp. 432ff. See here Fig. 19 (Figures 2 and 3 in the original text).

much earlier. There is a sharp rise in temperature as a result of the constantly growing multiplication rate, and as the heat rises the probability of fission is reduced—since the neutrons move more quickly and therefore spend a shorter time in the vicinity of a uranium nucleus. Thus the rising temperature results in a reduction in the number of 'marriages' and thus of multiplication. For this reason, at a certain temperature the multiplication of neutrons will just compensate the absorption.

The layered arrangement described will thus stabilize of its own accord at a specific temperature. As soon as energy is removed from the machine, it cools and multiplication resumes, and the energy that had been removed is replaced again by the energy of fission. Practically speaking, the engine stays at the same constant temperature.

Thus we have an engine that is possibly suited to heating a steam turbine; and in the course of time it can make all of its great energies available to such a heat-engine. We can therefore imagine the practical applicability of such engines in vehicles, particularly ships, that would gain a huge cruising radius from the large energy reserve of a relatively small amount of uranium. The fact that the engine does not combust oxygen would be of particular advantage for submarines.[13]

Following an idea by von Weizsäcker, once such an engine is operational, the matter of producing an explosive also takes on new significance.[14] As the uranium converts in the engine, a new substance is formed (an element of atomic number 94) that is most probably an explosive of the same inconceivable power as pure $^{235}_{92}U$. However, this substance can be produced much more easily than $^{235}_{92}U$ from uranium, since it can be isolated chemically from uranium.

As stated earlier, first it had to be determined through experiment whether a combination of uranium and a moderator could be found in which the chain reaction can take place. But even if such a combination is found, a large amount of this mixture will have to be accumulated to permit a chain reaction to really start up, because the loss of neutrons escaping at the surface of smaller quantities will always be larger than the multiplication within. For this reason experiments with very small amounts of substances are unsuited from the outset in deciding on

[13]See also Feitl's article on submarine engineering, doc. 100. An unsigned typewritten document among the Döpel Papers, Leipzig, dated 'summer 1942, Leipzig' goes into more detail: "*Military application:* Energy source for bigger combat vehicles (above all for ships, whose radius of action can be increased about a hundred-fold; the engine is too big for aircraft and tanks, the engine's intense radiation is an obstacle here as well, which requires careful radiation protection measures)." See Christian Kleint's contribution in Geyer et al. (Eds.) [1993], p. 149, See also footnote 16 below.

[14]Carl Friedrich von → Weizsäcker's idea first appeared in his internal secret and unpublished research report: 'Eine Möglichkeit der Energiegewinnung aus U 238', dated July 7, 1940, *Kernphysikalische Forschungsberichte* **G-59**; see the copy at the American Institute of Physics or the Kernforschungszentrum Karlsruhe; also in Irving (Ed.) [undated], nos. 29-451ff. Heisenberg was thus fully aware that a nuclear explosive, plutonium, could be produced, but he did not actually pursue this line of research in contrast to the Allies.

the suitability of mixtures for chain reactions. Without the generous support of the Army Ordnance Office for the research, in the form of materials, radioactive specimens and funding, it would certainly not have been possible to make any progress at all here. But it is still not possible to induce a chain reaction even with the quite large quantitites of heavy water, for example, that have been available up to now. For this reason we must still address briefly the question of how we can recognize whether the 'birth rate' already outweighs the 'death rate' in the chosen mixture in a model experiment.

To resolve this question, an appropriate neutron preparation with an already known neutron emission rate per sec[ond] is added to the mixture. If the amount of neutrons that escape at the surface of the mixture is larger than what the added preparation contributes, then one can conlude that multiplication outweighs absorption, and that therefore a suitable mixture has been found.

Experiments performed in the last year in Leipzig have demonstrated that a particular mixture of heavy water and uranium actually has the desired properties.[15] However, the 'birth rate' surplus over the 'death rate' is still so minimal in these experiments that just the additional minimal absorption from the mounting apparatus used there removes the surplus again. But later the mounting material will not be necessary or can be replaced with something else.

It can be concluded, with what confidence we can gain from laboratory experiments for large-scale trials, that the experiments clearly support the feasibility of building a machine of the type specified, in which uranium and a moderator are layered.

The results to date can be summarized as follows:

1) Energy generation from uranium fission is undoubtedly possible, provided the enrichment of isotope $^{235}_{92}U$ is successful. **Isolating** $^{235}_{92}U$ would lead to an explosive of unimaginable potency.[16]

2) Common uranium can also be exploited to generate energy when layered with heavy water. In a layered arrangement these materials can transfer their great energy reserves over a period of time to a heat-engine. It thus provides a means of storing very large amounts of energy that are technically measurable in relatively small quantities of substances. Once in operation, the machine can also lead to the production of an incredibly powerful explosive. In addition, it has good

[15]For overviews of Heisenberg's Leipzig experiments L 1–L 4 performed together with Robert → Döpel and Klara → Döpel see, e.g., doc. 115, footnote 47, Heisenberg & Wirtz [1948], pp. 149ff.; Heisenberg [1989], Ser. A, part II, pp. 419ff.; Kleint [1986]; and Kleint & Wiemers (Eds.) [1993], pp. 15ff.

[16]Note that in 1942 Heisenberg continues to underscore the possible military applications of the 'uranium machine' in 1942. His later postwar review articles omit this aspect: See here doc. 115 and Heisenberg & Wirtz [1948]. Potential military applications of nuclear fission were emphasized further in E. Schumann's talk at this meeting on Feb. 26, 1942, entitled 'Nuclear physics as a weapon' ('Kernphysik als Waffe').

potential for a lot of other scientifically and technically important applications that are, however, outside of the scope of this report.

Herewith, Fig. [18] from the slides that are probably already in the possession of the A[rmy] O[rdnance] O[ffice].[17]

96 Georg Dietrich Graue: Letter to Albert Vögler [March 13, 1942]

Source: Archive of the Max Planck Society, Berlin: I, Rep. 1A, No. 1652, main files. Typed signed letter on letterhead: 'Nationalsozialistische Deutsche Arbeiterpartei, Gauleitung, Berlin. Amt: N S D-Dozentenbund, Der Dozentenführer der Freien Forschungsinstitute, Berlin-Dahlem, Faradayweg 4–6', and addressed: 'An den Herrn Präsidenten der Kaiser Wilhelm-Gesellschaft, Herrn Dr. Vögler, Berlin C2, Schloß; Portal III'. File reference: 'Dr.Gr./Ha.'.[1]

Esteemed President,[2]

During one of my visits with you, we once discussed the deplorable pamphlet *Jewish and Non-Jewish Physics.*[3] It is my pleasure to be able to bring a review to your attention today, a transcript of which I am enclosing.[4] From it you can see that the battle against the completely impossible and unscientific argumentation is in full swing.

With best regards and Heil Hitler!

Yours very sincerely, Graue.[5]

[17]According to Helmut Rechenberg, editor of Heisenberg [1989], this last sentence is a handwritten note by Heisenberg, and the figures (numbered 1–3 in the original) are identical with Report No. 18, 'Die Energiegewinnung aus der Atomkernspaltung' in the same volume (Ser. A, part II), on pp. 570–575 (Fig. 1, p. 571; and Figs. 2–3, p. 573) which are reproduced here as Fig. 18.

[1]Administrative notations and other insignificant information on the letterhead are omitted.

[2]Albert → Vögler was president of the → *KWG* from 1941 to 1945.

[3]This pamphlet, Müller (Ed.) [1941]a, was edited by A. → Sommerfeld's successor to the chair of theoretical physics in Munich, Wilhelm → Müller. The draft version of the review was entitled as it appears in the text: 'Jüdische und nicht jüdische Physik'; the title of the published version was changed to 'Jewish and German Physics'. For other opinions on the pamphlet, see also docs. 85 and 92.

[4]For the enclosed review by W. → Weizel, see doc. 89.

[5]Georg Dietrich Graue (1903–) started his studies in chemistry at the Berlin Friedrich-Wilhelm University in 1925. From May 1928 to May 1930 he worked in Otto → Hahn's department at the → *KWIC*. In 1931 he submitted his Ph.D. thesis on surface formation and surface changes in thorium and iron gels and brines. He was then employed in Hahn's department at the *KWIC* until 1945. Because of his National Socialist involvement, representing the Nazi University Lecturers League (→ *NSDDB*) for independent research institutes (*Dozentenbundsführer der freien Forschungsinstitute*), he was not re-employed at this institute after its reestablishment in 1948.

97 Albert Vögler: Foreword to 'Yearbook of the Kaiser Wilhelm Society for the Advancement of the Sciences [after March 31, 1942]

Source: Albert Vögler, Foreword to: *Jahrbuch 1942 der Kaiser-Wilhelm Gesellschaft zur Förderung der Wissenschaften, Berlin,* Ernst Telschow, (Ed.), pp. 5–6. Signed published document.[1]

> There is no better way for a nation to publicize itself around the world and to present itself as a leading upholder of civilization than for it to extend human knowledge, find new means of employment and improve the health of present and future generations. That is why leadership in the area of the sciences is not only of hypothetical value; it is also of eminent importance nationally and politically. That an economic value can also be added to these goes without saying.

These words were taken from the petition in which Adolf von Harnack endorsed the foundation of the Kaiser Wilhelm Society in 1910, right in the lap of peace.[[2]] Since then the German nation has come to know war and a wartime economy. It seems almost superfluous to indicate the importance of the sciences to warfare. The development of the modern weapon, the production of synthetic raw materials, the physical maintenance and medical care of men in combat, all rest on the results of scientific research.

The institutes of the Kaiser Wilhelm Societies are prominently and helpfully engaged in all these areas. Cooperation on a large scale has been established. All research is inspired by only one goal: the quickest evaluation of findings for warfare.

The towering importance of independent research remains alongside this teamwork, which is determined by the war.[[3]] Developments in all scientific disciplines reveal that it was always the feat of individual scientists that marked out science's route, often for decades to come, until a new peak performance determined its direction and content. A[lbert]. Vögler

[1]Since Albert → Vögler's foreword is not dated, we can only assume it was written somewhere in the second half of 1942 (see → Telschow's foreword two years earlier, doc. 81) and certainly not before the end of the period covered by the yearbook: April 1, 1941–March 31, 1942.

[2]The Estonian-born Protestant theologian and church historian [Karl Gustav] Adolf von Harnack (1851–1930) was noted for his vocal opposition to the annexation policy in the first World War in favor of domestic reforms. He served as president of the → *KWG* 1911–29. On Harnack and the foundation of the Society see, e.g., Vierhaus & vom Brocke (Eds.) [1990], pp. 26–142.

[3]Among the topics covered in the yearbook, Konrad Meyer's talk on planning and reconstruction in the occupied territories in the East: 'Planung und Aufbau in den eingegliederten Ostgebieten', pp. 250–273, reflects most clearly the emphasis on war issues.

98 The Führer's Decree on the Reich Research Council [June 9, 1942]

Source: 'Erlaß des Führers über den Reichsforschungsrat. Vom 9. Juni 1942.' *Reichsgesetzblatt* (Berlin), June 15, 1942, Part I, No. 64, p. 389.

The necessity to develop all available resources most efficiently in the interest of the State dictates that a concerted effort be made not only in peace but primarily also in wartime to engage scientific research and to orient it towards the desired targets.

I therefore appoint Reich Marshal Hermann Göring to form a **Reich Research Council** for this purpose as a separate legal entity, to himself assume the presidency of the same, and to provide it with a charter.

Leading men of science working in teams must make a priority research that contributes to war combat within their fields of specialty.

The former Reich Research Council under the Reich Minister of the Sciences, Education, and Culture is to be incorporated within the new organization.[1]

The funds needed for research purposes are to be secured out of the Reich budget where they cannot be raised in the form of supporting contributions by research interest groups.

Führer Headquarters, the 9th of June, 1942.

The *Führer* Adolf Hitler

Reich Minister and Chief of the Reich Chancellery Dr. Lammers[2]

---

[1]The *Reichsforschungsrat* (→ *RFR*) was created on Mar. 16, 1937 by Reich Education Minister → Rust (see here doc. 52) and was now being reorganized when it became evident that it was not fulfilling its purpose of centralizing research and taking the place of the *Deutsche Forschungsgemeinschaft* (→ *DFG*). See, e.g., Goudsmit's account of the role of the 'State Research Council', doc. 111. On the persistent malfunctioning of this monitoring agency see, e.g., Ludwig [1974], chap. 6, and Simon [1948], chap. 6. On the science policy of the National Socialist government see also the Introduction, sec. 3.2, and references there.

[2]Hans Heinrich Lammers (1879–1962) was chief of the Reich Chancellery and A. → Hitler's closest legal adviser throughout the Nazi regime. He studied law in Breslau and Heidelberg and was appointed judge of the Beuthen lower district court (*Amtsgericht*) in 1912, becoming ministerial councillor at the → *RIM* in 1912. In 1933, he was named undersecretary and chief of the Chancellery, as well as member of the *Akademie für Deutsches Recht* and Prussian councillor of state. In 1937 he became Reich minister without portfolio; and on Nov. 30, 1939, Cabinet Minister (*Ministerialrat*) of Defense. In 1940 he was promoted to *SS-Obergruppenführer*, and chaired the cabinet meetings in Hitler's absence in January 1943, gaining considerable influence as a result. Lammers was one of Hitler's three proxies appointed to co-sign all orders from 1943 onwards; he was arrested on Hitler's orders as a result of intrigues by a rival top official. He was imprisoned by the Allies after the war; and in 1949 he was charged for crimes against peace and for having supported the leadership's anti-Semitic measures by giving them legal sanction. Sentenced to 20 years in prison, he was released early, however, in 1952. See Wistrich [1982], p. 216; Führerlexikon [1934], p. 267.

99 Hermann Göring et al.: Record of a Conference Regarding the Reich Research Council, July 6, 1942

Source: Bundesarchiv, Militärarchiv Freiburg, and National Archives, Washington: Milch Documents, Vol. 58, excerpts on pp. 3,663–3,664 and 3,707–3,708. Also in Irving (Ed.) [undated], nos. 29 1023–29 1025. Typed transcript of stenographic notes from a meeting on Monday, July 6, 1942 (11 A.M.–1:45 P.M.).

Participants: Reich Marschal [Hermann Göring[1]]; General Field Marshal [Erhard] Milch;[[2]] Reich Marshal [Albert] Speer;[[3]]; Reich Marshal [Walther] Funk;[[4]] Field Marshal Reich Minister [Wilhelm] Ohnesorge;[[5]] President [Max] Herz;[[6]] State Secretary [Paul] Körner;[[7]] Assistant Secretary [Werner] Gabel;[[8]] Prof. [Eduard]

---

[1]H. → Göring was the chairman of this meeting at the Reich Aviation Ministry (→ *RLM*) on the recently reorganized Reich Research Council (→ *RFR*) (cf. doc. 98). It was hoped that as its new president he could control German research with the same efficiency and discipline as he did the aviation sector. Göring also controlled the *RLM*'s research centralizing agency *Forschungsführung*, created a month earlier in June 1942.

[2]E. Milch (1892–1972) was general field marshal of the *Luftwaffe* and Göring's deputy. After serving in World War I in the Air Force, he followed a career in civil aviation and was director of the finance division of *Lufthansa* 1926–33. In 1933 Göring appointed him state secretary at the *RLM* and also chief of armaments for the *Luftwaffe*, which he was instrumental in building into a powerful force. In 1937 he was appointed vice-president of the German Academy of Aviation Research. In 1938 Milch was promoted to colonel general and in 1940 to general field marshal, also serving as air inspector general (*Luftzeugmeister*) 1941–44. In 1943 Milch attempted unsuccessfully to warn Göring of the danger posed by American aircraft production. In 1947 he was sentenced to life imprisonment but was released in 1954. Milch was himself a 'half-Jew'.

[3]At this time A. → Speer was Minister of Armaments and War Production and leader of the party's main office for technology and was also interested in the German uranium project.

[4]W. Funk (1890–1960) was appointed plenipotentiary of the war economy in 1938. After studying law, economics and philosophy at Berlin Univ., he became editor of the *Berliner Börsenzeitung* in 1920. He joined the party in 1931, becoming Hitler's personal economics adviser and chief of the party office for private economics. Funk proposed the public works program to stimulate the economic depression in Germany. He was elected to parliament in July 1933 and appointed press chief of the Reich government and undersecretary of the Reich Ministry for Public Information and Propaganda under J. → Goebbels. 1938 he was named Economics Minister and soon afterwards president of the *Reichsbank*. At the Nuremberg trials in 1945 he was sentenced to life imprisonment, but was released in 1957 due to ill health.

[5]The engineer W. → Ohnesorge, who had studied physics under P. → Lenard, took over leadership of the Reich Postal Ministry (→ *RPM*) in February 1937.

[6]The cardiologist M. Herz (1865–?) had been president of the Viennese Society for Physical Medicine since the 1930's. Upon completing his studies in internal medicine at Vienna University, Herz worked at the general hospital 1890–93 and became unpaid lecturer at the university in 1895. He founded the Viennese Medical Club in 1892, which was later renamed the Society for Internal Medicine.

[7]P. Körner (1893–?) was appointed state secretary for the Prussian Ministry on April 20, 1933. After holding managerial positions in industry he became Göring's personal consultant at the Prussian Interior Ministry in February 1933. He had been a party member since 1926.

[8]The food chemist *Ministerial Dirigent* W. Gabel (1904–1982) studied chemistry at the

Houdremont;[9] Colonel General [Fritz] Fromm;[10] State Minister [Hans-Joachim] Riecke for State Secretary Backe;[11] State Secretary [Leonardo] Conti;[12] Admiral [Karl] Witzell;[13] Prof. [Rudolf] Mentzel;[14] Prof. [Karl] Brandt;[15] Prof.

Karlsruhe Polytechnic and the Universities of Giessen and Marburg 1922–32, obtaining his doctorate at the latter in 1932 under K. Brand. In 1940 he submitted his habilitation thesis on lead levels in the liver and became lecturer at the Hanover Veterinarian School in 1943 and director of the Hanover municipal institution for chemical analysis. He later became ministerial councillor at the Federal Ministry of the Interior in Bonn and in 1952 was named honorary professor at Bonn University.

[9]The Luxemburg-born E. Houdremont (1896–?) took his doctorate in engineering following completion of his studies at the Berlin Polytechnic and was teaching assistant at the ferrous metallurgy department there 1919–21. He then moved into industry in 1921 and was employed at → *Krupp* in middle management, where he was promoted to director of operations in 1932. Houdremont specialized in steel research and development, also exploring aviation applications. He was appointed titular professor at the Aachen Polytechnic in the 1930's.

[10]*Generaloberst* F. Fromm (1888–1945) was armaments chief at that time. On Fromm see also footnote 1 of doc. 88.

[11]H.-J. Riecke (1899–?) earned his degree in agriculture in 1925 at Leipzig and after various consultancy positions was appointed state minister for Lippe in Detmold on May 22, 1933. Herbert Backe (1896–1947) studied at Göttingen and was teaching assistant at the Hanover Polytechnic 1923–24, becoming general councillor on agriculture and state secretary in the Reich Nutrition Ministry in Berlin in 1934.

[12]L. Conti (1900–1945) was chief physician (i.e. surgeon general) and one of the old guard of the Nazi Party, joining the *SA* in 1923. He built up its Sanitation Corps and also founded the National Socialist Doctors League in Berlin. In 1932 he was a member of the Prussian diet and in 1939 was appointed Reich Leader of Public Health and Reich and Prussian State Secretary for Health and was responsible for the murder of large numbers of mentally impaired German citizens. Conti was elected to parliament in 1941 and promoted to lieutenant general (*SS-Gruppenführer*). He committed suicide in custody in 1945.

[13]General-Admiral K. Witzell had been head of the Navy's Ordnance Office since the mid-1930's. Together with Milch from the Air Force and von Loeb from the Army, Witzell was a member of the Armaments Committee (*Rüstungsrat*) established on May 6, 1942 under the chairmanship of Reich Minister Speer. The committee's function was to improve coordination of research conducted at the ordnance offices of the three Armed Forces through the creation of a centralized planning agency.

[14]The Nazi science policy-maker R. → Mentzel had been appointed professor of military chemistry at the Berlin Polytechnic's defense technology department, and was also second vice-president of the → *KWG*.

[15]Karl Brandt (1904–1948) became one of Hitler's personal physicians in 1934. He started his medical studies in 1922 at Jena, Freiburg (Breisgau) and Berlin. After entering the *NSDAP* on 1932 and switching membership from the *SA* to the *SS* his career took off; a lower officer in the *SS* in 1934, he was promoted to battalion leader in 1937. In September 1939 he was appointed together with Philipp Bouhler to organize and carry out the euthanasia program for the 'extermination of unfit life'. In 1942 Brandt was named general health commissioner, thus coordinator of all the experiments, including his own, conducted on concentration camp prisoners. By Führer decree in 1943 he became a central figure in control of medical care and supplies as well as the coordination of medical research. In 1947 he was convicted to death at Nuremberg for his involvement in human experiments and hanged in the following year. His name appears last in a *RFR* membership list of 14, which according to Ludwig [1974], p. 235, was compiled in order of importance.

[Adolf] Baeumker;[16] Prof. [Ernst] Wagemann;[17] General Director [Albert] Vögler;[18] Reich Marshal [Alfred] Rosenberg;[19] 3 parliamentary stenographers.

[...] I would like to say one thing right at the start. What the *Führer* rejects is regimentation *per se* of science, done on principle: Yes, this product may well be very valuable, extremely valuable and it would take us very far. But we cannot exploit it, simply because the man is married to a Jewess or because he is a half-Jew.—You are shaking your head, Schultze![20] Your Lecturers League is the worst of all at it.[21] I just wanted to say this in passing. It is the living terror at universities in this respect.—But I only wanted to say: Precisely this has to

[16]Ministerial Councillor Adolf Georg Heinrich Baeumker (1891–1976) had headed the *RLM*'s research department since 1935. In World War I he had been an officer in the flying corps. 1920–22 he worked in weapons and equipment inspection at the → *RWM*, serving as official expert in the Troops Office there 1924–26. 1926–27 Baeumker was the Reich delegate on aviation issues at the League of Nations' disarmament conference in Geneva and was official expert on technological developments in civil aviation at the Reich Transportation Ministry 1927–32. In 1933 the office was incorporated into the Reich Aviation Commissioners Office and Baeumker became head of aviation research. With the establishment of the *RLM* he remained head of the Technology Office, retaining management positions there until 1945. In June 1942, he was on a four-person committee controlling the newly founded '*Fofü*' (*Forschungsführung des Reichsministers der Luftfahrt und Oberbefehlshabers der Luftwaffe*) together with L. → Prandtl as chairman, the engineers Walter Georgii and Friedrich Seewald): cf., e.g., Ludwig [1974], p. 237, and Stuchtey et al. (Eds.) [1938], pp. 42–43.

[17]The Chilean-born economist Ernst Friedrich Wagemann (1884–?) habilitated at Berlin University in 1914 and was official expert at the War Nutrition Office in Berlin 1916–18. In the following year he was appointed associate professor at Berlin University and served simultaneously on the State and Government Economics Committee at the Prussian Agricultural Ministry and was also councillor at the Reich Economics Ministry. In 1923 Wagemann presided over the Reich Statistics Offices, also becoming administrator of the Reich elections. In 1925 he was appointed director of the Institute for Cyclical Research. See Ludwig [1974], p. 100, and Albert Wissler: *Ernst Wagemann. Begründer der empirischen Konjunkturforschung in Deutschland*, Berlin: Duncker & Humblot, 1954.

[18]A. → Vögler was director of *Vereinigte Stahlwerke*, one of the leading steel manufacturers in Germany. He was also president of the *KWG* from 1940 to 1945.

[19]A. → Rosenberg was the leading National Socialist ideologue and in charge of the Foreign Affairs Office.

[20]Walther Schultze (1894–?) had been appointed head of the → University Lecturers League (→ *NSDDB*) in 1935. After acquiring a degree in medicine he served as a physician in the → *SA*. In 1933 he was state commissioner and physical education leader at the Bavarian State Ministry as well as president of the State Academy of Medicine in Munich. In 1934 Schultze was named honorary professor at Munich Univ. He was a dogmatic party hard-liner, insisting on consistent adherence to the anti-Semitic laws. Göring, on the other hand, like many other Nazi officials had reverted to a more pragmatic course in the face of the growing number of military set-backs and the realization of the necessity for strong research and development.

[21]The Lecturers League and the Student League (→ *NSDStB*) implemented the anti-Semitic laws in academia with fanatic dedication, monitoring professorship appointments and contributing signficantly to the politicization of universities. See sec. 3.4 of the Introduction, as well as Kelly [1973]. Cf. in particular doc. 11 as well as docs. 14, 20 and, e.g., Walk (Ed.) [1981].

be avoided. I have just presented this personally to the *Führer*. We have now hitched up a Jew in Vienna for another two years and another one in the field of photography, because they have those particular things we need and which would absolutely take us a step forward at this moment. In this situation it would be madness to say: He has to go! Granted, he was a very great scientist, a fantastic mind, but he has a Jewish wife and cannot be at university, etc. The *Führer* has granted exceptions in this case in the field of the arts, all the way down to the operetta, in order to preserve it.[22] He will approve and sanction exceptions even more readily where really very big research projects or scientists are involved. Really, we ought to be punishable for rejecting a man of the highest intellectual capacity in the field of science for such reasons or if we refused him the opportunity to conduct research because, as I said, he has a Sarah[23] or is perhaps a quarter-Jew or whatever else.[24] This cannot be; instead it has to be laid out before the *Führer*, as in so many other fields: This man has the following accomplishments to his name, the following usefulness! And then the *Führer* was always the first to say: Quite right, so be it, an exception must be made here! I wanted to ask the Party Comrades in particular to think of this and to facilitate, not complicate my work in this area. The prerequisite is obviously always that we get an actual advantage out of it. It should not become a passport for any man connected to the Jews through marriage who happens to be or to have been at university or who had once done some little trifle there; rather, I want to see it applied to those persons who can really achieve something for us in exchange in this war. On the other hand, we naturally cannot be so indecent as to accept the accomplishments and the work of such a man and let him help us, and then kick him in the pants at the end of the war because he is married to a Jewess. Of course we will not make the two Jews I have mentioned into university professors, now or after the war for this reason, even though only a minor field is involved; but we will compensate the service in the way always valued most highly by the

[22] A smoothly running entertainment industry reinforcing the illusion of a sound world formed an integral part of Propaganda Minister Goebbel's plan. The popular Hungarian-born operetta composer in this period Franz (Christian) Lehár (1870–1948) implied in the text had been married to a Jewish Viennese woman, Sophie Paschkis (1878–1947), since Feb. 20, 1924. According to a personal evaluation drawn up by the local Nazi cultural association in Halle dated Nov. 27, 1934, the fact that his closest friends and all his librettists were Jews still could not justify any restriction in Lehár performances. Hitler even personally awarded the Goethe Arts and Science Medal on Apr. 30, 1940 to the composer on his 80th birthday as a master of 'German operetta'. See Poliakov & Wulf (Eds.) [1959/83], pp. 321–322.

[23] In 1935 all female citizens of Jewish extraction were required to add Sarah to their official name as a means of distinguishing them as Jews. Göring's formulation here is deliberately intended to temper the official line, which considered such mixed marriages as 'racially unclean'.

[24] This realization obviously was very late in coming, since the majority of the most talented scientists affected by the anti-Semitic legislation were already in detention or had emigrated in the first years of the Nazi regime. The crystallographer Paul → Ewald, for example, whose wife was Jewish, resigned his rector's post at the Stuttgart Polytechnic in 1933, and finally decided to emigrate in 1937 after the issuance of the 1935 Nuremberg Laws.

Jews, namely, with a check, and then send them packing.[25] Then the matter is settled. But right now we need such people scattered here and there in science.

Now please, announce your wish to speak.[...[26]]

I wanted to ask the three Armed Forces branches[27] anyway again to reconsider in the next 2–3 days the special assignments and special constructions underway, so that each Armed Forces branch can say from his point of view how best to conceive it. As, this is the most secret of secret matters, of course, and some kind of agreement has to be reached.

Other things also require tremendous secrecy and must be provided with the strictest secrecy conditions even among the scientists. Take the area of such burning interest, atomic fission, for example.[28] I would like to see the utmost secrecy kept in this area even within scientist circles, and this is a bit difficult. The scientist has led a life of individuality for too long in this regard. It can make you sick to read that a conference had been held on this or that physical or chemical topic in New York or in London—I am speaking about the time before the war—and you see with what urgency he disperses his results, as if he could not hold it in his bladder any longer.[29] Everything was wonderful, everyone was informed. Only those who would have been interested in finding an application for it usually never found out about it. This is because first of all they cannot read the journals that scientists have published; in any case, I am too stupid for that. So many formulas pop up in there that you cannot pick your way through. As a result we who are interested in such things usually heard nothing about it. But conversely our colleagues in England, France and America were completely informed about what kind of an egg their colleague in Germany had hatched. ... Here also we will have to proceed according to the *Führer*'s maxim—the famous three points: Only say to whom, when and where. That's self-evident.

---

[25]Here Göring is using the derogatory anti-Semitic image of the money-grabbing *homo oeconomicus*, whose services may be exploited and then disposed of as quickly as possible.

[26]The chairman's invitation was followed by remarks by Rust, Schultze, Vögler, Conti, Riecke, Witzell, Funk, Klopfer and Wagemann omitted in this excerpt.

[27]These included the Army (*Heer*), the Air Force (*Luftwaffe*) and the Navy (*Marine*). As Reich Field Marshal, Göring was simultaneously Commander in Chief of the Air Force.

[28]Göring is referring to the German uranium project (*Uranverein*) which had been in operation since 1939. Various research teams explored the possible applications of nuclear fission, initially under the supervision of the Army Ordnance Office (→ *HWA*). Until 1942 this research had as its purpose the development of an atomic bomb as well as a slow-burning reactor until 1942, and then in the middle of that year was confined to the development of a mini reactor, the 'uranium machine' (once the *HWA* had relinquished this research to the *RFR*, due to its longterm nature): See Walker [1989]a, Irving [1967] as well as here the Introduction, sec. 5.4, pp. lxxxiiif., and docs. 103, 104, and 115 where Heisenberg omits any allusion to the military applications he had stressed previously at least until 1942.

[29]This sentence illustrates Göring's calculated use of obscene imagery to contest the commonly held assumption of the supranationality of science (equating urine with the product of such research).

100 Hans Feitl: War Physics. Specific Weight 'Zero'. Submarine Diving Conditions [May 30, 1943]

Source: Hans Feitl, 'Physik des Krieges. Das Spezifische Gewicht "Null". Voraussetzungen des U-Boot-Tauchens.' *Das Reich*, No. 22, May 30, 1943, p. 10.

Fig. 20: 'A demonstration model made of plexiglas with which the students can observe closely individual submersion processes, such as, flooding, trimming, and adjustment. PK. photo: War correspondent Feitl (Military Photography Service)'

A submerged submarine has practically the specific weight of zero, that is, it has been equalized to the surrounding seawater through trimming and adjustment. On the 'flood' command the air vents and flood valves are opened, whereupon the air that is in the second hull that encases the pressure hull escapes, and seawater flows in through the flood valves. As a result of the excess weight the vessel sinks downward and then levels off at a particular depth. Any imbalance fore or aft, which can possibly occur as a result of a few more men being at the bow or at the stern than foreseen at the moment of diving, is equalized through trimming. The Chief Engineer can determine this imbalance exactly with the instruments in the control room. He then orders that an amount of water be pumped into the lighter section of the boat to compensate for the overweight in the boat's heavier section. Only then is the vessel in a balanced horizontal position.

Once the submarine is suspended at the same specific weight as the surrounding seawater, it can then maneuver with the slightest propeller action; it can be held in the horizontal position, lifted, or submerged with the aid of the bow and stern rudders. Considering that a periscope may only extend a few centimeters above the surface of the water so as to be as invisible as possible to the opponent, it becomes clear how important trimming and adjustment are and what kind of experience the Chief Engineer must have in order to keep a submarine weighing from 250,000 kg to over 1,000,000 kg exactly at periscope depth.[1]

At the submarine school's physics laboratory, students are shown with a glass model how this maneuver works. For this purpose the liquid in the ballast and trimming tanks is dyed different colors to mark the maneuver more clearly. If the vessel is submerged only in sweet water, then it would suffice to make a single weight equalization determination. At each dive it would then always only be necessary to adjust the same amount of water, in order to maintain the submarine at the desired state of neutral buoyancy. It is different in seawater, since the specific weight varies constantly with depth, the structure and vegetation on the ocean floor, the region, the temperature, the season, and currents. At a depth of 30 meters there is already a pressure of 3 atmospheres. This means there is a pressure of 30 kilograms on each cubic centimeter of a surface and a proportionate increase in the specific weight of the water in relation to the same water at the surface. The nature of the vegetation on the ocean floor, the sun and the season are again of great significance to the lives of many trillions of minute living creatures: microorganisms. The specific weight is likewise affected by their presence. Heat from the sun results in an increase in the salinity, with a related rise in the specific weight. The salt content is increased additionally by the excrements of fish and plant life. The specific weight also increases during and following a major storm due to sand disturbance. Of course, the latter facts are particularly noticible above all in shallow waters.

If the water's specific weight changes by only a one-thousandth part of a gram, for a mid-sized submarine weighing 500 tons this means that the specific weight of the vessel must also be altered by a thousandth part [of a gram], but this already corresponds to a weight difference of over 500 kg. To bring a submarine to a state of neutral buoyancy, however, it must be balanced out exactly to within 5 kilograms using the ballast tank. An alteration of the seawater's specific weight by only a 1 millionth would already exceed this weight a hundredfold.

In practice, the specific weight of seawater is determined with a density gauge. Seawater from outside is let into a glass sampling tube, and a density indicator, similar to those for determining the density of gasoline or alcohol, indicates to an accuracy of one thousandth [of a gram] the density of each seawater sample. The specific weight obtained in this way is then converted into the weight of the vessel and the necessary amount of water is admitted accordingly into the

[1]When World War II began, the German Navy had only about twenty 500-ton, and ten 750-ton U-boats capable of traveling in the open sea. This fleet, together with additional submarines produced during the war, was quite effective in attacking and sinking Allied ships in the first two years of the war. In June 1942, an average of 42 U-boats operating at sea sank 143 ships. But with the introduction of new radar and microwave technology, a dramatic increase in German submarine losses resulted. Thanks to the Allies' specialized anti-submarine bombers and destroyers, the total number of U-boat sinkings rose from 85 in 1942 to a total of 237 in 1943 and 241 in 1944, with the corresponding total weight of Allied shipping falling from 8,245 tons in 1942 to 3,611 tons in 1943 and down to 1,422 tons in 1944; all data from Baxter [1947], pp. 37, 40, 48; cf. also the Introduction, pp. lxxxf.

trimming tank.

Ocean currents are another natural phenomenon that presents exceptional navigational challenges for submarine pilots. There are surface current s and undercurrents, and when they meet they vary greatly in direction and speed. In operations close to the coast, the currents of large rivers must be taken into account. For instance, the brackish water of the Amazon extends almost 500 nautical miles (about 900 km) away from land. At the same time, the specific weight of the water also changes here at the transition from seawater to the brackish river waters.

The submarine war, which has spread to the seven seas, demands from our crews seamanlike and technical achievements that until recently our enemies still considered impossible.[2]

101 Paul Karlson: War Physics. Dropping Bombs. Wind Tunnel Tests [May 30, 1943]

Source: Paul Karlson, 'Physik des Krieges. Bomben im Fall. Untersuchungen im Windkanal.' *Das Reich*, No. 22, May 30, 1943, p. 10.[1]

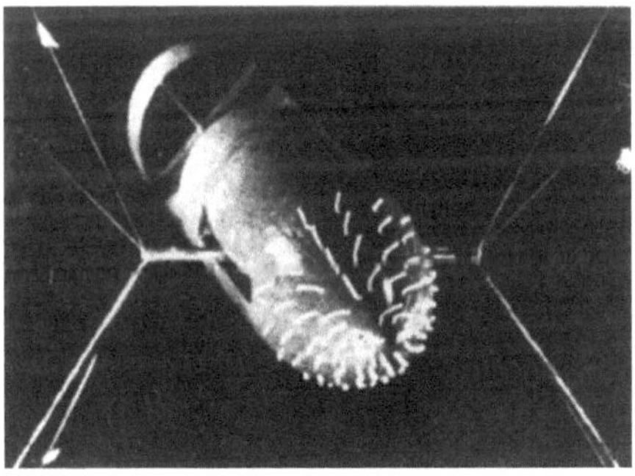

Fig. 21: 'The airflow around the tip of a bomb, which is crucial to the bomb's stable flight, is made visible with glued-on strands of wool. Photo: ZWB Archive'

[2]Contrary to the impression given in this article, by the summer of 1942 Admiral Dönitz had already secretly given up the submarine war for lost to Allied technical superiority, but the propaganda continued unabated. See, e.g., Reinhard Hardegen: *"Auf Gefechtstationen!" U-Boote im Einsatz gegen England und Amerika*, Leipzig: Boreas-Verlag, 1943. For a description of the 'seesaw of submarine warfare' from the Allied perspective see, e.g., Baxter [1947], chap. III, as well as Janet Marily Manson: *Unrestricted Submarine Warfare. The Change in U.S. Policy and German-American Relations, 1939–1941*, Pullman, Wash.: Washington State Press, 1987; Jürgen Rohwer: *Axis Submarine Successes, 1939–1945*, Annapolis: Naval Institute Press, 1983. The application of nuclear engines in submarines was only a matter of speculation at the time: See, e.g., doc. 95, footnote 13.

[1]P. → Karlson's article directly follows H. → Feitl's, doc. 100, with both carrying the main headline 'War Physics'.

"People who advocate dropping bombs from an airplane do not know that the effect of one or even several projectiles is equal to zero. Contrary to the common view, I maintain that a bomb attack can never bring down a courageous opponent, no matter how fiercely it is carried out." That is what Captain Ferber says, one of the very first flight pioneers in aviation.[2] "In the summer of 1918 .... there were so few casualties caused by air strikes that they were balanced out by the lower percentage of daily traffic accidents. (It will be a completely different matter when in a single night half of the inhabitants of a city lie buried under the rubble of their houses ...". This is the opinion of military writers who supported the offensive bombing war in 1938.)

We all know it: The ghastly utopias have become a reality; war has crossed the frontlines and reaches deep into the hinterland. In some areas of our Reich it has become as normal to women amd children as it has to soldiers at the front—with one difference, of course: The soldier fights and strikes back. The civilian population, however, which is only capable of passive resistance, dampens the impact of enemy attack as far as is humanly possible—but the German Air Force is its last hope.

The communiqués of the Armed Forces High Command inform us daily of the military operations. Behind these reports are the armaments industry's relentless efforts. Behind this again, is the indefatigable activity of the Air Force development and testing departments, as well as the efforts of research laboratories and institutes.[3]

Considering that a bomb must be dropped at an altitude of 5,000 m about 3,000 m ahead of the target, that it is then en route for almost half a minute, that an error of a half second during the release already amounts to a difference of 50–80 m at the target, that a target the size of a battleship only looks like a pencil stroke of a thumbnail's length, that even a light crosswind can displace the bomb from its flight path, and that many targets are also mobile as well and change their course in unforeseeable ways as the bomb is falling—considering all this, it will certainly seem surprising that a pilot can hit a target as accurately as he in fact can. Added to these difficulties, which are unavoidable by their very nature, are the various quirks of the object itself. As a result of careful, time-consuming trials bombs now drop the moment the bombardier presses the release button of his

[2][Louis] Ferdinand Ferber (1862–1909) studied at the École polytechnique in Lyons 1882–84, became artillery lieutenant in 1884 and was promoted to captain in 1893. He was professor at the École d'application de Fontainebleau 1897–1900. Ferber emphasized future civilian uses of aviation research, in which he had been actively involved since 1893, over its military potential. His notices to the Académie des Sciences on the theory of airplanes and helicopters led to his call at the École de l'aérostation de Chalais-Meudon in 1904 but his research was aborted for lack of funds. His research led to many improvements to the Wright brothers' airplane design. Ferber died in 1909 in a fatal flight test in a Voisin biplane. He was the author of many works on aviation development.

[3]For a description of the contemporary Allied research and development and on aerial warfare and bombing in particular see, e.g., Baxter [1947], chapter VI.

bomb-sight. They used to sometimes leave late and had landed correspondingly too far; or they had even gotten jammed—and maybe only finally bothered to fall above their own airfield. The bomb ballistics specialist is also familiar with 'runaway' bombs, which take a completely different flight path to those that fall in the prescribed manner. Thanks to the results of precise studies and trials, new bomb models no longer fall into a spin en route, in which case, rather than hitting the ground head on and getting stuck in the earth, they ricochet back upwards in hot pursuit of their own airplane and finally detonate far away from the target.

Bombs dropped at targets at sea are supposed to detonate underwater close to the ship's hull, so that the water pressure crushes its steel casing. But quite frequently they leap out of the water again and detonate in midair. Torpedoes show the troublesome characteristic of crashing headlong into the ocean floor or of getting stuck there; or else, they bounce back up out of the water, somersault in the air, glance off the water's surface again, and snap in two in the process.[4]

Mines are another story with their thin casings and often touchy fuses. Their shells can shatter upon impact with rock, cement, or steel. The explosive charges then explode harmlessly, instead of detonating in the proper way. Previously, a very substantial portion of mines were duds; but it should not be forgotten that the fuses are expected to function with amazing precision. For certain purposes the fuse must be a real 'touch-me-not' when put in place, while beforehand it must withstand rough shipment conditions. For example, in winter combat even heavy bombs must detonate if possible already at first contact with the snow and not only after penetrating innocuously into the deep deadening blanket. Bombs dropped on enemy landing craft must also detonate themselves when they are still above the ocean surface otherwise they only produce a steep and impressive fountain, which nevertheless poses no danger to the landing craft. In other cases the bomb must only explode after piercing the enemy ship's armor-plating—but not, as also happens, only after it had cut clean through the whole ship and is in deep, open waters again.

Thus the bomb and fuse design engineer certainly has to satisfy a lot of requirements that are in part very contradictory. Today all the resources of modern aeronautical research are used to meet them. First and foremost, of course, a very modest aid: paper and pencil. Experiments at the theoretical level [*Gedankenversuch*], or careful calculation and reflection, can save an infinite amount of work. Today we can calculate accurately the flight path of any bomb under all conditions in advance. The probability of strikes and, for example, the chances of hitting a particular target from a given approach with at least a fragment, can also be predicted quite well. Thus already at this stage major preselections can be made.

But naturally, practical tests are indispensible; and here the wind tunnel, the aerodynamicist's wonderful tool, comes into its own. A descending bomb passes

[4] On contemporary German torpedo technology, see in particular Rössler [1984].

through stationary air at high speed. If a rapidly moving stream of air is blown on to a stationary bomb, it amounts to the same thing. Thus the optimum shape of a bomb can be determined from a 'bomb drop on the ground'—or rather, the best compromise can be found; since, the technical considerations of airflow are not the only factors involved in bomb design. Many other kinds of requirements must be met: Often the explosive must be stored in an awkward position, in order to accommodate its center of mass; its sturdiness must be taken into account; the armaments industry demands a shape that is easily produced in quantity; the aircraft engineer sees to it that the bomb can be stowed well in the fuselage; finally, the troops reject any bomb, no matter how well conceived, that is not equal to the rough conditions of combat: To name only one example, a tail unit that bends easily is unacceptable—because bent tail units lead to unwanted course deviations. Once the bomb is dropped, from then on it is at the mercy of the forces acting upon it; there is no driver to bring it back on track. Wind tunnel experiments have taught us today how to stabilize the bombs reliably; that is, we have learnt how to make them fly accurately and correctly.[5] The troublesome spinning can also be analyzed today in special tunnels and can be eliminated through the appropriate measures. In high-speed tunnels the conditions to which bombs are subjected when they are dropped from very great heights—6,000 m and more—were analyzed, as well as their speed, which then exceeds that of sound. Finally, the crucial seconds or fractions of a second were examined during which the bomb ejects from the speeding airplane. Slow-motion cameras allow this important process to be observed from one one-thousandth of a second to the next; and our scientists and testers have explained beyond doubt why some bombs slide out of their bays at a shallow angle, while others leave too steeply. We now know why some get stuck despite all the measures taken in their design, and what causes fuse duds.

Of course, the wind tunnel cannot do everything; it must be complemented by tests out in the open. Since it is difficult to observe an actual bomb drop, bombs are now being launched: The canon is an important tool in modern bomb ballistics. With it the bomb's free flight and the manner of its arrival at the target can be observed at an exactly determined time and from a stationary place, without always necessitating the use of an airplane.

[5]See here Fig. 21 for a large wind tunnel. In Germany L. → Prandtl led in the development of wind tunnels at the *Aerodynamische Versuchsanstalt* in Göttingen in close cooperation with the *Heereswaffenamt* (→ *HWA*) and particularly with the German Air Force research divisions, which also conducted research of their own (see also Fig. 12, p. 189). The *Luftfahrt-Forschungsanstalt* at Braunschweig, for instance, operated a wind tunnel 8 meters in diameter. Cf., e.g., Trischler [1992]a, part II, on the expansion of aeronautic research during the Nazi period.

102 Carl Ramsauer: The Key Position of Physics in Science, Technology and Armament [June 18, 1943]

Source: C. Ramsauer, 'Die Schlüsselstellung der Physik für Naturwissenschaft, Technik und Rüstung.'* *Die Naturwissenschaften*, Volume 31, Issue 25/26, June 18, 1943, pp. 285–288.[1]

If physics research and education is to be encouraged to the degree that meets the actual importance of physics as a whole or if German physics is to be assessed accurately as a factor of military strength, we must first understand the precise nature of physics. Physics should not only be considered as an isolated science, but should be seen simultaneously in the light of its general importance, which extends well beyond the bounds of an individual scientific field. This will be demonstrated in the following text, which will address both experts and the general public. As a result, many known and obvious things must be mentioned; but the resulting **overall** picture should be new and of some value.

The main task of physics consists in investigating the **fundamental laws** of all existing things and of all events. Consequently, physics occupies a decidedly key position, since obviously every natural science and every technology is ultimately built upon these fundamental laws.

How this key position makes itself felt is evident most clearly and completely in the example of electrical engineering. The creation of the galvanic cell by the physicist Volta and the subsequent physical discoveries it enabled, such as the deflection of the magnetic needle, together introduce direct-current engineering, and thus, the first branch of electrical engineering.[2] This development soon reaches its natural limits, however, despite the great importance of telegraphy. Then the physicist Faraday gives a fresh and powerful stimulus with the discovery of induction.[3] The rapid development of alternating-current technology follows, until the sum of all possibilities contained in the law of induction is reached. Then physics again provides a fresh impetus: Hertz's discovery of electrical waves and

* [Footnote 1), p. 285, in original text:] "Expanded version of a lecture held at the Lilienthal Society for Aeronautical Research on the 6th of August, 1942."

[1]C. → Ramsauer's article Ramsauer [1944] is also based on the same lecture held at the *Lilienthal-Gesellschaft für Luftfahrtforschung*. This research organization emerged out of the aeronautical associations *Wissenschaftliche Gesellschaft für Luftfahrt* and *Vereinigung für Luftfahrtforschung* on June 12, 1936. Its presidency was held jointly by L. → Prandtl, C. → Bosch and A. Baeumker (on Baeumker see footnote 16 of doc. 99), and H. → Göring was its powerful patron: See Trischler [1992]a, pp. 211f., 236ff., (Ed.) [1992]b, pp. 149ff.

[2]The Italian physicist Alessandro Giuseppe Antonio Anastasio Volta (1745–1827) invented the so-called voltaic pile in 1800, a battery that produces electric current through chemical action. Many new effects were discovered from this invention, such as electrolysis, the production of heat from currents, and its influence on a magnetic needle. The latter effect, electromagnetism, was discovered by the Danish physicist and chemist Hans Christian Ørsted (1777–1851) in 1820.

[3]The English chemist and physicist Michael Faraday (1791–1867) made his discovery of electromagnetic induction, the basis of all electromotors and electric generators, in 1831.

the technological repercussions of this new idea by Marconi and his successors.[4] As the limits of development approach here as well, the new impetus comes from the research results of the physicist Lenard: the technical application of steering electrons with electrical fields,[5] which is the guiding principle behind amplifier valves. This impetus promoted the light-current engineering of the time to enormously important new areas.

The crucial point of this example is not the indisputable but nevertheless banal observation that electrical engineering grew out of physics, but rather the following fundamental fact: Each new technical development only goes as far as the impetus it received from physics takes it. The technical specialty itself is not capable of making fundamental advances from within its own field but must receive these advances again and again from physics.

These relations exist—albeit not so transparently—also in those cases involving a science or technology that initially did not arise out of physics but out of the necessities of practical life. Thus, astronomy emerges to begin with as an observational science, with Kepler's laws as the climax; and only later does it receive the more far-reaching impetuses from physics: Galileo's laws of mechanics and Newton's laws of motion, which are built upon them, provide the basis for astronomy as a science. Spectrum analysis opens up an entirely new field of astronomical research; and modern knowledge of the relation between energy and matter makes it possible for the first time to arrive at an overall understanding of the universe.

Chemistry is in a very similar position. In the beginning, chemistry is half empiricism, half mysticism. It evolves into a system of empirical knowledge; but it only gains its scientific foundation through the physical research of atoms, ions, and electrons. But even on this basis it cannot yet cope with its main problems, such as the fact that the atomic weights are not whole numbers, or such as the system of elements in general. Then physics arrives with the new major impetus: The discovery of radioactivity provides the genesis of the elements. The discovery of isotopes and of the atomic number places the concept of atomic weights in a completely new light.[6] Right now we are again experiencing an enormous leap forwards in physics, which will also set chemistry on completely new foundations, or rather, will create an entirely new chemistry. This new development is atomic fission. Just as the old chemistry breaks the molecule down into its

[4]On Heinrich Hertz's discovery of electromagnetic waves see footnote 5 in doc. 31. The Italian physicist Marchese Guglielmo Marconi (1874–1937) invented the wireless signalling system in 1894, using an induction coil and a spark discharger to transmit radio waves.

[5]Philipp → Lenard's paper on cathode rays originates from his time as Heinrich Hertz's assistant in Bonn around 1886. Hertz discovered that these 'rays' could penetrate thin metal foils, and Lenard used this for the construction of what became known as 'Lenard windows', which enabled further experimentation on cathode rays: J. J. Thomson then demonstrated these rays to be streams of negatively charged particles.

[6]For the early history on generating new elements through neutron bombardment of uranium see, e.g., Herrmann [1964], [1990], and Menke & Herrmann [1971].

atoms, or constructs it out of its atoms, the new chemistry will take apart the basic components of the atomic nucleus or will put these basic components back together again. The ancient alchemic problem of converting metals, upon which chemistry had labored fruitlessly for centuries, is thus solved through an impetus from physics.

Physics's key position becomes most obvious when we examine more closely the process of new scientific or technical fields emerging from physics as a whole. Electrical engineering had become an independent discipline already many decades ago. In contrast, other fields are still in various stages of development. Audio engineering is now in the process of disengaging itself from physics and of becoming an independent technical science. Electron optics is still in the first stages of development; and its technical potential can only be surmised. On the other hand, other areas, such as for example, refrigeration engineering, have already become totally independent and hardly expect continued stimulation from physics any more. Finally, there are also a set of special fields of application closely associated with physics. Three groups stand out in particular: military physics, the physico-medical field, and manufacturing control in the chemical industry.

All these relationships become particularly clear when summarized in a diagram (see Fig. [22]).[7] In the center we have basic physical research. Then follows a ring-shaped area, general experimental and theoretical physics, of which the separate branches of mechanics, acoustics, thermodynamics, optics, electrical science, magnetism, and atomic physics are not specifically marked in the figure.

Fig. 22: The 'Ramsauer medusa' illustrating the role of physics in science. (Labels from the center outwards: Basic research. General physics. Audio engineering, electrical engineering, chemistry, refrigeration engineering. Special Applications (Armed Forces, manufacturing control, medicine, others). Electron optics.)

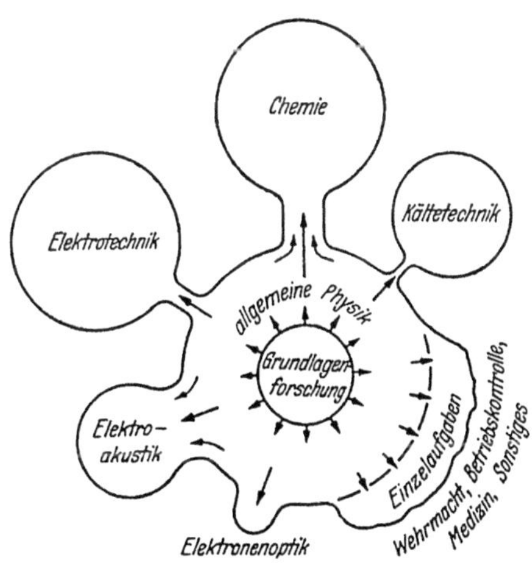

[7]When this figure (Fig. 1 in the original), known as the 'Ramsauer medusa' (*Ramsauer-Qualle*), was republished in an anthology in 1949, and again in 1979 in the *Physikalische Blätter* **35**, p. 664, the caption was slightly altered: Under 'Special applications', the entry 'Armed Forces' (*Wehrmacht*) was simply omitted; see Schmithals [1980].

General physics is in a constant state of growth, partly through fresh impetuses from basic research, partly through additional individual developments from within. The boundaries of the entire field of physics are defined by the material a modern physics professorship presently covers. The intellectual potential contained within these boundaries operates further outwards into the field of engineering, in that its own advances open up new technical branches. All the various phases of this development are represented. Electron optics is at the very beginning stage. Audio engineering has already achieved a certain degree of self-sufficiency but is still closely linked to general physics. Electrical engineering and chemistry are independent disciplines of technical science; but they have important connections to general physics, in that electrical engineering is stimulated further by advances in physics in the area of gaseous discharge, and chemistry, by the advances in atomic and nuclear physics. Finally, refrigeration engineering has only narrow ties to general physics, out of which it had originally developed on the basis of the physical laws [of thermodynamics]; since, it has already assimilated everything that it could use from physics and hardly expects any more progress out of it. The cross-section of ties between physics and technical specialties outlined here with the aim of characterizing the intensity of this relationship, nevertheless only reflects the present state of affairs. The moment we discover how to produce electrical current directly from coal, for example, or as soon as the temperature limit of superconductivity is shifted substantially upwards, then a greater influx of stimuli will pour from physics into electrical engineering. It would be the same in the case of refrigeration engineering, if the production of low temperatures by means of magnetic processes could be made practicable.

This picture is not yet complete until we answer the question, when the development of a technical science away from physics can be considered concluded, i.e., when the relevant technical science has become an independent discipline. This final stage has been reached when the following preconditions are met: – The field's practical applications must have become sufficiently important; – the material covered by the field, i.e., the necessity for specialized knowledge and experience, must have expanded so much that it can no longer be acquired in conjunction with general physics, but rather completely occupies a man's life-work; – the first generation that had developed this branch of physics into a technical science must have died out. Then, as a consequence of these three preconditions and as an external sign of the technical independence of this field, it follows finally: – The new technical science must have become officially recognized by the State and be adequately established as an academic subject, both through staffing and materially at universities or polytechnics.[8]

The whole model only reflects a small segment of the manifold ties between physics and a great variety of areas of technology and the natural sciences. In fact, the number of fields that are emerging or that have emerged from physics,

[8]Universities and polytechnics in Germany were and still are government-run institutions.

as well as the number of fields that are completely or at least to a large part indebted to physics for their scientific basis, are much larger than in the model. Entire groups of scientific or technical fields are sometimes involved. Next to the large electrical engineering group with all its branches, there is, for example, the mechanics group: applied mechanics, the study of oscillations, hydrodynamics, aerodynamics, ballistics; and the astronomical and geophysical group: astronomy, astrophysics, geophysics, meteorology.

Rather than listing all the individual fields of this type, it seems more appropriate to demonstrate in a practical example the quantitative importance of physics within technology as a whole. We would like to show, by means of the course catalog of the *Technische Hochschule Berlin* for [the academic year] 1939/40, what a prominent role physics plays in the curriculum of a large polytechnic and hence in science and technology on the whole. To this end we group the number of lectures and practice sessions of the course catalog according to the following criteria, whereby for brevity's sake, the exercises are always included under the label 'lecture' and are counted as such.

1. Physics lectures for physicists. Total 29.

2. Physics lectures for non-physicists (engineers, chemists, etc.). Total 22.

3. Lectures in fields that owe their existence to physics or would otherwise only exist as a crude empiricism and that are indebted to physics alone for their scientific foundation as well as for their additional scientific impetuses. The following groups belong here: – Mechanics group: Mechanics, the study of oscillation, hydrodynamics, aerodynamics, ballistics. – Heat technology group: Steam-engine building,* thermodynamics, theory of heat, refrigeration engineering. – Electrical engineering group: Heavy current engineering, light-current engineering, high-frequency engineering, telephony and telegraphy, etc. – Technical physics group: Engineering acoustics, applied optics, optical engineering, microscopy, radiography. – Physico-chemical group: Physical chemistry, electro-chemistry, photochemistry, photography, colloid chemistry. Total 327.

4. Lectures in fields that had received their deeper foundations and their decisive stimuli entirely or at least mainly from physics, though they could exist as a systematic science without physics. Included here are: Inorganic chemistry, metallography, mineralogy, astronomy, astrophysics, geophysics, meteorology. Total 64.

5. Lectures in fields that do not have as much basis in physics as the fields under 3 and 4 but that are more or less closely connected to physics. Included here are: the analytical, organic, and technological branches of chemistry; the

* [Original footnote 1), p. 287:] "This does not refer so much to the important but nevertheless somewhat out-dated fact that the steam engine is a discovery in physics, but above all to the fact that the whole modern field of steam-engine building first became a scientific technology through thermodynamics in physics, and that the development of the steam turbine is based essentially on the physical branch of fluid dynamics.—By the way, only the systematic lectures are counted here, not the practical sessions, altogether about 20."

various branches of statics and strength analysis of materials as well as dynamics; finally, fields that—while lacking physical content of their own—are devoted to a substantial degree to the implementation of physical instruments, such as, general metrology, photogrammetry, geodesy, the science of mine surveying. Total: 134.

6. Lectures in technical fields that have little or nothing to do with physics directly. Included here are: – Mathematics; – architecture, civil engineering; – general machine building, ship building and marine engineering, aircraft construction; – mining and metallurgical engineering—insofar as the individual lectures in this category do not also fall under numbers 3, 4, and 5. Total 579.

7. Lectures in non-technical fields. Included here are: Economics and jurisprudence, the fine arts, philosophy, languages, etc. Total 119.

Disregarding the non-technical lectures under point 7, which with their total of 119 make up a scant 10% of the total number of lectures, and comparing the totals of the individual categories against the total number of all **technical** lectures of [about] 1160, the following synopsis is thus obtained:

The number of actual physics lectures for physicists and non-physicists amounts to 4.4%; but **not even remotely** does this portray the real importance of physics. It increases to 33.3%, if all fields that owe their existence to physics are included, and to 38% when we include additional fields that would mostly remain mere empiricism without physics; and finally, it increases to about 50% when fields that still have noticeable ties to physics are added.

But even the remaining 50% are by no means uninfluenced by physics. Aside from the small unavoidable links—consider the physical instruments commonly used in **every** subfield of engineering, and the more minor individual problems, such as, for example, room acoustics—these lectures obviously significantly affect developments in engine construction, aircraft design, etc., which are based upon physical knowledge. Alternatively, they affect the manufacture of needed systems that appear in the wake of physical inventions, such as, the entire network, carriage, and fuse systems of steam and electrical railways.

All in all, the role of physics in present-day technical science is therefore so prominent, that it would be a **huge** mistake to assess physics only as an individual science and not in relation to its key position in technical science as a whole. Besides, its role has not diminished since to insignificance; rather it will influence future technology to at least the same degree that it has influenced it in the past and the present, but probably even more so. Thus, physics is one of the most essential foundations of our economy and our military strength. This applies especially to all competition with other nations on the economic and military levels. A finished technology can be copied without too much difficulty. Creating new technology and new science, however, is tied to the intellectual potential of the race. The State must also make allowance for the key position of physics in its decisions on funding physical research or on the optimal organization of physical education.[9]

103 Abraham Esau: Memorandum on the Budget of the Plenipotentiary of Nuclear Physics Research [November 19, 1943]

Source: Bundesarchiv Koblenz: Reichsforschungsrat R 26 III; also in Irving (Ed.) [undated], No. 29-1090. Typed signed document with handwritten postscript on letterhead: 'Der Reichsmarschall des Großdeutschen Reiches. Präsident des Reichsforschungsrates – Der Bevollmächtigte für kernphysikalische Forschung,' File ref.: 'Bb.-Nr. 349/43g.RFR'. Sender's address: 'Berlin-Chlbg., Werner-Siemensstr. 8-12', addressed: 'An den Leiter des Geschäftsführenden Beirates des Reichsforschungsrates Herrn Ministerialdirektor Prof. Dr. Mentzel, Berlin-Steglitz, Grunewaldstr. 35'. Handwritten administrative notes omitted. Stamped 'secret'.

Re.: Appropriation of budgeted funds.

Per the Reich Marshal's letter of Apr. 30, 1943, RM 2 000 000.– were granted to me for the 43/44 fiscal year.[1] Of this sum approximately RM 450 000.– has been spent up to now and another RM 450 000.– has been committed to longterm assignments. Out of this total an amount of RM 800 000.– is earmarked for the establishment of a production plant in Germany due to unforeseeable enemy action on factory plants in Norway,[2] an amount which was not taken into consideration in my projections for the running fiscal year.

As I was forced by these special circumstances to disburse RM 2 000 000.– prematurely and am no longer in the position to grant urgently needed projects without exceeding the approved budgetary funds, I request permission to make advance withdrawals totalling RM 1 000 000.– to be booked to the coming fiscal year.

[9]Ramsauer was interested in the acceleration of education in physics, including shortening the basic curriculum at universities which spanned 11–12 terms, through the introduction of a special degree which at that time was only available at polytechnics to students who had completed 7 semesters of study: *Diplom-Physiker*, which is approximately equivalent to a Bachelor of Science. See his guide to physical studies: Ramsauer [1938]; Orthmann [1939], Wien [1944], p. 89; and footnote 35 of doc. 110.

[1]Abraham → Esau was appointed Plenipotentiary of Nuclear Physics on Dec. 8, 1942 by *Reichsmarschall* H. → Göring, who had been been commissioned to oversee the Reich Research Council (→ *RFR*) in 1942.

[2]In February 1943 members of the Norwegian resistance managed to destroy most of the heavy water stocks at the Norsk Hydro electrolytic plant, at that time the only operational industrial production plant for this crucial moderator substance used in nuclear reactor research. A subsequent Royal Airforce bombing raid in November 1943 halted production altogether and the remaining heavy water supplies were eliminated in another act of sabotage when a ferry loaded for delivery to Germany was sunk in February 1944. See also doc. 115, footnotes 39 and 64. See Irving [1967], pp. 126ff., and Walker [1989]a, pp. 137ff., on the subsequent discussions on the construction of a German factory. The → *IG Farben*'s ammonia production plant in Merseburg (*Leuna Werke*) was commissioned to take up this production under the direction of Paul Herold.

It should be noted in this regard that disbursements in this fiscal year are expected not to exceed RM 1 000 000.– in total.[3] A Esau

Prof. Gerlach requests at a private meeting on Feb. 22, 1944, allocation of the funds applied for by Prof. Esau as his predecessor for nuclear physics research.[4] Although the previously appropriated sum is only expended in part, it is committed in its entirety, so that presently no additional research can be initiated.

104 Kurt Diebner: Listing of Nuclear Research Commissions Enclosed with a Letter to the President of the Reich Research Council [April 18, 1944]

Source: Bundesarchiv Koblenz: Reichsforschungsrat: R 26 III; also in Irving (Ed.) [undated], Nos. 29-1110, 29-1111. Typed signed enclosure letter on letterhead: 'Reichsforschungsrat. Der Bevollmächtigte des Reichsmarschalls für Kernphysik,' Bb.-Nr. 24/44g R RFR D/W. Preprinted sender's address: 'Berlin-Dahlem, Boltzmannstr. 20', addressed: 'An den Herrn Präsidenten des Reichsforschungsrates, Berlin-Steglitz, Grunewaldstr. 35'.[1] Handwritten administrative notes omitted. The attached typed list bears the secrecy stamp: 'Geheime Reichssache'.

On the instruction of Professor Gerlach you find enclosed a listing of the research commissions granted between Apr. 1, 1943 and Mar. 31, 1944. The commissions granted from Apr. 1, 1944 onwards will be reported monthly.[2]
By instruction Diebner (Govt. Planning Officer)

Commissions in Nuclear Physics Research

No.	Topic	Person Responsible	Priority Level	Secrecy Specif-ication
1.	Research with the Paris cycloton	Phys. Inst. at Univ. of Cologne, Lec. Prof. Riezler	SS	open
2.	"	Phys. Inst. at Darmstadt Polytech., Lec. Dr. Maurer	SS	open

[3]For an itemization of the nuclear research projects actually commissioned between April 1943 and March 1944, see doc. 104.

[4]At the close of 1943 Esau was replaced by Walther → Gerlach.

[1]Hermann → Göring was appointed head of the Reich Research Council (→ *RFR*) at its reorganization on July 6, 1942 (cf. doc. 99).

[2]In the following list the *DE* rating (*Dringlichkeitsentwicklung*) was the highest priority level for urgent development projects. For the historical context of the German uranium project which was under Kurt → Diebner's control until 1942 through the → *HWA* and thereafter under the supervision of the 'Reich Marshal's Plenipotentiary of Nuclear Physics' appointed by the → *Reichsforschungsrat*, then held by W. → Gerlach since the close of 1943, see in particular Walker [1989]a and Irving [1967], and the Introduction, sec. 5.4.

3.	"	Phys.Chem. Inst. at Munich Univ., Lec. Dr. Starke	SS	open
4.	Power generation from nuclear processes	Phys.Techn. Reichsanstalt (PTR), Councillor of State Prof. A. Esau	SS, for special purposes DE	secret, partly top secret
5.	"	KWI of Physics, Berlin-Dahlem, Prof. Heisenberg	SS	"
6.	"	KWI of Medical Research, Heidelberg, Prof. Bothe	SS	"
7.	"	1st to 3rd Phys. Inst. at Vienna Univ., Prof. Stetter	SS	"
8.	Theor. research on power generation from nuclear processes	Inst. for theor. Phys. at Strasbourg Univ., Prof. v. Weizsäcker	SS	secret
9.	Thermodynamic analyses of uranium machine, corrosion tests and investig. of physical properties of uranium	Phys. Techn. Reichsanst. (PTR), Councillor of State Prof. Esau	SS or DE	secret, partly top secret
10.	Investig. on the spectrum of U metal, esp. also on its Zeeman effect	Phys. Inst. at Tübingen Univ., Prof. Back	SS	
11.	Isotope separation	Inst. for phys. Chemistry at Hamburg Univ., Prof. Harteck	SS, for special purposes DE	secret, partly top secret
12.	Isotope separation	Inst. for phys. Chemistry and Electrochemistry, Kiel, Lec. Dr. Martin	SS	secret, partly top secret
13.	"	KWI of Chemistry, Berlin, Dr. Klemm	SS, for special purposes DE	secret
14.	Development of catalysts for H-D exchange	Phys.Chem. Inst. at Leipzig Univ., Prof. Bonhoeffer	SS	secret
15.	Preparation of gaseous uranium compounds for isotope separation	Danzig Polytechnic, Prof. Albers	SS	secret
16.	"	Chem. Inst. at Bonn Univ., Prof. Schmitz-Dumont	SS	secret
17.	Construction and development of a mass spectrograph	II. Phys. Inst. at Göttingen Univ., Prof. Kopfermann	SS	open, partly secret
18.	Development of a mass spectrograph	Deutsche Reichspost, Ministerial Councillor Gerwig	SS	open, partly secret
19.	Manufacture of an isotope sluice	Bamag-Meguin Co., Berlin	DE	secret
20.	Fission products in nuclear processes	KWI of Chemistry, Berlin-Dahlem, Prof. Hahn	SS	secret
21.	Investigations in nuclear physics	Phys. Inst. at Leipzig Univ., Prof. Döpel	SS	open, partly secret
22.	Investig. on natural occurences of element 85	Inst. for Radium Research, Vienna, Dr. Berta Karlik	SS	
23.	3 Samples of corrosion resistant metal sheets	Auer Company, Berlin	DE	secret

24.	Investig. on biological effects of radiation	Genet. Dept. at the KWI of Biology, Berlin-Buch, Lec. Dr. Zimmer	SS	secret
25.	" (Radiation protection)	KWI of Biophysics, Frankfurt/Main, Prof. Rajewsky	SS	secret.

Tab. 7: Commissions in nuclear physics, April 1943–March 1944

105 A[braham]. Esau [July 16, 1944]

Source: rh, 'A. Esau', *Das Reich. Deutsche Wochenzeitung*, Berlin, No. 29, July 16, 1944, p. 1.

The Councillor of State, Professor A. Esau, member of many academies and scientific organizations, and since the beginning of the war President of the Reich Physical and Technical Institute,[1] is descended from an ancient peasant family from Tiegenhagen, a small village in the Vistula delta,[2] where he was born in 1884. He attended Slaby's lectures;[3] later he studied under Max Wien, becoming his and teaching assistant.[4] In 1912 he took over the direction of the receiving laboratory of the wireless telegraphy company (*Telefunken*);[5] and in 1914 he embarked for the German colony Togo, where he was to direct the reception experiments of the first wireless high-power station.[6] But the declaration of war disrupted the preliminary experiments. At that time Esau destroyed the station before the eyes of the advancing French.[7] He was taken prisoner and abducted, suffering brutal hardship traveling by foot 800 kilometers into the interior of the

[1]The abbreviation of Abraham → Esau's given name is unusual for such a eulogy; clearly to avoid any suspicion of Jewish ancestry (Esau himself was a Mennonite.) He was president of the *Physikalisch-Technische Reichsanstalt* (→ *PTR*) from 1939 to 1945.

[2]A river in Poland flowing northward from the Carpathian Mountains near the Eastern border of the Czech Republic past Warsaw into the Baltic near Danzig.

[3]Adolf [Carl Heinrich] Slaby (1849–1913) took his doctorate in 1873 in Jena and became a teacher at the Potsdam Vocational School (*Gewerbeschule Potsdam*) in the same year. In 1876 he became unsalaried lecturer there and was full professor at the Polytechnic in Charlottenburg (Berlin) from 1882 to 1912. His specialties were telegraphy and electrical engineering.

[4]Max → Wien was professor of physics at the University of Jena between 1911 and 1935, specializing in experimental research of electrolysis and the development of high-frequency technology. Cf. Esau [1938].

[5]A major Berlin firm founded in 1903 by Count Georg von Arco (1869–1940). Esau published his research on topics related to atmospheric influence on wireless telegraphy in the *Physikalische Zeitschrift* 1911–12, and from 1920 on in the *Jahrbuch der drahtlosen Telegraphie.*

[6]This West African region was a German colony from 1885 until 1914, when Germany renounced her sovereignty over Togoland at the Treaty of Versailles and its administration was handed over to the British and French jointly.

[7]French colonial troops invaded Togoland from Dahomey on Aug. 7, 1914, and together with the British troops from the Golden Coast, quickly obtained an unconditional surrender from the Germans on August 26.

country. After the war he returned to *Telefunken*. In 1925 he followed a call to Jena as professor of technical physics.[8] He was rector of the local university from 1932–35; in 1933 he became Thuringian Councillor of State; in 1934 grant commissioner of the Carl Zeiss Foundation;[9] and in 1939 President of the Reich Physical and Technical Institute.

Fig. 23: [Abraham Esau]
'Private photo'

Esau is one of the few men in Germany to have not only experienced every step in the development of wireless telegraphy, but also to have been most actively involved in shaping it. Esau began when wireless telegraphy began. It started with waves of a kilometer's length. Only with time did they learn how to produce, transmit, and receive shorter waves reliably as well. Soon after the war oversea communications gained a new waveband as amateurs throughout the world bridged entire continents and finally even spanned the globe with midget transmitters [*Liliputsender*]. One segment just beyond these waves, however, was still largely avoided, because it seemed to present endless difficulties.[10]

Esau, who at that time had just taken up his professorship in Jena, forged ahead into this realm of ultra-shortwaves. Using exemplary theoretical considerations, he researched their properties in classic series of experiments and made them subservient to technology.[11] Thus he made that wave band accessible without which the televisor would have remained utopian even in the future and which

[8]His research during this period covered, for example, the influence of electromagnetic fields on ferromagnetic materials and related topics.

[9]The *Carl-Zeiss-Stiftung* was founded on May 19, 1889 following the death of Carl Zeiss (1816–1888) as a legal successor to the former → *Zeiss* optical works. His close friend Ernst Abbe (1840–1905), who owned 45% of the profits made by the optical works, signed the foundation charter, which transformed the big firm into an organization independent of private entrepreneurs, yet without turning it into a state-owned company: Cf., e.g., Schomerus [1952] and Hermann [1989], pp. 101ff.

[10]Cf., for instance, Esau's books: *Die Entwicklung der deutschen drahtlosen Nachrichtentechnik*, Jena, 1934; *Weltnachrichtenverkehr und Weltnachrichtenmonopole*, Jena, 1932.

[11]Cf., e.g., Esau's book *Elektrische Wellen im Zentimetergebiet*, Berlin, 1940, as well as his contributions in *Jahrbuch der Deutschen Luftfahrtforschung* special issue 1938, pp. 227–230; see also, 'Eigentümlichkeiten und Anwendungsmöglichkeiten kurzer elektrischer Wellen', *Elektrische Nachrichtentechnik* **2** [1925], and (together with W. M. Hahnemann) 'Electrical

has gained huge importance particularly during the war in other fields as well. It was he who had initiated the use of these waves in medical science; and he who had created the technical prerequisites for the development of ultra-shortwave therapy and had thus entrusted to doctors a unique tool against numerous, in part very serious, sicknesses.[12] Many purulent illnesses are cured today with ultra-shortwaves, not only externally, but also in the interior of the body—cold sicknesses and rheumatic illnesses, diseases of the circulatory organs, frostbite, and many others. Besides, the effective range of ultra-shortwaves can be narrowly confined also in the interior of the body; and its healing thermal effect can be dosed from the tolerance limit down to well below the limit of perceptibility. The ultra-shortwave has also stood the test outstandingly as a means of inducing artificial fever against a series of very unresponsive diseases.

The man who accomplished this exploratory work never came forward publicly; he always went silently about his business. Only once, half a year ago, a few words of his were heard in a nationwide broadcast of the German Radio Network [*deutscher Rundfunk*]: "We technicians do not believe in miracles; we believe in success only as the fruit of indefatigable, unerring effort." This statement is characteristic of a man who had earned his successes through hard labor. But this should not be misinterpreted as the statement of an unimaginative practitioner who dispenses with intuition. Since, Esau himself was constantly drawn towards the uncharted spots in the worldview of physics and electrical engineering, probing into the regions of the unexplored and the unknown. As recently as in the mid-thirties he undertook a second thrust forward into undiscovered wave territory, which led to waves of a centimeter's length.[13]

When standing face to face with him, the answer to the question of his personality seems to be simple and convincing. He is a stocky man with a solid peasant's skull; his accent reveals his East Prussian origin. The tough peasant's strength, a patience which can rise to the point of stubbornness—these give the face its features. But radiating from the person as a whole is the kindness and modesty of a man who knows too much and has accomplished too much to want too much.[14]

The present has again brought Esau an overabundance of the hardest tasks. As President of the Reich Physical and Technical Institute he directs all of its laboratories. Today its activities cater exclusively to the demands of the war. The first World War has already shown that it is crucial to engage not only the

---

waves of about 3 meters: Their propagation and use', *Proceedings of the Institute of Radio Engineers* **18** [1931].

[12]Cf., e.g., Esau's semipopular paper 'Temperaturverteilung in geschichteten biologischen Geweben', *Naturwissenschaften* **24** [1936], pp. 520f., and **26** [1938], pp. 477f., as well as Trendelenburg [1975], pp. 252f., on industrial research on ultrawave therapy.

[13]Cf. the texts mentioned in footnote 11 above and the Introduction, especially footnote 285 for other references on radar research, then the new crucial military technology.

[14]This portrait is the robust and down-to-earth Nazi ideal of a scientist.

most courageous soldiers but also the best and most modern weapons. Warfare is developing more and more into a conflict between engineers. We know today that high-frequency engineering is not just a supplementary aid for all the branches of the military services, but a weapon of decisive importance. Esau is responsible for guiding all high-frequency research, defining its goals, and making decisions of great consequence.[15] At sixty-one he bears the greatest responsibilities but is also at the peak of his abilities.

106 Heinrich Himmler: Letter to Albert Speer [August 3, 1944]

Source: Bundesarchiv Koblenz: Reichsforschungsrat, R 26 III, Sheets 142–143. Letter addressed: 'An den Reichsminister für Rüstung und Kriegsproduktion, Pg. Speer, Berlin W. 8, Leipziger Platz'. Salutation: 'Lieber Parteigenosse Speer!'. Reprinted in: Swinne [1992], footnote 10, pp. 194–195.

Dear Comrade Speer,[1]

Convinced that particularly in the present war situation our defense and military research must be brought to the highest possible level of performance, I issued as Commander of the Reserve Army about 2 weeks ago the order that under all circumstances scientists engaged in militarily crucial defense research remain in reserved occupation![2]

I now note with some dismay an edict that your technical office placed before you and that states among other things:

> "All research and development projects are to be discontinued unless my written special authorization to continue work has been applied for by the 31st of August 1944."[3]

[15]On Jan. 1, 1944 Esau became head of the Reich Research Council's (→ *RFR*) section that coordinated research in high-frequency technology, which was of direct relevance to radar technology and thus to anti-aircraft weaponry: Cf. also doc. 106. He was relieved of his functions as H. → Göring's plenipotentiary for nuclear research in December 1943. Cf. also the documents relating to Esau's replacement in 1943 by Gerlach as leader of the *RFR*'s physics division in Irving (Ed.) [undated], nos. 29-1068ff.; and Walker [1989]a, pp. 129ff. However, as comparative studies on radar research have shown, the Anglosaxon short-wave radar programs, particularly the one at the MIT Radiation Lab, far surpassed German advances. According to Guerlac [1987], radar research ultimately decided the outcome of World War II, giving the Allies control of the skies above Germany well in advance of the deployment of ground troops.

[1]This letter by the head of the → *SS* and (since 1943) Reich Interior Minister H. → Himmler to the head of the Armaments and War Production Ministry (→ *RMBM*), A. → Speer, illustrates the hectic efforts to coordinate research and to translate them into weapons production as the war situation became increasingly critical.

[2]Military deferrals were granted to civilian scientists working on high-priority '*uk*' (*unabkömmlich*) research projects.

[3]Speer's edict was based on the *Führer*'s order on the concentration development: 'Zur Konzentration der Entwicklung' dated June 19, 1944. According to W. Osenberg's letter to Curt Bley of July 5, 1949 (Irving (Ed.) [undated], no. 31-1082), Speer attempted to gain full

I share with you the view that it is expedient to concentrate our development projects on crucial weapons.[4] I also consider it necessary that research agencies receive exact information on which of the developments of the coming six months in technical warfare seem to claim particular priority. I consider it positively disastrous, however, that now large numbers of research projects are to be discontinued by Development Commissioners, because the officials in charge simply cannot keep track of their future potential and results, much as they would like to.

We should really have learnt from bitter experience in this war. As you know, following the France campaign, the decree was issued to halt research and development assignments that did not bear fruit within a fixed period.[5]

At that time a range of research projects were stopped, among others also in the field of high-frequency engineering. We had to pay dearly for this with the enemy's increasingly potent terrorization by air.[6] Are we now really to repeat this war's most disastrous of bad decisions again? Much as we would like, no one can say which technical means of combat our enemy will use to attack us in the future; they will not halt their own military research, that is absolutely certain.

Dear Comrade Speer, I will make you the following concrete proposal:

1. The Ordnance Departments of the separate military services are as a matter of fact heavily overstaffed. They are hampered by far too much specialization, by far too much bureaucracy, and by excessive conflicting secrecy regulations. I consider an internal reform of these Ordnance Departments not only appropriate, but plainly necessary.

I myself would like to begin with the Army Ordnance Office.

In this way, a considerable number of persons will actually become available for other assignments—which is obviously what was at issue for your expert officials.

2. I consider it imperative that your Development Commissioners sit down together with the Special Section heads [*Fachspartenleiter*] and authorized agents of the Reich Research Council and with the appropriate division heads of the

---

control over German research by interpreting 'development' to also include scientific research.

[4]This reflects the commonly held desperate hope that new weapons systems would be able to turn the tide of the war, which was increasingly threatening the fatherland. A. → Hitler nourished this morale-boosting belief with promises of using a 'wonder weapon' (*Wunderwaffe*) in the last resort.

[5]This decree was issued at various places in early February 1940 when the 'lightning war's' (*Blitzkrieg*) quick successes made a German victory seem imminent. It only allowed the continuation of projects likely to bear fruit in 1940 or in early 1941. See, e.g., the letter from E. Schumann to P. Harteck, Dec. 5, 1941, Irving (Ed.) [undated], no. 25-687, and Walker [1995], pp. 192–194; cf. Ludwig [1974], pp. 231f., Trischler [1992]a, pp. 244f.

[6]As a consequence, this research area was reactivated in 1943 with the official nomination of A. → Esau as H→ Göring's Plenipotentiary of High-frequency Engineering and Radar in the → *RFR* on Jan. 1, 1944.

Ordnance Departments of the separate military services.[7] In this way research would be informed about which lines of weapons development have gained particular priority.[8] Contact can continue to be maintained here through regular communication of the problems.

3. The Special Section heads or authorized agents of the Reich Research Council and the respective heads of the Defense Weaponry Offices [*Wehrwaffenämter*] must then assume the responsibility of directing research. Requiring that the thousands of research projects seek approval at one centralized location will unleash a senseless paper war. Moreover, it seems to me to be unjustifiable practically, because unfortunately, there are such geniuses capable of assessing such broad sectors of research in order to predict to which research developments the direction of enemy combat will still compel us in this war.

I would be grateful to you, dear Comrade Speer, if you would inform me of the steps you will take.

Heil Hitler Yours Himmler

107 Walther Gerlach: Letter to Reich Party Leader Martin Bormann [December 16, 1944]

Source: Bundesarchiv Koblenz: Files of the Reichsforschungsrat; also in Irving (Ed.) [undated], Nos. 29-1156, 29-1157. Carbon copy of typed signed letter, file ref.: 'RFR 66/44 Ge/Gud'. Sender's address: 'B[er]l[i]n.-Dahlem, Boltzmannstr. 20', addressed to: 'Herrn Reichsleiter Martin Bormann, Berlin, Wilhelmstr. 64'.[1] Classified as top secret, stamped 'Geheime Reichssache'. Reproduced with the kind permission of Dr. med. Helgard Krietsch (Munich).

Very esteemed Reich Party Leader,

In order to secure the indispensable staff for research essential to the war effort, on Sep. 3, 1944 you directed a telex message to all regional leaders according to which the enlisted scientists and technicians whom Profesor Osenberg of the Reich Research Council's planning office has secured for operation 'Research' are to be

[7]The *RFR* (originally founded in 1937 and revitalized in 1942) was unsuccessful in its concerted effort to coordinate research under its ægis, through research performed at the *HWA* and other military research divisions.

[8]The system of evaluating priority was quite bizarre by 1944: Due to continually shrinking supplies, it did not suffice to attain normal priority to receive needed materials; thus new categories were constantly being introduced. The '*DE*' (*Dringlichkeitsentwicklung*) was the highest priority rating, which Speer assigned to the construction of the → *KWIP* buildings, for instance. Cf., e.g., Macrakis [1993], p. 174.

[1]This carbon copy was prepared by Walther → Gerlach for the Berlin regional party leader Georg Dietrich Graue (on Graue see footnote 5 of doc. 96): 'An Herrn Dr. Graue' is typed at the top left corner followed by the note: 'This carbon copy (Headquarters classified matter) [g[eheime] K[omman]dos[ache].] is only intended for you personally. In negotiations with local authorities I ask you please not to use the letter itself, but to refer to the fact that the Plenipotentiary of Nuclear Physics has already contacted Reich Leader Bormann personally'. Administrative notes are omitted.

discharged from any special deployment recruitments.[2]

It goes without saying that the entire male staff at the research institutes have registered themselves for the People's Army.[3] A difficulty has arisen here for the research institutes transferred to the Württemberg-Hohenzollern region (esp. Hechingen, Haigerloch, Tai[l]fingen)[4] insofar as the majority of the men working there were enlisted during the 1st mobilization. Since for the reasons mentioned, the possibility must be considered that detachments of the People's Army be deployed outside of the locality, the local party authorities have privately granted the director of the local experimental stations a *temporary* discharge from deployment outside of the locality.

A removal of even only a small portion of the staff, which is already reduced to the absolute essentials, is tantamount to halting the research currently underway. But precisely these efforts are among the most important research and development projects within my sphere.[5] Thus I must take care that under all circumstances these projects continue without interruption. You are doubtlessly aware that these researches could become unexpectedly of crucial significance to the war. You are also aware that in America the greatest effort is being made in this area; I am convinced, however, that at the moment we still have a considerable advantage over America, as far as both research and development are concerned, even though we are using a negligibly small percentage against manpower employed in America.[6]

---

[2]This operation, *Aktion Osenberg*, was directed by the machine-tool and hard-solid tooling specialist Werner Osenberg (1900–), head of the Reich Research Council's (→ *RFR*) planning office between 1943–45. Osenberg was simultaneously professor at the Hanover Polytechnic (since 1938) and head of the Four-Year Plan's Institute of Production Engineering Procedures between 1941–45. Through his efforts in December 1943, about 5,000 scientists were recalled from the front, and in August 1944 another 10,000 able-bodied men already actively engaged in research were exempted from military service. Bormann's circular (file ref. no. 217/44) was transmitted to all regional party leaders (*Gauleiter*) on Sep. 4, 1944. In 1954 Osenberg became director of the Institute of Production Engineering and Machine Tooling. Cf. also Ludwig [1974], pp. 243ff.

[3]The *Volkssturm* which Hitler had called up on Sep. 25, 1944, was made up of all remaining unenlisted able-bodied men (loosely defined) between the ages of 16 and 60. Although these units were supposed only to be deployed as local defense militia, many were sent to the receding frontlines.

[4]Due to Allied bomb raids most of the research institutes were removed to rural areas. In 1943 the → *KWIP* was transferred to Hechingen, and the *Uranverein*'s last experiments were conducted in nearby Haigerloch (cf., e.g., Walker [1989]a, pp. 151ff, Irving [1967], pp. 238ff., and Wirtz et al. [1990]). The → *KWIC* was relocated to a textile factory building in Tailfingen.

[5]By pointing to the potentially significant military applications of nuclear research both Gerlach and his predecessor, Abraham → Esau, succeeded in obtaining the highest priority ratings (*SS* and *DE*) for their projects, even in the final years of the Nazi regime: See, e.g., doc. 103.

[6]Gerlach was thus informed about the existence of the American atomic bomb project evidently through intelligence sources, which failed to indicate, however, how far advanced it was. Gerlach's genuine surprise at the news of the dropping of the first American bomb over

Therefore, owing to the extreme urgency I take the liberty to turn to you directly, esteemed Reich Leader, with the request that you have Regional Leader Murr[7] in Stuttgart notified, who is informed personally about our research projects and has already provided us with every kind of assistance and support during the relocation, to the effect that the gentlemen at the transferred Kaiser Wilhelm Institutes and research offices of the Plenipotentiary of Nuclear Physics may only be enlisted in formations of the People's Army that are not intended for special deployments.[8]

I have sent a transcription of this letter to the president of the Reich Research Council and to Prof. Osenberg.[9]

Heil Hitler!

Yours very truly (Prof. Walther Gerlach)

Fig. 24: Walther Gerlach und Kurt Diebner as Farm Hall internees, England 1945

Japan on Aug. 6, 1945 (cf. Frank (Ed.) [1993], pp. 70ff.) indicates that this was more than simply an effort to protect the projects under his control. For a comparison of the total sums spent on German and American nuclear research up to 1945, see the Introduction, secs. 4.6 and 5.4 as well as doc. 103, and references there.

[7]Wilhelm Murr (1888–?) had been regional party leader since 1928. After entering the → *NSDAP* in 1923, Murr became local and regional propaganda head as well as representative in the state and national parliaments. In March 1933 he advanced to *Staatspräsident* and interior and economics minister, becoming governor of the state of Württemberg in May 1933.

[8]With this letter Gerlach wanted to protect his scientists from being deployed in the Third Reich's final desperate military campaigns.

[9]H. → Göring presided over the *RFR* since its reorganization in June 1942 (cf. doc. 98).

Part V
The Legacy of National Socialism

108 Lise Meitner: Letter to Otto Hahn [June 27, 1945]

Source: Archive of the Max Planck Society, Berlin: Carbon copy of letter. Original stenographic draft at Churchill College Archive, Cambridge, England. Reprinted in Krafft [1981], pp. 181–182 (see also footnotes 46–48 there); Deichmann [1992], pp. 317–318.

Dear Otto,[1]

Your last letter is dated the 25th of March. You can imagine how anxious I am to receive news about all of you. I have followed the events in the English war news reports very closely, and I believe I may assume that the area in which you and your family and Laue are was occupied without any fighting.[2] So I hope with all my heart that you all did not have to suffer personally. Naturally, it will be very difficult for you now, but that was unavoidable, of course. On the other hand, I am very worried about the Plancks, since there has been bitter fighting in their area.[3] Do you know anything about them and the Berlin friends? An American will take this letter along.[4] He will pick it up soon, and so I am writing in great haste; and yet, I have so much to say that is close to my heart. Please keep this in mind, and please read it with confidence in my unshakable friendship.

I have written to you very many letters in my thoughts these months past, because it was clear to me that even people like you and Laue had not grasped the real situation. I noticed this so clearly among other things when Laue wrote to me on the occasion of Wettstein's death[5] that his death was a loss also in the wider

[1]A handwritten note on the carbon copy indicates that this letter never arrived. L. → Meitner worked on nuclear physics and O. → Hahn on nuclear chemistry at the → *KWIC* in Dahlem (Berlin) and they collaborated on radioactivity research there until Meitner was forced to flee from Germany to Sweden in 1938. On their early collaboration see, e.g., Krafft [1981], chap. 3; Ernst (Ed.) [1992].

[2]The Hahns lived in the Dahlem suburb of Berlin at 48 Altenstern St. from mid-March 1929 to 1944, ten minutes walk away from the *KWIC.* Max von → Laue's family lived in the Zehlendorf suburb of Berlin at 17 Albertin St.

[3]Max → Planck's house, 21 Wangenheim St. in the Grunewald suburb of Berlin, was destroyed during an Allied air raid on Feb. 15–16, 1944. Planck stayed subsequently on a farm on the left bank of the Elbe at Rogätz, interrupted by a short stay at Amorbach, Lower Franconia, where he was operated on (he was suffering from spinal arthritis) by the well-known physician Sauerbruch in May 1944. For a description of Planck's experiences at that time and the fighting around Rogrätz in the Magdeburg government district, see Hermann [1973], pp. 105–116, and the correspondence between Planck, his wife Marga, and von Laue, DMM, call no. 1964-6. On May 16, 1945, American officers, including the astronomer Gerard P. Kuiper, helpfully drove Planck to relatives in Göttingen. On Marga Planck, see doc. 120, footnote 14.

[4]According to Krafft [1981], footnote 47: 'June 1945 (Dr. Berg)' is written in Meitner's hand above the shorthand notes.

[5]The biologist Fritz [Friedrich] von Wettstein Ritter von Westersheim (1895–1945) was assis-

sense, because with his diplomatic talents W[ettstein] could have been very useful at the end of the war. How could a man who had never objected to the crimes of recent years be of use to Germany? This is, of course, Germany's misfortune, the fact that all of you had lost your standard of justice and fairness. You had told me yourself in March 1938 that Hörlein[6] had said to you that terrible things were being done to the Jews.[7] So, he knew about all the crimes that were being planned and later executed and was a member of the Party despite it; and you regarded him as a very decent person—also despite it—and allowed yourself to be influenced by him in your behavior towards your best friend.[8]

You all have worked for Nazi Germany as well and never have even tried to put up a passive resistance either. Certainly, to buy off your consciences you have helped a person in distress here and there, but have allowed millions of innocent people to be slaughtered without making the least protest.

I must write this to you, because for your sake and for Germany's so much depends on your understanding what you have allowed to happen. Here in neutral Sweden there have been discussions already long before the end of the war about what should be done with German intellectuals after the war had ended. What might the English and Americans in particular be thinking about this? I and many others as well think that one solution for you would be that you make an open statement that you are all aware that through your passivity you had taken on a shared responsibility for what has happened and that you feel the need to help in making amends—as far as the past can be made good. But many think

tant at the Kaiser Wilhelm Institute of Biology in Berlin 1919–25. He submitted his habilitation to the University of Berlin and became full professor there in 1925, later at the University of Munich in 1931. In 1934, he was appointed director of the Kaiser Wilhelm Institute of Biology, succeeding Carl Correns (1864–1933) and was additionally head of the → *KWG*'s biology section 1936–42. Von Wettstein died unexpectedly on February 12, 1945 of pneumonia, which he had caught during an exhausting train trip from Innsbruck to Berlin. Lise Meitner knew Wettstein and his Austrian father and relatives. For references to biographical accounts on Wettstein see, e.g., Engel [1984], pp. 192 and 404, footnote 10; on his research see Deichmann [1992].

[6]The chemist Philipp Heinrich Hörlein (1882–1954) specialized in the synthesis of pharmaceuticals and medical chemistry and pharmacology. He submitted his Ph.D. thesis to the University of Jena in 1903 and was employed at the *Bayer* chemical company from 1909, later at the → *IG-Farben* trust, where he became executive board member in 1931. Hörlein was also treasurer of the *KWG* and lectured at the Medical Academy in Düusseldorf from 1931, retiring in 1949. *IG Farben* profited a lot from Jewish slave labor in special concentration camps, such as the *IG Buna* plant at Auschwitz and in many other smaller production units. On *IG-Farben*'s involvement in the Nazi crimes against humanity, which led to one of the biggest war-crimes trials after the end of war see, e.g., Hayes [1987], part V, Lifton [1986], pp. 18, 152ff., 187f., 291.

[7]In a letter to Martha Krause of Jan. 15, 1948, Meitner quoted a comment Max → Planck had made to her in Sweden in 1943: "[']terrible things ought to happen to us, we have done terrible things.['] He said 'us' and 'we' even though he himself had tried personally to fight against it."; Meitner papers at Churchill College, Cambridge, England.

[8]Meitner probably refers here to herself.

it is too late for this. These people say you had betrayed first your friends, then your men and children by allowing them to put their lives at stake in a criminal war, and finally that you had also even betrayed Germany, because when the war was already completely hopeless you did not even fight against the senseless destruction of Germany.

This sounds merciless and yet, believe me, it is out of the truest friendship that I write all this to you. You really cannot expect the rest of the world to pity Germany. The inconceivable atrocities at the concentration camps that have been reported these days surpasses everything we had feared. When I heard a very factual report on English radio by Englishmen and Americans on [Bergen]-Belsen and Buchenwald, I began to cry out loud and could not sleep the whole night long.[9] You should have seen the people who came here from out of the camps. Someone should force a man like Heisenberg and many million others to look at these camps and at the martyred people. His appearance in Denmark in 1941 is unforgettable.[10]

You might remember, while I was still in Germany (and today I know that it was not only stupid but very unfair of me not to have gone away immediately),[11] I often said to you: Things will not get better in Germany as long as only we have the sleepless nights and not all of you. But you did not have any sleepless nights. You did not want to see; it was too inconvenient. I could prove it to you with many examples, big and small. Please believe me that all that I write here is an attempt to help you all.

With very affectionate greetings to everyone. Yours Lise.

109 Alvin M. Weinberg & Lothar W. Nordheim: Memorandum [November 8, 1945]

Source: Technical Information Service of the Atomic Energy Commission in Oak Ridge, Tennessee (unclassified, Jan. 6, 1950); also in Irving (Ed.) [undated], nos. 31 1182-31 1184; also

[9]Bergen-Belsen and Buchenwald were two concentration camps in Germany. The former was in operation between April 1943 and Apr. 15, 1945, initially as a holding camp for Jews; from March 1944 on prisoners unfit for exploitation as laborers were moved there from other camps. Buchenwald held political prisoners and was in operation from July 15, 1937 to Apr. 11, 1945. Several tens of thousands lost their lives in these two camps. The total death toll at 'extermination' camps including Auschwitz reached 3 million, and if you include ghetto victims and mass shootings, the total loss of human life comes to 5 million. Cf., e.g., Hilberg [1961], appendix.

[10]During a visit to Denmark (then under German occupation) in September 1941 for a lecture series on astrophysics at a German propaganda institute, Werner → Heisenberg could not understand the Danes' increased sensitivity against the German consulate's 'cultural propaganda' efforts; see in particular, Walker [1992] and Cassidy [1992]a, pp. 436ff. Cassidy also gives a synopsis of the situation German physicists were facing in Nazi Germany and Heisenberg's solution: Cassidy [1992]b.

[11]Meitner repeats this point in her letter to Hahn in 1948: See doc. 120.

appended to: *Kernphysikalische Forschungsberichte*, **G-371**, Kernforschungszentrum Karlsruhe. Typed transcription of co-signed document with 'Monsanto Chemical Company, Clinton Laboratories' typed at the head of the first page. The appended list is as follows: 'Distribution: 1. A.H. Compton, 2. A.V. Peterson, 3. M.D. Whitaker, 4. J.R. Coe, 5. R.S. Mulliken, 6. K.D. Nichols, 7. T.R. Hogness, 8. F. Daniels, 9. E.P. Wigner, 10. P. Morrison, 11. L.W. Nordheim, 12. A.M. Weinberg, 13–18. Central Files, 19. Reading Files'. The German quotes in the text are translated here. Published by kind permission of Prof. A. M. Weinberg, Oak Ridge, Tennessee.

To: A. H. Compton[1]
From: A. H. Weinberg[2] and L. W. Nordheim[3]

---

[1]Arthur Holly Compton (1892–1962) graduated at Princeton Univ. in 1916, taught at various universities in the USA and England, and was appointed full professor of physics at the University of Chicago in 1923. His experimental demonstration of an increase in wavelength of scattered X rays, known as the 'Compton effect', showed that x-rays behave like particles in collisions with electrons; in 1927 he received the Nobel Prize for this discovery. Compton was chairman of the National Academy of Sciences Committee on Uranium, and upon America's entry into World War II, he headed the Manhattan District of the Corps of Engineers' metallurgical laboratory at the Univ. of Chicago, where he organized research on methods of isolating fissionable plutonium. Under his directorship Enrico Fermi produced a self-sustaining nuclear chain reaction in Dec. 1942. Compton also had a major influence in establishing the Argonne National Laboratory and the Clinton Engineer Works at Oak Ridge, Tennessee, as well as the plutonium production reactor establishment at Hanford. On Compton's career during the war see, in particular, his autobiographical account: *Atomic Quest. A Personal Narrative*, New York: Oxford Univ. Press, 1956.

[2]Alvin Martin Weinberg (born 1915) submitted his thesis at the Univ. of Chicago in 1939, worked as assistant in mathematical biophysics 1939–41, and then became researcher at the famous Chicago metallurgical laboratory 1941–45, where he took part in American atomic bomb research. From 1945–48 Weinberg worked at the Clinton Labs, Tennessee, and in 1948 became director of the physics division at the Oak Ridge National Laboratory, was research director 1949–55, and from 1955 director there. He was also member of the US Air Force scientific advisory board 1956–59, member of the presidential scientific advisory committee 1960–62, and has been appointed to several National Academy of Science Committees since. See also Weinberg's recent autobiography: *The First Nuclear Era: The Life and Times of a Technological Fixer*, New York: AIP Press, 1994.

[3]Lothar Wolfgang Nordheim (1899–1988) wrote his doctoral thesis on quantum-theoretic perturbation theory of degenerate cases in 1923 at Göttingen University, was assistant to the mathematician David Hilbert 1922–26, and after his habilitation in 1928 was lecturer there from 1928–33, at the same time serving as assistant to Max → Born. Nordheim emigrated first to Paris, where he worked at the Institute Henri Poincaré Oct. 1933–Aug. 1934, then at the Teyler's Stichting Foundation in Haarlem, The Netherlands Sep. 1934–Jan. 1935, moving subsequently to the USA, supported by a Rockefeller Foundation grant and by other emigré organizations, where he was visiting professor of physics at Purdue Univ. 1935–37, and professor at Duke Univ. 1937–43. 1943–47 Nordheim worked at the Clinton Laboratories, Oak Ridge, as section chief, and 1945–47 as director of the physics division, mainly working on reactor and neutron physics. He returned to Duke Univ. 1947–56, was visiting professor at Heidelberg and Karlsruhe Univs. 1949, and then 1956–59 was a physicist at the general and atomic division of the John Jay Hopkins Lab. for Pure and Applied Science, General Dynamics Corp., San Diego. 1960–68, Nordheim was chairman of the theoretical physics department and senior research adviser, since 1968 consultant for the same company. He was also member of the advisory committee on reactor physics from 1962.

We have just had an opportunity to read a few of the German *Kernphysikalische Forschungsberichte.*[4] We are writing in order to correct what we believe to be a very prevalent misconception concerning the state of the art as known to the Germans in 1945. We will proceed by posing a number of relevant questions and then answering them insofar as we can from the few reports we have been allowed to see. Presumably when more reports are made available we will be able to document our statements more fully.

I. Did the Germans know the correct lattice dimensions for a P-9U system?[5]

Via the grapevine we had heard rumors that the Germans were experimenting with plate lattices far too rich in U. Apparently these rumors were based on very early reports which are not yet available to us. At present, however, the answer to the above question is an unequivocal yes. The March 1944 *Forschungsberichte* contain a description of experiments on various lattice arrangements performed by Bothe and Funfer.[6] The experiments are integral ones in which the strength of a source is measured with and without the lattice in place. The main conclusion drawn from these experiments is "a combination of 20 cm D_2O and 1 cm U-metal of density 18 (would) probably be the most efficient... . Previously a larger U requirement had clearly been expected."[7] This conclusion is exactly the same as that reached by us, on the basis of calculations in August 1943 (CP-923). The German work apparently was done at the same time as ours.

Plates seem to have been preferred because they were most convenient for experiments. The advantage of cubes was recognized as early as June 1943 (Hocker), and the use of cylinders had been suggested on technical grounds.[8]

II. Did the Germans know the critical dimensions of the P-9 machine?

We have not had access to the reports in which critical size calculations are made. However, there are repeated references, in the reports available to us, of

---

[4]Distribution of these top-secret internal reports (→ *KPFB*) was restricted to the members of the *Uranverein* and selected members of the → *Heereswaffenamt.* According to W. → Walcher (quoted in Heisenberg [1989], Vol. A II, p. 377, footnote 3), even the authors of these reports were forbidden to keep copies.

[5]P-9 was the American code name for deuterium (heavy water). The 'P-9U system' thus refers to a lattice arrangement of uranium cubes with deuterium as the moderator as described, e.g., in Heisenberg & Wirtz [1948], pp. 149f., 151ff.; cf. also Walker [1989]a, pp. 99–104.

[6]See W. → Bothe's and E. Fünfer's report 'Schichtenversuche mit Variation der U- und D_2O-Dicke', dated Dec. 6, 1943 (*KPFB* **G-206**; also in Irving (Ed.) [undated], no. 30 437); Bothe and Erwin Fünfer (1911–) described experiments with a small pile with octagonal layers of uranium in an aluminium cylinder 48 cm long and 45 cm in diameter, carried out at the Kaiser Wilhelm Institute for Medical Research in Heidelberg; cf. also David & Warheit (Eds.) [1952], pp. 31 and 53, for a table of the dimensions analyzed in the 10 different tests.

[7]Ibid.

[8]See the paper by C.F. von Weizsäcker's collaborator at the Strasbourg Univ. department for theoretical physics, Karl Heinz → Höcker: 'Über die Abmessungen von Uran und schwerem Wasser in einer Kugelstrukturmaschine', dated June 23, 1943, *KPFB* **G-222**, see also Irving (Ed.) [undated], no. 30 414 and David & Warheit (Eds.) [1952], pp. 23 and 54.

about 4 tons as the required amount of P-9. This figure is essentially correct.[9]

The Laplacians, measured by the Germans are of the order 1000×10^{-6} cm^{-2}. This value is in excellent agreement with ours. It indicates, and this is important, that the U-metal used by them was about as pure as ours.

III. What was the state of German theory of the chain reaction?

Here we are badly hampered by the unavailability of the reports. What we do have shows:

(1.) Calculation of optimal lattice dimensions was understood and followed pretty much the same lines as ours. The calculated results on P-9 spheres agree well with ours.

(2.) The group model for reflector calculations was introduced in early 1944.[10] This was a little later than the time we began to use it extensively.

(3.) Generally we would say their approach was in no wise inferior to ours; in some respects it was superior.

IV. Why didn't the Germans succeed in establishing a chain reaction with P-9?

The answer is simple; they did not have sufficient P-9. The latest reference is to a 1.5 ton P-9 experiment.[11] According to our estimates, with the volume ratio they used (20:1), they would have needed somewhat less than 4 tons.

V. Are there any 'scientific secrets' concerning the design of the chain reaction which the Germans do not seem to have understood?

From the general state of the art as deduced from the few reports we have seen, we would say their understanding of the principles is comparable to ours. The only non-engineering 'secrets' we can think of which might affect the design of a chain reaction is the poisoning by Xe^{135}, and possibly, the properties of Pu^{240}.[12]

---

[9] Cf., e.g., Serber [1943] for the contemporary crude but essentially correct methods of obtaining the critical mass for an atomic bomb.

[10] See F. Bopp & E. Fischer: 'Einfluß des Rückstreumantels auf die Neutronenausbeute des U-Brenners', dated Jan. 10, 1944 *KPFB* **B-249**, where it is estimated that with up to 25% of the material can be economized using an efficient carbon reflector. See also Irving (Ed.) [undated], no. 30 418 and David & Warheit (Eds.) [1952], p. 36.

[11] For a description of the last experiment (B 7) performed in the → *KWIP*'s bunker laboratory in Dahlem (Berlin), and the last one for which we have an official report in the secret *KPFB* series, which preceded the very last experiment (B 8) at Haigerloch, see Bopp, Bothe, Fischer, Fünfer, Heisenberg, Ritter & Wirtz: 'Bericht über einen Versuch mit 1,5 to D_2O und U und 40 cm Kohlerückstreumantel', dated Jan. 3, 1945, *KPFB* **G-300**; see also Irving (Ed.) [undated], no. 29 656, as well as Heisenberg & Wirtz [1948], pp. 152–158, and here doc. 115, p. 376.

[12] The 'poisoning' in uranium machines used to breed plutonium refers to the generation of the isotope Xe 135 (from the inert gas xenon) as a by-product of the radioactive decay process with disturbing absorption characteristics. This by-product was detected in an experimental pile in Hanford in 1944; see, e.g., Kevles [1987], pp. 328f., Brown & MacDonald [1977], pp. 329f. German researchers had been familiar with it since 1942: See in particular, Wilhelm Groth: 'Trennung der Uranisotope nach dem Ultrazentrifugenverfahren. I. Anreicherung der

VI. What bearing does this have on publication of the parts of the PPR[13] dealing with principles of the chain reaction?

The Germans knew how to design a lattice which will work. From the practical standpoint this is all that matters. The details of elegant perturbation theory or transport theory (which would be contained in Vol. III) or the details of heat transfer calculations (Vol. IV) would tell them nothing essential to the determination of lattice dimensions. They already know how to calculate the optimum dimensions.

A question of ethics is raised by the existence of the German reports. In many cases useful information is contained therein. It is certainly extraordinary in a scientific treatise, to attribute a given result to an American author without at the same time giving due credit to his German counterpart who is known to have also done the work. Such a situation will arise for example, in Vol. III in the discussion of the multi-group methods where the Germans have duplicated our work.

VII. What bearing does this have on the general question of our 'secrets'?

On this we can presume to speak only as individuals.

The general impression from the German reports is that they were on the right track and that their thinking and developments paralleled ours to a surprising extent. The fact that they did not achieve the chain reaction is primarily due to their lack of sufficient amounts of heavy water.

In one of the reports a vivid description is given of the German efforts in this respect. The heavy water factories in Norway were designed for a capacity of 3–4 tons a year and were successfully operating during part of 1942 and 1943. This capacity would have been sufficient for the construction of a pile. However, the production was interrupted by sabotage and finally the main factory was destroyed by a bombing attack.[14] Toward the end of 1944 plans were made to initiate production of heavy water in Germany and to use enriched uranium in order to reduce the material requirements.[15]

Xenonisotope in einer einstufigen Ultrazentrifuge', dated Jan. 27, 1942, *KPFB* **G-146**; cf. also Irving (Ed.) [undated], no. 30 251.

[13]This abbreviation refers to the planned official history of the American nuclear research during World War II, which was completed in 1947 but remained classified material until it was finally published in 1961; cf. Brown & MacDonald [1977] for long excerpts from this material, and Hawkins et al. [1983] for the publication; cf. also Gowing [1964] for the British perspective, and Rhodes [1986] for a comparative account.

[14]Weinberg and Nordheim may be referring to *KPFB* **G-268**, an undated compilation of material on heavy water production by staff members at the → *IG-Farben* patent office; cf. also H. Jensen's analysis in 1943 of deuterium production via electrolysis in Norway, *KPFB* **G-226**, and Paul Harteck's earlier survey: 'Die Gewinnung von schwerem Wasser', *KPFB* **G-154**, dated Feb. 26, 1942. On the sabotage and later bombing of the Norwegian chemical factory → *Norsk Hydro* see, e.g., Irving [1967], pp. 126ff, 152f.

[15]See P. Harteck's and Hans E. Suess's paper *KPFB* **G-261** [1944] on deuterium production using the Clusius-Linde process in combination with hydrogen regeneration, in which the authors project increasing the production level to 3 tons a years within the next 18 months.

It is also fairly clear that the total German effort was on a very considerably smaller scale than the American effort.[16] This may be due to the strained German economy or to the less favorable attitude of their government. The fact remains that an independent group of scientists, of much smaller size than ours, operating under much more adverse conditions achieved so much.

We must proceed therefore on the basis that anyone knowing what is in the German reports can establish a chain reaction, provided he has sufficient materials. The Smyth report will give additional very helpful hints.[17] The time when others can establish a chain reaction is therefore no longer a matter of scientific research but mostly a matter of procurement. The policies of our authorities must, it seems to us, be formulated with a clear realization of these facts.

A.M. Weinberg,[18] L.W. Nordheim.

110 Wolfgang Finkelnburg: The Fight against Party Physics [c. 1946]

Source: Heisenberg Papers, Munich. Typed document: 'Der Kampf gegen die Partei-Physik'. A handwritten note appears in English at the top of the page: '(? 1946) to be published anony[m]ously in *Phy[si]k[alische]. Blä[tter]* H[eisenberg] N[achlaß].'[1]

Shortly after 1933 the Party started an aggressive campaign against German theoretical physics, which was alleged to be contaminated by Jews [*verjudet*]. Its public champions were the two Nobel laureates Stark and Lenard along with their students and followers.[2] This action received strong support from the Reich leadership of the National Socialist University Lecturers Association, whose physics consultant was Bühl,[3] as well as from the Party Chancellery and from

Cf. also Walker [1989]a, pp. 118–121, 137–146, on the strong support by *IG-Farben* and other companies such as *Linde* and *BAMAG*. On the contemporary American technology see, e.g., Brown & MacDonald [1977], pp. 142–146.

[16]For comparative personnel and budget figures, see the Introduction, p. lxix; cf. also doc. 103, and Kevles [1987], pp. 324ff.

[17]See Smyth [1945]; cf. with the later histories mentioned in footnote 13 above.

[18]In a letter of June 16, 1995 to the editor, Prof. Alwin M. Weinberg added the following afterthought: "In reviewing the memorandum today, I realize that we failed to notice a major oversight in the German effort: Heisenberg and his colleagues overlooked completely the delayed neutrons, which of course are essential for the control of a reactor. Had the Haigerloch experiment gone critical, this oversight would have been corrected—yet the fact that such an error was made in Heisenberg's analyses almost suggests to me that he did not take the matter as seriously as, say, Fermi, Wigner, and Szilard."

[1]W. → Finkelnburg's report was apparently never published, however.

[2]On Johannes → Stark and Philipp → Lenard's activities see, e.g., Beyerchen [1977], chaps. 5–6, and chap. 7 on 'Aryan physics', as well as, e.g., Lenard's foreword to his physics textbook, doc. 39, or the article by the → *NSDStB* activist W. → Menzel, followed by Werner → Heisenberg's and Stark's replies, docs. 42–44.

[3]Alfons → Bühl was a pupil of Lenard and a physics consultant for the → *NSDDB*.

the Reich Education Ministry (Undersecretary Mentzel and Senior Executive Adviser Dr. Führer).[4] Following the publication of the famous polemical article in the *Schwarzes Corps* carrying Stark's supporting signature,[5] in which Planck, Sommerfeld and Heisenberg were visciously attacked as 'white Jews', the journal *Zeitschrift für die gesamte Naturwissenschaft*[6] in particular, which is promoted by the Reich Student leadership, then took up the fight around 1937. Thüring, Hugo Dingler, Wilhelm Müller and others assailed especially the theory of relativity and quantum physics in this publication—never using factual arguments, only political and personal ones.[7] Whoever came to these theories' defense at a university lecturers' camp[8] (such as Mr. von Weizsäcker)[9] was labeled politically suspect and ran the risk of not getting a lectureship. Whoever covered relativity theory in their lectures was looked upon as a bad German and no longer came into consideration for professorial appointments; and they could be openly insulted, as Bühl did publicly against Heisenberg on the occasion of the inauguration of the Heidelberg Lenard Institute.[10] Even Tomaschek made himself available to this group, giving talks against relativity theory in the science student camps; and in the editions of his physics textbook that were published during those years, at least he did not dare to call things openly by their proper name.[11] This public boycott of theoretical physics slowly did begin to influence the students, who became insecure and began to have doubts. At the same time, the Party clique just mentioned filled more and more professorial chairs with to-

[4]*Ministerialdirektor* Rudolf → Mentzel and *Oberregierungsrat* Wilhelm → Führer were both political hardliners and supporters of the 'Aryan physics' movement.

[5]For these *Schwarzes Korps* articles see here docs. 55 and 56.

[6]Max → Planck and Arnold → Sommerfeld were attacked as important representatives of the 1915–16 version of quantum theory and Werner Heisenberg for his 1925 matrix theory of quantum mechanics. The → *Zeitschrift für die gesamte Naturwissenschaft*, edited by the → *NSDStB*, was controlled by Lenard's and Stark's supporters, who vehemently rejected both versions of quantum theory as formalistic.

[7]Besides Bruno → Thüring, who was one of the main editors of the journal, Hugo Dingler (see doc. 83, footnotes 19–20), Wilhelm → Müller, and Ludwig → Glaser were among its most vociferous contributors of inflammatory anti-Semitic articles attacking relativity theory and quantum physics; see, e.g., docs. 77 and 83.

[8]The National Socialist Lecturers League organized these camps [*Dozentenlager*], and participation in them was obligatory to qualify as a university teacher.

[9]C.F. → von Weizsäcker took part in the first German mathematician's *Reichslager* of July 1–3, 1938, for instance, in the youth hostel at Utzdorf near Bernau; see the report by Johannes Juilfs [1939], particularly pp. 119–121.

[10]Bühl's comments were published in 'Philipp Lenard und die deutsche Naturforschung' in Weigel (Ed.) [1937], pp. 5–19. See also Stark's speech of Dec. 13, 1935 at the same event, doc. 40.

[11]Rudolf → Tomaschek, a student of Lenard's, was full professor at Munich 1939–45. From 1933 onwards he revised the 8th and subsequent editions of the textbook by Ernst Grimsehl: *Lehrbuch der Physik. Zum Gebrauch beim Unterricht neben akademischen Vorlesungen und zum Selbststudium*, Leipzig & Berlin: Teubner. It was so widely used that it appeared in a 13th edition in 1943.

tally incompetent candidates [*Nichtskönner*], who called themselves 'Aryan physicists' [*deutsche Physiker*]: Primarily, the Sommerfeld chair with Wilhelm Müller (Aachen);[12] Gaede's in Karlsruhe with Bühl;[13] the chair for theoretical physics in Heidelberg with Wesch;[14] Regener's chair with Ferdinand Schmidt;[15] while the Berlin chair for theoretical physics remained unoccupied for years;[16] and the Charlottenburg professorship was even eliminated.[17]

Though the great majority of physicists were unanimously outraged by all these machinations, they did not speak out publically and their indignation did not reach the places of influence. Jordan alone made himself very clear in his various books and articles,[18] and later Weizel as well in a review of a book by Wilhelm Müller in the *Zeitschrift für technische Physik*.[19]

When I was urged in the summer of 1940 to assume the acting direction of the University Lecturers in Darmstadt on a temporary basis, I attached the condition that I would be able to make a stand openly against this physics policy of the Party in Munich. At the beginning of August in an initial discussion with the Reich University Lecturers leadership in Munich, I spoke quite frankly and explained that more than 95% of German physicists were unanimously opposed

---

[12]On the appointment of W. → Müller as A. → Sommerfeld's successor in 1939 see, e.g., doc. 85, no. 4 on p. 265.

[13]Wolfgang Gaede (1878–1945) was a gifted inventor of various airpumps and meters; and his rotating mercury airpump, which he invented in 1905 while still a teaching assistant in Freiburg, Breisgau, is named after him. He was appointed full professor at the Polytechnic in Karlsruhe in 1919 and continued to invent various instruments. He accepted the Dudell Medal in London in 1933 and was therefore dismissed as politically unreliable by the National Socialists in 1934 and was replaced by Lenard's pupil A. Bühl.

[14]Ludwig → Wesch was indoctrinated with the anti-Semitic and antitheoretical intellectual climate at Heidelberg as a student of A. F. Schmidt (see the following footnote). He was appointed full professor of technical physics in 1943, which he retained until the end of the war, in 1945.

[15]Erich → Regener, whose specialty was the physics of the stratosphere, was placed in temporary retirement in 1937 because he was married to a Jew. He could resume his research on cosmic ray physics in 1938 in Friedrichshafen (Lake Constance) on a semi-private and independent basis thanks to support by the → *KWG*. August Ferdinand → Schmidt was a former student of Lenard's. He succeeded Regener in Stuttgart in 1940.

[16]This chair was last occupied by Max → Planck and Erwin → Schrödinger; H. A. → Stuart was commissioned to lecture on theoretical physics 1936–39 at the Berlin Friedrich Wilhelms University, and Wilhelm → Westphal continued to teach theoretical physics from 1939, but in a lower ranking position, and the chair remained vacant.

[17]Richard → Becker had been full professor there since 1926 as head of the newly founded theoretical physics department. He was forced to transfer to Göttingen in 1936 on the basis of the newly issued law (see doc. 36), Westphal was appointed substitute head of the physics department there in 1935, but the chair was eventually eliminated. The department of military science (*Fakultät für Wehrwissenschaft*) at the Polytechnic was expanded in its place; cf., e.g., Cassidy [1979], p. 379.

[18]Pascual → Jordan continued to mention the theory of relativity and quantum mechanics in his semipopular accounts on the progress in modern physics; cf., however, doc. 87.

[19]Walter → Weizel published his book review in 1942; see doc. 89.

on this issue against the remaining 5% or so, and that with the exception of Tomascheck [sic] and Messrs. Stark and Lenard—the last two of whom can no longer be taken seriously—the latter group is made up entirely of embittered incompetents. I stated that the University Lecturers League's policy must be changed and that Bühl should no longer be the sole authority, since then German physics would go to the dogs. At first they responded to me very negatively, saying it was impossible that my assertions were correct, since none of the many full professors of physics with whom they had dealt had ever said such a thing to the Reich Leadership of the University Lecturers Association (!). But then finally they suggested that the question be examined at a debate in Munich, to which I was to appear with five supporters of my 'line' and to which Bühl would also be invited to appear with 5 supporters.

After lengthy delaying tactics by the opposing side in an attempt to obstruct the conference, and after extensive, in part very embittered, correspondence between me and the Reich University Lecturers leadership, the debate was then finally convened for the 15th of November, 1940, in Munich. I invited to this event Messrs. von Weizsäcker and Scherzer to represent pure theory, Mr. Joos as a specialist in both theory and experiment, Mr. Kopfermann as spokesman for experimental physics, and Mr. Heckmann as a specialist in the general theory of relativity.[20] Mr. Bühl appeared with his teaching assistant Volkmann, Mr. Thüring, Wilhelm Müller, Tomaschek, and Wesch;[21] Mr. Stuart also took part at the invitation of the Reich University Lecturers leadership.[22] Prof. Borger of the Reich University Lecturers leadership presided over the debate and confined himself to guiding the discussion.[23] In partly very heated debate, all the

[20]The theoretician Carl Friedrich von Weizsäcker studied under Heisenberg, taught at Berlin University from 1937 and was Meitner's assistant for a half year, and working at the Berlin → *KWIP* from 1936; Otto → Scherzer, a student of Sommerfeld, worked at the → *AEG* research lab under Ramsauer 1932–33 particularly on electron microscopy, then became associate professor of theoretical physics, and during World War II was employed at the German Navy's communications research lab (*Nachrichtenmittel Versuchskommando der Kriegsmarine*); Georg → Joos was full professor of experimental physics at Göttingen University and became head of the research division of the → *Zeiss* optical company in 1941; Hans → Kopfermann was a nuclear physicist who taught at the University of Kiel from 1937; Otto Heckmann (1901–1983) was an astronomer and cosmologist, who taught at the Göttingen University from 1927, and was full professor at the University of Hamburg and director of its observatory 1941-62.

[21]Harald Friedrich August Volkmann (born 1905) attended the Universities of Göttingen, Heidelberg, Jena and Königsberg 1924–30, where he wrote his doctoral thesis under R. → Gans in 1930, habilitated in 1935, was teaching assistant 1929–36, and was promoted to university lecturer there in 1936. He transferred to the Karlsruhe Polytechnic in the following year and was employed there as a lecturer until 1943 when he was appointed supernumerary professor. In 1948 he was employed as a scientist at the *Carl Zeiss* company in Oberkochen. Volkmann worked in the fields of electron microscopy and fluid structure.

[22]Herbert → Stuart was full professor of theoretical physics at the Dresden Polytechnic from 1939 to 1945. He initiated together with W. → Orthmann the exclusion of Jewish members of the German Physical Society (→ *DPG*) in late 1938.

[23]Gustav Borger (1899–1938) was a physician and full professor of pathology at the Univer-

opposition's statements, which attempted to reject theoretical physics, relativity theory, and quantum physics, were defeated; and it was determined unanimously that these three fields were necessary for the whole of physics.[24] Thüring and W. Müller chose not to reappear in the afternoon, though, after the morning's discussion had collapsed, and therefore had no part in the decision.[25]

In the following period we circulated widely the decision reached at the 'Munich Synod' [*Münchner Religionsgespräche*], as this debate soon came to be called; and we applied it wherever it was important to promote a reasonable approach to physics at Party agencies. This debate, which I had prompted, had a number of direct consequences: To begin with, when he noticed our resistance, Mr. Tomaschek carefully tried to distance himself from his original friends; while on the other side, Lenard accused his student Bühl of betrayal to the good cause, because he had not only sanctioned the Munich declaration, but had even helped to formulate it. The Reich University Lecturers leadership lost some of its confidence in Bühl as a result of his obvious defeat and in the following period also turned to other physicists for suggestions for physics professorship appointments, etc., so that the one-sided policy against a reasonable approach to physics gradually lost ground. However, very soon the opposing side broke the 'truce' regarding public polemics that had been made at the debate; since Thüring and W. Müller did not honor it, and Dr. Führer of the Ministry continued to use all means to agitate against us. The younger spokesmen on our side then suffered considerably as a result of their brave support. The appointments of Messrs. v. Weizsäcker and myself at Strasbourg University, which were at that time still undecided, were delayed for almost two years with open reference to our actions against 'German physics'; and we were then also treated unfairly besides (associate professorships instead of the planned full professorships). It is at least very probable that the delays in Mr. Heckmann's appointment in Hamburg also were caused by his having defended the cause of sensible physics.[26]

It had been agreed in Munich to continue the debate in a larger group of about 30 physicists, so I kept in contact with the Reich University Lecturers leadership despite the agitation against me. Prof. Ramsauer also encouraged me in this, who having been elected President of the German Physical Society in the autumn of 1940, appointed me Vice-President because of my activity against Party

sity of Munich. Cf. Dietmar Eckert, *Personalbibliographien der Professoren und Dozenten der Pathologie an der Maximilians-Universität München im ungefähren Zeitraum von 1870–1945*, Ph.D. thesis, University of Erlangen (Nuremberg), 1971, pp. 21–25.

[24]For a fuller account of the outcome of this debate see doc. 93.

[25]According to Beyerchen [1977], p. 178, Borger was so disgusted with his political colleagues that he lectured them during the lunch break, whereupon Müller and Thüring indignantly left the meeting.

[26]See Renneberg & Hentschel [1995] on Heckmann's Hamburg appointment; although the Hamburg faculty favored Heckmann, the appointment was delayed several years through politically motivated interventions at various levels of the Nazi bureaucracy.

physics.[27] I accepted this nomination despite great misgivings, after Messrs. von Laue, Gerlach, Joos, and Heisenberg, had urged me to do so as well.[28]

Then Prof. Ramsauer and I prepared a detailed memorandum against the abuses in the field of physics, which was subsequently sent officially by Mr. Ramsauer as President of the *DPG* to Reich Minister Rust, along with numerous supporting attachments.[29] As expected, there was no response; and the only result was a significant loss of favor for Mr. Ramsauer. But he did not let himself be deterred by this from speaking publicly about these problems (at the Aviation Academy, for example);[30] and he also sent our memorandum to various influential posts in business and in the Armed Forces, where it was received with great interest and had an impact as well.[31]

After endless delays and an additional preparatory discussion in Munich, which Mr. Ramsauer also attended, the 'Physics Camp' then took place from the 1st–3rd of November, 1942 in Seefeld in Tyrol.[32] In three days of extremely thorough reports and discussions on the issues of relativity theory and quantum physics, the meeting concluded with a one-hundred percent defeat for the few 'German physicists' who had had the courage to attend.

In an effort to regain the German Physical Society's proper place in science and in an effort to involve it in the consultations on physics matters, the 'Information Agency of German Physicists' was then created.[33] Its director Mr. Brüche[34] developed a bustling operation out of it with his *Physikalische Blätter*, which was sympathetic to our aims. Furthermore, as President of the *DPG*, Mr. Ramsauer took part in the consultations on the regulations on curricula and examinations for *Diplom* physicists.[35] Finally, following up on the efforts we had

---

[27]Carl → Ramsauer mentions this in his letter to Ludwig → Prandtl of June 4, 1941; see doc. 86.

[28]Max → von Laue, Georg Joos, Walther → Gerlach, and W. Heisenberg all actively fought the influence of the 'Aryan physics' movement on science policy.

[29]See docs. 90–93.

[30]This is an allusion to Ramsauer [1943].

[31]Judging from Ramsauer's, Joos's, and Prandtl's contacts, they probably sent it to important companies that produced, military equipment among other things, such as the *Zeiss* optical company and → *AEG*. H. → Göring who was the influential head of the Air Force within the Third Reich polycracy, probably was also contacted.

[32]On this physicists' retreat or conference (*Physikerlager*), see Goudsmit [1947], pp. 153f.; Beyerchen [1977], pp. 176–179; Walker [1989]b, pp. 69–70.

[33]In 1944 this agency (*Informationsstelle deutscher Physiker*) organized the publication of the journal later called → *Physikalische Blätter*. Immediately after the war, in 1946, 'New' was added to its title, but it returned to its former name in 1948. On the history of *DPG*'s journal see, e.g., Rechenberg [1994]a.

[34]Ernst → Brüche studied electrotechnics and took his doctorate under Ramsauer in 1926; he worked from 1928 to 1945 at the *AEG* research labs.

[35]This is also mentioned in Ramsauer's letter to Prandtl, doc. 88. In an effort to shorten education in physics at universities, the academic degree *Diplomphysiker* was proposed in 1937 and introduced in the early 1940's, requiring a thesis but involving significantly less research than a doctoral thesis. In 1938 this was only possible at polytechnics; see also footnote 9 of

already started with the memorandum, Mr. Ramsauer drafted a '*DPG* Program' with my support, which reflected the German physics community's wishes on how their field was to be represented particularly at universities; and it was discussed in August of 1944 at the meeting of the board of directors with the presidents of the District Association [*Gauverein*] in Bad Eilsen.[[36]]

The end of the war prevented all these efforts, as was the case with so many other things, from coming to a successful conclusion. But I believe that the German physics community has a right to know how the board of directors of the *DPG* had done everything within its power in the years following the last physics conference of 1940, which required a lot of courage, despite all the difficulties, to confront Party and Ministry for the sake of a true, decent, and scientific physics and to prevent worse from happening than had already happened. I believe that this struggle against Party physics can easily be considered an honorable page in the history of real German physics, because though few had led it actively, nevertheless, the overwhelming majority of physicists had supported it both practically and morally.[[37]]

111 Samuel A. Goudsmit: War Physics in Germany [January 1946]

Source: S.A. Goudsmit,* 'War Physics in Germany'. *The Review of Scientific Instruments*, Volume 17, Number 1, January 1946,[1] Announcements, pp. 49–52.

What happened to the German scientific war effort can be summarized as follows:

1. The esteem for bona fide science and scientists declined rapidly during the Nazi regime, while crackpots and second raters rose in power.

2. The German Government expected the war to be won in a short time and initially did not mobilize science for the war effort.

3. The broad fundamental training of German engineers eliminated to some extent the need for academic scientists' participation in wartime applied science.

doc. 102.

[36]On the role of the German Physical Society after the end of World War II in both East and West Germany see, e.g., Wilhelm Walcher and others in: Mayer-Kuckuck (Ed.) [1995], pp. F107–F182.

[37]This is a rather generous view of the performance of German physicists as leaders. For a more severe judgment see, for example, Lise Meitner's letter to Hahn, doc. 108.

* [Original footnote *:] "Dr. Goudsmit has spent most of the last two years in Europe, as scientific chief of the so-called Alsos Mission. This group, reporting to the War Department, was charged with obtaining information of military significance concerning enemy science."

[1]The journal prescribed that articles be submitted "not later than the twelfth of the month preceding the date of issue". For S. → Goudsmit's own description of the Alsos Mission, see doc. 116. See also the book review Morrison [1947]. For more on Goudsmit, Werner → Heisenberg and the German nuclear physics community, see docs. 117ff., and Walker [1990]b. The portrait photo of the author is omitted here; see, however, Fig. 26, doc. 116.

At the start of the war scientists of draft age were inducted into the armed forces. The German Navy, Air Force, Army Ordnance, Postal Department, and major industries had competent research staffs, primarily engineers but there was no coordination or cooperation among them.[2] The industrial research establishments had suffered least from Nazi influence and were occasionally able to employ scientists who were politically or racially undesirable in government laboratories.

It was mainly in these industrial research establishments, and in government-sponsored laboratories (such as that at Peenemunde) directed to a specific engineering goal, that the several real accomplishments of the Germans were made.[3] V-1 and V-2, the high speed jet fighter, and several other weapons were pioneered by the Germans.[4] But their development program lacked coordination and failed to fit realistic military needs.

Academic scientists were not in the picture. They offered their services to the various government agencies, but these were hardly ever accepted. University physicists continued their peacetime experiments but soon found that necessary materials and equipment could not be obtained without priorities. They then attempted to convince the authorities that their fundamental research work was important to the war effort, or at least essential for the dominating role Germany was to play after the war had been won. Again there was no coordination, though some research workers found a ready sponsor in the German Air Force, and others got support from the Army or the Postal Department, which was headed by the gullible minister Ohne[s]orge.[5]

---

[2]The Reich Research Council (→ *RFR*) was founded in 1937 and reorganized in 1942 (see docs. 52 and 98) in and effort to coordinate the different research divisions in the → *HWA*, → *RPM*, → *RLM* and in the military; but it never succeeded in overcoming the conflicting competencies of the National Socialist polycracy. Cf., e.g., Ludwig [1974], chap. 6, and the Introduction, p. xxi, and sec. 3.2.

[3]Peenemünde at the Baltic Sea was the principal center for the Army's long-range rocket research and development, installed at a cost of about 120 million dollars: Cf., e.g., Neufeld [1995]. At its peak the center employed 2,200 scientists and technicians before it was heavily bombed in Allied air raids. It was relocated in August 1943 to an underground facility in the Harz mountains called *Mittelwerk* where concentration camp prisoners were used as forced labor. See also Bode & Kaiser. For a broader overview of German weapons research, see, e.g., Simon [1948].

[4]These weapons of reprisal (*Vergeltungswaffen* or *Wunderwaffen*) were the Nazi regime's weapons of last resort. The V-1 was a pilotless jet-propelled airplane and the V-2 a ballistic missile. On German jet propulsion and supersonics research see, e.g., Simon [1948], chaps. IXf.; on the physics and design of the V-2 rocket, which was launched against London in the final years of the war, and its later application in high-altitude research, see Jetter [1948]. Ulrich Albrecht discusses Nazi German aerodynamics research and development with its gigantomanic tendencies and its desperate move towards suicide technology in the final years of the war in: Renneberg & Walker (Eds.) [1994], pp. 88–125. See also footnote 3 of doc. 112.

[5]For instance, M. → von Ardenne's research on nuclear physics and high-frequency technology at the Nuclear Physics Institute in East Lichterfelde (Berlin) was financed by the Postal Ministry (→ *RPM*), headed by Wilhelm → Ohnesorge, who was otherwise one of the least influential Nazi officials.

The net result was that several physicists were able to continue their work on a larger scale than before the war, at least with more money and materials for building and research, but often lacking assistants. It was said that "the war had been put in the service of science instead of science in the service of the war." Most of the research work was far remote from anything we would have called war work. In a few cases only were the immediate applications fairly obvious, such as transmission of the atmosphere for infrared and ultraviolet, a study of materials for screens of cathode-ray tubes, a thorough study of the carbon arc for searchlights and as a source of ultraviolet.[6]

The only non-military physics laboratories engaged in actual war research were those of some of the large industries. In these places the scientists were at least acquainted with the finished apparatus for which their work was intended. But even they never were sufficiently informed about its performance in actual military operations.

Nuclear physics was in the foreground from the very start. Using the promise of a uranium bomb and uranium energy as a selling point, some physicists succeeded in getting rather peripheral experiments in nuclear physics sponsored by the Air Force and the Postal Department. At the same time more serious attempts to harness uranium energy were undertaken by two competing groups, one under Heisenberg at the Kaiser Wilhelm Institute for Physics, the other under Professor Erich Schumann at the Army Ordnance Research Department.[7] Schumann is a second-rate physicist who held a very high position in Army research. Even *Gestapo* reports on him agree that he had neither the competence nor the personal qualities for this job.

Most of the academic research came under the Ministry of Education, where it was managed by the State Research Council. This council was the direct successor of the '*Deutsche Notgemeinschaft*' which was a great influence in furthering the brilliant achievements of German science in the twenties.[8] The president of this State Research Council was Professor Dr. Rudolph Mentzel, a brigadier general

[6]W. → Finkelnburg and M. → Steenbeck worked on carbon arc searchlights, which became important as the Anglo-american aerial raids increased. They were primarily concerned with reducing disturbing light scattering not focussed in the beam itself. Cf., e.g., Steenbeck [1977], p. 122; W. Finkelnburg, *The High Current Carbon Arc*, Berlin: Office of Military Government for Germany (US) (=FIAT final report, No. 1052), 1947. On the infrared absorption of the Earth's atmosphere and its relation to atmospheric visibility conditions relevant to aviation, see secs. 2.1.3f. in Fritz Möller's report in: Ratje Mügge (Ed.) *Meteorology and Physics of the Atmosphere*, Wiesbaden, Klemm, 1948 (= FIAT Reports, Vol. 19), pp. 62–72.

[7]For a detailed account of the researches by W . → Heisenberg's team at the → *KWIP* and K. → Diebner's team at the *HWA* (where E. → Schumann was one of the supervisors), which coordinated all research in this sector between 1939 and 1942, see Walker [1989]a.

[8]The → *Notgemeinschaft* had been under the supervision of the → *RIM* since its foundation in 1920, but was placed under the → *REM* in 1934. Its new president J. → Stark then changed its name to *Deutsche Forschungsgemeinschaft*. Contrary to Goudsmit's assertion, the *DFG* and the Reich Research Council (*RFR*) co-existed after the latter's foundation in 1937. See also footnote 2 above.

in the *SS*, also a second-rate scientist and an inferior administrator. He escaped from Berlin before it fell, taking with him all the special food packages which were under his control for distribution among deserving research workers; he has not been found yet.[9]

The enormous lack of coordination and the need for war research were finally felt when the war began to show signs of becoming a prolonged affair instead of a short blitz, and especially when the U-boat campaign appeared to be a fiasco.[10] Thus, early in 1943, a Hitler decree took the State Research Council out of the Ministry of Education and made it directly responsible to *Reichsmarschall* Goering, who became its president in name.[11] Unfortunately, the incompetent Mentzel remained its actual leader. While various branches of science were in general represented by really outstanding men who were given much freedom and power, fields of research of special importance were represented separately on the council. For example, there was a representative for physics, Walther Gerlach, and a chief for nuclear physics, first Abraham Esau, later also Gerlach. Their titles in German were '*Fachspartenleiter für Physik*' and '*Bevollma[e]chtigter des Reichsmarschalls für Kernphysik.*'[12]

The State Research Council was further encumbered by a presidential committee of lethargic Nazi big shots. It never functioned effectively as a coordinating agency; the council as such met only once, and its president, Goering, never appeared.[13] However, all research projects undertaken by academic institutions and by some industries were sponsored after 1943 only by the State Research Council, and it was up to Gerlach to make decisions regarding support for research programs in physics. The uranium project of Army ordnance was also turned over, with most of its personnel (excepting Schumann), to the Research Council.[14] Upon Gerlach fell the impossible task of reconciling this group under Diebner with the competing group at the Kaiser Wilhelm Institute under

[9]Rudolf → Mentzel was detained soon afterwards at an American prisoner-of-war camp for three years. For his conception of research in Germany, see Mentzel [1940].

[10]Due to the development of an effective means to localize German submarines, the initial naval threat to the Allies was diminished, and in the 1940's Allied naval and air forces began to destroy German submarines in great numbers.

[11]The reorganisation of the *RFR* installing Hermann → Göring as its new president actually took place in June 1942, see doc. 98.

[12]Walther → Gerlach succeeded Abraham → Esau as head of the physics section in January 1944: See footnote 15 of doc. 105.

[13]This severe judgment on the effectiveness of the *RFR* even under the directorship of H. → Göring is shared by other contemporaries, such as, e.g., Simon [1948], chap. 6, as well as historians of science such as, e.g., Ludwig [1974], chap. 6.

[14]The Army Ordnance Office began withdrawing from the German uranium project at the beginning of 1942 after it had become clear at conferences in Berlin at the end of January (cf. doc. 95) that it would be of no immediate military relevance. The *RFR* took over its direction at the beginning of July and the *KWG* resumed its authority over the project, but on July 9, 1942 the *RFR*, and with it the 'uranium project', was subordinated under Reich Marshal Göring. W. Gerlach became Göring's plenipotentiary of nuclear physics in December 1943.

Heisenberg.[15] Gerlach had only moderate success.

Though Gerlach, who had the full support of all good physicists, succeeded in bringing more unity in physics research, one cannot say that this research made a significant contribution to the German war effort. There were various reasons for this. It has been mentioned above that academic scientists were not in the confidence of the Armed Forces and were hardly aware of their real needs. A sharp distinction was made between research and development. The 'pure' scientists refused to engage in development, which was usually left to the more practical engineers. Whereas Goering was in charge of 'research', Speer, the minister of armament production, was responsible for 'development.'[16] This distinction gave rise to a lot of difficulties. It helped to keep the real scientific brains far away from any of the practical applications of their work. Finally, the main motive behind Gerlach's policies was the firm belief that Germany could win the peace if she again dominated the world of science, irrespective of the outcome of the military struggle.

Fig. 25: 'Hitler presenting an award to Speer, Armament Minister in charge of German wartime scientific development.'

The results of research in physics were reported to Gerlach. A great deal of it was published in a 'restricted' supplement to the *Physikalische Zeitschrift*, called *Berichte für Physik*. Only five numbers of this journal appeared during

[15]Kurt → Diebner was the designated head of the Nuclear Research Council (*Kernforschungsrat*) under and head of the nuclear research units of the *HWA* in Gottow and the *RFR* in Stadtilm. Early in 1940 the → *KWIP* in Dahlem (Berlin) was also annexed by the *HWA* for the duration of the war, Diebner also became its managing director; cf., e.g., doc. 79, and the Introduction, pp. lxxxiiif.

[16]Hitler's favored architect, Albert → Speer, was appointed on the *RFR* in 1942 and acquired additional power by becoming Armaments and War Production Minister in 1943. Speer attributed the Research Council's failure to the strained war economy and to Hitler's lack of appreciation for the importance of fundamental research. For Speer's retrospective account of his activities regarding the *Uranverein*, see Speer [1976]b, pp. 235–243.

the war. The work on nuclear physics was published in 'secret' and 'top secret' reports which had only a limited distribution among the members of the '*Uran Verein.*'[17]

Physicists in Germany were clearly aware of the fact that their field had lagged behind Allied achievements ever since the Nazis rose to power. Therefore, fundamental research was now pushed ahead. A vigorous but futile campaign was fought by Gerlach to rid physics of men who were appointed merely on the basis of their Nazi party affiliations without respect to their competence. Progress in physics suffered because modern developments were considered subversive by Nazi scientists such as Müller, the totally ignorant and bigoted successor to Sommerfeld in Munich, or Lenard and his equally misguided disciple, Wesch, at Heidelberg.[18] The Nazi *Bund* of University teachers arranged a meeting where twenty-five of the most influential physicists undertook to decide whether the theory of relativity was a Jewish concoction or good Aryan science. Several were flatly against it, but the diplomatic son of the diplomat von Weizsäcker succeeded in formulating a favorable statement regarding Einstein's work which proved acceptable to the most rabid anti-Semites.[19]

The better physicists had to waste part of their time on several non-productive activities. The crystallographer Von Schiebold sold Milch of the Air Force the idea that it was possible to build an x-ray searchlight which would cause burns on the crews of Allied bombers at altitudes up to 30,000 feet![20] The source of the x-rays was to be a betatron.[21] Not only was he promised money, but also

[17]For a list of 222 of these internal reports called *Kernphysikalische Forschungsberichte* (→ *KPFB*) on progress in nuclear physics with restricted distribution among the members of the research teams, the *HWA*, and the National Socialist bureaucracy, see Diebner (alias Tautorus) [1956]. For a more complete listing, see David & Warheit (Eds.) [1952]. For overviews, see also Bothe & Flügge (Eds.) [1948]. Heisenberg also refers to these *Berichte für Physik* in his overview article on nuclear physics in Germany, doc. 115.

[18]On the appointment of Wilhelm → Müller as successor to the theoretical physicist Arnold → Sommerfeld at the Maximilians University in Munich, see doc. 85, No. 4 on p. 265, and Cassidy [1992]a, chap. 18. Müller, Philipp → Lenard, and Ludwig → Wesch were advocates of 'Aryan' physics; see, e.g., Finkelnburg's account of the struggle against 'Party physics', doc. 110.

[19]On this meeting in Munich in 1940, see here docs. 93 and 110; at this event, Carl Friedrich → von Weizsäcker had the support of opponents of the 'Aryan physics' movement, such as G. → Joos and W. → Finkelnburg.

[20]The mineralogist and materials science expert Ernst von Schiebold (1894–1963) specialized in X-ray analysis of various materials. He was associate professor at the University of Leipzig 1926–41, and at the Dresden Polytechnic from 1941–45 and was simultaneously department head at the German Office of Materials Testing in Leipzig. Later, in 1954 von Schiebold became full professor and director of the Institute of Materials Science at the Technical University of Heavy Machine Construction in Magdeburg. Erhard Milch was *Generalinspekteur* of the Air Force from 1938 to January 1945 and Field Marshal of the Air Force from July 19, 1940; holding other important positions at the → *RLM* since 1933. On Milch see footnote 2 of doc. 99 and, e.g., Wistrich [1982].

[21]On the development of the betatron, invented in Germany by Max Steenbeck in 1935 and independently in the USA by D. W. Kerst in 1940 see, e.g., Trendelenburg [1975], pp. 68–70,

first priority for a high tension apparatus and other equipment which was badly needed for research elsewhere. Heisenberg, Gerlach, and others succeeded after much trouble in stopping this absurd project. Physicists were also ordered to investigate whether explosives could be ignited from a large distance by means of two infra-red rays intersecting at the explosive charge under a prescribed angle.

The picture is definitely incomplete without mentioning the interference of the *Gestapo*.[22] Every laboratory and every meeting had at least one *Gestapo* spy, who sent reports to the 'cultural' section of the *Gestapo* about disagreements among scientific workers and their relations to the authorities, about difficulties which delayed progress, about evaluation of results, and occasionally also about the political reliability of the men involved. Late in 1944 local *Gestapo* heads sent inquiries to various laboratories requesting information about what stood in the way of a more rapid advance in their research progress, whether it was red tape, lack of materials, lack of personnel, or lack of funds. At that time the *SS* and *Gestapo* were trying to take charge and save whatever could still be saved. In some cases they evacuated scientists and their equipment without advance notice from the war zone to the Bavarian redoubt. After the middle of 1943 the key scientific laboratories had been evacuated from the larger cities and dispersed in small villages, where they were housed in rather primitive quarters such as schoolhouses, a few rooms in minor factories, breweries, etc.

There was no lack of money. The State Research Council was given 50,000,000 marks for 1944, but expended only half of it. The principal projects in that year were the uranium project and high-frequency development under Abraham Esau, each with a budget of about 4,000,000 marks.[23] The high-frequency work was mainly in the hands of engineers, involving only a few physicists. The system of priorities (these were more essential than money) was in a state of confusion for a long while. High priorities could be assigned only by the Speer ministry, and were assigned much too freely. Nobody dared to take the responsibility to say "No" or to discontinue an apparently useless project. Speer tried, too late, to take drastic steps restricting research and development to the more important and promising projects. Even then, the State Research Council violently opposed Speer's intentions on the grounds that research was their and not Speer's business.[24]

At the end of the war, German physicists were confident that they had saved German research from annihilation. They believed that they were ahead of the

---

and Steenbeck [1977], pp. 84ff, 118.

[22]On the activities of the political secret police → *Gestapo* see, e.g., Gellately [1990], or Delarue [1964/79].

[23]Esau became head of the unit for high-frequency research on Jan. 1, 1944 after being replaced by Gerlach as coordinator of nuclear research at the end of 1943. Goudsmit's figures are approximately correct. According to Gerlach's letter to R. Mentzel of May 26, 1944, the *Uranverein*'s precise budget was, for instance, 2.56 million RM in 1943/44, and 3.65 million RM in 1944/45: Irving (Ed.) [undated], no. 29 1113. Cf. also doc. 103.

[24]See, e.g., H. → Himmler's protestations in a letter to Speer, doc. 106.

Allies in the uranium problem. They attempted to hide most of their more significant work but were willing to assist the Allies if offered suitable opportunities to rebuild their laboratories and return to work. Not until the seventh of August, 1945, did the German physicists realize that Germany had lost the scientific war, as well as the military one.[25]

112 Samuel A. Goudsmit: German Scientists in Army Employment I—The Case Analyzed [February 1947]

Source: Samuel A. Goudsmit, 'German Scientists in Army Employment. I—The Case Analyzed' in Letters to the Editor, *Bulletin of the Atomic Scientists*, Volume 3, No. 2, February, 1947, pp. 64, 67.[1] Reprinted by permission, © Education Foundation for Nuclear Science, Chicago.

A War Department release concerning the importation of enemy scientists for employment in military, academic and industrial research over here, has stirred up a lot of comment.[2] The opponents of such a plan point out that most of these enemy scientists were strong supporters of non-democratic doctrines and at heart are still ardent Nazis.

What to do with enemy scientists is unfortunately a more complex problem than opponents of the plan seem to realize. Let us make an attempt to analyze the situation objectively.

American and British scientists have played a prominent role in the war. The result was the recognition that science and scientists are essential to the welfare and strength of a country, in peace time as well as in an emergency. It is immaterial in this respect whether the scientists are working on military problems, on pure research, or on industrial applications.

From this point of view the finding of a number of good enemy scientists is

[25]Goudsmit had access to the secretly recorded conversations of a select group of German physicists interned at Farm Hall, England, by the Allies from July to December, 1945. On this day these leading German nuclear experts first became aware of the superiority of America in their field as they listened to a BBC radio broadcast reporting on the atomic bomb drop over Hiroshima on August 6. The official English translation of the transcripts are now finally open to the public and are published in: Frank (Ed.) [1993]; see especially pp. 70ff. See also footnote 25 of doc. 116.

[1]For a response to this article, see the following document.

[2]According to official figures provided in December 1952, 642 German jet and rocket propulsion specialists were invited to work in the United States soon after World War II (Trischler [1992]a, pp. 288f.). This program, known as 'Operation Paperclip' was arranged by the American secret service and other military agencies, often in violation of U.S. immigration law, which prohibits the immigration of former members of the → *NSDAP* or other Nazi organizations. It was later revealed that the United States even falsified its files or made conscious omissions in its scramble for capable foreign scientists to employ in U.S. weapons development programs. Through 'Operation Paperclip' even war criminals were employed at American and British agencies after 1945. On Operation Paperclip, see Lasby [1971], Gimbel [1986], Bower [1990], and Hunt [1991].

equivalent, militarily, to the finding of a few up-to-date enemy war ships; economically, to finding some enemy gold caches; industrially, to finding a new source of raw materials.

Some of these scientists, especially the technicians, have some actual contributions to make. There are a few fields of technology in which the enemy has gone further than we have. The best known example is the development of the V-2 rocket.[3] It is clear that the men who are responsible for this instrument may well be useful on future problems in this field.[4]

It is natural for the military to be concerned lest such valuable resources might be used eventually against us. The only absolute way to settle this minor and many major problems is the abolition of war. Neither political leaders nor the popular masses have made much progress towards this goal. As one-sided unpreparedness cannot prevent involvement in a war, we are, paradoxically, forced to consider the possibility of an armed conflict while sincerely working towards world peace.

In our dealings with these enemy scientists we must first of all exclude any undemocratic procedures. Keeping them segregated in camps or restricting them to small regions in or outside the territorial United States can be done for a limited period only.

It has been proposed to leave these men in their own place and by rigid control keep them from working on problems of military significance. Such a plan will work for a short time only. German and Japanese scientists are not allowed to work on rockets, jet propulsion, or atom bomb research.[5] As a result, experts in these fields move sooner or later to countries where they are not restricted in their research.[6] It must also be stressed that it is getting more difficult to draw the line between harmless research and research of use in warfare. No matter how strict the controls are now, they will lose their strength in a few years. It is furthermore evident that the defeated countries will make special efforts to gain leadership in fields of science, which is cheaper than maintaining an army.

It appears therefore that our interests are best served by using the skill and

[3]On the V-2 rocket and German rocket research see, e.g., M. J. Neufeld in: Renneberg & Walker (Eds.) [1994], pp. 51–71; Neufeld [1995], Bode & Kaiser [1995]. See also footnote 3 of doc. 111.

[4]Both the American and Soviet space programs, for example, benefited greatly from this influx of highly skilled specialists. See again Bower [1990]. On the corresponding transfer of scientists to the USSR, see Albrecht et al. [1992]; Hunt [1991], Ciesla [1993].

[5]These restrictions were imposed by Allied Control Law 25, issued April 29, 1946, prohibiting all research in applied nuclear physics. This law required further that all planned research be cleared in advance with a science officer, and that all results and publications be submitted to that officer for review. Allied Control Law 25 was finally relaxed in 1949, but experimental applied nuclear research was not permitted until 1955. Cf., e.g., Cassidy [1992]a, pp. 530f., and references there.

[6]Paul → Harteck, for instance, moved permanently to the United States in 1951 and became professor at Rensselaer Polytechnical Institute in Troy, New York.

potential knowledge of these scientists ourselves. It is known that our allies are doing the same with scientists found in their zone of occupation and those willing to move there. England, Russia and France have the geographical advantage of being near to the occupied territory, nevertheless in many instances the scientists are being moved to the controlling country.[7]

So far we have discussed this problem in a rather cold materialistic way as if these scientists and their knowledge were mere inanimate commodities. If we next consider the human side of the question a number of serious difficulties arise.

It cannot be denied that the majority of these enemy scientists consciously supported an aggressive war against us. Those who opposed the excesses of the Nazi regime, were nevertheless in agreement with its policy of an imperialistic Germany, ruling the world.[8] I know of only very, very few who clearly saw the German errors and acted accordingly. Security 'screening' will remove only the actual war criminals, members of the *SS* and the *Gestapo*.

It is thus, in most cases, morally wrong for our scientists to collaborate with these imported colleagues. However, moral issues seem to have very little practical effect and only a negligible number will refuse to work with any one of them. It is sad indeed to observe that the few surviving victims of Nazism are mentally and morally starving in Displaced Persons Camps, while these "*Heil*" shouting scientists are offered privileged positions in our country.

We must expect that these German scientists, once they are here, will continually try to defend the actions of Germany before and during the war. Their background and education will have supplied them with the necessary arguments. The rocket specialists have at times tried to convince some of us that they worked on rockets only for the purpose of scientific research.[9] The German atomic scientists, who failed in their attempt to make a bomb, planned to deny that they ever wanted to make one.[10]

[7] This is particularly true of the German Eastern sector: Many nuclear physicists, among whom some were victims of Nazi persecution, such as former Communists, and others collaborators of the Nazi regime with no prospects in the West, decided to work for the Soviets for several years. These physicists included Gustav → Hertz, Manfred → von Ardenne, and Peter → Thiessen. See Albrecht et al. [1992], as well as for autobiographical accounts: von Ardenne [1972], Steenbeck [1977], Riehl [1988]. On the postwar plundering of technology see also Gimbel [1990].

[8] Goudsmit may be alluding to a comment by Heisenberg, who though certainly not a radical Nazi propagandist, naively understood himself to be a goodwill ambassador. During one of his lectures in Copenhagen in September 1941, he purportedly said he would certainly prefer Europe under German rather than Russian control. During a trip to Holland in October 1943 he also reportedly told colleagues: "Europe under German leadership might be the lesser evil" against Russian domination. Cf., e.g., Walker [1992], pp. 365, 381. See also footnote 10 of doc. 108 here.

[9] An example is Werner von Braun (1912–1977), a prominent researcher in the American rocket and space program that culminated in the first manned flight to the moon in July 1969.

[10] Transcripts from the Farm Hall tapes record discussions between the interned German physicists, including on how they should respond to questions about their nuclear research motives. See Frank (Ed.) [1993], pp. 102ff., 175.

The belief of their superiority is widespread among the German scientists and their selection to work with us will strengthen that belief. Until Hiroshima woke them up German atomic scientists were absolutely convinced that their work was ahead of that of the Allies.[11] In reality they had hardly passed the initial stages of research in spite of an early start of their project.[12]

It is thus obvious that the majority of the imported German scientists will ascribe their selection to their superior achievements. Unfortunately there are many among us who share this opinion with the Germans and are likely to praise highly anything they see or hear about them.

In reality the Germans are not at all superior and in many phases lag far behind American achievements. If we had made the serious mistake of putting our principal technical efforts upon super rockets instead of radar and nuclear research, we too would have produced a V-2-like weapon and probably lost the war. The Germans had surpassed us in work on supersonics, jet propulsion and U-boat development.[13] In some technical fields isolation caused their developments to be along different lines from ours, giving uninitiated the impression of superiority. In a secret report (April, 1943) the president of the German physical society calls the attention of the Government to the fact that "physics is a weapon ... of probably decisive importance in the struggle of nations" and that "since the first world war Germany has lost the lead in this field to America." ...[14]

An improper evaluation of the scientific value of these Germans by unqualified

[11]The Americans dropped the first atomic bomb ever used in warfare over the Japanese seaport Hiroshima on August 6, 1945. The second atomic bomb dropped over Nagasaki three days later brought about the surrender of the Japanese on August 14, thus bringing World War II to an end. (On the political pressures leading to the deployment of a second bomb see also Goldberg [1995].) On the German physicists' changing perception of where their country stood in their field see also, doc. 111, footnote 25, and doc. 116, footnote 25.

[12]On the groping beginnings of Allied research with the investigations of the feasibility of atomic weapons by the Maud Committee in England, established in 1940, and the American counterpart which became known from June 1942 on as the Manhattan Project see, e.g., Gowing & Arnold [1979], chap. 3 and on its rapid expansion see also Goldberg [1995]. The German nuclear research had begun much earlier in September 1939 (not including earlier uncoordinated preliminary research work). See Smyth [1948] as well as Heisenberg's overview article on German nuclear research from 1938 to the end of the war, doc. 115, and Weinberg's and Nordheim's postwar assessment from an examination of the *KPFB* reports, doc. 109.

[13]On supersonics and jet propulsion research in Germany see, e.g., footnotes 3f. of doc. 111. On submarine technology see, e.g., Richard Lakowski *U-Boote: Zur Geschichte einer Waffengattung der Seestreitkräfte*, Berlin: Militärverlag, 1987.; Erich Grœner, Dieter Jung, et al., *Die deutschen Kriegsschiffe: 1815–1945*, Munich: Graetz, Vol. 3. Cf. also Baxter [1947] on the Allied weapons technology, especially pp. 37ff., 170 on their final dominance in submarine development. On submarine warfare, see also footnote 2, doc. 100.

[14]Carl → Ramsauer was president of the *Deutsche Physikalische Gesellschaft* (→ *DPG*)from 1941 to 1945. He addressed the issue of American superiority in physics in 1942 in a letter to Minister → Rust; see doc. 90 and its enclosure entitled 'American Physics Outdoes German Physics', doc. 91. See also doc. 102 and Ramsauer [1944].

persons may cause very serious drawbacks, far overshadowing the presumed gains to be obtained from having these men work in this country. It would indeed be unfortunate if for example all our military rocket research or ballistics were permanently dominated by German personnel. Misplaced hero worship may give these men better positions and their opinions more weight than they deserve.

On the other hand, if these imported scientists are gradually absorbed in places in accordance with their true ability and personality, they can fill a need. Among competent American colleagues they will eventually find their proper level. Most of them will turn out to be merely mediocre. The better ones will have a chance to progress. The mediocre ones will continue to preach German superiority and how much better they did things in Germany than in this country. But on the whole they will be harmless.

The problem as a whole shows clearly that scientists and science have lost much of their freedom of action. If this trend goes much further, they may even lose their freedom of thought, thus killing scientific progress and starting a modern 'dark age' era.

113 Hans A. Bethe & Henri S. Sack: German Scientists in Army Employment II—A Protest [February 1947]

Source: Hans A. Bethe and Henri S. Sack, 'German Scientists in Army Employment. II—A Protest' in Letters to the Editor, *Bulletin of the Atomic Scientists*, Volume 3, No. 2, February 1947, pp. 65, 67.[1] Reprinted by permission, © Educational Foundation for Nuclear Science, Chicago.

A few weeks ago, the Army revealed another of its secrets: The employment, in this country, of a considerable number of German scientists and technicians. It has been known for some time that the Armed Forces brought to this country German specialists in order to acquaint the Services with the developments in Germany in the field of jet- and rocket-propulsion, in particular to help in the reconstruction of German equipment.[2]

We may regret that there is a need for such a collaboration, but, as it is the

[1]This joint letter to the editor by Hans Albrecht → Bethe and Henri Samuel Sack (1903–1972) is in response to Samuel A. → Goudsmit's letter in the same issue, see doc. 112. H. S. Sack was Swiss-born and became a naturalized U.S. citizen in 1933. After acquiring his *Diplom* in Zurich in 1925, Sack took his doctorate in experimental physics in 1927 and worked as assistant in physics at Zurich University 1925–27, then in Leipzig 1927–33. He also served as assistant editor of the *Physikalische Zeitschrift* 1929–33. Sack received a fellowship in Brussels 1933–35, was assistant of physical chemistry 1935–37, and chief scientist 1937–40; 1940–44 he was resident associate in applied physics at Cornell University, then became associate professor of engineering physics 1944, and full professor there in 1949. Sack worked on applied physics, supersonics, dielectric properties, infrared spectroscopy, electric and magnetic birefringency, analog computers and the elastic properties of solids.

[2]See footnote 2 of doc. 112.

Army's responsibility to keep pace with the modern technical developments, one cannot raise serious objections against it. However, the reports that appeared in the press at the occasion of the official release create an uneasy feeling that the deal was not as straightforward as could be wished. Timid and vague hints concerning possible citizenship, preferred treatment of their families, permanent employment in Universities or industrial laboratories, awaken suspicion that far-reaching concessions have been made in order to gain this collaboration, and that the Armed Forces plan to make use of the services of these scientists beyond the stage of merely bringing our Services 'up to date.' It is therefore not astonishing that already protests have been voiced, the most significant of them in the form of telegrams to President Truman and the Secretaries of State and of War, by the "Council Against Intolerance." [3]

It is difficult to get an exact picture of the situation. Only very few facts have been officially and clearly reported. But according to rumors, little attempt is made to train American personnel so as to be able to eventually replace the Germans. It also seems that some are already employed in new research projects, and that at least one University was offered the "assistance" of a German scientist, (which offer was declined by the University staff). It is furthermore contested that these men are in basic knowledge or ability superior to American scientists, and it is rumored that some have only been taken in order to prevent their employment by other countries. Their usefulness for the Army seems to lie primarily in their intensive specialization in restricted fields, which have not during the war received as much attention in this country as in Germany.

It is, however, definite, that promises for permanent stay and employment, as well as for preferred treatment of the families have been given. From some reports, one could gain the impression that these scientists waited eagerly for the Americans and were only too thankful to be allowed to continue their scientific work under the Army's sponsorship. This makes good reading, but it is probably nearer to the truth that they joined the Americans for fear of the Russians and in the hope of material advantages. It is astonishing that they were allowed by us to make conditions regarding their employment.

The question then arises: Is it wise, from a long-range point of view, or even compatible with our moral standards, to make this bargain? Would it not have been better to restrict definitely the stay of these scientists in this country to the

[3]Under Harry S. Truman (1884–1972), who acceded to the presidency in April 1945 and became the 33rd President of the United States until 1953, George Catlett Marshall (1880–1959), known particularly for his economic aid program for Europe, the 'Marshall Plan', was U.S. Secretary of State 1947–49. Kenneth Claiborne Royall (1894–1971) held the office of Undersecretary of War in 1945–47 and became Secretary of War in 1947 but was succeeded in the same year by the former Secretary of the Navy James V. Forrestal (1892–1949), who held the office until 1949. The 'Council Against Intolerance in America', located in New York City, advocated in particular the elimination of racial discrimination and presented its annual Thomas Jefferson awards to individuals for their contributions toward the 'advancement of democracy'.

absolutely necessary minimum? For it must be borne in mind, that many of them, probably the majority, are die-hard Nazis, or, at least worked whole-heartedly with the Nazis; otherwise they would not have held their high posts so vital for the Nazi war machine. Some German scientists are known to have opposed the Nazis and to have refused unrestricted collaboration; if these scientists wish to come to this country, they will be heartily welcomed by their American colleagues. However, no anti-Nazi action or attitude seems to be connected with the names released in the different press-reports.

Is it possible that the State Department does not object to their becoming citizens? Does it believe that they will turn into good democrats just by receiving favors from us? And if their stay in this country really reforms them, would this not be one reason more for sending them back to Germany so that they can lead their people towards democracy?

In the reports it is mentioned that the presence of these men saved millions of dollars. Does this imply that permanent residence and citizenship can be bought? Is it not more desirable to admit those people who have been longing for freedom and are willing to defend it, than those who, according to press release would work for any government. Hundreds of billions of dollars were spent for a war fought to keep this country free from Nazi-oppression and Nazi-ideology. It does not seem logical now to admit Nazis to important national position in order to save a few hundred million dollars.

These scientists are now working on highly secret projects. Can the Army put any trust in them? Can they be relied on in case of emergency? Even the insurance of screening by the Army can not dissipate the fear that while working for us, they will have in mind the interests of a nationalistic Germany, and that at the critical hour their holding of important positions in our defense-organization might prove fatal. Furthermore, in their strongly indoctrinated hate against the Russians,[4] they may contribute to increase the divergency between Russia and the U.S.A. Can we count on them as having peace as the aim of their efforts?

The admission of these Germans to our country creates another danger, particularly if they be allowed to take employment in industries or Universities. It must be feared that willingly or unwillingly they will be advocates of Nazi methods and ideology. Has this war been fought in order to have Nazi ideology creep into our educational and scientific institutions by the back door? Every effort should be made to prevent exposing American youth in such a direct way to Nazi influence. As far as the staff members are concerned, the mere presence of these Nazis in the laboratories will antagonize many of them and poison the atmosphere of friendly cooperation.

If the Army believes that there are foreign scientists whose unrestricted presence would be a positive asset to this country, then it should have them screened

[4]The Nazis had capitalized on anti-Russian sentiments and fear of the 'Communist threat', which has a long history in Germany including the cold war.

not only by an Army organization and according to Army standards, but also by a civilian group, among which the American scientist should have a decisive voice and which should consult with persons who have known these men previously. Army screening is more concerned with finding out whether there is an immediate danger in their employment, while the civilian screening will scrutinize more thoroughly their former connection with the Nazi regime, and the sincerity of their newly acquired democratic convictions.

Finally, should we not also consider what impression our actions will make on other peoples? These men who had a preferred standing under the Nazis are again offered a preferred treatment. Must that not hurt any German democratic feeling still alive, or now being born? The exaggerated appraisal of their scientific achievements and abilities will foster the self-consciousness and vanity of the Germans, so strongly emphasized by the Nazi regime.

We must admit that more scientists could be used in this country, but we must ask ourselves: Do we want science at any price? If the American scientists believe in freedom and peace and in their responsibility towards humanity, then they have to show concern about the conditions under which the Army is employing German scientists. Before all, they must request that the mystery shrouding this affair be lifted, that an exact account be given of what has been promised and to whom. It is not wholesome to have rumors going around. If it should turn out that these rumors are ill founded, then the public should be assured so by being given the facts. If, however, there is some truth in them, then the American scientists should join individually, and as a group, in a strong protest.[5]

114 Max Planck: My Audience with Adolf Hitler [May 6, 1947]

Source: Max Planck, 'Mein Besuch bei Adolf Hitler', *Physikalische Blätter*, Volume 3 [1947], p. 143, report dated May 6, 1947. Reprinted in Albrecht [1993], pp. 41–42; and elsewhere.

Following Hitler's seizure of power, I had the responsibility as president of the Kaiser Wilhelm Society of paying my respects to the *Führer*.[1] I believed I should

[5]Despite this appeal, 'Operation Paperclip' (see footnote 2, doc. 112) continued into the 1950's, organizing the employment of hundreds of German scientists (some of them former members of the *NSDAP, SA* or *SS*) in American weapons development programs.

[1]Max → Planck became president of the *Kaiser-Wilhelm-Gesellschaft* (→ *KWG*) in 1930; after Hitler came to power, he was re-elected on June 28, 1933 and remained in office until April 1, 1936. In this period he was instrumental in the *KWG*'s policy of self-realignment; cf., e.g., here doc. 38. See Albrecht & Hermann in: Vierhaus & vom Brocke (Eds.) [1990], pp. 356–382, especially p. 369; Cassidy [1992]a, pp. 306–307, and particularly, Albrecht [1993], pp. 44ff. Planck's visit took place on May 16, 1933 at 11 o'clock. Contrary to what is suggested, Planck and Glum had specifically arranged the meeting: C.f., e.g., Planck's letter to Hitler of May 2, 1933, in which he used the opportunity of thanking the *Führer* for his 75th birthday wishes (on April 23) to ask him for an appointment soon "on the current situation and future plans

take this opportunity to put in a favorable word for my Jewish colleague Fritz Haber,[2] without whose invention of the process for producing ammonia from nitrogen in air the previous war would have been lost from the start.[3] Hitler answered me with these words: "I have nothing against Jews as such. But Jews are all Communists, and it is the latter who are my enemies; it is against them that my fight is directed."[4] I commented that there are different types of Jews, both worthy and worthless ones to humanity, with old families of the highest German culture among the former; and when I suggested that a distinction would have to be made between them after all,[5] he replied: "That's not right. A Jew is a Jew; all Jews stick together like burrs. Where there is one Jew, all kinds of other Jews gather right away. It would have been the duty of the Jews themselves to draw a dividing line between the various types. They did not do this; and that is why I must act against all Jews equally." He ignored my comment that forcing worthy Jews to emigrate would be equivalent to mutilating ourselves outright, because we direly need their scientific work and their efforts would otherwise go primarily to the benefit of foreign countries.[6] Instead, he broke out in generalities and

---

of the Society"; and the *Generaldirektor* of the *KWG*, Friedrich Glum's letter of May 5, 1933 to State Secretary Hans Heinrich Lammers. For background information, see Albrecht [1993], p. 47, where both letters are quoted.

[2]Fritz → Haber (1868–1934) was director of the Kaiser Wilhelm Institute of Physical Chemistry (→ *KWIPC*) from 1911. He resigned from his position on April 30, 1933 (cf. doc. 15) in protest against the new government's requirement that his Jewish colleagues at the institute be dismissed. Haber died in exile in Switzerland in 1934. See doc. 31 for von Laue's obituary on Haber; for a description of the memorial session in honor of Fritz Haber, organized by Planck in January 1935, see footnote 3 there, and Vierhaus & vom Brocke (Eds.) [1990], pp. 372–374.

[3]During World War I, Haber organized the mass production of ammonia using the Haber-Bosch process, which yielded 25 million tons per month in 1916. Without the synthetic production of ammonia, the German Army would have run out of explosives already in 1915, and the German population would have starved from a shortage of fertilizer. Haber also played an active part in developing poison gas, which was first used by the German Army in Ypern in April 1915; cf. Fritz Stern in: Vierhaus & vom Brocke (Eds.) [1990], pp. 529–532.

[4]Hitler's answer demonstrates his main rhethorical technique: the use of succinct oversimplifications and assuming popularly held prejudices peppered with cue words, such as 'my fight' (*mein Kampf*), which immediately brings to mind his autobiography of that title.

[5]This retrospective account was written at the request of the editor of the *Physikalische Blätter*, Ernst → Brüche. A letter by Planck's wife Marga to the editor dated May 7, 1947, indicates not only her involvement in the exact wording of this text (because of Planck's sick condition at the time) but simultaneously challenges the general assumption that the text was a verbatim account of the visit: "Looking over the interview again one part has given us pause, namely with regard to the 'Eastern Jews'. You obviously cannot say that all Eastern Jews are worthless, and my husband might justifiably be seriously charged for it. Would you please change the section."

[6]On Planck's own values marked by loyalty to his fatherland and a strong national pride see, e.g., Bleuel [1968], Döring [1974], Heilbron [1986]b, Vogel [1991], Albrecht [1993], pp. 50f., Mehrtens in Meinel & Voswinckel (Eds.) [1994]. For a statistical analysis of the actual contributions to physical research of German-speaking emigrés after 1933, see Fischer [1988].

finally ended with: "People say I suffer occasionally from nervous debility.[7] This is slander. I have nerves of steel." As he said this he pounded his knee, spoke faster and faster, and whipped himself up into such a fury that I was left with no other choice than to remain silent and to take my leave.[8]

115 Werner Heisenberg: Research in Germany on the Technical Application of Atomic Energy [August 16, 1947]

Source: Werner Heisenberg, 'Research in Germany on the Technical Application of Atomic Energy',*[1] *Nature*, Volume 160, No. 4059, August 16, 1947, pp. 211–215.

Even ten years ago, physicists were well aware that the utilization of atomic energy could not be realized without a fundamental extension of scientific knowledge. In spite of the remarkable progress in experimental nuclear physics which followed the introduction of high-voltage equipment and the invention of the

[7]Cf., e.g., Waite [1977] for a psychoanalytical biography of Adolf → Hitler.

[8]The fact that this account was published soon after the end of World War II and was reprinted several times since without any comment on the factual collaboration of the Kaiser Wilhelm Society with the Nazi regime has caused Albrecht [1993] to suspect that Brüche had used the then aging Planck in the ongoing effort to put the actions of the *DPG* and the *KWG* during the National Socialist period in a better light (cf. Brüche [1946]a). Planck's distinction between 'worthy' and 'worthless' Jews above, for example, shows on the contrary how pervasive anti-Semitism, an integral part of the Nazi ideology, actually was even among leaders of the German intelligentsia. Thus this text cannot simply be considered an authentic documentation of Planck's 'heroic' intervention for a select few of his persecuted colleagues. It is also an illustration of the selective, even distorting use of historical events for the purposes of the politics of the day. This is equally true of the earlier accounts of Planck's visit, such as in W. → Heisenberg's letter to M. → Born (here doc. 22). In it Heisenberg only reported that Planck had been assured that the government would not impede science beyond the Civil Service Law provisions. This more positive version better suited Heisenberg's own motives of trying to dissuade Born from emigrating. It is therefore by no means clear that it is any more justified to grant greater historical weight to Heisenberg's contemporary report as is suggested in Albrecht [1993], p. 47.

* [Original footnote *:] "This article is a slightly abridged translation of a paper appearing in *Die Naturwissenschaften.*"

[1]W. → Heisenberg's paper referred to in the preceding original footnote * is entitled: 'Über die Arbeiten zur technischen Ausnutzung der Atomkernenergie in Deutschland', *Die Naturwissenschaften* **33** [1946], pp. 325–329. Contemporaries from the Allied countries disputed the impartiality of this survey of nuclear research in Germany. S. → Goudsmit thought the author had pulled the events out of their historical context and had artificially severed them from their military origin in an attempt to portray pure fundamental research; see, e.g., doc. 116 and Walker [1989]a, pp. 210f., for a historical discussion of this issue. At the end of the paper appeared the following statement: "Publication of results for which no source is cited was prohibited during the War." The relevant reports published in the series of secret reports on German nuclear research, *Kernphysikalische Forschungsberichte* (→ *KPFB*), are indicated. On this publication see also doc. 109, footnote 4.

cyclotron,[2] no physical phenomenon was known, even as late as 1937, which offered the remotest possibility of exploiting the enormous quantities of energy lying latent in atomic nuclei.

It was the discovery of the fission of uranium by Hahn and Strassmann†[3] in December 1938—in other words, the fact that the uranium nucleus can be split into two fragments of comparable mass when bombarded by neutrons—which brought the actual utilization of atomic energy within reach. Following this discovery, Joliot and his co-workers§[4] succeeded in proving, in the spring of 1939, that in the act of fission the uranium nucleus itself emits several neutrons, thus making a chain reaction fundamentally possible. Thereafter the possibility of nuclear chain reactions was eagerly debated among physicists, particularly in the United States; in Germany it was discussed by Flügge‡[5] in *Die Naturwissenschaften* in the summer of 1939. Meitner and Frisch¶[6] had already directed attention to the enormous quantities of energy set free in the fission process.

Public interest in the problems of atomic physics was negligibly small in Germany between the years 1933 and 1939, in comparison with that shown in other countries, notably the United States, Britain and France. Thus, while in America, previous to 1939, a whole series of modern research laboratories equipped with high-voltage plants and cyclotrons was springing up, in Germany there were only two adequately equipped laboratories; and these were not supported by the State, but sponsored by a private body, the *Kaiser Wilhelm Gesellschaft.* These two institutes were the Kaiser Wilhelm Institutes at Heidelberg and Berlin-Dahlem; each possessed a small high-voltage set suitable for nuclear research.[7] A cyclotron for such work was altogether lacking—the Heidelberg cyclotron, again built entirely by private funds, and mainly designed for medical investigations, was started as late as 1938, and could not be tested out before 1944.[8] Only with the outbreak of war did the awakened interest of the authorities allow of more

[2]On the development of the cyclotron or particle accelerator, see footnote 320, doc. 74.

† [Original footnote 1:] "*Naturwiss.*, **27**, 11 (1939)."

[3]Otto → Hahn's and Fritz → Straßmann's article referred to in footnote † above is: Hahn & Straßmann [1939]a. Cf. also O. Hahn's Nobel lecture delivered on Dec. 13, 1946, for his own historical account of the discovery of uranium fission: Hahn [1946/48].

§ [Original footnote 2:] "*Nature*, **143**, 470 (1939)."

[4]For the citation in footnote §, see the paper by Hans von Halban, Frédéric Joliot & Lew Kowarski: 'Liberation of neutrons in the nuclear explosion of uranium', *Nature* **143** [1939], pp. 470–471.

‡ [Original footnote 3:] "*Naturwiss.*, **27**, 402 (1939)."

[5]Footnote ‡ refers to Siegfried → Flügge's article: Flügge [1939]; see also doc. 74 for a semipopular version of this article published on Aug. 15, 1939.

¶ [Original footnote 4:] "*Nature*, **143**, 239 (1939)."

[6]The article by Lise → Meitner & O. R. → Frisch cited in footnote ¶ is: Meitner & Frisch [1939].

[7]For a description of the high-voltage facility at the → *KWIP* see doc. 54.

[8]On the first European cyclotrons see footnote 7 of doc. 74.

extended facilities for nuclear research.[9]

The following report deals with those particular investigations which had for their purpose the technical utilization of atomic energy. The many purely scientific problems which arose in more or less close connexion with the technical problem will not be discussed here; they will be dealt with in a forthcoming FIAT** review by Flügge and Bothe.[10] I may, however, mention the extensive chemical investigations of Hahn and his co-workers on the fission products of uranium, carried out throughout the War in the Kaiser Wilhelm Institute for Chemistry, the great majority of which have been published.[11]

Almost simultaneously with the outbreak of war, news reached Germany that funds were being allocated by the American military authorities for research in atomic energy.††[12] In view of the possibility that England and the United

[9]The → *REM* had actually already been informed about the possible military applications of nuclear energy in a short letter by Georg → Joos dated April 22, 1939. Seven days later, on April 29, 1939, a group of physicists, including Joos, W. → Bothe, P. → Debye, R. → Döpel, H. → Geiger, W. → Gentner, W. → Hanle, and G. → Hoffmann, met for a conference concerning the feasibility of a self-sustaining nuclear chain reaction at the *REM*, organized by Wilhelm Dames. Although research was commenced in Göttingen, this informal working group, later called the 'first *Uranverein*', did not get very far due to the outbreak of World War II; and Joos, Hanle, and Reinhold → Mannkopff were drafted into the Army. Afterwards, the military took the initiative (see footnote 13 below).

** [Original footnote *:] "FIAT Reviews of German Science, 1939–1946: a series of authoritative accounts of the progress made in both natural and applied science in Germany during the War. These reviews have been sponsored jointly by British, American and French FIAT (Field Intelligence Agency (Technical)), and a limited edition is expected to be ready for distribution before the end of 1947."

[10]On the above reference in footnote **, see Flügge & Bothe (Eds.) [1948], a two-volume survey of more than 400 pages on nuclear physics and cosmic ray physics in Germany between 1939 and 1946. Possible military applications are not mentioned at all in it. For a critique of this overview report see, e.g., Walker [1989]a, pp. 210f.

[11]For a full bibliography of the published work by Otto Hahn and his collaborators, in particular F. Straßmann and J. → Mattauch, see Hahn [1975], pp. 287–304, esp. pp. 298–301, for the papers published between 1939 and 1945. See also O. Hahn's reports: 'Bericht über die Arbeiten des Kaiser-Wilhelm-Instituts für Chemie über 'Präparat 38'', *KPFB* **G-34** (dated Dec. 10, 1940); 'Die Spaltung des Urankerns', *KPFB* **G-150** (dated Feb. 26, 1942); 'Künstliche Atomumwandlung und Atomkernspaltung', *KPFB* **G-216** (dated May, 5, 1943); and O. Hahn & F. Straßmann, 'Zur Folge nach der Entstehung des 2,3-Tage-Isotops des Elements 93 aus Uran', *KPFB* **G-151** (dated Feb. 27, 1942).

†† [Original footnote †:] "Cf. Smyth Report, p. 27: Initial approaches to the Government. In actual fact, U.S. Government funds were first used at the turn of the year 1939–40 (p. 28), whereas the first discussions between men of science and the American navy took place as early as March 1939."

[12]For the full reference in footnote ††, see the official report: Smyth [1945]. Henry DeWolf Smyth (1898–1986) was then chairman of the Department of Physics at Princeton University and consultant to the Manhattan District U.S. Corps of Engineers. According to the titlepage he published the report at the request of Major General L. R. Groves, who also provided its foreword (on Groves, see footnote 5 of doc. 116). On the progress of atomic weapons research in the USA see, e.g, Rhodes [1986], or the preliminary discussions between Enrico Fermi, two civil

States might undertake the development of atomic weapons, the German Army Ordnance Office created a special research group, under Schumann, whose task it was to examine the possibilities of the technical exploitation of atomic energy.[13] As early as September 1939 a number of nuclear physicists and experts in related fields were assigned to this problem, under the administrative responsibility of Diebner.[14] I should mention the names of Bothe, Clusius, Döpel, Geiger, Hahn, Harteck, Joos and von Weizsäcker among those so employed.[15] At Schumann's behest, the *Kaiser Wilhelm Institut für Physik* in Berlin-Dahlem was nominated as the scientific centre of the new research project.[16] The Institute came accordingly under the administration of the German Army Ordnance Office; a step which disregarded the rights of the *Kaiser Wilhelm Gesellschaft*, and so led to the departure of its director, Debye, who as a Dutch citizen could not continue to serve under the ægis of a German war department.[17]

---

scientists from the Naval Research Laboratory, and Rear Admiral Stanford Caldwell Hooper (1884–1955), an expert in Naval radio communications who had been instrumental in the lab's completion, as well as officers of the Ordnance Bureau of the U.S. Army.

[13]The *Heereswaffenamt* (→ *HWA*) had perhaps noted a *New York Times* report on the American Physical Society's spring meeting on Apr. 29, 1939, at which the (controversial) prospects were discussed of making atomic weapons capable of destroying New York City in a single blow. At the same time, a letter by Paul → Harteck and Wilhelm → Groth to the Ministry of Defense (→ *RWM*) reported the possibility of developing a nuclear explosive. The head of the *HWA*, General → Becker, passed this information on to the research department head Erich → Schumann.

[14]Kurt → Diebner had been working in nuclear physics since 1931 as assistant to Gerhard Hoffmann in Halle before changing to the *RWM* in 1934. He became consultant on nuclear physics and radiation issues to the *HWA* and established good relations with E. Schumann. His selection as head of the *HWA*'s *Uranverein* was thus not surprising.

[15]Walther Bothe, Hans Geiger, Paul Harteck and Georg Joos were all members of the 'second *Uranverein*' created in early September 1939 under the *HWA*. Other scientists drafted in September 1939 to serve in this secret research unit were: S. Flügge, G. Hoffmann, J. Mattauch, and G. → Stetter. After their first meeting on Sep. 16, 1939, the following scientists were also called into the research unit: Klaus → Clusius, Robert Döpel, Otto Hahn, W. Heisenberg, G. Joos, and C. F. → von Weizsäcker. On the research performed by the *Uranverein* see, e.g., Walker [1989]a, Karl → Wirtz's article: 'Papers on the uranium project (1939–1945)', in Heisenberg [1989], Vol. A II, pp. 365ff., Irving [1967]; cf. also the concise chronology of the German uranium project in: Rechenberg [1988]b.

[16]However, the original plan to confine all research to this one center failed; it was shared among the following physical laboratories: The physico-chemical department at Hamburg University (Harteck and collaborators), the physical institute at Leipzig University (Döpel and Heisenberg), the *Heeresforschungsstelle* Gottow or Kummersdorf (Diebner), the physics department of the Kaiser Wilhelm Institute of Medical Physics in Heidelberg (Bothe), and the *KWIP* in Dahlem (Berlin) (Fritz → Bopp, Diebner, von Weizsäcker, K. Wirtz). Some of the chemical research was conducted at the Kaiser Wilhelm Institute of Chemistry (→ *KWIC*) in Berlin (Hahn).

[17]On the annexation of the *KWIP* by the *HWA*, see here doc. 79; Peter Debye declined assuming German citizenship, a prerequisite to directing the German *Uranverein*). He was granted leave of absence and took up a guest professorship at Cornell University in Ithaca, USA, in January 1940, and Diebner became acting director. Through the support of von Weizsäcker

As a result of the first conferences in the autumn of 1939, it was clear that there were two lines of attack possible in the exploitation of nuclear energy.[18] One could attempt the separation of the rare isotope U(235) from ordinary uranium. This isotope, following theoretical arguments due to Bohr,[19] must be immediately applicable either to the controlled production of energy, using primarily the slow-neutron reaction, or directly as an explosive in bombs, using the fast-neutron reaction; the separation of U(235) was, however, a problem which made the greatest possible demands on engineering technique. Secondly, one could mix ordinary uranium with a substance which would slow down the neutrons produced in the nuclear fission without absorbing them. These slow neutrons give rise preferentially to the fission of U(235) and thus maintain the chain reaction. A rapid deceleration of the neutrons is required, in order that the region of resonance absorption by U(238) should be rapidly traversed; for if absorbed they are lost to the chain reaction. The advantage of this arrangement is that the chain reaction can be controlled through the heat developed thereby, so that energy can be abstracted in amounts sufficient for technical applications.

Thus two lines of purely scientific investigation were marked out: first, to develop refined methods for the separation of isotopes; secondly, by measurement of the effective cross-sections of a range of possible substances, to determine whether the alternative line of attack was at all practicable. Harteck pointed out, as early as the autumn of 1939, that it might be advantageous, in regard to the second scheme, to have the so-called moderator physically segregated from the uranium.[20] This suggestion gave rise to theoretical investigations as to whether, with the effective cross-sections of such moderator substances as were known at the time, an arrangement having a homogeneous mixture of uranium and moderator, or one with a local separation (for example, in layers), led to the more favorable production of energy. A tentative theoretical investigation made by Heisenberg, in December 1939, led to the result that while ordinary water was unsuitable as a moderator, it should be possible with heavy water (D_2O) or very pure carbon to produce energy in positive amount, provided the moderator

and Wirtz, Heisenberg visited the Berlin institute regularly to consult with his colleagues, and was named acting director at the institute in the beginning of July, 1942.

[18]This first 'uranium conference' took place on Sep. 16, 1939, and was opened by Diebner's superior, H. Basche. According to E. → Bagge, its purpose was "to make all preparations to be able to answer beyond doubt the question of whether generating nuclear energy was feasible. It would certainly be very nice if it were possible to acquire a new source of energy, it would also very probably have military importance; a negative answer would be just as important, since we could be sure that the enemy would also not be able to make use of it." Rechenberg [1988]b, p. 457.

[19]This refers to Niels Bohr's paper co-authored with John Archibald Wheeler: 'The mechanism of nuclear fission', *Physical Review* **56** [1939], pp. 426–450.

[20]Harteck's suggestion is already mentioned in Heisenberg's first secret report of Dec. 6, 1939 (quoted in footnote 21 below), p. 18/391. Cf. also Harteck's letter to Diebner, dated Oct. 9, 1939 and his 'working program' dated Nov. 23, 1939, both in Irving (Ed.) [undated], nos. 29 617ff.

and uranium were arranged in layers.[21] This arrangement, however, demanded the highest degree of purity of the substances involved. At the same time, it was evident that a certain minimum size of apparatus was necessary for the production of energy.[22] Nevertheless, with a small set it is still possible to determine whether there would be a production of energy if the apparatus were suitably enlarged. Thus if we feed such a small plant with neutrons from an internal source, more neutrons must escape from the surface than are supplied by the source, if the layer arrangement is favourable to energy production; if unfavourable, then fewer neutrons escape than are supplied by the source. These small model plants, which are continuously fed from a neutron source, are called 'neutron-injected piles'. The ratio k of the number of neutrons escaping from the pile to that fed in by the source can be used to characterize the pile. If $k \leq 1$, the arrangement is unsuitable for the production of energy; if $k > 1$, energy will be produced on enlarging the pile.[23]

In 1940, measurements of the most important effective cross-sections were carried out, especially by Bothe and his collaborators, and by Döpel and Heisenberg.[24] At the same time, investigations on the masses and energies of the fission

[21]On the technical points of the foregoing two paragraphs, see W. Heisenberg, 'Die Möglichkeiten der technischen Energiegewinnung aus der Uranspaltung', *KPFB* **G-39** (undated, but according to later listings, completed Dec. 6, 1939), reprinted and annotated together with the other *KPFB* reports, co-authored by Heisenberg, K. Wirtz, and H. Rechenberg in Heisenberg [1989], Vol. A II, pp. 378ff.

[22]This volume minimization also had military importance, aside from the general shortage of purified radioactive materials: See Heisenberg's 1939 report (quoted in the preceding footnote), p. 24/396: "The enrichment of U^{235}_{92} is the only method by which the machine can be reduced in size against 1 cubic meter. It is furthermore the only method by which an explosive can be produced which supercedes the explosiveness of the most powerful conventional explosive by several powers of a ten." (See the Introduction, p. lxxxii.) Note that while present in the original document, this hint at a military application for 'uranium machines' is completely absent in this postwar article.

[23]German scientists usually spoke of these devices in terms of a 'uranium combuster' (*Uranbrenner*) or a 'uranium machine' (*Uranmaschine*), but in America and England they were usually referred to as 'uranium piles', as Heisenberg indicates. In the original papers the directly observable 'coefficient of neutron production' $\overline{\nu}$ is used instead of the multiplication factor k, which is related to k via $\overline{\nu} = (k-1)/l$, with l denoting the lifetime of the neutron. According to Wirtz in Heisenberg [1989], Vol. A II, p. 366, in Heisenberg's two pioneering papers on the theory of the uranium pile of late 1939 and February 1940, "all crucial concepts of current reactor theory already appear, e.g., critical size, exponential increase of the neutron density for $k > 1$, temperature dependence of k".

[24]See, e.g., W. Bothe, 'Die Diffusionslänge für thermische Neutronen', *KPFB* **G-12** (June 7, 1940); R. Döpel, Klara → Döpel & W. Heisenberg: 'Bestimmung der Diffusionslänge thermischer Neutronen in schwerem Wasser', *KPFB* **G-23**, (Aug. 7, 1940), and 'Bestimmung der Diffusionslänge thermischer Neutronen in Präparat 38', *KPFB* **G-22** (Dec. 5, 1940), the latter two reprinted in Heisenberg [1989], Vol. A II, pp. 419ff., and 427ff. See also the published papers by W. Bothe and P. → Jensen, cited in footnote 43 below, and by P. Jensen: 'Die Bremsung von Neutronen in Kohlenstoff, Wasser und schwerem Wasser', *Zeitschrift für Physik* **122** [1945], pp. 756–768 (submitted on Jan. 21, 1944, but written in July 1940 according to footnote 1 of

products were being pursued by Jentschke and Prankl§§[25] and by Flammersfeld, P. Jensen and Gentner,‡‡[26] and on the spectrum of the neutrons produced by Kirchner[27] and von Droste and by Bothe and Gentner.¶¶[28] The theory of neutron absorption in the U(238) resonance line was laid down by Flügge and Heisenberg.[29] On the technical side the following results were the most important: the absorption cross-section of heavy water proved to be so low that this substance was certainly usable in the construction of a uranium pile (Döpel and Heisenberg).[30] The work of von Droste on large quantities of sodium uranate,

---

this paper).

§§ [Original footnote 5:] "*Z. Phys.*, **119**, 696 (1942)."

[25]The report cited in footnote §§ by W. → Jentschke & F. Prankl: 'Energien und Massen der Urankernbruchstücke', *KPFB* **G-44** (Aug. 1940), was later published in the *Zeitschrift für Physik* **110** [1942], pp. 696–712 (submitted on June 27, 1942, but completed in August 1940 according to footnote * of this paper). See also footnote 1 there for references to other papers by these two authors, who estimated that the average kinetic energy of both fission products together amounted to 150 MeV, ranging between 120 and 190 MeV.

‡‡ [Original footnote 6:] "*Z. Phys.*, **120**, 450 (1943)."

[26]Footnote ‡‡ refers to the report by P. Flammersfeld, P. Jensen & W. Gentner: 'Die Energietönung der Uranspaltung' and 'Die Aufteilungsverhältnisse und Energietönung bei der Uranspaltung', *KPFB* **G-25** and *KPFB* **G-26** (May 24, and Sep. 24, 1940), the latter published in *Zeitschrift für Physik* **120** [1943], pp. 450–467, submitted on Sep. 4, 1942. These authors determined the total energy of the fission products as ranging between 120 and 180 MeV with energy and mass relations between 0.4:1 and nearly 1:1, peaking at a mass relation for the fission products of 96:140 at 151 MeV.

[27]Fritz [Franz] Kirchner (1896–1967) specialized in X-ray physics. He studied at the Munich and Jena Universities, took his degree at Jena in 1921 under Max Wien, and habilitated in 1925, receiving a teaching assignment in theoretical physics in 1930. 1931 Kirchner was appointed associate professor at the University of Munich and in 1934 at the University of Leipzig.

¶¶ [Original footnote 7:] "*Z. Phys.*, **119**, 568 (1942)."

[28]For the paper cited in footnote ¶¶, see W. Bothe & W. Gentner, 'Die Energiegrenze der Spaltungsneutronen aus Uran', *Zeitschrift für Physik* **119** [1942], pp. 568–574 (submitted on Apr. 30, 1942, but written in May 1940 according to the concluding sentence on p. 574 of this paper). These authors measured the spectrum of neutrons produced in fission processes of uranium to up to 10 MeV. G. → von Droste's independent work on the same topic was referred to in footnote 2 of this paper with the note, "private communication"; it was apparently not published. See also the report by W. Bothe: 'Die Diffusionslänge für thermische Neutronen in Kohle', *KPFB* **G-12** (June 7, 1940), and W. Bothe & W. Gentner, 'Die Energie der Spaltungsneutronen aus Uran', *KPFB* **G-17** (May 9, 1940). See also G. von Droste, 'Über den Spaltprozeß bei Präparat 38', *KPFB* **G-78** (dated 1941).

[29]See S. Flügge's three papers: 'Berechnung des Bruchteils von Neutronen, der in einem Uran-Wasserstoffgemisch in der Resonanzlinie absorbiert wird', unpublished research reports, dated 1939 and 1940, quoted and commented on in Bothe & Flügge (Eds.) [1948], part I, pp. 213ff. For Heisenberg's work see: 'Bericht über die Möglichkeit technischer Energiegewinnung aus der Uranspaltung I' (section I.2) and II (section II.2.), *KPFB* **G-39** and **G-40**, issued on Dec. 6, 1939 and on Feb. 29, 1940, respectively. See also Heisenberg & Wirtz [1948], sec. 7.1.1, pp. 143ff., and von Droste in Bothe & Flügge (Eds.) [1948], pp. 203ff., for further comments.

[30]R. Döpel, K. Döpel & W. Heisenberg, 'Bestimmung der Diffusionslänge thermischer Neutronen in schwerem Wasser', *KPFB* **G-23** (Aug. 7, 1940), reprinted in Heisenberg [1989], Vol. A II, pp. 419ff., esp. pp. 10/425f.

and of Harteck and the Hamburg group—Groth, [J.] H. Jensen, Knauer and Süss—on U_3O_8 in solid carbon dioxide, furnished the first criteria for the distribution of neutron density in certain arrangements of uranium and moderator.[31] In the autumn of 1940, the first pile, constructed of layers of U_3O_8 and light paraffin, was built at Berlin-Dahlem and its characteristics measured (Wirtz, Fischer, Bopp).[32] This model pile gave as expected, $k < 1$, that is, the arrangement was not suitable for the production of energy. Nevertheless, it yielded valuable data for further piles using alternative layers of U_3O_8 and heavy water.[33]

In the summer of 1940, von Weizsäcker pointed out that a uranium pile, besides generating the fission products of uranium, will constantly produce the uranium isotope U(239) and its transformation series; and that theoretically these transformation products should show the same properties as U(235) in regard to neutron fission.[34] It remained thereby an open question whether β-decay ended at element No. 93 or 94 or even later; for, since no cyclotron was available in Germany, these elements could not be prepared in sufficient quantity for the examination either of their nuclear properties or chemical characteristics. Nevertheless, it appeared likely from von Weizsäcker's work, that an energy-producing pile might be used for the production of an atomic explosive, even though the practical details involved remained uncertain.[35] In fact, this method had been

[31]See G. Freiherr von Droste, 'Der Spaltquerschnitt von Uran für thermische Neutronen', and 'Die Vermehrung der Neutronen an 2 t Uran', unpublished research reports, dated 1941 and June 1940 respectively, quoted and commented on by von Droste in Bothe & Flügge (Eds.) [1948], part I, pp. 197ff.; and Droste's report in *KPFB*: 'Bericht über einen Versuch mit 2 t Natriumuranat', *KPFB* **G-24** (dated Sep. 24, 1940); and the report by P. Harteck, J. → Jensen, Friedrich → Knauer and Hans → Suess: 'Über die Bremsung, die Diffusion und den Einfang von Neutronen in fester Kohlensäure und über ihren Einfang in Uran', *KPFB* **G-36** (dated Aug. 19, 1940). Cf. also Harteck's correspondence in 1940, Irving (Ed.) [undated], nos. 29 625ff.

[32]In this experiment B 1, a multiplication factor of $k = 0.92$ was reached, corresponding to a still negative neutron production coefficient of –950. See the report: 'Bericht über die Versuche mit Schichtenanordnungen von Präparat 38 und Paraffin am Kaiser-Wilhelm-Institut für Physik in Berlin-Dahlem', dated March 1941, attributed to Heisenberg in Heisenberg [1989], Vol. A II, pp. 432ff., with the comment: "The report had been written in part by Wirtz, Fischer and Bopp". Section 2b of this report describes the new external laboratory connecting the *KWIP* and the Kaiser Wilhelm Institute of Biology, in which this experiment was conducted with no less than 6 tons of uranium. See W. Heisenberg, 'Bericht über Versuche mit Schichtenanordnungen von Präparat 38 und Paraffin am Kaiser-Wilhelm-Institut für Physik in Berlin-Dahlem', *KPFB* **G-93** (dated May 1941); the reports by F. Bopp, Erich → Fischer, W. Heisenberg, C. F. von Weizsäcker & Karl Wirtz: 'Untersuchungen mit neuen Schichtenanordnungen aus U-Metall und Paraffin', *KPFB* **G-127** (dated March 1942) and 'Messungen an Schichtenanordnungen aus 38-Metall und Paraffin', *KPFB* **G-162** (dated Oct. 30, 1942); see also E. Fischer, 'Bestimmung des Absorptionsquerschnittes von Uran für langsame Neutronen', *KPFB* **G-79** (dated June 26, 1941); and C. F. von Weizsäcker, 'Bemerkungen zur Berechnung von Schichtenanordnungen', *KPFB* **G-122** (dated Aug. 2, 1941).

[33]Cf. Heisenberg [1989], Vol. A II, sections 3–5, pp. 438ff.

[34]See C. F. von Weizsäcker, 'Eine Möglichkeit der Energiegewinnung aus Uran 238', *KPFB* **G-59** (dated July 17, 1940).

[35]Besides the report mentioned in the preceding footnote, see also C. F. von Weizsäcker, Paul

employed on the grand scale in America. The American piles deliver as a transformation product of U(239) the element plutonium $^{293}_{94}$Pu, which is used in the manufacture of atomic bombs.[36]

Important technical problems immediately arose out of these scientific investigations. The production of U_3O_8 of the highest purity was assigned to the *Auer-Gesellschaft* by the German Army Ordnance Service.[37] The casting of the corresponding metal powder was afterwards allotted to *Degussa* in Frankfurt.[38] The production of heavy water, which was obviously of the greatest importance for the construction of a uranium pile, was planned at the *Norsk Hydro* factory at Rjukan in Norway.[39] Harteck, in conjunction with Süss, [J.] H. Jensen and Wirtz, developed a number of projects which resulted in an increase of heavy water production at Rjukan far beyond the former output of 10–20 litres a month.[40] Moreover, Harteck and Clusius put forward detailed plans for the production of heavy water in Germany.[41] The improvements in the *Norsk Hydro* factory fi-

→ Müller & Karl-Heinz → Höcker, 'Berechnung der Energieerzeugung in der Uranmaschine', *KPFB* **G-60** (dated Feb. 26, 1940).

36On progress in uranium pile development and plutonium production in the U.S. from May 1942 onwards at Stagg Field, Chicago (headed by Enrico Fermi), see Rhodes [1986], pp. 428ff.

37The head of the research division of the → *Auer* Company, Nikolaus → Riehl, contacted the *HWA* upon becoming aware of the potential applications of nuclear energy through Flügge's article (doc. 74); he realized the profit-making possibilities for his company, which specialized in the fields of radioactivity, luminescence, and uranium and thorium products.

38The precious metals dealer and processing company *Deutsche Gold- und Silber-Scheideanstalt* (*Degussa*, formerly called Roessler and founded in 1873) is one of the largest in Germany. It is also involved in the chemical and pharmaceutical industries.

39In 1940 the → *Norsk Hydro* factory was the only site where commercial production of heavy water (deuterium) was undertaken, actually only as a by-product of electrolytic nitrogen and hydrogen. The Norwegian company started this branch of production using a process discovered in 1932 recommended by Karl Friedrich → Bonhoeffer.

40As a first step, in the spring of 1941 heavy water production was increased by installing additional electrolysis units and optimizing the units used to concentrate the deuterium: See Karl Wirtz, 'Bericht II. Eine 10stufige Elektrolysieranlage zur Gewinnung von schwerem Wasser', *KPFB* **G-61** (dated Jan. 19, 1940); P. Harteck, 'Die Produktion von schwerem Wasser', *KPFB* **G-86** (dated December 1941); P. Harteck & J. Jensen, 'Der Thermodiffusionseffekt im Zusammenspiel mit der Konvektion durch mechanisch bewegte Wände und Vergleich mit der Thermosiphonwirkung', *KPFB* **G-89** (dated Feb. 18, 1941); P. Harteck, 'Die Gewinnung von schwerem Wasser', *KPFB* **G-154** (dated Feb. 26, 1942); K. Wirtz, 'Die elektrolytische Schwerwassergewinnung in Norwegen', *KPFB* **G-198** (dated Feb. 26–28, 1942). On the earlier German output see Erika Cremer, 'Untersuchungen des Schwerwassergehaltes einiger technischer Elektrolyseure in Deutschland', *KPFB* **G-21** (dated June 24, 1940).

41In autumn 1941 Harteck and Suess proposed additionally increasing the deuterium production through a hydrogen, water vapor and deuterium exchange described by Harteck in Bothe & Flügge (Eds.) [1948], pp. 182. Harteck and Clusius also corresponded with Heisenberg about alternative ways of producing deuterium in Germany where the supply of electricity was not as abundant as in Rjukan with its hydroelectric power station. Cf., e.g., Harteck's letter to Heisenberg, Jan. 1, 1940 and Harteck's progress report of April 17, 1941, in which he reports a price of 1 RM per gram (!) of deuterium in Germany, and continues: "Besides—as has been stressed repeatedly—from the point of view of power generation, it is out of the question to produce

nally increased production in the summer of 1942 to about 200 litres per month. Further, steps were taken by order of the *Heereswaffenamt* for the production of very pure carbon. The attempts to exceed the degree of purity afforded by the best technical electrographite failed, however for the time being.

The most important progress in the uranium project was achieved during the year 1941. Initially some negative results were recorded. Thus, the enrichment of U(235) by the Clusius–Dickel thermal diffusion method, using gaseous UF_6, proved impossible (Fleischmann, Harteck and Groth).[42] The absorption cross-section for neutrons of the highest purity electrographite was determined in the Kaiser Wilhelm Institute at Heidelberg (Bothe and Jensen)***[43] and the behaviour of pure carbon estimated from the results. It appeared, according to the information then available, that even the purest possible carbon was unsuitable for the construction of a uranium pile; whereas, as is well known, carbon has been used in the United States with complete success.[44] Whether the Heidelberg conclusions were falsified by insufficient consideration of the chemical impurities present in commercial graphite (for example, hydrogen or nitrogen), or by deficiencies in the theory, can scarcely be assessed for the moment. In any event, the Heidelberg experiments on graphite and beryllium (Fünfer and

heavy water on a large scale through electrolysis in Germany." See also his letter to the *HWA*, dated Jan. 24, 1940, according to which 100,000 tons of coal would be needed to produce one ton of deuterium via electrolysis. See the protocol of the meeting between Bonhoeffer, Harteck, Clusius, Suess and others at the *HWA* on Nov. 11, 1941 for a full list of alternative methods of producing heavy water, such as through the exchange of water and hydrogen at two different temperatures, using low-pressure columns, or via rectification of fluid hydrogen. (Irving (Ed.) [undated], nos. 29 625, 29 628, 29 662ff., and 29 668ff.)

[42]For a description of the Clusius-Dickel thermal diffusion method see, e.g., Clusius [1943], Walker [1989]a, pp. 28ff. Cf. Rudolf → Fleischmann's report: 'Ein mögliches Verfahren zur Isotopentrennung von Uran', *KPFB* **G-27** (dated July 3, 1940); P. Harteck, 'Die Trennung der Uranisotope', *KPFB* **G-88** (dated December 1941); W. Groth, 'Stand der Arbeiten zur Trennung der Isotope des Präparats 38', *KPFB* **G-83** (dated December 1941).

*** [Original footnote 8:] "*Z. Phys.*, **122**, 749 (1944)."

[43]The reference in footnote *** is: W. Bothe & P. Jensen, 'Die Absorption thermischer Neutronen in Kohlenstoff', *Zeitschrift für Physik* **122** [1945], pp. 749–755 (submitted on Jan. 21, 1944, but completed in January 1941 according to footnote 1 of this paper). The absorption coefficient σ_a and diffusion length l were determined in this paper as $\sigma_a = (6.4 \pm 1) \cdot 10^{-27}\text{cm}^2$, and $l = (67 \pm 4)\ \text{g/cm}^2$. Bothe used the purest electrographite available in Germany, produced by → *Siemens*; but later chemical analyses of this substance by Hanle showed that even this electrographite was contaminated with boron and cadmium, both efficient absorbers of slow neutrons, which explains why Bothe's determination of the carbon absorption coefficient was too high: Cf., e.g., Walker [1989]a, pp. 26f. See also W. Bothe & P. Jensen, 'Die Absorption thermischer Neutronen in Elektrographit', *KPFB* **G-71** (Jan. 20, 1941).

[44]The crucialness of this initial erroneous determination of the carbon absorption coefficient implied by Heisenberg in this and in comparable texts, is qualified and sometimes disputed in the secondary literature. According to Walker [1989]a, p. 27, the *HWA* decided not to use carbon because the cost of very pure carbon was too high under the prevailing conditions and because Heisenberg's analysis had shown that uranium piles using carbon as the moderator would need much more uranium and moderator than piles using a deuterium moderator.

Bothe)†††[45] made it clear, in connexion with later experiments in the Berlin-Dahlem Institute, that both pure carbon and pure beryllium were highly suitable for use as an outer cover for a uranium pile, inasmuch as their low absorption cross-section and high reflecting power restrict the spread of the neutrons escaping from the pile, thus reducing its minimum dimensions.[46]

In the summer of 1941, 150 litres of heavy water were available for the first experiments on a neutron-injected pile built up of uranium and heavy water (Döpel and Heisenberg, Leipzig).[47] The uranium and heavy water were arranged in alternate spherical layers with the neutron source at the centre. The oxide U_3O_8 which was first employed produced only a slight increase in the number of escaping neutrons, which could scarcely be considered a clear proof that $k > 1$. The use of pure uranium metal, however, gave such a decided improvement that no further doubt of a real increase in the number of neutrons ($k > 1$) was possible (about February or March 1942). Here then was the proof that the technical utilization of atomic energy was possible, and that the mere enlargement of the Leipzig apparatus must furnish an energy-producing uranium pile.[48]

At the same time, important administrative changes were taking place. At a meeting held in the building of the *Reichsforschungsrat* in Berlin, on February 26, 1942, the results to date were reported to the Minister of Education, Rust, and

††† [Original footnote 5:] "*Z. Phys.*, **119**, 696 (1942)."

45The paper in footnote ††† was by Jentschke and Prankl (fully cited in footnote 25 above).

46See the report by Erwin Fünfer & Walther Bothe: 'Absorption thermischer Neutronen und die Vermehrung schneller Neutronen in Beryllium', *KPFB* **G-81** (Oct. 10, 1941); cf. their article: 'Die Wechselwirkung von Neutronen und γ-Strahlen mit Beryllium', *Zeitschrift für Physik* **122** [1945], pp. 769–777 (submitted on Jan. 21, 1944, but completed in March 1942 according to footnote 1 of this paper).

47See R. Döpel, K. Döpel & W. Heisenberg, 'Versuche mit Schichtenanordnungen von D_2O und 38', *KPFB* **G-75** (dated Oct. 28, 1941 according to later listings) reprinted in Heisenberg [1989], Vol. A II, pp. 481ff. See also the later account of the Leipzig experiments L 1–L 4 in Heisenberg & Wirtz[1948], sec. 7.1.2, pp. 149ff., Kleint [1986], and Kleint & Wiemers (Eds.) [1993], pp. 15ff. The L 1 experiment used 150 liters of deuterium and two concentric spherical 3.5-cm thick layers of uranium.

48See W. Heisenberg, 'Über die Möglichkeit der Energieerzeugung mit Hilfe des Isotops 238', unpublished and undated research report; according to the postscript it was received on Nov. 19, 1941, reprinted in Heisenberg [1989], Vol. A II, pp. 499ff.; and R. Döpel, K. Döpel & W. Heisenberg, 'Der experimentelle Nachweis der effektiven Neutronenvermehrung in einem Kugel Schichten-System aus D_2O und Uran-Metall', undated and unpublished research report with cover letter dated July 1942, reprinted in Heisenberg [1989], Vol. A II, pp. 536ff. According to the editor's footnote 1 on p. 536 of this reprint, Döpel later gave the date of the experimental proof of neutron multiplication as April 1942. See also the reports about the Leipzig experiments: R. Döpel, K. Döpel & W. Heisenberg, 'Die Neutronenvermehrung in einem D_2O-38 Metall-schichtensystem', and 'Die Neutronenvermehrung in einem 38-Metall durch rasche Neutronen', unpublished and undated research reports, presented at the Berlin meeting of Feb. 26–28, 1942, and reprinted in Heisenberg [1989], Vol., A II, pp. 526ff. Cf. again the published account of the Leipzig experiments L 1–L 4 in Heisenberg & Wirtz [1948], sec. 7.1.2, pp. 149ff., Kleint [1986], and Kleint & Wiemers (Eds.) [1993].

several directors of war research.[49] The uranium project was transferred from the *Heereswaffenamt* to the control of the *Reichsforschungsrat*; and the then president of the *Physikalisch-Technische Reichsanstalt*, Esau, was made responsible for the project. On June 6, 1942, there was a second meeting at Harnack House in Berlin, when the results of the uranium project were reported to Speer, as Minister for War Production, and to the armament staff.[50]

The facts reported were as follows: definite proof had been obtained that the technical utilization of atomic energy in a uranium pile was possible. Moreover, it was to be expected on theoretical grounds that an explosive for atomic bombs could be produced in such a pile. Investigation of the technical sides of the atomic bomb problem—for example, of the so-called critical size—was, however, not undertaken. More weight was given to the fact that the energy developed in a uranium pile could be used as a prime mover, since this aim appeared to be capable of achievement more easily and with less outlay. As to the separation of the uranium isotopes, no method was known which would have allowed of the production of an atomic explosive without an enormous and therefore impossible technical equipment. Incidentally, the use of protoactinium as an atomic explosive was also considered, since its nucleus is fissionable by neutrons with energies down to 10^5 eV, with the consequent possibility of a fast chain reaction.[51] It was, however, considered to be impracticable to prepare the necessary quantities of the element.

Following this meeting, which was decisive for the future of the project, Speer ruled that the work was to go forward as before on a comparatively small scale. Thus the only goal attainable was the development of a uranium pile producing energy as a prime mover—in fact, future work was directed entirely towards this one aim. Again during the summer of 1942, discussions were held with heat ex-

---

[49] At this secret meeting at the → *KWG*'s *Harnack-Haus* in Dahlem (Berlin) on Feb. 26–28, 1942, Bagge, Bothe, Clusius, Döpel, Hahn, Harteck, Heisenberg, von Weizsäcker, and Wirtz delivered technical reports on the progress of their research (see the previous footnote as well as the progress report on the Berlin experiments B 3 and B 4 in: F. Bopp, E. Fischer, W. Heisenberg, C.F. von Weizsäcker & K. Wirtz, 'Untersuchungen mit neuen Schichtenanordnungen aus U-Metall und Paraffin', undated and unpublished secret research report, presented at the Berlin 1942 meeting, and reprinted in Heisenberg [1989], Vol. A II, pp. 522ff. At the same time, another meeting was held at the *Haus der Deutschen Forschung*, with less technical talks by Bothe, Clusius, A. → Esau, Geiger, Hahn, Harteck, Heisenberg, and Schumann. For Heisenberg's talk on nuclear energy at the latter meeting, see here doc. 95.

[50] This transfer took place in July 1942 after another meeting of some of the *Uranverein* members (Diebner, Hahn, Harteck, Heisenberg and Wirtz) with the Reich Minister of Armament and Munition Albert → Speer and Field Marshal E. Milch of the German Air Force, as well as other high ranking military officials on June 4 (!), 1942. On this meeting, at which Heisenberg allegedly specified that a nuclear explosive would be the size of a pineapple see, e.g., Irving [1967], p. 109, and Wirtz in Heisenberg [1989], Vol. A II, p. 368. When the *RFR* was placed under H. → Göring on July 9, 1942, the whole uranium project thus came under his control. In December 1942 Göring appointed Abraham Esau as his plenipotentiary of nuclear physics.

[51] Protactinium, or protoactinium, has 91 protons, thus one less than uranium.

perts on the technical problems of heat transfer from the uranium to the working material (that is, water or steam). Technical experts from the Navy attended the meeting with a view to the possible use of a uranium power unit in warships.[52] The Kaiser Wilhelm Institute of Physics was restored to the *Kaiser Wilhelm Gesellschaft*, with the author as director.[53] In preparation for investigations on the larger uranium piles planned in the Institute, a spacious underground laboratory was added (Wirtz).[54]

About this time, however, the strain of the War on the already overloaded German industry was making itself felt. Uranium and uranium slugs were produced in such small quantities that deliveries were late and the larger-scale experiments were repeatedly postponed. Nevertheless, important progress was made. As early as 1941, a research group at the *Heereswaffenamt* (Diebner, Pose,[55] Czulius) had made measurements on a large pile built up as a lattice of uranium cubes in a paraffin matrix;[56] the subsequent theoretical investigation (Höcker) demonstrated that the lattice construction could in certain circumstances show advantages over the layer arrangement.[57] An experiment made by this group with a model pile of uranium cubes in D_2O ice did, in fact, yield a larger increase in the number of neutrons than the Leipzig pile. In a later experiment, using 500 litres of heavy water, a further increase in the number of neutrons was recorded. Measurements made in the Heidelberg Institute with a small model pile defined

[52]See also footnote 13 of doc. 95.

[53]The *KWIP*, which had been temporarily handed over to the Kaiser Wilhelm Society (→ *KWG*) for the duration of the war (see the contract, doc. 79), was thus returned much sooner than planned. This legal transfer and the appointment of Heisenberg as acting director of the *KWIP* (since Debye was still officially its director on leave) took place in July 1942.

[54]On the experiments conducted at the *KWIP* in Dahlem (Berlin) between 1940 and early 1945, see again Heisenberg & Wirtz [1948], sec. 7.1.2, esp. p. 157, on the underground bunker laboratory, completed in 1943 for experiments B 6_{a-d} and B 7.

[55]Heinz → Pose started working at the *KWIP* in 1940, then was employed two years later at the *Physikalisch-Technische Reichsanstalt* (→ *PTR*) in Berlin.

[56]In this so-called **G 1** experiment performed at the *HWA Versuchsstelle* in Gottow, lattices of 6,800 uranium oxide cubes 9.7 cm in length were used weighing a total of approx. 25 tons. They were emersed in paraffin, each pair of uranium oxide blocks being separated by a distance of 2 cm. The complete team, consisting of F. Berkei, W. Borrmann, W. Czulius, K. Diebner, Georg Hartwig, K. H. → Höcker, W. Herrmann, H. Pose, and W. Rexer, co-authored the report: 'Bericht über einen Würfelversuch mit Uranoxyd und Paraffin', *KPFB* **G-125** (dated to before Nov. 26, 1942); cf. also Heisenberg in Bothe & Flügge (Eds.) [1948], pp. 3–4. Cf. the later progress reports by K. Diebner, G. Hartwig, W. Herrmann, H. Westmeyer, W. Czulius, F. Berkei & Karl-Heinz Höcker: 'Vorläufige Mitteilung über einen Versuch mit Uranwürfeln und schwerem Eis als Bremssubstanz', *KPFB* **G-211** (dated April 1943), and 'Bericht über einen Versuch mit Würfeln aus Uran-Metall und schwerem Eis', *KPFB* **G-212** (dated July 1943).

[57]For a list of K.-H. Höcker's many theoretical papers on uranium machines, see Heisenberg & Wirtz [1948], footnote 10 of sec. 7.1.1. See in particular K.-H. Höcker, 'Auswertung der Würfelversuche mit Uranoxyd und Paraffin in der Versuchsstelle Gottow des Heereswaffenamts', *KPFB* **G-164** (dated Nov. 26, 1942); 'Über die Anordnung von Uran und Streusubstanz in der U-Maschine', *KPFB* **G-218** (dated Jan. 25, 1943); 'Über die Abmessungen von Uran und schwerem Wasser in einer Kugelstrukturmaschine', unpublished research report dated 1943.

the relation between the increase in the number of neutrons on one hand, and the thickness of the layers on the other; while experiments undertaken by Bothe and Flammersfeld at Heidelberg, and by Stetter and Lintner in Vienna, threw new light on the fission processes occurring in U(238).[58] A theoretical investigation by Bothe stressed the importance of the 'stopping distance' (*Bremslänge*) for the minimum size of a self-sustaining pile.[59]

In preparation for further experiments with larger quantities of heavy water and uranium metal, the *Kaiser Wilhelm Institut für Physik* in Berlin began a study of the effect of graphite and water as an outer cover for the pile. Resonance absorption in uranium had been studied by Volz and Haxel and by Sauerwein.[60] Further, the absorption cross-sections of a series of different substances were measured by Ramm, as well as by Volz and Haxel at the Berlin

---

[58]See the reports by W. Bothe & A. Flammersfeld: 'Über die Vermehrung thermischer und schneller Neutronen in Uran', unpublished (dated 1941), 'Die Wirkungsquerschnitte von 38 für thermische Neutronen aus Diffusionsmessungen', *KPFB* **G-67** (dated Jan. 20, 1941), 'Resonanzeinfang an einer Uranoberfläche', *KPFB* **G-68** (dated May 8, 1941), 'Versuche mit einer Schichtenanordnung von Wasser und Präp. 38', *KPFB* **G-74** (dated Apr. 28, 1941), 'Messungen an einem Gemisch von 38-Oxyd und -Wasser; der Vermehrungsfaktor *K* und der Resonanzeinfang *w*', *KPFB* **G-69** (dated May 26, 1941), and 'Die Neutronenvermehrung bei schnellen und langsamen Neutronen in 38 und die Diffusionslänge in 38 Metall und Wasser', *KPFB* **G-70** (dated July 11, 1941). The unpublished report by G. → Stetter and K. → Lintner is: 'Schnelle Neutronen im Uran. Der Zuwachs durch den Spaltprozeß und der Abfall durch unelastische Streuung', (dated Nov. 7, 1941); G. Stetter, 'Schnelle Neutronen im Uran', *KPFB* **G-190**–**G-193**, and **G-380** (1942). Cf. the commentary in Heisenberg & Wirtz [1948], sec. 7.1.1. p. 144., von Droste in Bothe & Flügge (Eds.) [1948], pp. 204–207.

[59]See Walther Bothe's three papers dealing with formulas for the spherically symmetric diffusion in scattering and absorbing media and the resulting density distribution: 'Einige Diffusionsprobleme', *Zeitschrift für Physik* **118** [1941], pp. 401–408, dated June 1941, submitted Aug. 25, 1941; 'Die Diffusion von einer Punktquelle aus', ibid. **119** [1942], pp. 493–497, dated Apr. 28, 1942, submitted Apr. 30, 1942; 'Die räumliche Verteilung abgebremster Neutronen', ibid. **122** [1945], pp. 648–748, dated Dec. 7, 1943, and submitted Dec. 30, 1943. See also his report: 'Einige Eigenschaften des U und der Bremsstoffe. Zusammenfassender Bericht über die Arbeiten', *KPFB* **G-66** (dated Mar. 28, 1941).

[60]See the paper by H. → Volz and O. → Haxel: 'Über die Absorption von Neutronen im Uran', *KPFB* **G-118**, commented on further by von Droste in Bothe & Flügge (Eds.) [1948], pp. 202f., and in Heisenberg & Wirtz [1948], sec. 7.1.1, p. 145, which both add references to two unpublished research papers co-authored by S. Flügge & K. Sauerwein: 'Untersuchungen über den Resonanzeinfang von Neutronen beim Uran', dated 1942. See, however, their later account: Flügge & Sauerwein in Bothe & Flügge (Eds.) [1948], part I, pp. 208ff; and K. Sauerwein & S. Flügge, 'Untersuchungen I und II über den Resonanzeinfang von Neutronen beim Uran', *KPFB* **G-185** (dated Jan. 28, 1942). See also the report by O. Haxel, Ernst Stuhlinger & H. Volz: 'Über die Absorption und Verlangsamung von Neutronen in Berylliumoxyd', *KPFB* **G-91** (dated Aug. 4, 1941); O. Haxel & H. Volz, 'Absorptionsquerschnitte für langsame Neutronen', *KPFB* **G-39**; H. Volz, 'Über die Absorption des Urans im Resonanzgebiet', *KPFB* **G-116** (dated 1941); H. Volz, 'Über die Geschwindigkeitsverteilung der Neutronen in einem Gemisch von schwerem Wasser und Uran', *KPFB* **G-117** (dated 1941); K. Sauerwein, 'Untersuchungen über den Resonanzeinfang von Neutronen bei Uran', *KPFB* **G-110** (dated November 1941).

Technical High School.[61] As regards the question of thermal stabilization of the energy production resulting from the temperature broadening of the resonance lines, experiments by Sauerwein and Ramm with the Berlin-Dahlem high-voltage plant were significant.[62]

In the spring of 1943, the *Norsk Hydro* electrolytic plant was put out of action in a Commando raid.[63] Its reconstruction was begun, but finally the responsible army command in Norway reported that effective protection of the plant, particularly against air raids, was impossible. In October 1943 the plant was completely destroyed in a heavy air raid.[64] Nevertheless, about two tons of heavy water were available in Germany at the time: a quantity which, according to our calculations, was just enough for the construction of an energy-producing pile. The *Reichsforschungsrat* had made no effective provision for the construction of a new heavy water factory in Germany, and the pilot plant at *I. G. Leuna* made slow progress.[65] It was proving, in fact, barely possible, in view of air raids and the overall strain on German production, to undertake such big building projects. The production of uranium slugs came to a temporary standstill after the raids on Frankfurt in the spring of 1944.

[61]The absorption cross-sections of cadmium, hydrogen and silver were determined by O. Haxel & H. Volz in: 'Über die Absorption von Neutronen in wässrigen Lösungen', *Zeitschrift für Physik* **120** [1943], pp. 493–507, and 'Anhang. Zur Berechnung der Neutronenabsorption im Kugelinnern', by H. Volz, ibid., pp. 507–512, submitted on Sep. 4, 1942 from the Polytechnic in Charlottenburg (Berlin); see also O. Haxel & H. Volz, 'Bestimmung von Absorptionsquerschnitten für langsame Neutronen. Methode I. Konzentrationsabhängigkeit', *KPFB* **G-37** (dated Dec. 17, 1940). W. Ramm determined the absorption cross-section in uranium for different temperatures in: 'Einfang thermischer Neutronen und Diffusionslänge in Uran bei verschiedenen Neutronentemperaturen', unpublished research report (dated November 1942,); commented on further by von Droste in Bothe & Flügge (Eds.) [1948], pp. 206f., and by Ramm himself in: Bothe & Flügge (Eds.) [1948], part I, pp. 129ff.

[62]On Sauerwein's and Ramm's researches, see the foregoing footnotes. On high-voltage plants in Germany, see F. Kirchner in Bothe & Flügge (Eds.) [1948], part II, pp. 24ff.

[63]On this successful sabotage mission carried out by Norwegian resistance fighters and supported by the British intelligence service on Feb. 27, 1943, destroying 900 kg of heavy water and shutting down the deuterium production unit for several months see, e.g., Irving [1967], or Walker [1989]a, pp. 118ff. See also here doc. 103.

[64]According to Walker [1989]a, p. 120, the Allied air raid on Nov. 16, 1943 failed to destroy any essential part of the heavy production unit; but the Norwegian director of the factory, supported by the local → *IG Farben* representative Bey, refused to continue to produce heavy water on the reasoning that the Allied forces would obviously take this as a threat and would react with air raids, putting at stake the other crucial production units at Rjukan as well.

[65]*IG Farben* had a small trial production branch at *IG Leuna*, headed by Dr. Bütefisch and Dr. Herold, which they intended to expand after the failure of the Rjukan production center. Cf., e.g., Walker [1989]a, pp. 120f., on the consultations of the *Uranverein* representatives Bonhoeffer, Diebner and Harteck with *IG Farben* managers, and the memorandum on the collaboration between the *HWA* and *IG Farben* on the production of heavy water in Germany, dated April 30, 1942. Later, patent quarrels arose between the participating researchers and the company based on this document; see Irving (Ed.) [undated], nos. 29 716f., 29 723f., and 29 757ff.

Even then, some progress was achieved by the Harteck-Groth-Beyerle group.[66] As early as 1942, this group had succeeded in developing an ultracentrifuge for the enrichment of the isotope U(235). It was planned to use uranium enriched with the rare isotope in the construction of improved uranium piles, possibly in conjunction with ordinary water.[67] At about this time, the direction of the uranium project was transferred from Esau to Gerlach.[68] Gerlach had taken over the physics section of the *Reichsforschungsrat*, and he strove to promote more particularly the scientific side of the uranium problem; and at that, not only the physical, but also the medical aspect. In connexion with the medical applications, the construction of a low-temperature pile, in liquid carbon dioxide, was undertaken on Harteck's suggestion. Such a pile, even of small dimensions, could be expected to yield profitable amounts of radioactive elements for tracer research, in view of the decreased absorption in the resonance lines at low temperatures.[69]

In the winter of 1943–44, a model pile of 1.5 tons of heavy water and about the same weight of uranium plates was constructed in the Dahlem air-raid shelter through the co-operative efforts of the Kaiser Wilhelm Institutes for Physics in Berlin and Heidelberg (Wirtz, Fischer, Bopp, P. Jensen, Ritter).[70] The number

---

[66]See the reports by W. Groth: 'Stand der Arbeiten zur Herstellung einer Ultrazentrifuge', *KPFB* **G-82** (dated Dec. 14, 1941), 'Trennung der Uranisotope nach dem Ultrazentrifugenverfahren. I. Anreicherung der Xenonisotope in einer einstufigen Ultrazentrifuge', *KPFB* **G-146** (dated Jan. 27, 1942), and 'Die Trennung der Uranisotope nach dem Trennrohr- und nach dem Ultrazentrifugenverfahren', *KPFB* **G-147** (dated Mar. 23, 1942); the report by W. Groth & Albert Suhr: 'Trennung der Uranisotope nach dem Ultrazentrifugenverfahren', *KPFB* **G-149** (dated Aug. 17, 1942); P. Harteck & J. Jensen, 'Berechnung des Trenneffektes und der Ausbeute verschiedener Zentrifugenanordnungen zur Erhöhung des Wirkungsgrades einer einzelnen Zentrifuge', *KPFB* **G-158** (dated February 1943); and the report by Konrad Beyerle: 'Die Gaszentrifugenanlage für den Reichsforschungsrat', *KPFB* **G-248** (dated Dec. 12, 1944). See also the later overview of their research achievements with ultracentrifuges in: Beyerle, Groth, Harteck & Jensen [1950], and the description in Walker [1989]a, pp. 121ff.

[67]See P. Harteck's letter to *HWA* dated June 26, 1942 on the importance of separating U^{235} for atomic explosives and the importance of concentrating the efforts of ultracentrifuge research: Irving (Ed.) [undated], nos. 29 719ff. See also Irving [1967], p. 145, for a photograph of the first ultracentrifuge.

[68]Esau had resigned as Göring's commissioner of nuclear research in a letter dated Oct. 28, 1943 and was replaced by Walther → Gerlach officially in December. Cf. Walker [1989]a, pp. 129ff., on the reasons for Esau's replacement; as well as footnote 15 of doc. 105.

[69]Medical applications of radioactive isotopes are described, e.g., in: G. Schubert, 'Radioaktive Isotope als Indikatoren in Biologie und Medizin', in Bothe & Flügge (Eds.) [1948], pp. 134–142.

[70]For reports on these last two experiments (B 6 and B 7) carried out in a special bunker lab at the Dahlem *KWIP* (Berlin), see W. Heisenberg, 'Bemerkungen zu dem geplanten halbtechnischen Versuch (B_6 und B_7) mit 1.5 to D_2O und 3 to 38-Metall', unpublished report (1942), reprinted in Heisenberg [1989], pp. 545ff. On B 7 in particular, see the report by F. Bopp, W. Bothe, Fischer, E. Fünfer, W. Heisenberg, O. Ritter & K. Wirtz: 'Bericht über einen Versuch mit 1,5 to D_2 und U und 40 cm Kohlerückstreumantel (B 7)', *KPFB* **G-300** (Jan. 3, 1945), reprinted in Heisenberg [1989], Vol. A II, pp. 595ff. Cf. also Heisenberg & Wirtz [1948], pp. 153ff.

of neutrons injected from the internal source was multiplied by the factor 3,[71] a performance approaching considerably nearer to what we called the *Labilitätspunkt*, at which the ratio k increases beyond limit and at which the uranium pile begins to radiate independently of the neutron source and thus to produce energy. The relation between neutron increase and layer thickness fulfilled expectations. Further, the stopping distances of the fission neutrons in carbon and heavy water were redetermined (Wirtz) and the former inexact measurements considerably improved.[72] These experiments were made in the air-raid shelter of the Institute during the heaviest air raids on Berlin, and were naturally to some extent hindered by the raids. On February 15, 1944, the *Kaiser Wilhelm Institut für Chemie* received a direct hit. In the meantime, the *Kaiser Wilhelm Institut für Physik* had been partly evacuated to Hechingen. On the instructions of Gerlach, a cellar cut out of the solid rock, situated in the village of Haigerloch, was equipped for the rebuilding of the uranium pile. It was not until February 1945, however, that the greater part of the necessary material (about 1.5 tons of heavy water, 1.5 tons of uranium, 10 tons of graphite, cadmium for the regulating rods, etc.) was finally assembled at Haigerloch, and a new pile, this time built up of uranium cubes in heavy water, with an outer cover of graphite, constructed (Wirtz, Fischer, Bopp, Jensen, Ritter).[73] A branch of the *Reichsforschungsrat* at Stadtilm was allotted the remaining quantity of heavy water and a great part of the available uranium. The Haigerloch pile yielded a sevenfold neutron increase. The material available at Haigerloch, however, was just insufficient to attain $k = \infty$. A relatively small amount of uranium would in all probability have sufficed; but it was no longer possible to obtain it, since transport from Berlin or Stadtilm could no longer reach Hechingen. On April 22, Haigerloch was occupied, and the material confiscated by the Americans.[74]

When we compare the German work reported here with the corresponding Anglo-American effort, so far as it has been made known, then the beginning of 1942 seems to be the turning point. Up to that time, both sides were dealing predominantly with the scientific problem as to whether nuclear energy could

[71]The neutron multiplication k-value reached in experiments B 6_b and B 7 was actually 1.08, corresponding to $\overline{\nu} = 266$.

[72]See the references to experiments B 6_b and B 7 in the preceding footnotes. The earlier measurements were done by Bothe and his colleagues in Heidelberg, cf. footnote 43 above.

[73]The report on this Haigerloch experiment (B 8), co-authored by F. Bopp, E. Fischer, W. Heisenberg, K. Wirtz (the last three from the *KWIP*) and by W. Bothe, P. Jensen & O. Ritter (from the physics section of the Heidelberg Kaiser Wilhelm Institute of Medicine): 'Bericht über den Versuch B_8 in Haigerloch', is apparently not preserved. See however, Heisenberg [1943]b, 3rd ed., pp. 161ff., and in particular Heisenberg & Wirtz [1948], sec. 7.1.3, for his detailed account of the construction of uranium cubes suspended on 40 chains of 9 cubes each and 38 chains of 8 cubes each. This final Haigerloch experiment B 8 determined k at 1.11, corresponding to $\overline{\nu} = 430$.

[74]On the Haigerloch redoubt see also K. Wirtz et al. [1990] as well as, e.g., Walker [1989]a, pp. 150ff., especially pp. 157ff., and doc. 116 or Goudsmit [1947] on the Alsos Mission.

be utilized at all, and what fundamental methods had to be employed to that end. Both sides had arrived almost simultaneously at very similar results, if one excludes the field of isotope separation, in which the Anglo-Americans had made much greater progress. Furthermore, in the United States far more attention had been given to laying the groundwork for subsequent full-scale development of the uranium project; so that the first self-supporting pile was functioning as early as December 1942.[[75]]

It remained to determine the technical sequel to these results. In the United States, the final decision was taken to go for the production of atomic bombs, with an outlay that must have amounted to a considerable fraction of the total American war expenditure; in Germany an attempt was made to solve the problem of the prime mover driven by nuclear energy, with an outlay of perhaps a thousandth part of the American. We have often been asked, not only by Germans but also by Britons and Americans, why Germany made no attempt to produce atomic bombs. The simplest answer one can give to this question is this: because the project could not have succeeded under German war conditions. It could not have succeeded on technical grounds alone: for even in America, with its much greater resources in scientific men, technicians and industrial potential, and with an economy undisturbed by enemy action, the bomb was not ready until after the conclusion of the war with Germany. In particular, a German atomic bomb project could not have succeeded because of the military situation. In 1942, German industry was already stretched to the limit, the German Army had suffered serious reverses in Russia in the winter of 1941–42, and enemy air superiority was beginning to make itself felt. The immediate production of armaments could be robbed neither of personnel nor of raw materials, nor could the enormous plants required have been effectively protected against air attack. Finally—and this is a most important fact—the undertaking could not even be initiated against the psychological background of the men responsible for German war policy. These men expected an early decision of the War, even in 1942, and any major project which did not promise quick returns was specifically forbidden. To obtain the necessary support, the experts would have been obliged to promise early results, knowing that these promises could not be kept. Faced with this situation, the experts did not attempt to advocate with the supreme command a great industrial effort for the production of atomic bombs.[[76]]

[75]On Enrico Fermi's successful uranium reactor design using a graphite moderater in Chicago, see e.g., Rhodes [1986], chap. 13, which also describes in detail the subsequent progress leading to the development of the atomic bomb in Los Alamos. See also Baxter [1947], part 6; Gowing & Arnold [1979], pp. 23ff., and Hawkins, Truslow & Smith [1983].

[76]This is very different to what Heisenberg said as an internee at Farm Hall, to the effect that he had not dared propose such a risky project to the German military and government. See Frank (Ed.) [1993], p. 76: "We wouldn't have had the moral courage to recommend to the government in the spring of 1942 that they should employ 120,000 men just to build that thing up."

From the very beginning, German physicists had consciously striven to keep control of the project and had used their influence as experts to direct the work into channels which have been mapped in the foregoing report.[77] In the upshot they were spared the decision as to whether or not they should aim at producing atomic bombs. The circumstances shaping policy in the critical year of 1942 guided their work automatically towards the problem of the utilization of nuclear energy in prime movers. To a German physicist, this task seemed important enough. The mere possibility of solving the problem had been rendered possible by the discovery of the German scientific workers Hahn and Strassmann and so we could feel satisfied with the hope that the important technical developments, with a peace-time application, which must eventually grow out of their discovery, would likewise find their beginning in Germany, and in due course bear fruit there.[78]

116 Samuel A. Goudsmit: Nazis' Atomic Secrets. The Chief of a Top-Secret U.S. Wartime Mission Tells How and Why German Science Failed in the International Race to Produce the Bomb [October 20, 1947]

Source: Samuel A. Goudsmit, 'Nazis' Atomic Secrets. The chief of a top-secret U.S. wartime mission tells how and why German science failed in the international race to produce the bomb.' *Life*, Volume 23, October 20, 1947, pp. 123–134 with 6 photos and 1 cartoon.

More than two years have passed since the war's end, but many of our scientific and military experts still believe that we were engaged in a desperate race with the Germans for the secret of the atomic bomb and that it was only by a miracle that we got there first.

The plain fact is that the Germans were nowhere near getting the secret of the atomic bomb. Months after our scientists had established, beyond a doubt, the feasibility of the atomic bomb, the Germans were still talking about the 'uranium problem' and the possibility of constructing a 'uranium machine'. They did not yet know how to produce a chain reaction in a uranium pile.[1] They did not know how to produce plutonium. They knew only, perhaps with typical German arrogance, that if they could not make an atomic bomb nobody else

[77]This postwar portrayal of the stance of German nuclear physicists is very controversial. The American historian Mark Walker argues that the decision rested rather on the Army Ordnance Office (*HWA*), noting that even physicists like C. F. von Weizsäcker had stated that their hope of retaining control of the fate and speed of their research had been naive; see also footnote 1 above.

[78]On the birth of this postwar myth at Farm Hall about the peaceful motives of German nuclear scientists, see Walker [1990]a, [1993]b as well as here doc. 116, pp. 389.

[1]See figures 27 and 28 here for photos of the installations and grounds (both on p. 123 of the original).

could either. That arrogance received a terrible jolt the day our atomic bomb fell on Hiroshima.[2]

Fig. 26: [Left] 'Chief 'target' of the secret U.S. scientific mission sent to Germany to discover her atomic-bomb progress was 45-year-old Werner Heisenberg.
A world-famous nuclear physicist, some of whose discoveries rank with Einstein's, Heisenberg was the man most feared by U.S. atomic-bomb specialists, who felt that he might be far ahead in development of the bomb. Heisenberg escaped once, was finally captured and is now interned in British-occupied Germany.'[3] [Right:] 'Top U.S. atomic sleuth overseas was Dutch-born, 45-year-old Samuel A. Goudsmit, who was picked because he knows all key European scientists. A naturalized U.S. citizen, Goudsmit taught physics at the University of Michigan before the war and is co-discoverer with George E. Uhlenbeck of the electron's spin. On these pages *Life* presents Goudsmit's own story of his mission. His book *Alsos* (Henry Schuman, Inc.; $3.50) will be published this week.'[4]

[2] The reaction of some leading German physicists to the atomic bomb on Aug. 6–7, 1945, is recorded in English translations of the Farm Hall tape transcripts in Frank (Ed.) [1993], pp. 70ff. See also footnote 11 of doc. 112.

[3] (Photo on p. 123 of the original.) When the 'Alsos' Mission arrived in Hechingen, Württemberg, Southwest Germany, where the Kaiser Wilhelm Institute of Physics (→ *KWIP*) had been relocated, Heisenberg had already left by bike. Goudsmit's team finally tracked him down at his summer house in Urfeld, Upper Bavaria. Heisenberg and several other atomic scientists were then interned at Farm Hall near Cambridge, England, from July to December 1945 when they were allowed to return to Göttingen in British-occupied Germany. See Frank (Ed.) [1993] and footnote 20 below.

[4] S. A. Goudsmit and George E. Uhlenbeck (1900–) published their concept of electron spin (corresponding classically to internal angular momentum) in November 1925. For the context of this discovery which explained some of the earlier puzzles of quantum theory including Pauli's exclusion principle see, e.g., Jammer [1966], pp. 133–153. Goudsmit borrowed heavily from his book Goudsmit [1947] in writing this essay.

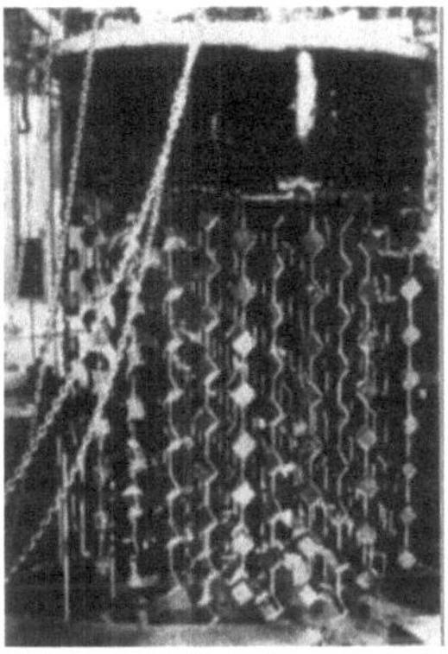

Fig. 27: 'This was Germany's 'A-bomb.' It turned out to be a harmless Uranium pile.' [The photo in the original article duplicates fig. 19 in doc. 95, p. 298. Above is the only photograph of the uranium lattice arrangement used during the last Haigerloch experiment in early 1945].

These are some of the facts that were established by the Alsos Mission. It was named Alsos, which is the Greek equivalent of 'grove,' after the head of the U.S. atomic-bomb project, Major General Leslie R. Groves.[5] Our job was to follow immediately in the wake of the invasion of Europe and even to penetrate into enemy territory on our own in order to determine precisely how much the Germans knew about the atomic bomb and how far they had progressed in its construction.

Alsos headquarters were set up in Paris the day the first Allied troops moved in.[6] We soon learned that a well-known German chemical concern had kept an office there throughout the Occupation and that their head representative had been chiefly interested in locating stocks of uranium.[7] We also learned, to our horror at the time, that this very same chemist had confiscated and sent to Germany all the thorium he could find in Paris.

This really scared us. Thorium, we knew, could be used in an A-bomb project when that project was in a well-advanced stage. Did this interest in thorium mean that the Germans were really ahead of us?

---

[5]The Alsos Mission was created in the fall of 1943 by the Army's G-2 intelligence department, Groves's Manhattan District, the U.S. Navy and Vannevar Bush's Office of Scientific Research and Development. For background information see also Walker [1989]a, pp. 212ff. Brigadier General Leslie Richard Groves (1896–1979), a graduate of West Point Military Academy and deputy director of U.S. Army construction projects, was appointed head of the 'Manhattan Engineer District' (as the American project was soon called) on Sep. 17, 1942, and he remained in this position until 1947. Cf., e.g., Rhodes [1986], pp. 424ff., Baxter [1947], p. 439, Groves's own account in Groves [1962], and Goldberg [1995].

[6]Paris was liberated on Aug. 25, 1944.

[7]N. → Riehl headed the research division of the → *Auer Gesellschaft*, which was responsible for the production of uranium in cubes and in the form of oxide and hexafluoride; cf., e.g., Riehl [1968] for the contemporary techniques of uranium enrichment.

The thorium mystery became an obsession; we had to find out more about it. No one in Paris had any idea what the Germans wanted it for. We finally located the German chemical company's Paris representative—one Herr Doktor Petersen—in a little town on the German-Belgian border which had just fallen into Allied hands.

When captured, Petersen had with him a suitcase full of documents. It was quite an event for us. Here was our own first real Alsos Mission prisoner. We had great hopes that we finally had found somebody who had inside information about the Germany uranium project. Thus far all we knew was that in Norway they had been working hard on the production of heavy water.[8] In addition, we had heard through neutral sources that the Kaiser Wilhelm Institute for Physics had been evacuated from Berlin to Hechingen, a little village in southern Germany near the family castle of the former *kaiser*.[9]

We transformed our hotel suite in Paris into a tribunal for the interrogation of our prize quarry. We put on our Sunday uniforms and Colonel Boris T. Pash, the energetic military head of our mission who had personally done the capturing, put on as many ribbons and medals as he could find.[10] Then Petersen was brought in. We placed him facing the window so that we could observe all his reactions and then proceeded to shout dozens of questions at him. But his answers were all unsatisfactory. Either he was hiding something or he really didn't know what it was all about.

The radioactive smile

Greatly disappointed, I began to study the numerous papers he had in his suitcase. This man Petersen, like most Germans, was so systematic that his papers revealed exactly where he had been and what he had done during the previous week. A streetcar ticket showed that he had been in Berlin only two weeks before. But what excited me the most was a hotel bill which showed that he had visited that little village of Hechingen near the *kaiser*'s castle. There was no doubt in my mind that Petersen, on his way from Paris to the factory near Berlin, had stopped off at the German uranium-research project. It seemed a tight case. It connected the theft of thorium from France with the secret project in Hechingen.

The next morning we pestered our prisoner with questions about Hechingen. He claimed he had gone there merely to visit his mother. A further investigation

[8]This Norwegian company → *Norsk Hydro* terminated its heavy-water production in 1943: See footnotes 39 and 64 of doc. 115.

[9]On the Hechingen redoubt in the region south of Stuttgart in the Swabian alps, in Württemberg, which was designated French-occupied territory see, e.g., Goudsmit [1947], pp. 98ff. The *KWIP* was relocated there in 1944 in a former textile factory building to escape the heavy bombardments in Berlin.

[10]Boris T. Pash (1892–1989) was a former high-school teacher, retrained by the FBI after becoming a security officer in the U.S. Army G-2 intelligence unit. On Pash, who later wrote his own account of the Alsos Mission, Pash [1969] see, e.g., Rhodes [1986], pp. 605–613.

showed that indeed his mother was living there.

We were not quite satisfied, however. We did some more interrogating and studied the papers more and more carefully. Finally, we pieced together the amazing story of the missing thorium.

The company officials, it seems, had been thinking about peacetime, when they could no longer make money by manufacturing gas masks, carbons for searchlights or other war materials. They didn't want the end of the war to find their business shrinking away. They realized that in peacetime the big money would be in cosmetics, and they had been impressed by American methods of advertising.

Now, as it happened, one of the officials had a certain patent on toothpaste—thoriated toothpaste, a toothpaste mixed with thorium oxide which was supposed to have the same effect as peroxide. The company was already dreaming of its advertising for the future. "Use toothpaste with thorium! Have sparkling, brilliant teeth—radioactive brilliance!" After all, America had its Bob Hope and irium toothpaste.[11] To be sure of success the firm had planned to monopolize all the thorium supply it could lay its hands on. That was the real reason it had stolen all the thorium from the French. It sounded incredible, but later information confirmed the story completely. And all the time we had been afraid that it was going to be used for an atomic bomb!

The thorium for toothpaste was not the only false scent that Alsos followed in its first days overseas. We had unqualified support from General Groves and War Department Intelligence, and the mission has since been cited "as one of the finest examples of cooperation of the scientists and the Armed Forces" in James Phinney Baxter's *Scientists Against Time*.[12] Nevertheless some of our early experiences abroad were downright ludicrous, chiefly because long-range communication with Washington presented certain difficulties.

Fig. 28: 'Atomic laboratory on the army proving grounds in a Berlin suburb was housed in this small wooden shack. Sign says, "Halt. Approach forbidden."[13]

[11]Bob [Leslie Townes] Hope (born 1904), the celebrated actor and TV star born in England. For a facsimile of a 'Doramed' advertisement, a product promoted by the *Auer* Company as "the radioactive, biologically effective toothpaste", see Irving [1967], p. 161.

[12]See Baxter [1947], p. 413.

[13](Photo on p. 126 of original.) It is not clear from Goudsmit's description where this laboratory was, but the Army Ordnance Office's (→ *HWA*) most important experimental testing grounds was located in Gottow near Kummersdorf; and from the extensive description of the

Here is an example, admittedly an extreme one, of what could happen. To help us in our search for uranium plants we shipped back from Germany a sample of water from the Rhine, which we wished Washington to test for radioactivity. At the same time, to please his colleagues at home, the Army major who was General Groves's representative on our mission added a fine bottle of liberated French wine and wrote on the label, "Test this for activity, too." It seemed a nice thing to do and a rather entertaining idea.

A few days later came one of those 'top-secret ACTION' radiograms with which we had all become familiar. "Water negative. Wine shows activity. Send more. Action." Ah, we thought, they were carrying our little joke further. But we were quite mistaken. Washington apparently could take no chance. Further radiograms indicated that they had actually measured the radioactivity of the wine instead of drinking it. Might it not mean that the Germans had a secret laboratory somewhere in the mountains where the grapes were grown?

At first I refused to look into this wine nonsense. But I lost the battle. I was forced to send one of my physicists, Major Russell A. Fisher, on a 10-day wild-goose chase.[14] Fortunately we knew where the wine came from: it was a bottle which Major Fisher had liberated when he came to France on the heels of the southern invasion. Now I had to send him all the way back to the Marseille region to find more of the same.

"Do a complete job." I said. "Don't be stingy. And above all, be sure that for every bottle of wine you locate, you secure a file copy for our office in Paris."

Major Fisher, accompanied by Captain Wallace Ryan, did a thorough job. They returned to Paris with a most representative supply of Rhône wines, samples of the grapes, the soil where they grew, the water of little rivers, wholesale samples, retail samples, all absolutely complete. Except for our file copies, all of this was sent on to Washington accompanied by Major Fisher's report.

I was later reprimanded, justly and fairly, I admit, by General Groves's scientific adviser for "playing practical jokes over 3,000 miles of ocean during wartime." As time went on, however, remote control of Alsos was reduced to a minimum.

---

site among the Osenberg Papers (Irving (Ed.) [undated], nos. 29 1134–29 1138), it was certainly much larger, accommodating a total of 125 employees in June 1944, including Kurt → Diebner and three other reseachers. This photograph thus illustrates Goudsmit's inclination to ridicule German research efforts.

[14]According to Goudsmit [1947], p. 19, Goudsmit chose his longtime acquaintance Prof. Russell Arden Fisher (born 1904) because he was a trained physicist. Fisher earned his Bachelor of Arts in Michigan in 1927, his Masters degree in 1929, and his Ph.D. in physics in 1931, working as a graduate at the Institute of Physics in Michigan. 1931–32 he transferred to Northwestern University and became assistant professor there in 1932 and associate professor 1937–42, later becoming full professor in 1946 and working at the chemistry department 1950–57. He served in the Army 1942–46 and was promoted to Lieutenant-colonel and awarded the bronze star medal. Fisher worked on atomic spectroscopy, nuclear moments, interferometry and nuclear spectroscopy.

The mysterious German scientist

One other occurrence helped clarify the problem of freedom of action for the Alsos Mission. At first it had been hard to convince our Military Intelligence colleagues that searching for scientists was not the same as searching for spies or criminals, that some were important and others were utterly unimportant. A list of 'suspects' had been furnished us. In military jargon they were called 'targets,' and our job in the normal police procedure would have been to investigate each one individually.

One day we received a dispatch from Washington about a mysterious German scientist who had visited the U.S. before the war. Instead of furnishing us with data about his scientific work, which would have told us everything or at least aroused our interest, the secret message told us about his beer-drinking habits, his opinion of American women, how he had had German measles in 1938, had a polyp in his right nostril and an atrophied left testicle.

After this memorable dossier even the military members of our mission treated such dispatches from Washington with some skepticism. Whenever Washington proposed that we go after someone we knew to be of no importance to us, we would file a propo[sal] philosophically, remarking that, after all, we hadn't even caught that other fellow yet.

How did we know who **was** important? Our 'secret' method of operation was like the so-called secret of the atomic bomb itself. An algebra problem can be a deep mystery to the uninitiated and duck soup to a high-school student. To an outsider a professor is a professor, but we knew that no one but Professor Werner Heisenberg of the Kaiser Wilhelm Institute could be the brains of a German uranium project. Every physicist in the world knew that.[15] In other words we knew who our chief targets in Germany were before we started. What we had to find out was how far they had advanced on their atomic-bomb project.

We continued to fear that German atomic-bomb development might prove superior to ours until we reached Strasbourg. Colonel Pash went forward as soon as that city was occupied and rounded up the scientists we wanted.[16] He found our 'targets,' who had all been attached to the University of Strasbourg's nuclear-physics laboratory, posing as medics in the famous Strasbourg Hospital.[17] He took the whole crew into custody, but they refused to talk. We found documents, however, which suddenly raised the curtain of secrecy for us. Here, in apparently harmless memoranda written by these German scientists, was hidden a wealth of secret information available to anyone who understood it. No, it was not in

[15] W. → Heisenberg had become director at the *KWIP* in Berlin and was thus the head of the Berlin section of the '*Uranverein*'; see footnotes 1 and 14ff. of doc. 115. Other important groups in Hamburg, Stadtilm, and Heidelberg were headed by P. → Harteck, K. → Diebner, and W. → Bothe. See figure 26 above.

[16] The Allied occupation of Strasbourg started on Nov. 15, 1944.

[17] C. F. → von Weizsäcker had taken up a professorship in theoretical physics at Strasbourg's physics institute in 1942.

code. The papers were not even secret. They were just the usual scientific gossip between colleagues, here and there a hint, nothing objectionable from a secrecy point of view—just ordinary notes such as we had all sent to our own colleagues when we were in the U.S. Obviously such memos would have told nothing to the uninitiated; on the other hand they were not meant to be read by Allied scientists.

We found the letterheads of some of the German secret atomic laboratories. Even the precise addresses and telephone numbers were given. We found a copy of a letter to "*Lieber* Werner," who was obviously Heisenberg, telling just what problems they were working on at Strasbourg. We found references to "special metal," which was obviously uranium. We learned that 'large-scale' experiments were being performed at an army proving ground near Berlin. We found parts of computations which clearly applied to the theory of a uranium pile.

Villages surrendering by telephone

These seemingly innocent papers gave us an over-all view of the whole German uranium project. They showed that the Germans had been unsuccessful in their attempts to separate U-235. They had probably started separation on a very small scale by means of a centrifuge and they were working hard to construct a uranium pile. But apparently they had only recently succeeded in manufacturing uranium.

Their computation showed that as of August 1944 their pile work was still in a very early stage. They had not yet succeeded in producing a chain reaction. Their preliminary experiments had not even given them hints about certain major difficulties they would have to overcome before the pile would work. In short they were about as far as we were in 1940, before we had begun any large-scale efforts on the atomic bomb at all.[18]

In spite of the decisive finds at Strasbourg, our main objective had not yet been reached. The center of atomic-bomb research was not yet in our hands. We still had to contact the brains of the German uranium project, the physicist, Werner Heisenberg. We knew that he and his group were working in a makeshift laboratory in Hechingen. Otto Hahn, the discoverer of atomic fission, and his men were in the nearby village of Tailfingen.[19] Many other important research laboratories had also been evacuated to that region south of Stuttgart in the state of Württemberg.

The official code name for our mission's thrust into this key area was "Op-

[18]For Heisenberg's view on how far German nuclear research had advanced, see doc. 115. Cf. doc. 109 for the Allied estimation on how far advanced German research was compared to the Americans at this time.

[19]For Otto → Hahn's account of his capture, see the chapter on his internment in his autobiography Hahn [1968/86], pp. 169ff. For photos of the Alsos Mission's arrival in Hechingen see, e.g., Irving [1967], p. 225.

eration Humbug." For this operation into territory only sparsely occupied by French colonial troops we had two armored cars, a caravan of jeeps and other vehicles. Cautiously they proceeded south along the eastern bank of the Rhine and then penetrated eastward to where the secret laboratories were. There was no opposition to the first Americans to enter this strategic area; a few shots were exchanged and some of the villages were summoned to surrender by telephone. It turned out to be probably the least dangerous of all Alsos advances.

Fig. 29: 'As prewar friends: Goudsmit (left), Heisenberg of Germany (center) and famed Enrico Fermi of the U.S. (second from right) met at Ann Arbor.'

Colonel Pash, who headed our military caravan, was accompanied by General Eugene L. Harrison, who was chief of Intelligence for the area. When our force reached Hechingen it proceeded at once to Heisenberg's laboratory, which was not hard to find. The colonel and the general entered Heisenberg's office; he was not there. But the first thing they saw, to the consternation of the general, was a photograph of Heisenberg and myself standing side by side. The photograph had been taken when Heisenberg was my guest at the University of Michigan in 1939.[[20]] Egged on by Colonel Pash, the general almost began to believe that I could not be trusted and that I had close contact with the enemy. At any rate I had to stand a lot of teasing about this incident afterward.

Next a thorough investigation of Heisenberg's laboratory began. For this final achievement of the mission some of General Groves's officers and some of our British colleagues had made a special trip to the scene.

I came too late to prevent some of our guests from blowing up the cave in which the uranium-pile research had been located. It was an utterly senseless

[20]Heisenberg had been in the United States to lecture in the summer of 1939. He visited Goudsmit at his home in Ann Arbor (Fig. 29 was on p. 131 of the original article) and met the Italian physicist Enrico Fermi there, who was teaching at the University of Michigan that summer and who later became famous for constructing the first self-sustaining nuclear generating reactor at the University of Chicago together with others for the Manhattan Project in December 1942. The two other persons on the photo are Dean Yokum (second from left) and Dean Kraus (right), both from the University of Michigan. See Goudsmit [1947], pp. 98f., 163, and for the photo, the page facing p. 124. On this visit see also footnote 3 of doc. 117.

thing to do but fortunately harmless since all equipment was first removed. It was the same reasoning which later caused the useless destruction of the Japanese cyclotrons. A cyclotron is not a vital part of an atomic bomb any more than a harmless physicist may be considered a potential saboteur about to blow up New York City.

Even after we knew how insignificant the work of the German physicists had been, we had orders to capture and intern them all. Consequently we picked up half a dozen at Hechingen, including the famous scientists Otto Hahn, Von Weizsäcker and Von Laue.[21] It was a sad caravan of command cars and jeeps that departed for Heidelberg with our 'prisoners.' Seeing them off, I could not help feeling a little like James Thurber's fierce female pointing her gun at the droopy professors in his cartoon, "Capture of Three Physics Professors."[22] It was so obvious that the whole German uranium setup was on a ludicrously small scale. Here was the central group of laboratories, and all it amounted to was a little underground cave, a wing of a small textile factory, a few rooms in an old brewery.

But we were still lacking Heisenberg, our chief target among the German physicists. A few days before we took over Hechingen he had left by bicycle to join his family in Bavaria, and that territory was still in German hands. He had left behind in Hechingen, however, the evidence of his activities. We discovered all the stuff that had been so carefully buried—on his instructions—2 tons of uranium, 2 tons of heavy water, 10 tons of carbon and the principal reports on the pile research.[23]

Fig. 30: 'Capture of three Physics Professors' by James Thurber was recalled by Goudsmit when he had to intern captured German scientists.' Copyright © 1943 James Thurber. Copyright © 1971 Rosemary A. Thurber. From *Men, Women and Dogs*, published by Harcourt Brace.

I was in Heidelberg when Colonel Pash brought in Werner Heisenberg, whom he had captured in Bavaria. I had known Heisenberg for many years. He had visited me in my native Holland as far back as 1925. He had been in the U.S.

[21]The three prisoners aside from Otto Hahn, C. F. von Weizsäcker and Max → von Laue, were Karl → Wirtz and two younger men who had done "novel research work on isotope separation" (Erich → Bagge was probably one of them). This selection puzzled von Weizsäcker; he reportedly exclaimed: "What kind of selection is this?". See Goudsmit [1947], p. 104.

[22]See figure 30 (on p. 131 of the original) for the cartoon by the American humorist, writer and artist James [Groves] Thurber (1894–1961).

[23]See the photo, Fig. 31.

several times and had spent summers with me at the University of Michigan. He is still the greatest German theoretical physicist and among the greatest in the world. His contributions to modern physics rank with those of Einstein.

I greeted my old friend and former colleague cordially. Purely on the impulse of the moment I said, "Wouldn't you want to come to America now and work with us?" But he was still too impressed by his own importance and that of his work, to which he ascribed his internment.

"No, I don't want to leave," he said. "Germany needs me. If American scientists wish to learn about the uranium problem, I shall be glad to show them the results of our researches if they come to my laboratory."

It was sad and ironic to hear him say this, when I was aware how much more we knew about the problem than he did. But I did not enlighten him. I merely thanked him for his offer and left him secure in the belief that his work was ahead of ours.

This was the end of our roundup of the German nuclear physicists. We had found all the key men on the project and all documents and materials. We had interned 14 of them. Four were already in the U.S. and the rest in England, where they were housed in an estate about 50 miles from London.[24] Just why these top German physicists were interned in England, I never understood. Perhaps our military experts did not know what to do with these scientists after we had found them and felt quite grateful when the British offered to take them over. As a result the majority of the best German scientific brains are now in the British zone of occupation. That there are still a few good nuclear physicists in the American zone is due to the fact that the Alsos Mission did not intern them all.

"The Nazis worked for peace"

It was in England on Aug. 6, 1945 at dinnertime that the 10 German physicists first heard the news of Hiroshima. Their first reaction was one of utter incredulity. "It can't be an atomic bomb," one of them said. "They may have some new explosive or an extra-large bomb they call 'atomic,' but it's certainly not what we would have called an atomic bomb. It has nothing to do with uranium."[25]

But at 9 o'clock came the detailed news broadcast. The impact on them was shattering. At one stroke all their self-confidence was gone and the belief in their own scientific superiority gave way to an intense feeling of despair and futility.

[24]Erich Bagge, Kurt Diebner, Walther → Gerlach, Otto Hahn, Paul Harteck, Werner Heisenberg, Horst → Korsching, Max von Laue, Carl Friedrich Freiherr von Weizsäcker, and Karl Wirtz were interned by the British at Farm Hall in Godmanchester near Cambridge, England, from July to December 1945.

[25]Goudsmit had access to the secret Farm Hall tape recordings (or their transcripts translated into English) to be able to give such a precise account. They were only recently published in the official English version: Frank (Ed.) [1993]. The German original tapes were apparently destroyed, and the German version, Hoffmann (Ed.) [1993], is a retranslation.

They spent hours discussing the bomb and trying to figure out its mechanism. But the radio, for all its details, had not given enough, and the German scientists still believed that what we had dropped on Hiroshima was a complete uranium pile. No wonder they were bewildered. As yet, no plane has ever been built that could do that. Even if there were such a plane, a uranium pile could never be a bomb. It could only fizz. But these German experts failed to realize this basic fact.

It was not until more than a full day after the first announcement of Hiroshima that Heisenberg began to understand how he and his colleagues had completely missed the basic principle of the atomic bomb. It was only then that he finally came to understand that we had used the uranium pile merely to produce material—plutonium—and out of this new substance had made the bomb.[26]

When Heisenberg called his colleagues together and explained to them what it was all about, some of the younger men hit upon a brilliant rationalization of their failure. They decided to turn their very failure to advantage by denying they had ever tried to make an atomic explosive. This, then, was to be the new theme song of German science: "Germany worked on the uranium problem for peaceful uses only; the Allies, for purposes of destruction." It was therefore no surprise, at least not to the members of the Alsos Mission, when, almost two years after Hiroshima, Heisenberg told an Associated Press reporter that he had been building up Germany's uranium pile "to create energy for machines and not for bombs." "As the world now knows," Heisenberg added innocently, "the explosive, plutonium, is produced in such a uranium pile."[27]

Heisenberg's statement is a beautiful example of how to mislead with half truths. It is true that the German scientists were working on a uranium machine and not the bomb, but it is true only because they failed to understand the difference between the machine and the bomb. The bomb is what they were after. And what the whole world now knows about plutonium, the German scientists did not know—until they were told about it after Hiroshima.

German science failed miserably in the atomic-bomb race because it was guilty of three major errors: 1) political bungling, 2) deterioration of interest in pure science and 3) complacency. These same mistakes are the principal ones that we too can make if we are not on guard against them.

In Germany the administrators of the uranium-research projects were selected because they had the confidence of the Nazi bosses. This is not likely to happen here. But the difficulties which arose when the Atomic Energy Commission was to

[26]Germans were aware of the possibility of producing an explosive, however; see doc. 95, p. 296. See also Frank (Ed.) [1993] for transcriptions of their conversations on Aug. 6–7, 1945, pp. 70ff. On the later emergence of the myth that the German nuclear scientists had never intended to develop a weapon, see footnote 78 of doc. 115, as well as Walker [1990]a, [1995], chap. 10.

[27]The only published statement by Heisenberg in the years 1945–48 listed in a comprehensive bibliography compiled by Cassidy & Baker (Eds.) [1984] is: Heisenberg [1946].

be confirmed show clearly how near we can come to making the Nazi mistake.[28]

German science was also severely handicapped by the Nazi dogma which insisted on exiling all scholars with a Jewish 'taint' and which made 'non-Aryan' sciences like modern physics unpopular.[29] In the U.S., on the other hand, there has been an increase in the number of physics students since the atomic bomb. Nevertheless as soon as the glamour of the new wears off, this number will in all likelihood decrease. A higher salary level and increased prestige for teachers, together with a popularization, on the high-school level, of the value of pure science would help to remedy this situation.[30]

We cannot go on forever living on borrowed scientific capital from Europe. At present the roster of some of our specialized scientific societies reads like the lineup of a Notre Dame football team. In the future we may not be able to import an Enrico Fermi, whose work was the key to our atomic bomb.[31] The sources from which Europe's great physicists came are dry now and we shall have to produce our own geniuses. We must convince our young people that new ideas are more important to their country and the world than new gadgets, even though the latter may bring in more immediate cash.

Perhaps the most dangerous mistake made by the German scientists was to assume the supremacy of German science. U.S. scientists, far from being conceited about their success in producing the atomic bomb, are impressed primarily by

[28]The confirmation proceedings on the Atomic Energy Commission were beset by problems in clearing security, especially concerning the leftist U.S. nuclear physicist J[ulius]. Robert Oppenheimer (1904–1967): In March 1947 the FBI also discovered that his brother Frank Friedman Oppenheimer had been a member of the Communist Party. Cf. Philip M. Stern, *The Oppenheimer Case: Security on Trial*, New York: Harper & Row, 1969, pp. 100–110, on this episode. Despite this J. Robert Oppenheimer became a member of the General Advisory Committee to the U.S. Atomic Energy Commission as well as its secretary. 1947–52 he was chairman of the General Advisory Committee to the U.S. Atomic Energy Council, and 1945–54 member of the Committee on Atomic Energy of the Research and Development Board. Goudsmit's concern about the general climate of research in the USA and the tight secrecy rules imposed by the military were not groundless, as is demonstrated by the famous Oppenheimer hearings before the American Energy Commission's Personal Security Board in 1954 in the McCarthy era. On this Commission see R. Hewlett & O. Andersson: *The New World. A History of the Atomic Energy Commission 1939/1946*, Berkeley: Univ. of California Press, 1962.

[29]For a survey of the quantitative and qualitative losses within German physics through emigration, see the Introduction, sec. 4, as well as sec. 5 on physics research in Germany.

[30]This sentence reveals Goudsmit's intention to influence postwar American science policy.with this article.

[31]Nevertheless, the USA continued to attract foreign scientists, though not always under such secret circumstances as in 'Operation Paperclip'. Backed by the U.S. president during the cold war era, it arranged for German scientists to work in the USA several of whom would not have obtained a visa otherwise because of their former involvement in Nazi organizations: Cf., e.g., Bower [1990]. The Italian nuclear physicist E. Fermi constructed nuclear reactors at Columbia University in New York in 1941, then after his transfer to the University of Chicago, the first self-sustaining uranium pile in December 1942. He joined the Los Alamos project in 1943 with several other exiled physicists; see, e.g., footnote 11 in doc. 74.

its ultimate simplicity. They believe that any other country that has the raw materials can make a bomb in a few years. The basic principles are well known to any student of modern physics, and the engineering difficulties may have various solutions, some perhaps even simpler than ours.

But if there is little complacency among U.S. scientists, the same thing cannot be said with equal confidence of the general public and certain of their representatives in the government. The belief is far too prevalent that we are way ahead of any possible competition because we have the atomic bomb. Many people have a childish notion of the atomic-bomb 'secret' which we must be careful not to give away. In their minds the secret consists of a formula or a diagram on a piece of paper, to be swallowed when a potential enemy approaches. Confident that we, at least, are safe from an attack by atomic bombs, they would fetter science lest the secret leak out and their night's sleep be severely disturbed. But too much secrecy stifles the progress of science. It also fetters the training of young scientists. If such false secrecy notions are permitted to become general, our scientific unpreparedness could let us in for another surprise attack. Compared with an atomic-bomb attack, Pearl Harbor would seem a mere broken window.

Fig. 31: 'Cubes of Uranium were dug up in a big field near a cave laboratory outside Hechingen. Here members of Alsos stack their first big haul on the grass.'[32]

[32]This photo (on p. 134 of the original) demonstrates the complete ignorance about the hazards of radioactive materials at that time. In the early postwar period U.S. troops were occasionally even commanded to approach the center of A-bomb test-sites soon after detonation to test atomic combat conditions. Concentrated medical observation of the survivers of Hiroshima and Nagasaki and other research yielded new information on the radiation dosage a person can tolerate in a year, monitored with portable dosimeters.

117 Max von Laue: The Wartime Activities of German Scientists [April 1948]

Source: Max von Laue, 'The Wartime Activities of German Scientists.' *Bulletin of the Atomic Scientists*, Volume 4, No. 4, April 1948, p. 103; reprinted in Henning [1992]. Reprinted by permission. [1]

In the *Bulletin of the Atomic Scientists* for December 1947 (p. 365) Philip Morrison, in a review of Goudsmit's book *Alsos*, writes as follows:[[2]]

"The documents cited in *Alsos* prove simply, that no different from their Allied counterparts, the German scientists worked for the military as best their circumstances allowed. But the difference, which it will be never possible to forgive, is that they worked for the cause of Himmler and Auschwitz, for the burners of books and the takers of hostages. The community of science will be long delayed in welcoming the armorers of the Nazis, even if their work was not successful. Men were able to remain aloof from the German war efforts, and brave and good men like Laue and Gentner could resist the Nazis even in the sphere of science. That is a story *Alsos* does not fully tell."

At the time of writing we have not yet seen the book itself. One concludes however from the above quotation that it is the reviewer who explicitly puts forward the monstrous suggestion that German scientists as a body worked for Himmler and Auschwitz. How far Morrison has suffered personally through both or either is unknown to us. We do know that Goudsmit lost not only father and mother, but many near relatives as well, in Auschwitz and other concentration camps.[[3]] We realize fully what unutterable pain the mere word Auschwitz must always evoke in him. But for that very reason we can recognize neither him, nor his reviewer Morrison, as capable of an unbiased judgment of the particular circumstances of the present case. Accordingly a few words in protest may be allowed us here.

---

[1]Max → von Laue is introduced to the journal readers as follows: 'The following communication has been received from Dr. von Laue, co-director of the *Kaiser Wilhelm Institute fuer Physik* (Max Planck Institute). Dr. von Laue is also engaged in the reorganization of the *Physikalisch-technische Reichsanstalt.* A reply from Dr. Morrison and an editorial comment follow.' For Philip Morrison's reply see doc. 118.

[2]For this review see Morrison [1947].

[3]His mother Marianne Gompers Goudsmit ran a millinary shop in The Hague and his father Isaac Goudsmit was a wholesale dealer in bathroom fixtures. At the request of the Dutch physicist Dirk Coster, von Laue wrote a brief letter of recommendation on their behalf but unfortunately forgot to sign it. Heisenberg apparently delayed too long in responding to Coster. His letter of support was dated Feb. 16, 1943: "As Goudsmit has always been very hospitable to us Germans even at a time when anti-Germany trends were clearly perceptible in America (I was his guest as early as in the summer of 1939), I would be very sorry if for reasons unknown to me difficulties arose for his parents in Holland." (On Heisenberg's visit see footnote 20 of doc. 116.) Goudsmit's parents both died on Feb. 11, 1943 (the 70th birthday of one of them) at Auschwitz. Irving (Ed.) [undated], no. 29 1051; Goudsmit Papers finding aid, American Institute of Physics, Niels Bohr Library, p. 4.

Ever since war between civilized states relapsed once more into the old barbaric 'total' war between peoples, it has been no easy matter for an isolated citizen of a warring nation to withdraw himself altogether from war service; indeed, it is relatively unimportant whether he puts his heart into such war service, or whether he is opposed to the methods of his government, or even whether he would depose it. If one or other among the German scientists found it possible during the war to avoid being drawn with his work into the maelstrom, it is not allowable to conclude that it was so for all.

The directors of the larger research institutes in particular were under the absolute necessity of putting the facilities of their institutes at least partially and formally at the service of the war effort. Open refusal on their part, immediately classable as 'sabotage', would have led inexorably to catastrophic consequences to themselves.[4] On the other hand, an (often fictitious) compliance with the demands of the armed forces had advantages which our opponents should recognize as legitimate. In particular one could in this way shield a quite considerable number of the younger specialists from a much more direct mobilization for war; and by this and several other means preserve throughout the war years the foundation on which now, **after** the war, we seek to build afresh.[5]

Sometimes too the possibility arose of protecting political suspects from concentration camps or worse, by assigning them research work of more or less 'military importance'. Our severer critics might perhaps inquire about such cases among non-Aryan Germans (there were a few such in Germany even during the war).[6] Or do they wish to declaim even in these cases about "armorers of Himmler and Auschwitz"?

Nevertheless: whosoever put his institute in this fashion at the disposal of the military authorities, put himself in an ambiguous position. That is the particular curse of such a time: and it visited not Germans only. Thus a foreign colleague,

---

[4]'Sabotage' allegations did play a key role and were taken very seriously, as is demonstrated in docs. 84–85. See also the Göttingen University Lecturers statement, doc. 11, where this accusation is directed against the representatives of 'Aryan' physics.

[5]Indeed, many research projects were successful in obtaining the desired *uk* priority designation (*unabkömmlich*), ranking it as militarily important and thus exempting the scientists from military service.

[6]One example is Gustav → Hertz, who was forced to leave his position at the Berlin Polytechnic because he was classified as a 'second degree part-Jew', but found refuge in industry as the director of the research lab at → *Siemens* in 1935. Under the protective wing of this big electronics company he could remain in Germany and continue his experimental research. Another is the case of the Jewish physicist Richard → Gans. After losing his professorship at Königsberg in 1935 he found temporary employment at → *AEG* from 1936–39 and later at *Telefunken* and other companies. When he was ordered by the 'Jewish Employment Office' (*Amt für Judeneinsatz*) to clear rubble from bombed buildings despite his 63 years of age, a group of physicists including F. → Houtermans, J. H. → Jensen, von Laue, W. → Gerlach, R. → Becker, and Heinz Schmellenmeier, arranged that he be employed in a research project to generate hard X-ray radiation, which they were able to portray successfully as important to the war effort. Cf. Swinne [1992], pp. 105–122.

who during the occupation of his fatherland by Hitler's troops had been particularly active in the underground movement, told me recently that the double dealing to which he was forced thereby lays him open even yet among many of his compatriots to the suspicion of collaboration. This example illustrates how careful one must be in passing judgment on events which took place under a tyranny.[7]

Part of the answer to the question which asks what German scientists were actually about during the war is given in the 50-odd volumes of the *Fiat* Reviews of German Science 1930–46, now in course of publication.[8] There one may find, not perhaps world shaking results, but **honest**, solid scientific investigation, following steadily in the steps of the preceding peacetime research, now once more resumed. Admittedly, the scientific journals lapsed, one after another, towards the end of the war; but not for lack of work submitted for publication, but because of paper shortages, bomb damage to printing houses, and other economic strictures. In the case of the *Zeitschrift für Physik*,[9] for instance, where I know the circumstances at first hand, there were 60 manuscripts awaiting publication at the end of the war; since when the editors could have accepted 86 more, which still deal, to a considerable extent, with work carried out during the war years. But nothing in their contents points to that fact; and they have verily nothing whatsoever to do with Himmler and Auschwitz.

What do the 'Atomic Scientists' aim at in their *Bulletin*? So far as I understand it, the attainment of a durable peace. But that high aim is ill served by articles such as the review from which we have quoted; they keep alive hate, on the elimination of which **everything** hinges. We recommend as the foundation of every utterance of peace politics, in great and small things alike, the words which Sophocles puts in the mouth of Antigone, citizeness of a victorious state: "To league with love not hatred was I born".[10]

---

[7]See also Lise → Meitner's comment on the responsibility of scientists in doc. 120.

[8]The FIAT Reviews of German Science 1930–46 were published in the late 1940's; see, e.g., Bothe & Flügge (Eds.) [1948].

[9]The → *Zeitschrift für Physik* was published by the German Physical Society (→ *DPG*).

[10]In Sophocles's play *Antigone*, Creon the king of Thebes denies his slain nephew Polyneices the rites of burial as revenge for having attacked the city to wrest control from his brother Eteocles, who was also slain in the battle. Antigone chooses to bury her brother in obedience to the higher law of the gods. The above refers to the line that appears in her courageous confrontation with the king (line 564).

118 Philip Morrison: A Reply to Dr. von Laue [April 1948]

Source: Philip Morrison,[1] 'A Reply to Dr. von Laue.' *Bulletin of the Atomic Scientists*, Volume 4, No. 4, April, 1948, p. 104. Reprinted by permission. © Educational Foundation for Nuclear Science, Chicago.[2]

It is with regret that I must take issue with the moving statement of Professor von Laue, and with his implied defense of those scientists who worked for the *Waffenamt*, the *Wehrmacht*, and the *Luftwaffe*.[3] I said in the review of which he writes that these men worked, not for Himmler, but for Himmler's cause, the victory of a National Socialist Germany. Where this is true, I feel that my indictment must still stand. Where laboratory work was in fact a cover for underground activity, where it was meant chiefly to save the enemies of the Nazis from imprisonment, or even when it was carried out in a way remote from the war, by men—perhaps they were the majority of *Zeitschrift* contributors—who tried "to remain aloof from the German war effort", no such criticism applies, nor was it stated or intended. I do not believe that there is an unwarranted subtlety in such a distinction.

One point remains at issue. What of those men, "directors of the larger research institutes in particular", says Professor von Laue, who complied with the requests of the Nazi armed forces? Perhaps their compliance was in some cases fictitious: here too there would be no issue. But many of the most able and distinguished men of German science, moved doubtless by sentiments of national loyalty, by traditional response to the authority over them, and by simple fear, worked for the advantage of the Nazi state. These men **were** in fact the armorers of the Nazis. Professor von Laue, as the world knows and admires, was not among them. It is not for the reviewer to judge how great was their peril; it is certainly not for him to imply that he could have been braver or wiser than they. But it was sentiments like theirs, weakness like theirs, and fear like theirs which helped bring Germans for a decade to be the slaves of an inhuman tyranny, which has

---

[1]Philip Morrison (1915–) took his doctorate in 1940 at the University of California at Berkeley. His thesis, written under the guidance of J. Robert Oppenheimer, was on gamma rays and energy fluctuations in an electromagnetic field. Soon afterwards he became a member of the Manhattan Project, first at the Metallurgical Laboratory of the University of Chicago, then at Los Alamos. After witnessing the destruction caused by atomic weapons (he took a reconnaissance flight above Hiroshima in August 1945), he turned to advocating international controls on atomic research. For more on Morrison, who later taught at Cornell University and MIT see, e.g., *Current Biography* 1981, pp. 308–311.

[2]This is in reply to the preceding document, doc. 117.

[3]The controversial issue of the collaboration of scientists with Nazi military research departments, the → *HWA*, the German Armed Forces and *Luftwaffe* despite their private rejection of the Nazi ideology see, e.g., Mehrtens in Meinel & Voswinckel (Eds.) [1994]. See also the editor's comment by Eugene Rabinowitch, directly following this article in the *Bulletin* on pp. 104–105, and Lise → Meitner's letter discussing the responsibility of scientists, doc. 120.

wrecked Europe, and in its day attacked the very name of culture. Are we to forget the tragic failure of those German men of learning?

Professor von Laue, whose outspoken opposition under Hitler was a token of his wisdom and integrity, begins his piece with a most uncharacteristic reference *ad hominem*. He wonders whether I suffered personally through Himmler and Auschwitz. I did not. But I do not see that it is fair or relevant to ask. I am of the opinion that it is not Professor Goudsmit who cannot be unbiased, not he who most surely should feel an unutterable pain when the word Auschwitz is mentioned, but many a famous German physicist in Göttingen today,[4] many a man of insight and of responsibility, who could live for a decade in the Third Reich, and never once risk his position of comfort and authority in real opposition to the men who could build that infamous place of death.

119 Samuel A. Goudsmit: Our Task in Germany [April 1948]

Source: Samuel A. Goudsmit, 'Our Task in Germany.' *Bulletin of the Atomic Scientists*, Volume 4, No. 4, April, 1948, p. 106. Reprinted by permission. © Educational Foundation for Nuclear Science, Chicago.[1]

Science has always benefited by world-wide cooperation. The work of each individual scientist depends upon knowing what others in his field have done or are doing. This pooling of knowledge has made research workers acutely aware of the futility of national barriers in the fields of science. It may well serve as an example of true international cooperation, based on a realistic approach to the advantages for all and not on idealistic pipe dreams. The continuous exchange of scientific information, though inspired by necessity, provides also a basis for better understanding and real friendship among men of science.

It is with this in mind that we must consider renewing our relations with

[4]The Göttingen physicists referred to here include W. → Heisenberg and Otto → Hahn. Heisenberg had been instrumental in the relocation of the Kaiser Wilhelm Institute of Physics (→ *KWIP*), which he directed 1946–58, to Göttingen, one of the few German university towns physically unscathed by the war. In the British occupied zone there were significantly less restrictions, and O. Hahn was thus able to assume the direction of the Kaiser Wilhelm Society (→ *KWG*) which was re-established in Göttingen and renamed the Max Planck Society.

[1]Introductory sentence to the article: 'Dr. Goudsmit is deeply convinced that only a more positive concern on the part of American scientists can encourage a belief in democracy among their German colleagues. These remarks were written independently of the discussion on pages 103–105 that stems from a review of ALSOS, Dr. Goudsmit's story of German Wartime Science. Formerly, a member of the Physics Departments at the University of Michigan and at Northwestern, Dr. Goudsmit has recently joined the staff of the Brookhaven National Laboratory.' See the exchange between Philip Morrison and M. von → Laue, docs. 117–118. S. → Goudsmit became full professor at the University of Michigan in 1932; his professorship at Northwestern University was between 1946–48, after which he took on a position as senior scientist at the Physics Department in Brookhaven in 1948.

German scientists. It would be understandable if many among us were reluctant to converse with our German colleagues again as if nothing had happened. They supported the Hitler crime against humanity, or at least did not do much against it at the earlier times when opposition was still possible. Many of these Germans did not see the evils of a totalitarian regime until its dogmas interfered with the teachings of science. They believe that they did their full share by merely having resisted this encroachment in their domain of learning.

The attitude of most of our colleagues over here is, however, not due to considerations of this kind, but to extreme indifference. While this avoids the necessity of deciding how to confront former enemies, it is a dangerous indication of complacency.

It is true that pure science in Germany under Hitler declined rapidly in quality and quantity and has little to offer compared to the leading role it played in the 1920's.[2] In fact the few worthwhile contributions came from scientists already prominent in that pre-Hitler period; almost no new talent was added under the Nazi regime. Though we can gain almost nothing from German scientific work and only little from a few technological developments,[3] an attitude of indifference on our part will be detrimental in the long run.

The United States' foreign policy should determine just how to deal with the problem of German science and scientists. In a democracy the foreign policy is meant to reflect the will of the people and thus American scientists should inspire their Government's handling of science in occupied territories. Our present unconcern has certainly no constructive influence.

The Kind of Program That Must Be Developed

As scientists we could help create a constructive policy of promoting democratic ideals among our German colleagues.

Due to our apathy the situation in German scientific circles seems to be deteriorating rapidly. It is almost impossible to obtain factual and reliable information from there, which in itself is a bad sign. Casual visitors, official or unofficial, cannot penetrate beneath the surface except for those few who knew the German colleagues well before the war. Communication is slow and inadequate. But the few facts that reach us paint a gloomy picture. Several foes of democracy or advocates of revenge have regained prominent educational positions.[4] "Those who

[2]On highlights of science during the Weimar period and the great importance of scientists of Jewish descent in these years see, in particular, Nachmansohn [1979].

[3]Goudsmit is downplaying the importance and usefulness of a great number of German engineers and scientists who were transferred to the USA soon after the end of World War II, sometimes even concealing evidence for their involvement into Nazi crimes. On Operation Paperclip and its Russian counterpart, see also the Introduction, p. xcii, and footnote 335 there.

[4]For example, the physicist Johannes → Juilfs, who had been an *SS* member and therefore was barred from a position in the German Civil Service after 1946, worked in industry for a while, but then gained academic positions at the Hanover Polytechnic, advancing to full

would have dared to speak up are silent forever. Those who still can speak up are again afraid to open their mouths," a letter informs me.

It is our duty to do whatever we can to improve this situation. What is obviously needed is a more active interest on the part of our scientists in the events that take place in German academic circles. Only then can we prevent, perhaps, a return to power of the undesirables.

Though it is too readily believed that Hitler was the only advocate of totalitarian excesses, it is nevertheless a fact that he had strong supporters among our German colleagues.[5] It is, of course, impossible with a few exceptions, to decide who were active Nazis and who were not. But fortunately there are many German scientists with some understanding of the meaning of democracy and its value. Moreover the problem of de-Nazification if soluble at all, is entirely an internal German affair.[6]

We need a positive program to guide events in German scientific circles in a more favorable direction. This means that we must morally support those German colleagues in whose integrity we have confidence. There are many of them. We do not have to agree with all their opinions and should make allowances for the disturbing circumstances under which they have lived and are still living. We must again communicate with them as in the days before Hitler. The exchange of scientific literature, now practically at a standstill due to our indifference, should be actively promoted.[7] German scientific research should again be helped by grants and supplies.[8] By thus supporting the right persons, institutions or pro-

---

professor in the 1950's through the support of Heisenberg and others, who certified that he had helped them in their fight against the *NSDAP*'s efforts to politically streamline physics. Cf., e.g., Walker [1989]a, p. 199. For another example, see footnote 332 of the Introduction.

[5]On the attitude of German professoriate towards Hitler, cf. the Introduction, p. xxiv and the secondary literature mentioned in footnote 39.

[6]The denazification procedure was far from straightforward and hardly successful. In October 1947, Military Governor Clay reluctantly agreed to accept the first amendment to the Liberation Law, which provided that the majority of low-level officials were exempted from the occupation ban through a 'financial absolution' and also allowed suspected Nazi activists to be charged as fellow travellers (i.e., nominal supporters). Then on Mar. 25, 1948 the second amendment allowed the conversion of all the 'incriminated' individuals (group 2) in accelerated proceedings to fellow travelers and limited the occupation ban to convicted major offenders: Vollnhals [1991], pp. 22f. for the abrupt change in American procedure around 1948.

[7]Before World War II, many institutions of higher learning exchanged publications, thus furthering the strong trend towards an internationalization of science. Interrupted by the Nazi regime, this tradition was resumed, however, soon after the founding of the Federal Republic of Germany in 1949. Moreover, a great many duplicate copies from Americal university and college libraries were sent to Germany as part of the Marshall Plan to rebuild Germany. On international scientific exchange and its implications see also Goudsmit [1946].

[8]American institutions such as the Rockefeller Foundation had supported European scientific research before the war to a considerable extent: cf., e.g., Macrakis [1986] for one example related to a physics institution. See also Fosdick [1989], docs. 30, 38, and the last paragraph of doc. 9. After the war they took up this task again; cf., e.g., the periodicals *Rockefeller Foundation Grants*: 1950-64, *Rockefeller Foundation Quarterly Report*: (since 1964), and the *Annual Report*

grams we might finally create an atmosphere of mutual understanding, beneficial to the spread of democratic ideals and practices.

We cannot wait much longer. Many foes of the Nazi regime have been bitterly disappointed by postwar events in Germany.[9] Former Nazis and Communists are quick at makin[g] use of this feeling of discontent. If we as American scientists take a greater interest in this problem, the proper U.S. Government agencies may adopt a policy of action more in line with our recommendations.

120 Lise Meitner: Letter to Otto Hahn [June 6, 1948]

Source: Max Planck Society Archives, Berlin; carbon copy at Churchill College, Cambridge, England. Published in: Krafft [1981], pp. 185–186 (see footnotes 50–51 there).

Dear Otto,

[...] Yesterday I heard here from a reliable source that Mattauch is coming to Sweden for 2 months, ostensibly to study [current scientific] literature, probably also to advise Siegbahn on the construction of a mass spectrograph.[1] I cannot quite understand Mattauch's attitude. Research material would surely be at least as much available to him through Scherrer as through Siegbahn, and the Swedish climate even in the summer is not particularly favorable to someone suffering from Tb.[2] The number of persons suffering from tuberculosis in Sweden is much higher than in countries in the West. But this is Mattauch's business, of course.

In any case, I do not think I can accept the position in Mainz.[3] I have few worries about the unfavorable living conditions but very serious misgivings about the mental attitude there. Whenever I would happen to be of a different opinion to my colleagues—outside the realm of physics—they would surely respond with

(1967–1985).

[9]For some examples, see, e.g., Gerhardt [1948] on the → *IG-Farben* trial, or von Laue [1949] on the brain drain of German scientists to the USA.

[1]Josef → Mattauch succeeded L. → Meitner at the Kaiser Wilhelm Institute of Chemistry (→ *KWIC* in Dahlem (Berlin) and became in addition associate professor of nuclear chemistry at the University of Berlin in 1940. In 1948, Mattauch had just become guest professor at the University of Berne. [Karl] Manne Georg Siegbahn (1886–1978) was a specialist in X-ray spectroscopy, professor at the University of Stockholm and first director of the Nobel Institute of Physics in Stockholm, where Lise Meitner had been working since her emigration to Sweden in 1938. On Siegbahn, see also footnote 5 in doc. 51, and Atterling's obituary in *Biographical Memoirs of Fellows of the Royal Society* **37** [1991], pp. 429–444.

[2]Meitner's close friend, Paul [Hermann] Scherrer (1890–1969), was full professor of experimental physics at the the Zurich Polytechnic 1920–60. Together with P. → Debye he developed X-ray analysis of polycrystalline materials. Later, he also worked on cosmic-ray and nuclear physics. Because Scherrer worked in Switzerland, which was neutral during World War II, his institute was fairly well-off compared to most institutions in Germany or in Allied countries.

[3]Mattauch and Fritz → Straßmann, who were both employed at the University of Mainz at that time, actively endorsed Meitner's appointment as head of the physics department there; cf., e.g., Krafft [1981], pp. 180ff., 399ff., and Kerner [1986], pp. 112f.

the words: Naturally she doesn't understand German circumstances, because she is an Austrian, or because she is of Jewish decent. I also mentioned these misgivings to Straßmann;[4] and he only answered by repeating his assertion of how indispensable I would be for the Institute. Therefore he did not dare to allay my doubts. This means that I would not be able to count on the younger colleagues' trust that I had once enjoyed and that in my opinion is always—and nowadays in particular—the most important basis for good collaboration. [...]

It would be a similar struggle to the one I had waged with very little success in the years [19]33–38—and it is very clear to me today that I committed a big moral injustice by not leaving in [19]33, since in the end, by staying I supported Hitlerism after all. These moral reservations do not exist today, of course; but with the general attitude there, my personal situation would not be very different to the one at that time, and I would not really win the confidence of my colleagues and could therefore not really be effective.

I regretted very much not having read Laue's article (I mean the article in the *Bulletin of the Atomic Scientists*)[5] before I came to Göttingen.[6] I would have liked to discuss the article with him, because from what I know of the mentality abroad, contrary to Laue's intention and hope, this article will be no use. There are, as you know, too many proofs of how many scientists went along with Hitler (with or without conviction). When Einstein warned against the fatal consequences of Hitlerism in 1933 while he was in America, he was immediately expulsed from the Academy of Sciences (before April 1933).[7] This surely could not have been carried out by a minority of the Academy. Almost simultaneously, Leux and another chemist submitted the proposal that Einstein be stripped of his honorary membership in the Chem[ical] Society, which also can only be decided by a majority of the board of trustees.[8] When Franck justified his resignation from his professorship in a newspaper article with the statement that his children were being barred from being able to "prove themselves worthy Germans", 42 university lecturers at Göttingen University published a proclamation against

[4] Straßmann had been working on Meitner's team at the *KWIC* since 1935 and replaced her in 1938. Soon afterwards, he and O. → Hahn discovered nuclear fission. He became full professor and director of the Institute of Inorganic Chemistry and Nuclear Chemistry at the University of Mainz in 1946. On Straßmann and his collaboration with Hahn and Meitner, see Krafft [1978]b, [1981].

[5] This is Max → von Laue's response to Morrison's review of Goudsmit [1947]; see here doc. 117. See also Laue's article: 'The World Needs New Ethical Standards' in *Bulletin of the Atomic Scientists* 4 [1948], pp. 337f.

[6] Meitner visited Göttingen in April 1948 to attend the session in commemoration of M. → Planck, who had died on Oct. 4, 1947.

[7] On A. → Einstein's withdrawal of his membership in the *Preußische Akademie der Wissenschaften*, see Kirsten & Treder (Eds.) [1979], and here doc. 6.

[8] The organic chemist Friedrich Hermann Leuchs (1879–1945) wrote his Ph.D. thesis under the guidance of Emil Fischer and submitted it to the University of Berlin in 1902. After his habilitation at the same university in 1910, he became associate professor in 1914, department head of the chemical institute in 1916, and was full professor at the University of Berlin 1935–45.

him, that "he is sabotaging the Third Reich".[9] I surely do not need to remind you of Bi[e]berbach's lecture that he held directly after the boycott of the 1st of April [1933] on Germanic and Jewish mathematics.[10] Furthermore, I recommend that Laue read the Berlin Academy's *Forschungen und Fortschritte* for the years 1939–45 and earlier and that he take into account the Professors' Tribute to Hitler and his intellectual position (did he have such a thing at all?).[11] Is it really justifiable to say that the majority of scientists were against Hitler from the beginning? [...] When Planck held the obsequies for Haber, Laue and Heubner were the only professors to dare to come to it:[12] At the same time the Chemical Society and the Glass Engineering Society [*Glastechnische Gesellschaft*] (others do not come definitely to mind and I therefore do not want to make any contentions) had forbidden their members from attending the ceremony.[13] Doesn't this all show that the subordination to the Hitler ideas was very prevalent and that the opposition—aside from the fact that it kept itself very much in the background publicly from the beginning—was a minority? I truly do not have the intention of saying unpleasant things with these observations, but I am afraid that with his inclination to defend everything that has happened—out of an understandable attachment to Germany—Laue is not helping Germany but risks achieving the opposite. Since I know that he is going to America very soon, I think it important to make him aware of this danger. That is why I would be very grateful to you if you would give him this letter to read.

Now I hope to hear good news about all of you soon, and I also hope that you understand all that I have written the way it was meant. [...]

Very affectionate greetings also to Edith, the Laues and to Marga Planck as well as to the Frasers,[14]

Yours Lise.

---

[9]For J. → Franck's resignation, see doc. 9, and the lecturers' protest, doc. 11.

[10]Ludwig Bieberbach (1886–1982), full professor of mathematics at the University of Berlin since 1921, was one of the main protagonists of 'Aryan mathematics'; cf., e.g., Bieberbach [1934]. On Bieberbach and 'Aryan mathematics', see also Lindner [1980], Mehrtens [1987], [1989].

[11]The periodical *Forschungen und Fortschritte* was a publication of the Reich Research Council (→ *RFR*) in Berlin. Meitner alludes to the following tribute to Hitler: *Bekenntnis der Professoren an den deutschen Universitäten und Hochschulen zu Adolf Hitler und dem nationalsozialstischen Staat*, Dresden, 1934, containing several hundred declarations of support by professors throughout Germany.

[12]For Laue's article commemorating Fritz → Haber, see doc. 29. Wolfgang [Otto Leonhard] Heubner (1877–1957) was a pharmacologist and full professor at the University of Berlin 1932–45 and at the *Freie Universität* in Berlin, 1949–53.

[13]According to Krafft [1981], footnote 51, the Reich Association of Technical and Scientific Occupations (*Reichsgemeinschaft der Technisch-Wissenschaftlichen Arbeit*) published a directive in a letter by its president dated Jan. 24, 1935, "regarding the commemoration ceremony for Fritz Haber which is fixed for the 29th of January of this year" forbidding all member organizations of the association to attend. A similar notice was sent out accordingly to the German Chemists Society (*Gesellschaft deutscher Chemiker*), for example, on Jan. 25, 1935; see also D. Hahn (Ed.) [1979], p. 140; Hahn [1960].

[14]On Edith Hahn see footnote 5 of doc. 5. Max von Laue and Magdalene Degen Laue (1891–

121 Walther Gerlach: Affidavit on Rudolf Mentzel [December 13, 1948]

Source: Deutsches Museum, Munich: Gerlach Papers, 'Evaluations 1946–1955' binder. Unsigned typed carbon copy. 'G/D' typed in right upper corner next to the date. Published with the kind permission of Dr. med. Helgard Krietsch (Munich).

Affidavit

In response to a request directed to me, I make the following declaration upon oath in support of Professor Rudolf Mentzel:[1]

The first closer contacts I had with Professor Mentzel that I can recall took place around 1936 on the occasion of my call to Berlin. I was aware that my departure would have been welcomed at that time from certain quarters.[2] But Professor Mentzel acknowledged my wish to remain in Munich for professional reasons and decided accordingly without attempting to exercise any pressure whatsoever on me.

In the years that followed as member of the Kaiser Wilhelm Society's board I met repeatedly with Professor Mentzel, since he also was on the same board.[3] At the meetings I cannot remember having ever heard him mention any political views.[4] On the contrary, he was greatly interested in the concerns of the Kaiser Wilhelm Society. I know from statements by Privy Councillor Bosch, president at that time, that he could work confidentially with Professor Mentzel, and I know that this was also the opinion of Privy Councillor Bosch's successor, Dr.

1961), who were married in 1910 in Munich, resided in Göttingen 1946–51, where the → *KWG*, renamed the Max Planck Society (*Max-Planck-Gesellschaft*), had relocated in 1948. Marga Planck (maiden name, Magda von Hoesslin, † 1949) was Max → Planck's second wife, whom he had married in 1911 after the death in 1909 of his first wife Marie Merck (married to Planck in 1887). The Plancks moved to relatives in Göttingen in 1945. See Meitner's correspondence after 1946 with the British physicist Ronald Fraser in Göttingen, who represented the Military Government from Oct. 1946: Meitner Papers, Churchill College, Cambridge.

[1]Rudolf → Mentzel had requested this affidavit in a letter dated Dec. 6, 1948, also on file at DMM, Gerlach Papers.

[2]Walther → Gerlach, then successor to Wilhelm Wien in the chair for experimental physics at Munich University since 1929, had thwarted attempts by Rudolf → Tomaschek and others to establish 'Aryan' physics at the department, and as chairman of the appointment committee had resolutely opposed Wilhelm → Müller's appointment to Arnold Sommerfeld's chair for theoretical physics: cf. doc. 84, footnote 9 and doc. 85, p. 265.

[3]Gerlach was elected to the supervisory board (*Senat*) of the → *KWG* by the general assembly and confirmed by the minister. Mentzel was also member of the society's advisory board from 1937 through a supplementary nomination: See Vierhaus & vom Brocke (Eds.) [1990], pp. 405f.

[4]With this phrasing Gerlach does not exclude that political reasons might nevertheless have played a certain role for Mentzel, although he was careful enough not to mention them at the *KWG* meeting, where they would not have been well received (see also footnote 12).

Vögler.[5] I particularly remember a private conversation in which I described to Dr. Vögler charges and complaints by colleagues about biased influences at universities. Dr. Vögler later included Professor Mentzel in these discussions; all the circumstances could be discussed frankly in his presence: – On the proposal to appoint Professor Heisenberg as director of the Kaiser Wilhelm Institute of Physics, Professor Mentzel backed the position of the Society's supervisory board fully, despite the ongoing and even public controversy about him.[6]

From the years 1938–1940 I recall numerous statements by Professor Mentzel in which he took a clear stance against the attacks by the NS University Lecturers League at universities. Professor Mentzel expressed to me his disapproval about the appointment of Professor Müller as successor to Privy Councillor Sommerfeld which had such unpleasant consequences for Munich. I remember almost verbatim a reply to my complaint about Müller's complete incompetency: The students should not put up with it; it should be shown with this case what the consequences are of the University Lecturers League's meddling. These statements by Professor Mentzel and the fact that in the end he was responsible for Professor Müller's appointment led me to the conclusion (and I still consider it correct today) that Professor Mentzel had to fight against a number of forces whose disastrous influence on science he was not in the position to avert.[7]

When Professor Mentzel invited me to take over the direction of the physical science section at the German Scientific Research Association (Reich Research Council),[8] I was astonished because Professor Mentzel knew very well that I was not a member of the *NSDAP* and that many charges and complaints about me had been filed which had, of course, also led to some investigations.[9] I was even more astonished that I was to take the position held until then by Councillor of State Esau.[10] Professor Mentzel replied to my reservations that he was only concerned about professional competence. Professor Mentzel did not follow through on a complaint against me submitted by an agency of the Security Service immediately after I assumed the position, and immediately informed me of it openly. In all

[5]Carl → Bosch was president of the society 1937–40, and Albert → Vögler 1941–45.

[6]On these controversies, see here docs. 42–44, 55ff., 62ff., and 73, as well as Cassidy [1992].

[7]Considering that Mentzel did not side with the party ideologues and the → *NSDDB*, but rather with pragmatists like Göring and many in the military who sought efficiency as a priority in science, Gerlach's statements above are not implausible. For Mentzel's own contemporary views on science policy as president of the *DFG*, cf. Mentzel [1940].

[8]This took place in November and December of 1943; his official appointment as head of the physics section of the Reich Research council (→ *RFR*), which was subordinate to the → *DFG*, was on Jan. 1, 1944. Cf., e.g., Walker [1989]a, chap. 4. On the *RFR*, founded in 1937 and reorganized in 1942, see the Introduction, sec. 5.1, and doc. 52.

[9]On Gerlach's conflicts with the Nazi establishment see Heinrich & Bachmann [1989], pp. 78ff.

[10]Abraham → Esau had been appointed the powerful position of coordinating German nuclear research on Dec. 8, 1942 but resigned the following year in a letter dated Oct. 28, 1943. Gerlach replaced him in December and was officially named Göring's plenipotentiary of nuclear physics on Jan. 1, 1944.

cases I am informed about, as managing director of the Reich Research Council Professor Mentzel sought professionals most respected by their colleagues and willing and able to assume the requisite administrative work, such as Sauerbruch for medicine, Kuhn for chemistry, Marx for electrical engineering, and Köster for physical metallurgy.[11]

As head of physical science at the German Scientific Research Association (Reich Research Council) I had frequent meetings with Professor Mentzel both on professional questions and personal affairs. Views on party membership or the like never arose on such occasions.[12] As head of physical science I was given complete freedom. Professor Mentzel did not influence the distribution of funds, the conferral of research projects or the approval of grants. The sums of money requested by me were granted and upon expenditure were deducted on the due dates indicated. When Professor Mentzel's signature was required for contracts, he countersigned on the basis of a statement that the matter was in proper order. I have no knowledge that Professor Mentzel supported enterprises against the advice of the specialists. It certainly did not happen in my sphere. On the contrary, Mentzel backed my denial of support for a research institute established by Dr. Ley,[13] which I had rejected because I had been given no information on it, likewise regarding support for fraudulent research ventures, for example.

Since I know Professor Mentzel's stance on many issues not only through

[11]The four individuals mentioned are the famous surgeon Ernst Ferdinand Sauerbruch (1875–1951), who practised in Munich from 1918 and in Berlin from 1928; Richard Kuhn (1900–1967), a student of R. Willstätter, from 1929 director of the chemistry department at the Kaiser Wilhelm Institute of Medical Research and honorary professor at Heidelberg University; Erwin Marx (1893–1980), professor of electrical engineering at the Braunschweig Polytechnic (cf. Maier [1993]); and Werner Köster (1896–1989), from 1934 full professor of physical metallurgy at the Stuttgart Polytechnic and director of the Kaiser Wilhelm Institute of Metallurgical Research.

[12]A former full professor of technical chemistry at the Berlin Polytechnic in Charlottenburg, Leo Ubbelohde (1876–1964) filed a complaint that Mentzel had given party members preferential treatment. As a result Mentzel's rating as fellow traveller (category 4) was retracted and denazification proceedings were initiated, eliciting this correspondence. While Gerlach's exonerating statement may well be true, it does not preclude that Mentzel made no politically motivated decisions. This affidavit thus suggests a much stronger implication without actually saying so—a typical tactic used in such personal evaluations which frequently had the effect of whitewashing past involvement and came to be known as *Persilscheine* after a brandname washing powder.

[13]The chemist Robert Ley (1890–1945) obtained a doctorate and passed the state examination for food chemists. His successful military career, starting as a volunteer in the artillery, then in 1917 in the Air Force led to his serving as lieutenant at Verdun. In 1921 he was employed as a chemist at → *IG Farben*. Ley joined the outlawed party in March 1924 and became regional leader of the Rhineland in 1925. On Jan. 1, 1928 he was sacked from *IG Farben* for his political activities, but elected to the Prussian Legislature in the same year and to the *Reichstag* in 1930. He was regional leader in the Cologne district 1931–34 and appointed to the Munich party headquarters in 1932. On May 2, 1933 Ley was placed in charge of coordinating the workers unions and became head of the German Workers Front. He committed suicide while in Allied custody.

our collaboration at the German Scientific Research Association but also on the supervisory board of the Kaiser Wilhelm Society and can attest at least in all these incidences that he relied on the best advice of specialists, I consider it out of the question that he supported experiments that could be decried as inhumane.[14] I was in constant close contact with many heads of the separate divisions at the Reich Research Council, particularly with the above-named gentlemen and know from them that Professor Mentzel always accepted the judgment of these specialty heads and did not demand more detailed specific justifications for their requests and applications.

Obviously, in discussions with Professor Mentzel there were also differences of opinion between us. But these were always debated in a factual manner, and very specifically without Professor Mentzel's making use of political points of view. From this I must conclude that Professor Mentzel exhibited the same comportment in all cases and did not allow himself to be guided by biased motives.

I declare in lieu of an oath that I am not related to Professor Mentzel either by blood or through marriage and that I was not a member of either the Party or one of its branches. My qualifications have been confirmed by the English military government under – 210/EDN/6 – of Feb. 8, 1946, by the American military government in Bavaria under – AG 210. I MGBAE – of Apr. 19, 1946.[15]

(Professor Walther Gerlach)

[14]Such experiments were particularly prevalent in the fields of biological and medical research during the Third Reich; see Müller-Hill [1985], and Deichmann [1992] on biological experiments conducted by *KWG* scientists, and Lifton [1986] on experiments, e.g., by the → *Ahnenerbe*, performed on concentration camp prisoners. Cf. also Poliakov & Wulf (Eds.) [1959/83]b, pp. 321f. on the use of such prisoners to assist in calculation work and other routine tasks in basic research.

[15]This closing paragraph was part of the prescribed procedure for such exoneration attestations. They could only be issued by persons who had themselves gone through the denazification procedure and had had no affiliation with the → *NSDAP*. On the denazification of Germany see in particular Niethammer [1982], Vollnhals [1991], and the Introduction, sec. 5.5.

Appendix

A Professional Institutions and Associations

AVA: Aerodynamische Versuchsanstalt (see under → *KWIS*)

DFG: Deutsche Forschungs-Gemeinschaft (previously → *NG*)
German Scientific Research Association. Full title: *Deutsche Gemeinschaft zur Erhaltung und Förderung der Forschung* (German Association for the Support and Advancement of Scientific Research). Successor organization to the → *NG*, which was renamed *DFG* unofficially since about 1929 and officially in 1937. During the terms of its presidents: J. → Stark (June 1934–36); R. → Mentzel (Nov. 1936–39) and A. → Esau (1939–45), the *DFG* also had a dominant influence on the research policy of the → *RFR*. It was funded by government grants in the millions and smaller contributions by the → *Stifterverband*. Refs.: Mentzel [1940], Stark [1943]c, Zierold [1968], Nipperdey & Schmugge [1970].

DGtP: Deutsche Gesellschaft für technische Physik
German Society of Technical Physics. Founded on June 6, 1919 by Georg Gehlhoff as an alternative to the → *DPG* with a total of 13 local associations and its own journal → *Zeitschrift für technische Physik*. Around 1924 the *DGtP* had approximately 3,000 members, thus somewhat more than the *DPG*, but membership fell by 1945 to around 1,500. Chairmen: G. Gehlhoff (1920–31); K. → Mey (1931–45). Refs.: Gehlhoff et al. [1920], Ludwig [1974], Richter [1977], Peschel (Ed.) [1991], chap. 1, Heinicke [1985], p. 43, Hoffmann & Swinne [1994].

DPG: Deutsche Physikalische Gesellschaft
German Physical Society. Founded in 1899 a national organization at to succeed the Berlin Physical Society, which dates back to 1845. The Society issued regular biweekly proceedings, reports (*Berichte*) on the same, as well as the journal: *Fortschritte der Physik* (since 1845). Other publication series of the *DPG* include: → *Verhandlungen der DPG*; → *Zeitschrift für Physik*; → *Physikalische Blätter*. Treasurers: Walter Schottky (1928–39); M. → Steenbeck (1939–44). Secretaries: Karl → Scheel (1918–36); Walter Grotrian (1936–45). Chairmen include: F. → Paschen (1925–27); M. → von Laue (1931–33); K. → Mey (1933–35); J. → Zenneck (1935–37 and 1939–40); P. → Debye (1937–39); C. → Ramsauer (1940–45). Ramsauer and his deputy, W. → Finkelnburg, steered a relatively independent course from the party line and against the *Deutsche Physik* movement, and were supported, among others, by M. → Wien and L. → Prandtl. However at the end of 1938, on the initiative of H. → Stuart and W. → Orthmann, all Jews were asked to withdraw their membership in the Society. Around 1918 membership totalled about 750 and in the 1930's about 1,400. The 'Max Planck Medal' of the *DPG* was awarded to, among others: M. von Laue (1932); W. → Heisenberg (1933); E. → Schrödinger (1937/38); P. → Jordan (1942); F. → Hund (1943); M. → Born (1948), L. → Meitner (1949); P. → Debye (1950), J. → Franck & G. → Hertz (1951); W. → Bothe (1953), W. → Pauli (1958). After the war the Society was initially re-established in the individual occupation zones and united in 1950 in the West. Refs.: Warburg [1925], Ramsauer [1947]a, Peschel (Ed.) [1991], chap. 1, Heinicke [1985] and Mayer-Kuckuck (Ed.) [1995].

HG: Helmholtz-Gesellschaft
Helmholtz Society. Founded in 1920 as an alternative to the → *NG* following an appeal to business by the physics and engineering research community. Contributors to the Society were mainly from the the coal, steel and chemical industries, as well as banks, and to a lesser degree

electric companies. As was the case with the *NG*, the Society was autonomously organized with advisory boards to decide on applications in physics and applied physics, particularly in materials testing. Around 1923 the governing board included: P. → Lenard; J. → Stark; W. → Wien. After 1933 the *HG* managed to remain relatively independent of the State. Refs.: Richter [1977].

KWG: Kaiser-Wilhelm-Gesellschaft zur Förderung der Wissenschaften e.V.
Kaiser Wilhelm Society for the Advancement of the Sciences, registered association (From Sep. 1946: *Max-Planck-Gesellschaft zur Förderung der Wissenschaften*). Founded in 1911, based upon the plans drawn up by von Althoff and Schmidt-Ott and on the instigation of Adolf von Harnack "to promote the sciences, especially by founding and maintaining scientific research institutions". Its establishment of research institutes that were independent of the State was most important (→ *KWIP*, → *KWIC*, → *KWIPC*, and the Aerodynamics Experimental Station (→ *KWIS*)) whose researches were conducted under the guidance of prominent directors (W. → Bothe, P. → Debye, A. → Einstein, F. → Haber, O. → Hahn, P. → Thiessen) and under the supervision of a professionally competent board of trustees. In June 1937 the *KWG* was formally reorganized, adopting new Articles implementing the *Führerprinzip* which primarily strengthened the role of its president but also ceded power to the → *REM*. Presidents of the *KWG*: Adolf von Harnack (raised to the nobility in 1914: 1911–29); M. → Planck (1930–37); C. → Bosch (1937–40); A. → Vögler (1941–45); M. Planck (Jul. 24, 1945–Mar. 31, 1946); O. Hahn (Apr. 1946–59). Supervisory board (*Senat*) composed of members from the fields of science, finance, industry and politics. The managing director Friedrich Glum and the President Planck initiated the voluntary realignment (*Selbstgleichschaltung*) of the Society using their own definition with the intention of thus circumventing more radical intervention by the Nazis shortly after their rise to power in 1933–34. After 1933: Dismissal of Jewish staff members at Institutes that received at least 50% of their financing from government funds. Altogether 71 scientists were lost as a result, 6 of whom were directors and the remainder department heads, staff scientists and assistants. The total budget of the *KWG* rose from 5.6 million in 1933 to 14.3 million reichsmarks in 1944, thus research at the *KWI*s expanded during the Nazi regime. However, this led to many of the Institutes being drawn into war preparations. Out of the 48 member scientists of the *KWG* holding director positions during the regime, approximately 40% were members of the → *NSDAP*, 3 of whom had joined before 1933 (including Gerhard Jander and P. → Thiessen), 3 in Mar. 1933, 1 in 1935, 1 in 1936, 7 in 1937, 1 in 1938 and 3 in 1940. 21 of the 48 party members belonged either to the → *NSDDB* or the → *NSLB*. The general secretary Ernst Telschow was also influential, becoming additionally national defense adviser to the Reich and commissioner of defense for the *KWI*s in 1939. He was also instrumental in the relocation of the Society's management from Berlin to Göttingen in Feb. 1945, as well as in its rebuilding after the war. In Sep. 1946 the *KWG* was succeeded by the Max Planck Society in Göttingen (initially just for the British occupation zone; notably E. Telschow, O. Hahn, W. → Gerlach and M. → von Laue as well as Colonel B. Blount participated), then established in Feb. 1948 in Göttingen as a bizonal "association of independent research institutes, that are not connected to the government or industry" and was eventually expanded to include all three Western zones. Refs.: Annual reports and lists of papers in the journal → *Die Naturwissenschaften* 1933–43; Festschrift [1961], Burchardt [1975], Lemmerich (Ed.) [1981], Henning & Kazemi [1988], Vierhaus & vom Brocke (Eds.) [1990], Guérout [1992], pp. 19–27, Albrecht [1993], pp. 48f., Hermann [1993], Macrakis [1993], Oexle [1994].

KWIC: Kaiser-Wilhelm Institut für Chemie
Kaiser Wilhelm Institute of Chemistry. Established in late 1911 through a contract between the chemical association *Verein Chemische Reichsanstalt* and the → *KWG*. It was funded almost entirely by monies from the Emil Fischer Association for the Advancement of Chemical

Research. Ernst Otto Beckmann was appointed its first director in Apr. 1912, who simultaneously directed the Inorganic and Physical Chemistry Division. Also included are the Divisions of Organic Chemistry (until 1916 under R. Willstätter) and Radioactivity (under O. → Hahn (chemistry) and L. → Meitner (physics)). O. Hahn was director of the Institute 1928–46. The board of trustees included: C. → Bosch (1921–40); M. → von Laue (starting 1921); W. → Nernst (1921–41); M. → Planck (1931–46). In 1932 the *KWIC* had 25 staff scientists. In 1938 L. Meitner was forced into exil. Feb. 1, 1939, J. → Mattauch succeeded her; and Nov. 1, 1943: he was promoted to deputy director of the Institute. Under Hahn's direction from 1939 on, experimental and theoretical research directly relevant to military applications of nuclear fission was conducted for the → *HWA*. During the last two years of war the *KWIC* was partially destroyed in two bomb raids (on Feb. 15 and Mar. 24, 1944), and then removed to buildings formerly part of a textile factory in Tailfingen. After the war the *KWIC* was reestablished in 1949 in the French occupation zone close to the University of Mainz. Refs.: Fischer & Beckmann [1913], Hartmann (Ed.) [1936], pp. 164–174, Meitner [1954], Hahn [1962]b, part VI, Burchardt [1975], pp. 95–98, Engel [1984], pp. 222ff., D. Hahn (Ed.) [1988], pp. 89–210, Johnson [1990], Weiss [1994], pp. 271ff.

KWIP: Kaiser-Wilhelm Institut für Physik
The Kaiser Wilhelm Institute of Physics. Founded at the proposal of F. → Haber, W. → Nernst, M. → Planck, Heinrich Rubens and Emil Warburg and with the financial support of the Koppel Endowment in Mar. 1914. Unlike other *KWI*s it did not conduct its own research initially, only financing physics projects at university departments. The director decided on the applications together with the board of trustees, which included: F. → Paschen, Nernst, Planck, M. → von Laue, Warburg, J. → Franck and E. → Schrödinger. Its building was finally completed in 1937 using funds donated by the Rockefeller Foundation. Directors: A. → Einstein (1914–22). (On Einstein's request, von Laue was appointed his deputy in 1922, though practically, Laue managed the *KWIP*.) P. → Debye (1935–39), who also conducted experiments in the Institute's low temperature laboratory on the magnetocaloric effect to attain the temperature absolute zero. In Oct. 1939, Debye was placed on leave of absence and he left for the United States at the beginning of 1940. Starting in the fall of 1939 the Institute was placed under the authority of the → *HWA* but was returned to the Society in Jan. 1942. It became the center of Army Ordnance sponsored nuclear energy research. K. → Diebner became the new administrative director to whom O. → Hahn and W. → Heisenberg were subordinated. Heisenberg then became acting institute director. In Oct. 1940 another laboratory was erected, the so-called 'Virus' building, where small-scale experiments were conducted on controlled chain reactions under the direction of K. → Wirtz. Around 1942 a dozen chemists and physicists conducted experiments on transuranic elements, fission products, isotope separation and the measurement of atomic constants. A research group led by C. F. → von Weizsäcker, which also included K.-H. → Höcker and P. → Müller, worked on the theory behind uranium reactors. In 1943 the Institute was moved to Hechingen in Southern Germany to avoid the escalating number of air raids. In 1945 the Institute was renamed *Max-Planck-Institut für Physik* and transferred to Göttingen. Refs.: Heisenberg [1971], Engel [1984], pp. 234ff., Macrakis [1986], Kant [1987], [1992] and in Geyer et al. (Eds.) [1993], pp. 152–158, Weiss [1994], pp. 275ff.

KWIPC: Kaiser-Wilhelm Institut für physikalische Chemie und Elektrochemie
Kaiser Wilhelm Institute of Physical Chemistry and Electrochemistry. Founded in 1911 in Dahlem (Berlin) with the financial support of the Koppel Endowment. Directors: F. → Haber (1911–33); Gerhard Jander as substitute (1933–35); P. → Thiessen (1935–45); K. F. → Bonhoeffer (1948–51). Under Jander and particularly under Thiessen this Institute was transformed into a model National Socialist establishment concentrating most of its research efforts on areas of military relevance. 19 of its staff members went into exile, the largest number of emigrants from a single institute. Apr. 17, 1933: Haber resigned in response to the Institute's

racially motivated dismissal of the department heads Herbert Freundlich and Michael Polányi. Member scientists included: R. → Ladenburg (1924–32); J. → Franck (1919–20, corresponding member 1926–28). After the war, Bonhoeffer became director of the Institute (1948–51), followed by M. → von Laue (1951–59). On the latter's suggestion the Institute renamed *Fritz Haber Institute of the Max Planck Society* in 1953. Ref.: P. Thiessen in Hartmann (Ed.) [1936], pp. 175–178, Harteck [1960], Burchardt [1975], pp. 98–100, Engel [1984], pp. 214ff., Chmiel et al. [1986], Vierhaus & vom Brocke (Eds.) [1990], p. 375, Stoltzenberg [1991], [1994], Macrakis [1993], pp. 67ff., 85ff.

KWIS: Kaiser Wilhelm Institut für Strömungsforschung
Kaiser Wilhelm Institute of Fluid Dynamics Research. Founded in 1925 in addition to the Aerodynamics Design Testing Station at Göttingen University (*Aerodynamische Versuchsanstalt* (*AVA*), formerly called *Modellversuchsanstalt für Aerodynamik*), founded in 1908. L. → Prandtl was director of the *KWIS* 1925–46. Albert Betz was deputy director and also from 1925: director of the *AVA*, which expanded greatly from 1934–36: From a total workforce of 80 in 1933, it grew to 230 in 1935, and to 450 in 1936. In 1937 it was severed from the *KWG* and renamed the → *KWG*'s Aerodynamics Experimental Station in 1937, with M. → Planck in the board of trustees. In 1937 alone, 4 million RM were spent on land acquisition and construction in Göttingen as well as Reyershausen in a shut down potash mine. In 1939 the *AVA* staff came to 700 and in the beginning of the 1940's approximately 800 persons designed bombs, rockets and aircraft using wind tunnels for the → *RLM*. A new wind tunnel was purchased in autumn 1936 for 2.5 million RM, and in 1935/36 a cold wind tunnel was installed for aerodynamic tests reaching -60°C and 0.1 atm pressure. Ballistic experiments were also carried out for the *HWA*. Refs.: A. Betz in Hartmann (Ed.) [1936], pp. 129–139. C. Tollmien in Becker et al. (Eds.) [1987]; Festschrift [1975], pp. 11–34, Lüst (Ed.) [1975], pp. 11–34, Rotta [1990], Trischler [1992]a, pp. 128ff., 188ff., 199ff., 290ff., (Ed.) [1992]b.

NG: Notgemeinschaft der Deutschen Wissenschaft (later → **DFG**)
Emergency Association of German Science. Founded on Oct. 30, 1920 on the initiative of F. → Schmidt-Ott and F. → Haber; politically realigned in 1934. In Oct. 1937 renamed *Deutsche Gemeinschaft zur Erhaltung und Förderung der Forschung* (German Association for the Support and Advancement of Scientific Research). In short: → *Deutsche Forschungs-Gemeinschaft*. Members of the *NG* included: all German universities and polytechnics, the 5 scientific academies, and the → *KWG*. Until 1934 the *NG* was under the supervision of the → *RIM* and thereafter under the → *REM*. Presidents of the *NG*: F. Schmidt-Ott (1920–34); J. → Stark (June 1934–36, appointed by B. → Rust); R. → Mentzel (1936–39); A. → Esau (1939–45). E. Wildhagen, who had already served as aide-de-camp to Schmidt-Ott was a departmental official (from 1932), managing director (from 1934), and vice-president (from 1936): M. → von Laue was Chairman of the Physics Commission from 1921. Applications for support submitted to the *NG* (for loans of instruments or for staff grants) were reviewed by expert panels (*Fachausschüsse*) of specialists in the relevant disciplines. Aside from the 21 panels, foreign grant monies were managed by the electrophysical and the Japan committees. The physics panel included M. von Laue, W. → Wien, G. → Mie, J. → Zenneck. The organization was officially renamed *DFG* in Oct. 1937, and in the same year the *REM* established the → *RFR* as a parallel organization to the society. In 1945 the *DFG* was no longer active; in 1949 it was refounded first as the *NG*, then from 1951 again bearing the name *DFG*. Refs.: Notgemeinschaft [1920], Zierold [1968], Nipperdey & Schmugge [1970], Richter [1971], [1972], Forman [1974], Schlicker [1979]a, Deichmann [1992], pp. 56f.

PTR: Physikalisch-Technische Reichsanstalt
Reich Physical and Technical Institute. Founded in 1887 as a new type of institution combining metrology with research in physics and technology with the goal of supervising and

directing calibration and establishing metrological standards. Werner von Siemens was instrumental in its establishment. Research emphasis included: Spectroscopy, photometry, electrical engineering, low-temperature physics. Until 1934 the *PTR* was under the supervision of the → *RIM*, then under the → *REM*. However, staffing was basically decided by the Institute's board of trustees, which included: H. → Konen and W. → Nernst around 1930, as well as A. → Einstein (1917–33), L. → Prandtl, M. → Planck, and industrialists from the companies → *Siemens*, → *Zeiss*, and → *Krupp*. Presidents: H. von Helmholtz (1887–92), F. W. Kohlrausch (1892–1905), W. Nernst (1922–24); F. → Paschen (1924–Apr. 1933), J. → Stark (May 1, 1933–39); A. → Esau (1939–45). From 1938: the addition of a vice-president, G. Moeller, who represented the growing influence of the → *HWA* on the *PTR*'s research, marking the move away from fundamental physics research and towards military applications. The Institute collaborated closely with the → *RLM*. 1933: more than 50% of the staff were members of the → *NSDAP*. In 1932 it had 292 employees and in 1937, 443 staff members. In the first war year, the staff was expanded to over 500 in 1942. 1934: Installation of a new acoustics laboratory headed by Martin Grützmacher, later also research into ultrasound. Max → von Laue was the *PTR*'s adviser on theoretical physics from around 1925 to Dec. 1933. His advisory contract with the *PTR* was terminated following his dispute with J. Stark at the Würzburg physics conference of Sep. 18, 1933. A decree by the Reich Chancellor finally brought about the dissolution the board in the academic year 1934/35, and the *Führerprinzip* was applied under Stark's presidency. However, Stark's overly ambitious plans to expand the *PTR* were not carried out. The last conference of the *PTR* before the end of the war took place in 1943 on nuclear physics. Esau, Beuthe, Bomke and others also worked at the *PTR* on the secret German uranium project. After 1945 the Institute was renamed *Physikalisch-Technische Bundesanstalt* (*PTB*): Federal Physical and Technical Institute. Collaborators included: O. Lummer; E. Pringsheim; E. Lau; W. → Bothe (from 1927); K. → Diebner (from 1934); O. → Frisch (1927–30); E. → Gehrcke (1901–46); H. → Geiger (from 1912); W. → Meissner (1908–33); I. → Noddack; K. → Scheel (1891–1931). Refs.: Stark [1937], Kausch [1937], Burchardt [1976], Lundgren et al. [1986], chap. 1, Lemmerich [1987], Bortfeld et al. (Eds.) [1987], Cahan [1989], Hoffmann [1993]a, Kern [1994].

Stifterverband der Notgemeinschaft der Deutschen Wissenschaft e.V.
Donor Federation of the Emergency Association of German Science, registered association. Founded on Dec. 14, 1920 to collect donations for the → *NG* or → *DFG* through its representatives in industry or through their political influence. Chairmen: Carl Friedrich von Siemens (1920–34); Duisberg (1934–35); F. → Schmidt-Ott (1935–45). Managing director 1920–45: R. Fellinger of the company → *Siemens*. The Federation's contributions were only small when compared to the government grants, not least because of the competing → *HG* founded by A. → Vögler in 1920 as well as the Society for the Advancement of German Industry (*Fördergesellschaft der Deutschen Industrie*, 1942). Refs.: Mentzel [1940], Zierold [1968], pp. 29ff., 181, 226f., Nipperdey & Schmugge [1970], Forman [1974], Richter [1977].

B State and Military Institutions

HWA: Heereswaffenamt
Army Ordnance Office. Central agency for the technical design and manufacture of weapons, equipment and ammunition, subordinate to the Army High Command (*Oberkommando des Heeres*: OKW) in Berlin. To mid-1943 it was located in the center of Berlin (*Bahnhof-Zoo*), when it was relocated in Sep. 1943 to Wünsdorf nd Wittenberg. The *HWA* monitored all scientific areas for potential military applications using its own experts and external consultants, and in

1926 already had in addition an unofficial research department, directed by E. → Schumann from 1934. Department head until 1938: General K. → Becker, who became head of the entire Office in 1938. Then headed by the Artillery General Emil Leeb (1881–?) Apr. 15, 1940–end of war, who had previously directed procurement (from 1933). Expert consultant on nuclear physics and radiation issues: K. → Diebner (from 1934). Consultant on chemical weapons: P. → Harteck (from 1937), who also made the *HWA* aware of the potential of creating a nuclear explosive for military purposes through a chain reaction in 1939. Thereafter an independent nuclear physics section was erected, headed by K. Diebner, who together with his colleague Dr. Zipprich, also consulted H. → Geiger and other German nuclear physicists (*Uranverein*). From 1939 the *HWA* distributed the necessary nuclear materials among the research institutes (i.e., uranium oxide and other fissionable materials), and assumed the supervision of nuclear research. It had its own nuclear test site in Gottow until 1945, directed by Diebner, and in 1942 about a half dozen physicists took nuclear constant and cross-section measurements. In 1942 supervision of the *Uranverein* was turned over to the → *RFR*. From around 1930 on, the *HWA* had, among others, a military rocket research center at West Kummersdorf near Berlin, an artillery testing site. In 1932, Arthur Rudolph became expert adviser on special weapons (*Referent 'Sondergerät'*) to the *HWA*, which coordinated further concentrated rocket research. Between 1940–44 it increasingly shed authority to the → *RMBM*. The *Chemisch-Technische Reichsanstalt* assisted the *HWA* on the physical and chemical final quality controls of newly developed weapons. As the war progressed, the Materials Testing Office in Berlin and polytechnics were also drawn into this area. Refs.: Simon [1948], Leeb [1958], Walker [1989].

PrKM: Preußisches Kultusministerium, officially: **Preußisches Ministerium für Wissenschaft, Kunst und Volksbildung**
Prussian Culture Ministry, officially: **Prussian Ministry of Science, Arts and Culture**. Ministers: F. → Schmidt-Ott (1917–19), succeeded by: Konrad Haenisch; Carl Heinrich → Becker, under whose direction the *PrKM* became the most important counterpart to the → *NG*. Other important functionaries: Undersecretary Krüss, 1920–25 head of the university division along with his successor, Prof. W. Richter. On Feb. 2, 1933 the nationalist Minister Kahler resigned the post he had been holding since Oct. 1932. B. → Rust first was named deputy to the Reich Commissioner of Prussia (on Feb. 4, 1933) in the areas of science, education and culture, and on Apr. 21, officially the Prussian Culture Minister. On Mar. 15, 1933, T. → Vahlen was appointed to the science subdivision, who also exercised considerable control over the Prussian Academy of Sciences. George Gerullis became head of the science department in May 1933. In 1934 the *PrKM* was incorporated into the → *REM*. Refs.: Grau et al. [1979], Fischer et al. (Eds.) [1994], Walker [1995], chaps. 4–5.

REM: Reichserziehungsministerium, officially: **Reichsministerium für Wissenschaft, Erziehung und Volksbildung**
Reich Education Ministry, officially: **Reich Ministry of Science, Education and Culture**. Successor to the former Prussian Ministry of Science, Art and Culture (*Preußisches Ministerium für Wissenschaft, Kunst und Volksbildung*), creating for the first time in German history a centralized and hierarchical institution in control of the Reich's education sector. In 1934 this ministry took over from the → *RIM* the supervision of colleges and universities as well as other research institutions like the → *PTR*. Minister from June 1, 1934: B. → Rust, who also assumed the administration of the culture division of the *RIM* on the same date. Head of the Science Office there (*Amt für Wissenschaft* or *Amt W*) (1934–37): T. → Vahlen (1937–39); Otto Wacker; from May 1939, R. → Mentzel. Among his subordinates: F. Bachér as head of the university division *W* 1, from 1934, on commission; from 1935 on a full-time basis. Also E. Wildhagen (1934); W. Zschintsch; W. Dames, W. → Führer and A. → Esau as heads of the physics section of the → *RFR*, as well as P. → Thiessen as part-time adviser. In 1939 G. → Joos and W. → Hanle brought the potential military applications of nuclear physics to the attention

of the *REM*. Refs.: Rantzau [1939], Seier [1964], Geuter [1992]b, pp. 105f., Deichmann [1992], pp. 56f., Macrakis [1993], pp. 76ff.

RFR: Reichsforschungsrat
Reich Research Council. Inaugurated by B. → Rust on Mar. 16, 1937 on the initiative of E. → Schumann as a coordinating agency in the → *REM* with the purpose of centralizing the planning of all basic and applied research, and expanding the influence of the military (excluding aeronautical research, which was kept beyond the grasp of the *RFR* and placed under H. → Göring's supervision). Not coincidentally, the general K. → Becker, head of the → *HWA*, was its President (1937–40). Vice-president: O. Wacker of the → *REM*. Actual direction of the Council was carried out by the president of the → *DFG* (from 1936), R. → Mentzel. The heads of the 13 special sections (later called *Fachspartenleiter*) decided on all applications for support. Following conferences on nuclear physics between Dec. 16, 1941 and Feb. 1942, the Armed Forces leadership decided to leave nuclear research in the hands of the *RFR*, since according to W. → Heisenberg, no direct importance to the war could be ascertained. The edict of June 9, 1942 moved the Council out of the jurisdiction of the *REM* and placed it under H. Göring's authority. The reorganization of the Council was made on the initiative of A. → Speer. The *RFR* maintained its own nuclear physics test site at Stadtilm in Thuringia, under the leadership of K. → Diebner. Other members of the Council were: E. Schumann (1939–45); P. → Thiessen, the head of the division of chemistry and organic materials; A. → Esau (Dec. 1942–the close of 1943 head of the physics division, then Reich plenipotentiary of the high-frequency physics working group (*A.G. Hochfrequenzphysik*) and W. → Gerlach (from the end of 1943 head of the physics division and the Reich Marshal's plenipotentiary for nuclear physics—replacing Esau); and O. → Scherzer (1944–45): head of radar direction finding. The *RFR* was never able to fulfill its aim of guiding research effectively (that is, free from administrative confusion and competency disputes), probably not least because of the unsuited placement and weak position of Rust and Mentzel against Göring and the Armed Forces. Werner Osenberg was head of the planning board in the *RFR* from June 29, 1943–45. In 1944 he initiated the recalling of around 5,000 scientists and engineers from the front to conduct research 'critical to the war effort'. Refs.: Mentzel [1940], Simon [1948], chap. 6, Zierold [1968], pp. 192, 215ff., 266ff., Ludwig [1974], pp. 237–270, Deichmann [1992], pp. 57ff., Trischler [1992]a, pp. 252ff., 260f.

RIM: Reichsinnenministerium
Reich Interior Ministry. The *RIM* supervised State institutions like the → *PTR* until 1934 and issued far-reaching laws which facilitated the political realignment of government agencies by the National Socialists and the elimination of their political opponents. It also negotiated with the → *KWG* on the support and funding of institutes as well as, i.e., the construction of the → *KWIP* after 1930. The ministers at the *RIM* were replaced frequently; these included: Koch (around 1920); von Kendell (1927–29); Carl Severing (around 1929); Wirth (from 1930); Wilhelm → Frick (from Jan. 30, 1933 to Aug. 1943). Other important officials: Privy Councillor Krüss, head of the science division (until 1920); ministerial councillor in the Army high command and senior leader in the *SS*, Dr. Richard Donnevert; Leist; Niessen; assistant secretaries Buttmann and Pallengahr (from 1927) as well as Undersecretary Hans Pfundtner. Refs.: Schmidt-Ott [1936], Pfundtner (Ed.) [1937], Hilberg [1961], p. 43, etc.

RKM: Reichskriegsministerium (see → *RWM*)

RLM: Reichsluftfahrtministerium
Reich Aviation Ministry. From its foundation in May 1933 under the control of H. → Göring (who had already been nominated Reich Commissioner for Air Traffic in Jan. 30, 1933), with

Field Marshal Erhard Milch as his deputy as well as Ministerial Councillor Adolf Baeumker heading the research department. The Air Force (*Luftwaffe*), since May 1935 also under Göring's control as its commander-in-chief, had about a dozen research institutes, including in particular the *Deutsche Versuchsanstalt für Luftfahrt* (German Aviation Testing Institute) in Adlershof (Berlin); *Deutsches Forschungsinstitut für Segelflug* (German Glider Research Institute), located in Darmstadt, and from 1942 Ainsing near Freilassing, at which Alexander Lippisch conducted his pioneering research on delta-wing airplanes; the radar research institute *Flugfunk-Forschungsinstitut Oberpfaffenhofen*, of which J. → Zenneck was board chairman; *Technische Akademie der Luftwaffe* (Technical Academy of the Air Force); the *Deutsche Forschungsanstalt für Luftfahrt* (German Aviation Research Institute), in Braunschweig; the *Deutsche Akademie der Luftfahrtforschung* (German Academy of Aviation Research) founded in 1936/36, and various research institutes in industry attached to it as well as the rocket experimentation site at West Peenemünde, where the pilotless plane carrying a load of explosives (V-1) and anti-aircraft missiles were developped. Important wind-tunnel experiments were also conducted for the Air Force at the *Aerodynamische Versuchsanstalt* (→ *KWIS*) of the → *KWG* in Göttingen under L. → Prandtl as well as at the research institute *Forschungsanstalt Graf Zeppelin* in Stuttgart. The *Luftfahrtforschungsanstalt* in Munich founded in early 1939 was originally envisioned to have gigantic dimensions which could however not be realized any more in the war-time economy. In June 1942 a Reich agency was created called *Forschungsführung* (*FoFü*) (Research Command) as a part of the *RLM*, which assumed Baeumker's responsibilities and was directed by L. Prandt, Seewald, Georgi and Baeumker. Although certainly far more effective than the rival → *RFR* in centralizing scientific research, the *Fofü* also suffered from the polycratic Nazi system and from growing material and staff shortages. Refs.: Simon [1948], Völker (Ed.) [1968], Zierold [1968], pp. 222f., 271, Ludwig [1974], Homze [1976], Trischler [1992]a, pp. 174ff., [1992]b, chap 3.

RMBM: Reichsministerium für Bewaffnung und Munition (from autumn 1943: **Reichsministerium für Rüstung und Kriegsproduktion**)
Reich Ministry of Armament and Ammunition, (or from autumn 1943: **Reich Ministry of Armament and War Production**). Created in Mar. 1940 on A. → Hitler's orders with the aim of improving logistics and weapons production for the 3 branches of the Armed Forces. From 1940 to the end of 1944 this ministry accumulated power from the Armed Forces and the Army. While the Army ceded authority to the Reichsminister from the beginning, the more politically powerful wings of the Armed Forces only did so much later: the Navy in May 1943, and the Air Force around Jan. 1944. the *RMBM* focussed on the main interests of the Army High Command (*OKW*), such as tanks, anti-tank guns and artillery, to which the necessary raw materials were supplied. Less prestigious matters were left to the → *HWA*, which also increasingly lost power to the *RMBM*. F. → Todt was the first War Production Minister, succeeded by A. → Speer, under whose direction the *OKW*'s Defense Economy and Armament Offices were also transferred into his authority, with the result that despite the escalation in aerial raids between 1942 to mid-1944, German armament production increased more than three-fold, and tank production even six-fold. In the summer of 1936, construction of the largest German research compound was started on the Baltic island of Peenemünde, involving about 10,000 construction workers until 1939. About 1,200 engineers and other scientists were engaged there in 1939 plus another 3,500 officers and military scientific personnel from 1940 on. In May 1940 Speer was responsible for further developments at Peenemünde. The *Wunderwaffe* or the weapons of reprisal (*Vergeltungswaffen*) were built there, principally, a ballistic missile (V-2). Following an Allied air raid in Aug. 1945, production of the V-2 rockets was transferred to mine shafts deep in the Harz Mountains, the *Dora-Mittelbau*, where 10,000 slave laborers worked and perished under appalling conditions. Refs.: Leeb [1958], Janssen [1968], Speer [1970], Bode & Kaiser [1995], Neufeld [1995], Zilbert [1981].

RPM: Reichspostministerium
Reich Postal Ministry. The engineer Wilhelm → Ohnesorge was Postal Minister from Feb. 1937. On Jan. 1, 1937, all research and development departments in the fields of television and high-frequency technology, cable transmission (wide-band cable development to increase transmission capacity), metrology and acoustics (microphone technology) were united under the *Forschungsanstalt der deutschen Reichspost*. Dept. VIII of the former *Reichspostzentralamt* formed the core of this new research establishment. The television specialist Dr. Banneitz was head of research, and the high-frequency engineer Dr. Vilbig was named his deputy. The *RPM* had its own 500,000 m^2 research site in Miersdorf outside of Berlin where development of a cyclotron was started in 1940 and completed by the end of the war. However, the installation was dismantled by the Soviets in May and June of 1945. The *RPM* also supported independent research, e.g., by M. von → Ardenne in Berlin-Lichterfelde on isotope separation, electron microscopy, and communications technology. In May 1942 the armed postal security service was subsumed in the → *SS* as one more step in the 'national socialization' of the *Deutsche Reichspost*. Refs.: Flanze [1937], N.N. [1942], H. Salow in Bothe & Flügge (Eds.) [1948], part II, pp. 32f., Zierold [1968], p. 261; Ardenne [1972], Lundgren et al. [1986], pp. 122–124, Leclerc [1988], Ueberschär [1989].

RWM or RKM: Reichswehrministerium, (from 1939:) **Reichskriegsministerium**
Reich Ministry of Defense, then **Reich Ministry of War**. The *RWM* supervised the → *HWA* which developed weapons systems along with the necessary basic and applied research. In the science section, E. → Schumann was head of the scientific central agency from 1929, from 1932 ministerial councillor, and 1939–45 assistant secretary. K. → Diebner was expert adviser in nuclear physics there from Dec. 1934, from 1945 senior executive planning officer. From 1939 the researches of the *Uranverein* were also coordinated there. Especially after 1939 the *RKM* and *HWA* research projects were given top priority, and the scientist positions involved were rated 'indispensable' or reserved (*uk*). Ref.: Leeb [1958], Völker (Ed.) [1968]

C National Socialist Organizations

Ahnenerbe der SS
Ancestral Heritage Foundation of the Defense Squadron through the Research and Education Association of the → *SS* (*SS Forschungs- und Lehrgemeinschaft*). Founded in 1935 as the 'Society for the Study of Intellectual History' (*Studiengesellschaft für Geistesgeschichte*) by H. → Himmler and the *SS*. Wolfram Sievers was its managing director. Initially, the Ancestral Heritage served primarily to promote folklore research (ancient and prehistory, genealogy, investigations into the origin of proper names, clan symbols, and the like). Its later activities included: Subsidizing archeological expeditions and anthropological research, for the most part based on racist premises, as well as increasingly, scientific research (in 1938, it had 5 science departments) which was presided by the curator and senior batallion officer of the *SS* Walther Wüst (1901–?), who was also full prof. at the Munich Univ. and had close ties to the → *KWG*. In addition it patronized pseudo-theories of its own, for example, the 'cosmic ice theory' (*Welteislehre*). But it also organized slave labor in research as well as subjecting concentration camp prisoners to fatal medical experiments, such as those conducted at the Institute of Practical Military Research (*Institut für wehrwissenschaftliche Zweckforschung*), which was founded on July 7, 1942 under the direction of Wolfram Sievers, and was subordinate to Himmler and the *SS*. Refs.: Goudsmit [1947], pp. 202ff., Mitscherlich & Mielke (Eds.) [1962], Kater [1974], Zierold [1968], pp. 187f., Deichmann [1992], pp. 199ff.

Gestapo: Geheime Staatspolizei

Secret State Police. 1933–45 the political secret police force of the Hitler dictatorship with a large network of spies. It made arbitrary arrests, used torture and murdered at will. As was the case with the → *SS*, the *Gestapo* was subordinate to the Security Headquarters of the Reich, *Reichssicherheitshauptamt* (*RSHA*). In 1933 H. → Göring became head of the *Gestapo*. It came under H. → Himmler's control in June 30, 1934 until 1936, during which time it became the feared executive of the Nazi State. In 1936 R. Heydrich took over, using it not only to spy on the population to get information on politically 'unreliable elements', but especially to deport politically undesirable social groups (i.e., leftists, Jews and Gypsies) to concentration camps, parallel to the activities of the → *SA* and *SS*. Refs.: Delarue [1964/79], Gellately [1990].

NSDAP: Nationalsozialistische Deutsche Arbeiterpartei

National Socialist German Workers Party. Founded 1919 in Munich as the German Workers Party (*Deutsche Arbeiterpartei*). 'National Socialist' was added to its title in Mar. 1920. Following the failed *putsch* of 1923 it was dissolved and refounded by A. → Hitler on Feb. 25, 1925. Its basic view of the world promoted the fascist one-party state Hitler had been advocating since 1921, with himself at the helm. Crucial points of the demagogic ideology include: anti-Semitism and anti-Communism; the 'stab in the back' myth (*Dolchstoßlegende*) of Germany's having been swindled into defeat in 1918; militarism and extreme nationalism; pseudo-Darwinian racist theories and the idea of a Nordic "master race". These elements were spread in particular in the party newspaper → *Völkischer Beobacher* but also in publications loyal to the regime: *Der Angriff*, → *Das Reich* and *Der Stürmer*. Hitler's autobiography *Mein Kampf* und A. → Rosenberg's book on the 20th century 'myth', *Mythos des 20. Jahrhunderts*, also promoted these ideas. The Scientific and Academic Division was led by von Bochmann from around 1932. Following the nomination of Hitler as Reich Chancellor on Jan. 30, 1933 and the subsequent seizure of power by the National Socialists after the burning of the *Reichstag* on Feb. 27, 1933 until the end of the war in 1945, an attempt was made to put this ideal into practice, through politically realigning (*Gleichschaltung*) all institutions and applying the authoritarian *Führerprinzip* to all subsidiary organizations within the party (such as the → *SA* and the → *SS*; the University Lecturers League (→ *NSDDB*); the Teachers League (→ *NSLB*) the Student League (→ *NSDStB*), the Hitler Youth, etc.), by means of political terrorism. From July 14, 1933 on, the Nazi Party was the only legal political party in Germany until 1945. From May 1, 1933 to the spring of 1937 new members were no longer admitted. Refs.: Hilberg [1961], pp. 37f., Orlow [1969], Bracher [1969], chap. II; Broszat [1969], chaps. 2–3 and 6, Broszat & Frei (Eds.) [1989], pp. 80–93, 194f. For statistics on party membership: Kater [1983], Ash & Geuter [1985], Benz [1990], pp. 29ff., Deichmann [1992], pp. 225ff.

NSDDB: Nationalsozialistischer Deutscher Dozentenbund

National Socialist German University Lecturers League. Founded in 1933. This League grew out of the professional branch for university teachers of the → *NSLB*: *Reichsfachschaft Hochschullehrer des NS-Lehrerbundes*, the main organization of faculty and teaching assistants at universities and colleges. Like the Nazi Party and its subordinate organizations, such as the *NSLB* (founded in 1927) this League was organized geographically into districts, each headed by a district leader (*Gauführer*), who were appointed by the Education Minister after consultation with the relevant Nazi authorities and were subordinate to the university rector of the area. The physician Walther Schultze (born 1894) became the national head of the *NSDDB* in 1935; the astronomer Wilhelm → Führer was the University Lecturers Leader (*Dozentenführer*) in Munich; A. → Bühl was the physics expert adviser. The director of the University Lecturers League in Darmstadt, W. → Finkelnburg, organized regular 'camp' meetings (*Dozentenlager*) in which the exchange of ideas at the professional level was connected to political indoctrination. Although membership was not obligatory, it was considered tactically advantageous, if not unavoidable, for career advancement: The district leadership's evaluations were decisive in the acceptance of *habilitation* (post-doctoral) theses, which is a prerequisite to attaining the rank of university lecturer. A new regulation was issued requiring all new university lecturers to demonstrate their "personal ability to educate students in the National Socialist spirit" during a six-week period at one of such official camps. While among the 71 members of the Uranium project about 55% were party members, average membership in the *NSDDB* was probably 75%. On Finkelnburg's initiative the League organized a physics meeting in 1940 which concluded with a vote against the 'Aryan Physics' movement and with official acknowledgment of the relativity and quantum theories. The 'physics camp' (*Physikerlager*) follow-up meeting of university lecturers in 1942 in Seefeld in Tyrol, Austria, confirmed this result. Refs.: Seier [1964], Kelly [1973], Snyder [1976], p. 315.

NSDStB: Nationalsozialistischer Deutscher Studentenbund
National Socialist German Student League. A nation-wide student association of the → *NSDAP* founded in Feb. 1926 in Schwabing (Munich), under the direction of the later head of the Party's Office for Race Policy (*Rassenpolitisches Amt*), Prof. Walter Groß. It had considerable influence at some universities, i.e., the Berlin Polytech. or the universities in Munich, even before the rise to power of the Nazi Party. At the student council elections of 1931/32 the *NSDStB* won as much as 60% of the votes. Heads of the League: Wilhelm Tempel; Oskar Stäbel; from 1928, Baldur von Schirach, until he became Reich Governor (*Reichsstatthalter*) of Vienna. The physics group leaders included: W. → Menzel; B. → Thüring; and F. Kubach. Refs.: supplement issue of *Volk im Werden* **3** [1935], pp. 64–107, Bleuel & Klinnert [1967], Seier [1964], Faust [1973]b, Giles [1985], Wengenroth [1993], pp. 219f.

NSLB: Nationalsozialistischer Lehrerbund
National Socialist Teachers League. Founded in 1927 arising from radical right-wing teachers organizations such as, e.g. the *Bund völkischer Lehrer*, *Junglehrerbund Baldur* and the *Deutschnationaler Lehrerbund.* Until 1932 the *NSLB* was mostly a political activist association for National Socialist elementary schoolteachers, and this reputation caused the other teachers associations not to acknowledge it immediately until after the Nazis' seizure of power 1933. The → RIM feared losing control over a substantial percentage of civil servants to the *NSLB*. In mid-1933 a cooperation contract was signed between the League and 48 teachers associations from all kinds of schools, resulting in the creation of the German Education Society *Deutsche Erziehungsgemeinschaft*, but at the end of the same year the German Philologists Federation *Deutscher Philologenverband* and the University Teachers Federation formed a rival organization. However, a growing membership as well as the influence of important → *NSDAP* officials strengthened the *NSLB*. For example, in 1935 the German Association for the Promotion of Mathematics and Science Instruction (*Deutscher Verein zur Förderung des mathematisch-naturwissenschaftlichen Unterrichts*) was subsumed in the League but it was only with the passage of the law on civil servant associations of May 27, 1935 that the remaining independent teachers associations were dissolved. The *NSLB* was thus able to claim 97% of all teachers as their members. Moreover, 32% were also Nazi party members (against 17% of Civil Service League members), 14% were even political leaders (7 district leaders and deputy district leaders, 78 were circuit leaders (*Kreisleiter*), and 2,668 were local group leaders), and 11% were members of the → *SA*. By 1939 around 215,000 League members had participated in political training camps, notably those not members of the party. Irratic management of the League led to its dissolution in 1943 "as a simplification measure determined by the war". Refs.: Eilers [1963], Brämer & Kremer (Eds.)[1980], pp. 25ff.

SA: Sturmabteilung

Storm Detachments. Founded in Aug. 1921 through the reorganization of the personal protection and propaganda force of the → *NSDAP*. It emerged out the Armed Forces' (*Wehrmacht*) 'gymnastics and sports division' founded in Nov. 1920 and expanded into a virtually independent defense unit at the beginning of 1923. It played a major role in the Munich *putsch* attempt of Nov. 8–9, 1923, where 1,500 of its members (out of a total of 4,000) were involved and was subsequently banned along with the other Nazi organizations. But it continued to exist underground and was reorganized by Franz von Pfeffer, who became its supreme command officer (*oberster SA Führer*) in 1926. Latest in 1926 it became the party's private army, initially under Klintsch in collaboration with Röhm, then under H. → Göring, and from Jan. 1931 under Röhm and Victor Lutze. The *SA* often recruited former soldiers. Like the → *SS*, its structure and operation was military-like; and even before Hitler's *putsch*, it orchestrated brawls in public halls and street battles against political opponents. After 1933 it carried out the arrest of these opponents, shooting or deporting them to concentration camps. Following the arrest and summary execution of Röhm and other leaders of the *SA* by the *SS* between June 30 and July 2, 1934, known as the 'Röhm affair', the Storm Detachments lost political importance. During World War II it was used as an auxilliary force to the *Wehrmacht*. Refs.: Höhne [1984/90], Longerich [1989], Delarue [1964/79], part II.

SS: Schutz-Staffel

Defense Squadron. Founded in Apr. 1925 as a combat force and terror organization of the → *NSDAP*, also known as the 'Blackshirts'. In contrast to the large scale of the → *SA*, the *SS* considered itself an élite and exclusive organization. Its leaders included from 1926 on Pfeffer, then Schreck, Berchthold and Heiden. From Jan. 6, 1929: H. → Himmler. Following the murder of heads of the competing *SA* organization, the *SS* became an independent organization on July 20, 1934 under Himmler's Reich Security Headquarters (*Reichssicherheitshauptamt*) (*RSHA*). Himmler in particular built the Defense Squadron into the third pillar of power in the National Socialist State, alongside of the party and the Armed Forces (*Wehrmacht*). The *SS* had a military structure and trained its members similarly. Aside from its primary tasks of protecting its own operations and of serving as A. → Hitler's bodyguards, it also systematically eliminated political opponents and 'undesirable' minorities like Jews, Gypsies and homosexuals. Subdivided into the general and the armed branches (*Waffen-SS*), the latter also served as combat troops during the second World War. The Defense Squadron's newspaper *Das Schwarze Korps* launched smear campaigns against individuals and attacked trends that went against their racially based nationalistic (*völkischen*) ideals (e.g., labeling W. → Heisenberg a 'White Jew' in 1937). German physicists who were members of the *SS* included: J. → Juilfs; R. → Mentzel and T. → Vahlen. Refs.: Kogon [1973], Höhne [1984/90], Lifton [1986], Deichmann [1992], chap. 8, Guérout [1992], pp. 114–118.

D Some Major Industrial Firms

AEG: Allgemeine Elektrizitäts-Gesellschaft

General Electric Company. German electrical combine with headquarters in Berlin and Frankfort-on-Main. Founded in 1883 by E. Rathenau. E. → Brüche was its long-standing general manager. In 1928/29 it established its own research institute at Reinickendorf (Berlin) initially employing 45 technicians and staff scientists. By 1939 the staff had increased to 365. Research conducted there included electron optics and high-tension generation: The → *KWIP* obtained a high-tension generator from the *AEG* after 1937. C. → Ramsauer was director of the research division (until 1945). 365 staff members there. Other employees included: Johannesson; and O. → Scherzer. Refs.: Ramsauer [1949]; Osietzki in Eckert & Osietzki [1988], p. 43.

Anschütz & Co.
Machine-tool and gyroscope manufacturer, which produced among other things, gyrocompasses for submarines and ships, as well as centrifuges. Founded in 1905 in Neumühlheim (Kiel) by Hermann Anschütz-Kaempfe (1872–1931), inventor of the gyrocompass. Konrad Beyerle was its longtime scientific director. After 1939 the *Anschütz* Company manufactured not only gyrocompasses for the Navy and Air Force, but also ultracentrifuges designed by P. → Harteck and his co-workers for the separation of uranium isotopes. In July 1944 the company's facilities were destroyed in an air raid and removed to Kandern (Freiburg). Refs.: Gerlach & Sommerfeld [1931], Walker [1989]a, pp. 82f., 123, 147f.

Auer-Gesellschaft
Auer Company. The company was originally owned by the Jewish banker Leopold Koppel (1854–1933) and was initially primarily a gaslight manufacturer under the name *Deutsche Gasglühlicht Auergesellschaft.* Research on rare earths, gas mantles, radioactivity and luminescence, and on uranium and thorium compounds was conducted there. The director of the corporation N. → Riehl became aware of the developments in nuclear physics and sought out contacts within the uranium project (*Uranverein*). In 1942 two or three scientists focussed their efforts on the problems of uranium production and enrichment. The firm's uranium oxide production took place in their Oranienburg plant near Berlin, where they exploited the labor of about 2,000 female concentration-camp prisoners from Sachsenhausen. The Auer Company provided the uranium sheets and cubes for model experiments on the 'uranium machine' at the → *KWIP* and elsewhere. Refs.: Riehl [1988], Walker [1989]a.

BASF: Badische Anilin- und Sodafabrik
Baden Aniline and Soda Factory. Founded in 1865 in Ludwigshafen on the Rhine, the chemical company developed rapidly in the 1870's, benefiting from the boom in synthetic dye production. In this century the company produced nitrogen in conjunction with *Bayer* and *Agfa.* C. → Bosch, who had been working there since 1899 on salt synthesis and after 1914 achieved (together with F. → Haber) the synthesis of ammonia from nitrogen and hydrogen at high pressure, was its longtime director. In 1912 a branch was opened in Oppau to develop the Haber and Bosch patents. In 1919 Bosch became chairman of the board of this firm. On Dec. 2, 1925 the company merged with the dye works *Hoechst* and other major firms in the chemical industry to form the → *IG Farben* syndicate. Refs.: Teltschik [1992].

GKSS: Gesellschaft zur Kernenergieverwertung in Schiffbau und Schiffahrt m.b.H.
Company for the Commercial Exploitation of Nuclear Energy in Ship Building and Shipping, Ltd. Founded on Apr. 18, 1956 with headquarters at Geesthacht near Hamburg, after the Paris agreements of 1955 lifted the ban on nuclear research in Germany. It developed nuclear powered engines for ships and submarines. The most famous product was the 'Otto Hahn', the only German ship propelled by nuclear energy. E. → Bagge became its general manager in 1956; K. → Diebner was on its supervisory board (see also E. → Fischer). Refs.: Bagge in: Bagge, Diebner & Jay [1957], pp. 73ff., Renneberg [1993].

Interessengemeinschaft Farbenindustrie Aktiengesellschaft **(IG Farben A.G.)**
Community of Interests of the Dye Industry Stock Company. A German chemical trust founded in Dec. 1925 in Frankfort-on-Main, including in particular the companies: *Hoechst*, the dye factory in Leverkusen and → *BASF*; chairman of the supervisory board (managing director): Carl Duisberg (until 1935); C. → Bosch (1935–40); longstanding member of the board: C. → Krauch, who simultaneously directed the Research and Development Department

at the Raw Materials and Exchange Office under H. → Göring. Supervisory board member: F. → Schmidt-Ott (1925–45). Through the privately organized Emil Fischer Association for the Advancement of Chemical Research, which was funded primarily by *IG Farben*, considerable influence could be exerted on the → *KWIC*, which also received most of its financial support from the Fischer Association. The *Farben* concern supported the efforts to achieve economic self-sufficiency in raw materials prescribed in the Four-Year Plan, since this goal depended on the development of substitute materials. Following the occupation of foreign territories, industrial plants in operation there, for example, → *Norsk Hydro*, were forcibly incorporated into the production plans of the *Farben* trust. When production was crippled, the ammonia plant in Merseburg (*Leuna Werke*) was commissioned under the leadership of Dr. Paul Herold to continue the production of heavy water. After the war, *IG Farben*'s capital assets were confiscated and the West German companies were again broken up into independent firms. Refs.: Gerhardt [1948], Borkin [1978], Schlicker [1979]b, Lifton [1986], pp. 18, 152ff., 187f., 291, Hayes [1987], Guérout [1992], pp. 119–128, Teltschik [1992].

Krupp
Steelworks, founded by Friedrich Krupp (1787–1826) in Essen in 1811 and expanded by his son Alfred Krupp (1812–1887) to the largest cast-steel and ordnance factory of its time in the world. The founder developed an efficient method of producing cast steel, and his son expanded the manufacturing business into a major industry in Germany and substantially improved the quality of the steel, which became known worldwide. He acquired mines and began producing parts for the railway and the military. Aside from making the business profitable, he also founded worker colonies guided by the motto: "the common good should be the purpose of work". His heir, Friedrich Alfred Krupp, moved the emphasis toward scientific research and expanded the business to include the production of raw iron, large ships and later military equipment like warships, submarines and tanks. He also continued improving working conditions for his employees well above the standards of the day. Alfred's son-in-law Gustav von Bohlen und Halbach (1870–1950) managed the firm until 1943, turning it into Germany's armaments smith for both World Wars. Alfried von Bohlen und Halbach (1907–1967) directed the company 1943–45, exploited the labor of Jews from Ausschwitz concentration camp where he had built a munition factory, and through an agreement with → Speer also used 45,000 Russian civilians and 120,000 prisoners in coal mines. Krupp was arrested by the Allies in 1945, convicted to 12 years imprisonment, but already released in 1951 due to an amnesty, and resumed direction of the family firm.

Norsk Hydro-Elektrisk Kvaelstoffaktieselskab
Norwegian Hydro-Electric Power Stock Company. Norwegian firm 110 kilometers west of Oslo that produced hydrogen at its water-powered electrolysis plants in Rjukan and Vemork, as well as heavy water D_2O as a by-product. During the German occupation, the firm became the main supplier of heavy water for German nuclear research. Its annual production rose from 20 liters to 1 ton. In 1941 a catalytic conversion reactor designed by P. → Harteck and H. → Suess was installed which was supposed to increase the annual production to 4–5 tons. In 1943 the electrolysis plant was bombed by the Allies and the remaining heavy water destroyed through sabotage. Refs.: Irving [1967], Kramish [1986], Walker [1989]a, Powers [1993], chap. 18.

Siemens Aktiengesellschaft
Siemens, Stock Company. The largest European firm in the electrical industry. Founded in 1847 in a new suburb of Berlin, *Siemensstadt*, by the inventor of the self-correcting dynamo and electrical pyrometers, (Ernst) Werner von Siemens (1816–1892, knighted in 1888) together with Johann Georg Halske (1814–1890), initially under the name *Siemens & Halske, A.G.* as a telegraph construction establishment; it became a limited company in 1890 and a stock company 1897. In 1873: establishment of the physical-technical research laboratory, one of the

earliest in the electrotechnical sector. Siemens was also instrumental in the foundation of the → *PTR* in 1886. The company was consolidated by the founder's youngest son, Carl Friedrich von Siemens (1872–1941), among others, and later merged with other companies such as the *Schuckert-Werke* and the *Werner-Werke*. C. F. von Siemens was chairman of the supervisory board of the → *Stifterverband* 1920–34 and second vice-president of the → *KWG* 1937–41. In 1924, the Siemens research lab had about 68 white-collar and 90 blue-collar workers. In 1929 the numbers had grown to 88 and 111, and in 1931 (due to the global economic crisis) fell again to 51 and 41 respectively. After 1935 the company expanded rapidly. In 1933 the research labs. were reorganized: metallography (under Masing) and development of metal alloys (Neumann) were shifted into the dept. of electrochemistry; theoretical magnetism (von Auvers, Kühlewein) and photocells and semi-conductors (Waibel) were incorporated into the central research lab. of the Werner Works, which had existed since 1920 (until 1929 headed by F. Lüschen, until 1935 headed by Hermann von Siemens, and since then by B. Pohlmann and H.F. Meyer). In 1929, 200 engineers and an approximately equal number of mechanicians were working there. In 1930 the number of engineers jumped to 350. At the research lab. of the *Siemens-Schuckert Werke*, which was headed by Reinhold Rüdenberg until 1935. One of the staff scientists, M. → Steenbeck, developed the first functioning betatron in the years 1933–35. His X-ray flash technique developed in 1938 led to new insights on highspeed processes which had applications in ballistics, for instance. In 1937, Ernst Ruska developed his electron microscope at the central lab of the *Werner-Werke*. When G. → Hertz lost his professorhip at the Berlin Polytech. in 1935, he became director of the newly established 2nd Research Lab at Siemens. Research lab I, headed by Prof. Gerdien, laid an emphasis on acoustics, general physics, mechanics, ballistics, light arc and electronics. Research lab II conducted research on gas discharge, electron physics, atomic and nuclear physics. Both laboratories coexisted until Gerdien's retirement, at which time Hertz assumed the direction of both. Research Lab II started developing one of the first European cyclotrons in 1938, but it only became operational at the end of the war and was promptly removed to the USSR. The longtime director of the Center for Scientific and Technical Research (*Zentralstelle für wissenschaftlich technische Forschungen*) Robert Fellinger (1920–45) was at the same time general manager of the → *Stifterverband*. Siemens was the main supplier for electrical appliances used in research, including, for example, the cyclotron magnets for W. → Bothe's institute, and the extra-high voltage system for the → *KWIP*. In 1940 Siemens supplied the electrographite (pure carbon) needed by Bothe in Heidelberg to measure the absorption coefficient of thermal neutrons. After World War II, G. Hertz left for Russia, and W. → Finkelnburg became head of the research department of gaseous discharge at the Erlangen research lab of the Siemens-Schuckert works, other divisions being metrology (R. Schade), solid state physics (H. Welker), inorganic chemistry (G. Iwantscheff) and organic chemistry (E. Kasten). In 1955, he also became head of a study group on reactor development, an independent research group since 1957. Refs.: Schmidt-Ott [1943], G. Siemens [1947/49], Trendelenburg [1975], Laitko et al. [1987], pp. 560f., Schubert [1987], Osietzky [1988], Eckert & Osietzky [1989], pp. 43ff., Osietzky in Renneberg & Walker (Eds.) [1994], pp. 262ff.

Zeisswerke Jena

Zeiss Works Jena. Founded in 1846 in Jena as a small manufacturer of microscopes and magnifying glasses. Collaboration with Ernst Abbe (1840–1905) from 1869 on converted the company into a world-famous producer of optical instruments, ranging from projectors to telescopes, prisms, theodolites, spectacle lenses, spectrographs, medical and other precision instruments. 1889/91 the reorganization of the company into the Carl Zeiss Foundation was complete. G. → Joos took over the direction of its large research departments in 1941. Emphasis turned toward military optics (i.e., binoculars, periscopes and infrared equipment), becoming the main supplier to the military which took up 82% of the company's sales in 1943. In 1943 the Institute of Microbiology was founded, in which production of penicillin was commenced in 1946. After the second World War the firm was split into the *VEB Jenaoptik* (German Democratic Republic) and the Zeiss Works in Heidenheim (Federal Republic of Germany). Refs.: Ley [1942], Brüche [1946], Münster [1946], Schomerus [1952], Hermann [1989].

E Journals and Periodicals

Kernphysikalische Forschungsberichte
'Research Reports in Nuclear Physics'. Internal progress reports of the *Uranverein*, the German uranium project initiated in 1939 under the → *HWA*. These top-secret reports had extremely limited distribution within the official research program; the authors were officially prohibited from keeping copies. At the end of the war the Allied 'Alsos Mission' sent the confiscated reports to the United States Atomic Energy Commission for evaluation. After 1971 they were declassified and returned to Germany. Today many are archived at the *Kernforschungszentrum Karlsruhe* and the American Institute of Physics, New York (see the listing in David & Wahrheit (Eds.) [1952], providing file reference nos. **G 1–G 395**). Refs.: Irving (Ed.) [undated], Diebner [1956], Rechenberg in Heisenberg [1989], Vol. A II, pp. 371–375, and the appendix to Walker [1989]a.

Die Naturwissenschaften
'The Natural Sciences'. Subtitle: 'Weekly Publication of the Advances in the Natural Sciences, Medicine and Technology' (*Wochenschrift für die Fortschritte der Naturwissenschaften, der Medizin und der Technik*). Publication of the → *KWG* (after 1945, the *Max-Planck-Gesellschaft*) together with the Society of German Scientists and Medical Doctors (*Gesellschaft Deutscher Naturforscher und Ärzte*). Founded in 1913. Place of publication: Berlin; J. Springer Publishers. Edited by Arnold Berliner (until 1935) together with Curt Thesing (until 1914), then with Aug. Püttner (until 1922). After 1938: P. → Rosbaud. After 1922 annual reports appeared in addition on 'Results in the Exact Sciences' (*Ergebnisse der exakten Wissenschaften*) until 1935, edited by A. Berliner. Refs.: Laue [1946], Holl [1992].

Physik in regelmäßigen Berichten
'Physics in Regular Reports'. Founded in 1933 as a supplement to the → *Zeitschrift für technische Physik* of the German Society of Technical Physics (→ *DGtP*); Place of publication: Leipzig; Edited by: C. → Ramsauer. It published longer review articles, similar to the *Reviews in Modern Physics.*

Physikalische Blätter
'Physical Pages'. Founded in 1943 by E. → Brüche, who was also its editor from 1944–72. The first issue in Jan. of that year was destroyed in an air raid before it could be distributed, but was reprinted and delivered in May 1944. Issued by the 'Informationsstelle Deutscher Physiker'. J. → Goebbels and A. → Speer assured its paper supply during the war years, and publication continued after 1945, initially under the name *Neue Physikalische Blätter* ('New Physical Pages') from 1946, but it readopted its former name in 1948. Since its reappearence in 1946, the official publication of the German Physical Society (→ *DPG*). Place of publication in 1944: Braunschweig (Vieweg publishers); in 1946: Stuttgart (Mittelbach), 1947-49: Karlsruhe (Volk und Zeit), and from 1949 on: Mosbach, then Weinheim (Physik-Verlag publishers). Refs.: Rechenberg [1994]a, Mayer-Kuckuck (Ed.) [1995], pp. F 139–F 142.

Physikalische Zeitschrift
'Physical Journal'. Founded in 1901. Place of publication: Leipzig, later Berlin. J. Springer publishers. After Volume 25 (1924) merged with J. → Stark's *Jahrbuch für Radioaktivität und*

Elektronik; the supplements *Reichsberichte für Physik* started to appear in 1944. Editors: E. Riecke and Th. Simon, later P. → Debye. Together with the traditional *Annalen der Physik*, this journal represented the more conservative elements within the German physics community. Most experimental and technical physicists submitted their papers to one of these publications, if not to more specialized journals, but not to the → *Zeitschrift für Physik*.

Das Reich. Deutsche Wochenzeitung
National Socialist propaganda weekly published in Berlin from Mar. 1940 to Apr. 16/22, 1945. Editor: Propaganda Minister J.→ Goebbels. With its good-quality illustrations and well-researched articles (albeit with an obvious bias in its war and politics reports) this publication catered to intellectuals and others not interested in the standard party organs.

Verhandlungen der Deutschen Physikalischen Gesellschaft
'Proceedings of the German Physical Society'. Published from 1899–1944, as the successor publication to the 'Proceedings of the Physical Society in Berlin' (*Verhandlungen der Physikalischen Gesellschaft zu Berlin*) (since 1882) and to the journal 'Advances in Physics' (*Fortschritte der Physik*) (since 1845). Main editors: Karl → Scheel (1902–36); Walter Grotrian (1937–44); Ernst → Brüche (1950–65).

Volk im Werden
'Nation in the Making'. Founded in 1933, edited by A. → Rosenberg und Ernst Krieck as a joint publication of the Reich Student League (→ *NSDStB*) and the University Lecturers League (→ *NSDDB*). Place of publication: Leipzig (Armanen Verlag publishers). It mostly contained articles by the Rosenberg faction within the Nazi polycracy, which later lost its influence in the political power struggles.

Völkischer Beobachter
'Nationalist Observer'. Blatantly anti-Semitic and chauvinistic propaganda newspaper of the → *NSDAP*. Main place of publication: Berlin. With editions for Berlin, Northern and Southern Germany and Munich. It was banned along with the Nazi Party between Nov. 1923 and Feb. 1925. Main editor: A. → Rosenberg.

Zeitschrift für die gesamte Naturwissenschaft
'Journal for all the Natural Sciences'. Founded in 1935. Place of publication: Braunschweig. Edited by the science section (*Fachgruppe*) of the Reich student leadership of the → *NSDStB*. Chief editors from 1937: Fritz Kubach; B. → Thüring; Ernst Bergdoldt. Unofficial mouthpiece of the supporters of the 'Aryan Physics' movement. Refs.: Beyerchen [1977], chap. 8, Deichmann [1992], p. 207.

Zeitschrift für Physik
'Journal of Physics'. Founded 1920. Published by the German Physical Society (→ *DPG*) in addition to their Proceedings. Places of Publication: Berlin and Braunschweig. Main editor from 1920: K. → Scheel. This journal published more progressive research articles such as, for example, most of the papers published on quantum mechanics. Refs.: Holl [1992].

Zeitschrift für technische Physik
'Journal of Technical Physics'. The publication of the German Society of Technical Physics (→ *DGtP*) founded in 1919. It catered toward industrial physicists and engineers. Place of publication: Leipzig. Editors: G. Gehlhoff, H. Rukop, as well as Wilhelm Hort, later C. → Ramsauer. It appeared between 1920 and 1943. Refs.: Gehlhoff et al. [1920], Hoffmann & Swinne [1994].

F Biographical Profiles

This listing includes prominent individuals, predominantly physicists, mentioned repeatedly in the preceding text. It is intended as an orientation guide for the reader and not as a 'who's who of the Third Reich'. For the latter see, e.g., Wistrich [1982], Führerlexicon [1934], or Snyder [1976]; and Benz & Graml (Eds.) [1988] for the Weimer period. The main sources for this section are *Poggendorffs Biographisches Handwörterbuch* and *Kürschners Gelehrtenkalender*, along with other standard biographical lexica, obituaries in the *Physikalische Blätter* and other journals. Additional references are indicated at the end of each entry roughly in chronological order.

The following translations are used in this appendix: *ordentlicher Professor*: (full) prof.; *ausserordentlicher*: associate, *planmäßiger*: regular, and *nichtplanmäßiger*: supernumerary professors; *(nicht)beamteter Professor*: (un)tenured prof. *(Privat)dozent*: (unpaid) university lecturer. *Assistent*: teaching assistant (approximately equivalent to assistant prof. in the US); *Hilfsassistent*: auxiliary aid/part-time assistant. *Technische Hochschule*: Polytechnic; *Technische Universität*: Technical University.

Ardenne, Manfred Baron von (1907–)
Autodidact and entrepreneur. Ardenne maintained a private high-frequency engineering laboratory (*Forschungsinstitut Manfred von Ardenne*) in Lichterfelde (Berlin) (1928–45) to conduct his own research on communications technology, notably television and electron microscopy, and in 1910 turning also to electronics. He was supported by the → *RPM*. Isotope separation research was also conducted at Bad Weißer Hirsch 1945–55: Employed in the USSR on nuclear physics as a research institute in Suchumi, thereafter working again at his own lab with a staff of 500 on electron beam technology, mass spectrography, and biophysics in Dresden. Refs.: Ardenne [1972].

Bagge, Erich Rudolf (born 1912)
Nuclear physicist. 1931–34/35: Studied physics at Munich Univ. and the Berlin Polytech.; 1935–38: Leipzig Univ. 1938: Doctorate under the guidance of W. → Heisenberg on the theory of nuclear forces. 1937–41: Heisenberg's assistant. From Sep. 1939: Collaboration in the uranium project of the → *HWA*. 1941: Habilitation. 1941–45: Scientific collaboration at the → *KWIP*, working on a new isotope separation process. 1945f.: Internment in France, Belgium, and England. 1946: Assistant at the Max Planck Institute of Physics in Göttingen. 1948/49: Regular associate prof., then regular associate prof. of physics at the University of Hamburg and head of the *Physikalisches Staatsinstitut*, Hamburg, working on cosmic radiation. 1956: Managing scientist at the → *GKSS* near Hamburg. 1957: Director of the department of pure and applied nuclear physics at Kiel Univ. Refs.: Bagge, Diebner & Jay [1957], Bagge [1985], Walker [1989]a, [1995], pp. 208–215, 229ff., Frank (Ed.) [1993], Hoffmann (Ed.) [1993]b, p. 61, Renneberg [1993], p. 148.

Becker, Carl Heinrich Emil (sometimes also Karl) (1879–1940)
Weapons engineer and artillery general. From 1898: Military engineer; 1901–03: Studies at Munich Artillery and Engineering School; 1906–11: studies at the Berlin Military Engineering

Academy, in particular, on ballistics under Carl Cranz. 1908–11: Teaching assistant at the Ballistics Lab. there. 1911–14: Member of the Artillery Examining Board. 1914–16: Command of a 42 cm mortar battery. 1917–19: Adviser of artillery ballistics at the "Weapons and Equipment Inspection" (the later → *HWA*). 1919–23: Studied chemistry and metallurgy; 1922: engineering degree with a thesis on cathodic change of phenol. From 1922: Adviser to the *HWA* inspections office. From 1932: Honorary prof. at the Berlin Polytech. From autumn 1933 full prof. of technical physics, from 1935 full prof. of defense technology, physics and ballistics as well as dean of the military technology department newly created in 1933 at the Polytech. in Charlottenburg (Berlin). On the supervisory board of the → *KWG*, and from 1935: the first active general to be a member of the Prussian Academy of Sciences. From Nov. 1933: head of the research department and the Weapons Testing Office, and from Feb. 4, 1938 head of the *HWA*. From 1937: also first president of the → *RFR*. 1940 suicide in the presence of the → *Gestapo*, which was covered up with a state funeral. Refs.: Stuchtey et al. (Eds.) [1938], pp. 44–46, K. Becker & E. Schumann in Donnevert (Ed.) [1938/39], Justrow [1940], Karl [1940], N.N. [1940]a, Thiessen [1940], Ludwig [1974], pp. 220ff.

Becker, Richard (1887–1955)
Theoretical physicist, specializing in thermodynamics and statistical mechanics. Studied zoology and physics in Hamburg, was assistant at the Hanover Polytech., and then went briefly into industry. During World War I: Worked on explosives development. Following his habilitation in 1922 at the Berlin Polytech., unpaid lecturer; and 1926: full prof. there as head of the newly founded theoretical physics department. 1936: Elimination of his theoretical physics chair and compulsory transfer to the chair in Göttingen, which had been vacant since M. → Born's emigration. In 1954: Elected chairman of the regional association of the *Physikalische Gesellschaften*. Refs.: Döring [1947], Leibfried [1955]b, Neumann [1955].

Bethe, Hans Albrecht (1906–)
Nuclear physicist. Physics studies at Munich Univ. (notably with A. → Sommerfeld as his teacher and doctoral adviser); 1928: doctorate and 1929–32: teaching assistant to Sommerfeld; 1930: Habilitation; 1930–33: unpaid lecturer, Munich Univ. 1932/33: Teaching assignment at Tübingen Univ. 1933: Emigration first to England, where he became lecturer at Manchester Univ.; 1934–35: Bristol. Then from 1935: in the USA: 1935–37: assistant prof.; from 1937: prof. at Cornell Univ. lab. of nuclear studies, Ithaca, NY. 1942–43: Staff member, Radiation Lab., Massachusetts Institute of Technology. 1943–46: Collaboration in the Manhattan Project as chief of the 'theoretical physics division' of the Los Alamos Atomic Science Lab., New Mexico. Since 1947 consultant of the US Atomic Energy Commission. 1967 Nobel Prize in physics for, among other things, the discovery of the Bethe-Weizsäcker cycle to explain the combustion of hydrogen in the sun. Refs.: Bernstein [1980], Eckert [1993], pp. 132f., 152–158, 189–195, 227–233.

Bloch, Felix (1905–1983)
Theoretical physicist. 1924–27: Studies at the Zurich Polytech.; 1927–28: Leipzig Univ. 1928: Doctorate at Leipzig under W. → Heisenberg with a thesis on electron waves in crystal lattices; 1931: habilitation on the theory of resonance and hysteresis of ferromagnetica; 1932: unpaid university lecturer of theoretical physics. 1934: Prof. of theoretical physics, Stanford University, California. After a brief half-year stay at Los Alamos, 1941–44: Investigator, Manhattan Project. 1944–45: Radio Research Lab, Harvard. 1954–55: Director, CERN. 1952 Nobel Prize in physics. Refs.: Eckert [1993], pp. 130f., 248.

Bonhoeffer, Carl-Friedrich (also Karl-Friedrich) (1899–1957)
Physical chemist; discoverer of parahydrogen; brother of the theologian and resistance fighter Dietrich Bonhoeffer, who was executed by the Nazis. 1919–22: Studies at Berlin and Tübingen

Univs. 1922: Berlin Univ. (under W. → Nernst). 1923–30: Teaching assistant at the → *KWIPC* where he performed experiments on parahydrogen and orthohydrogen together with P. → Harteck. 1930–33: Full prof. of physical chemistry, Frankfort Univ.; research on reaction kinetics. 1934–45: Full prof. and colleague of W. → Heisenberg at Leipzig Univ. where he worked in electrochemistry and nuclear physics. K. → Wirtz was one of his assistants, who studied heavy water (D_2O). On Bonhoeffer's suggestion, the company → *Norsk Hydro* took up the production of heavy water for German nuclear research. He had considerable influence through his brother-in-law Hans von Dohnanyi, who worked at the Foreign/Defense Office, in obtaining military deferrals for members of the staff. 1947–49: Full prof., Berlin Univ.; 1948–51: Director of the → *KWIPC* (after 1949 renamed the Max Planck Institute of Physical Chemistry in Göttingen). Refs: Jaenicke [1957], Walker [1989]a.

Bopp, Friedrich (Fritz) Arnold (1909–1987)
Theoretical physicist. 1929–34: Studied physics at Frankfort-on-Main and Göttingen Univs. Wrote his masters thesis in 1933 under H. → Weyl. 1934: Teaching assistant at Göttingen; 1937 doctorate on Compton scattering under F. → Sauter. 1936–41: Teaching assistant, Breslau Univ.; 1941: habilitation under Erwin Fues with a study on a consistent field-theory of the electron. 1941–47: Staff scientist at the → *KWIP* in Berlin, later in Hechingen; 1946–47: Teaching assignment at Tübingen Univ. 1947–50: Regular associate prof., and 1950: full prof. of theoretical physics at Munich Univ.; worked on quantum field theory. 1954: Member of the board of trustees of the Institute. Co-signer of the 'Göttingen Manifesto' against rearming Germany with atomic weapons. 1964–65: President of the → *DPG*. Refs.: Rechenberg [1980], Kastrup [1988], Walker [1989]a, pp. 84, 126, 150, 186f.

Born, Max (1882–1970)
Theoretical physicist. 1900–06: Studies in Breslau, Heidelberg, Zurich and Göttingen; 1907: doctorate at Göttingen Univ. (with Minkowski's guidance). Early articles on relativity theory. 1915–19: Associate prof., Berlin Univ. 1915–23: Important contributions in the physics of crystals. 1919: Exchanged posts with M. → von Laue as full prof. of theoretical physics at Frankfort Univ. 1921: Full prof. and director of the theoretical physics department, Göttingen Univ. 1925: Develops statistical interpretation of quantum mechanics and (together with W. → Heisenberg and P. → Jordan) the algebraic description of problems in quantum mechanics, and later the 'Born approximation' for equations that cannot be solved exactly in quantum theory. Until 1933: Prof. of theoretical physics, Göttingen Univ. 1933: Emigration to England, continuation of work on electron field theory. 1954: Nobel Prize (together with W. → Bothe) for work on quantum mechanics. 1933–35: Stokes Lecturer in applied mathematics, Cambridge Univ. 1936–52: Tait professor of natural philosophy, Edinburgh Univ. 1939: British citizenship. 1954: Returned to Germany following his retirement. Co-signer of the Göttingen manifesto against rearming Germany with atomic weapons. Refs.: Born [1968], Lemmerich (Ed.) [1982], Holl [1992].

Bosch, Carl (1874–1940)
Physical chemist. 1894–98: Studies in chemistry, mechanical and metallurgical engineering at the Berlin Polytech. and chemistry at Leipzig Univ. 1898: Doctorate under the guidance of J. Wislicenus at the Karlsruhe Polytech. Apr. 15, 1899: Engaged by → *BASF*, where he succeeded in converting to industrial scale the process F. → Haber had developed in 1909 for the catalytic synthesis of ammonia from its elements hydrogen and nitrogen. From 1914: member of the executive board of *BASF*. He was awarded the 1931 Nobel Prize in chemistry jointly with Friedrich Bergius (1884–1949), for the invention and development of chemical high-pressure methods, including the catalytic coal hydrogenation process that produces oil and synthetic rubber. 1935–40: Chairman of the board of directors of → *IG Farben*. Longtime member of the

governing board of the → *KWIC*. 1937–40: President of the → *KWG*. Refs.: P. Müller [1940], N.N. [1940]b, Telschow [1953], Holdermann [1954], Kerstein [1970], Lenoir [1992].

Bothe, Walther Wilhelm Georg Franz (1891–1957)
Experimental physicist. 1908–12: Studies at Berlin Univ; 1914: doctorate (under the guidance of M. → Planck); from 1913: Planck's teaching assistant. 1920: privy councillor and 1927: member of the → *PTR*. 1925: Unpaid university lecturer, 1929: associate prof., Berlin Univ. 1930: Full prof. and director of the physics department, Giessen Univ. From 1932: Full prof. and director of the physics and radiology department, Heidelberg Univ. 1934: Loss of full professorship, but remained honorary prof. and director of the *KWI* of Medical Research in Heidelberg (which he had been holding since 1932), where he and his assistant W. → Gentner constructed the first German cyclotron. From 1939: Collaboration with R. → Fleischmann, W. Gentner, E. → Fünfer and others (in 1942, 6 physicists altogether) on the experimental determination of atomic constants, energy distribution in fission fragments and cross-sections for the → *HWA*'s nuclear energy project. His results on the absorption of neutrons in electrographite (pure carbon), which contradicted W. → Hanle's in Göttingen, were pivotal in the decision to favor heavy water as the moderator in uranium reactors. 1946: Regained professorship at Heidelberg Univ. 1949: Return to operation of the cyclotron that had been built in 1939 but seized by the Allies after 1945. 1954: Nobel Prize award in physics (together with M. → Born) for his atomic energy research. Refs.: R. Fleischmann in Treue & Hildebrandt (Eds.) [1987], Walker [1989]a, [1995], pp. 150, 165, 225, Hoffmann [1991], Osietzky and Weiss in Renneberg & Walker (Eds.) [1994], pp. 263ff., 271ff.

Brüche, Ernst Carl Reinhold (1900–1985)
Electrical engineer. 1919–24: Studies at Danzig Polytech.; 1926: doctorate there (under C. → Ramsauer). From 1920: Teaching assistant at the physics department, 1927: habilitation, and until 1933: unpaid university lecturer of experimental and technical physics, Danzig Polytech, where he worked on the measurement of scattering cross-sections of molecular gases. 1928–45: Head of the physics laboratories at → *AEG* in Reinickendorf (Berlin), where he concentrated mostly on geometrical electron optics and developed an electrostatic electron microscope. 1941: Award of Leibniz Medal. From 1944: Founding editor of the journal → *Physikalische Blätter*. 1946–51: Head scientist of the *Süddeutsches Laboratorium*, Mosbach; 1948: Managing director of the company *Physik-GmbH* in Mosbach; After 1951: at the *Physikalisches Laboratorium*, Mosbach. Refs.: Heinicke [1985], Rechenberg [1994]a, pp. 21–23, Mayer-Kuckuck (Ed.) [1995], pp. F 139–F 142.

Bühl, Alfons (1900–?)
Physicist. 1919–25: Studies at Berlin and Heidelberg Univs. 1925: Doctorate at Heidelberg (pupil of P. → Lenard) and teaching assistant. 1928: Teaching assistant at Freiburg (Breisgau); from 1929: unpaid university lecturer in physics. 1931–33: Lectureship at the physics department of the Zurich Polytech. 1933: Directorship of the physics department of the Karlsruhe Polytech. (to replace Wolfgang Gaede, who had been determined 'politically unreliable'). 1936: Untenured associate prof., and 1937–45: full prof., Karlsruhe Polytech. Physics adviser of the → *NSDDB*. Refs.: Walker [1995], pp. 30f.

Clusius, Klaus Paul Alfred (1903–1963)
Physical chemist. 1922–26: Studies at Breslau Polytech.; 1926: doctorate (under A. Eucken) on specific heat of solids at low temperatures. 1926–29 Eucken's teaching assistant. 1929/30: Foreign stays at Oxford and Leiden. 1931: habilitation and teaching assistant of physical chemistry, Göttingen, Univ. 1934: Regular associate prof., Würzburg Univ. From 1936: Full prof., Munich Univ. where he conducted major experiments on heavy water (D_2O) and developed an isotope separation tube in 1938 together with Gerhard Dickel that uses thermodiffusion. 1939:

P. → Harteck, R. → Fleischmann, W. → Groth and others initiated tests with the tube on uranium hexafluoride, the most important starting substance for isotope enrichment but which is unaffected by thermal diffusion. 1942: Employment of about 4 physical chemists on isotope separation and heavy water production problems. From 1947: Full prof. of physical chemistry, Zurich Univ. Ref.: Eggert [1963], Walker [1989]a, [1995], p. 165.

Debye, Peter Joseph Wilhelm (1884–1966) [Debije, Petrus Josephus Wilhelmus]
Theoretical physicist and physical chemist. 1901–05: Studies in electrotechnics at the Aachen Polytech. 1905: Doctorate on light pressure at Munich Univ. (under A. → Sommerfeld). From 1906: Teaching assistent of theoretical physics, and 1910: habilitation and unpaid lecturer at Munich Univ. 1911: Associate prof. of theoretical physics, Zurich Univ. 1912: Full prof. of theoretical physics, Utrecht Univ. From 1914: Full prof., Göttingen Univ., later head of the department of theoretical physics. From 1920: Full prof., Zurich Polytech. From 1927: Full prof. of experimental physics and later director of that department; also researched crystalline structure using X rays. 1937–39: Chairman of the → *DPG*. From autumn 1935: Director of → *KWIP* in Dahlem (Berlin). 1936: Nobel Prize in chemistry for his contribution to the theory of molecular structure through his investigations on dipole moment as well as X-ray and electron interference in gases. 1939: Forced to choose between adopting German citizenship, to permit his participation in militarily important research, and resignation of his directorship at the → *KWG*. 1940: Compromise solution by the → *REM*: approval of leave of absence, 1940–42: guest professorship at Cornell Univ. while still on Society's payroll without returning to Germany. Replaced by W. → Heisenberg as administrative director of the *KWIP*. 1948–52: Full prof. of chemistry and department head at Cornell; research on macro-molecules and polymers. 1952–66: Emeritus. Refs.: Falkenhagen [1966], Busch [1985], von Meyenn in Treue & Hildebrandt (Eds.) [1987], Walker [1989]a, pp. 19, 59f., 86.

Diebner, Kurt (1905–1964)
Nuclear physicist. From 1925: Physics studies at Halle and Innsbruck. 1931: Doctoral thesis on column ionization of individual α rays under the guidance of G. → Hoffmann in Halle; from 1931 Hoffmann's teaching assistant. Papers on nuclear disintegration and construction of a high-tension facility for proton acceleration. From 1934: Part-time employee at the → *PTR* in Berlin; adviser to the → *RWM*, and military adviser to the → *HWA* on nuclear physics. 1945: promotion to senior government building officer. Army specialist on nuclear physics and explosives; from 1939: director of a newly created Nuclear Research Council (*Kernforschungsrat*) under General C. → Becker (*HWA*), with participation in the uranium project (*Uranverein*) and managing director of the → *KWIP*. Until 1945: Director of the testing site of the *HWA* in Gottow, where a half-dozen physicists measured nuclear constants, and the experimental station of the → *RFR* in Stadtilm, where experiments on neutron multiplication with uranium cubes and heavy water were being conducted with the goal of designing a functioning nuclear reactor. 1946: Internment in England. Later employed in industry (development of electronic measuring instruments): From 1947/48: Director and joint-owner of the instrument manufacturer *DURAG-Apparatebau GmbH* in Hamburg. From 1956: Member of the supervisory board of → *GKSS*, and from 1957: lecturer at the state School of Naval Engineers in Flensburg. Refs.: Diebner [1955], Bagge, Diebner & Jay [1957], pp. 9–82, Walker [1989]a, [1995], Frank (Ed.) [1993], Hoffmann [1993], p. 63, Renneberg [1993], pp. 148ff.

Döpel, Klara (Minna) Renate (née Mannß) (1900–1945)
Experimental physicist and feminist. 1922/23–34: Law studies at Munich Univ. 1934: Married to R. → Döpel. Change to physics studies at Würzburg. 1938: Move to Leipzig following R. Döpel's appointment there. Published together with R. Döpel and W. → Heisenberg secret *Deutsche Forschungsberichte* reports on the progress of the 'uranium machine' experiments at

Leipzig. Apr. 6, 1945: Died in an air raid on the physics building. Refs.: Kleint & Wiemers (Eds.) [1993], pp. 82–84.

Döpel, Georg **Robert** (1895–1982)
Experimental physicist, specializing in nuclear physics and electromagnetism. 1919–24: Studies at Leipzig, Jena and Munich Univs. 1924: Doctorate (under W. Wien) at Munich. Then teaching assistant at Göttingen Univ. under R. W. Pohl working at the private lab. of Baron R. von Hirsch zu Planegg, on canal rays together with J. → Stark. Since 1929: Teaching assistant, and since 1932: unpaid lecturer at Würzburg Univ. 1939–45: Regular associate prof. in the civil service, Leipzig Univ.; colleague of W. → Heisenberg; change of focus to nuclear physics. Aug. 1940: Proved the viability of heavy water as a moderator substance. From June 1940 to summer 1942: Supplemented K. → Wirtz's experiments in Berlin with independent spherical, layered 'uranium machine' model. 1945–58: Employed in the USSR, then prof. at the electrical engineering college at Ilmenau, *Hochschule für Elektrotechnik*, where he conducted spectroscopic analyses of the mechanicsm of gaseous discharge. Refs.: Hanle [1983], Kleint [1986], Walker [1989]a, Kleint & Wiemers (Eds.) [1993], pp. 53–84.

Droste zu Vischering-Padberg, Gottfried Baron von (1908–)
Physical chemist. 1926–33: Studies at the Univs. of Heidelberg, Munich and Berlin. 1933: Doctorate at Berlin (under L. → Meitner). 1933–42: Scientific assistant at the → *KWIC*; work on nuclear chemistry. From 1933: Member of the → *SA*, and from 1937: Member of the → *NSDAP*. 1942–44: Strasbourg Univ. 1944–45: Collaborator in the physics department of the *KWI* of Physical Medicine, Heidelberg. 1946–51: Physics department, Heidelberg Univ. From 1951: Employed at the → *PTR*, renamed *Physikalisch-Technische Bundesanstalt*, Braunschweig. Refs.: Walker [1989]a, pp. 197f.

Einstein, Albert (1879–1955)
Theoretical physicist. From 1896: Studies at the Zurich Polytech.; 1900: graduation (*Diplom*). 1902–09: On staff of Berne Patent Office. Development of special (1905) and general (1915f.) theories of relativity. 1905: Dissertation on the determination of molecular dimensions; mathematical calculation of Brownian motion and the implications of the photoelectric effect for quantum theory, for which he was awarded the 1921 Nobel Prize in physics. 1908: Habilitation at Berne Univ.; associate prof., Zurich Univ. 1911: Full prof., Prague Univ. 1912: Full prof., Zurich Polytech. 1914–33: Regular full-time member of the Prussian Academy of Sciences, Berlin; simultaneously 1917–33: Director of the → *KWIP* in Dahlem (Berlin) (active director only until 1922). 1922–24: Member of the supervisory board of the → *KWG*, as well as on the board of trustees of the Potsdam Astrophysical Observatory and of the → *PTR* (to 1933). 1933: Research visit to the USA, from which he never returned upon Hitler's rise to power. Emigration to the USA and honorary prof. at the Institute for Advanced Study, Princeton, where he worked in increasing isolation on unified field theory. Refs.: Einstein [1933], Reichinstein [1935], Hermann [1966], Clark [1971], Herneck [1976], Kleinert & Schönbeck [1978], Kirsten & Treder (Eds.) [1979], Kleinert [1979], Pais [1982], Elton [1986], H. J. Treder and M. von Laue in Treue & Hildebrandt (Eds.) [1987], Guérout [1992], pp. 44–54, etc.

Esau, Abraham (1884–1955)
Technical physicist. 1902–07: Studies at Berlin Univ. and Danzig Polytech. 1906–09: Teaching assistant of M. → Wien at Danzig Polytech.; 1908: doctoral thesis at Berlin Univ. 1909–10: Volunteer at the radio transmission division of the Berlin telegraph battalion. 1910–12: teaching assistant at Halle Univ. 1914: active service in Togo where he became a prisoner of war of the French, returning to Germany 1918. 1912–25: Laboratory chairman of the Wireless Telegraphy Society (*Gesellschaft für drahtlose Telegraphie*) in Berlin and pioneer of very high frequency waves (VHF) used in radar, radio and television (at *Telefunken*). 1921–22: Foreign stays in

Argentina and Brazil. From 1925: President of the *Deutscher Telefunken Verband.* 1925–27: Associate prof. of technical physics; 1927–39: full prof. of technical physics and director of the technical physics department, Jena Univ.; and 1932–35 and 1937: Rector there. Oct. 1933: Thuringian state councillor (*Staatsrat*), thus having direct access to → Hitler. 1939–45: Full prof., Berlin Polytech. and president of the → *PTR*. At the same time, visiting lecturer at the *Technische Akademie Bergisch-Land* and until the end of 1943 head of the physics special division of the → *RFR* of which he had been member since 1937. Apr. 29, 1939: Organized the inaugural meeting of the uranium project (*Uranverein*). Dec. 8, 1942: Appointment as H. → Göring's plenipotentiary of nuclear physics, with which he controlled nuclear research in Germany; Oct. 28, 1943: submitted letter of resignation; Dec. 1943: replaced by W. → Gerlach. Jan. 1, 1944: Appointment as plenipotentiary of high-frequency engineering and radar. After the war until 1948: Prisoner of the Dutch. From 1949: Visiting prof. of short-wave technology at the Aachen Polytech. From 1953: also head of the Institute of High-Frequency Engineering of the German Aeronautical Research Institute in Mülheim (Ruhr). After 1954: Honorary doctorate, Freiburg Univ. for his research on diathermy. Refs.: Stuchtey et al. (Eds.) [1938], p. 61, Rindfleisch [1955], Schröter [1959], Brandt [1962], p. 27, Walker [1989]a, [1995], pp. 138ff., 165.

Ewald, Peter Paul (1888–1985)
Physicist, specialist in X-ray analysis of crystalline structures. Chemistry studies at Cambridge, mathematics and physics studies at Göttingen and Munich Univs. 1912: Doctoral thesis on dispersion and double refraction of electron grids at Munich Univ. under the guidance of A. → Sommerfeld. Teaching assistant, and from 1918: unpaid lecturer of theoretical physics, Munich Univ.; work on the physics of interferences of X rays in crystals and its applications in crystallography. From 1921: Associate prof. of theoretical physics, Stuttgart Polytech.; from 1922: full prof. there, and 1933: rector. Apr. 20, 1933: Voluntary withdrawal from post as rector in protest against the new Laws for the Restoration of the Professional Civil Service. 1937: Emigration to England; until 1939: fellow at Cambridge Univ. 1939–49: Lecturer, later prof. of mathematics and physics, The Queen's University, Belfast, Northern Ireland. 1948–59: Editor of *Acta Crystallographica.* 1949: Move to the USA, and until 1959: at the Brooklyn Polytech.al Institute. Recipient of the 1978 Max-Planck Medal of the → *DPG*. Ref.: Hildebrandt [1985], [1993], Walker [1989]a, pp. 42, 83f., 170.

Finkelnburg, Wolfgang Karl Ernst (1905–1967)
Experimental physicist. 1924–28: Studies at Tübingen and Bonn Univs. 1928: Doctorate under the guidance of H. → Konen at Bonn; 1928: Konen's teaching assistant. 1931: Teaching assistant at the theoretical physics department, and 1932: unpaid university lecturer at the Karlsruhe Polytech. 1933/34 fellowship to work with R.A. Millikan in the USA on continuous spectra. 1936: Unpaid university lecturer, and from 1936: associate prof. at the Darmstadt Polytech. 1942–45: Associate prof. and director of the physics department, Strasbourg Univ.; work on the high-temperature carbon arc. Also from 1941: Vice president of the → *DPG* and influential member of the → *NSDDB*. He organized the 'Munich Synod' in 1941 between the supporters and opponents of the 'Aryan Physics' movement, which ended with the physics community's official rejection of the latter and acknowledgment of the validity of the theories of relativity and quantum mechanics as a part of German science. 1946–52: Guest lecturer at the Catholic Univ. of America, and scientific consultant of the Research and Development Lab. at Fort Belvoir, Virginia. From 1952: member of the research dept., and from 1955: head of the dept. for reactor development. From 1963: general manager of the *Siemens-Schuckert* plant. From 1955: also honorary prof. of nuclear physics at Erlangen Univ. and 1966–67: president of the → *DPG*. Ref.: Brüche [1967], Trendelenburg [1975], pp. 260f., Walker [1989]a, pp. 58, 67ff., 74ff.

Fischer, Erich Horst (1910–1969)
Experimental physicist. 1929–35: Studies at Bonn, Munich and Berlin Univs. 1935: Doctorate at Berlin under W. → Nernst and A. Deubner. 1935–36: Teaching assistant of W. Friedrich, Berlin Univ. 1937: Assistant at the → *KWIP*, Berlin, under P. → Debye and W. → Heisenberg. 1939: Habilitation, and 1942: Lecturer at Berlin Univ.; member of the German *Uranverein*, for which he determined the rate of neutron multiplication in heterogeneous uranium-moderator combinations. After 1945: helped to rebuild the Hechingen branch of the *KWIP*. In 1948: transferred to Tübingen Univ. as lecturer of experimental physics, and 1950: supernumerary prof. there. 1951: Full prof., Ankara Univ. From 1956: employed at the → *GKSS* firm, conducting materials research for the construction of reactors. Ref.: Bagge [1969], Walker [1989]a, pp. 84, 125, 150, [1995], pp. 75, 87f., 95ff., 121.

Flammersfeld, Arnold Rudolf Karl (1913–)
Experimental physicist. 1931–37: Studies in physics, Berlin Univ., and 1938: doctorate there (student of L. → Meitner). From 1937: Meitner's assistant at the → *KWIC*. 1939–41: Staff scientist at the physics department of the *KWI* of Medical Research, Heidelberg with W. → Bothe. From 1941: Employed at the *KWIC*. 1947: Habilitation at Tübingen Univ., and from 1948: unpaid lecturer there. Also worked on the installation of the belt generator in Tailfingen. From 1949: Unpaid lecturer of physics, Mainz Univ. From 1954: Full prof. Göttingen Univ. Refs.: Weiss [1994], p. 289.

Fleischmann, Arthur Wilhelm **Rudolf** (1903–)
Experimental nuclear physicist. 1922–26: Studies at Erlangen and Munich Univs.; 1929: doctorate (under B. Gudden). 1931: Teaching assistant at physics department I, Göttingen Univ. 1932–34: Teaching assistant to W. → Bothe, Heidelberg Univ.; 1934–41: Collaboration with Bothe at the *KWI* of Medical Research, Heidelberg. 1937: Habilitation at Heidelberg Univ., and from 1938: unpaid university lecturer there. From 1939: Testing of the Clusius-Dickel isotope separation process on uranium hexafluoride. 1941–44: Regular associate prof., Strasbourg Univ. 1947-53: Full prof., Hamburg Univ. Since 1953: Full prof., Erlangen Univ. Refs.: Walker [1989]a, pp. 31f., 67, 154ff.

Flügge, Siegfried (1912–)
Theoretical physicist. 1929–33: Studies at the Dresden Polytech. and Göttingen Univ., and 1933: Doctorate at the latter. 1933–35: Teaching assistant, Frankfort Univ. 1936–37: Teaching assistant, Leipzig Univ. (under W. → Heisenberg). 1937–42: Assistant of O. → Hahn at the → *KWIC* (as successor to the emigré M. Delbrück). 1938: Habilitation at the Munich Polytech., and in the winter semester of 1940: Lectured there on theoretical physics on the initiative of R. → Tomaschek, despite W. → Müller's objections. From 1941: Lecturer at Berlin Univ. 1939: Publication of two influential articles on the exploitation of nuclear energy, and from 1939: Collaboration with C. F. → von Weizsäcker and F. → Houtermans on the theoretical basis of uranium reactors and extended Bohr's and Wheeler's theory of nuclear fission to the transuranic elements. 1944: Associate prof., Königsberg Univ. 1945–47: Transfer to Göttingen Univ. From 1947: Marburg Univ. 1949–50: Visiting prof. at Madison, Wisc., and 1953: at the Carnegie Institute of Technology, Pittsburgh; later at Freiburg Univ. Refs.: Walker [1989]a, Wengenroth [1993], pp. 246, 249; Weiss [1994], p. 274.

Franck, James (1882–1964)
Experimental physicist. Studies in Berlin; 1906: doctorate there (under E. Warburg). 1911: Start of electron collision experiments, which since 1913 have been considered a clear verification of N. Bohr's and A. Sommerfeld's atomic theory. From 1918: Department head and member of the board of trustees of the → *KWIPC*. 1925: (together with G. → Hertz) Nobel Prize in physics. 1920–34: Full prof. of experimental physics, Göttingen Univ., and director of physics

department II there. 1919–20: Member scientist at the *KWIPC* and then 1926–38: corresponding member. Also longstanding member of the Electrophysics Committee of the → *NG* as well as 1931–33: Board of trustees member of the → *KWIP*. 1933: Voluntary resignation of his professorship and exile: 1934–35: Guest prof., Copenhagen Univ.; 1935–38: Full prof., Baltimore Univ.; 1938–47: Full prof., chemistry department, Chicago Univ. Collaboration at the Chicago branch of the Manhattan Project; according to Franck's report of June 1945, he and six other physicists involved in research leading to the development of the American atomic bomb advised the U.S. government not to use the bomb against the Japanese (Franck report). From 1947: Emeritus. Special interest in photosynthesis. Refs.: Kroebel [1964], Westphal [1964], Engelmann [1964], Franck [1964], pp. 324–334, N.N.[1964], Hertz [1965], Kuhn [1965], Lemmerich (Ed.) [1982].

Frick, Wilhelm (1877–1946)
National Socialist official. 1896–1901: Law studies at Göttingen, Munich, Berlin and Heidelberg Univs., doctorate at the latter in 1901. 1900–03: Bavarian law trainee in Kaiserslautern. 1904–24: Administrative assistant of the Upper Bavarian government, and public prosecuter [*Amtsanwalt*] at the Munich police headquarters, and 1907–17: District assessor in Pirmasens. 1917–23: Government assessor and senior official at the Munich police headquarters. 1919–25: Head of the political police department. In 1920's: A. → Hitler's contact at police headquarters. Arrested for his involvement in the *putsch* attempt of Nov. 9, 1923 and condemned to 15 month's imprisonment, but one of the first Nazi representatives to be released prematurely. Elected to parliament in 1924. 1926–30: Employed at the Munich *Oberversicherungsamt*. 1928: Parliamentary party leader of the → *NSDAP*. Jan. 23, 1930–Apr. 1 1931: Thuringian Interior Minister; Jan. 30, 1933–Aug. 1943: Minister of the → *RIM*. Outlawed Communist and Socialist parliamentary opposition. Apr. 7, 1933: Issued Law for the Restoration of the Professional Civil Service. From May 1, 1934: also head of the Prussian Interior Ministry. Aug. 23, 1943–end of the war: Reich Protector of Bohemia-Moravia. Sep. 15, 1935: Drew up discriminatory Nuremburg Laws on citizenship and race. Oct. 16, 1946: Execution following his conviction at the Nuremburg Trials. Refs.: Führerlexikon [1934], Snyder [1976], Wistrich [1982].

Frisch, Otto Robert (1904–1979)
Nuclear physicist. 1922–26: Mathematics and physics studies at Vienna Univ., and 1926: doctorate there. 1927–30: Staff scientist at the → *PTR*, Berlin. 1930–33: O. Stern's teaching assistant at the physical chemistry department, Hamburg Univ. 1933: Following the Nazi government take-over, forced exile: 1933–34: London (Blackett's assistant); 1934–39: Copenhagen (collaboration with N. Bohr on nuclear physics). Beginning 1939: calculated together with his aunt L. → Meitner the amount of energy released during nuclear fission (published in *Nature*). Jan. 16, 1939: Determined the recoil energy of fission products, which experimentally confirmed nuclear fission. 1939–40: Birmingham (Oliphant); 1940–43: Liverpool; 1943–46: Los Alamos, where he determined the critical mass of spontaneous fission. 1946–47: Harwell. From 1947: Prof. at Trinity College, Cambridge Univ. Refs.: Frisch [1979], [1981]b, Lemmerich [1980], [1988].

Führer, Wilhelm August Ludwig (1904–1974)
Astronomer and National Socialist official. From 1925: Studies in astronomy, mathematics, physics and chemistry at Frankfort Univ. From 1927: Studies at Kiel Univ., and 1933: doctoral thesis there on color equivalence of 51 subpolar stars. 1930–33: Engaged as editor of the journal *Astronomische Nachrichten*. 1933–36: auxiliary aid and assistant at the Munich Observatory. 1934–36: Head of the lecturer's organization *Dozentenschaft* at Munich Univ. From July 1935: Leader of the → *NSDDB*; 1936–39: Munich district leader of the Lecturer's League, and from Oct. 1936: In the university affairs division of the Bavarian Ministry of Culture. From July 1939:

Senior official adviser at the → *REM*, until Oct. 1940 additionally at the → *HWA*. 1936–41: → *SS* Junior company commander; 1941–43: Senior company commander, → *Ahnenerbe* Office of the *SS*; 1942–43: Junior company commander of the armed branch, *Waffen-SS*; 1943–45: Company commander-in-chief of the *SS*, and senior company commander *Waffen-SS*. 1945–48: Interned. 1949–50: Rated as lessor offender and fellow traveller. Refs.: Litten [1992].

Gans, Richard Martin (1880–1954)
Experimental physicist. Studied physics 1898–99: at Hanover Univ., 1899–1901: at Strasbourg, and wrote his doctoral thesis under H. Weber on induction in rotation conductors. From 1901: Teaching assistant at Heidelberg, from 1902: at Tübingen, 1903: habilitation there and from 1903: unpaid university lecturer. 1911: unpaid university lecturer at Strasbourg, 1912: prof. and director of the physics department, La Plata Univ., Argentina, Research in electrodynamics, magnetism and optics. 1925–35: Full prof. and director of the Institute of Physics II, Königsberg Univ., 1936–39 technical science consultant to → *AEG* in Berlin, 1943–45 on the staff of the → *RFR* in a project mainly devised to exempt him from service in a work camp. 1946: Prof. of theoretical physics, Munich Univ. 1947–51: Full prof., La Plata Univ., from 1951: full prof. of physics, Buenos Aires Univ. Refs: Mrowka [1954], Swinne [1992].

Gehrcke, Ernst J. L. (1878–1960)
Experimental physicist. Student of E. Warburg. 1903: Developed together with O. Lummer (1860–1925) a precision interferometer or plate (*Lummer-Gehrcke-Platte*). 1908: Discovery together with O. Reichenheim of anode rays. 1901–46: Employed at the → *PTR*. From 1921: Senior official adviser; from 1926: director of its optical labs. 1927: Developed (together with E. Lau) the multiplex interferometric spectroscope for precision resolution of spectral-line structures. From 1921: Associate prof., Berlin Univ. 1922–25: Member of the board of trustees of the Potsdam Astrophysical Observatory. Critic of relativity theory. After the war: Employed at → *Zeiss* in Jena; helped establish the physiological optics department at Jena Univ. 1949: Moved to West Berlin and engaged at the Materials and Products Testing Office. Then appointed to the German Office of Weights and Measures, where he was serving as independent scientist in 1958. Refs.: N.N. [1958], Goenner [1993], pp. 114f.

Geiger, Hans (1882–1945)
Experimental physicist. Studies at Erlangen, Munich and Manchester Univs. 1906: Doctorate at Erlangen. 1907–12: Lecturer at the physics department, Manchester Univ. as Rutherford's assistant collaborating on the scattering of alpha particles that confirmed Rutherford's model of the atom. From 1912: Member of the → *PTR*; 1913: development of the demand meter and head of the radium research laboratory there. 1925–29: Full prof. of experimental physics and director of the physics department, Kiel Univ., and 1928: development there (together with Walter Müller) of the Gehrcke-Müller counter tube for measurements in nuclear physics. 1929–36: Full prof. of physics, Tübingen Univ. From 1936: Full prof. and department head at the Polytech. in Charlottenburg (Berlin) (successor to Gustav → Hertz). Refs.: Stuhlinger [1946], Cassidy [1979], O. Haxel in Treue & Hildebrandt (Eds.) [1987], Lemmerich [1988], pp. 95ff., Swinne [1988], Walker [1995], pp. 129, 154, Mayer-Kuckuck [1995].

Gentner, Wolfgang (1906–1980)
Experimental physicist. 1925–30: Studies at Erlangen and Frankfort Univs. 1930: Doctorate at Frankfort (under the guidance of F. Dessauer). 1932: Auxiliary aid at the physical foundations of medicine department, Frankfort Univ. 1933: Scholarship at the Curie Radium Institut Lab., Sorbonne. 1936–45: Assistant at the physics department of the *KWI* of Medical Research, Heidelberg (under W. → Bothe). Work on the nuclear photoeffect. Member of the National Socialist Flying Corps. 1937: Habilitation, Frankfort Univ., then university lecturer, 1938–39: fellowship at Berkeley where he studied the physics and technology of cyclotrons. From 1941:

teaching assignment. 1946: Full prof., Freiburg Univ. where he worked on nuclear and cosmic-ray physics. From 1955: Director of the synchro-cyclotron department of the European nuclear physics organization (CERN), and 1971–74 chairman of the CERN board. In 1965: Officer in the French Legion of Honor, from 1975: Member of the board of governors at the Weizmann Institute, Israel. From 1972: Vice-president of the Max Planck Society (successor to → *KWG*), from 1974: member of the *Ordre pour le Mérite*. Ref.: Citron [1980], Walker [1989]a, pp. 44, 125, 157, 181, 197.

Gerlach, Walther (1889–1979)
Experimental physicist specializing in atomic physics and quantum theory. 1908–11: Studies at Tübingen Univ. 1912: Doctoral thesis on the measurement of radiation under the guidance of F. → Paschen. From 1911: Teaching assistent, 1916: habilitation, then unpaid lecturer at Tübingen Univ. From 1917: Unpaid lecturer at Göttingen Univ. 1919–20: Head of the physical laboratory of the Eberfeld dye factory (formerly *Bayer-Werke*). 1920: Teaching assistant and university lecturer, and 1921: associate prof., Frankfort-on-Main Univ., where he discovered jointly with Otto Stern in Nov. 1921 space quantization in a magnetic field (Stern-Gerlach effect). From 1925: Full prof., Tübingen Univ. (successor to Paschen). 1929–May 1945: Full prof., Munich Univ. (successor to W. Wien). From Apr. 1935: Chairman of the committee to appoint A. → Sommerfeld's successor. From 1937: Member of the supervisory board of the → *KWG*. From Nov. 1939: Member of the Comerlin working group on ship degaussing and torpedo physics. Jan. 1, 1944: Officially named head of the physics section of the → *RFR* and plenipotentiary of nuclear physics (replacing A. → Esau). In Apr. 1944: Founded the *Reichsberichte für Physik*, official reports appearing as a supplement to the → *Physikalische Zeitschrift*. From May 1945: Interned in France and Belgium; July 1945–Jan. 1946 at Farm Hall, England. After 1946: Influential official of the *KWG* and its successor *Max-Planck-Gesellschaft*, 1948–51: rector of the Univ. of Munich. 1949–51: Founding president of the Fraunhofer Society. From 1946: Visiting prof., Bonn Univ.; 1948–57: Full prof. of experimental physics and director of the physics department, Munich Univ.; member of the Göttingen, Halle and Munich academies of science, member of the *Ordre pour le Mérite* and recipient of high awards. 1949–61: Vice-president of the → *DFG*, 1956–57: chairman of the → *DPG* and co-signer of the 'Göttingen Manifesto' against the rearmament of the Federal Republic of Germany with atomic weapons in 1957. Refs.: Stierstadt [1980], Nida-Rümelin (Ed.) [1982], Heinrich & Bachmann (Eds.) [1989], Walker [1989]a, pp. 129–148, 151, 157–163, 205, 208, [1995], Frank (Ed.) [1993], Hoffmann (Ed.) [1993]b, p. 65.

Glaser, Ludwig (1889–?)
Applied physicist. Studies at the Polytech. and Univ. in Berlin and at Imperial College of Science, London. Doctorate in engineering; worked subsequently as engineer at → *Krupp*. 1915–22: Editor of *Glasers Annalen für Gewerbe und Bauwesen*. 1921: Unpaid lecturer at Würzburg Univ. Before 1931: Associate prof.: Protegé of J. → Stark and P. → Lenard. Convinced antirelativist and opponent of Bohr's and Sommerfeld's quantum theories. Feb. 1932: entered Nazi Party. After 1933: Active proponent of the 'Aryan Physics' movement, flagrant anti-Semite. After 1939: W. → Müller's teaching assistant, Munich Univ. Refs.: Goenner [1993], pp. 118f., Walker [1995], pp. 11, 55–57.

Goebbels, Paul **Joseph** (1897–1945)
National Socialist politician. From 1922: Member of the → *NSDAP*; 1927–35: Editor of his weekly pro-Nazi newspaper *Der Angriff*. District leader (*Gauleiter*) of Brandenburg (Berlin) and 1929: Reich Propaganda Leader. 1932: Organized A. → Hitler's two presidency campaigns. Mar. 13, 1933: appointed Reich Minister of Public Information and Propaganda, in which capacity he attempted to influence all fields of culture (7 sections: propaganda, radio, press, film, theater, music, and fine arts). 1938: Responsible for the 'Night of Broken Glass' (*Reichskristallnacht*).

After the failed assassination attempt on Hitler in July 1944: appointed Plenipotentiary of the Mobilization for Total War, and assigned by Hitler as his successor as Reich Chancellor. 1945: Suicide together with his entire family. Refs.: Fest [1976], Vol. 2, p. 580, Walker [1995], etc.

Göring, Hermann Wilhelm (1893–1946)
National Socialist politician. From 1922: Member of the → *NSDAP*. Dec. 1922: appointed supreme commander of the → *SA*. Involved in the 1923 *putsch* in Munich, exile until 1926. 1928 and 1930: Elected to parliament. 1932: Parliament president. 1933: Briefly, Prussian Prime Minister, Reich Commissioner for Air, and Interior Minister, possibly the man behind the arson of the parliament building. June 30, 1934: Orchestrated the purge of the *SA*. From May 1935: General and commander in chief of the Air Force. From 1936: Commissioner of the Four-Year Plan, which led to the massive rearmament of Germany. End Aug. 1939: Chairman of the Reich Council for National Defense and designated → Hitler's chosen successor. June 19, 1940: Appointed Reich Marshal; from 1942: commissioned additionally with the direction of the → *RFR*. Commanded the *Blitzkrieg*. 1945: Removed from his posts by Hitler as a result of the failure of the German air defense and thrown out of the party. 1946: Forestalled his Nuremberg death sentence through suicide. Refs.: Fraenkel & Manvell [196.], Kube [1986], Trischler [1992]a, pp. 174ff., etc.

Goudsmit, Samuel Abraham (1902–1978)
Dutch-American atomic physicist. 1919–26: Studies, Leiden Univ., and 1927: doctorate there (under P. Ehrenfest). Research on the Zeeman effect; 1925: Discovery of electron spin (together with G. Uhlenbeck). 1927–28: Instructor, 1928–32: assistant prof., and 1932–36: full prof., Michigan Univ. 1941–46: simultaneously member of the Radiation Lab., Massachusetts Institute of Technology. Feb. 11, 1943: Death of both his parents at Auschwitz. 1944/45: Chief scientific officer of the Allies' 'Alsos Mission' to track down the physicists who had collaborated in the German uranium project. 1946–48: Full prof., Northwestern Univ. From 1948: Senior scientist, and 1952–60: Chairman of the physics department, Brookhaven National Lab. Refs.: Walker [1989]a, [1990]b.

Groth, Wilhelm (1904–1977)
Physical chemist. 1922–27: Studies at the Munich Polytech. and Univ. and Tübingen Univ. 1927: Doctorate (under W. → Gerlach). 1927: Teaching assistant, Hanover Polytech. From 1932: Assistant to O. Stern and P. → Harteck in Hamburg. End of 1938: Habilitation following complicated negotiations with the Hamburg district university lecturers leadership. End of 1939: Tested simultaneously with R. → Fleischmann the Clusius-Dickel isotope separation process on uranium hexafluoride, with negative result. Autumn of 1941: Beginning of construction of an ultracentrifuge (together with Harteck and Suhr) for the enrichment of uranium 235. 1943: Successful enrichment to about 5%, however, large-scale production was not yet possible due to technical difficulties. After 1945: Appointed supernumerary associate prof. of physical chemistry at Hamburg Univ., with the permission of the British occupation government; from 1948: regular associate prof. there. From 1950: Full prof., Bonn Univ. Refs.: Walker [1989]a.

Haber, Fritz (1868–1934)
Physical chemist. 1886–91: Studies at Berlin and Heidelberg Univs. and at the Berlin Polytech. 1891: Doctorate (under C. Liebermann). From 1894: Karlsruhe Polytech.; 1894-96: teaching assistant; from 1896: unpaid university lecturer of technical chemistry, from 1898: associate prof., and from 1906: full prof. of physical chemistry, Berlin Univ. In World War I: mainly responsible for the use of poison gas at the Western front. 1919: Nobel Prize award in chemistry for the year 1918, for his collaboration with C. → Bosch in developing the process for producing synthetic ammonia from hydrogen and nitrogen using high pressure. 1920: Together with F. →

Schmidt-Ott, founding father of the → *NG*, and longstanding vice-president and representative in its Japan Committee. 1917–33: Member of the board of trustees of the → *KWIP*. 1933: Resignation as director of the → *KWIPC*, which post he had held since 1911, to protest the dismissal of his colleagues for "reasons of race", including the department heads Herbert Freundlich and Michael Polányi. 1933: Emigration to Cambridge, England. Jan. 29, 1934: Death in Switzerland. Jan. 29, 1935: A commemoration event organized by M. → Planck, M. → von Laue and others was held in Dahlem (Berlin) despite an official order forbidding it. Refs.: Berl [1928] including a bibliography, Jaenicke [1935], von Laue et al. [1953], Harteck [1960], pp. 464–465, Jaenicke [1966], Goran [1967], Zierold [1968], pp. 9ff., 50, 124ff., 151f.; Chmiel et al. [1986], pp. 7–43, A. Becker et al. in: Treue et al.(Eds.) [1987], F. Stern in: Vierhaus & v.Brocke (Eds.) [1990], Lenoir [1992], Macrakis [1993], pp. 32ff., 68ff., 85ff., Stoltzenberg [1994].

Hahn, Otto (1879–1968)
Physical chemist. Studies at Munich and Marburg Univs.; 1901 doctorate at Marburg. From 1910: Associate prof., Berlin Univ., and from 1924: also director of physics department II there. From 1912: Head of the radioactivity department of the → *KWIC*, and from 1926: provisional director there; from 1928: director. Member of the Japan Committee of the → *NG*. From 1933: Substitute director of the → *KWIPC*, also in Dahlem (Berlin). 1934: Resignation of his professorship at Berlin Univ. in response to F. → Haber's departure. End of 1938: Discovery together with F. → Straßmann of the nuclear fission of uranium through purely chemical methods, which his former collaborator L. → Meitner provided with a correct theoretical explanation (together with O. → Frisch) at the beginning of 1939. From 1939: Hahn and Meitner's successor J. → Mattauch laid great emphasis on nuclear physics research at the Institute. After the outbreak of war: Several reports on the contributions of his Institute to the nuclear energy project toward realizing the → *HWA*'s military goals. In Nov. 1945: Award of the 1944 Nobel Prize in chemistry. Apr. 1946–59: President of the → *KWG*, later renamed Max Planck Society. Refs.: Laue [1939], Bagge, Diebner & Jay [1957], Hahn [1962]b, [1968], pp. 337–450, [1968/86], [1975], Gerlach [1968], Hartmann [1968], Karlik [1970], Berminger [1974], Spence [1977], E. H. Berninger in Treue & Hildebrandt (Eds.) [1987], D. Hahn (Ed.) [1988], Lemmerich [1988], pp. 35ff., Walker [1989]a, [1995], Frank (Ed.) [1993], Oexle [1994], Weiss [1994].

Hamel, Georg Karl Wilhelm (1877–1954)
Mathematical physicist. 1895–1901: Studies at Aachen, Berlin, and Göttingen Univs., and 1901: doctorate at the latter (under Hilbert). 1905–12: Prof. of mechanics, Brünn Polytech. 1912–19: Full prof., Aachen Polytech. 1919–45: Full prof. of mathematic and mechanics, Berlin Polytech. After 1933: Supported the 'Aryan Mathematics' movement. 1946–49: Full prof., Berlin Polytech., and also 1946–47: Guest prof., Tübingen Univ., and 1947–48: Guest prof., Berlin Univ. From 1948: Lectureship, Munich Polytech.

Hanle, Wilhelm (1901–1993)
Experimental physicist. 1919–24: Studies at Heidelberg and Göttingen Univs. 1924: Doctorate at the latter (under J. → Franck). 1924: Teaching assistant, Göttingen Univ; 1925: Tübingen Univ. 1926–29: Halle Univ., from 1927: unpaid university lecturer. From 1929: Associate prof. and head of the physics department, Jena Univ. 1937–41: Göttingen Univ. Collaborator of G. → Joos. Indicated boron and cadmium impurities in electrographite, which are potent absorbers of thermal neutrons, thus invalidating W. → Bothe's results in Heidelberg. Aug. 1939: Left German uranium project developing purified carbon as a moderator in reactors, which was in its initial stages in Göttingen, to perform his military service. From 1941: Full prof. of experimental physics, Giessen Univ. Refs.: Walker [1989]a, pp. 17, 26, Rechenberg [1993].

Harteck, Paul (1902–1985)
Physical chemist. 1921–24: Studies in chemistry at Vienna and Berlin Univs., and 1926: doctorate at the latter (under M. Bodenstein). 1926–28 A. Eucken's teaching assistant at Breslau Polytech. 1928/33: Employed at the → *KWIPC*, where he performed experiments together with K.-F. → Bonhoeffer on parahydrogen and orthohydrogen. 1931: Habilitation at Berlin Univ. 1933/34: Research visit with E. Rutherford in Cambridge, England. 1934–45: Full prof. and director of the physical chemistry department, Hamburg Univ. From 1937: Adviser to the → *HWA* on chemical explosives. Investigated ways of refining crude oil to produce gasoline using the thermal cracking method. Apr. 1939: Contacted the → *RKM*, together with his teaching assistant W. → Groth about the potential military applications of nuclear chain reactions ('nuclear explosives'). Member of the *Uranverein*, and became the driving force behind the nuclear energy project. From 1939: Research for the *HWA*, together with his colleagues J. → Jensen, W. Groth, Friedrich → Knauer and his department as a whole, with emphasis on uranium isotope separation. From 1940: Together with his colleague H. → Suess, focussed on heavy water as a neutron moderator. 1941: Construction of a conversion unit that produced heavy water catalytically for → *Norsk Hydro*, but failed in their attempt to introduce the process of electrolysis in Germany and its territories. 1942: Circumvented appointment in Russia with the help of W. → Heisenberg, among others. Feb. 1943: Suggested, together with Jensen, a new type of centrifuge construction for isotope separation, which the → *Anschütz* Company adopts. 1946–50: Again director of the physical chemistry department, Hamburg Univ. From 1951: Resident prof., Rensselaer Polytechnical Institute, Troy, N.Y., USA. Refs.: Bagge, Diebner & Jay [1957], pp. 19ff., 35ff., 54, Ermenc (Ed.) [1989], pp. 77–160, Walker [1989]a, [1995], Frank (Ed.) [1993], Hoffmann (Ed.) [1993], p. 69.

Haxel, Otto (born 1909)
Nuclear physicist. 1927–33: Studies at the Munich Polytech., and 1933: doctorate (under H. → Geiger). 1933–36: Geiger's teaching assistant. 1936: Habilitation at Tübingen Univ. 1936: Teaching assistant, Berlin Polytech., and 1939: university lecturer there. In 1940's: Worked on resonance absorption and determined neutron absorption cross-sections for the *Uranverein*. 1946–50: Assistant to W. → Heisenberg at the Max Planck Institute of Physics, Göttingen. 1949: Supernumerary prof., Göttingen Univ. Since 1950: Full prof. of physics, Heidelberg Univ.

Heisenberg, Werner (1901–1976)
Theoretical physicist. 1920–23: Studies in physics at Göttingen and Munich Univs. 1923: Doctorate at Munich (under A. → Sommerfeld), 1924: Habilitation at Göttingen. 1924–27: Unpaid university lecturer there (1926: in Copenhagen). 1927: Full prof., Leipzig Univ., where he became head of the theoretical physics department. One of his students was C. F. → von Weizsäcker. 1932: Nobel Prize in physics, and 1933: Max Planck Medal, for his papers on quantum mechanics. Also from 1939: scientific adviser at the → *KWIP*. From 1941: Acting institute director, also full prof., Berlin Univ. 1937: Approached H. → Himmler to protest a smear campaign against him conducted by a leader of the 'Aryan Physics' movement in an *SS* newspaper. 1939: Restoration of his reputation, as well as that of modern physics, helped along by appeals by L. → Prandtl, M. → Wien, and C. → Ramsauer. Despite being the candidate openly favored by the Munich physics faculty to succeed Sommerfeld, lost theoretical physics chair to the professionally unsuitable aerodynamics specialist W. → Müller for political reasons. From 1940: Occupied with the theory of chain reactions and uranium reactors; prepared several reports on the state of nuclear research for the → *HWA*. July 5, 1942: Appointed acting director of the *KWIP*. Beginning of 1945: Research removed to the Haigerloch redoubt in Württemberg, where experiments were conducted in a last failed attempt to bring about a working chain reaction. Otherwise devoted a large part of his time to cosmic radiation and the foundations of elementary particle physics. 1946–58: Director of the Max Planck Institute of Physics, Göttingen; from

Mar. 1948: First president of the newly founded *Deutscher Forschungsrat*, which later merged with the → *DFG*. 1955-75: President of the grant organization *A. von Humboldt-Stiftung*. From 1958: Prof., Munich Univ. 1960–70: Director of the Max Planck Institute of Physics and Astrophysics in Munich. Refs.: N.N. [1944], Dolch [1955], Hermann [1976], [1988], Cassidy & Baker [1984] for a complete bibliography, E. Heisenberg [1984], Walker [1989]a, [1990], [1995], chaps. 6–9, Eckert [1990], Cassidy [1992], Guérout [1992], pp. 62–69, Rechenberg (Ed.) [1992]a, Geyer et al. (Eds.) [1993], Powers [1993].

Hertz, Gustav Ludwig (1887–1975)
Experimental physicist. 1906–11: Studies at Göttingen, Munich and Berlin Univs. 1911: Doctorate at Berlin Univ. (under H. Rubens). 1917: Unpaid lecturer, Berlin Univ. Since 1920: Employed at the Philipps lightbulb industrial laboratory at Eindhoven. 1925: Nobel Prize in physics (together with J. → Franck) for their experiments on inelastic electron collision conducted between 1911–13. 1925: Appointment to Halle Univ. 1928: Prof. of experimental physics, Berlin Polytech. 1932: Developed an isotope separation method via diffusion in a gaseous state. End of 1934: Forced to leave Berlin Polytech., because classified as a "2nd degree part-Jew"; but 1935: became director of research lab. II at → *Siemens*. From the winter term 1935/36: called again as honorary prof. at the same Berlin Polytech. department. 1935–45: Continued his work on atomic physics and ultra-sound waves, but discontinued his work on isotope separation, upon which he had concentrated previously. 1945–55: Took up research in Russia. 1955: Full prof., Leipzig Univ. and from 1955–67, he was president of the Physical Society of the former German Democratic Republic, from 1967–75 honorary chairman. Refs.: Steenbeck [1977], pp. 82f., Cassidy [1979], Kuczera [1985], G. Richter in Treue & Hildebrandt (Eds.) [1987], Walker [1989]a, pp. 4, 31, 34, 183, Osietzky in Renneberg & Walker (Eds.) [1994], pp. 262f.

Himmler, Heinrich (1900–1945)
National Socialist politician. After World War I, agricultural studies at the Munich Polytech. Became business manager of the → *NSDAP*. 1923: Took part in the Munich *putsch* attempt. 1925: District leader of Lower Bavaria, and 1926: of Upper Bavaria. 1925–30: Acting propaganda leader. Jan. 6, 1929: Appointed by A. → Hitler to head the → *SS*. 1931–33: Organized secret civilian security organization, *Sicherheitsdienst* (*SD*) under Heydrich. 1933: Named chief of police at Munich. Apr. 20, 1934: H. → Göring appointed him acting chief of the → *Gestapo*. June 30, 1934: purge of the left wing of the Nazi movement, subsequently subordinating the → *SA* to the *SS*. June 17, 1936: *Reichsführer* of the *SS* and head of the *Gestapo*. 1939: Attacks against W. → Heisenberg and theoretical physics as 'Jewish in spirit' in an *SS*-controlled newspaper are stopped on Himmler's orders. Oct. 1939: Appointed by Hitler as Commissar for the Consolidation of German Nationhood, in which post he implemented his ruthless policy of eliminating 'non-Aryan' elements in the extended German Reich. From 1943: also Reich Interior Minister. Initiator of the → *Ahnenerbe* of the *SS*. Following the failed assassination attempt on Hitler in July 1944, the secret service is also integrated into Himmler's Security Service (*SD*) under Walter Schellenberg. July 21, 1944: Named supreme commander of the Peoples Army (*Volkssturm*). May 23, 1945: Suicide while in the custody of the Allies. Refs.: Höhne [1984/90], chap. 2, Schröder [1993], pp. 335ff.

Hitler, Adolf (1889–1945)
Austrian-born National Socialist politician and Reich Chancellor. Initially part-time construction worker and artist, influenced by anti-Semitic groups in Vienna. From 1919: Built up the → *NSDAP*. From 1921: Party chairman. Feb. 26, 1924: Condemned to 5 years imprisonment following the Munich *putsch* attempt on Nov. 8, 1923, but granted amnesty in Dec. 1924. 1925: Refounding of the then illegal Nazi Party and expansion of its subdivisions, in particular the → *SA* and the → *SS*, whose activities contributed to the destabilization of the Weimar Republic.

Mar. 13, 1932: Lost a close election campaign for the presidency against Hindenburg. Jan. 31, 1933: Appointed Reich Chancellor by Hindenburg. Following Hindenburg's death 1934: 'Lifetime *Führer* and Reich Chancellor', upon whose person the entire government was centered. 1938: Sole commander of the Armed Forces. Sep. 1, 1939: Triggered World War II with the invasion of Poland, following the reincorporation of the Rhineland and the Saar region, and the annexation of Austria, Czechoslovakia and the Memel region in NW Lithuania. 1945: Suicide as the Allied forces were approaching the *Führerbunker* in Berlin. Refs.: Stark [1930]c, Broszat [1969], [1977], Fest [1976] Waite [1977], Fox [1979], Baynes [1983], Kershaw [1985], Albrecht [1993], Walker [1995], etc.

Höcker, Karl-Heinz Friedrich (1915–)
Theoretical physicist. 1935–40: Studies at Marburg and Berlin Univs. 1940: Doctorate at Berlin Univ. (under C. F. → von Weizsäcker). After 1939: Collaborated with C. F. von Weizsäcker and P. → Müller at the → *KWIP* on the theory behind 'uranium machines', initially on horizontal and spherical layer arrangements. 1942: Assistant at the *KWIP*, and then drafted to the Eastern front. Von Weizsäcker was only able to delay his departure; but he returned in the same year for health reasons and became von Weizsäcker's assistant at Strasbourg Univ. 1942: Analysis of the interior geometry of uranium reactors concluding with the choice of a lattice arrangement. 1948: Supernumerary lecturer, and 1955: supernumerary prof. of theoretical physics and nucleonics, Stuttgart Univ. Refs.: Walker [1989]a.

Hoffmann, Gerhard (1880–1945)
Nuclear physicist. Studies at Göttingen, Leipzig and Bonn Univs. 1906: Doctorate at Bonn Univ. (under W. Kaufmann). 1908: Kaufmann's teaching assistant at Königsberg Univ., and 1911: Habilitation there in pure and applied physics. 1917: Associate prof. Work on precision measurement of radioactivity and cosmic rays. 1928–37: Prof., Halle Univ. (as successor to G. → Hertz). 1937–45: Full prof. of experimental physics, Leipzig Univ. (succeeding P. → Debye), specializing there in nuclear physics. Refs.: Osietzky in Renneberg & Walker (Eds.) [1994], p. 262.

Houtermans, Fritz Georg (1903–1966)
Atomic and nuclear physicist. 1921–27 Studies at Göttingen Univ. (student of J. → Franck); habilitation under the guidance of G. → Hertz at the Berlin Polytech. 1935–37: Director of the Ukrainian Physics Institute at Char'kov, then implicated in a political trial in the Soviet Union. 1939: Return to Germany, made possible by the Hitler-Stalin Pact, but subjected to brutal interrogation by the → *Gestapo* as a presumed Soviet spy. From 1940: Employed at M. → von Ardenne's semi-private Institute of Nuclear Physics, where he showed the substitutability of elements of atomic mass 93 or above (neptunium or plutonium) for uranium as nuclear fuel, and continued to work on nuclear fission and fusion until the end of the war. 1941: Figured out the theory behind uranium reactors independently of the *Uranverein*. 1942: Probable author of the telegram to E. P. Wigner warning the USA of the imminent construction of a German atomic bomb. From 1945: Max Planck Institute of Physics, Göttingen. From 1952: Prof., Berne Univ. Refs.: Kramish [1986], chap. 26.

Hund, Friedrich (born 1896)
Theoretical physicist. 1915–16: Studies at Marburg Univ., and 1919–22: Göttingen Univ. 1922: Doctorate (under the guidance of M. → Born). 1927: Associate prof., and 1928: full prof., Rostock Univ. 1929–46: Full prof. of mathematical physics, Leipzig Univ. (colleague of W. → Heisenberg). 1943: Award of Max Planck Medal by the → *DPG*. 1946–51: Full prof., Jena Univ. 1951–56: Full prof. of theoretical physics, Frankfort Univ. Since 1956: Göttingen Univ. Refs.: Rechenberg [1988]c, Hentschel & Tobies [1996].

Jensen, Johannes **Hans** Daniel (1907–1973)
Nuclear physicist. 1926–31: Studies at Freiburg (Breisgau) and Hamburg Univs. 1932: Doctorate at Hamburg (under W. Lenz), and 1936: Habilitation. From 1933: Member of the → *NSDDB*, from 1934: member of the → *NSLB*, and from 1937: member of the → *NSDAP*. 1937: Unpaid lecturer, Hamburg Univ., and collaboration with P. → Harteck. 1940: In Heidelberg (together with A. → Flammersfeld and W. → Gentner) measurements of cross-sections and energy distribution of fission fragments. 1941: Associate prof. of theoretical physics within the civil service at the Hanover Polytech. 1943: Together with Harteck development of double centrifuges for the separation of uranic isotopes. 1946: Full prof., Hanover Polytech. 1947: Honorary prof., Hamburg Univ. Since 1949: Full prof., Heidelberg Univ. Guest prof. at 1950: Wisconsin Univ.; 1952: Institute for Advanced Study, Princeton; and 1953 at Indiana Univ. and Caltech. 1963: Nobel Prize for his shell theory of the atomic nucleus (jointly with M. M. Goeppert-Mayer and E. P. Wigner). Refs.: Walker [1989]a.

Jensen, Peter Herbert (1913–1955)
Experimental nuclear physicist. 1932–38: Studies at Göttingen and Freiburg (Breisgau) Univs. 1938: Doctorate at Göttingen Univ. (under G. → Joos). 1938: Volunteer at the physics department of the *KWI* of Medical Research at Heidelberg, and 1939–46: teaching assistant there under W. → Bothe. 1943: habilitation on nuclear cross sections of neutron scattering experiments at the Univ. of Heidelberg. 1946–53: W. → Gentner's teaching assistant, and 1953–54: senior assistant there. 1947: University lecturer. 1951: Supernumerary prof. at Freiburg Univ. (Breisgau) where he focussed on the installation of a van-de-Graaff generator for experiments in nuclear physics. 1954: Department director at the high-voltage section of the Max Planck Institute of Chemistry in Mainz and supernumerary prof., Mainz Univ. Refs.: Gentner [1956], Weiss [1994], p. 289.

Jentschke, Willibald Karl (born 1911)
Experimental nuclear physicist. 1930–36: Studies at Vienna Univ., and 1935: Doctorate there (under G. → Stetter). 1937–42: Stetter's teaching assistant, Vienna Univ. 1939: Measurements (together with Friedrich Prankl) on the energy distribution in fission fragments. 1942–45: Lecturer at Vienna Univ. 1946–47: Innsbruck Univ. 1947–48: Physicist of the Air Material Command, Wright Field, Ohio. 1950–55: Resident assistant prof., and 1955: Resident prof., Department of Physics, Illinois Univ., Urbana. Since 1956: Full prof., Hamburg Univ. and one of the initiators of the construction of the German electron synchrotron (*DESY*) in Hamburg. Refs.: Walker [1989]a, pp. 22, 124, 180.

Joos, Georg Jakob Christof (1894–1959)
Physicist. Studies at the Stuttgart Polytech. and Tübingen Univ. 1920: Doctorate under the guidance of C. Füchtbauer. 1921: Teaching assistant of J. → Zenneck. 1922: Unpaid university lecturer, Munich Polytech. End of 1923: Entered → *NSDAP*, subsequently inactive membership. 1924: Associate prof. with a lectureship in electron and relativity theory. 1928: Full prof. of theoretical physics, Jena Univ. (successor to Felix Auerbach). End of 1920's: Involved in precision experiments at the labs. of the → *Zeisswerke*, reproducing the Michelson-Morley experiments with increasingly refined equipment, and arriving at the same negative results. Before end of 1929: Member of the first *Uranverein*. 1935: Compulsory transfer to replace J. → Franck as prof. of experimental physics at the physics department II, Göttingen Univ. Apr. 1939: Together with W. → Hanle contacted the physics expert at the → *REM*, Dames, about the potential military applications of 'uranium machine' research. Aug. 1939: Conscription into military, resulting in withdrawal from the German uranium project. 1941: Departure from academia because of the increased constraints set by the → *NSDDB*; chief physicist at the optical manufacturer

Zeisswerke, and from 1941–45: In the management there; also honorary prof., Jena Univ. From Sep. 1, 1946: Full prof. of experimental physics at the Munich Polytech. (succeeding R. → Tomaschek, who had been suspended). 1947–49: Visiting prof., Boston Univ. Refs.: Meissner [1959], Becker, Dahms & Wegeler (Eds.) [1987], pp. 386–393, Walker [1989]a, pp. 17, 26, 74f., Wohofsky in Magnus (Ed.) [1993], p. 105.

Jordan, Pascual Ernst (1902–1980)
Theoretical physicist specializing in quantum mechanics and relativity theory. 1921–24: Studies at Hanover and Göttingen Univs. 1924: Doctorate, Göttingen Univ. (student of M. → Born). 1926: Unpaid lecturer, Göttingen Univ. Developed the algebraic variant of quantum mechanics (together with M. Born and W. → Heisenberg). From 1927: Unpaid lecturer, then prof., Hamburg Univ. From May 1, 1933: Member of the → *NSDAP*. Nov. 1933: Member of an → *SA* unit. 1935–44: Full prof. of theoretical physics, Rostock Univ. 1944–45: Associate prof., Berlin Univ. From 1939: Worked additionally in an air defense research institute near Bremen for the *Luftwaffe*, and then 1943: in Peenemünde on rocket development. 1947–53: Visiting prof., and from 1953: full prof., Hamburg Univ. 1942: Award of the Max Planck Medal of the → *DPG*, and 1955: the Gauß Medal. 1957–61: Member of parliament for the conservative Christian Democratic Union party. Refs.: Kramish [1986], chap. 25, Beyler [1994], Wise in Renneberg & Walker (Eds.) [1994], Walker [1995].

Juilfs, Johannes Wilhelm Heinrich (1911–)
Theoretical physicist. 1930–38: Academic studies. 1938: Doctoral student of W. Kohlhörster. 1938–45: Scientific assistant; 1941–45: Teaching assignment and substitute for the chair of theoretical physics, Rostock Univ. Mar. 30, 1945: Habilitation, Berlin Univ. Member of the → *SS*. Supported W. → Heisenberg in his negotiations with the *SS* in 1939, and was present at the → *NSDDB* meeting in Munich in 1940. After 1945: Denazification court trial, and initially barred from returning to a university career. 1948–50: Principal of an adult education school in Helmstedt. From 1950: Head of the physical department of the Textile Research Institute (*Textilforschungsanstalt*) at Krefeld. 1955: Appointment as temporary lecturer of theoretical physics at the Hanover Polytech., then since 1958: full prof. there. Refs.: Walker [1989]a, pp. 65, 70, 198f., [1995], pp. 203, 267.

Knauer, Friedrich Wilhelm Karl (1897–)
Physical chemist. 1918–24: Studies at Göttingen and Hanover Univs. 1923: Doctorate in engineering, Hanover Univ. (student of Beckmann and W. Kohlrausch). 1933: Unpaid university lecturer, and 1939: scientific assistant at the department of physical chemistry, Hamburg Univ., and then supernumerary prof. of physics there. In 1940's: Member of P. → Harteck's group at Hamburg Univ. conducting research on neutron diffusion and capture in uranium for the *Uranverein.*

Konen, Heinrich Matthias (1874–1948)
Spectroscopist. 1893–98: Studies at Bonn Univ., and 1897: doctorate there. 1899: Teaching assistant, and from 1902: unpaid lecturer, Bonn Univ., as well as for a time part-time teacher at the Bonn highschool (*Gymnasium*). 1905–12: Associate prof. of theoretical physics, and 1919–20: full prof., Münster Univ. 1920–1933: Until his forced retirement, full prof. of physics, Bonn Univ. (as successor to H. Kayser). Involved in the founding and organization of the → *NG* (thanks to his good relations with → Schmidt-Ott), and longstanding member of its main committee (1920 initially as 'representative of the occupied areas'). Until 1933: On the → *KWG* supervisory board, and influential member of the board of trustees of the → *PTR*, of the chemical and technical research institute (*Chemisch-Technische Reichsanstalt*) and on the board of the *Deutsches Museum* in Munich. 1933–45: Adviser in industry, especially for

the company *Troisdorfer Werke* (*Dynamit Nobel AG*). After 1945: Rector of Bonn Univ., then headed the Culture Ministry of North Rhine-Westphalia and in this position brought about the refounding of the *NG*. Refs.: Gerlach [1949].

Kopfermann, Hans (1895–1963)
Atomic and nuclear physicist. 1925: Doctorate, Göttingen Univ. under J. → Franck, then scientist at the → *KWIPC*; worked on dispersion and stimulated emission. Thereafter assistant to G. → Hertz at the Polytech. in Charlottenburg (Berlin). 1932: Unpaid lecturer, Berlin Univ. 1937: Full prof., Kiel Univ. 1940: His standard text on nuclear moment *Kernmomente* influenced a whole generation of nuclear physicists. 1941: Named dean against his will by the rector of Kiel Univ., thus pressuring him to join the → *NSDAP*. 1942: Full prof., Göttingen Univ. on the chair of experimental physics formerly held by J. Franck succeeded by G. → Joos. Around 1942: A couple of physicists worked under his direction on isotope separation. 1953: Full prof. and head of the department of physics I, Heidelberg Univ. (successor to W. → Bothe). Refs.: Becker, Dahms & Wegeler (Eds.) [1987], pp. 393–396, Walker [1989]a, pp. 29, 33, 53, 134, 158, Dreisigacker [1995], p. 54.

Korsching, Horst (1912–)
From 1932: Studies in physics at Berlin University. From 1935: on the scientific staff at the → *KWIP*. 1938: Doctoral thesis under H. Schüler. Research conducted for the *Uranverein* included the determination of nuclear moment and work on thermodiffusion (current between two differently tempered plates). July–Dec. 1945: Interned at Farm Hall in England with other prominant German nuclear physicists. 1946: Returned to Göttingen to work as staff scientist at the Max Planck Institute. 1958: Moved with the institute to Munich. Refs.: Bagge, Diebner & Jay [1957], Walker [1989]a, Walker [1995], pp. 208f., 220, 235, 240, Frank (Ed.) [1993].

Krauch, Carl (also 'Karl') (1887–1968)
Chemist. 1906–12: Studies at Giessen and Heidelberg Univs. 1912: Doctorate at Heidelberg (under T. Curtius). 1911–12: Unpaid teaching assistant, Heidelberg Univ. (under R. Stallé). From 1912: Employed at → *BASF*, (later → *IG Farben*); longstanding member of the board and general committee as well as chairman of the supervisory board, 1940–45. From 1936: Head of the Research and Development Department of the *Amt für Deutsche Roh- und Werkstoffe* (from Dec. 1939, head of the renamed *Reichsamtes für Wirtschaftsausbau*, established in 1936 as part of the Four-Year Plan), which was also nicknamed 'Office for the Expansion of *IG-Farben*' ("Amt für IG-Farben Ausbau") because of Krauch's dual function. From 1937: Member of the → *NSDAP*. 1938–45: Plenipotentiary of Special Issues in Chemical Production and member of the board of the → *RFR*. 1939: Additionally, honorary prof., Berlin Univ. 1948: Implicated in the *IG Farben* trial and sentenced to 6 years in prison. Refs.: Eichholtz [1969], pp. 184–187, Borkin [1978], chaps. 3, 6–8, Schlicker [1979]b.

Ladenburg, Rudolf (1882–1952)
Atomic physicist. 1900–06: Studies at Heidelberg, Breslau and Munich Univs.; Doctorate at Munich (under Röntgen). Unpaid university lecturer; from 1921 associate prof., Breslau Univ. 1924: Appointment, Berlin Univ. and scientific member → *KWIPC*. As early as 1932: Exile to USA; 1932–50: Brackett Research Prof., Princeton Univ. When the emigration wave started in Apr. 1933, Ladenburg was the principal coordinator in the job placement of exiled physicists in USA. From 1950: Retirement at Princeton.

von Laue, Max Theodor Felix (1879–1960)
Theoretical physicist. From 1899: Studies at Strasbourg, Göttingen and Munich Univs. 1903: Doctoral thesis, Berlin Univ. (student of M. → Planck) on 'Interference on plane-parallel plates'. From 1905: Planck's teaching assistant, research on optics, quantum theory and relativity theory

(the latter of which he was one of the first consider seriously). From 1906: Unpaid lecturer, Berlin Univ., and 1909–12: in Munich. 1914: Award of 1914 Nobel Prize in physics for his successful demonstration of the wave-like nature of X rays and his explanation of interference patterns of X-ray beams passing through crystals observed by P. Knipping and W. Friedrich. 1912–14: Zurich Univ.; 1914–19: Prof. of theoretical physics, Frankfort Univ. 1919–43: Full prof. of theoretical physics, Berlin Univ. and organizer of the weekly Berlin physics colloquium. His students included: Fritz London, Kleve, Kohler, Beck, Päsler, Möglich and Szilard. From 1921: Regular member of the Prussian Academy of Sciences, chairman of the physics special committee of the → *NG* and member of the electrophysical committee. From 1922: Member of the board of trustees of the Postdam Astrophysical Observatory. About 1925–Dec. 1933: Adviser to the → *PTR* in particular on superconductivity, and 1922–1946/48: substitute director of the → *KWIP*, where he had already been a member of its board of trustees since 1917. 1932: Award of Max Planck Medal. 1931–33: Chairman of the → *DPG*. At the opening ceremony of the 1933 physics conference in Würzburg, he daringly compared A. → Einstein to Galileo; in the same year he blocked J. → Stark membership in the Prussian Academy of Sciences. 1934: Together with M. Planck the driving force behind an officially disapproved meeting in commemoration of F. → Haber. 1933–45: With these and other occasions, Laue publicized his opposition to the National Socialist science policy. 1943: Followed the *KWIP* to Hechingen and 1946: to Göttingen, where he returned following his internment in England in the same year. 1951–59: Director of the Berlin → *KWIPC* formerly under Haber, until retirement. Major role in post-war *DPG*, in the reconstruction of the *Physikalisch-Technische Bundesanstalt* and the → *DFG*. Refs.: Planck [1929], [1939], Leibfried [1955], Bagge, Diebner & Jay [1957], Brill et al. [1959], Päsler [1960], Meissner [1960], Franck [1960], Ewald [1960], [1962], Rosbaud [1960], Herneck [1979], Hermann [1982], G. Hildebrandt in Treue & Hildebrandt (Eds.) [1987], Walker [1989]a, [1995], Henning [1992], Frank (Ed.) [1993], Hildebrandt [1993].

Lenard, Philipp (1862–1947)
Experimental physicist. 1883–86: Studies at Heidelberg (attending lectures by Bunsen) and Berlin (attending lectures by Helmholtz). 1886: Doctoral thesis: 'On the oscillations of falling drops', Heidelberg Univ.; then Quincke's teaching assistant. Research on phosphorescence and hydroelectricity, and from 1891: on cathode rays (under H. Hertz). Following Hertz's death: Edition of the collected papers of Heinrich Hertz. 1894–95: Appointment to Breslau, where he left after one year due to unsatisfactory working conditions. 1895: Unpaid university lecturer at the Aachen Polytech. under Wüllner. 1896: Associate prof., Heidelberg Univ. 1898: Full prof. and director of the physical laboratories, Kiel Univ. 1905: Nobel Prize for his work on cathode rays and on the photoelectric effect, which A. → Einstein explained theoretically in 1905. 1907: Successor to Quincke, Heidelberg Univ., and director of the physical and radiological laboratories there. 1909: Wrote memorandum at A. von Harnack's request that played a major part in the founding of the → *KWG* and the → *KWIP*. 1914: Established a new laboratory at Heidelberg, 1935: renamed the 'Philipp Lenard Laboratory'. In World War I, revealed a strong nationalistic streak which grew during the Weimar Republic and acquired anti-Semitic elements. 1920: The most influential proponent against relativity theory at the Bad Nauheim science conference, and from 1924: one of the earliest supporters of A. → Hitler among scientists, to whom he was later physics adviser. Together with J. → Stark one of the leaders of the 'Aryan Physics' movement, which he propagated in a four-volume work published in 1936 and 1937. After the war, not arrested because of his age (he was 83). Refs.: Bühl [1937], Weigel (Ed.) [1937], Kubach [1937/38], [1942], Becker [1942], Brüche [1947]a, Wolf [1962], Beyerchen [1977], sec. 5; Kleinert & Schönbeck [1978], Kleinert [1980], Neumann & Putlitz [1986], Wolgast [1986], Walker [1989]a, pp. 2, 61–80, 199f., [1995], Guérout [1992], Kleinert in Olff-Nathan [1993], pp. 149–166.

Lintner, Karl (1917–)
Austrian nuclear physicist. 1936–40: Studies at Vienna Univ., and 1940: doctorate there under G. → Stetter. 1941: Stetter's teaching assistant. 1945: Assistant at physics department II at Vienna Univ. under Haschek, Przibram and Schmid. 1949: Unpaid university lecturer there. Around 1959: Titular associate prof. at Vienna Univ.

Mannkopff, Reinhold (1894–1978)
Experimental physicist. 1913–14 and 1919–25: Studied physics at Freiburg (Breisgau), Berlin and Göttingen Univs. 1926: Doctorate under J. → Franck on atomic scattering in sodium vapor. 1927–30: Franck's teaching assistant. 1926–27: Assistant to C. Runge and work using Rowland gratings. 1929: switch to mineralogy: 1930–33: Assistant to V. M. Goldschmidt with whom he worked on quantitative applications of spectroscopy for chemical analysis of the earth's crust. 1934: University lecturer, and from 1939: supernumerary prof., Göttingen Univ. After World War II, secretary of the Northwest German branch of the → *DPG* for over 20 years. Refs.: Kohler [1978].

Mattauch, Josef (1895–1976)
Nuclear physicist, expert in mass spectroscopy. 1928: Unpaid university lecturer, and 1937: associate prof., Vienna Univ. Feb. 1, 1939: appointed successor to L. → Meitner at the → *KWIC* in Dahlem (Berlin). From 1940: also supernumerary prof., Berlin Univ. Nov. 1, 1943: Appointed deputy director of the *KWIC*, for which he acquired a high-voltage generator and a pressure belt generator for performing artificial atomic transformations. From 1947: Staff scientist at the Max Planck Institute of Chemistry, Mainz. 1948: Visiting prof., Tübingen Univ.; from 1948: at Berne Univ.; and from 1952: honorary prof., Mainz Univ. Ref.: Weiss [1994], pp. 273f.

Meissner, Friedrich (Fritz) **Walther** (1882–1974)
Low-temperature physicist and mechanical engineer. 1901–06: Studies at the Polytech. and Univ. in Berlin. 1907: Doctorate at Berlin Univ. on radiation pressure. 1908: Assistant at the → *PTR*, first at the pyrometry laboratory, then from 1913: work in low-temperature physics. 1925: Succeeded in liquefying helium, and 1926: head of the low-temperature laboratory there as well as senior official adviser. 1933: Discovery together with Robert Ochsenfeld of a new effect in superconductivity. Since 1934: Full prof., Munich Polytech. Refs.: W. Buckel in Treue & Hildebrandt (Eds.) [1987], Lemmerich [1987], pp. 126ff., Andres in Magnus (Ed.) [1993], p. 104.

Meitner, Lise (1878–1968)
Nuclear physicist. 1901–05: Studies at Vienna Univ., and 1906: doctoral thesis there on thermal conductivity in inhomogenous bodies (under L. Boltzmann), and work with Stefan Meyer on radioactive radiation. 1907: Studies under M. → Planck in Berlin. 1909: Finally received permission to matriculate as a women in Prussia. From 1912: Prussia's first female teaching assistant. Collaboration with O. → Hahn at the chemistry department investigating radioactive radiation; unpaid lecturer of physics at the chemistry department of the Friedrich Wilhelm University, Berlin. From 1913: Staff scientist at the newly established → *KWIC*, where she verified pair formation of energy-rich γ rays and, 1909: radioactive recoil. Worked jointly with Hahn on problems in radiochemistry and discovered with him protactinium. 1918: Establishment of her own physical radioactivity department. 1922: Habilitation, and 1926: associate prof. of experimental nuclear physics (untenured). 1933: Stripped of title and permission to teach as a Jew, but allowed to continue to work at the *KWIC*. 1938: Emigrated to Holland, Denmark and then Sweden, since her foreigner status, which had provided some protection, disappeared upon Austria's annexation in Mar. 1938. 1939: Calculated jointly with her nephew O. → Frisch the energy released from a uranium nucleus during fission using the liquid-drop model, for which she

was awarded together with Hahn 1949: Golden Max Planck Medal; 1955: Otto Hahn Prize; and 1966: Fermi Prize of the Atomic Energy Commission. From 1947: Prof. at the Nobel Institute of Physics at the Stockholm Polytech. 1946: Acquired Swedish citizenship. 1960: Joined her nephew in Cambridge, England. Meitner was member or corresponding member of the Berlin, Halle, Göttingen and Stockholm Academies of Science and member of the *Ordre pour le Mérite*. Refs.: Meitner [1954], [1963]; Hahn [1963], [1975], pp. 75–130, Straßmann [1968], Frisch [1970], Krafft [1978], [1981], chap. 2, [1988], Frisch [1981]a, Kerner [1986], [1987], E. H. Berninger in Treue & Hildebrandt (Eds.) [1987], Lemmerich [1988], pp. 30ff., [1993], Sime [1990], [in press].

Mentzel, Rudolf (1900–?)
Chemist and National Socialist science policy-maker holding a golden party badge because of his involvement in the Kapp *putsch* in 1923. District head of the → *NSDAP* in Göttingen, of which he had been a member since 1922, and Senior Battalion Commander, then Brigadier of the → *SS*. Studies in chemistry at Göttingen Univ., 1925: Doctoral thesis at Göttingen on 'Stereoisomerism and transformation of b-substituted decahydro-naphthalenes'. 1933: Habilitation on 'military chemistry' ('*Wehrchemie*') with a paper on a top-secret topic not disclosed to the department. From 1934: Expert adviser on science at the → *REM*; and supernumerary prof. at the Berlin Polytech. Shared with P. → Thiessen the flat that had previously been made available to F. → Haber for business purposes on Faradayweg in Dahlem. From 1939: Undersecretary of the Science Office (*Amt für Wissenschaft*) of the *REM*. J. → Stark's rival and from Dec. 1936: his successor as president of the → *DFG*. 1935: Full prof. of military chemistry at the defense technology department of the Berlin Polytech. (on the instigation of B. → Rust, who also appointed him commissioner in the managing committee of the → *KWG* with the support of P. Thiessen). From 1937: On the advisory board of the → *KWG*, and 1940–44: Second Vice President of the *KWG*. After the war: 3-year internment by the Americans. Refs.: Zierold [1968], pp. 190ff., 213ff., Becker, Dahms & Wegeler (Eds.) [1987], p. 57, Walker [1989]a, [1995], Deichmann [1992], p. 357, Macrakis [1993], pp. 79f.

Mey, Karl (1879–1945?)
Technical physicist. 1897–1902: Studies physics and mathematics at Berlin and temporary supervisor of Warburg's physics practical session. Nov. 1902: Doctoral thesis on the cathode gradient of alkali metals at Berlin Univ., supervised by Emil Warburg. Employed by the *Militärversuchsamt* Tegel, then at the → *AEG* where he specialized on the research and development of incandescent bulbs and became head of the *AEG* light bulb factory in 1909. From 1914–17, he served in the infantry at the Western Front. After the merger of the light bulb factories *AEG*, → *Siemens & Halske* and → *Auer* Company to the *Osram AG* in Berlin, Mey supervised the whole research and development branch. 1931–45: President of the → *DGtP* and 1933–35: simultaneously of the → *DPG*. Furthermore from circa 1931: vice-chairman of the *Deutsche Glastechnische Gesellschaft*. In May 1945: arrested as 'leading military industrialist' and deported to the Soviet Union. Refs.: Rukop [1939], Hoffmann & Swinne [1994].

Mie, Gustav Adolf Feodor Wilhelm Ludwig (1868–1957)
Physicist. 1886–91: Studies at Rostock and Heidelberg Univs. 1891: state teaching examination in mathematics and physics, and doctorate in the same year at Heidelberg. 1889: Teaching assistant at the mineralogy department, Heidelberg Univ. 1892: Teaching assistant, physics department, Karlsruhe Polytech., 1897: habilitation and unpaid lecturer there, working on electrodynamics. From 1902: Associate prof. of physics, Greifswald Univ., and 1905–16: full prof. there, working on ether theory and from 1912 on a field theory of matter. 1917–24: Full prof. of experimental physics, Halle Univ. 1924–36: Full prof. of experimental physics, Freiburg Univ. Longstanding member of the physics special committee of the → *NG*, then the → *DFG*. Since 1936: Emeritus. Refs.: Kast [1957], Rechenberg [1994]b.

Müller, Wilhelm Carl Gottlieb (1880–1968)
Specialist in aerodynamics and fluid dynamics, applied mathematics and mechanics. 1921: Unpaid university lecturer, Hamburg Polytech. 1928: Associate prof., Hanover Polytech., then in the same year at the German Polytech. in Prague, where he lectured in particular on flight mechanics, 1929–34. 1934–39: Full prof., Aachen Polytech. Dec. 1, 1939–54: Successor to A. → Sommerfeld in the chair for theoretical physics, Munich Univ., despite the physics department's declared preference for W. → Heisenberg. Retained his post to his retirement. Refs.: Eckert et al. (Eds.) [1984], pp. 150–159, Beyerchen [1977], chap. 8, Walker [1989]a, pp. 66f., 71, 79, [1995], pp. 51–57, Litten [in preparation].

Nernst, Hermann **Walther** (1864–1941)
Physical chemist. 1883–87: Studies at Zurich, Berlin, Graz and Würzburg Univs. 1887: Doctorate at Würzburg. From 1894: Full prof., Göttingen Univ. From 1905: Full prof. of physical chemistry, Berlin Univ., as well as 1922–24: Director of the → *PTR* and longtime director of the physical chemistry department at Berlin Univ. 1905–06: Discovery of the Third Law of Thermodynamics, according to which the temperature of a body can approach absolute zero but never reach it. 1920: Award of Nobel Prize in chemistry for his discovery. Longstanding member of the → *NG*'s electrophysical committee. 1917–31: Member of the board of the → *KWIP*. 1921–41: On the governing board of the → *KWIC*, and from 1922: on the board of the Potsdam Astrophysical Observatory. He spent his last years in seclusion on his landowner's estate Zibelle in Oberlausitz, subject to pressure, because his two daughters had married Jews. Refs.: Harteck [1960], pp. 463–464, Mendelssohn [1973], E. Cremer in Treue & Hildebrandt (Eds.) [1987], Bartel [1989].

Noddack, Ida Eva (née Tacke) (1896–1978)
Chemist. 1915–21: Studies at the Polytech. in Charlottenburg (Berlin). 1921: Doctorate in engineering under D. Holde. 1924–35: Visiting researcher at the → *PTR*. 1925: Discovery jointly with W. → Noddack of the element rhenium and possibly also of masurium. 1934: Hypothesized that heavy nuclear cores could explode into a number of lighter fragments when bombarded, but this idea was ignored until 1939. 1935–45: Scientist at the department of physical chemistry at Freiburg (Breisgau) and Strasbourg Univs.; pro-Nazi involvement. From 1956: Staff scientist at the State Research Institute of Geochemistry at Bamberg. Refs.: Noddack [1934], [1939], U. Kern in Treue & Hildebrandt (Eds.) [1987], Lemmerich [1988], pp. 126–127, Assche [1989].

Noddack, Walter Karl Friedrich (1893–1960)
Photochemist. 1912–14 and 1919–20: Studies in physics at Berlin Univ. Senior official adviser. Member and chairman of the chemical and photochemical laboratory at the → *PTR*. 1925: There discovery jointly with Ida → Noddack of the elements masurium und rhenium. From 1935: Full prof. of physical chemistry at Freiburg (Breisgau) Univ. 1941–45: Transfer to Strasbourg Univ. 1946–56: Full prof. and director of chemistry department at the Bamberg University of Philosophy and Theology. From 1956: Director of the State Research Institute of Geochemistry, Bamberg. Refs.: Bayer & Remy [1963], Lemmerich [1987], pp. 63f., [1988], p. 126.

Ohnesorge, Wilhelm (1872–1962)
Signal engineer and science policy-maker. 1897: State examinations; 1898–1901: academic studies in mathematics and physics under P. → Lenard in Kiel, later in Berlin under Emil Warburg. 1892: first patent application, 1900: Employed at the Telegraph Research Office in Berlin then, 1902–14: at Postal Headquarters in Berlin. 1912/14: Ohnesorge first employed loudspeakers to improve the radio broadcasting from the Berlin opera house. During the war 1914–Jan. 1915: communications specialist at the Army High Command, then 1915–18: head of telegraph man-

agement of the Main Headquarters. He succeeded in establishing the first telephone link between Mezières (France) and Constantinople using the four-wire long-distance switching technique and was awarded the Iron Cross. 1919: Senior manager in Dortmund postal administration. 1920: Entered the → *NSDAP* (membership card no. 42), founding the local branch in Dortmund. 1924: In senior postal management in Berlin; 1929: appointed director of the Tempelhof (Berlin) branch of the postal administration, and honorary doctorate from Stuttgart Polytech. Mar. 1, 1933: appointed State Secretary of the → *RPM* and named head in Feb. 1937. Ohnesorge's research specialties included telecommunications, microphone and speaker technology and he was instrumental in improving the Berlin telephone network. He was a member of the committee on wave research and of the Deutsches Museum; from 1933: president of the Federation of German Electrical Technicians (*Verband deutscher Elektrotechniker*) and member of the board of the Association of German Engineers (*Verein Deutscher Ingenieure*). In May 1934 he committed the board members to adopt Hitler's approach as a guideline. On May 1, 1942, Ohnesorge was named 'Pioneer of Work' by Hitler. Refs.: N.N. [1942], Leclerc [1988], Ueberschär [1989], Hoppe in Böndel et al. [1995], pp. 53–88.

Orthmann, Wilhelm (1901–1945)
Studies at Halle and Berlin Univs. 1926: Doctorate at Berlin under P. Pringsheim on resonance lines. Then teaching assistant to Pringsheim and W. → Nernst, as well as L. → Meitner. 1931: Habilitation on the dielectric constants of electrolytes. From 1931: Unpaid university lecturer of physics, Berlin Univ. 1938: Untenured associate prof., 1940: associate prof., and 1942: full prof. and director of the physico-technical department of the Berlin industrial college *Wirtschafts-Hochschule*. End of 1938: Submitted a petition to the → *DPG* jointly with H. → Stuart for the expulsion of all Jewish members. Both were engaged in physical studies reform. During the war: Employment at the → *RLM*, and finally scientific adviser in the development of anti-aircraft artillery. Refs.: Hettner [1947].

Paschen, Louis Carl Heinz **Friedrich** (1865–1947)
Experimental physicist, specializing in spectroscopy. 1884–88: Studies at Strasbourg and Berlin Univs. 1888: Doctorate at Strasbourg, then teaching assistant at the physics institute of the Münster Academy. From 1891: Teaching assistant of physics, Hanover Polytech., and from 1895: University lecturer of physics and photography there. 1901: Full prof. of physics, Tübingen Univ. His measurements of the helium spectrum was generally taken as an excellent verification of Sommerfeld's fine-structure theory. From 1925: Honorary prof., Berlin Univ. Also 1924–33: President of the → *PTR*, and 1925–27: Chairman of the → *DPG*. From 1925: On the board of trustees of the → *KWIP*. Refs.: Schüler [1947], Swinne [1989].

Pauli, Wolfgang (1900–1958)
Theoretical physicist. Son of the biochemist and physician Wolfgang Joseph Pauli. 1918: Started his studies at Munich Univ., 1921: Doctorate there (under the guidance of A. → Sommerfeld) on the hydrogen molecule ion. 1921–22: Research at Göttingen Univ. 1922/23: Research stay in Copenhagen (visiting N. Bohr) and in Göttingen (visiting M. → Born). 1924: Habilitation and unpaid lectureship at Hamburg Univ. 1925: Formulation of his exclusion principle which states that two electrons of the same atomic orbit cannot have the same quantum state. From 1928: Full prof., Zurich Polytech. 1930: Awarded the Lorentz Medal. 1931: Postulated the existence of the neutrino on the basis of the conservation of energy and momentum in elementary particle reactions, which could later be verified. 1935/36, 1939–46 and 1949/50: Visiting prof. at the Institute for Advanced Study, Princeton, where he was on leave 1940–45, becoming subsequently permanent member. 1945: the Nobel Prize in physics, and 1958: the Max Planck Medal of the → *DPG*. Refs.: Peierls [1959], Scherrer [1959], Meyenn et al. (Eds.) [1978–93].

Planck, Max (1858–1947)
Theoretical physicist. 1874–78: Studies at the Munich and Berlin Univs. (attending lectures by Kirchhoff and Helmholtz). 1878: Doctoral thesis at Munich Univ.: 'On the second law of thermodynamics'. 1880: Habilitation, Munich Univ. 1885: Associate prof. of mathematical physics, Kiel Univ. From 1889: Prof., Berlin Univ. (as successor to G. Kirchhoff). From 1894: Member of the Prussian Academy of Sciences; and 1912–38: its permanent secretary (pressured by the National Socialists to resign). 1899: Discovery of the quantum of action *h*; 1900: Founder of quantum theory through his articles on radiation field statistics, for which he was awarded the 1918 Nobel Prize in physics in 1919. Member of the Japan committee and chairman of the electrophysical committee of the → *NG*. 1917–46: member of the board of trustees of the → *KWIP*; and 1931–36: on the governing board of the → *KWIC*. In addition, 1926–30: second secretary, and 1930–37: president of the → *KWG*, playing a major part in establishing the society's policy of political realignment; thereafter honorary member of the supervisory board. May 16, 1933: Audience with A. → Hitler in which he made a futile attempt to dissuade him from his racist science policy, but also indicated the society's willingness to cooperate with the new rulers. His son Erwin (born 1893), formerly undersecretary at the Reich Chancellery in 1932, but who had resigned upon Hitler's ascension to power, was arrested on July 23, 1944 following the failed plot against Hitler's life, and executed on Jan. 23, 1945. July 24, 1945–Mar. 31, 1946: Managing president of the *KWG*; from Apr. 1946: honorary president. Refs.: Ka. [1943], Laue [1947]a, Planck [1948], pp. 125–126, 133–180, Frühauf et al. [1959], Hermann [1973], Heilbron [1986]b, Walker [1989]a, [1995], Albrecht [1993], Macrakis [1993], pp. 57ff., 97ff., Oexle [1994].

Pose, Heinz Ferdinand Hermann (1905–)
Nuclear physicist, also worked on radioactivity. Studies at Göttingen and Halle Univs. 1928: Doctorate at Halle. 1930: Unpaid lecturer, Halle Univ. 1934–33: Teaching assignment there. 1938: Untenured associate prof., then 1939: supernumerary prof. at Halle. 1940: Engaged at the → *KWIP* in Dahlem (Berlin) then 1942: at the → *PTR* in Berlin. 1945: Research in the Soviet Union. 1957: Appointment to a professorship with a full teaching load at the Dresden Polytech. Refs.: Walker [1989]a, pp. 102, 125, 183ff.

Prandtl, Ludwig (1875–1953)
Aeronautical engineer. 1894–98: Studies in mechanical engineering at the Munich Polytech., then part-time assistant to A. Föppl. 1900: Doctorate at Munich Univ. on the torsional instability of beams. From 1900: Employed as engineer by the machine manufacturer *Maschinenfabrik Augsburg-Nürnberg* in Nuremberg. 1901–04: Prof. of mechanics, Hanover Polytech. 1904–07: Associate prof. of applied physics, and 1907–46: Full prof. of mechanics, Göttingen Univ. and 1907–25: director of the Aerodynamics Design Testing Station, Göttingen Univ., renamed at the end of 1918: *Aerodynamische Versuchsanstalt* and from 1925: associated with the new → *KWIS*; 1925–46: director of the latter, which greatly expanded between 1934–36. 1941: Involved in C. → Ramsauer's petition to B. → Rust. 1942: Head of the → *RLM*'s Reich research planning office *Forschungsführung* and thus head of Air Force research, also figuring prominently in the Aircraft Research Academy. 1946: Emeritus. Refs.: Stuchtey et al. (Eds.) [1938], pp. 39–41, Busemann [1959], Tollmien in Becker et al. (Eds.) [1987], Walker [1989]a, pp. 63ff., 73ff., 80; Rotta [1990], Trischler (Ed.) [1992]b, esp. pp. 144f., Vogel-Prandtl [1993].

Ramsauer, Carl Wilhelm (1879-1955)
Industrial physicist. 1897–1907: Studies at Munich, Tübingen, Berlin, Kiel, London and Breslau Univs. 1903–07: Assistant at the Imperial Torpedo Laboratory (*Kaiserliches Torpedo-Laboratorium*). 1907–09: Teaching assistant to P. → Lenard at the physics department, Heidelberg Univ.; research there on the quantum effect of the transparency of nobel gases to slow electrons, later named after him (also known as the Ramsauer-Townsend effect). 1909–21: Staff

scientist, Radiological Institute, Heidelberg. During World War I, artillery officer. From 1921: Full prof., Danzig Polytech. 1928–45: Director of the research division of → *AEG*. Simultaneously, 1931–45: Honorary prof., Berlin Polytech. From 1937: Chairman of the Berlin section of the → *DPG*, and from 1940–45: general chairman of the society. Editor of the journals → *Zeitschrift für technische Physik* and → *Physik in regelmäßigen Berichten*. Jan. 20, 1942: On Felix Klein's initiative with the support of L. → Prandtl, submission of a petition to the → *REM* on the atrocious state of physical instruction in Germany, which Ramsauer concludes is the result of its politicization. From 1945: Full prof. and director of the physics department, Berlin Polytech. Refs.: Ramsauer [1949], Brüche [1956], Boeters [1979], Schmithals [1980], Gobrecht in Treue & Hildebrandt (Eds.) [1987], Walker [1989]a, [1995], Macrakis [1993], p. 95, Soon Im [1995].

Regener, Erich (1881–1955)
Experimental physicist. 1900–05: Studies at Berlin Univ.; 1905: doctoral thesis on the effect of short-wave radiation on gaseous bodies (under Emil Warburg). 1920: Appointment at the Stuttgart Polytech. 1931–33: Balloon experiments on cosmic rays and ozone. Autumn 1937: Placed in 'temporary retirement' (*Wartestand*) after 17 years in the civil service because of his marriage to a Jew. Dec. 1937: The → *KWG* established the 'Prof. Regener Private Research Station' in a one-family house belonging to Marianne von Weizsäcker at Lake Constance where he could continue his research on the physics of the stratosphere. 1946: Resumption of his post in Stuttgart, and vice-president of *KWG*. Refs.: Ehmert [1955], Walker [1995], pp. 129f.

Riehl, Nikolaus (1901–1990)
Industrial physicist specializing in applications of radioactive materials. 1920–27: Studies at St. Petersburg Polytech.al Institute and Berlin Univ. 1927: Doctorate at Berlin Univ. (under O. → Hahn and L. → Meitner) on Geiger-Müller counters for β-spectrometry. Specialist in the field of luminescence. From 1927: Staff scientist in the radiology department, and from 1937: head of the optical engineering department and 1939–45: director of the scientific headquarters of the company → *Auer-Gesellschaft*, in whose interest he developed contacts in 1939 with the *Uranverein* and the → *HWA* in the hope of developing new lucrative products from nuclear energy research. 1945–55: Head of the group of German scientists in the Soviet Union, from 1950 head of a radiological department in Sungul in the Urals. 1955–57: Directed jointly with H. Maier-Leibnitz the construction of the first German postwar research nuclear reactor at Garching (Munich). 1957–69: Full prof., Munich Polytech. Ref.: Riehl [1988], Walker [1989]a, Becker & Mader [1990].

Ritter, Oskar (1913–)
Experimental physicist. Studies at Würzburg and Leipzig Univs., notably under R. → Döpel. 1943: Doctoral thesis in nuclear physics in Leipzig with Döpel's guidance. Co-authored studies together with Döpel for the German uranium project. Refs.: Kleint & Wiemers [1993], p. 72.

Rosbaud, Paul (1896–1963)
Metallurgical physicist and science editor. Studies at the Darmstadt Polytech. 1926: Doctorate in engineering, Berlin Polytech. 1926–30: Staff scientist at the *KWI* of Silicate Research in Berlin. One of the few German physicists to actively oppose the Nazi regime. His Jewish wife and daughter managed to escape to England. Editor of many scientific journals, including *Metallwirtschaft*, *Spectrochimica Acta*, as well as → *Die Naturwissenschaften* following Berliner's dismissal. 1930–45: Supervision of scientific publications by the Julius Springer publishing house. Close contacts with many prominent physicists, including O. → Hahn and L. → Meitner. 1945–52: Employed at Butterworth-Springer in London. 1952–56: Science director at Pergamon Press. 1956–63: Leading position at Interscience Publ. and adviser to other publish-

ers. Refs.: Heitler [1963], N.N. [1963], Kramish [1986], Heinrich & Bachmann [1989], pp. 91f., Walker [1989]a, pp. 72f., 140.

Rosenberg, Alfred (1893–1946)
National Socialist ideologue. From 1934: Führer's Commissioner on the General Ideological Schooling and Education of the Nazi Party. From 1921: Main editor of the Nazi newspaper → *Völkischer Beobachter*; editor of → *Volk im Werden* and author of a book on the '20th century myth': *Mythos des 20. Jahrhunderts* (1930). Leading ideologue of the → *NSDAP*; 1933–45: direction of its foreign affairs office, representing the party. July 17, 1941: Appointed Minister of the Eastern Occupied Territories where he promoted the 'Germanization' of the Eastern peoples and exploited slave labor. 1946: In the Nuremberg trials, death sentence and execution on Oct. 16, 1946 as a major war criminal. Refs.: Bollmus [1970], Snyder [1976], Wistrich [1982], Walker [1995], pp. 38, 47, 159.

Rust, Bernhard Karl Josef (1883–1945)
National Socialist politician. Studies in literature, philosophy and classical philology at the Univs. of Munich, Berlin, Göttingen, and Halle. 1908: passed state examinations for secondary teaching in Halle. 1909–31: Senior teacher at the *Realgymnasium* in Hanover. He served for four years in World War I and was decorated. From 1925: District leader of Braunschweig, South Hanover. 1930: Elected to parliament as a Nazi delegate. Feb. 1933–34: Prussian Minister of Science, Art and Culture, and May 1934–45: Reich Minister in the → *REM*, in which capacity he supervised state institutions like the → *PTR*. Initiated the formation of the → *RFR* to centralize research. 1941: Received C. → Ramsauer's memorandum on the deteriorating state of physics instruction in Germany. May 1945: Suicide. Refs: Führerlexikon [1934], Stuchtey et al. (Eds.) [1938], p. 32, Wistrich [1982], Walker [1995].

Sauter, Fritz Eduard Josef Maria (1906–1983)
Austrian theoretical physicist, specializing in electrodynamics and solid-state physics. 1924–28: Studies in mathematics and physics, Innsbruck Univ., and 1928: doctorate there (under A. Mar.) on Kirchhoff's theory of diffraction. 1931–34: Teaching assistant to R. → Becker, Berlin Polytech. in Charlottenburg, and 1933: university lecturer there; work on Dirac's theory and atomic physics. 1934: Lectureship, Göttingen Univ. 1936: Teaching assignment and acting director, and from 1939: Director of the theoretical physics department and full prof., Königsberg Univ. 1939: Membership in the → *NSDAP*. 1942–45: Full prof., Munich Univ. 1950–51: Teaching assignment and substitute director, theoretical physics department, Hanover Polytech. 1951–52: Teaching assignment, Göttingen and Bamberg Univs. 1952–71: Full prof. and director of the theoretical physics department, Cologne Univ. Ref.: Mühlschlegel [1983], Walker [1995], p. 129.

Scheel, Karl Friedrich Franz Christian (1866–1936)
Physicist. 1885–90: Studies at Rostock and Berlin Univs. 1890: Doctoral thesis on water expansion at Berlin Univ. then part-time assistant at the imperial bureau of standards: *Kaiserliche Normal-Aichungs-Commission.* From 1891: Employed at the → *PTR*, 1904–31: member, and finally Senior Executive Officer and head of Department IIIb there. 1899–1918: Editor of the journal *Fortschritte der Physik* and the semi-monthly bibliographic section of the journal *Physikalische Berichte.* From 1902: editor of the Proceedings (*Verhandlungen*) of the → *DPG*, and from 1920: the Society's journal → *Zeitschrift für Physik*; also 1926–33: the physics textbook: *Handbuch der Physik*, together with H. → Geiger. Refs.: Grotrian [1936], Heinicke [1985], F. W. Seeman in Treue & Hildebrandt (Eds.) [1987], Brüche [1987], Mayer-Kuckuck (Ed.) [1995], pp. F 135–F 138.

Scherzer, Otto (1909–1982)
Specialist in electron microscopy. 1927–31: Studies in physics, Munich Polytech. and Univ. 1931: Doctorate under A. → Sommerfeld, Munich Univ. on the quantum theory of bremsstrahlung. 1932–33: C. → Ramsauer's assistant at the → *AEG* research laboratory in Reinickendorf (Berlin) where he specialized in electron optics research. 1934: Habilitation, then teaching assignment in physics, Munich Univ. 1935 substitute member, from 1936: Associate prof. of physics and director of the theoretical physics department, Darmstadt Polytech. 1939–45: Communications research headquarters of the German Navy (*Nachrichtenmittel-Versuchskommando der Kriegsmarine*), work on radar. 1944–45: Head of radar direction finding research (*Arbeitsbereich Funkmeßtechnik*) of the → *RFR*. 1954: Full prof., Darmstadt Polytech. where he helped found the Society for Heavy Ion Research. Ref.: Beck [1983].

Schmidt, August **Ferdinand** (1889–1960)
1907–13: Studies in physics in Giessen and Heidelberg. 1913: Doctorate (under P. → Lenard) at Heidelberg, and from 1913: Lenard's teaching assistant. 1921: Staff scientist and 1922: unpaid university lecturer, Heidelberg. 1926: Teaching assignment in radiology, and 1927: supernumerary associate prof. with teaching assignment, Heidelberg Univ., and 1937: regular associate prof. 1940: Full prof. Stuttgart Polytech.

Schmidt-Ott, Friedrich (1860–1956)
Science policy-maker. Law degree. 1888: At the → *PrKM*, and for many years assistant to Friedrich Althoff; from 1903: head of the arts division. 1917: Minister of Culture. 1911–19: On the boards of trustees of all Kaiser Wilhelm Institutes. 1920–37: Second vice-president (practically, acting vice-president) of the → *KWG*, then honorary member of the supervisory board of that society. 1920–34: President of the → *NG*; 1934–45: Chairman of the → *Stifterverband*. Refs.: Schmitt-Ott [1952], Laue [1956], Zierold [1968], pp. 8ff., 40ff., 109ff., 150ff., Walker [1995], pp. 21, 25.

Schrödinger, Erwin (1887–1961)
Theoretical physicist. 1906–10: Studies at Vienna Univ., 1910: doctorate there, 1911: teaching assistant at the physics department, and 1914: habilitation. 1920: Teaching assistant and unpaid lecturer at Jena Univ. (under M. → Wien), then prof. at the Stuttgart Polytech. 1921: Full prof., Breslau Univ. From the end of 1921: Full prof., Zurich Univ. Co-founder of modern quantum mechanics through his development of de Broglie's ideas of 1925 of the wave-like character of matter. 1933: Awarded the Nobel Prize in physics together with Paul Dirac, 1938: Max Planck Medal of the → *DPG*, along with many other distinctions for his pioneering work in quantum mechanics. 1927–33: Full prof., Berlin Univ. (as successor to M. → Planck). 1931–33: On the board of trustees of the → *KWIP*. Autumn 1933: Exile, despite the fact that the new Laws for the Restoration of the Professional Civil Service did not yet apply to him. 1933–36: Fellow of Magdalen College, Oxford. 1936–38: Full prof., Graz Univ. 1939: Foundation Francqui, Belgium. From 1940: Senior prof., Institute for Advanced Study, Dublin. His book *What is Life?*, published in 1944 had a significant impact on biological studies. From 1956: Full prof., Vienna Univ. Refs.: N.N. [1938]b, Born [1961], Hoffmann [1984], Hoch & Yoxen [1987], Moore [1989].

Schumann, Erich (1898–?)
Physicist and systematic musicologist. 1922: Physicist at the → *RWM*, from 1926: Upon passing his state examinations, civil service candidate (*Referendar*), from 1929: head of the Central Science Office and from 1932: ministerial councillor there. 1929: Habilitation on acoustics and appointed lecturer of systematic musicology (acoustics), Berlin Univ., from 1931: prof. of experimental and theoretical physics, and from 1933: full prof. of applied physics and systematic

musicology. 1933: Membership in the → *NSDAP.* 1933–45: Director of the newly established Physics Department II, Berlin Univ., which was commissioned by the High Command of the Armed Forces (*OKW*) with physical research projects which were funded with subsidies of 120,000–150,000 RM annually. Also 1934–45: Head of the research department of the → *HWA* and assistant secretary in the Science Department of the *RKM*, then 1938–45: of the *OKW.* Like R. → Mentzel, a direct rival of J. → Stark. Autumn 1938: Additional appointment as full prof. of ballistics and military technology at the Berlin Polytech. 1939–45: On the → *RFR*, 1942–45: plenipotentiary for the physics of explosives. Again from 1951: Director of the Helmholtz Institute of Sound Psychology and Medical Acoustics in Berlin. Refs.: Zierold [1968], pp. 191ff., Ludwig [1974], pp. 205f., Walker [1989]a, [1995], Macrakis [1993], p. 79.

Sommerfeld, Arnold Johannes Wilhelm (1868–1951)
Theoretical physicist. 1886–91: Studies at Königsberg, and 1891: doctorate there (under F. Lindemann). 1897–1900: Prof. of mathematics, Clausthal Mining Academy. 1900–06: Prof. of mechanics, Aachen Polytech., working on bremsstrahlung. 1906–38: Full prof. of theoretical physics, Munich Univ., and until 1940: directorship of the physics department. 1915: Calculated the relativistic corrections to Bohr's atomic model. 1919–20: Chairman of the → *DPG.* Author of the most important textbook on atomic physics for the Weimar period: *Atombau und Spektrallinien.* From 1938: Emeritus. His successor was W. → Müller. Students included: P. → Debye, W. → Heisenberg, W. → Pauli and G. Wentzel. Refs.: Planck [1938]a, Joos [1951], Jordan [1951], Laue [1951], Scherzer [1951], Born [1952], Hermann [1967], Benz [1975], Eckert et al. (Eds.) [1984], Walker [1989]a, [1995], Eckert [1993].

Speer, Albert (1905–1981)
Architect, than National Socialist politician. Studies at the Karlsruhe, Munich and Berlin Polytechs., followed by a teaching assistant position in Berlin. 1931: → *NSDAP* membership. 1932: Entry into the → *SS.* Appointed section leader in the German Labor Front and to staff of the *Führer*'s deputy. 1937: General building inspector of Berlin. 1941: Elected delegate in parliament. From 1942: Acquired a number of posts, succeeding F. → Todt as Minister at the → *RMBM* in Feb.: Member of the Central Planning Bureau, general inspector for water and energy, head of the Nazi Automobile Corps as well as leader of the Party's Main Office for Technology (*Hauptamt für Technik*) and of the *NS Bund Deutscher Technik.* In May he was put in charge of the central planning of raw materials, and in June he was also made responsible for the development of armaments and building construction, as well as energy and highway construction in the Eastern territories. As Armaments Minister he revealed a great interest in nuclear physics and particularly in the uranium project, which explains his backing of → *KWG* from 1942 on in its power struggle with the → *RFR*, where he also became presiding councillor in 1942. From 1943: Expanded powers as Reich Minister of Armament and War Production, setting the priority of militarily relevant research projects. Mar. 1945: Disobeyed A. → Hitler's destruction order. 1946: Sentenced to 20 years in prison in the Nuremberg trials. 1966: Released from Spandau. Refs.: Speer [1942], [1970], [1976], Milward [1965], chap. IV, Janssen [1968], Zierold [1968], p. 263, Ludwig [1970], Zilbert [1981], Osietzky in Renneberg & Walker (Eds.) [1994].

Stark, Johannes (1874–1957)
Experimental physicist. 1894–97: Studies in physics, chemistry and mineralogy at Munich Univ., obtaining his doctorate there in 1897, and becoming Lommel's teaching assistant. in the same year. 1900: Habilitation, then 1900–06: unpaid lecturer, Göttingen Univ., where he demonstrated the validity of the Doppler Law in the laboratory using canal rays in 1905. 1904: Founded the journal *Jahrbuch der Radioaktivität und Elektronik.* Apr. 1906–09: Associate prof., Hanover Polytech. Apr. 1909–17: Full prof., Aachen Polytech., where he discovered the influence of elec-

tric fields on spectral lines in late 1913 (the 'Stark effect'). 1913: Awarded the Vahlbruch Prize of the Göttingen Academy of Science, and 1919: the Nobel Prize in physics for his discovery of the splitting of spectral lines in an electric field. Apr. 1917–20: Full prof., Greifswald Univ. 1920–22: Full prof., Würzburg Univ. From the spring of 1921: Private business (porcelain factory in the Upper Palatinate). Around 1927: Research in the private laboratory of Baron R. von Hirsch zu Planegg on the polarization of canal rays; from spring 1929: in his private laboratory (together with T. → Vahlen as his assistant, among others). May 1933: Appointed President of the → *PTR*, against the unanimous advice by the consulted specialists. His nomination as regular member of the Prussian Academy of Sciences was blocked, however, by M. → von Laue, to which he responded by promptly sacking him from his advisory position at the *PTR*. 1934–36: President of the → *DFG*. Stark was, with P. → Lenard, the main advocate of 'Aryan Physics', which was based on racist and nationalistic lines. After 1945: One of the few German physicists to stand before a denazification tribunal. Initially categorized as a major offender, sentencing him to 4 years forced labor, but in an appeal it was reduced to a 1,000 Mark fine. Refs.: Beuthe [1939], Kuhn [1957], Zierold [1968], pp. 173–212, Stark [1987]; H. Hermann [1966], [1967], [1975], Beyerchen [1977], chap. 6, [1980], Kleinert [1980], [1983], Walker [1989]a, [1995] chaps 2–3, Guérout [1992], pp. 55–61, Kleinert in Olff-Nathan (Ed.) [1993], pp. 149–166, Schröder [1993],

Starke, Kurt (1911–)
Radiochemist. 1931–36: Studies at Berlin Univ., and 1937: doctorate there (under O. → Hahn). From 1937: Assistant at the → *KWIC*, working on problems of isotope separation and uranium enrichment. Discovered the transuranic element neptunium (atomic number 93). 1941: Assistant at the → *KWIPC*. 1942: Work on the enrichment of uranium isotope U 239 and its decay product. 1943: Habilitation, Munich Univ. 1944: Assistant at the *KWI* of Medical Research (in the physics division). From 1948: Various professorships abroad. Since 1959: Associate prof. and director of the nuclear chemistry department, Marburg Univ. Refs.: Walker [1989]a, pp. 23, 42ff., 126f., 181, 211, [1995], p. 217.

Steenbeck, Max Christian Theodor (1904–1981)
Technical physicist. 1922–27: Physics and chemistry studies (under W. Kossel) at Kiel, 1927/28: submission of doctoral thesis on x-rays. 1927–45: Technical physicist at the → *Siemens-Schuckert* Works in Berlin. 1934: Laboratory director there, submitting his patent on the betatron in the same year. 1943: Appointed technical director of the static converter plant at *Siemens*, conducting research in gas-discharge physics. 1945: Brief internment. 1946–55: Research in the USSR, developing gas centrifuge for the Soviet atomic program. 1956: Prof. of plasma physics at Jena Univ., 1956–59: director of the Institute for Magnetic Materials, and 1958–69: director of the German Academy of Science Institute for Magnetohydrodynamics in Jena. 1957–63: Head of the Technological Science Bureau on Reactor Construction in Berlin. 1962–64: Vice-president, and 1965: president of the German Academy of Science. 1970: President of the East German Committee on European Security, 1978: honorary president of the East German Research Council. Refs.: Steenbeck [1977].

Stetter, Georg Karl Friedrich (1895–1988)
Nuclear physicist. 1914 and 1919–22: Studies at Vienna Univ., 1922: doctorate there and then teaching assistant at the physics department II. 1928: Habilitation, and 1934: titulary associate prof., Vienna Univ. From 1939: Member of the *Uranverein*, for which he prepared reports together with K. → Lintner on the inelastic dispersion of rapidly moving neutrons around 1941. Around 1942: Directed a half dozen physicists and physical chemists in measuring atomic constants, cross sections and transuranic elements. From 1953: Full prof., Vienna Univ.

Straßmann, Fritz (1902–1980)
Chemist. 1920–29: Studies at Hanover Polytech., and 1929: doctorate there (under H. Brauner), then teaching assistant at the physical chemistry department at the Hanover Polytech. and recipient of a scholarship from the → *NG*. 1935: Staff scientist, and 1938: (together with W. → Mattauch) successor to L. → Meitner at the → *KWIC*. End of 1938: Chemical verification jointly with O. → Hahn of the fission of uranium, revealing barium as a decay product. 1945–53: Director of the chemistry department and university lecturer, and from 1946: full prof. and director of the department of inorganic chemistry and nuclear chemistry, Mainz Univ. Refs.: Krafft [1981], E. H. Berninger in Treue & Hildebrandt (Eds.) [1987], Lemmerich [1988], pp. 129ff., Walker [1989]a.

Stuart, Herbert Arthur (1899–1974)
Experimental physicist. 1920–25: Studied physics, chemistry and mathematics at the Würzburg and Göttingen Univs. 1925: Doctoral thesis (under J. → Franck) in Göttingen on resonance fluorescence of mercury vapor. Transfer to the physics departments at Hamburg (O. → Stern), and Königsberg Univ. (R. → Gans); 1928: Habilitation on the temperature dependency of dielectric constants in gases and vapors. From 1928: unpaid university lecturer, and work on the Kerr effect and light scattering. 1930: Temporary scholarship from the Rockefeller Foundation to the Univ. of California, Berkeley. Specialization in molecular structure research. From 1935: untenured associate prof. and substitute director at Königsberg. 1936: Substituted the theoretical physics chair at Berlin Univ.; Stuart and W. → Orthmann moved the → *DPG* to exclude its last remaining Jewish members. 1939–45: Full prof. and director of the experimental physics department, Dresden Polytech. where he began studying the viscosity and light scattering of macromolecules. 1948–55: Physics adviser to the *Bayer* Company at Leverkusen. From 1955: Associate prof. of chemical physics and head of the highpolymer physics laboratory, Mainz Univ. Refs.: Peterlin [1959], Fischer [1974].

Suess, Hans Eduard (1909–)
Physical chemist. 1929–36: Studies Vienna Univ., and 1936: doctorate there (under Philip Groß). From 1937: Co-worker of P. → Harteck in Hamburg, starting as part-time scientist. After 1939: Involved in the construction there of the first layered arrangement of uranium trioxide and solid carbonic acid in a model 'uranium machine'. 1940: Habilitation. At this time also shared with Harteck an interest in analyzing heavy water as a moderator, constructing an isotope separation unit for the company → *Norsk Hydro*, which went into operation in 1941. 1941–48: University lecturer, and 1948–50: supernumerary prof., Hamburg Univ. From 1950: Visiting prof. in the USA, including Chicago (1950–51), Univ. of California, La Jolla (since 1955). Refs.: Walker [1989]a.

Thiessen, Peter Adolf (1899-1990)
Physical chemist. 1919–23: Studies at the Breslau, Freiburg, Greifswald, and Göttingen Univs. 1923: Doctorate at Göttingen (under R. Zsigmondy). 1923: Supernumerary assistant, and 1924–30: regular teaching assistant there. From 1925: Member of the → *NSDAP*. 1926: Unpaid lecturer, Göttingen Univ. 1930: Head of the department of inorganic chemistry, Göttingen Univ. 1932: Untenured associate prof., Göttingen Univ. From 1933: Department head of the → *KWIPC* in Berlin. 1935: Full prof. of chemistry, Münster Univ. 1935–45: Full prof., Berlin Univ. and director of the *KWIPC*. Main adviser and confidant of R. → Mentzel, head of the chemistry and organic materials section of the → *RFR*, with whom he also shared the flat that had previously been made available to F. → Haber for business purposes on Faradayweg in Dahlem. Thiessen had Communist contacts even before the end of the war. 1945–56: researched in the Soviet Union, where he received high Soviet decorations and the Stalin Prize. Upon his return, 1956: appointed head of the Physical Chemistry Institute of the East-German Academy

of Sciences, full prof. at Humboldt Univ., Berlin and from 1957: chairman of the Research Council of the German Democratic Republic. From 1960: East German councillor of state. Refs.: Chmiel et al. [1986], pp. 44ff., Walker [1989]a, pp. 135, 184, Macrakis [1993], pp. 88ff., 195ff.

Thüring, Bruno Jakob (1905–1989)
Astronomer. 1924–28: Studies at Bamberg College and Erlangen and Munich Univs. While a student, organized the antirelativist declaration by the → *NSDStB* and was politically active. 1928: Doctorate (under A. Wilkens) at Munich, and 1928–33: Wilkens's teaching assistant. 1930: Membership in the → *NSDAP*, and Mar. 1933: in the → *SA*. 1933: Teaching assistant, Breslau Univ. Observatory, and 1934: at the observatory of the Land of Baden in Heidelberg; 1935: Habilitation there. From 1935: Observer at the Munich Observatory. 1937–40: Lecturer at Munich Univ., leading position there in the *NSDStB*, and co-editor of the journal → *Zeitschrift für die gesamte Naturwissenschaft.* Jan. 1941: Denounced the director Alexander Wilkens, which led to his dismissal. In the same month became full prof. of astronomy and director of the observatory, Vienna Univ. Lost his position after the war and worked as a private scholar (computer specialist). 1949: Rated in his denazification trial as lesser offender, and 1950: as fellow traveller. Refs.: Litten [1992], pp. 148–151, 256, Walker [1995], pp. 50–56, 147.

Todt, Fritz (1891–1942)
Engineer. 1910–11: Volunteer service in a field artillery regiment. 1911–14: Studies at the Munich Polytech., and 1918–20: at the Karlsruhe Polytech, submitting his doctoral thesis on road construction technique at Munich. During World War I: In the artillery on the Western front, from Jan. 1916: flying observer, wounded in air battle; decorated with the Iron Cross. 1919–20: Construction engineer, 1920–21: building foreman; also temporarily assistant at a university laboratory. 1921–27: Building supervisor at private road construction companies, then advancing to company manager; and 1927–33: technical director. Jan. 5, 1923: Entered the → *NSDAP*; Oct. 31, 1931: Membership in the → *SA*, becoming unit leader in the staff of the High Command. From 1933: General road inspector and head of the *Organisation Todt*, which was assigned major building contracts for the military, including the *Westwall* fortifications at the Western front in 1938, as well as the national highway system (*Reichsautobahn*), which facilitated strategic East-West circulation. Also designated leader of the Four-Year Plan and Director of the Engineering Headquarters and since its foundation in Nov. 1933, vice-president of the League of German Technicians (*NS Bund Deutscher Technik*), advancing to president Nov. 26, 1934. 1940–42: Reich Minister for Armament and Munitions (at the → *RMBM*). Feb. 8, 1942: Died suddenly in an airplane accident and was succeeded by A. → Speer. Refs.: Führerlexikon [1934], Schrimf [1940], Speer [1942], Milward [1965], chap. III, Ludwig [1974], Zilbert [1981], pp. 86ff.

Tomaschek, Rudolf (1895–1966)
Experimental physicist. 1913–18: Studies at the German Univ., Prague. 1918: Doctorate under P. → Lenard, Heidelberg Univ. 1934: Director of the physics department, Dresden Polytech. 1939–45: Full prof., Munich Polytech. and director of the physics department. 1945: Suspended from office. Replaced by G. → Joos. 1948–54: Employed at the ADC Research Centre, Kirklington Hall near Newark, England. From 1954: Retired in Breitbrunn-Chiemsee. Research work on gravitation measurement, phosphorescence and fluorescence. Refs.: Wengenroth [1993], pp. 243f., Walker [1995], pp. 30, 42, 52.

Vahlen, Karl **Theodor** (1869–1945)
Mathematician. 1889–93: Studies in Berlin, and 1893: doctorate there. From 1893: Unpaid lecturer of mathematics, Königsberg Univ. 1911: Full prof., Greifswald Univ. From 1924: First → *NSDAP* Pomeranian district leader. 1927: Dismissed from chair for having the national flag (black, red and gold) removed from the university building. F. → Schmidt-Ott then increased

his grant, which he had been receiving since 1922 for his work for the German Navy. This was terminated by the → *RIM* in 1929. Worked briefly in J. → Stark's private laboratory as his assistant. 1930: Lecturer of mathematics, Vienna Polytech. After 1933: Strong advocate of the 'Aryan Mathematics' movement; full prof. of mathematics, Berlin Univ., as successor to the expulsed R. von Mises. 1933–37: 3rd vice-president of the → *KWG*. From Mar. 15, 1933: In the Science Department of the → *PrKM*. From May 1934: Assistant Secretary and head of the Science Office at the → *REM*. In → *SS* he held the rank of *SS-Oberführer*. 1934–37: Full prof., Berlin Univ., from 1937: emeritus. 1938: Appointed president of the Prussian Academy of Sciences. Refs.: Siegmund-Schultze [1984], Macrakis [1993], pp. 78f., Walker [1995], chaps. 4–5.

Vögler, Albert (1877–1945)
Industrialist and science politician. Studies in mechanical engineering, Karlsruhe Polytech. 1900–02: Employed at the machine factory *Baum AG* in Herne. From 1902–05: Engineer at the mine *Georgsmarienhütte* in Osnabrück. From Sep. 1905: Senior engineer, and from 1906: director of the *Dortmunder Union*. From 1915: General manager of the German and Luxemburg Mining and Ironworks Corp. in Bochum. From 1926: Deputy chairman of the supervisory board, and from 1940: chairman of the supervisory board (general manager) of Germany's dominant steel producer, *Vereinigte Stahlwerke*. From 1919: Representative of the *Deutsche Volkspartei*. From 1933: Representative of the → *NSDAP* in the *Reichstag*. 1920: Vögler founded the → *HG*. From 1942: Vice president of the *Fördergemeinschaft der Deutschen Industrie* (Society for the Promotion of German Industry), formed as a second attempt to offset the influence of the → *NG*. 1941–45: President of the → *KWG*. Apr. 14, 1945: Suicide at Wittenbräucke near Herdecke. Refs.: Pudor [1955], pp. 120–123, 129, Walker [1989]a.

Volz, Helmut (1911–)
Experimental nuclear physicist. 1929–33: Studies at Tübingen and Munich Univs. 1935: Doctorate at Tübingen under Hans → Geiger, and from 1937–44: teaching assistant to Geiger. Involved in the nuclear research conducted at Berlin for the *Uranverein* on neutron absorption in uranium. 1943–44: University lecturer at the Polytech. in Charlottenburg (Berlin). 1944: Associate prof., Erlangen Univ., and 1958: full prof. there.

Walcher, Wilhelm (1910–)
Experimental physicist. 1929–35: Studies at the Munich and Berlin Polytechnics. 1933–37: Teaching assistant (to G. → Hertz, H. → Kopfermann, W. → Westphal and H. → Geiger) in Berlin. 1933: On the advice of Hertz he became a member of the Nazi Automotive Corps (*Kraftfahrer Korps*). 1937: Doctorate at the Berlin Polytech. (under Kopfermann). 1937–42: Kopfermann's teaching assistant, Kiel Univ., where he developed a mass spectrograph in 1937 both for isotope separation and for the determination of the degree of enrichment of uranium samples. 1940: He worked within the *Uranverein* on the development of two mass spectrometers to determine the composition of isotope mixtures and for neutron-spin analyses. Due to shortages, the project was not completed by the end of the war. 1942: Habilitation thesis rejected due to 'political unreliability'. Through the intercession of Kopfermann at Berlin it was eventually accepted. 1942–47: Lecturer, Göttingen Univ. Since 1948: Full prof. of experimental physics, Marburg Univ. Refs.: Becker et al. (Eds.) [1987], pp. 393, 396, Walker [1989]a.

Weizel, Walter Friedrich Karl (1901–1982)
Theoretical physicist. 1918–25: Studied chemistry in Göttingen, Munich and Heidelberg. Doctorate in physical chemistry in Heidelberg under the guidance of M. Trautz. Changed from chemistry to physics to work on the quantum mechanics of molecules as a fellow of the → *NG* at the Univ. of Rostock. Habilitation in 1929, then brief employment at → *BASF* at Lud-

wigshafen; Rockefeller fellowship to Chicago for 1931. From late 1931: Full prof. of theoretical physics, Karlsruhe Polytech. Forced retirement for opposition to National Socialists, but call to a full professorship at Bonn Univ. in 1936. After World War II, focussed on the physics of gas discharge, but also involved in the establishment of the Jülich nuclear research facility, and worked at the spectroscopy department in Dortmund and the Duisburg inland shipbuilding department. After the war, also state representative of the Social Democratic Party. Ref.: Rollnik [1983].

von Weizsäcker, Carl Friedrich Baron (born 1912)
Theoretical physicist. Son of the Secretary of State of the Foreign Ministry, Ernst von Weizsäcker. 1929–33: Studies at Berlin, Göttingen and Leipzig Univs. Student notably of W. → Heisenberg; 1933: Doctoral thesis with Heisenberg's guidance on 'The passage of rapidly moving corpuscular radiation through ferromagnetic material.' 1934: Heisenberg's teaching assistant, Leipzig Univ. 1936: Habilitation thesis: 'Spin-dependence of nuclear forces'. From 1937: Lecturer, Berlin Univ., and 1936/37–42: employed at the → *KWIP* (initially to substitute Delbrück). From 1939: Resident theoretician there together with S. → Flügge, F. → Houtermans and others on the theory of nuclear fission, 'uranium machines' and cosmology. 1938: Suggested "element transmutations within the interior of the stars" as a stellar energy source (fusion of hydrogen nuclei to helium nuclei through the carbon cycle). 1940: Pointed to the importance of plutonium (atomic number 94) as a nuclear fuel and explosive. 1942: Research assignment in Russia avoided with the help of Heisenberg. 1942–44: Prof. of theoretical physics at the newly established Strasbourg Univ., after overcoming substantial political hurdles; also corresponding member of the *KWIP*. A subsequent call to the Berlin Univ. as associate prof. of physics fell through though, because of a "lack of National Socialist political awareness". 1946–57: Honorary prof., Göttingen Univ., and department head of the Max Planck Institute of Physics, Göttingen. 1957–69: Prof. of philosophy, Hamburg Univ. 1969–80: Head of a Max Planck Institute established to research the conditions essential for the presence of biological life. 1980: Emeritus. Refs.: Bagge, Diebner & Jay [1957], Ackermann et al. (Eds.) [1989], Walker [1989]a, [1995], Frank (Ed.) [1993], Hoffmann (Ed.) [1993], pp. 77, 331ff.

Wesch, Ludwig (1909–)
Physicist. 1927–32: Studies Munich and Heidelberg Univs. (also attending lectures by P. → Lenard). 1931: Doctoral thesis (under the guidance of F. → Schmidt) on 'Discoloration and phosphorescence from cathode rays and high frequency waves'. 1934: Teaching assignment, and 1937: university lecturer, then with Lenard's support and thanks to his membership in the → *SS*, associate prof. of theoretical physics, Heidelberg Univ. 1943–45: Full prof. of technical physics, Heidelberg Univ. Represented the 'Aryan Physics' movement at the 'Munich Synod' on the foundations of physics, organized by the → *NSDDB* on Nov. 15, 1940. Refs.: Wolgast [1986], p. 156, Walker [1995], pp. 30, 42.

Westphal, Wilhelm Heinrich (1882–1978)
Physicist. 1902–08: Studies at Bonn, Munich, Stuttgart and Berlin Univs. 1908: Doctoral thesis at Berlin (under Arthur Wehnelt) on measurements of potential in Wehnelt cylinders. Then H. Rubens's teaching assistant and research on thermal radiation and the physics of gaseous discharge. 1913: Habilitation, then unpaid university lecturer, 1918: titular prof. at the Friedrich Wilhelm Univ., Berlin. 1922–24: Expert adviser to the → *PrKM* additionally; and 1925/26: physics teacher at the *Landschulheim* in Salem. From 1928: Head physics demonstrator, and 1934–45: associate prof., at the Berlin Polytech., where he and G. → Hertz together replaced F. Kurlbaum. From 1935: Substitute head of the physics department, Berlin Polytech. Simultaneously from 1939: associate prof., Berlin Univ., where the chair for theoretical physics had not been adequately filled since the departure of E. → Schrödinger. Additionally, a successful

textbook author and editor of Vols. 12 and 17 of the *Handbuch der Physik*, the *Physikalisches Wörterbuch*, and of the series *Die Wissenschaft*. 1955: Emeritus associate prof. of the Berlin Technical University. Refs.: Krebs [1952], Westphal [1955], Brüche [1957], [1962], Westphal [1972], Päsler [1977].

Weyland, Paul (1888–1972)
Engineer and radical political activist. Around 1920: Active member of the *Deutschnationale Volkspartei*, an ultra-conservative nationalistic party in existence 1919–35. 1921: Editor of the anti-Semitic periodical *Völkische Monatshefte* and attempted to establish even more extreme-right factions, including the professional association for the protection of 'pure' science: *Arbeitsgemeinschaft deutscher Naturforscher zur Erhaltung reiner Wissenschaft e. V.*, which organized lectures against relativity in 1920 in Berlin. Emigrated later to the USA, became an American citizen and an FBI informant. After World War II, returned to West Germany. Refs.: Hentschel [1990]b, Goenner [1993], pp. 120f., Kleinert [1993].

Wien, Max Carl Werner (1866–1938)
Physicist. Studies at Königsberg (Prussia), Freiburg (Breisgau) and Berlin Univs. (also attended lectures by H. von Helmholtz). 1888: Doctoral thesis on acoustics in Berlin under the guidance of Helmholtz and A. Kundt, then one year of military service followed by his return to Berlin. From 1892: C. Röntgen's teaching assistant at Würzburg, where he did research on electrodynamics and wireless telegraphy. 1893: Habilitation on a new kind of induction balance at Würzburg. 1898: Associate prof., Aachen Polytech., where he worked on acoustics and technical applications of alternating current circuits. 1904: Full prof. and director of the physics department of the newly founded Polytech. in Danzig. 1911–35: Full prof. of physics, Jena Univ., where he continued his work on electrolysis and high frequency engineering. 1918–19 and 1924–25: Chairman of the → *DPG*. 1934 and 1936: preparation of memoranda regarding the decline of physics education since the rise to power of the National Socialists. Refs.: Esau [1938], G. Hoffmann [1938], Lemmerich [1987], pp. 138ff., Wittig [1988], D. Hoffmann [1989].

Wirtz, Karl Eugen Julius (1910–1994)
Nuclear physicist and nuclear chemist. 1929–34: Studies in physics, chemistry and mathematics at Bonn, Freiburg and Breslau Univs. 1934: Doctorate at Breslau (under C. Schäfer), then A. Smekal's teaching assistant at Halle. 1935–37: Teaching assistant of C.-F. → Bonhoeffer at Leipzig Univ. Member of the → *NSLB* but not of the → *NSDAP*. 1938: Habilitation thesis at Berlin Univ. on electrochemical foundations of electrolytic heavy water production. From 1937: Staff scientist at the → *KWIP* (working with P. → Debye and W. → Heisenberg), where he worked 1940: on 'uranium machine' horizontal layer designs (together with F. → Bopp and E. → Fischer). From 1941: Unpaid lecturer, Berlin Univ. 1944: Appointed head of the experimental department of the *KWIP*, which was removed to Hechingen due to the increasing number of air raids. Following his internment at Farm Hall, from 1946: at the Max Planck Institute of Physics in Göttingen, 1948–57: also associate prof., Göttingen Univ. From 1950: Scientific member of the Max Planck Society, the successor institution to the → *KWG*. 1957–79: Full prof. of physical foundations of reactor technology, at the Karlsruhe Polytech. and director of the Institute of Neutron Physics and Reactor Technology at the Nuclear Research Center in Karlsruhe founded in 1957. 1965–67: Chairman of the scientific council of the Karlsruhe Center for Nuclear Research. 1966–68: Executive vice-president of the European Atomic Energy Society and consultant of the West German Government in affairs related to the Nonproliferation Treaty. 1972–77: member of the presiding committee of the German Atomic Forum *Deutsches Atomforum*, and 1974–76: Dean of the faculty of mechanical engineering at the Karlsruhe Polytech. Refs.: Walker [1989]a, [1995], Frank (Ed.) [1993], Hoffmann (Ed.) [1993], p. 79, Keßler [1994].

Zenneck, Jonathan Adolf Wilhelm (1871–1959)
Technical physicist. 1889–94: Studies in mathematics, science and zoology at London and Tübingen. 1894: Doctorate in Tübingen (under von Eimer). 1895: Teaching assistant at the physics department and from 1901: unpaid lecturer, Strasbourg Univ. 1905: Associate prof., Danzig Polytech., worked notably on wireless telegraphy. 1906: Full prof., Braunschweig Polytech., 1911: Danzig. During World War I, captain of the first naval batallion in Flanders. 1913–39: Full prof. of experimental physics, Munich Polytech., and regular member of the Bavarian Academy of Sciences. Also from 1933/34: Director of the *Deutsches Museum* in Munich. 1935–37 and 1939-40: Chairman of the → *DPG* and longstanding member of the physics committee of the → *DFG*. Also from 1936: chairman of the board of the aviation research center of the → *RLM* in Oberpfaffenhofen. 1939: Retired. 1926: Award of the Heinrich Hertz Medal. P. → Lenard's student R. → Tomaschek became his successor at the Munich Polytech. 1950–51: Chairman of the regional branch of the *Physikalische Gesellschaften*. Refs.: Joos [1959], Dieminger [1961].

Bibliography

Abbreviations:

AJP: American Journal of Physics, Lancaster
BMFRS: Biographical Memoirs of Fellows of the Royal Society, London
BT: Berliner Tageblatt, Berlin
BWG: Berichte zur Wissenschaftsgeschichte, Weinheim
DSB: Dictionary of Scientific Biography, New York (Gillispie, Ed.)
FuF: Forschungen und Fortschritte, Berlin
HSPS: Historical Studies in the Physical Sciences, Baltimore
KPFB: Kernphysikalische Forschungsberichte (see Appendix E)
Natw.: Die Naturwissenschaften, Berlin, Heidelberg
NDB: Neue Deutsche Biographie, Munich
NSM: Nationalsozialistische Monatshefte, Munich
NTM: Schriftenreihe f. Geschichte d. Naturwissenschaften, Technik & Medizin, Leipzig
NYT: The New York Times, New York
PB: Physikalische Blätter, Weinheim (Deutsche Physikalische Gesellschaft)
PZ: Physikalische Zeitschrift, Berlin
VB: Völkischer Beobachter, Berlin/Munich
VDPG: Verhandlungen der Deutschen Physikalischen Gesellschaft, Berlin/Weinheim
VHZG: Vierteljahreshefte für Zeitgeschichte, Munich
ZgN: Zeitschrift für die gesamte Naturwissenschaft, Braunschweig
ZgSS: Zeitschrift für das gesamte Schieß- und Sprengstoffwesen, Munich
ZtP: Zeitschrift für technische Physik, Leipzig

The entries in this bibliography are ordered chronologically under the first author's name. Accents on vowels are ignored in the alphabetical listing. The texts selected for the anthology are generally omitted as well as specific references appearing in full in the footnotes.

References

[1] Ackermann, Peter, Wolfgang Eisenberg, Helge Herwig & Karlheinz Kannegießer (Eds.) [1989] *Erfahrung des Denkens – Wahrnehmung des Ganzen. Carl Friedrich von Weizsäcker als Physiker und Philosoph*, Berlin: Akademie Verlag.

[2] Adam, Uwe Dietrich [1977] *Hochschule und Nationalsozialismus: Die Universität Tübingen im 3. Reich*, Tübingen: Mohr.

[3] Albrecht, Helmuth [1993] ''Max Planck. Mein Besuch bei Adolf Hitler' – Anmerkungen zum Wert einer historischen Quelle', in: Albrecht (Ed.) [1993], pp. 41–63.

[4] —— (Ed.) [1993] *Naturwissenschaft und Technik in der Geschichte. 25 Jahre Lehrstuhl für Geschichte der Naturwissenschaft und Technik am Historischen Institut der Universität Stuttgart*, Stuttgart: GNT-Verlag.

[5] Albrecht, Ulrich et al. [1992] *Die Spezialisten. Deutsche Naturwissenschaftler und Techniker in der Sowjetunion nach 1945*, Berlin: Dietz Verlag.

[6] Aly, Hermann [1936] 'Hat Deutschland zu wenig Ingenieure? Betrachtungen zum Nachwuchs des Ingenieurberufes', *Deutsche Technik* **4**, pp. 432–433.

[7] Ardenne, Manfred von [1972] *Ein glückliches Leben für Technik und Forschung. Autobiographie*, a) 1st ed., Berlin: Verlag der Nation; b) 10th expanded ed. under the title *Die Erinnerungen*, Munich: Harbig, 1990.

[8] Arendt, Hannah [1977] *Eichmann in Jerusalem. A Report on the Banality of Evil*, rev. ed., Viking Penguin.

[9] Ash, Mitchell G. & Ulfried Geuter [1985] 'NSDAP-Mitgliedschaft und Universitätskarriere in der Psychologie', in: C. F. Graumann (Ed.) *Psychologie im Nationalsozialismus*, Berlin: Springer.

[10] Ash, Mitchell G. [1995] 'Wissenschaftswandel in Zeiten politischer Umwälzungen: Entwicklungen, Verwicklungen, Abwicklungen', *NTM*, new ser. **3**, pp. 1–21.

[11] Assche, Pieter H. M. van [1989] 'The first discovered fission product hampered the discovery of fission', *International Conference on Fifty Years Research in Nuclear Fission*, Berlin: Hahn-Meitner Institute, p. 14.

[12] Bagge, Erich, Kurt Diebner & Kenneth Jay [1957] *Von der Uranspaltung bis Calder Hall*, Hamburg: Rowohlt (RDE Ser. No. 41), esp. Bagge & Diebner: 'Zur Entwicklung der Kernenergieverwertung in Deutschland', pp. 9–80.

[13] Bagge, Erich [1969] 'Erich Fischer †', *PB* **25**, p. 515.

[14] —— [1985] 'Fünfzig Jahre als Physiker in Deutschland', *Fusion* **6**, No. 6, pp. 26–34.

[15] Banse, Ewald [1933] *Wehrwissenschaft. Einführung in eine neue nationale Wissenschaft*, Leipzig: Armanen-Verlag.

[16] Bartel, Hans-Georg [1989] *Walther Nernst*, Leipzig: Teubner.

[17] Bauman, Zygmunt [1989] *Modernity and the Holocaust*, Oxford: Polity Press.

[18] Bäumer(-Schleinkofer), Anne [1990] *NS-Biologie*, Stuttgart: Wissenschaftliche Verlagsges.

[19] Bavink, Bernhard [1933] 'Die Naturwissenschaften im Dritten Reich', *Unsere Welt* **25**, pp. 225–236.

[20] —— [1934]a 'Rasse und Kultur', *Unsere Welt* **26**, pp. 98–114, 161–190.

[21] —— [1934]b Reply to Bieberbach [1934]a, *Unsere Welt* **26**, pp. 342–348.

[22] —— [1938] 'Vom Sinn und Ethos der Wissenschaft (Zur 95. Versammlung deutscher Naturforscher und Ärzte in Stuttgart)', *Unsere Welt* **30**, pp. 241–265.

[23] Baxter, James Phinney III [1947] *Scientists against Time* (1st ed. Nov. 1946) reprinted Boston: Little, Brown & Co.

[24] Bayer, Otto & Heinrich Remy [1963] 'Walter Noddack. 1893–1960', *Chemische Berichte* **96**, pp. xxvii–lii.

[25] Baynes, Norman Hepburn (Ed.) [1942] *Adolf Hitler: The Speeches of Adolf Hitler, April 1922–August 1939*, London (New York: H. Fertig, 1969; New York: Gordon, 1981).

[26] Beck, Friedrich [1983] 'Otto Scherzer: Wegbereiter der Elektronenoptik' *PB* **39**, p. 50.

[27] Becker, August (Ed.) [1936] *Naturforschung im Aufbruch. Reden und Vorträge zur Einweihung des Philipp-Lenard Instituts der Universität Heidelberg am 13. und 14. XII. 1935*, Munich, J.F. Lehmanns.

[28] —— [1942]a 'Das Philipp Lenard-Institut zur 80. Geburtstag seines Meisters', *ZgN* **8**, pp. 139–143.

[29] —— [1942]b 'Philipp Lenard und seine Schule', *ZgN* **8**, pp. 143–147.

[30] Becker, Heinrich, Hans-Joachim Dahms & Cornelia Wegeler (Eds.) [1987] *Die Universität Göttingen unter dem Nationalsozialismus. Das verdrängte Kapitel ihrer 250 jährigen Geschichte*, Munich: Saur.

[31] Becker, Karl [1939] 'Wehrtechnik' in: R. Donnevert (Ed.) *Wehrmacht und Partei*, Leipzig: Barth, pp. 49–132.

[32] Becker, L. & L. Mader [1990] 'In memoriam Nikolaus Riehl', *PB* **46**, p. 450.

[33] Bentwich, Norman [1953] *The International Problem of Refugees*, Geneva (Geneva Special Studies, 6,5).

[34] —— [1953] *The Rescue and Achievement of Refugee Scholars. The Story of Displaced Scholars and Scientists, 1933–1952*, The Hague: Nijhoff.

[35] Benz, Ulrich [1975] *Arnold Sommerfeld. Lehrer und Forscher an der Schwelle zum Atomzeitalter 1868–1951*, Stuttgart: Wissenschaftliche Verlagsgesellschaft.

[36] Benz, Ulrich & Hermann Graml (Eds.) [1988] *Biographisches Lexikon zur Weimarer Republik*, Munich, Beck.

[37] Benz, Ulrich [1990] *Herrschaft und Gesellschaft im nationalsozialistischen Staat*, Frankfurt (Main): Fischer.

[38] Berl, E. [1928] 'Fritz Haber zum 60. Geburtstag', *Zeitschrift für Elektrochemie* **34**, pp. 797–803.

[39] Berninger, Ernst [1974] *Otto Hahn in Selbstzeugnissen und Bilddokumenten*, Reinbek: Rowohlt (2nd ed. 1979).

[40] Bernstein, Jeremy [1980] *Hans Bethe, Prophet of Energy*, New York: Basic Books.

[41] Bethge, Klaus & Horst Klein (Eds.) [1989] *Physiker und Astronomen in Frankfurt*, Neuwied: Metzner.

[42] Betz, Albert (Ed.) [1948] *Hydro- and Aerodynamics*, Wiesbaden: Klemm (FIAT Review of German Science 1939–1945, Vol. 11).

[43] Beuthe, H. [1939] 'Johannes Stark', *PZ* **40**, pp. 321–322.

[44] Beyerchen, Alan D. [1977] *Scientists under Hitler: Politics and the Physics Community in the Third Reich*, New Haven/London: Yale Univ. Press, (see also the reviews by Kleinert [1979]b and Fölsing [1981]).

[45] —— [1980] 'Der Kampf um die Besetzung der Lehrstühle für Physik im NS-Staat', in: Heinemann (Ed.) [1980], pp. 77–86.

[46] —— [1983] 'Anti-intellectualism and the cultural decapitation of Germany under the Nazis', in: Jackman & Borden (Eds.) [1983], pp. 29–44.

[47] —— [1992] 'What we now know about Nazism and science', *Social Research* **59**, 515–641.

[48] Beyerle, Konrad, Wilhelm Groth, Paul Harteck & Johannes Jensen [1950] *Über Gaszentrifugen: Anreicherung der Xenon-, Krypton- und der Selen-Isotope nach dem Zentrifugenverfahren*, Weinheim: Verlag Chemie.

[49] Beyler, Richard [1994] *From Positivism to Organizism: Pascual Jordan's Interpretation of Modern Physics in Cultural Context*, Ann Arbor, Mich.: U.M.I. Dissertation Services No. 9500016 (Ph.D. thesis, Harvard Univ.).

[50] Bie, Oskar, Werner Kraus, Andreas Lazlo, Alexander Moissi, Max Reinhard, Johanna Terwin, Hella Thimig & Stefan Zweig [1920] 'Eine Sympathiekundgebung für Professor Einstein', *Neue Freie Presse*, Sep. 1, No. 20120, and *BT* **49**, Aug. 31, No. 409 (A), in response to Einstein [1920].

[51] Bieberbach, L. [1934]a 'Persönlichkeitsstruktur und mathematisches Schaffen', a) *Unterrichtsblätter für Mathematik und Naturwissenschaft* **40**, pp. 236–243; b) condensed in: *FuF* **10**, pp. 235–237 (see also Bavink [1934]b).

[52] —— [1934]c 'Stilarten mathematischen Schaffens', *Sitzungsberichte der Preußischen Akademie der Wissenschaften*, math. phys. Klasse, pp. 352–360.

[53] Bieberstein, Johannes von [1965] 'Die geistige SA rückt ein', *Politikon* **9**, pp. 11–13.

[54] Blackett, Sir Basil et al. [1934] 'A call to scientists', *The New Republic* **80**, pp. 76–77.

[55] Bleuel, Hans Peter [1968] *Deutschlands Bekenner. Professoren zwischen Kaiserreich und Diktatur*, Bonn: Scherz (reprinted New York: Arno, 1977).

[56] Bleuel, Hans Peter & Ernst Klinnert [1967] *Deutsche Studenten auf dem Weg ins Dritten Reich. Ideologien, Programme, Aktionen 1918–1935*, Gütersloh: Mohn.

[57] Blumenfeld, Kurt [1962] *Erlebte Judenfrage. Ein Vierteljahrhundert Deutscher Zionismus*, Stuttgart: Deutsche Verlagsanstalt.

[58] Boberach, Heinz (Ed.) [1984] *Meldungen aus dem Reich: Die geheimen Lageberichte des Sicherheitsdienstes der SS, 1938–1945*, Herrsching: Pawlak.

[59] Bode, Volkhard & Gerhard Kaiser [1995] *Raketenspuren. Peenemünde 1936–1994*, Berlin: Links.

[60] Boelcke, Willi A. [1969] *Deutschlands Rüstung im zweiten Weltkrieg: Hitlers Konferenzen mit Albert Speer 1942–1945*, Frankfurt: Athenaion.

[61] Boeters, Karl Ernst [1979] 'In memoriam Carl Ramsauer (1879–1955)', *PB* **35**, 664–666.

[62] Böhles, Hans-Jürgen et al. [1982] *Frontabschnitt Hochschule. Die Gießener Universität im Nationalsozialismus*, Giessen: Anabas & Fochs.

[63] Bollmus, Reinhard [1970] *Das Amt Rosenberg und seine Gegner. Zum Machtkampf im nationalsozialistischen Herrschaftssystem*, Stuttgart: Deutsche Verlagsanstalt.

[64] Böndel, Dirk et al. [1995] *Ich diente nur die Technik. Sieben Karrieren zwischen 1940 und 1950*, Berlin: Nicolai.

[65] Borkin, Joseph [1978] *The Crime and Punishment of I.G.-Farben*, New York: Free Press.

[66] Born, Max [1952] 'Arnold Johannes Wilhelm Sommerfeld', *Obituary Notices of the Fellows of the Royal Society* **8**, pp. 275–296.

[67] —— [1961] 'Erwin Schrödinger †', *PB* **17**, pp. 85–87.

[68] —— [1968] *My Life and My Views*, a) New York: Scribners. b) Slightly altered German translation by H. Degner et al.: *Mein Leben. Die Erinnerungen eines Nobelpreisträgers*, Munich: Nymphenburger, 1969.

[69] Bortfeld, J., W. Hauser & Helmut Rechenberg (Eds.) [1987] *Forschen – Messen – Prüfen. 100 Jahre Physikalisch-Technische Reichsanstalt/Bundesanstalt 1887–1987*, Weinheim: Physik-Verlag.

[70] Bothe, Walther [1943] 'Die Forschungsmittel der Kernphysik', *Schriften der Deutschen Akademie der Luftfahrtforschung Berlin*, pp. 37–59 (= *KPFB* **G-205**).

[71] Bothe, W. & Siegfried Flügge (Eds.) [1948] *Nuclear Physics and Cosmic Rays*, Wiesbaden: Klemm (FIAT Review of German Science 1939–1945, Vol. 13: part 1; Vol. 14: part 2).

[72] Bower, Tom [1990] *Blind Eye to Murder: Britain, America, and the Purging of Nazi-Germany—A Pledge Betrayed*, London: Deutsch.

[73] Bracher, Karl Dietrich [1969] *Die deutsche Diktatur. Entstehung, Struktur, Folgen des Nationalsozialismus*, Cologne: Kiepenheuer & Witsch a) 1st ed., 1969; b) 7th ed., 1993.

[74] Bracher, Karl D., Manfred Funke & Hans-Adolf Jakobsen (Eds.) [1986] *Nationalsozialistische Diktatur 1933–1945. Eine Bilanz*, Bonn: Bundeszentrale für politische Bildung.

[75] Brämer, Rainer & Armin Kremer [1980] *Physikunterricht im Dritten Reich*, Marburg: Soznat.

[76] Brämer, Rainer (Ed.) [1983] *Naturwissenschaft im NS-Staat*, Marburg: Soznat.

[77] —— [1986] 'Heimliche Komplizen? Zur Rolle der Naturwissenschaften im Dritten Reich', in: *Aus Politik und Zeitgeschichte*, supplement to the weekly *Das Parlament* **B 12/86**, Mar. 22, pp. 15–30.

[78] Brandt, Leo [1962] *Forschen und Gestalten. Reden und Aufsätze*, Cologne/Opladen: Westdeutscher Verlag.

[79] Braun, Hans-Joachim [1992] 'Aero-engine production in the Third Reich', *History of Technology* **14**, pp. 1–15.

[80] Brill, R., O. Hahn, W. Brandt, W. Hartke, P. P. Ewald, P. ten Bruggencate, W. Meißner, G. Schenk, R. Vieweg, F. Trendelenburg, F. Laves, K. Schäfer, J. Schormüller [1959] 'Feier-Stunde zu Ehren von Professor Dr. Max von Laue', *Mitteilungen der Max-Planck-Gesellschaft* **6** pp. 323–366.

[81] Broszat, Martin [1969] *Der Staat Hitlers. Grundlegung und Verfassung seiner inneren Entwicklung*, Munich: Deutscher Taschenbuch Verlag (No. 4009).

[82] —— [1977] 'Hitler und die Genese der "Endlösung": Aus Anlaß der Thesen von David Irving', *VHZG* **25**, pp. 739–775.

[83] —— [1986] *Nach Hitler. Der schwierige Umgang mit der Geschichte*, Munich: Oldenbourg.

[84] Broszat, Martin & Norbert Frei (Eds.) [1989] *Das Dritte Reich im Überblick. Chronik, Ereignisse, Zusammenhänge*, Munich/Zurich: Piper.

[85] Broszat, Martin & Klaus Schwabe (Eds.) [1989] *Die deutschen Eliten und der Weg in den Zweiten Weltkrieg*, Munich: Beck (Beck Ser. No. 401).

[86] Brown, Anthony Cave & Charles B. MacDonald [1977] *The Secret History of the Atomic Bomb*, New York: The Dial Press.

[87] Brüche, Ernst [1946]a "Deutsche Physik' und die deutschen Physiker', *PB* **2**, pp. 232–236.

[88] —— [1946]b '100 Jahre Carl Zeiss Jena', *PB* **2**, pp. 193–200, 223-224.

[89] —— [1947]a 'Philipp Lenard', *PB* **3**, p. 161.

[90] [——] [1947]b 'NS-Physiker', *PB* **3**, p. 168 (comment on Stark [1947]).

[91] —— [1956] 'Abschied von Carl Ramsauer', *PB* **12**, pp. 49–54.

[92] —— [1957] 'Wilhelm Westphal. 75 Jahre', *PB* **13**, pp. 131–132.

[93] —— [1962] 'Wilhelm Westphal zum 80. Geburtstag', *PB* **18**, pp. 127–130.

[94] —— [1967] 'Wolfgang Finkelnburg †' *PB* **23**, pp. 520–521.

[95] —— [1987] 'Scheel, Karl', in: Hermann (Ed.) [1987], pp. 341–342.

[96] ten Bruggencate, Paul (Ed.) [1948] *Astronomy, Astrophysics and Cosmogony*, Wiesbaden: Klemm (FIAT Review of German Science 1939–1946, Vol. 20).

[97] Bühl, Alfons [1937] 'Philipp Lenard und die deutsche Naturforschung', in: Weigel (Ed.) [1937], pp. 5–19.

[98] —— [1939] Review of Lenard [1936]a, *ZgN* **5**, pp. 151–153.

[99] Burchardt, Lothar [1975] *Wissenschaftspolitik im Wilhelminischen Deutschland. Vorgeschichte, Gründung und Aufbau der Kaiser-Wilhelm-Gesellschaft zur Förderung der Wissenschaften*, Göttingen: Vandenhoek & Ruprecht.

[100] —— [1976] 'Der Weg zur PTR', *PB* **32**, No. 7, pp. 289–297.

[101] Busch, Georg [1985] 'Peter Debye (1884–1966). Werden und Wirken eines großen Naturforschers', *Vierteljahresschrift der Naturforschenden Gesellschaft in Zürich* **130**, pp. 19–34.

[102] Buselmeier, Karin, Dietrich Harth & Christian Jansen (Eds.) [1985] *Auch eine Geschichte der Universität Heidelberg*, Mannheim: Edition Quadrat.

[103] Busemann, A. [1959] 'Ludwig Prandtl 1875–1953', *BMFRS* **5**, pp. 193–205.

[104] Cahan, David [1989] *An Institute for an Empire. The Physikalisch-Technische Reichsanstalt, 1871–1918*, Cambridge: Cambridge Univ. Press.

[105] Carroll, P. Thomas [1983] 'Immigrants in American Chemistry', in: Jackman & Borden (Eds.) [1983], pp. 189–203.

[106] Casimir, Hendrik Brugt Gerhard [1983] *Haphazard Reality? Half a Century of Science*, New York: Harper & Row.

[107] Cassidy, David C. [1979] 'Gustav Hertz, Hans Geiger und das Physikalische Institute der Technischen Hochschule Berlin in den Jahren 1933–1945', in: Rürup (Ed.) [1979], pp. 373–387.

[108] Cassidy, David C. & Martha Baker (Eds.) [1984] *Werner Heisenberg: A Bibliography of His Non-Technical Writings*, Berkeley: Office History of Science and Technology.

[109] —— [1992]a *Uncertainty: The Life and Science of Werner Heisenberg*, San Francisco: Freeman.

[110] —— [1992]b 'Heisenberg, German Science, and the Third Reich', *Social Research* **59**, pp. 643–661.

[111] Chmiel, Gerd, et al. [1986] *'...im Frieden der Menschheit, im Krieg dem Vaterland...', 75 Jahre Fritz Haber Institut der Max Planck Gesellschaft. Bemerkungen zur Geschichte und Gegenwart.* Berlin: Öffentliche Tarif-Vereinigung.

[112] Ciesla, Burghard [1993] 'Der Spezialistentransfer in die UdSSR und seine Auswirkungen in der SBZ und DDR', *Aus Politik und Zeitgeschichte*, supplement to the weekly: *Das Parlament*, December, No. B49–50, pp. 24–31.

[113] Citron, A. [1980] 'In memoriam Wolfgang Gentner', *PB* **36**, pp. 358–359 (see also 320).

[114] Clark, W. Ronald [1971] *Einstein. The Life and Times*, New York: World Publ. Co.; reprinted London: Hodder & Stoughton, 1973.

[115] Clusius, Klaus [1943] 'Isotopentrennung' *Schriften der Deutschen Akademie der Luftfahrtforschung Berlin* (Probleme der Kernphysik) (= *KPFB* **G-207**), pp. 61–78.

[116] —— (Ed.) [1948] *Physical Chemistry*, Wiesbaden: Klemm (FIAT Review of German Science 1939–1945, Vol. 30).

[117] Cooper, Ray M. (Ed.) [1992] *Refugee Scholars: Conversations with Tess [Esther] Simpson*, Leeds: Moorland Books.

[118] Dahms, Hans-Joachim [1986] 'Verluste durch Emigration. Die Auswirkungen der nationalsozialistischen "Säuberungen" an der Universität Göttingen. Eine Fallstudie', *Exilforschung: Ein internationales Jahrbuch* **4**, p. 163.

[119] David, Lore R. & I. A. Warheit (Eds.) [1952] *German Reports on Atomic Energy. A Bibiliography of Unclassified Literature* (Atomic Energy Commission Critical Review Series. Technical Information Division, TID 3030, June 6, 1952).

[120] Deichmann, Ute [1992] *Biologen unter Hitler*, Frankfurt (Main): Campus.

[121] Delarue, Jacques [1964/79] *Geschichte der Gestapo* a) Düsseldorf: Droste Verlag, 1964; b) reprinted Königstein: Athenäum, 1979.

[122] Diebner, Kurt [1955] 'Der deutsche Forscheranteil', *Die Zeit*, Aug. 18.

[123] —— (alias Werner Tautorus) [1956] 'Die deutschen Geheimarbeiten zur Kernenergieverwertung während des zweiten Weltkrieges 1939–1945', *Atomkernenergie* **1**, pp. 368–370, 423–425 (cf. also David & Warheit (Eds.) [1952]).

[124] Dieminger, Walter [1961] 'Jonathan Zenneck', *Deutsches Museum. Abhandlungen und Berichte* **29**, pp. 1–44.

[125] Dingler, Hugo [1934] 'Zur Philosophie des Dritten Reiches', *Zeitschrift für Deutschkunde* (48th ser. of *Zeitschrift für den deutschen Unterricht*), pp. 609–622.

[126] —— [1937/38] 'Die Physik des 20. Jahrhunderts. Eine prinzipielle Auseinandersetzung. (Zu einem Buche von P. Jordan)', *ZgN* **3**, pp. 321–335 (review of Jordan [1936/49]).

[127] —— [1938] *Die Methode der Physik*, Munich: Reinhardt (see also reviews by Steck [1938/39], Weizsäcker [1939]a,b).

[128] —— [1938/39] 'Zur Entstehung der sogen. modernen theoretischen Physik', *ZgN* **4**, pp. 329–341.

[129] —— [1940] 'Methode der Physik. Zu einem Aufsatz von Karl Friedrich von Weizsäcker', *ZgN* **6**, pp. 75–88 (reply to Weizsäcker [1939]a).

[130] —— [1942] 'Philipp Lenard und die Prinzipien der Wissenschaft', *ZgN* **8**, pp. 115–117.

[131] —— [1943] 'Über den Kern einer fruchtbaren Diskussion über die 'moderne theoretische Physik' ', *ZgN* **9**, pp. 212–221 (reply to Heisenberg [1943]a).

[132] Dolch, Heimo [1955] 'Werner Heisenberg', in: Hans Schwerte & Wilhelm Spengler (Eds.) *Forscher und Wissenschaftler im heutigen Europa*, Oldenburg: Stalling, pp. 94–102.

[133] Donnevert, Richard [1939] 'Partei und Wehrmacht', in: R. Donnevert (Ed.) *Wehrmacht und Partei*, 2nd ed. (1st ed. 1938) Leipzig: Barth, pp. 1–31.

[134] Döring, Herbert [1974] 'Deutsche Professoren zwischen Kaiserreich und Drittem Reich', *Neue Politische Literatur* **19**, pp. 340–352.

[135] Döring, W. [1947] 'Richard Becker 60 Jahre', *PB* **3**, p. 393.

[136] Dreisigacker, E. [1995] 'Gedenkkolloquium für Hans Kopfermann', *PB* **51**, p. 527.

[137] Dreyer, Ernst Adolf (Ed.) [1934] *Deutsche Kultur im Neuen Reich. Wesen, Aufgabe und Ziel der Reichskulturkammer unter Mitarbeit der Präsidenten und Präsidialratsmitglieder der Kammern*, Berlin: Schlieffen-Verlag.

[138] Duggan, Stephen & Betty Drury [1948] *The Rescue of Science and Learning. The Story of the Emergency Committee in Aid of Displaced Foreign Scholars*, New York: Macmillan Co.

[139] Ebert, Hans [1979] 'Die Technische Hochschule Berlin und der Nationalsozialismus: Politische "Gleichschaltung" und rassistische "Säuberungen", in: Rürup (Ed.) [1979], pp. 455–468.

[140] Eckert, Michael, Willibald Pricha, Helmut Schubert & Gisela Torkar (Eds.) [1984] *Geheimrat Sommerfeld – Theoretischer Physiker: Eine Dokumentation aus seinem Nachlaß*, Munich: Deutsches Museum.

[141] Eckert, M. [1985] 'Die 'Deutsche Physik' und das Deutsche Museum', *PB* **41**, pp. 87–92.

[142] Eckert, Michael & Maria Osietzki [1989] *Wissenschaft für Macht und Markt. Kernforschung und Mikroelektronik in der Bundesrepublik Deutschland*, Munich: Beck.

[143] Eckert, M. [1990] 'Primacy doomed to failure: Heisenberg's role as scientific adviser for nuclear policy in the Federal Republic of Germany', *HSPS* **21,1**, pp. 29–58.

[144] —— [1993] *Die Atomphysiker. Eine Geschichte der theoretischen Physik am Beispiel der Sommerfeldschule*, Braunschweig: Vieweg.

[145] Eggert, J. [1963] 'Klaus Clusius †', *PB* **19**, p. 319.

[146] Ehmert, A. [1955] 'Erich Regener †', *PB* **11**, pp. 174–176.

[147] Eichholtz, Dietrich & Wolfgang Schumann (Eds.) [1969] *Anatomie des Krieges. Neue Dokumente über die Rolle des deutschen Monopol-Kapitals bei der Vorbereitung und Durchführung des Zweiten Weltkrieges*, Berlin: Deutscher Verlag der Wissenschaft.

[148] Eilers, Rolf [1963] *Die nationalsozialistische Schulpolitik. Eine Studie zur Funktion der Erziehung im totalitären Staat*, Cologne: Westdeutscher Verlag.

[149] Einstein, Albert [1918]a 'Bemerkung zu Gehrckes Notiz: Über den Äther', *VDPG* **20**, p. 261 (see also Gehrcke [1920]a).

[150] —— [1918]b 'Dialog über Einwände gegen die Relativitätstheorie', ibid., pp. 697–702 (reply to Gehrcke [1913], and Lenard [1918/21]; see also Gehrcke [1919]).

[151] —— [1920] 'Meine Antwort. Über die antirelativitätstheoretische G.m.b.H.', *BT* **49**, Aug. 27, p. 1 (reprinted in: Sugimoto [1989], p. 66; see also Bie et al. [1920], Gehrcke [1920]c).

[152] [——] [1933]a 'Professor Einstein and the Prussian Academy', *Science* **77**, No. 1997, 346.

[153] [——] [1933]b 'Professor Einstein', *Science* **77**, No. 2002, p. 444.

[154] Elton, Lewis [1986] 'Einstein, general relativity and the German press', *ISIS* **77**, 95–103.

[155] Engel, Michael [1984] *Geschichte Dahlems*, Berlin: Spitz.

[156] [Engelmann, B.] [1964] 'Der Franck-Report', *PB* **20**, pp. 329–334.

[157] Erickson, Robert P. [1985] *Theologians under Hitler*, New York: Yale Univ. Press.

[158] Ermenc, Joseph J. (Ed.) [1989] *Atomic Bomb Scientists: Memoirs, 1939–1945. Interview with Werner Karl Heisenberg, Paul Harteck, Lew Kowarski, Leslie R. Groves, Aristid von Grosse, C. E. Larson*, Westport, Conn.: Meckler.

[159] Ernst, Sabine (Ed.) [1992] *Lise Meitner an Otto Hahn. Briefe aus den Jahren 1912 bis 1924*, Stuttgart: Wissenschaftliche Verlagsgesellschaft.

[160] Esau, Abraham [1938] 'Gedankenworte auf Max Wien', *VDPG* (3) **19**, pp. 118–121.

[161] —— (Ed.) [1943] 'Probleme der Kernphysik', *Schriften der Deutschen Akademie der Luftfahrtforschung* (= *KPFB* **G-213**f., **G-216**ff.).

[162] Euler, Hans & Werner Heisenberg [1938] 'Theoretische Gesichtspunkte zur Deutung der kosmischen Strahlung', *Ergebnisse der exakten Naturwissenschaften* **17**, pp. 1–69.

[163] Evola, Julius [1940]a 'Über das Problem der arischen Naturwissenschaft', *ZgN* **6**, 161–172.

[164] —— [1940]b 'Die Juden und die Mathematik', *NSM*, Feb., (119) **11**, pp. 81–87.

[165] Ewald, Peter Paul [1960] 'Max von Laue 1879–1960', *Memoirs of Fellows of the Royal Society London* **6**, pp. 135–156.

[166] —— (Ed.) [1962] *Fifty Years of X-ray Diffraction. Dedicated to the International Union of Crystallography on the Occasion of the Commemoration Meeting in Munich, July 1962*, Utrecht: A. Oosthoek.

[167] Falkenhagen, Hans [1966] 'In memoriam Peter J. W. Debye', *PB* **22**, p. 578.

[168] Faust, Anselm [1971] *Der Nationalsozialistische Deutsche Studentenbund: Studente und Nationalsozialismus in der Weimarer Republik* (Thesis), a) Düsseldorf: Schwann, b) reprinted, 1973.

[169] Feigl, Georg & Georg Hamel [1939] 'Ausbildungsplan für Lehramtsanwärter in den Fächern reine Mathematik, angewandte Mathematik und Physik', *Deutsche Mathematik* **4**, pp. 98–108.

[170] Ferber, Christian von [1956] *Die Entwicklung des Lehrkörpers in deutschen Universitäten und Hochschulen, 1865–1954*, Göttingen: Vandenhoek & Ruprecht.

[171] Fermi, Laura [1954] *Atoms in the Family. My Life with Enrico Fermi*, Chicago: Chicago Univ. Press.

[172] —— [1968] *Illustrious Immigrants. The Intellectual Migration from Europe 1930–1941*, Chicago: Chicago Univ. Press.

[173] Fest, Joachim C. [1976] *Hitler. Eine Biographie*, Vol. 1: *Der Aufstieg*, Vol. 2: *Der Führer* Frankfurt/Berlin: Ullstein (1st ed., 1973).

[174] Festschrift [1961] *50 Jahre Kaiser-Wilhelm-Gesellschaft und Max-Planck-Gesellschaft zur Förderung der Wissenschaften, 1911–1961, Beiträge und Dokumente*, Göttingen.

[175] Festschrift [1975] *Max-Planck-Institut für Strömungsforschung Göttingen: 1925–1975. Festschrift zum 50-jährigen Bestehen des Instituts*, Göttingen: Hubert & Co.

[176] Finkelnburg, Wolfgang [1944]a 'Wesen und Bedeutung der Physik', *PB* **1**, pp. 27–29 (excerpt from Finkelnburg's *Physik* (Wissenschaftskunde in Einzeldarlegung)).

[177] —— [1944]b 'Physik hoher Temperaturen', *Natw.* **32**, pp. 105–111.

[178] Fischer, E. W. [1974] 'Herbert Arthur Stuart 1899–1974', *PB* **30**, pp. 510–511.

[179] Fischer, Emil & Ernst Beckmann [1913] *Das Kaiser-Wilhelm-Institut für Chemie [in] Berlin-Dahlem*, Braunschweig: Vieweg.

[180] Fischer, H. J. [1937/38] 'Völkische Bedingtheit von Mathematik und Physik', *ZgN* **3**, pp. 422–426.

[181] Fischer, Klaus [1987] 'Vom Wissenschaftstransfer zur Kontextanalyse – oder: Wie schreibt man die Geschichte der Wissenschaftsemigration?' in: Rainer Erb & Michael Schmidt (Eds.) *Antisemitismus und jüdische Geschichte*, Berlin: Wissenschaftlicher Autorenverlag, pp. 267–293.

[182] —— [1988] 'Der quantitative Beitrag der nach 1933 emigrierten Naturwissenschaftler zur deutschsprachigen physikalischen Forschung', *BWG* **11**, pp. 83–104.

[183] —— [1991]a 'Die Emigration deutschsprachiger Physiker nach 1933: Strukturen und Wirkungen', Strauss et al. (Eds.) [1991], pp. 27–72.

[184] —— [1991]b 'Die Emigration von Wissenschaftlern nach 1933', *VHZG* **39**, pp. 535–549.

[185] —— [1993] *Changing Landscapes of Nuclear Physics. A Scientiometric Study on the Social and Cognitive Position of German-Speaking Emigrants within the Nuclear Physics Community, 1921–1947*, Berlin: Springer.

[186] Fischer, [Ernst] Peter [1985] *Licht und Leben. Ein Bericht über Max Delbrück, der Wegbereiter der Molekularbiologie*, Constance: Universitäsverlag Konstanz.

[187] Fischer, Wolfram, Klaus Hierholzer, Michael Hubenstorf, Peter T. Walther & Rolf Winau (Eds.) [1994] *Exodus von Wissenschaften aus Berlin: Fragestellungen – Ergebnisse – Desiderate: Entwicklungen vor und nach 1933*, Berlin/New York: de Gruyter.

[188] Fisher, David. E. [1988] *A Race on the Edge of Time. Rader—The Decisive Weapon of World War II*, New York: McGraw-Hill.

[189] Flanze, G. [1937] 'Die Forschungsanstalt der Deutschen Reichspost', *Telegraphen, Fernsprech- und Funktechnik* **26**, pp. 97–98.

[190] Flechtner, Hans-Joachim [1942] *Atomzertrümmerung. Zauberei? Alchemie? Wissenschaft!*, Berlin: Wilhelm Limpert Verlag.

[191] Fleming, Donald [1969] 'Émigré physicists and the biological revolution', in: Fleming & Bailyn (Eds.) [1969] pp. 152–189.

[192] Fleming, Donald & Bernard Bailyn (Eds.) [1969] *The Intellectual Migration: Europe and America, 1930–1960*, Cambridge, Mass.: Harvard Univ. Press.

[193] Flügge, Siegfried [1939] 'Kann der Energieinhalt der Atomkerne technisch nutzbar gemacht werden?' *Natw.* **27**, pp. 402–410.

[194] Fölsing, Albrecht [1981] 'Naturwissenschaft im Dritten Reich – Verluste, die unersetzlich sind. Selbstgleichschaltung und Flucht in den Professionalismus', *Die Zeit*, Mar. 27, No. 14, p. 151.

[195] Forman, Paul [1971] 'Weimar culture, causality and quantum theory, 1918–1927', *HSPS* **3**, pp. 1–115.

[196] —— [1973] 'Scientific internationalism and the Weimar physicists: The ideology and its manipulation in Germany after World War I', *ISIS* **64**, pp. 151–180.

[197] —— [1974] 'The financial support and political alignment of physicists in Weimar Germany', *Minerva* **12**, pp. 39–66.

[198] Fosdick, Raymond Blaine [1989] *The Story of the Rockefeller Foundation*, New Brunswick, NJ: Transaction Publications (reprint of 1952 ed.).

[199] Fox, John P. [1979] 'Adolf Hitler: The continuing debate', *International Affairs* (Oxford) **55**, pp. 252–264.

[200] Fraenkel, Heinrich & Roger Mauvell [1962] *Göring*, New York: Simon & Schuster.

[201] Franck, James [1960] 'Max von Laue (1879–1960)', *Yearbook of the American Philosophical Society* (Philadelphia), pp. 155–159.

[202] [——] [1964] Special issue, *PB* **20**, esp. pp. 324–334.

[203] Frank, Charles (Ed.) [1993] *Operation Epsilon. The Farm Hall Transcripts*, Bristol/Philadelphia: Institute of Physics Publ.

[204] Frank, Philipp [1947] *Albert Einstein. His Life and Times*, New York: Knopf.

[205] Frank, Walter [1936] 'Die "graue Eminenz"' der Notgemeinschaft. Auf dem Wege zum 30. Januar der Wissenschaft', *Westfälische Landeszeitung – Rote Erde* (Dortmund) Friday, June 19, 2nd edition, **164**, pp. 3–4.

[206] Freise, Gerda [1983] 'Autonomie und Anpassung – Das Selbstverständnis von Naturwissenschaftlern im Nationalsozialismus', in: Brämer (Ed.) [1983], pp. 31–58.

[207] French, Anthony Philip (Ed.) [1980] *Einstein: A Centenary Volume*, Cambridge: Harvard Univ. Press.

[208] Frisch, Otto Robert [1939] 'Physical evidence for the division of heavy nuclei under neutron bombardment', *Nature* **143**, p. 276.

[209] —— [1970] 'Lise Meitner, 1878–1968', *BMFRS*, Vol. 16, pp. 404–420.

[210] —— [1979] *What Little I Remember*, Cambridge: Cambridge Univ. Press.

[211] —— [1981]a 'Lise Meitner' *DSB*, Vol. 9, pp. 260–263.

[212] —— [1981]b *Woran ich mich erinnere. Physik und Physiker meiner Zeit*, Stuttgart: Akademische Verlagsgesellschaft.

[213] Frühauf, H., M. von Laue, G. Hertz & W. Heisenberg [1959] *Max Planck zum Gedenken, herausgegeben von der Deutschen Akademie der Wissenschaften zu Berlin*, Berlin: Akademieverlag.

[214] Führerlexikon [1934] *Das Deutsche Führerlexikon 1934–1935*, Berlin: O. Stollberg, 1934.

[215] Füllgraf, Georges & Annegret Falter (Eds.) [1990] *Wissenschaft in der Verantwortung. Möglichkeiten der institutionellen Steuerung*, Frankfurt/M.: Campus.

[216] Funke, Manfred [1986] 'Universität und Zeitgeist im Dritten Reich. Eine Betrachtung zum politischen Verhalten von Gelehrten', *Das Parlament. Aus Politik und Zeitgeschichte* supplement, Mar. 22, B12/1986, pp. 3–14.

[217] Gay, Peter [1968] *Weimar Culture: The Outsider as Insider*, New York: Harper & Row.

[218] Gehlhoff, G., H. Rukop & W. Hort [1920] 'Einführung und Aufruf zur Gründung der Deutschen Gesellschaft für technische Physik', *ZtP* **1**, pp. 1–6.

[219] Gehrcke, Ernst [1913] 'Die gegen die Relativitätstheorie erhobenen Einwände', *Natw.* **1**, pp. 62–66 (see also Einstein [1918]b and Gehrcke [1919]).

[220] —— [1916] 'Zur Kritik und Geschichte der neueren Gravitationstheorien', *Annalen der Physik* (4) **51**, pp. 119–124 (see also Laue [1917]).

[221] —— [1919] 'Berichtigung zum Dialog über die Relativitätstheorie', *Natw.* **7**, pp. 147–148 (reply to Einstein [1918]b).

[222] —— [1920]a 'Zur Diskussion über den Äther', *Zeitschrift der Physik* **2**, pp. 67–68 (reply to Einstein [1918]a).

[223] —— [1920]b *Die Relativitätstheorie, eine wissenschaftliche Massensuggestion. Gemeinverständlich dargestellt*, Leipzig: Köhler (Schriften aus dem Verlag der Arbeitsgemeinschaft deutscher Naturforscher zur Erhaltung reiner Wissenschaft, Vol. 1; reprinted in: Gehrcke [1924]b, pp. 54–68).

[224] —— [1920]c Reply to Einstein [1920], *BT* **49**, Aug. 31, No. 409 (A).

[225] —— [1924]a *Die Massensuggestion der Relativitätstheorie. Kultur-historisch-psychologische Dokumente*, Berlin: Meußer.

[226] —— [1924]b *Kritik der Relativitätstheorie. Gesammelte Schriften über absolute und relative Bewegung*, Berlin: Meußer.

[227] Gellately, Robert [1990] *The Gestapo and German Society. Enforcing Racial Policy, 1933–1945*, Oxford: Clarendon.

[228] Gentner, W. [1956] 'Peter Jensen †', *PB* **12**, pp. 125–126.

[229] Gerhardt, Ulrich [1948] 'Das Nürnberger Urteil im Chemieprozeß', *PB* **4**, pp. 429–432.

[230] Gerlach, Walther & Arnold Sommerfeld [1931] 'Hermann Anschütz-Kaempfe', *Natw.* **10**, pp. 666–669.

[231] Gerlach, Walther [1936] 'Theorie und Experiment in der exakten Naturwissenschaft', *Natw.* **24**, pp. 721–741.

[232] —— [1949] 'Heinrich M. Konen †', *FuF* **25**, pp. 117–118; see also *PB* **5**, p. 226.

[233] —— [1963] 'Münchener Erinnerungen', *PB* **19**, pp. 97–103.

[234] —— [1968] 'Otto Hahn zum Gedächnis', *PB* **24**, pp. 337–338.

[235] Gerstengarbe, Sybille [1994] 'Die erste Entlassungswelle von Hochschullehrern deutscher Hochschulen aufgrund des Gesetzes zur Wiederherstellung der Berufsbeamtentums vom 7.4.1933', *BWG* **17**, pp. 17–39.

[236] Geuter, Alfred [1992]a *The Professionalization of Psychology in Nazi Germany*, Cambridge: Cambridge Univ. Press.; b) Translation by R. Holmes of: *Die Professionalisierung der Psychologie im Nationalsozialismus* Frankfurt (Main): Suhrkamp, 1984.

[237] Geyer, Bodo, Herwig & H. Rechenberg (Eds.) [1993] *Werner Heisenberg – Physiker und Philosoph*, Heidelberg: Spektrum.

[238] Giles, Geoffrey J. [1985] *Students and National Socialism in Germany*, Princeton: Princeton Univ. Press.

[239] Gimbel, John [1986] 'U.S. policy and German scientists', *Political Science Quarterly* **101**, pp. 433–451.

[240] —— [1990] *Science, Technology and Reparations. Exploitation and Plunder in Postwar Germany*, Stanford: Stanford Univ. Press.

[241] Glaser, Ludwig C. [1920] 'Zur Erörterung über die Relativitätstheorie', *Tägliche Rundschau* (Berlin), Aug. 16, No. 180, (reprinted in: Weyland [1920]b pp. 29–30).

[242] Glum, Friedrich J. [1962] *Der Nationalsozialismus: Werden und Vergehen*, Munich: Beck.

[243] —— [1964] *Zwischen Wissenschaft, Wirtschaft und Politik. Erlebtes und Erdachtes in 4 Reichen*, Bonn: Bouvier.

[244] Goebbels, Joseph [1934] 'Die deutsche Kultur vor neuem Anfang (Opening speech of the Reich Chamber of Culture on Nov. 15, 1933), in: Dreyer (Ed.) [1934].

[245] Goenner, Hubert [1993] 'The reaction to relativity theory and the anti-Einstein campaign in Germany in 1920', *Science in Context* **6** (1), pp. 107–136.

[246] —— [1994] 'The reaction to relativity theory in Germany, III: "A hundred authors against Einstein" ', in: John D. North (Ed.) *The Attraction of Gravitation*, Basel: Birkhäuser, pp. 248–273.

[247] Golczewski, Frank [1988] *Kölner Universitätslehrer und der Nationalsozialismus. Personengeschichtliche Ansätze*, Cologne: Böhlau (2nd ed., 1989).

[248] Goldberg, Stanley & Thomas Powers [1992] 'Declassified files reopen "Nazi bomb" debate', *Bulletin of the Atomic Scientists* **48** (7), pp. 32–40.

[249] Goldberg, Stanley [1995] 'Der Endspurt', *PB* **51**, pp. 827–831.

[250] —— [in press] *Fighting to the Finish: The Private Wars of Leslie Groves*, Steerforth Press.

[251] Goran, Morris [1939] 'Swastika science', *The Nation* **148**, pp. 641–643.

[252] —— [1967] *The Story of Fritz Haber*, Norman: Univ. of Oklahoma Press.

[253] Göring, Hermann [1937] Foreword: *Jahrbuch der deutschen Luftfahrtforschung*, **1**, p. i.

[254] Götz von Olenhusen, Albrecht [1966] 'Die 'nichtarischen' Studenten an den deutschen Hochschulen. Zur nationalsozialistischen Rassenpolitik 1933–1945', *VHZG* **14**, pp. 175–206.

[255] Goubau, Georg & Jonathan Zenneck (Eds.) [1948] *Electronics incl. Fundamental Emission Phenomena*, Wiesbaden: Klemm (FIAT Review of German Science 1936–1946, Vols. 15, 16).

[256] Goudsmit, Samuel A. [1946] 'Wissenschaft oder Geheimhaltung', *PB* **2**, pp. 203–207.

[257] —— [1947] *Alsos*, New York: Schuman (also reprinted with introduction by R. V. Jones as Vol. 1 of *The History of Modern Physics, 1800–1990*, Los Angeles/San Francisco: Tomash, 1983; see also the review by Morrison [1947]).

[258] Gowing, Margaret M. [1964] *Britain and Atomic Energy 1939–1945*, London: Macmillan.

[259] Gowing, Margaret M. & Lorna Arnold [1979] *The Atomic Bomb*, London: Butterworths.

[260] Grab, Walter (Ed.) [1986] 'Juden in der deutschen Wissenschaft', *Jahrbuch des Instituts für Deutsche Geschichte, Tel Aviv*, suppl. issue 10 (esp. Shulamit Volkov: 'The social origins of success in science: Jews in 19th Century Germany', pp. 179–184).

[261] Grau, Conrad, Wolfgang Schlicker & L. Zeil [1979] *Die Berliner Akademie der Wissenschaften in der Zeit des Imperialismus*, Part III: *Die Jahre der faschistischen Diktatur*, Berlin: Akademie-Verlag.

[262] Grothusen, Klaus-Detlef (Ed.) [1989] *Der Scurla-Bericht: Die Tätigkeit deutscher Hochschullehrer an türkischen wissenschaftlichen Hochschulen*, Frankfurt (Main) (1st ed.: Dagyeli, 1987).

[263] [Grotrian, Walter] [1936] 'Karl Scheel †. Trauerrede, gehalten am 13. November 1936 im Krematorium Berlin-Wilmersdorf', *VDPG* (3) **17**, pp. 32–35.

[264] Groves, Leslie [1962] *Now It Can be Told. The Story of the Manhattan Project*, a) New York: Harper; b) reprinted New York: Da Capo Press, 1975, 1983.

[265] Grundmann, Siegfried [1965] 'Zum Boykott der deutschen Wissenschaft nach dem ersten Weltkrieg', *Wissenschaftliche Zeitschrift der Technischen Universität Dresden* **14**, pp. 799–806.

[266] —— [1967] 'Das moralische Antlitz der Anti-Einstein-Liga', *Wissenschaftliche Zeitschrift der Technischen Universität Dresden* **16**, pp. 1623–1626.

[267] Grüttner, Michael [1995] *Studenten im Dritten Reich*, Paderborn/Munich: Schöningh.

[268] Guerlac, Henry E. [1987] *Radar in World War II*, New York: Am. Institute of Physics.

[269] Guérout, Serge [1992] *Science et Politique sous le Troisième Reich*, Paris: Marketing.

[270] Gumbel, Emil Julius [1938] 'Arische Naturwissenschaft?', in: Gumbel (Ed.) *Freie Wissenschaft. Ein Sammelbuch aus der deutschen Emigration*, Strasbourg, pp. 246–262.

[271] Günther, Erich [1933] 'Die Bedeutung des Physikunterrichts für die Erziehung zur Wehrhaftigkeit', *Unterrichtsblätter für Mathematik und Naturwissenschaften* **39** (7) pp. 230–234.

[272] —— [1936] *Wehrphysik – Ein Handbuch für Lehrer*, Frankfurt: Diesterweg.

[273] Haber, Fritz [1924] *Fünf Vorträge aus den Jahren 1920–23: Über die Darstellung des Ammoniaks aus Stickstoff und Wasserstoff – Die Chemie im Kriege – Das Zeitalter der Chemie. Neue Arbeitsweise –Zur Geschichte des Gaskrieges*, Berlin: Springer.

[274] Haberditzl, Werner [1963] 'Der Widerstand deutscher Naturforscher gegen die 'Deutsche Physik' und andere faschistische Zerrbilder', *NTM*, 1st supplement, pp. 320–326.

[275] Hahn, Dietrich (Ed.) [1979] *Otto Hahn – Begründer des Atomzeitalters. Eine Biographie in Bildern und Dokumenten*, Munich: List.

[276] —— (Ed.) [1988] *Otto Hahn, Leben – Werk in Texten und Bildern*, Frankfurt: Insel.

[277] Hahn, Otto & Fritz Straßmann [1939]a 'Über den Nachweis und das Verhalten der bei der Bestrahlung des Urans mittels Neutronen entstehenden Erdalkalimetalle', *Natw.* **27**, pp. 11–15.

[278] —— [1939]b 'Nachweis der Entstehung aktiver Bariumisotope aus Uran und Thorium durch Neutronenbestrahlung; Nachweis weiterer aktiver Bruchstücke bei der Kernspaltung', *Natw.* **27**, pp. 89–95.

[279] Hahn, Otto [1943] 'Künstliche Atomumwandlungen und die Spaltung des Urans', *Schriften der Akademie der Luftfahrtforschung* (Probleme der Kernphysik: *KPFB* **G-216**), pp. 19–27.

[280] —— [1946/48] 'Von den natürlichen Umwandlungen des Urans zu seiner künstlichen Zerspaltung' (Nobel lecture, Dec. 3, 1946), *Les Prix Nobel en 1946*, Stockholm: Norstedt, 1948, pp. 167–183.

[281] Hahn, Otto & F. H. Rein [1947] 'Gelehrtenexport nach Amerika', a) in: *Göttingen Univ.-Zeitung*, Feb. 21; b) entitled: 'Einladung nach USA' in *PB* **3**, pp. 33–35.

[282] Hahn, Otto [1960] 'Zur Erinnerung an die Haber-Gedächtnisfeier vor 25 Jahren, am 29. Januar 1935, im Harneck-Haus in Berlin-Dahlem', *Mitteilungen der Max Planck-Gesellschaft*, pp. 3–13.

[283] —— [1962]a 'Die "falschen" Trans-Urane. Zur Geschichte eines wissenschaftlichen Irrtums', *Naturwissenschaftliche Rundschau* **15**, pp. 43–47.

[284] —— [1962]b *Vom Radiothor zur Uranspaltung. Eine wissenschaftliche Selbstbiographie*, Braunschweig: Vieweg. (Revised ed. 1988).

[285] —— [1963] 'Lise Meitner, 85 Jahre', *Natw.* **50**, pp. 653–654.

[286] [——] [1968] Special issue *PB* **24**, especially pp. 337–450.

[287] —— [1968/86] *Mein Leben*, Munich: F. Bruckmann. Revised edition: *Mein Leben. Die Erinnerungen des großen Atomforschers und Humanisten*, Munich: Piper, 1986.

[288] —— [1975] *Erlebnisse und Erkenntnisse*, Düsseldorf/Vienna: Econ.

[289] Haldane, J. B. S. [1934] 'The attitude of the German government towards science', *Nature* **133**, p. 726 (reply to Stark [1934]b).

[290] Hallows, Ralph Watson [1946] *Radar. Radiolocation Simply Explained*, London: Chapman & Hale.

[291] Hamel, Georg [1933] 'Die Mathematik im Dritten Reich', *Unterrichtsblätter für Mathematik und Naturwissenschaften* **39**, pp. 306–309.

[292] Hammerstein, Notker [1989] 'Zur Geschichte des philosophischen Seminars der Johann Wolfgang Goethe-Universität während des Dritten Reichs', *Hessisches Jahrbuch für Landesgeschichte* **39**, pp. 271–310.

[293] —— [1995] *Antisemitismus und deutsche Universitäten 1871–1933*, Frankfurt (Main): Campus.

[294] Hanle, Wilhelm [1983] 'Nachruf auf Robert Döpel', *PB* **39**, p. 104.

[295] Hanle, Wilhelm & Erich R. Bagge [1982] '40 Jahre Nutzung der Kernenergie', *Atomkernenergie* **40**, pp. 3–6.

[296] Hansen, Max (Ed.) [1948]a *General Metallurgy*, Wiesbaden: Klemm (FIAT Review of German Science 1939–1946, Vol. 31)

[297] —— (Ed.) [1948]b *Non-Ferrous Metallurgy* Wiesbaden: Klemm (FIAT Review of German Science 1939–1946, Vols. 32, 33).

[298] Harbordt, Steffen (Ed.) [1983] *Wissenschaft und Nationalsozialismus*, Berlin: Technische Universität Berlin (especially pp. 101ff.).

[299] Harteck, Paul [1960] 'Physical Chemists in Berlin 1919–1933', *Journal of Chemical Education* **37**, pp. 462–466.

[300] Hartmann, Hans [1968] 'Otto Hahn als Mensch und Forscher. Erinnerungen an einen großen Wissenschaftler', *PB* **24**, pp. 442–450.

[301] Hartmann, Max (Ed.) [1936] *25 Jahre Kaiser-Wilhelm-Gesellschaft zur Förderung der Wissenschaften*, Vol. 2: *Naturwissenschaften*, Berlin: Springer.

[302] Hartshorne, Edward Yarnall, Jr. [1937] *The German Universities and National Socialism*, a) London: Allan & Unwin b) reprinted: New York: AMS Press, 1981.

[303] —— [1938] *The German Universities and the Government*, Chicago: Univ. of Chicago.

[304] Hawkins, David, Edith C. Truslow & Ralph Carlisle Smith [1983] *Project Y. The Los Alamos Story*, Part 1: 'Toward Trinity' by D. Hawkins; Part 2: 'Beyond Trinity' by E. C. Truslow & R. C. Smith. Los Angeles: Tomash Publ.

[305] Hayes, Peter [1987] *Industry and Ideology: IG Farben in the Nazi Era*, Cambridge: Cambridge Univ. Press.

[306] Heckmann, Otto [1942] *Theorien der Kosmologie*, Berlin: Springer.

[307] —— [1976] *Sterne, Kosmos, Weltmodelle. Erlebte Kosmologie*, Munich/Zurich: Piper.

[308] Heiber, Helmut [1991/92] *Universität unterm Hakenkreuz*, Part I: *Der Professor im Dritten Reich. Bilder aus der akademischen Provinz*, (1991) and Part II: *Die Kapitulation der hohen Schulen*, Munich: Saur (1992).

[309] Heilbron, John L. [1986]a 'The first European cyclotrons', *Rivista de la Storia Scienza* **3** (1), pp. 1–44.

[310] —— [1986]b *The Dilemmas of an Upright Man: Max Planck as Spokesman for German Science*, Berkeley: Univ. of California Press.

[311] Heilbron, John L. & Robert W. Seidel [1989] *Lawrence and His Laboratory. A History of the Lawrence Berkeley Laboratory*, Vol. 1, Berkeley: Univ. of California Press.

[312] Heinemann, Manfred (Ed.) [1980] *Erziehung und Schulung im Dritten Reich.* Part 2: *Hochschule, Erwachsenenbildung*, Stuttgart.

[313] Heinicke, Wolfgang [1985] 'Daten aus der Geschichte', *VDPG* (6) **20**, pp. 4–84.

[314] Heinrich, Rudolf & Hans-Reinhardt Bachmann [1989] 'Walther Gerlach: Freiheit und Verantwortung des Physikers', *Kultur und Technik* **13**, issue 3.

[315] —— (Eds.) [1989] *Walther Gerlach. Physiker – Lehrer – Organisator*, Munich: Deutsches Museum.

[316] Heisenberg, Elisabeth [1984] *Inner Code: Recollections of a Life with Werner Heisenberg*, a) translated by S. Cappellani and C. Moris, Boston: Birkhäuser; b) paperback reprint of German original (1980), Munich: Piper, 1991.

[317] Heisenberg, Werner [1933] 'Probleme der modernen Physik', a) *Deutsche Optische Wochenschrift* **18**, pp. 27–28; b) in: *BT* **62**, No. 530, Nov. 10 (ed. A), 1st suppl.

[318] —— [1934] 'Wandlungen der Grundlagen der exakten Naturwissenschaft in jüngster Zeit', *Natw.* **22**, pp. 669–675.

[319] —— [1939] 'Die Entwicklung der jüngsten Zweige der Atomphysik in Deutschland', *FuF* **15**, pp. 241–244 (reprinted in: Heisenberg [1989], Ser. C, Vol. I, pp. 142–145 and in: Rechenberg (Ed.) [1992]a, pp. 57–67).

[320] —— [1943]a 'Die Bewertung der 'modernen theoretischen Physik' ', *ZgN* **9**, pp. 201–212 (see also reply by Dingler [1943]).

[321] —— [1943]b *Die Physik der Atomkerne*, Braunschweig: Vieweg, 1st ed. (3rd ed., 1949: Die Wissenschaft, No. 100; also reprinted in: Heisenberg [1989], Ser. C, Vol. IV.)

[322] —— [1945/92] 'Vortrag über die Uranbombe' *PB* **48**, pp. 995–1001.

[323] —— [1946] 'Das Deutsche Atombomben-Geheimnis', *Die Welt* **1**, No. 39, Aug. 13, p. 3.

[324] —— [1948]a 'Wer weiß was rechtes wird', *Die Welt* **3**, No. 143, Dec. 9, 1948, p. 3.

[325] —— [1948]b 'Nazi spurned idea of an atomic bomb', *NYT*, Dec. 28, 1948, p. 10.

[326] Heisenberg, Werner & Karl Wirtz [1948] 'Großversuche zur Vorbereitung der Konstruktion eines Uranbrenners' in: Bothe & Flügge (Eds.) [1948], part II, pp. 142–165.

[327] Heisenberg, Werner [1949] 'German atomic research', *NYT*, Jan. 30, 1949, sec. 4, p. 8.

[328] —— [1967]a 'Das Dritte Reich versuchte nicht, die Atombombe zu bauen', *Frankfurter Allgemeine Zeitung*, No. 286, Saturday, Dec. 9.

[329] —— [1967]b 'Gott sei Dank, wir konnten sie nicht bauen', *Der Spiegel* **21**, No. 28, 79–87.

[330] —— [1969] *Der Teil und das Ganze*, Munich: Piper.

[331] —— [1971] 'Das Kaiser-Wilhelm-Institut für Physik, Geschichte eines Instituts', *Jahrbuch der Max Planck-Gesellschaft zur Förderung der Wissenschaften*, pp. 46–89.

[332] —— [1989] *Collected Works*, Ser. A: *Original Scientific Papers*, Ser. A, esp. Vol. II, and Ser. C, esp. Vols. I, IV, Berlin: Springer.

[333] Heitler, W. [1963] 'Dr. Paul Rosbaud', *Nature* **197**, No. 4872, Mar. 16, p. 1051.

[334] Henning, Eckart & Marion Kazemi [1988] *Chronik der Kaiser-Wilhelm-Gesellschaft zur Förderung der Wissenschaften*, Berlin (Veröffentlichungen aus dem Archiv zur Geschichte der Max-Planck-Gesellschaft, 1).

[335] Henning, Eckart [1992] 'Der Nachlaß Max von Laues' *PB* **48**, pp. 938–940.

[336] Hentschel, Klaus [1989] 'The rise of antisemitic arguments against Einstein's theories of relativity since 1920', *Abstracts of the XVIIIth International Congress of History of Science*, Hamburg/Munich. Section A22, Abstract No. 20.

[337] —— [1990]a *Interpretationen und Fehlinterpretationen der speziellen und der allgemeinen Relativitätstheorie durch Zeitgenossen Albert Einsteins*, Basel: Birkhäuser.

[338] —— [1990]b 'A postscript on Einstein and the FBI', *ISIS* **81**, pp. 279–280.

[339] —— [1993] 'Bernhard Bavink (1879–1947), der Weg eines Naturphilosophen vom deutschnationalen Sympathisanten der NS-Bewegung bis zum unbequemen Non-Konformisten', *Sudhoffs Archiv* **77**, pp. 1–32.

[340] Hentschel, Klaus & Renate Tobies [1996] Interview with Prof. Friedrich Hund in: *NTM*, new ser. **4**, January issue.

[341] Hentschel, Klaus [1996] *The Einstein Tower*, Stanford: Stanford Univ. Press (in press).

[342] Herf, Jeffry [1984] *Reactionary Modernism: Technology, Culture, and Politics in Weimar and the Third Reich*, Cambridge: Cambridge Univ. Press.

[343] Hermann, Armin [1966] 'Albert Einstein und Johannes Stark – Briefwechsel und Verhältnis der beiden Nobelpreisträger', *Sudhoffs Archiv* **50**, pp. 267–285.

[344] —— [1967] 'Die frühe Diskussion zwischen Stark und Sommerfeld über die Quantenhypothese', *Centaurus* **12**, pp. 38–59.

[345] —— (Ed.) [1968] *Albert Einstein/Arnold Sommerfeld Briefwechsel. Sechzig Briefe aus dem goldenen Zeitalter der modernen Physik*, Basel/Stuttgart: Schwabe.

[346] —— [1973] *Max Planck in Selbstzeugnissen und Bilddokumenten*, Reinbek: Rowohlt.

[347] —— [1975] 'Stark, Johannes', *DSB* **12**, pp. 613–616.

[348] —— [1976] *Werner Heisenberg in Selbstzeugnissen und Bilddokumenten*, Reinbek: Rowohlt.

[349] —— [1977] 'Der Kampf um die Relativitätstheorie', *Bild der Wissenschaft* **14**, 109–116.

[350] —— [1982] 'Laue, Max v.' *NDB* **13**, pp. 702–705.

[351] —— [1988] 'Heisenberg und das deutsche Atomprojekt', *Bild der Wissenschaft* **25** (10), pp. 139–145.

[352] —— [1989] *Nur der Name war geblieben: Die abenteuerliche Geschichte der Firma Carl Zeiss*, Stuttgart: Deutsche Verlagsanstalt (3rd ed., 1991).

[353] —— [1993] 'Science under foreign rule – Policy of the Allies in Germany, 1945–49' in: Fritz Krafft & Christoph J. Scriba (Eds.) *XVIIIth International Congress of History of Science Hamburg-Munich, 1989*, Stuttgart: Steiner, pp. 75–86.

[354] Herneck, Friedrich [1976] *Einstein u. sein Weltbild. Aufsätze u. Vorträge*, Berlin: Morgen.

[355] —— [1979] *Max von Laue*, Leipzig: Teubner.

[356] Herrmann, Günther [1964] '25 Jahre Kernspaltung. Methoden und Ergebnisse chemischer Studien. Entdeckungsgeschichte, Spaltprodukte', *Radiochimica Acta* **3**, pp. 169–184.

[357] —— [1990] 'Vor fünf Jahrzehnten: Von den 'Transuranen' zur Kernspaltung', *Angewandte Chemie* **102**, pp. 469–496.

[358] Hertz, Gustav [1965] 'James Franck, † 21.5.1964', *Annalen der Physik* (7) **15**, pp. 1–4.

[359] Hettner, G. [1947] 'Wilhelm Orthmann †', *PB* **3**, p. 160.

[360] Hiedemann, Egon [1941] 'Die deutsche Physik in Straßburg', *Straßburger Monatshefte* **5**, Nov. issue, pp. 735–737.

[361] Hilberg, Raul [1961] *The Destruction of the European Jews*, Chicago: Quadrangle Books.

[362] Hildebrandt, G. [1985] 'Zum Tode von Paul Peter Ewald', *PB* **41**, pp. 412–413.

[363] —— [1993] 'The discovery of the diffraction of x-rays in crystals—A historical review', *Crystal Research and Technology* **28**, pp. 747–766.

[364] Hill, A. V. [1933] 'International status and obligations of science', *Nature* **132**, 952–954.

[365] —— [1934] Reply to Stark [1934]a, *Nature* **133**, p. 290.

[366] Hillers, Wilhelm [1935] 'Vorschläge zur Verdeutschung physikalischer Fachausdrücke', *Unterrichtsblätter für Mathematik und Naturwissenschaften* **41**, pp. 304–311.

[367] Hirsch, Walter [1975] 'The autonomy of science in totalitarian societies. The case of Nazi Germany', in: Karin D. Knorr, Hermann Strasser, & Hans G. Zilian (Eds.) *Determinants and Controls of Scientific Development*, Dordrecht/Boston: Reidel, p. 359.

[368] Hirschfeld, Gerhard [1988] 'The defence of learning and science'. Der Academic Assistance Council in Großbritannien und die wissenschaftliche Emigration aus Nazi-Deutschland', in: Koebner et al. (Eds.) [1988], pp. 28–43.

[369] —— [1992] 'Zuflucht in Großbritannien. Zur Emigration deutschsprachiger Prager Wissenschaftler nach 1938', in: Peter Becker & Peter Heumos, *Drehscheibe Prag. Zur deutschen Emigration in der Tschechoslowakei, 1933–39*, Munich: Oldenbourg, pp. 75–86.

[370] Hitler, Adolf [1934] 'Der deutschen Kultur Zukunftsaufgabe', in: Dreyer (Ed.) [1934], pp. 9–19.

[371] Hoch, Paul K. [1983] 'The reception of Central European refugee physicists of the 1930's: USSR, UK, USA', *Annals of Science* **40**, pp. 217–246.

[372] —— [1987] 'Migration and the generation of new scientific ideas', *Minerva* **25**, 209–237.

[373] Hoch, Paul K. & E. J. Yoxen [1987] 'Schrödinger at Oxford: A hypothetical national cultural synthesis which failed', *Annals of Science* **44**, pp. 593–616.

[374] Hoddeson, Lillian, Paul W. Hendriksen, Roger A. Meade & Catherine Westfall [1993] *Critical Assembly: A Technical History of Los Alamos during the Oppenheimer Years, 1943–1945*, Cambridge: Cambridge Univ. Press.

[375] Hofer, Walther (Ed.) [1957] *Der Nationalsozialismus. Dokumente 1933–1945*, Frankfurt (Main): Fischer

[376] Hoffmann, Dieter [1984] *Erwin Schrödinger*, Leipzig: Teubner.

[377] —— [1987] 'Hans Euler (1909–1941): Vom begabten Schüler Heisenbergs zum Opfer des Zweiten Weltkrieges – Zur Lebensgeschichte eines Talents', in: *Physikalische Gesellschaft der DDR, Arbeitsgruppe Physikgeschichte, Vorträge*, Berlin, pp. 31–40.

[378] —— [1989] 'Die Physikdenkschriften von 1934/36 und zur Situation der Physik im faschistischen Deutschland', *Akademie der Wissenschaften, Kolloquien*, (E. Berlin) issue 68, pp. 185–211.

[379] —— [1991] 'Zum 100. Geburtstag von Walther Bothe: Ein Physiker per se', *Wissenschaft und Fortschritt* **41**, pp. 162–165.

[380] —— [1992] 'Farm-Hall-Tonbänder. Operation Epsilon', *PB* **48**, pp. 989–993.

[381] —— [1993]a 'Nationalsozialistische Gleichschaltung und Tendenzen militärischer Forschungsorientierung an der Physikalisch-Technischen Reichsanstalt im Dritten Reich', in: Albrecht (Ed.) [1993], pp. 121–131.

[382] —— (Ed.) [1993]b *Operation Epsilon. Die Farm-Hall-Protokolle oder die Angst der Alliierten vor der deutschen Atombombe*, retransl. by Wilfried Sczepan. Berlin: Rowohlt.

[383] Hoffmann, Dieter & Edgar Swinne [1994] *Über die Geschichte der "Technischen Physik" in Deutschland und den Begründer ihrer wissenschaftlichen Gesellschaft Georg Gehlhoff*, Berlin: ERS Verlag.

[384] Hoffmann, Gerhard [1938] 'Nachruf auf Max Wien', *Berichte der Sächsischen Akademie der Wissenschaft* **90**, pp. 225–234.

[385] —— [1940] 'Methoden und Ergebnisse neuerer kernphysikalischer Forschung', *PZ* **41**, pp. 514–522.

[386] Höhne, Heinz [1984/90] *Der Orden unter dem Totenkopf. Die Geschichte der SS*, a) Munich: Bertelsmann Verlag, 1984; b) Bindlach: Gondrom, 1990.

[387] Holdermann, Karl [1954] *Im Banne der Chemie. Carl Bosch, Leben und Werk*, Düsseldorf: Econ.

[388] Holl, Frank [1992] *Produktion und Distribution wissenschaftlicher Literatur 1913–1933. Der Physiker Max Born und sein Verleger Ferdinand Springer.* (Thesis, Munich University; to be published by *Archiv zur Geschichte des Buchwesens.*)

[389] Holton, Gerald [1983] 'The migration of physicists to the United States', in: Jackman & Borden (Eds.) [1983], pp. 169–188.

[390] Homze, Edward L. [1976] *Arming the Luftwaffe. The Reich Air Ministry and the German Aircraft Production, 1919–1939*, Lincoln: Univ. of Nebraska Press.

[391] Hooker, Ruth H. [1935] 'A study of scientific periodicals', *Review of Scientific Instruments*, new ser. **6**, pp. 333–338.

[392] Howse, Derek [1993] *Radar at Sea. The Royal Navy in World War 2*, London: Macmillan.

[393] Huber, Hans et al. (Eds.) [1942] *Die Deutsche Hochschulverwaltung. Sammlung der das Hochschulwesen betreffenden Gesetze, Verordnungen und Erlasse*, Vol. 1, Berlin.

[394] Hunt, Linda [1991] *Secret Agenda. The United States Government, Nazi Scientists, and Project Paperclip, 1945 to 1990*, New York: St. Martin.

[395] Hüttenberger, Peter [1976] 'Nationalsozialistische Polykratie', *Geschichte und Gesellschaft* (Göttingen), No. 2, pp. 417–442.

[396] Irving, David [1967] *The Virus House. Germany's Atomic Research and Allied Counter Measures*, London: Kimber.

[397] —— [1968] *The German Atomic Bomb: The History of Nuclear Research in Nazi Germany*, New York: Simon & Schuster.

[398] —— (Ed.) [undated] *Records of the German Atomic Research Programme, 1938–1945*, compiled by D. Irving, microfilms: DJ–29, DJ–30, DJ–31 (copies at the Deutsches Museum, Munich).

[399] Jackman, Jarrell C. & Carla M. Borden (Eds.) [1983] *The Muses Flee Hitler: Cultural Transfer and Adaptation 1930–1945*, Washington, DC: Smithsonian Institution Press.

[400] Jaenicke, Erna & Johannes Jaenicke [1966] 'Haber, Fritz Jacob', *NDB*, **7**, pp. 386–389.

[401] Jaenicke, Johannes [1935] 'Habers Forschung über das Goldvorkommen im Meerwasser', *Natw.* **23**, pp. 57–63.

[402] Jaenicke, W. [1957] 'Karl-Friedrich Bonhoeffer †', *PB* **13**, pp. 369–370.

[403] Jaensch, Erich Rudolf [1931] 'Über die Grundlagen der menschlichen Erkenntnis (Über den Aufbau der Wahrnehmungswelt und die Grundlagen der menschlichen Erkenntnis) (Offprint of *Zeitschrift für Psychologie*, **120**, pp. 46–125).

[404] Jammer, Max [1966] *The Conceptual Development of Quantum Mechanics*, New York: McGraw-Hill (Reprint: American Institute of Physics, 1989).

[405] Jansen, Christian [1992] *Professoren und Politik. Politisches Denken und Handeln der Heidelberger Hochschullehrer 1914–1935*, Göttingen: Vandenhoek & Ruprecht.

[406] Janssen, Gregor [1968] *Das Ministerium Speer. Deutschlands Rüstung im Krieg*, Frankfurt (Main): Ullstein.

[407] Jarausch, Konrad Hugo [1984] *Deutsche Studenten 1800–1970*, Frankfurt: Suhrkamp.

[408] Jensen, Hermann [1992] 'Warum Hitler die Bombe nicht baute. Eine neue These zum Gespräch zwischen Bohr und Heisenberg 1941 und den welthistorischen Folgen', *Die Zeit*, No. 18, April 24, p. 41 (See also Rechenberg [1992]b, Weizsäcker [1992]).

[409] Jetter, U. [1946] 'Was wurde über die Atombombe mitgeteilt?', *PB* **2**, pp. 6–8.

[410] —— [1948] 'Höhenforschung mit Raketen' *PB* **4**, pp. 93–99.

[411] Johnson, Brian [1978] *The Secret War*, London: BBC.

[412] Johnson, Jeffrey Allan [1990] *The Kaiser's Chemists. Science and Modernization in Imperial Germany*, Chapel Hill: Univ. of North Carolina Press.

[413] Jones, Bessie Zaban [1984] 'To the rescue of the learned: The asylum fellowship plan at Harvard 1938–1940', *Harvard Library Bulletin* **32**, pp. 204–238.

[414] Joos, Georg (Ed.) [1948] *Physics of Solids*, Wiesbaden: Klemm (FIAT Review of German Science 1939–1945, Vols. 8, 9).

[415] —— [1951] 'Arnold Sommerfeld', *Sitzungsberichte der Bayerischen Akademie der Wissenschaften, München*, math. phys. Klasse, pp. 152–159.

[416] —— [1959] 'Jonathan Zenneck †', *PB* **15**, pp. 271–272.

[417] Jordan, Pascual [1935] *Physikalisches Denken in der neuen Zeit*, Hamburg: Hanseatische Verlagsanstalt.

[418] —— [1936]a 'Gibt es eine 'Krise' der modernen physikalischen Forschung?', *Die Tatwelt* **12**, pp. 59–68.

[419] —— [1936]b 'Positivistische Bemerkungen über die parapsychologischen Erscheinungen', *Zentralblatt für Psychotherapie* **9**, pp. 1–17.

[420] —— [1936/49] *Die Physik des 20. Jahrhunderts*, Braunschweig: Vieweg (Die Wissenschaft, No. 88) a) 1st ed. 1936; b) 7th ed. 1949; see also review by Dingler [1937/38].

[421] —— [1938] 'Die Verstärkertheorie der Organismen in ihrem gegenwärtigen Stand', *Natw.* **26**, pp. 537–545.

[422] —— [1944]a 'Physik im Vormarsch', *Das Reich*, Jan 2, No. 1, pp. 9–10.

[423] —— [1944]b 'Quantenphysik und Biologie', *Natw.* **32**, pp. 309–316.

[424] —— [1951] 'Nachruf auf Arnold Sommerfeld', *Jahrbuch der Akademie der Wissenschaften und der Literatur, Mainz*, pp. 143–154.

[425] —— [1957] 'Atom-Angst – gefährliche Ratgeber. Pascual Jordan antwortet auf das Göttinger Manifest', *Das Fenster* **6**, pp. 4–5, 7.

[426] Juilfs, Johannes [1939] 'Das erste deutsche Mathematikerlager', *Deutsche Mathematik* **4**, pp. 109–140.

[427] Jungk, Robert [1958] *Brighter Than a Thousand Suns. A Personal History of the Atomic Scientists*, (transl. by James Cleugh), New York: Harcourt Brace.

[428] Justrow [1940] 'Prof. Dr. K. Becker, General der Artillerie, 60 Jahre', *ZgSS* **35**, No. 1, pp. 1–2, 121–122.

[429] Ka. [1943] 'Max Planck', *Das Reich*, No. 15, Apr. 11, p. 1.

[430] Kaempffert, Waldemar [1948] 'Nazis spurned idea of an atomic bomb', *NYT*, Dec. 28.

[431] Kamp, Norbert, Peter Haase, Gerhart W. Rathenau & Friedrich Hund [1983] *James Franck und Max Born in Göttingen. Reden zur akademischen Gedenkfeier am 10.11.1982*, Göttingen: Vandenhoek & Ruprecht.

[432] Kampe, Norbert [1985] 'Jews and antisemitism at universities in Imperial Germany I', *Yearbook of the Leo Baeck Institute* **30**, pp. 357–394.

[433] —— [1988] *Studenten und "Judenfrage" im Deutschen Kaiserreich: Die Entstehung und akademische Trägerschuld des Antisemitismus*, Göttingen: Vandenhoek & Ruprecht.

[434] Kampmann, Wanda [1979] *Deutsche und Juden. Die Geschichte der Juden in Deutschland vom Mittelalter bis zum Beginn des ersten Weltkrieges*, Frankfurt (Main): Fischer.

[435] Kangro, Hans [1976] *Early History of Planck's Radiation Law*, London: Taylor & Francis.

[436] Kant, Horst [1987] 'Peter Debye und die Entwicklung des Kaiser-Wilhelm-Instituts für Physik', in: *Physikalische Gesellschaft der DDR, Arbeitsgruppe Physikgeschichte, Vorträge, Berlin*, pp. 91–103.

[437] —— [1992] 'Institutsgründung in schwieriger Zeit. 75 Jahre Kaiser-Wilhelm/Max-Planck-Institut für Physik', *PB* **48**, pp. 1031–1033.

[438] Kappler, Eugen (Ed.) [1948] *Physics of Liquids and Gases*, Wiesbaden: Klemm (FIAT Review of German Science 1939–1945, Vol. 10).

[439] Karl, Colonel [1940] 'General der Artillerie Prof. Dr. phil. h.c. Dr. ing. Becker', *ZgSS* **35** (6), pp. 121–123.

[440] Karlik, Berta [1970] 'Otto Hahn', *Almanach der Österreichischen Akademie der Wissenschaften für 1969* (obituary, vol. published 1970), pp. 266–271.

[441] Karlson, Paul [1943] 'Präzision als Aufgabe', *Das Reich*, No. 11, Mar. 14, pp. 9–10.

[442] Karlson, Peter [1990] *Adolf Butenandt. Biochemiker, Hormonforscher, Wissenschaftspolitiker*, Stuttgart: Wissenschaftliche Verlagsgesellschaft.

[443] Kast, W. [1957] 'Gustav Mie †', *PB* **13**, pp. 129–130.

[444] Kastrup, Hans A. [1988] 'Nachruf auf Fritz Bopp † 1987', *PB* **44**, pp. 77–78.

[445] Kater, Michael Hans [1974] *Das Ahnenerbe. Die Forschungs- und Lehrgemeinschaft in der SS. Organisationsgeschichte von 1935 bis 1945*, Stuttgart: Deutsche Verlagsanstalt.

[446] —— [1975] *Studentenschaft und Rechtsradikalismus in Deutschland 1918–1933. Eine sozialgeschichtliche Studie zur Bildungskrise in der Weimarer Republik*, Hamburg.

[447] —— [1981] 'Die nationalsozialistische Machtergreifung an allen deutschen Hochschulen. Zum politischen Verhalten akademischer Lehrer bis 1939', in: Hans Johann Vogel, Helmut Simon & Adalbert Podlech (Eds.) *Die Freiheit der Anderen. Festschrift für Martin Hirsch*, Baden-Baden: Nemos, pp. 49–75.

[448] —— [1983] *The Nazi Party. A Social Profile of Members and Leaders 1919–1945*, Oxford: Basil Blackwell.

[449] —— [1989] *Doctors under Hitler*, Chapel Hill/London: Univ. of North Carolina Press.

[450] Kausch, Dr. [1937] 'Die Physikalisch-Technische Reichsanstalt: Großzügige Forschung', *Braunschweiger Neueste Nachrichten – Braunschweigische Landeszeitung*, Wed., Nov. 22, No. 272, p. 5.

[451] Kbr. [1921] 'Eine gewichtige Stimme gegen die Relativitätstheorie', *Naturwissenschaftliche Wochenschrift* **36** (new ser. **20**), p. 551 (comment on Lenard [1918/21]).

[452] Kelly, Reece Conn [1973] *National Socialism and German University Teachers: The NSDAP's Efforts to Create a National Socialist Professoriate and Scholarship*, Ph.D. thesis: Univ. of Washington.

[453] Kern, Ulrich [1994] *Forschung und Präzisionsmessung: Die Physikalisch-Technische Reichsanstalt zwischen 1918 und 1945*, Weinheim: VCH.

[454] Kerner, Charlotte [1986] *Lise, Atomphysikerin. Die Lebensgeschichte der Lise Meitner*, Weinheim/Basel: Beltz (6th ed. 1990).

[455] Kershaw, Ian [1985] *The Nazi Dictatorship. Problems and Perspectives of Interpretation*, London: Arnold, a) 1st ed., 1985; b) 2nd ed., 1989; c) 3rd ed., 1993.

[456] Kerstein, Günter [1970] 'Bosch, Carl', *DSB*, **2**, pp. 323–324.

[457] Keßler, G. [1994] 'Zum Gedenken an Karl Wirtz', *PB* **50**, p. 867.

[458] Kevles, Daniel J. [1987] *The Physicists. The History of a Scientific Community in Modern America*, London/Cambridge, Mass.: Harvard Univ. Press.

[459] Kirsten, Christa & Hans-Jürgen Treder (Eds.) [1979] *Albert Einstein in Berlin 1913–1933*, 2 Vols., Berlin: Akademie-Verlag.

[460] Klamper, Elisabeth (Ed.) [1995] *Die Macht der Bilder. Antisemitische Vorurteile und Mythen*, Vienna: Jewish Museum.

[461] Kleinert, Andreas [1978] 'Von der Science Allemande zur deutschen Physik. Nationalsozialismus und moderne Naturwissenschaft in Frankreich und Deutschland zwischen 1914 und 1940', *Francia* **6**, pp. 509–525.

[462] Kleinert, Andreas & C. Schönbeck [1978] 'Lenard und Einstein. Ihr Briefwechsel und ihr Verhältnis vor der Nauheimer Diskussion von 1920', *Gesnerus* **35**, pp. 318–333.

[463] Kleinert, Andreas [1979]a 'Nationalistische und antisemitische Ressentiments von Wissenschaftlern gegen Einstein', in: Nelkowski et al. (Eds.), [1979], pp. 501–516.

[464] —— [1979]b Review of Beyerchen [1977], *ISIS* **70**, pp. 156–157.

[465] —— [1980] 'Lenard, Stark und die Kaiser-Wilhelm-Gesellschaft', *PB* **36**, pp. 35–43.

[466] —— [1983] 'Das Spruchkammerverfahren gegen Johannes Stark', *Sudhoffs Archiv* **67**, pp. 13–24.

[467] —— [1993] 'Paul Weyland, der Berliner Einstein-Töter', in: H. Albrecht (Ed.) [1993], pp. 198–232.

[468] Kleint, Christian [1986] 'Aus der Geschichte der Leipziger Uranmaschinenversuche', *Kernenergie* **29**, pp. 245–251.

[469] Kleint, Christian & Gerald Wiemers (Eds.) [1993] *Werner Heisenberg in Leipzig 1927–1942*, Berlin: Akademie Verlag.

[470] Klemperer, Victor [1975] *LTI. Lingua Tertii Imperii. Die Sprache des Dritten Reiches*, Leipzig: Reclam (reprint of 1945 ed.).

[471] Koebner, Thomas, Wulf Köpke, Claus-Dieter Krohn, Sigrid Schneider & Lieselotte Maas (Eds.) [1988] *Exilforschung. Ein Internationales Jahrbuch*, Vol. 6: *Vertreibung der Wissenschaften und andere Themen.*

[472] Kogon, Eugen [1973] *The Theory and Practice of Hell: The German Concentration Camps and the System Behind Them*, transl. by Heinz Norden of 1946 German ed., New York: Octagon Books.

[473] Kohler, Max [1978] 'Nachruf auf Reinhold Mannkopf', *PB* **34**, pp. 436–437.

[474] Kopfermann, Hans (Ed.) [1948] *Physics of the Electron Shells*, Wiesbaden: Klemm (FIAT Review of German Science 1939–1945, Vol. 12).

[475] Krafft, Fritz [1978]a 'Lise Meitner und ihre Zeit – Zum hundertsten Geburtstag der bedeutenden Naturwissenschaftlerin', *Angewandte Chemie*, **90**, pp. 876–892.

[476] —— [1978]b 'Vierzig Jahre Uranspaltung. Historische Betrachtungen zum Forscherteam Hahn-Meitner-Straßmann', *Frankfurt Allgemeine Zeitung*. Dec. 28, science supplement, No. 288, pp. I–II.

[477] —— [1981] *Im Schatten der Sensation. Leben und Wirken von Fritz Straßmann dargestellt nach Dokumenten und Aufzeichnungen*, Weinheim: Verlag Chemie.

[478] —— [1988] *Lise Meitner: Eine Biographie*, Berlin: Berichte des Hahn-Meitner-Instituts.

[479] Kramish, Arnold [1986] *The Griffin*, Boston: Houghton Mifflin Co.

[480] Krauch, Carl [1937] 'Forschung und Entwicklung. Aufgaben und Arbeiten des Amtes für deutsche Roh- und Werkstoffe', in: *Der Vierjahresplan*, **1**, pp. 261–263.

[481] Krause, Eckart, Ludwig Huber & Holger Fischer (Eds.) [1991] *Hochschulalltag im 'Dritten Reich'. Die Hamburger Universität 1933-1945*, Berlin/Hamburg: Reimer (3 Vols.).

[482] Krbek, Franz von [1942]a 'Die neue Welt der Physik', *Das Reich*, No. 8, Feb. 22, p. 13.

[483] —— [1942]b 'Verwandlung der Energie', *Das Reich*, No. 47, Nov. 22, p. 9.

[484] Krebs, K. [1952] 'Wilhelm Westphal 70 Jahre', *PB* **8**, pp. 131–132.

[485] Krieck, Ernst [1936] 'Der Wandel der Wissenschaftsidee und des Wissenschaftssystems im Bereich der nationalsozialistischen Weltanschauung', *Volk im Werden* **4**, pp. 378–381.

[486] —— [1940] 'Krisis der Physik', *Volk im Werden* **8**, pp. 55–62 (see reply by Müller [1940]b).

[487] Krockow, Christian Graf von [1983] '1933 – Bücher im Feuer: Selbstzerstörung des Geistes', *Spektrum der Wissenschaft*, issue 5, pp. 60–69.

[488] Kroebel, W. [1964] 'Zum Tode von James Franck', *Natw.* **51**, pp. 421–423.

[489] Kröner, Peter [1983] *Vor fünfzig Jahren. Zur Emigration deutschsprachiger Wissenschaftler 1933–1939*, Münster: Gesellschaft für Wissenschaftsgeschichte.

[490] Kubach, Fritz [1936] Review of Lenard [1936]a, *Volk im Werden* **4**, pp. 414–416.

[491] —— [1937/38]a Review of Lenard [1929]c, *ZgN* **3**, pp. 94–95.

[492] —— [1937/38]b 'Geheimrat Lenards 75. Geburtstag', *ZgN* **3**, pp. 164–165.

[493] —— [1942] 'Philipp Lenard – der deutsche Naturforscher', *ZgN* **8**, pp. 98–100.

[494] Kube, Alfred [1986] *Pour le Mérite und Hakenkreuz. Hermann Göring im Dritten Reich*, Munich/Vienna: Oldenbourg.

[495] Kuczera, Joseph [1985] *Gustav Hertz*, Leipzig: Teubner.

[496] Kuhn, H. G. [1965] 'James Franck, 1882-1964' *BMFRSL* **11**, pp. 53–74.

[497] Kuhn, K. [1957] 'Johannes Stark †', *PB* **13**, pp. 370–371.

[498] Kuhn, Thomas [1987] *Black-Body Theory and the Quantum Discontinuity, 1894–1912*, Chicago: Chicago Univ. Press.

[499] Kühnl, Reinhard [1990] *Faschismustheorien. Ein Leitfaden*, Heilbron: Distel-Verlag.

[500] Kuiper, Gerard Peter [1946] 'German astronomy during the war', *Popular Astronomy* **54**, pp. 263–287.

[501] Laitko, Hubert, Eginhard Fabian, Wolfgang Girnus, Conrad Grau, Dieter Hoffmann, Horst Kant, Wolfgang Schlicker & Annette Vogt [1987] *Wissenschaft in Berlin: Von den Anfängen bis zum Neubeginn nach 1945*, Berlin: Dietz, esp. pp. 502–593.

[502] Lasby, Clarence G. [1971] *Project Paperclip. German Scientists and the Cold War*, New York: Athenäum.

[503] Laue, Max von [1917] 'Die Fortpflanzungsgeschwindigkeit der Gravitation. Bemerkungen zur gleichnamigen Abhandlung von Paul Gerber', *Annalen der Physik* (4) **53**, pp. 214–216 (reply to Gehrcke [1916]).

[504] —— [1920] 'Zur Erörterung über die Relativitätstheorie', a) *Tägliche Rundschau*, Aug. 11, No. 175 (A); b) reprinted in: Weyland [1920]b, pp. 25–27, 30–31 (see Glaser [1920]).

[505] Laue, Max von, Walther Nernst & Heinrich Rubens [1920] Reply to Weyland [1920]c in: *Kreuz-Zeitung, Tägliche Rundschau, BT*, Aug. 26 and *Freiheit*, Aug. 27.

[506] Laue, Max von [1923] Review of Stark [1922] in: *Natw.* **11**, pp. 29–30.

[507] —— [1936] 'Deutsche Physik', *Frankfurter Zeitung*, **80** No. 110/111, Saturday, Feb. 29, p. 10 (review of Lenard [1936]a).

[508] —— [1937] 'Experimentelle und theoretische Physik', *Zeitschrift für den physikalischen und chemischen Unterricht* **50**, pp. 164–167.

[509] —— [1939] 'Otto Hahn zum sechzigsten Geburtstag', *Natw.* **27** (10), Mar. 10.

[510] —— [1946] 'Arnold Berliner (26.12.1862–22.3.1942)', *Natw.* **33**, pp. 257–258.

[511] —— [1947]a 'Max Planck †', *PB* **3**, pp. 249–252, 325, 393–394.

[512] —— [1947]b 'Bemerkung zu der vorstehenden Veröffentlichung von J. Stark', *PB* **3**, pp. 272–273 (reply to Stark [1947]).

[513] —— [1947]c 'Die Kriegstätigkeit der deutschen Physiker', *PB* **3**, pp. 424–425.

[514] —— [1949] 'A report on the state of physics in Germany', *AJP*, **17**, pp. 137–141.

[515] —— [1951] 'Sommerfelds Lebenswerk', *Natw.* **38**, pp. 513–518.

[516] Laue, Max von et al. [1953] 'Ansprachen am 9. Dez. 1952 zur Enthüllung der Haber-Gedenktafel', *Zeitschrift für Elektrochemie* **57**, p. 128.

[517] Laue, Max von [1956] 'Gedenkworte für Friedrich Schmidt-Ott', in: von Laue [1961], Vol. 3, pp. 238–242.

[518] —— [1961] *Gesammelte Arbeiten*, Braunschweig: Vieweg,

[519] Laurence, William L. [1940] 'The atom gives up', *Saturday Evening Post* (Philadelphia), Sep. 7, pp. 12–13, 60–63.

[520] Leclerc, Herbert [1988] 'Dr. Ohnesorge und die deutsche Reichspost', *Archiv für deutsche Postgeschichte* **2**, pp. 120ff.

[521] Leeb, Emil [1958] *Aus der Rüstung des Dritten Reiches (Das Heereswaffenamt 1938–1945)*, Frankfurt (Main): Mittler.

[522] Lehmanns [1940] *50 Jahre Lehmanns Verlag, 1890–1940*, Munich/Berlin: J. F. Lehmanns.

[523] —— [1965] *75 Jahre J. F. Lehmanns Verlag München, 1890–1965*, Munich: Markus.

[524] Leibfried, Günther [1955]a 'Max von Laue', in: Hans Schwerte & Wilhelm Spengler (Eds.) *Forscher und Wissenschaftler im heutigen Europa*, Oldenburg: G. Stalling, pp. 56–62.

[525] —— [1955]b 'Richard Becker †', *PB* **11**, pp. 319–320; see also pp. 176f.

[526] Lemmerich, Jost [1980] 'Nachruf für Otto Robert Frisch', *PB* **36**, p. 43.

[527] —— (Ed.) [1981] *Dokumente zur Gründung der Kaiser-Wilhelm-Gesellschaft und der Max-Planck-Gesellschaft zur Förderung der Wissenschaften. A Review of the History of Physics in Berlin (–1933)*, Amsterdam: Reidel.

[528] —— (Ed.) [1982] *Max Born, James Franck: Physiker in ihrer Zeit. Der Luxus des Gewissens*, Berlin: Catalog No. 17 of Staatsbibliothek Stiftung Preußischer Kulturbesitz.

[529] —— [1987] *Maß und Messen. Ausstellung aus Anlaß der Gründung der Physikalisch Technischen Reichsanstalt am 28. März 1887*, Braunschweig/Berlin: Physikalisch Technische Bundesanstalt.

[530] —— [1988] *Die Geschichte der Entdeckung der Kernspaltung*, Berlin: Technische Univ.

[531] —— [1993] 'Der Nachlaß von Lise Meitner in Cambridge', *Naturwissenschaftliche Rundschau* **46**, pp. 482–484.

[532] Lenard, Philipp [1910] 'Über Äther und Materie', a) *Sitzungsberichte der Akademie der Wissenschaften, Heidelberg*, 1910, 16th paper; b) (separatum) Heidelberg: Winter.

[533] —— [1918/21] *Über Relativitätsprinzip, Äther, Gravitation*, a) *Jahrbuch der Radioaktivität und Elektronik* **15**, pp. 117–136; b) (offprint) Leipzig, 1st ed. 1918; c) 3rd ed. 1921: *Mit einem Zusatz, betreffend die Bad Nauheimer Diskussion* (see Kbr. [1921]).

[534] —— [1920/21] 'Über Äther und Uräther', a) *Jahrbuch der Radioaktivität und Elektronik* **17**, pp. 307–355, b) (offprint) Leipzig: Hirzel.

[535] —— [1921] Introduction to Johann Soldner: 'Über die Ablenkung eines Lichtstrahls von seiner geradlinigen Bewegung durch die Attraktion eines Weltkörpers, an welchem er nahe vorbei geht' (excerpts from 1801 paper in *Astronomisches Jahrbuch für das Jahr 1804*, Berlin, 1801, pp. 161–172), *Annalen der Physik* (4) **65**, pp. 593–600.

[536] —— [1929] *Große Naturforscher. Eine Geschichte der Naturforschung in Lebensbeschreibungen*, Munich: J. E. Lehmann, a) 1st ed. 1929; b) 2nd ed. 1930; c) 3rd ed. 1937; d) 4th ed. 1941 (see reviews by May [1942], Kubach [1937/38]a).

[537] —— [1934] 'Wissenschaft, Volk und Rasse. Zum Präsident Johannes Starks 60. Geburtstag', *Volk und Rasse. Illustrierte Monatsschrift für deutsches Volkstum*, **5**, 131–132.

[538] —— [1936]a *Deutsche Physik*, Munich: J. F. Lehmann (4 Vols.), 1st ed. 1936; b) 2nd ed. of Vol. 1 1939 (see also reviews by Tomaschek [1937/38], Bühl [1939], Kubach [1936], Laue [1936]).

[539] —— [1936]c 'Gedanken zu deutscher Naturwissenschaft', *Volk im Werden* (Leipzig) **4**, pp. 381–383.

[540] Lenoir, Timothy [1992] 'Practical reason and the construction of knowledge: The life-world of Haber-Bosch' in: Ernan McMullin (Ed.) *The Social Dimensions of Science*, Notre Dame, Ill., pp. 158–197.

[541] Leske, Monika [1990] *Philosophen im Dritten Reich*, Berlin: Dietz.

[542] Ley, Robert [1942] 'Die letzten 10 Prozent', *Zeiss-Werkzeitung* N.S. **17**, pp. 87–88.

[543] Lilge, Frederic [1948] *The Abuse of Learning: The Failure of the German University*, New York: Macmillan

[544] Lifton, Robert Jay [1986] *The Nazi Doctors. Medical Killing and the Psychology of Genocide*, New York: Basic Books.

[545] Lindner, Helmut [1980] 'Deutsche und gegentypische Mathematik. Zur Begründung einer 'arteigenen' Mathematik im 'Dritten Reich' durch Ludwig Bieberbach', in: Mehrtens & Richter (Eds.) [1980], pp. 88–115.

[546] Litten, Freddy [1992] *Astronomie in Bayern, 1914–1945*, Stuttgart: Steiner.

[547] —— [1994]a 'Die Carathéodory Nachfolge in München 1938–1944', *Centaurus* **37**, 154–172.

[548] —— [1994]b 'Oskar Perron – Ein Beispiel für Zivilcourage im Dritten Reich', *Mitteilungen der Deutschen Mathematiker-Vereinigung*, No. 3, pp. 11–12.

[549] Loewenstein, Karl [1939] *Hitler's Germany: The Nazi Background to War*, New York: Macmillan.

[550] Longerich, Peter [1989] *Die braunen Bataillons: Geschichte der SA*, Munich: Beck.

[551] Lorenz, Charlotte [1943] *Zehnjahresstatistik des Hochschulbesuchs und der Abschlußprüfungen* (2 vols.), Berlin: Verlag für Sozialpolitik, Wirtschaft und Statistik.

[552] Ludwig, Karl-Heinz [1970] 'Die wohlreflektierten 'Erinnerungen' des Albert Speer. Einige kuriose Bemerkungen zur Funktion des Architekten, des Ingenieurs und der Technik im Dritten Reich', *Geschichte in Wissenschaft und Unterricht* **21**, pp. 695–708.

[553] —— [1974] *Techniker und Ingenieure im Dritten Reich*, Düsseldorf: Athenäum (unaltered reprint: Droste, 1979).

[554] Lundgren, Peter (Ed.) [1985] *Wissenschaft im Dritten Reich*, Frankfurt: Suhrkamp.

[555] Lundgren, Peter, Bernd Horn, Wolfgang Krohn, Günter Küppers & Rainer Paslack [1986] *Staatliche Forschung in Deutschland 1870–1980*, Frankfurt: Campus.

[556] Lüst, Reimar (Ed.) [1975] *Max-Planck-Institut für Strömungsforschung Göttingen, 1925–1975*, Göttingen: Max Planck Institut für Strömungsforschung.

[557] Macrakis, Kristie [1986] 'Wissenschaftsförderung durch die Rockefeller-Stiftung im 'Dritten Reich'', *Geschichte und Gesellschaft* **12**, pp. 348–379.

[558] —— [1993] *Surviving the Swastika: The Kaiser-Wilhelm Society 1933–1945*, New York: Oxford Univ. Press.

[559] Magnus, Kurt (Ed.) [1993] *125 Jahre Technische Universität München, 1868–1993*, Munich: Technische Universität.

[560] Maier, Helmut (Ed.) [1966] *Die deutsche Universität im Dritten Reich*, Munich: Piper (esp. the contributions by Maier, Pascher and Roegele).

[561] —— [1993] *Erwin Marx (1893–1980). Ingenieurwissenschaftler in Braunschweig, und die Forschung und Entwicklung auf dem Gebiet der elektrischen Energieübertragung auf weite Entfernungen zwischen 1918 und 1950*, Stuttgart: GNT-Verlag.

[562] Mard, Arthur [1943] 'Die kleinste Länge. Die Physik an der Grenze des Meßbaren', *Das Reich*, No. 46, Nov. 14, p. 10.

[563] Marx, Fritz Morstein [1936] *Government in the Third Reich*, New York: McGraw-Hill (2nd ed., 1937).

[564] Mason, Brian [1992] *Victor Moritz Goldschmidt. Father of Modern Geochemistry*, San Antonio: Geochemical Society, USA.

[565] May, Eduard [1937] 'Die Bedeutung der modernen Physik für die Theorie der Erkenntnis', in: Grete Hermann et al., *Die Bedeutung der modernen Physik für die Theorie der Erkenntnis*, Leipzig: Hirzel, 1937, pp. 47–154.

[566] —— [1940] 'Über die Anfänge der Elektrik', *ZgN* **6**, pp. 217–242.

[567] —— [1941]a *Am Abgrund des Relativismus*, Berlin: Lüttke (see review: Thüring [1941]b).

[568] —— [1941]b 'Dingler und die Überwindung des Relativismus', *ZgN* **7**, pp. 137–149.

[569] —— [1941/42] 'Relativismus und seine Überwindung', *Gesetzgebung und Literatur* (Berlin) **22**, p. 102.

[570] —— [1942] Review of Lenard [1929]d, *ZgN* **8**, pp. 153–154.

[571] Mayer-Kuckuck, Theo (Ed.) [1995] 'Festschrift 150 Jahre Deutsche Physikalische Gesellschaft', *PB* **51**, No. 1, pp. F3–F238.

[572] Mayer-Kuckuck, Theo [1995] 'Gedenken an Hans Geiger', *PB* **51**, p. 1154.

[573] Mehra, Jagdish & Helmut Rechenberg [1982] *The Historical Development of Quantum Theory*, New York: Springer.

[574] Mehrtens, Herbert [1979] 'Die Naturwissenschaften im Nationalsozialismus', in: R. Rürup (Ed.) [1979], pp. 427–443.

[575] Mehrtens, Herbert & Richter, Steffen (Eds.) [1980] *Naturwissenschaft, Technik und NS-Ideologie*, Frankfurt (Main): Suhrkamp (STW No. 303).

[576] Mehrtens, Herbert [1987]a 'The social system of mathematics and National Socialism: A survey', *Social Inquiry* **57**, pp. 159–182.

[577] —— [1987]b 'Ludwig Bieberbach and 'Deutsche Mathematik' in: Esther R. Phillips (Ed.) *Studies in the History of Mathematics* (Studies in Mathematics, No. 26), Washington, D.C.: Mathematical Association of America, pp. 195–241.

[578] —— [1989] 'Mathematics in the Third Reich: Resistance, adaptation and collaboration of a scientific discipline', in: R. P. W. Visser et al. (Eds.) *New Trends in the History of Science. Proceedings of a Conference Held at the University of Utrecht*, Amsterdam, 1989, pp. 141–166 (see comment by Freudenthal, ibid, pp. 167–170).

[579] —— [1990] *Moderne Sprache Mathematik*, Frankfurt (Main): Suhrkamp.

[580] Meinel, Christoph & Voswinckel, Peter (Eds.) [1994] *Medizin, Naturwissenschaft, Technik und Nationalsozialismus. Kontinuitäten und Diskontinuitäten*, Stuttgart: GNT.

[581] Meissner, W. [1959] 'Georg Joos †', *PB* **15**, pp. 320–321.

[582] —— [1960] 'Max v. Laue als Wissenschaftler und als Mensch', *Sitzungsberichte der bayerischen Akademie der Wissenschaften, München*, pp. 101–121.

[583] Meitner, Lise & Otto R. Frisch [1939] 'Desintegration of uranium by neutrons: A new type of nuclear reaction', *Nature* **143**, pp. 239–240.

[584] Meitner, Lise [1954] 'Einige Erinnerungen an das Kaiser-Wilhelm-Institut für Chemie in Dahlem', *Natw.* **41** (5), pp. 97–99.

[585] —— [1963] 'Wege und Irrwege zur Kernenergie', *Natw.* **16**, pp. 167–169.

[586] Mendelssohn, Kurt [1973] *The World of Walther Nernst. The Rise and Fall of German Science*, London: Macmillan.

[587] Menke, H. & G. Herrmann [1971] 'Was waren die "Transurane" der dreißiger Jahre in Wirklichkeit?', *Radiochimica Acta* **16**, pp. 119–132.

[588] Mentzel, Rudolf [1940] 'Zwanzig Jahre deutsche Forschung', in: *Aus der Arbeit des Stifterverbandes der Deutschen Forschungsgemeinschaft. Zur 20. Wiederkehr des Geburtstages des Stiftverbandes, 14. Dez. 1940*, pp. 1–15.

[589] Meyenn, Karl von et al. (Eds.) [1978–93] *Wolfgang Pauli. Wissenschaftlicher Briefwechsel*, Berlin: Springer. Vol. 1 (–1929), 1978; Vol. 2 (1930–39), 1985; Vol. 3 (1940–49), 1993.

[590] Michaelis, Anthony R. & Roswitha Schmid [1983] *Wissenschaft in Deutschland. Niedergang und neuer Aufstieg*, Stuttgart: Wissenschaftliche Verlagsgesellschaft.

[591] Milward, Alan Steele [1965] *The German Economy at War*, London: Athlone Press.

[592] —— [1977] *War, Economy and Society, 1939–1945*, Berkeley.

[593] Mitscherlich, Alexander & Fred Mielke (Eds.) [1962] *The Death Doctors*, London: Elek Books. Translation by James Cleugh of *Medizin ohne Menschlichkeit. Dokumente des Nürnberger Ärzteprozesses*, 1960, Frankfurt: Fischer.

[594] Möller, Horst [1984] 'Wissenschaft in der Emigration – quantitative und geographische Aspekte', *BWG* **7**, pp. 1–9.

[595] Mommsen, Hans [1966] *Beamtentum im Dritten Reich, mit ausgewälten Quellen zur nationalsozialistischen Beamtenpolitik*, Stuttgart: Deutscher Verlagsanstalt.

[596] —— [1991] *Der Nationalsozialismus und die deutsche Gesellschaft*, edited by Lutz Niethammer and Bernd Weisbrod, Reinbeck: Rowohlt.

[597] Moore, Walter [1989] *Schrödinger. Life and Thought*, Cambridge: Cambridge Univ. Press.

[598] Moraw, Peter [1982] *Kleine Geschichte der Universität Gießen: 1607–1982*, Giessen: Ferber.

[599] Morrison, Philip [1947] 'Alsos: The story of German science', *Bulletin of the Atomic Scientists*, **3**, No. 12, December, pp. 354, 365 (Review of Goudsmit [1947]).

[600] Moseley, Russell [1977] 'Tadpoles and frogs: Some aspects of the professionalization of British physics, 1870–1939', *Social Studies of Science* **7**, pp. 423–446.

[601] Mrowka, B. [1954] 'Richard Gans †', *PB* **10**, pp. 512–513.

[602] Mühlschlegel, B. [1983] 'Nachruf auf Fritz Sauter', *PB* **39**, p. 350.

[603] Müller, Gerhard [1978] *Ernst Krieck und die nationalsozialistische Wissenschaftsreform. Motive und Tendenzen einer Wissenschaftslehre und Hochschulreform im Dritten Reich*, Weinheim/Basel: Beltz.

[604] Müller, Hans [1961] 'Die pseudoreligiöse Charakter der nationalsozialistischen Weltanschauung', *Geschichte in Wissenschaft und Unterricht* **12**, pp. 337–352.

[605] Muller, Jerry Z. [1986] 'Enttäuschung und Zweideutigkeit: Zur Geschichte rechter Sozialwissenschaftler im 'Dritten Reich'', *Geschichte und Gesellschaft* **12**, pp. 289–316.

[606] Müller, Paul [1940] 'Carl Bosch †', *ZgSS* **35** (6), pp. 123–124.

[607] Müller, Wilhelm [1936] *Judentum und Wissenschaft*, Leipzig: Fritsch.

[608] —— [1939] 'Jüdischer Geist in der Physik', *ZgN* **5**, pp. 162–175.

[609] —— [1940]a 'Über Zehnders Darstellung des physikalischen Weltganzen', *ZgN* **6**, 134–135.

[610] —— [1940]b 'Zur 'Krisis der Physik' ', *ZgN* **6**, pp. 321–322 (reply to Krieck [1940]).

[611] —— (Ed.) [1941]a *Jüdische und Deutsche Physik. Vorträge zur Eröffnung des Kolloquiums für theoretische Physik an der Universität Munich*, Leipzig: Helingsche Verlagsanstalt (see also review by Weizel [1942]).

[612] —— [1941]b 'Dinglers Bedeutung für die Physik', *ZgN* **7**, pp. 150–156.

[613] —— [1944] *Kampf in der Physik*, Leipzig: Helingsche Verlagsanstalt.

[614] Müller-Hill, Benno [1985] *Murderous Science: Elimination by Scientific Selection of Jews, Gypsies and Others, Germany 1933–1945*, New York: Oxford Univ. Press. Translation by G. R. Frazer of: *Tödliche Wissenschaft. Die Aussonderung von Juden, Zigeunern und Geisteskranken 1933–1945*, Reinbek: Rowohlt (Rororo Aktuell No. 5349A), 1984.

[615] Münster, Clemens [1946] 'Fabrik im Kriege', *Frankfurter Hefte*, **1** (5), Aug., pp. 54–67.

[616] Mußgnug, Dorothee [1988] *Die vertriebenen Heidelberger Dozenten. Zur Geschichte der Ruprecht-Karls-Universität nach 1933*, Heidelberg: Winter.

[617] N.N. [1920]a 'Der Kampf um Einstein', *Vorwärts*, c. Aug. 25, reprinted in: Weyland [1920]b, pp. 7–8.

[618] N.N. [1920]b 'Um Einstein', *BT* **49**, Aug. 31, No. 409 (A), (contains declaration by Bie et al. [1920] and Gehrcke [1920]c).

[619] N.N. [1922] 'May Jews go to college?', *The Nation* **114**, p. 708.

[620] N.N. [1933]a 'Die deutsche Geisteswelt für Liste 1. Erklärung von 300 deutschen Universitäts- und Hochschullehrern', *VB*, 62nd, Munich ed., Mar. 3, suppl.

[621] N.N. [1933]b 'Das neue Beamtengesetz', *Göttinger Tageblatt*, Apr. 7, p. 1.

[622] N.N. [1933]c 'Das neue Studentenrecht', *Vossische Zeitung*, Thursday, Apr. 13, No. 176, pp. 1–2.

[623] N.N. [1933]d 'Zur Studentenschaft Boykott der Vorlesungen und Übungen von Hochschullehrern', *Vossische Zeitung*, Apr. 28, evening ed., p. 2.

[624] N.N. [1933]e 'Nazi "purge" of the universities. A long list of dismissals', *The Manchester Guardian Weekly*, Fri., May 19, p. 399.

[625] N.N. [1933]f 'The academic situation in Germany and the American Association of University Professors', *School and Society* **37** (10), No. 963, June 10, p. 742.

[626] N.N. [1933]g 'Psychopath Einstein', *VB* **46**, No. 256, Wed., Sep. 9, 1st. suppl., p. 2, col. 1.

[627] N.N. [1934] *Bekenntnis der Professoren an den deutschen Universitäten und Hochschulen zu Adolf Hitler und dem nationalsozialistischen Staat. Überreicht vom "NS-Lehrerbund"/Sachsen*, Dresden.

[628] N.N. [1936]a 'Philipp-Lenard-Institut at Heidelberg. Ceremonial dedication', *Nature* **137**, Jan. 18, pp. 93–94.

[629] N.N. [1936]b 'German science goose-steps', *NYT*, Mar. 12, p. 20.

[630] N.N. [1938]a 'The 'Jewish' spirit in science' and 'Prof. Stark's views', *Nature* **141**, Apr. 30, p. 778 (reply to Stark [1938]).

[631] N.N. [1938]b 'Prof. E. Schrödinger and the University of Graz', *Nature* **141**, p. 929.

[632] N.N. [1938]c 'Intellectual freedom', *School and Society* **48**, No. 1251, Dec. 17, p. 787.

[633] N.N. [1938]d 'Science and the dictator nation', *School and Society* **48**, No. 1251, Dec. 17, pp. 800–802.

[634] N.N. [1939] *Deutsche Wissenschaft. Arbeit und Aufgabe. Festgabe für Adolf Hitler zum 50. Geburtstag*, Leipzig: Hirzel.

[635] N.N. [1940]a 'General der Artillerie Prof. Dr. Ing. Karl Becker †', *Deutsche Technik* **8**, supplement: *Mitteilungen des Hauptamptes für Technik und des NSBDT* **65**, 181–182.

[636] N.N. [1940]b 'Geheimrat Professor Dr. Carl Bosch', *Jahrbuch der Kaiser-Wilhelm-Gesellschaft zur Förderung der Wissenschaften*, Berlin, p. 5.

[637] N.N. [1942] 'Reichsminister Ohnesorge 70 Jahre', *Elektrotechnische Zeitschrift* **63**, p. 249.

[638] N.N. [1944] 'Werner Heisenberg', *Das Reich*, No. 20, May 14, p. 1.

[639] N.N. [1946] 'Ein Brief nach Frankreich', *PB* **2**, pp. 8–11.

[640] N.N. (Ed.) [1949] *Statistisches Handbuch von Deutschland 1928–1944*, published by the American Occupation Zone State Commission in Munich (Länderrat des Amerikanischen Besatzungsgebietes), Munich: Franz Ehrenwirth.

[641] N.N. [1958] 'Ernst Gehrcke 80 Jahre alt', *Tagesspiegel* (Berlin) No. 3892, July 1, col. 2.

[642] N.N. [1963] 'Paul Rosbaud', *Physics Today* **16**, June issue, p. 105.

[643] N.N. [1964] 'Der Franck-Report', *PB* **20**, pp. 329–334.

[644] Nachmansohn, David [1979] *German-Jewish Pioneers in Science 1900–1933. Highlights in Atomic Physics, Chemistry, and Biochemistry*, a) New York: Springer; b) Revised ed. in German translation: *Die große Ära der Wissenschaft in Deutschland 1900 bis 1933. Jüdische und nichtjüdische Pioniere in der Atomphysik, Chemie und Biochemie* (ed. by Roswitha Schmid), Stuttgart: Wissenschaftliche Verlagsgesellschaft, 1988.

[645] Nagel, Brigitte [1991] *Die Welteislehre. Ihre Geschichte und ihre Rolle im Dritten Reich*, Stuttgart: GNT-Verlag (see also Schröder [1991]).

[646] Nathan, Otto & Heinz Norden (Eds.) [1960] *Einstein on Peace*, New York, 1960.

[647] Nelkowski, H., Armin Hermann, H. Poser, R. Schrader & R. Seiler (Eds.) [1979] *Einstein Symposium Berlin*, Berlin: Springer (Lecture Notes in Physics, Vol. 100).

[648] Neufeld, Michael J. [1995] *The Rocket and the Reich: Peenemünde and the Coming of the Ballistic Missile Era*, New York: Free Press.

[649] Neumann, Franz Leopold [1942] *Behemoth. The Structure and Practice of National Socialism* a) Oxford: Oxford Univ. Press; b) 2nd revised and expanded ed., 1944; c) reprinted New York, 1963.

[650] Neumann, Kurt [1955] 'Richard Becker †' *PB* **11**, p. 320.

[651] Neumann, Reinhard & Gisbert Freiherr zu Putlitz [1986] 'Philipp Lenard 1862–1947', *Semper Apertus. Sechshundert Jahre Ruprecht-Karls-Universität Heidelberg 1386–1986. Festschrift in 6 Bänden*, **3**, pp. 376–405.

[652] Neumark, Fritz [1980] *Zuflucht am Bosporus. Deutsche Gelehrte, Politiker und Künstler in der Emigration 1933–1953*, Frankfurt (Main): Knecht.

[653] Nida-Rümelin, Margret [1982] *Bibliographie Walther Gerlach. Veröffentlichungen von 1912–1979*, Munich: Forschungsinstitut des Deutschen Museums für die Geschichte der Naturwissenschaft und der Technik.

[654] Niemeier, Georg [1941] 'Von der wehrhaften Bedeutung naturwissenschaftlicher Grundlagenforschung', *Straßburger Monatshefte* **5** (11), pp. 721–734.

[655] Niethammer, Lutz [1982] *Die Mitläuferfabrik. Die Entnazifizierung am Beispiel Bayerns*, 2nd ed., Bonn: Dietz (1st ed. under different title, 1972).

[656] Nipperdey, Thomas & Ludwig Schmugge [1970] *50 Jahre Forschungsförderung in Deutschland: Ein Abriss der Geschichte der Deutschen Forschungsgemeinschaft. 1920–1970*, Bonn: Deutsche Forschungsgemeinschaft.

[657] Noakes, Jeremy & Geoffrey Pridham (Eds.) [1990] *Nazism 1919–1945. A Documentary Reader*, a) 3 vols., Exeter: Univ. of Exeter, 1983/1988; b) paperback edition in 2 vols., New York: Schocken Books, 1990.

[658] Noddack, Ida [1934] 'Das periodische System der Elemente und seine Lücken', *Angewandte Chemie* **47**, pp. 301–305.

[659] —— [1939] 'Bemerkung zu den Untersuchungen von O. Hahn, L. Meitner und F. Straßmann über die Produkte, die bei der Bestrahlung von Uran mit Neutronen entstehen', *Natw.* **27**, pp. 212–213.

[660] Nolte, Ernst [1965] 'Zur Typologie des Verhaltens der Hochschullehrer im Dritten Reich', *Aus Politik und Geschichte* **15** (64/65), pp. 3–14.

[661] [Notgemeinschaft] [1920] 'Zur Notgemeinschaft der deutschen Wissenschaft', *Internationale Monatsschrift für Wissenschaft, Kunst und Technik* **15**, issues 1–4.

[662] Notgemeinschaft deutscher Wissenschaftler im Ausland (Ed.) [1936] *List of Displaced German Scholars*, London: Notgemeinschaft (reprinted Stockholm, 1978; and in: Strauss, Buddensieg & Düwell (Eds.) [1987]).

[663] Oexle, Otto Gerhard [1994] 'Wie in Göttingen das Max-Planck-Gesellschaft entstand', *Max-Planck-Gesellschaft Jahrbuch*, pp. 43–60.

[664] O'Flaherty, James C. [1992] 'Werner Heisenberg on the Nazi revolution: Three hitherto unpublished letters', *Journal for the History of Ideas* **53**, pp. 487–494.

[665] Olff-Nathan, Josiane (Ed.) [1993] *La Science sous le Troisième Reich*, Paris: Le Seuil.

[666] Orlow, Dietrich [1969] *The History of the Nazi Party 1919–1933*, Pittsburgh: Pittsburgh Univ. Press.

[667] Orthmann, Wilhelm [1939] 'Über die Ausbildung des Physikers', and 'Bemerkung zum Frauenstudium', *Deutsche Mathematik* **4**, pp. 117–126 (see also Stuart [1939], Feigl & Hamel [1939]).

[668] Osietzki, Maria [1984] *Wissenschaftsorganisation und Restauration. Der Aufbau außeruniversitärer Forschungseinrichtungen und die Gründung des west-deutschen Staates, 1945–1952*, Cologne/Vienna: Bohlau.

[669] —— [1988] 'Kernphysikalische Großgeräte. Zwischen naturwissenschaftender Forschung, Industrie und Politik. Zur Entwicklung der ersten deutschen Teilchenbeschleuniger bei Siemens 1933–1945', *Technikgeschichte* **55**, pp. 25–46.

[670] —— [1993] 'Die Technisierung der Strahlenforschung und Hochenergiephysik als Durchsetzungsstrategie', *BWG* **16**, pp. 203–215.

[671] Overesch, Manfred & Saal (Eds.) [1991] *Das III. Reich 1933–1945. Eine Tageschronik der Politik, Wirtschaft, Kultur* (2 vols.), Augsburg: Weltbild Verlag (original ed.: Düsseldorf: Droste, 1982/83).

[672] P.F.F. [1936] 'The ceremonial dedication of the Philipp-Lenard-Institut at Heidelberg, *Nature* **137**, p. 233.

[673] P.K. [1940] 'Uran 235: Der Treibstoff der Zukunft?', *Die Koralle* **8**, issue 46.

[674] Pais, Abraham [1982] *Subtle is the Lord – The Science and the Life of Albert Einstein*, Oxford: Oxford Univ. Press.

[675] —— [1991] *Niels Bohr's Times in Physics, Philosophy, and Polity*, Oxford: Clarendon.

[676] Pash, Boris T. [1969] *The Alsos Mission*, New York: Award House.

[677] Päsler, Max [1960] 'Leben und wissenschaftliches Werk Max von Laues', *PB* **16**, 552–567.

[678] —— [1977] 'Ein Senior der Physik. Wilhelm Westphal 95 Jahre alt', *PB* **33**, pp. 123–124.

[679] Pauwels, Jacques R. [1984] *Women Nazis and Female University Students in the Third Reich 1933–1945*, New York: Westport, Conn.: Greenwood.

[680] Peierls, R. E. [1959] 'Wolfgang Ernst Pauli 1900–1950', *BMFRS* **5**, pp. 175–192.

[681] Perron, Oskar [1946] 'Verfälschung der Wissenschaften', *Neue Zeitung* (Munich), Oct. 7, 1946, p. 1.

[682] Peschel, I. (Ed.) [1991] *Physik-Handbuch - Daten, Fakten, Adressen*, Bad Honnef: DPG.

[683] Peterlin, Anton [1959] 'Herbert Stuart 60 Jahre', *PB* **15**, p. 224.

[684] Petzina, Dietmar [1968] *Autarkiepolitik im Dritten Reich: Der nationalsozialistische Vierjahresplan*, Stuttgart: Deutsche Verlagsanstalt.

[685] Pfundtner, Hans (Ed.) [1937] *Dr. Wilhelm Frick und sein Ministerium*, Munich: Eher.

[686] Pincussen, L. [1920] 'Bad Nauheim, 86. Versammlung Deutscher Naturforscher und Ärzte, 19.–25. IX. 1920', *Deutsche Medizinische Wochenschrift* **43**, p. 1212.

[687] Pinl, Max [1969–76] 'Kollegen in einer dunklen Zeit' *Jahresbericht der Deutschen Mathematiker Vereinigung*, Part 1: 1969, pp. 167–228; Part 2: 1971, pp. 165–189; Part 3: 1972, pp. 153–208; Part 4: 1974, pp. 166–208; postscript: 1976, pp. 161–164.

[688] Pinl, Max & Furtmüller, Lux [1973] 'Mathematicians under Hitler', *Leo Baeck Institute Yearbook* **18**, pp. 129–182.

[689] Planck, Max [1920] 'Die Entstehung und bisherige Entwicklung der Quantentheorie', *Les Prix Nobel en 1919–1920*, Stockholm: Norstedt, 1922, pp. 1–14 (Nobel prize lecture, June 2, 1920).

[690] —— [1929] 'Max von Laue. Zum 9. Oktober 1929', *Natw.* **17**, pp. 787–788.

[691] —— [1933]a 'Überproduktion von wissenschaftlichen Ideen', *BT*, Sunday ed., June 18, No. 281, 1st suppl.

[692] —— [1933]b *Wege zur physikalischen Erkenntnis. Reden und Vorträge*, Leipzig, 1st ed.; 4th ed., 1944.

[693] —— [1935] *Die Physik im Kampf um die Weltanschauung. Vortrag, gehalten am 6. März 1935 im Harnack-Haus, Berlin-Dahlem*, Leipzig: Barth (see review by Sommerfeld [1936]).

[694] —— [1938]a 'Arnold Sommerfeld zum siebzigsten Geburtstag', *Natw.* **26**, pp. 777–779.

[695] [——] [1938]b 'Feier des 80. Geburtstages des Ehrenmitgliedes der Deutschen Physikalischen Gesellschaft Max Planck', *VDPG* (3) **19**, pp. 58–76.

[696] —— [1939] 'Max von Laue. Zum 9. Oktober 1939', *Natw.* **27**, pp. 665–666.

[697] —— [1942] 'Warum kann Wissenschaft nicht populär sein?', *Das Reich*, Aug. 9, No. 32, p.9.

[698] [——] [1948] Special issue *PB* **4**, esp. pp. 125–1267, 133–180.

[699] —— [1958] *Physikalische Abhandlungen und Vorträge*, Vols. I–III, Braunschweig: Vieweg.

[700] Poliakov, Léon [1955/68] *Histoire de l'antisémitisme*, Vol. 1: *Du Christ aux Juifs de cour*, 1955; Vol. 2: *Liberté de l'Ésprit (de Mahomet aux marranes)*, 1961; Vol. 3: *De Voltaire à Wagner*, 1968.

[701] Poliakov, Léon & Joseph Wulf (Eds.) [1955/83] *Das Dritte Reich und die Juden*, a) 1st ed. Berlin: Arani; b) reprinted, Frankfurt (Main): Ullstein, chap. 3.

[702] —— (Eds.) [1959/83] *Das Dritte Reich und seine Denker*, a) 1st ed. Berlin: Arani; b) reprinted, Frankfurt (Main): Ullstein, chap. 6.

[703] Poliakov, Léon [1968] *Le Développement de l'antisémitisme en Europe aux temps modernes (1700–1850)*, thesis, Paris.

[704] —— [1974-85] *The History of Anti-Semitism*, in 4 vols. from Roman times to 1933, translated by R. Howard, N. Gerardi, M. Kochan and G. Klim. London: Rentledge & Kegan Paul; New York: Vanguard.

[705] Pollard, Ernest Charles & William Lee Davidson [1942] *Applied Nuclear Physics*, New York: Wiley.

[706] Popplow, Ulrich [1977] 'Die Machtergreifung in Augenzeugenberichten Göttingen, 1932–1935', in: *Göttinger Jahrbuch*, pp. 157–186.

[707] Powers, Thomas [1993] *Heisenberg's War: The Secret History of the German Bomb*, New York: Knopf.

[708] Proctor, Robert [1988] *Race Hygiene. Medicine Under the Nazis*, Cambridge, Mass.: Harvard Univ. Press.

[709] Pross, Helge [1955] *Die deutsche akademische Emigration nach den Vereinigten Staaten 1933–1941*, Berlin: Duncker & Humblot.

[710] Quetsch, Cäcilie [1960] *Die zahlenmäßige Entwicklung des Hochschulbesuchs in den letzten fünfzig Jahren*, Berlin: Springer.

[711] Rajewsky, E. & M. Schön (Eds.) [1948] *Biophysics I and II*, Wiesbaden: Klemm (FIAT Review of German Science, Vols. 21, 22).

[712] Ramsauer, Carl [1938] 'Der Physiker. I. Aufgaben und Anforderungen', in: *Die Akademischen Berufe*, Berlin, 1st ed. 1938, issue No. 15; 2nd ed. 1940.

[713] —— [1943] 'Über Leistung und Organisation des angelsächsischen Physik mit Ausblicken auf die deutsche Physik (Lecture held at the 2nd science section meeting of the regular members of the Akademie der Luftfahrtforschung on April 2, 1943) *KPFB* **G-241**.

[714] —— [1944] 'Die Schlüsselstellung der Physik', *PB* **1**, pp. 30–33.

[715] —— [1947]a 'Zur Geschichte der Deutschen Physikalischen Gesellschaft in der Hitlerzeit', *PB* **3**, pp. 110–114.

[716] —— [1947]b 'Eingabe an Rust', *PB* **3**, pp. 43–46.

[717] —— [1949] 'Meine physikalischen Erinnerungen', in: *Physik – Technik – Pädagogik. Erfahrungen und Erinnerungen*, Karlsruhe: G. Braun, pp. 99–130.

[718] Rantzau, Otto Graf zu [1939] *Das Reichsministerium für Wissenschaft, Erziehung und Volksbildung*, Berlin: Junker & Dünnhaupt.

[719] Rasch, Manfred [1989] *Geschichte des Kaiser-Wilhelm Institute für Kohlenforschung in Mülheim an der Ruhr, 1913–1943*, Weinheim: VCH.

[720] Rechenberg, Helmut [1980] 'Für Fritz Bopp zum 70. Geburtstag', *PB* **36**, p. 18.

[721] —— [1988]a 'Vor fünfzig Jahren', *PB* **44**, p. 418.

[722] —— [1988]b 'Transurane, Uranspaltung und das deutsche Uranprojekt', *PB*, **44**, 453–459.

[723] —— [1988]c 'Quantenmechanik im Aufbruch – Friedrich Hund berichtet aus seinem Leben' (Interview, Göttingen) Institut für den wissenschaftlichen Film, Göttingen, No. G 239.

[724] —— (Ed.) [1992]a *Werner Heisenberg: Deutsche und jüdische Physik*, Munich: Piper.

[725] —— [1992]b "Nicht die Spur von Landesverrat", *Die Zeit*, No. 24, June 5, p. 84 (reply to Jensen [1992]; see also Weizsäcker [1992]).

[726] —— [1993] 'Nachruf auf Wilhelm Hanle', *PB*, **49**, p. 678.

[727] —— [1994]a '50 Jahre Physikalische Blätter', *PB* **50**, pp. 21–28.

[728] —— [1994]b 'Mie, Gustav' *NDB* **17**, pp. 465–466.

[729] Reichinstein, David [1935] *Albert Einstein. Ein Bild seines Lebens und seiner Weltanschauung*, Prague: Stella, a) 1st ed. 1934; b) 2nd and 3th ed. 1935.

[730] Reid, Constance [1976] *Courant in Göttingen and New York. Stay of an Improbable Mathematician*, Berlin: Springer.

[731] Reingold, Nathan [1983] 'Refugee mathematicians in the United States, 1933–41: Reception and reaction', in: Jackman & Borden (Eds.) [1983], pp. 205–232.

[732] Renneberg, Monika [1993] 'Wissenschaft, Wirtschaft und Politik – Gründung und Aufbau des GKSS-Forschungszentrum in Geesthacht', *NTM*, N.S. **3**, pp. 147–159.

[733] Renneberg, Monika & Mark Walker (Eds.) [1994] *Science, Technology and National Socialism*, Cambridge: Cambridge Univ. Press.

[734] Renneberg, Monika & Klaus Hentschel [1995] 'Anatomie einer akademischen Karriere im Nationalsozialismus. Der Astronom Otto Heckmann', *VHZG* **43**, Issue 4.

[735] Requard, Friedrich [1937/38] 'Rassisch-völkische Gegebenheit und exakte Naturwissenschaft', *ZgN* **3**, pp. 193–201.

[736] —— [1938/39] 'Wissenschaftliche Strenge und Rasse', *ZgN* **4**, pp. 342–353.

[737] —— [1940] 'Physik und Erbcharakter', *ZgN* **6**, pp. 172–184.

[738] Reuter, Frank [1971] *Funkmeß. Die Entwicklung und der Einsatz des Radar-Verfahrens in Deutschland bis zum Ende des Zweiten Weltkrieges*, Opladen: Westdeutscher Verlag.

[739] Rhodes, Richard [1986] *The Making of the Atomic Bomb*, New York: Simon & Schuster.

[740] Richter, Steffen [1971] 'Das Wirken der Notgemeinschaft der Deutschen Wissenschaft: Erläutert am Beispiel der Relativitätstheorie in Deutschland, 1920–1930', *PB* **27**, pp. 497–504.

[741] —— [1972] 'Forschungsförderung in Deutschland 1920–1936. Dargestellt am Beispiel der Notgemeinschaft der Deutschen Wissenschaft und ihrem Wirken für das Fach Physik', *Technikgeschichte in Einzeldarstellungen* **23**.

[742] —— [1973] 'Die Kämpfe innerhalb der Physik in Deutschland nach dem 1. Weltkrieg. Einige Auswirkungen auf die Wissenschaftsorganisation', *Sudhoffs Archiv* **57**, pp. 195–207.

[743] —— [1977] 'Physik und Gesellschaft', *PB* **33**, pp. 49–57.

[744] —— [1978/79] 'Physiker im Dritten Reich', *Jahrbuch der Technischen Hochschule Darmstadt*, pp. 103–113.

[745] —— [1980] 'Die 'Deutsche Physik'', in: Mehrtens & Richter (Eds.) [1980], pp. 116–141.

[746] Rider, Robin E. [1984] 'Alarm and opportunity: Emigration of mathematicians and physicists to Britain and the United States, 1933–1945', *HSPS* **15** (1), pp. 107–176.

[747] Riehl, Nikolaus [1988] *Zehn Jahre im goldenen Käfig. Erlebnisse beim Aufbau der sowjetischen Uran-Industrie*, Stuttgart: Riederer.

[748] Rindfleisch, Hans [1955] 'Abraham Esau †', *PB* **11**, pp. 181f., 318–319.

[749] Ringer, Fritz K. [1969] *The Decline of the German Mandarins: The German Academic Community*, Cambridge: Harvard Univ. Press.

[750] —— [1983] 'Inflation, antisemitism and the German academic community of the Weimar period', *Leo Baeck Institute Yearbook* **28**, pp. 3–10.

[751] Ritchie, James MacPherson [1983] *German Literature under National Socialism*, London: C. Helm.

[752] Roeder, Werner & Herbert A. Strauss (Eds.) [1980/83] *International Biographical Dictionary of Central European Emigrés 1933–1945* (ed. by Institut für Zeitgeschichte, Munich and Research Foundation for Jewish Emigration, New York, 1980 and 1983); esp. Vol. II: *The Arts, Sciences, and Literature*, 1983, Munich: Saur.

[753] Rollnik, H. [1983] 'Zum Tode von Walter Weizel' *PB* **39**, p. 51.

[754] Rosbaud, Paul [1948] 'Secret mission', *Times Literary Supplement* (London), June 5, p. 320 (rev. of Goudsmit [1947]).

[755] —— [1959] 'Brighter than a thousand suns', *Discovery*, Mar., pp. 96–97 (rev. of Jungk [1958]).

[756] —— [1960] 'Prof. Max von Laue, For. mem. Acad.', *Nature* **187**, Apr. 27, pp. 738, 905.

[757] Rosenberg, Alfred [1934] 'Freiheit der Wissenschaft', *Münchner Neueste Nachrichten*, No. 306, Fri, Nov. 9, (Speech held on Nov. 7, 1934 at the opening of the winter semester at Munich Univ. See also *Münchner Zeitung*, No. 312, Nov. 9, p. 3).

[758] —— [1937] 'Rosenberg über die Freiheit der Forschung. Eine parteiamtliche Stellungnahme', *Deutsche Allgemeine Zeitung* (Berlin) **76**, Thurs., Dec. 9, greater Berlin morning ed., No. 574, p. 2.

[759] Rosenow, Ulf [1987] 'Die Göttinger Physik unter dem Nationalsozialismus', in: Becker, Dahms & Wegeler (Eds.) [1987], pp. 374–409.

[760] Rössler, Eberhard [1984] *Die Torpedos der deutschen U-Boote: Entwicklung, Herstellung und Eigenschaften der deutschen Marine-Torpedos*, Herford: Köhler.

[761] Rössler, Mechtild [1990] *Wissenschaft und Lebensraum. Geographische Ostforschung im Nationalsozialismus*, Hamburg: Reimer.

[762] Rotta, Julius C. [1990] *Die Aerodynamische Versuchsanstalt in Göttingen, ein Werk Ludwig Prandtls*, Göttingen: Vandenhoek & Ruprecht.

[763] —— (Ed.) [1990] *Dokumente zur Geschichte der Aerodynamischen Versuchsanstalt in Göttingen, 1907–1925*, Göttingen (DLR-Mitt. 90-05).

[764] Rüdin, Ernst [1937] 'Der Deutsche Forscher spricht. IX. Das Wohl des Volkes', *Illustrierter Beobachter* **12**, issue 4, pp. 114, 116, 118.

[765] Rukop, H. [1939] 'Dr. Karl Mey zum 60. Geburtstag', *ZtP* **20**, pp. 65–66.

[766] Rürup, Reinhard (Ed.) [1979] *Wissenschaft und Gesellschaft: Beiträge zur Geschichte der Technischen Universität Berlin 1879–1979*, Berlin: Springer, 1979 (2 Vols.).

[767] Rutherford, Lord of Nelson [1934] 'Exiles in British sanctuary', *Science* **79**, June 15, pp. 533–534.

[768] —— [1935] 'The scholar in exile. Help for German refugees. An academic problem', *The Times* (London), Sat., Mar. 9, Royal ed., No. 47,008, pp. 13–14.

[769] Schappacher, Norbert & Kneser [1990] 'Fachverband – Institut – Staat. Streiflichter auf das Verhältnis von Mathematik zu Gesellschaft und Politik in Deutschland, unter besonderer Berücksichtigung der Zeit des Nationalsozialismus 1890–1990', in: J. Fischer et al. (Eds.) *Festschrift zum Jubiläum der Deutschen Mathematiker-Vereinigung*, Braunschweig: Vieweg, pp. 1–82.

[770] Schappacher, Norbert & Erhart Scholz (Eds.) [1992] 'Oswald Teichmüller – Leben und Werk', *Jahresbericht der Deutschen Mathematiker-Vereinigung* **94**, pp. 1–39.

[771] Scherrer, P. [1959] 'Wolfgang Pauli †', *PB* **15**, pp. 34–35.

[772] Scherzer, Otto [1951] 'Arnold Sommerfeld †', *PB* **7**, pp. 222–224.

[773] —— [1965] 'Physik im totalitären Staat', in: Andreas Flitner (Ed.) *Deutsches Geistesleben und Nationalsozialismus*, Tübingen: Wunderlich, pp. 47–58.

[774] Schlicker, Wolfgang [1979]a 'Konzeptionen und Aktionen bürgerlicher deutscher Wissenschaftspolitik. Zum gesellschaftlichen Stellenwert der Forschung nach 1918 und zur Gründung der Notgemeinschaft der Deutschen Wissenschaft', *Zeitschrift für Geschichtswissenschaft* **27**, pp. 423ff.

[775] —— [1979]b 'Carl Krauch, die IG Farben und die Forschungslenkung im faschistischen Deutschland', *Bulletin des Arbeitskreises 'Zweiter Weltkrieg'*, No. 1/2.

[776] —— [1983] 'Physiker im faschistischen Deutschland. Zum Geschehen um eine naturwissenschaftlich-technische Grundlagendisziplin', *Jahrbuch für Geschichte* **27**, pp. 109–142.

[777] Schmithals, F. [1980] 'Carl Ramsauer und das Dritte Reich', *PB* **36**, p. 345.

[778] Schmidt-Ott, Friedrich [1936] 'Max Donnevert †', *FuF* **12**, Mar. 1, p. 96.

[779] —— [1943] 'Robert Fellinger siebzig Jahre alt', *FuF* **19**, p. 212.

[780] —— [1952] *Erlebtes und Erstrebtes 1800–1950*, Wiesbaden: Steiner.

[781] Schnauber, Cornelius [1972] *Wie Hitler sprach und schrieb. Zur Psychologie und Prosodik*, Frankfurt, Athenäum.

[782] Schoenbaum, David [1967] *Hitler's Social Revolution*, Garden City, NY: Doubleday.

[783] Scholz, Heinrich [1943] 'Forschung und Symbol', *Das Reich* No. 31, Aug. 1, p. 14 (review of Planck: *Wege zur physikalischen Erkenntnis*, Leipzig: Hirzel, 1943, 3rd ed.).

[784] —— [1944] 'Vorträge zur Atomphysik', *Das Reich*, No. 11, Mar. 12, p. 11 (review of Heisenberg [1943]b).

[785] Schomerus, Friedrich [1952] *Geschichte des Jenaer Zeisswerkes 1846–1946*, Stuttgart: Piscator.

[786] Schöne, Albrecht [1983] *Göttinger Bücherverbrennung* (Göttingen Universitätsreden, No. 70), Göttingen: Vandenhoek & Ruprecht.

[787] Schottländer, Rudolf [1979] 'Antisemitische Hochschulpolitik: Zur Lage an der Technischen Hochschule Berlin 1933/34', in: Rürup (Ed.) [1979], pp. 445–453.

[788] —— [1988] *Verfolgte Berliner Wissenschaft. Ein Gedenkwerk*, Berlin: Edition Hentrich.

[789] Schrimpf [1940] 'Dr. Todt Reichsminister für Bewaffnung und Munition', *ZgSS* **35**, 73–74.

[790] Schröder, Reinald [1991] 'Welteislehre, Ahnenerbe und 'Weiße Juden', *Die Zeit*, No. 17, Apr. 19, pp. 41–43 (review of Nagel [1991]).

[791] —— [1993] 'Die 'schöne deutsche Physik' von Gustav Hertz und der 'weiße Jude' Heisenberg – Johannes Starks ideologischer Antisemitismus', in: Albrecht (Ed.) [1993] pp. 327–341.

[792] Schröter, Fritz [1959] 'Esau, Abraham', *NDB* **4**, pp. 640–641.

[793] Schubert, Helmut [1987] 'Industrielaboratorien für Wissenschaftstransfer. Aufbau und Entwicklung der Siemens-Forschung bis zum Ende des zweiten Weltkriegs anhand von Beispielen aus der Halbleiterforschung', *Centaurus* **30**, pp. 245–292.

[794] Schüler, Hermann [1947] 'Friedrich Paschen', *PB* **3**, pp. 232–233.

[795] Schüller, Hermann [1943] 'Kosmos Atom' *Das Reich*, No. 37, Sep. 12, pp. 9–10.

[796] Schwabe, Klaus (Ed.) [1988] *Deutsche Hochschullehrer als Elite 1815–1945. Büdinger Forschungen zur Sozialgeschichte*, Boppard am Rhein: Boldt.

[797] Scriba, Christoph J. (Ed.) [1995] *Die Elite der Nation im Dritten Reich – Das Verhältnis von Akademien und ihrem wissenschaftlichem Umfeld zum Nationalsozialismus*, Halle: Leopoldina.

[798] Seier, Hellmut [1964] 'Der Rektor als Führer. Zur Hochschulpolitik des Reichserziehungsministeriums 1934–1945', *VHZG* **12**, pp. 105–146.

[799] —— [1976] 'Niveaukritik und partielle Opposition. Zur Lage an den deutschen Hochschulen 1939/40', *Archiv für Kulturgeschichte* **58**, pp. 227–246.

[800] —— [1984] 'Universität und Hochschulpolitik im nationalsozialistischen Staat', in: Klaus Molettke (Ed.) *Der Nationalsozialismus an der Macht*, Göttingen: Vandenhoek & Ruprecht, pp. 143–165.

[801] —— [1988] 'Die Hochschullehrerschaft im Dritten Reich', in: Schwabe (Ed.) [1988], pp. 247–295.

[802] Serber, Robert [1943] *The Los Alamos Primer: The First Lectures on How to Build an Atomic Bomb* (originally delivered April 1943), edited by Richard Rhodes, Berkeley: Univ. of California Press.

[803] Siegele-Wenschkewitz, Leonore & Gerda Stuchlik (Eds.) [1990] *Hochschule und Nationalsozialismus. Wissenschaftsgeschichte und Wissenschaftsbereich als Thema der Zeitgeschichte*, Frankfurt/M.: Haag & Heschen Verlag.

[804] Siegmund-Schultze, Reinhard [1984] 'Theodor Vahlen – Zum Schuldanteil eines deutschen Mathematikers am faschistischen Mißbrauch der Wissenschaft', *NTM* **21**, pp. 17–32.

[805] —— [1986] 'Faschistische Pläne zur "Neuordnung" der europäischen Wissenschaft. Das Beispiel Mathematik', *NTM* **23**, pp. 1–17.

[806] —— [1989] 'Zur Sozialgeschichte der Mathematik an der Berliner Universität im Faschismus', *NTM* **26** (1), pp. 49–68.

[807] —— [1993] *Mathematische Berichterstattung in Hitlerdeutschland: der Niedergang des Jahrbuchs über die Fortschritte der Mathematik*, Göttingen: Vandenhoek & Ruprecht.

[808] Sime, Ruth Lewine [1990] 'Lise Meitner's escape from Germany', *AJP* **38**, pp. 262–267.

[809] —— [in press] Biography of Lise Meitner, Berkeley: Univ. of California Press.

[810] Simon, Leslie E. [1948] *German Research in World War II. An Analysis of the Conduct of Research*, New York: John Wiley; London: Chapman & Hall.

[811] Simonsohn, G. [1992] 'Physiker in Deutschland 1933–1945', *PB* **48**, pp. 23–28.

[812] Smyth, Henry Dewolf [1945] *Atomic Energy for Military Purposes. The Official Report on the Development of the Atomic Bomb Under the Auspices of the US Government 1940–1945*, Princeton: Princeton Univ. Press.

[813] Snyder, Louis L. [1976] *Encyclopedia of the Third Reich*, London: Robert Hale.

[814] Sommerfeld, Arnold [1930] 'Über Anschaulichkeit in der modernen Physik', *Unterrichtsblätter für Mathematik und Naturwissenschaft* **36**, pp. 161–167 (see reply by Stark [1930]b).

[815] —— [1936] Review of Planck [1935], *PZ* **37**, p. 144.

[816] Sontheimer, Kurt [1962] *Antidemokratisches Denken in der Weimarer Republik*, Munich: Nymphenburger, a) 1st ed. 1962; b) revised ed. 1968.

[817] Soon Im, Gyeong [1995] 'The formation and development of the Ramsauer effect', *HSPS* **25** (2), pp. 269–300.

[818] Sopade (Ed.) [1980] *Deutschland-Berichte der Sozialdemokratischen Partei Deutschlands 1934–1940*, Frankfurt (Main): Zweitausendeins; Salzhausen: Nettelbeck.

[819] Speer, Albert [1942] 'Männer der deutschen Technik!' *Elektrotechnische Zeitschrift* **63**, p. 105; see also pp. 31, 394.

[820] —— [1970] *Inside the Third Reich: Memoirs*, translated by R. & C. Winston, New York: Macmillan.

[821] —— [1976] *Spandau: The Secret Diaries*, translation of the 1975 original by R. & C. Winston, New York: Macmillan; London: Collins.

[822] Spence, R. [1977] 'Otto Hahn 1879–1968', *BMFRS* **16**, pp. 278–313.

[823] Stadler, Friedrich (Ed.) [1987/89] *Vertriebene Vernunft. Emigration und Exil österreichischer Wissenschaft*, Vienna/Munich: Jugend und Volk (Vol. 1: 1987; Vol. 2: 1989).

[824] Stark, Gary D. [1976/77] 'Der Verleger als Kulturunternehmer: Der J. F. Lehmanns Verlag und Rassenkunde in der Weimarer Republik', *Archiv für Geschichte des Buchwesens* **16**, pp. 291–318.

[825] Stark, Johannes [1920] 'Änderungen der Struktur und des Spektrums chemischer Atome' (Nobel prize lecture June 3, 1920), a) *Les Prix Nobel en 1919-1920*, Stockholm: Norstedt, 1922, pp. 1–10; see also *Physikalischer Nobelpreis für 1919*, ibid., pp. 25–28. b) in Engl. trans.: 'Structural and spectral changes of chemical atoms', in: *Nobel Lectures in Physics, 1901–1921*, Amsterdam: Elsevier, pp. 427–435.

[826] —— [1922] *Die gegenwärtige Krisis in der deutschen Physik*, Leipzig: Barth (esp. pp. 7–17 and 24–32. See also review by Laue [1923]).

[827] —— [1927] *Die Axialität der Lichtemission und Atomstruktur*, Berlin: Seydel.

[828] —— [1928] *Atomstruktur und Atombindung*, Berlin: Seydel.

[829] —— [1930]a 'Die Verjudung der deutschen Hochschule', *NSM* **1**, pp. 360–370.

[830] —— [1930]b 'Ueber den Dogmatismus moderner Theorien in der Physik', *Unterrichtsblätter für Mathematik und Naturwissenschaft* **36**, pp. 305–309 (reply to Sommerfeld [1930]).

[831] —— [1930]c *Adolf Hitlers Ziele und Persönlichkeit*, Munich.

[832] —— [1931] *Fortschritte und Probleme der Atomforschung*, Leipzig: Barth.

[833] —— [1934]a 'International status and obligations of science', *Nature* **133**, p. 290 (reply to Hill [1933]; see also Hill [1934]).

[834] —— [1934]b 'The attitude of the German government towards science', *Nature* **133**, p. 614 (see also reply by Haldane [1934]).

[835] —— [1934]c *Adolf Hitler und die deutsche Forschung. Ansprachen auf der Versammlung der Deutschen Forschungsgemeinschaft in Hannover*, Berlin.

[836] —— [1934]d *Nationalsozialismus und Wissenschaft*, Munich: Eher.

[837] —— [1937] *Forschung und Prüfung. 50 Jahre Physikalisch-Technischer Reichsanstalt*, Leipzig: Hirzel.

[838] —— [1938] 'The pragmatic and dogmatic spirit in physics', *Nature* **141**, p. 770 (see also reply N.N. [1938]).

[839] —— [1939] 'Zur Neuordnung des physikalischen Unterrichts', *Unterrichtsblätter für Mathematik und Naturwissenschaften* **45**, pp. 81–83.

[840] Stark, Johannes & Wilhelm Müller (Eds.) [1941] *Jüdische und deutsche Physik. Vorträge zur Eröffnung des Kolloquiums für theoretische Physik an der Universität München*, Leipzig, pp. 21–56.

[841] —— [1942] 'Philipp Lenard als Vorbild', *ZgN* **8**, pp. 100–101.

[842] —— [1947] 'Zu den Kämpfen in der Physik während der Hitler-Zeit', *PB* **3**, pp. 271–272 (see also reply by Laue [1947]b and Brüche [1947]b).

[843] —— [1987] *Erinnerungen eines Deutschen Naturforschers*, (originally written 1945) ed. by A. Kleinert, Mannheim: Bionomica.

[844] Steck, M.[1938/39] Review of Dingler [1938], *ZgN* **4**, pp. 118–121.

[845] Steenbeck, Max [1977] *Impulse und Wirkungen. Schritte auf meinem Lebensweg*, Berlin: Verlag der Nation.

[846] Steinberg, Michael Stephen [1973] *Sabers and Brown Shirts: The German Student's Path to National Socialism, 1918–1935*, Chicago/London: Univ. of Chicago.

[847] Steinmetz, Max et al.[1958] *Geschichte der Universität Jena 1548/58–1958*, Jena: Fischer.

[848] Stern, Fritz [1972]a *The Failure of Illiberalism. Essays on the Political Culture of Modern Germany*, New York: Knopf.

[849] —— [1972]b 'The political consequences of the unpolitical German', in: Stern [1972]a, pp. 3–25.

[850] —— [1982] 'Einstein's Germany', in: Gerald Holton & Y. Elkana (Eds.) *Albert Einstein: Historical and Cultural Perspectives*, Princeton: Princeton Univ. Press, pp. 319–343.

[851] —— [1986] 'Einstein and Germany', *Physics Today*, Feb., pp. 40–49.

[852] Stierstadt, Klaus [1980] 'Walther Gerlach †', *PB* **36**, pp. 18–19.

[853] Stoltzenberg, Dietrich J. [1991] 'Beitrag zur Geschichte des Kaiser Wilhelm-Instituts für Physikalische Chemie und Elektrochemie', *BWG* **14**, pp. 15–23.

[854] —— [1994] *Fritz Haber: Chemiker, Nobelpreisträger, Deutscher Jude: Eine Biographie*, Weinheim: VCH.

[855] Strätz, Hans-Wolfgang [1968] 'Die studentische "Aktion wider den undeutschen Geist" im Frühjahr 1933', *VHZG* **16**, pp. 347–372.

[856] Straßmann, Fritz [1968] 'Lise Meitner †', *PB* **24**, pp. 508–509.

[857] Strauss, Herbert A., Tilmann Buddensieg & Kurt Düwell (Eds.) [1987] *Emigration. Deutsche Wissenschaftler nach 1933. Entlassung und Vertreibung*, Berlin: Technische Universität Berlin (includes a reprint of the List of Displaced German Scholars (1936, 1937) and its supplement issued by the Notgemeinschaft Deutscher Wissenschaftler im Ausland [1936]).

[858] Strauss, Herbert A., Klaus Fischer, Christhard Hoffmann & Alfons Söllner (Eds.) [1991] *Die Emigration der Wissenschaften nach 1933. Disziplinengeschichtliche Studien*, Munich: Saur.

[859] Stuart, Herbert Arthur [1939] 'Bedeutung der Physik und Aufgaben des Physikers', *Deutsche Mathematik* **4**, pp. 116–117 (see Orthmann [1939] and Feigl & Hamel [1939]).

[860] Stuchtey, Karl, Hans Wegener & Walter Boje (Eds.) [1938] *Deutsche Akademie der Luftfahrtforschung. Jahrbuch 1937/38*, Munich/Berlin: Oldenbourg.

[861] Stuckart, Wilhelm & Hans Globke [1936] *Kommentare zur deutschen Rassengesetzgebung*, Vol. 1, Munich: Beck.

[862] Stuewer, Roger H. [1984] 'Nuclear physicists in a new world. The emigrés of the 1930's in America', *BWG* **7**, pp. 23–40.

[863] Stuhlinger, Ernst [1946] 'Hans Geiger †', *Zeitschrift für Naturforschung* **1**, pp. 50–52.

[864] Sugimoto, Kenji [1989] *Albert Einstein. A Photographic Biography*, New York: Schocken.

[865] Swinne, Edgar [1988] *Hans Geiger - Spuren aus einem Leben für die Physik*, a) Berlin: David; b) 2nd expanded ed., 1990.

[866] —— [1989] *Friedrich Paschen als Hochschullehrer*, Berlin: DAVID Verlagsgesellschaft.

[867] —— [1992] *Richard Gans. Hochschullehrer in Deutschland und Argentinien*, Berlin: ERS.

[868] Tarter, Donald E. [1992] 'Peenemünde and Los Alamos. Two studies', *History of Technology* **14**, pp. 150–170.

[869] Teichmann, Horst [1942] 'Philipp Lenard und die theoretische Physik', *ZgN* **8**, 137–139.

[870] Telschow, Ernst [1953] 'Carl Bosch 27.8.1874–26.4.1940', *Mitteilungen der Max-Planck-Gesellschaft*, May, issue 6, pp. 2–6.

[871] Teltschik, Walter [1992] *Geschichte der deutschen Großchemie. Entwicklung und Einfluß in Staat und Gesellschaft*, Weinheim: VCH.

[872] Thiessen, Peter A. [1935] 'Forschung und Wirtschaft (Vortrag auf einer Veranstaltung der 'Wirtschaftsgruppe Nichteisenmetall-Industrie' in Würzburg am 22.11.35)', *Metall-Wirtschaft, Wissenschaft, Technik* **14**, No. 50, Dec. 13, pp. 1007–1009.

[873] —— [1940] 'Wehrchemie in Deutschland. Karl Becker zum Gedächtnis', *Angewandte Chemie* **53**, pp. 377–378.

[874] Thüring, Bruno [1936]a 'Deutscher Geist in der exakten Naturwissenschaft', *Deutsche Mathematik* **1**, pp. 10–11.

[875] —— [1936]b 'Kepler–Newton–Einstein—ein Vergleich', ibid., pp. 705–711.

[876] —— [1937/38] 'Physik und Astronomie in jüdischen Händen', *ZgN* **3**, pp. 55–70.

[877] —— [1941]a 'Hugo Dinglers Werk, ein Kampfruf und Aufschwung deutscher Wissenschaft', *ZgN* **7**, pp. 130–137.

[878] —— [1941]b Review of May [1941]a, *ZgN* **7**, pp. 318–319.

[879] —— [1941/43] *A. Einsteins Umsturzversuch der Physik und seine inneren Möglichkeiten und Ursachen*, Berlin: Lüttke, a) 1941; b) separatum: *Forschungen zur Judenfrage* **4**, pp. 134–162; see also review by J. Stobbe in: *ZgN* **8** [1942], pp. 306–307.

[880] Titze, Harmuth [1989] 'Hochschulen', in: *Handbuch der Deutschen Bildungsgeschichte*, Vol. 5: *1918–1945. Die Weimarer Republik und die nationalsozialistische Diktatur*, Dieter Langenwiesche & Heinz-Elmar Tenarth (Eds.), Munich, pp. 209–240.

[881] Tobien, Waldemar [1938] *Der Einstein-Putsch als Werkzeug zur Verewigung der Jahweherrschaft. Seelenkundlich und naturwissenschaftlich gesehen*, Mülheim-Ruhr: Fabri.

[882] Toepler, Max [1941] 'Gedanken über eine natürliche, deutsche Physik', *ZgN* **7**, 257–268.

[883] Tolischus, Otto D. [1936] 'Reich scientists uphold freedom', *NYT*, Vol. 85, No. 28,477, Sunday, June 12, sec. 1, p. 31.

[884] Tomaschek, Rudolf [1924] 'Über das Verhalten des Lichtes außerirdischer Quellen', *Annalen der Physik* (4) **73**, pp. 105–126.

[885] —— [1933] *Lehrbuch der Physik. Zum Gebrauch beim Unterricht neben die der Vorlesungen und zum Selbststudium*, Leipzig/Berlin: Teubner.

[886] —— [1937/38] Review of Lenard [1936]a, *ZgN* **3**, pp. 95–99.

[887] —— [1942] 'Lenards Äthervorstellung', *ZgN* **8**, pp. 117–136.

[888] Trendelenburg, Ferdinand [1975] 'Aus der Geschichte der Forschung im Hause Siemens', *Technikgeschichte in Einzeldarstellungen* **31**.

[889] Trenkle, Fritz [1986/87] *Die deutschen Funkmeßverfahren bis 1945*, Heidelberg: Hüthig-Verlag (2nd ed., 1987).

[890] Treue, Wilhelm & Gerhard Hildebrandt (Eds.) [1987] *Berlinische Lebensbilder: Naturwissenschaftler*, Berlin: Colloquium Verlag.

[891] Trischler, Helmuth [1992]a *Luft- und Raumfahrtforschung in Deutschland 1900–1970. Politische Geschichte einer Wissenschaft*, Frankfurt (Main)/New York: Campus

[892] —— (Ed.) [1992]b *Dokumente zur Luft- und Raumfahrtforschung in Deutschland 1900–1970* (Source book to Trischler [1992]a, on the Third Reich), issued by Deutsche Forschungsanstalt für Luft- und Raumfahrt: Cologne: DLR Mitteilung 92-08.

[893] Tschirner, Martina & Heinz-Werner Göbel (Eds.) [1990] *Wissenschaft im Krieg – Krieg in der Wissenschaft*, Marburg: Philipps-Univ. (Schriftenreihe des Arbeitskreises Marburger Wissenschaftler für Friedens- und Abrüstungsforschung, AMW).

[894] Ueberschär, Gerd [1989] 'Die Deutsche Reichspost im zweiten Weltkrieg' in: Wolfgang Lotz (Ed.) *Deutsche Postgeschichte*, Berlin: Nicolai, 1989, pp. 289–320.

[895] Uhlig, Ralph (Ed.) [1991] *Vertriebene Wissenschaftler der Christian-Albrechts-Universität zu Kiel nach 1933: Zur Geschichte der Christian-Albrechts-Universität im Nationalsozialismus*, Frankfurt (Main): Lang.

[896] Vahlen, Theodor [1942] 'Die Paradoxien der relativen Mechanik', *Deutsche Mathematik*, 3rd Suppl.

[897] Vierhaus, Rudolf & Bernhard vom Brocke (Eds.) [1990] *Forschung im Spannungsfeld von Politik und Gesellschaft: Geschichte und Struktur der Kaiser-Wilhelm/Max-Planck-Gesellschaft*, Stuttgart: Deutsche Verlagsanstalt.

[898] Vogel, Barbara [1991] 'Anpassung und Widerstand. Das Verhältnis Hamburger Hochschullehrer zum Staat 1919-1945', in: Krause et al. (Eds.) [1991], Vol. 1, pp. 3–89.

[899] Vogel-Prandtl, Johanna [1993] *Ludwig Prandtl. Ein Lebensbild. Erinnerungen, Dokumente*, Göttingen: Max-Planck-Institut für Strömungsforschung.

[900] Voigt, Johannes [1981] *Universität Stuttgart. Phasen ihrer Geschichte*, Stuttgart: Wittwer.

[901] Völker, Karl-Heinz (Ed.) [1968] *Dokumente und Dokumentarphotos zur Geschichte der deutschen Luftwaffe. Aus den Geheimakten des Reichswehrministeriums (1919–1933) und des Reichsluftfahrtministeriums (1933–1939)*, Stuttgart: DVA.

[902] Volkov, Shulamit [1987] 'Soziale Ursachen des Erfolgs in der Wissenschaft. Juden im Kaiserreich', *Historische Zeitschrift* **245**, pp. 315–342.

[903] Vollnhals, Clemens [1991] *Entnazifizierung. Politische Säuberung und Rehabilitierung in den vier Besatzungszonen 1945–1949*, Munich: Deutscher Taschenbuch Verlag.

[904] Waite, Robert [1977] *Adolf Hitler. The Psychopathic God*, New York: Basic Books.

[905] Walk, Josef (Ed.)[1981] *Das Sonderrecht für die Juden im NS-Staat. Eine Sammlung der gesetzlichen Maßnahmen und Richtlinien - Inhalt und Bedeutung*, Heidelberg: Müller.

[906] Walker, Mark [1989]a *German National Socialism and the Quest for Nuclear Power. 1939–1949*, Cambridge: Cambridge Univ. Press

[907] —— [1989]b 'National socialism and German physics', *Journal of Contemporary History* **24**, pp. 63–89.

[908] —— [1990]a 'Legenden und die deutsche Atombombe', *VHZG* **38**, pp. 45–74.

[909] —— [1990]b 'Heisenberg, Goudsmit and the German Atomic Bomb', *Physics Today* **43** (1), pp. 52–60.

[910] —— [1992] 'Physics and propaganda: Werner Heisenberg's foreign lectures under National Socialism', *HSPS* **22**, pp. 339–389.

[911] —— [1993]a 'La Physique sous le troisième Reich', *La Recherche* **262**, Feb., pp. 152–159.

[912] —— [1993]b 'Selbstreflexionen deutscher Atomphysiker', *VHZG* **41**, pp. 520–542.

[913] —— [1995] *Nazi Science. Myth, Truth and the German Atomic Bomb*, New York: Plenum.

[914] Wallach, Curt [1946] 'Völkische Wissenschaft - Deutsche Physik', *Deutsche Rundschau* **69** (5), pp. 126–141.

[915] Walle, Heinrich [1982] 'Deutsche jüdische Soldaten 1914–1945', in: Militärgeschichtliches Forschungsamt (Ed.) *Deutsche Jüdische Soldaten 1914–1945*, pp. 14–84.

[916] Warburg, Emil [1925] 'Zur Geschichte der Physikalischen Gesellschaft', *Natw.* **13**, pp. 35–39 (see also Goldstein, ibid, pp. 39ff.).

[917] Watzlawek, A. [1942]a 'Das Zyklotron', *Elektrotechnische Zeitschrift* **63**, pp. 319–326.

[918] —— [1942]b 'Beschleunigung von Elektronen durch magnetische Induktion', *Elektrotechnische Zeitschrift* **63**, pp. 591–592.

[919] Wehler, Hans-Ulrich [1980] 'Psychoanalysis and history', *Social Research* **47**, pp. 531ff.

[920] Weigel, Rudolf G.(Ed.) [1937] *Philipp Lenard, der Vorkämpfer der deutschen Physik*, Karlsruhe, Müller (= Karlsruher Akademischen Reden, No. 17).

[921] Weiner, Charles [1969] 'A new site for the seminar: The refugees and American physics in the Thirties', in: Fleming & Bailyn (Eds.) [1969], pp. 190–234.

[922] Weingart, Peter, Jürgen Kroll & Kurt Bayertz [1988] *Rasse, Blut und Gene. Geschichte der Eugenik und Rassenhygiene in Deutschland*, Frankfurt (Main): Suhrkamp.

[923] Weinreich, Max [1946] *Hitlers Professors: The Part of Scholarship in Germany's Crimes against the Jewish People*, New York: Yiddish Scientific Institute-YIVO.

[924] Weiss, Burghard [1994] 'The 'Minerva' project. The accelerator laboratory of the Kaiser Wilhelm Institute/Max Planck Institute of Chemistry: Continuity in fundamental research', in: Renneberg & Walker (Eds.) [1994], pp. 271–290.

[925] Weisskopf, Victor [1991] *The Joy of Insight. Passions of a Physicist*, New York: Basic Books (also in German transl. *Mein Leben*, Munich: Scherz, 1991).

[926] Weizel, W. [1942] Review of Müller (Ed.) [1941]a, *ZtP* **23**, p. 25.

[927] Weizsäcker, Carl Friedrich von [1939]a 'Die Physik der Gegenwart und das physikalische Weltbild', *Deutsche Mathematik* **4**, pp. 119–121 (see also Dingler [1940]).

[928] —— [1939]b 'Methode der Physik', *Die Tatwelt* **15**, pp. 97–106; c) abridged in: *PZ* **40**, p. 141 (review by Dingler [1938]).

[929] —— [1942] 'Das Atom in Physik, Chemie u. Philosophie', *Das Reich*, No. 24, June 14, p. 9.

[930] —— [1944] *Physique atomique et philosophie. Conférence prononcée à la Maison de la Chimie le 25 Novembre 1943*, Cahiers de l'Institut allemand, VII, Paris: Sorlot.

[931] —— [1991] 'Ich gebe zu, ich war verrückt', *Der Spiegel* **45**, No. 17, pp. 227–238.

[932] —— [1992] 'Spaziergang an der langen Linie', *Die Zeit*, No. 24, June 5, p. 84 (reply to Jensen [1992]; see also Rechenberg [1992]b).

[933] Welch, David [1983] *Propaganda and the German Cinema 1933–1945*, Oxford: Clarendon.

[934] Wendehorst, Alfred [1993] *Geschichte der Friedrich-Alexander-Universität Erlangen-Nürnberg 1743–1993*, Munich: Beck (esp. pp. 179-216).

[935] Wendel, Günter [1963] 'Der Kampf deutscher Naturwissenschaftler gegen die Vorbereitung eines Atomkrieges', *NTM* Supplement issue, pp. 327–342.

[936] Wengenroth, Ulrich [1993] 'Zwischen Aufruhr und Diktatur. Die Technische Hochschule 1918–1945', in: U. Wengenroth (Ed.) *Technische Universität München. Annäherungen an ihre Geschichte*, Munich: Technische Universität, pp. 215–260.

[937] Wennemuth, Udo [1994] *Wissenschaftsorganisation und Wissenschaftsförderung in Baden. Die Heidelberger Akademie der Wissenschaften 1909-1949*, Heidelberg: Winter.

[938] Wesch, Ludwig [1939] 'Die Zukunft der deutschen Physik (Eine Betrachtung zur Schulphysik)' *Der Deutsche Erzieher. Reichszeitung des NSLB*, pp. 27–29.

[939] —— [1942] 'Lenards Werk – Vorbild zukünftiger Forschung', *ZgN* **8**, pp. 101–114.

[940] Westphal, Wilhelm [1943] 'Probleme der Physik', *Das Reich*, No. 41, Oct. 10, pp. 9–10.

[941] —— [1945] 'Physik begründet die Technik', *Das Reich*, No. 6, Feb. 11, pp. 5–6.

[942] —— [1955] 'Das Physikalische Institut der Techn. Universität Berlin', *PB* **11**, 554–558.

[943] —— [1964] 'James Franck †', *PB* **20**, pp. 324–328.

[944] —— [1972] '68 Jahre als Physiker in Berlin', *PB* **28**, pp. 258–265.

[945] Weyl, Hermann [1920] 'Die Diskussion über die Relativitätstheorie auf der Naturforscherversammlung', *Umschau* **24**, pp. 609–611.

[946] Weyland, Paul [1920]a 'Einsteins Relativitätstheorie – eine wissenschaftliche Massensuggestion', *Tägliche Rundschau*, Aug. 6, (A), No. 171 (reprinted in: Weyland [1920]b, pp. 21–24, see also Laue et al. [1920] and Glaser [1920]).

[947] —— [1920]b *Betrachtungen über Einsteins Relativitätstheorie und die Art ihrer Einführung*, Leipzig/Berlin: Köhler (Schriften aus dem Verlage der Arbeitsgemeinschaft Deutscher Naturforscher zur Erhaltung reiner Wissenschaft, Vol. 2).

[948] —— [1920]c 'Die Naturforschertagung in Nauheim. Die Erdrosselung der Einstein-Gegner', *Deutsche Zeitung. Unabhängiges Tagesblatt für nationale Politik*, Sep. 26, First Suppl.

[949] —— [1920]d Reply to Laue et al. [1920], *Tägliche Rundschau* (Berlin), Aug. 11.

[950] —— (Ed.) [1921] *Deutsch-Völkische Monatshefte*, (only one issue appeared).

[951] Wheaton, Bruce R. [1983] *The Tiger and the Shark. Empirical Roots of Wave-Particle Dualism*, Cambridge: Cambridge Univ. Press.

[952] Whittaker, Edmund [1951/53] *A History of the Theories of Aether & Electricity* in 2 vols. a) London/New York: Nelson & Sons (2 vols.); b) reprint, New York: Dover, 1989.

[953] Widmann, Horst [1973] *Exil und Bildungshilfe in die Türkei nach 1933*, Berlin/Frankfurt.

[954] Wien, Max [1944] 'Der Nachwuchs der Physiker', *PB* **1** (88/89), pp. 55–57 (excerpts from Wien's 1937 memorandum).

[955] Wininger, Salomon [1921/35] *Große jüdische National-Biographie, mit mehr als 12 000 Lebensbeschreibungen namhafter jüdischen Männer und Frauen aller Zeiten und Länder*, Vols. 1–7, Cernauti: Arta.

[956] Winkler, Robert [1936] 'Neue Wissenschaft', *Volk im Werden* **4** (7), pp. 14–23.

[957] Wirtz, Karl et al. [1990] *Atom-Museum Haigerloch*, Haigerloch: Stadtverwaltung.

[958] Wistrich, Robert [1982] *Who's Who in Nazi Germany*, London: Weidenfeld & Nicolsen.

[959] Wittig, J. [1988] 'Zum Leben und Wirken von Max Wien', *Feingerätetechnik* **37**, pp. 490, 516–519.

[960] Wohlfarth, Horst (Ed.) [1979] *40 Jahre Kernspaltung. Eine Einführung in die Originalliteratur*, Darmstadt: Wissenschaftliche Buchgesellschaft.

[961] Wolf, Franz [1962] 'Zur 100. Wiederkehr des Geburtstages von Philipp Lenard', *Natw.* **49**, pp. 245–247.

[962] Wolff, Stefan [1991] 'Das ungarische Phänomen – ein Fallbeispiel zur Emigrationsforschung', *Wissenschaftliches Jahrbuch des Deutschen Museums* (Munich), 228–245.

[963] —— [1993] 'Vertreibung und Emigration in der Physik', *Physik in unserer Zeit* **24**, pp. 267–273.

[964] Wolgast, Eike [1986] *Die Universität Heidelberg 1386–1986*, Berlin: Springer.

[965] Wolters, Gereon [1992] 'Opportunismus als Naturanlage: Hugo Dingler und das Dritte Reich', in: Peter Janich (Ed.) *Entwicklung der methodischen Philosophie*, Frankfurt (Main): Suhrkamp.

[966] Zenneck, Jonathan et al. [1940] 'Deutsche Physikertagung Kriegsjahr 1940, veranstaltet von der Deutschen Physikalischen Gesellschaft und der Deutschen Gesellschaft für technische Physik am 1. und 2. September 1940 in Berlin', *VDPG* (3) **21**, pp. 31–50.

[967] Zierold, Kurt [1968] *Forschungsförderung in drei Epochen*, Wiesbaden: Steiner.

[968] Zilbert, Edward R. [1981] *Albert Speer and the Nazi Ministry of Arms. Economic Institutions and Industrial Production in the German War Economy*, Rutherford: Farleigh Dickinson Univ. Press.

[969] Zneimer, Richard [1978] 'The Nazis and the professors: Social origin, professional mobility and political involvement of the Frankfurt faculty, 1933–1939', *Journal of Social History* **12**, pp. 147–158.

Name Index

This index lists the persons mentioned in the introduction, the main texts and the annotation, omitting authors of later secondary literature. Entries in the biographical section (Appendix F) are indicated. Firstname initials are provided to distinguish between otherwise identical entries. Uncapitalized articles, contractions or particles attached to names as well as accented vowels are ignored in the alphabetical listing ('v.' abbreviates 'von'). Italicized numerals indicate biographical information on the relevant pages.

Editor's Errata List

Below please find a few important corrections and updates (many being year of death of physicists featured in the biographical profiles, Appendix F).

p. xv, line 9: nuclear physicist → quantum physicist
p. lviii, line 12: 14.3% → 9.2%
p. lxxxi, lines 24–25: 1939 → 1938
p. 95: the number 1 in the second-to-last row (bottom right) has to be shifted one field to the left
p. 139, footnote 6: Doc. 21 → Doc. 31
p. 294: delete the last sentence in the caption starting with 'Originally'
p. 329, line 10 should read: there are no such geniuses ...
p. 331: the two images have to be switched
p. IV in the Appendix under NG, line 2: "In Oct. 1937" → "In 1929"
p. XVIII: Bagge (1912–1996)
p. XXV: Flammersfeld (1913–2001); Fleischmann (1903–2002)
p. XXIX, line 16: Manwell [196.] → Mauwell [1962]
p. XXXI: Haxel (1909–1998)
p. XXXIII: Hund (1896–1997)
p. XXXIX: Mentzel (1900–1987)
p. XLIV entry Sauter, line 7: Munich Univ. → Munich Polytechnic
p. L: Walcher (1910–2005)

Since this commented anthology first appeared in 1996 in the Birkhäuser series *Science Networks*, the historiography of the sciences under national socialism has produced many new and important studies on all kinds of aspects, none of which can be cited in this unmodified photostatic reprint. However, I would like to mention the following more recent publications of special pertinence to the themes discussed in this anthology:

- The multi-volume series of studies on the history of the former *Kaiser Wilhelm Society*, produced by a presidential commission of the *Max Planck Society*, active from 1997 to 2008; see http://www.mpiwg-berlin.mpg.de/KWG/publications
- An anthology of papers on *The German Physical Society in the Third Reich*, Dieter Hoffmann & Mark Walker, eds., forthcoming at Cambridge Univ. Press
- The history of the DFG 1920–70, edited by Karin Orth et al., Wiley 2011
- An exemplary study of university history during the 'Third Reich': *"Kämpferische Wissenschaft". Studien zur Universität Jena im Nationalsozialismus* (edited by Uwe Hossfeld et al., Böhlau 2003)
- New biographies such as Jost Lemmerich on James Franck (Stanford 2011)
- My own follow-up study on the mentality of German physicists after 1945, appeared in 2007 at Oxford Univ. Press under the title *The Mental Aftermath*.

MIX
Papier aus verantwortungsvollen Quellen
Paper from responsible sources
FSC® C105338

If you have any concerns about our products,
you can contact us on
ProductSafety@springernature.com

In case Publisher is established outside the EU,
the EU authorized representative is:
Springer Nature Customer Service Center GmbH
Europaplatz 3, 69115 Heidelberg, Germany

Printed by Libri Plureos GmbH
in Hamburg, Germany